ELECTRONIC PACKAGING AND INTERCONNECTION HANDBOOK

Electronic Packaging and Interconnection Series

Charles M. Harper, Series Advisor

To order or receive additional information on these or any other McGraw-Hill titles, please call 1-800-822-8158 in the United States. In other countries, contact your local McGraw-Hill representative.   **WM16XXA**

ELECTRONIC PACKAGING AND INTERCONNECTION HANDBOOK

Charles A. Harper Editor-in-Chief

Technology Seminars, Inc., Lutherville, Maryland

Second Edition

McGRAW-HILL

New York   San Francisco   Washington, D.C.   Auckland   Bogotá
Caracas   Lisbon   London   Madrid   Mexico City   Milan
Montreal   New Delhi   San Juan   Singapore
Sydney   Tokyo   Toronto

Library of Congress Cataloging-in-Publication Data

Electronic packaging and interconnection handbook / Charles A. Harper,
   editor-in-chief. — 2nd ed.
       p.   cm. — (Electrical packaging and interconnection series)
   Includes bibliographical references and index.
   ISBN 0-07-026694-8 (hc)
   1. Electronic packaging—Handbooks, manuals, etc. I. Harper,
Charles A. II. Series.
TK7870.15.E42   1997
621.381'046—dc20   96-31294
                                               CIP

McGraw-Hill

A Division of The McGraw-Hill Companies

3   4   5   6   7   8   9   0   DOC/DOC   9   0   1   0   9   8   7

ISBN 0-07-026694-8

*The sponsoring editor for this book was Stephen S. Chapman and the
production supervisor was Pamela Pelton. It was set in Times by J. K. Eckert
& Company, Inc.*

Printed and bound by R. R. Donnelley & Sons Company.

This book is printed on acid-free paper.

CONTENTS

Chapter 4. Wiring and Cabling for Electronic Packaging   4.1

Chapter 5. Solder Technologies for Electronic Packaging   5.1

Part 2   Interconnection Technologies

Chapter 6. Packaging and Interconnection of Integrated Circuits   6.1

Part 3 System Packaging Technologies

Chapter 11. Packaging of High-Speed and Microwave Electronic Systems   11.1

Chapter 12. Packaging of High-Voltage Systems   12.1

PREFACE

It is a pleasure to present to my reading audience this second edition of *Electronic Packaging and Interconnection Handbook*. It will probably be remembered by my established audience that this handbook is, in fact, a descendent of my original *Handbook of Electronic Packaging*, which was the first complete handbook in this field. In the interim, there have been many books on specific subjects in the electronic packaging field, but this handbook remains the only comprehensive handbook covering all of the disciplines and technology levels in electronic packaging. As readers will note, I have had the good fortune to gather together a group of outstanding chapter authors, each with many years of experience in his or her field. Together, they cover the multidisciplinary technology of electronic packaging as perhaps no other group could do. Hence, I would like to give special credit to these authors in this preface.

The technology of electronic packaging, often understandably described as the technology of electronic interconnections, remains a dynamic technology. Since it is a multidisciplinary technology, defining electronic packaging is often difficult. Nonetheless, definitions and terminology are sufficiently important to have led me to include a separate statement following this preface, entitled Electronic Packaging Defined. With respect to the dynamics of electronic packaging, trends toward higher performance and higher density electronic packages continue to accelerate. Electronic packaging was, until recently, the afterthought of electronic system designers. Today it is widely accepted that proper electronic packaging fundamentals and practices must be considered in the initial system design phases. In fact, electronic packaging is often the critical limiting factor in the success of modern electronic systems. Much of the great potential of modern semiconductor and electronic system technologies can be seriously degraded, or even negated, without proper electronic packaging. The purpose of this *Electronic Packaging and Interconnection Handbook* is to make available the fundamentals and practices necessary for achieving success in any of the disciplines that form the technology of electronic packaging.

This thoroughly revised and updated edition continues to cover all of the disciplines in the multidisciplinary field of electronic packaging. Chapters 1 through 5 cover the fundamental technologies of electronic packaging. These chapters deal with materials for electronic packaging, thermal management, connectors and interconnection technology, wiring and cabling for electronic packaging and, finally, soldering and solder-paste technology. The second group of chapters, namely Chapters 6 through 10, cover the interconnection technologies of electronic packaging. These chapters discuss packaging and interconnection of integrated-circuit and semiconductor devices, hybrid microelectronic and multichip module packaging, rigid and flexible printed-wiring boards, surface mount technology and advanced electronic packaging and interconnection technologies such as ball grid array, flip chip, tape automated bonding, and chip-scale packaging. The last two chapters, 11 and 12, cover system packaging technologies for the important categories of high-speed and microwave systems, and high-voltage systems.

It is my hope, and the hope of the chapter authors, that this new edition of *Electronic Packaging and Interconnection Handbook* will serve all of our readers very well. Any comments will be appreciated.

Charles A. Harper
Technology Seminars, Inc.
Lutherville, Maryland

ELECTRONIC PACKAGING DEFINED

Electronic packaging is a multidisciplinary technology. This creates a problem when a definition is to be found that will be acceptable to all those involved with this technology. While task forces have been formed to define electronic packaging, with resulting definitions ranging from one line to one page, perhaps is has been best and most simply defined in a recent dictionary—the only dictionary devoted to this technology.[*] This dictionary contains 235 pages of multidisciplinary terms, including many pages of acronyms and abbreviations, which are often so troublesome. In this dictionary the author, Martin B. Miller, defines electronic packaging as *the combination of engineering and manufacturing technologies required to convert an electronic circuit into a manufactured assembly.* In fact, the engineering technologies include electrical, mechanical, thermal, chemical, materials, components, and others. Therefore, since none of these engineering groups is fluent in the terminologies of all the others, a definition of electronic packaging may depend on which group defines the term. With the increasing criticality of electronic packaging (as discussed in the Preface), a common understanding of the various languages, and hence a common definition of electronic packaging, is becoming more and more important. Again, perhaps the definition by Miller is broad enough to define electronic packaging. It is my hope that this definition, coupled with the multidisciplinary presentations in this handbook, will serve well the interests and needs of our readers.

Charles A. Harper

---

*Martin B. Miller, *Electronic Packaging, Microelectronics and Interconnection Dictionary,* McGraw-Hill Companies, New York, NY.

CONTRIBUTORS

Frank E. Altoz, Consultant (Chap. 2)

John K. Bonner, JPL, California Institute of Technology (Chap. 9)

Victor J. Brzozowski, Northrop Grumman Corporation (Chap. 8)

Charles Cohn, Lucent Technologies (Chap. 6)

Edward J. Croop, Technology Diversified Services (Chap. 4)

William J. Dunbar, Consultant (Chap. 12)

Jennie S. Hwang, H-Technologies Group, Inc. (Chap. 5)

Stephen G. Konsowski, Technology Seminars, Inc. (Chap. 11)

Robert S. Mroczkowski, AMP, Incorporated (Chap. 3)

Ronald N. Sampson, Technology Seminars, Inc. (Chap. 1)

Donald P. Schnorr, Consultant (Chap. 10)

Jerry E. Sergent, BBS Power Mod (Chap. 7)

Ming T. Shih, Lucent Technologies (Chap. 6)

P · A · R · T · 1

FUNDAMENTAL TECHNOLOGIES

CHAPTER 1

MATERIALS FOR ELECTRONIC PACKAGING

Ronald N. Sampson
Technology Seminars, Inc.
Lutherville, Maryland

Douglas M. Mattox
Department of Ceramic Engineering
University of Missouri–Rolla, Rolla, Missouri

1.1   INTRODUCTION

The materials used in the packaging of electronic devices play key roles in the proper functioning and useful life of the packaged assembly. The most obvious function is the conduction of signals through the circuit contributed by metals in the form of wires, contacts, foils, plating, and solders. Of equal and often greater importance is the electrical insulation function, which prevents the loss of the signal currents and confines them to the desired paths. The insulation systems exist in a variety of forms, including liquids, solids, and gases, and they determine the life of the devices. Other materials perform structural roles and support the circuit physically. Finally, there are materials whose function it is only to protect the circuit from the environment—moisture, contamination, heat, and radiation. No wonder then that these packaging materials are critical to the function, efficiency, and life of electronic circuits.

This section discusses these materials and their properties and provides the designer with data and performance information needed in modern electronic packaging design.

1.1.1   Classification of Materials

It is possible to classify packaging materials in many different ways, but a system relating to volume resistivity is preferred by packaging engineers. Figure 1.1 shows the materials most commonly used in electronic applications in relation to their insulating (10^{19} to 10^6 $\Omega \cdot$ cm), semiconducting (10^6 to 10^{-3} $\Omega \cdot$ cm), or conducting (10^{-3} to 10^{-6} $\Omega \cdot$ cm) resistivity properties.

Insulation materials may be organic or inorganic. Organic packaging materials are those based on molecules consisting of carbon-to-carbon chains [C—C—C]. Chain modi-

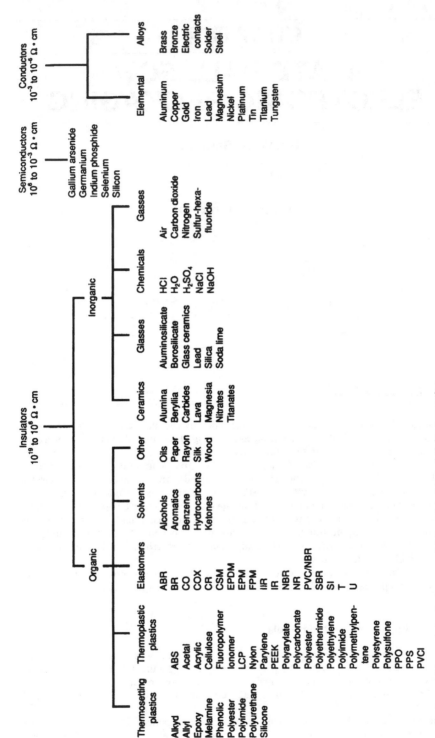

FIGURE 1.1 Electronic packaging materials as related to resistivity properties.

1.2

fications containing oxygen, sulfur, nitrogen, chlorine, fluorine, and other atoms are common. When the chains are straight, the molecules are called aliphatic. When the molecules consist of rings, they are called aromatic. The most common ring structure is based on the benzene molecule C_6H_6, which is usually represented by ⬡ . An exception to the carbon chain polymer structure is the family of silicones, which is based on a structure consisting of silicon and oxygen [Si—O—Si—O].

The most common insulators are polymers or plastics; they consist of very large molecules. A plastic is either thermosetting or thermoplastic. Thermosetting means that the plastic reacts chemically while being formed, resulting in a product that is infusible and not liquefied by further heating. Epoxy-glass circuit boards are an example. Thermoplastic polymers are reacted by the manufacturer and can be liquefied and solidified by heat repeatedly. A good example is candle wax. Elastomers (rubbers) can be either thermosetting or thermoplastic and are polymers that can be greatly elongated by stress.

Inorganic materials usually consist of much smaller molecules than plastics. Thousands of inorganic materials exist, but the packaging engineer is concerned with such chemicals as sodium chloride (NaCI), water (H_2O), hydrochloric acid (HCl), sodium hydroxide (NaOH), and others. Ceramics are also inorganic materials and are critical for electronics applications. They are most often oxides of metals, but more recently carbides and nitrides are finding special uses. Common electronic ceramics include silica (SiO_2), alumina (Al_2O_3), beryllia (BeO), and various glasses.

One other special inorganic insulation is gas. Used most often in higher-voltage applications, such gases include sulfur hexafluoride (SF_6), dry nitrogen (N_2), and carbon dioxide (CO_2).

Semiconducting materials have resistivity properties that are intermediate between insulations and conductors (10^{-3} to 10^6 $\Omega \cdot$ cm) and comprise the active components of electronics. Familiar semiconductors include silicon, germanium, and gallium arsenide.

Conductors are most commonly metals. They have resistivity values from 10^{-6} to 10^{-3} $\Omega \cdot$ cm. Most metals are alloyed in mixtures of two or more metallic elements. Those of most interest to electronic packaging engineers are copper, aluminum, gold, solder, and platinum.

1.2 POLYMERS

Plastics and polymers (see also Ref. 16) are used extensively in electronic circuits. They comprise a large list of materials and are available in liquids, films, coatings, shapes, plates, rods, adhesives, foams, laminates, and composite structures. A key consideration for the electronic packaging engineer is whether the plastic is thermoplastic or thermosetting, for this influences the effects of temperature, many of the electrical and physical properties, and how the material is fabricated.

1. *Thermosetting plastics.* Thermosetting plastics are only partially reacted or polymerized at the time of manufacture, forming a resin. Full polymerization is accomplished by the person who forms the final shape. The polymerization process is called cross-linking, curing, hardening, or vulcanization. This chemical reaction is most often caused by applying heat to the plastic. A coreacting, or initiating, material is frequently required. When this material reacts into the plastic and remains a part of the chain, it is called a hardener or copolymer. When it only provides chemical reactivity and remains outside of the plastic molecule, it is known as a catalyst, activator, or initiator. The chemical reaction occurring as the plastic cross-links usually produces heat (exothermic reaction). The heat involved can be considerable, and internal temperatures can rise to unacceptable levels if consideration is not given to removing or controlling the heat. Cracking of cast parts is often caused by this thermal effect.

The cross-linking reaction is also accompanied by a reduction in volume (shrinkage) of 1 to 10 percent. Polymerization shrinkage control is a factor in the design of cast and embedded parts, and materials such as fillers, reinforcing fibers, lubricants, and pigments are usually added to thermosetting plastics to control shrinkage. Since the reaction of the plastic goes to completion, scrap material cannot be added back into the virgin compound.

2. *Thermoplastic plastics.* Thermoplastic plastics are fully reacted or polymerized at the time of manufacture. They can be liquefied and hardened by heat repeatedly, analogous to the freezing and melting of ice. In the melt form (the viscosity may be extremely high), the plastic may be forced into a mold or through a forming die by pressure. Cooling then preserves the desired shape. Scrap can be reground, added to virgin materials, and reused. Thermal aging of the material over many repeated cycles causes degradation of the properties of the plastic. Materials such as fillers, reinforcements, thermal stabilizers, ultraviolet light stabilizers, and pigments are usually combined with the plastic to impart specific properties.

Thermoplastics have higher creep, more sensitivity to chemicals, and less tolerance to high temperatures than do thermosets. A loose grouping called *engineering plastics* comprises thermoplastics with high service temperatures, which permit their use in applications formerly reserved for thermoset materials, ceramics, and metals.

The following sections describe those plastics used by electronic packaging engineers and designers.

1.2.1 Thermosetting Plastics

Full descriptions of all thermosetting plastics and their properties are given in the *Modern Plastics Encyclopedia.*[1] The following discussions describe the unique features for the thermosetting plastics used in electronic applications.

Thermosetting plastics are fabricated by casting, compression molding, filament winding, laminating, pultrusion, reaction injection molding, and transfer molding. A description of these processes is given later in this chapter. Fillers and reinforcements are usually combined with thermosetting materials and are also described later.

1.2.1.1 Alkyds.   Alkyds are described in Sec. 1.2.1.5.

1.2.1.2 Allyls.   Allyl-based polymers are used for their dimensional stability, thermal resistance, moisture resistance, and electrical properties. Commercial compounds are based on diallyl phthalate (DAP) or diallyl isophthalate (DAIP). DAP materials are generally used as molded shapes. They can be compression molded, transfer molded, or injection molded. The strength of parts made from the latter two methods is less than that of compression molded parts. In all cases, the resins are reinforced with glass, mineral, or synthetic fibers.

The superior performance of DAP at high humidity is shown in Fig. 1.2. Properties for short fiberglass-filled DAP are given in Table 1.1. DAP compounds are defined in MIL-M14G in the following categories:

Type GDI-30	Glass filled, general purpose
Type GDI-30F	Glass filled, flame retardant
Type MDG	Mineral filled, low shrinkage
Type SDG	Glass filled, low losses

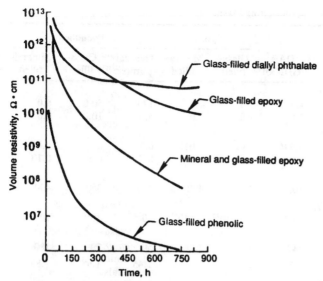

FIGURE 1.2 Effect of 95% humidity on resistivity of several polymer-filler systems.

Type SDG-F	SDG with flame retardancy
Type SDI-5	Acrylic fiber filled, high electric strength
Type SDI-30	Polyethylene terephthalate fiber filled, low loss, and high impact strength

DAP materials are used in connectors, chip carriers, and terminal boards for their stability and electrical performance. They are often used in switches because they have good arc resistance. Combined with glass fabric, they can be laminated and used as radomes and ducts.

1.2.1.3 Epoxies. In electronic applications, epoxies are selected for their toughness, processibility, rigidity, chemical resistance, low shrinkage, and adhesion.

Epoxies are used in all aspects of electronic manufacturing. The presence of an epoxide ring (oxirane) $CH\overset{O}{\diagup\!\diagdown}CH_2$ defines an epoxy. Because so many starting compounds can be used to react epoxies, there are many different resins, but most commonly epoxies are formed by reacting bisphenol A with epichlorohydrin.

In many cases, bromine is reacted into the phenol ring section of the molecule to add flame retardance. Multifunctional epoxies have more than two epoxide rings in the molecule. This basic resin must be cured or cross-linked by adding a coreactant or catalyst. Many different compounds can react with either the epoxide ring or the hydroxyl group (—OH); the most common curing agents are shown in Table 1.2.

Epoxies can be made with any viscosity from liquid to solid, they can be flexible or rigid, they can be filled with mineral fillers or fibers or left unfilled, and they can react slowly or rapidly at both high and low temperatures. Epoxies can be cast as a liquid into a mold to produce a shape or to embed a circuit. More detail is given later in this chapter.

TABLE 1.1 Properties of Thermosetting Plastics

		Epoxy		Phenolics		
	DAP (GDI-30)	Glass-filled	Mineral-filled	General-purpose	Glass-filled	Mineral-filled
Dielectric constant, D-150						
60 Hz	4.2	5.0	4.0	12.0	50.0	6.0
10^6 Hz	3.5	4.6	5.0	6.0	10.0	6.0
Dissipation factor, D-150						
60 Hz	0.004	0.01	0.01	0.3	0.3	0.07
10^6 Hz	0.01	0.01	0.01	0.7	0.8	0.10
Dielectric strength, D-149; V/mil	400	360	400	400	350	400
Volume resistivity, D-257; $\Omega \cdot$ cm	10^{13}	3.8×10^{15}	9×10^{15}	10^{13}	10^{13}	10^{14}
Arc resistance, D-495; seconds	140	140	180	50	70	180
Specific gravity, D-792	1.7	1.8	2.1	1.45	1.95	1.83
Water absorption, D-570; % 24 h	<0.2	0.2	0.04	0.7	0.5	0.5
Heat deflection temperature, D-648; at 264 lb/in^2, °F	500	400	250	340	400	500
Tensile strength, D-638; lb/in^2	10,000	30,000	15,000	10,000	7000	11,000
Impact strength (Izod), D-256; ft · lb/in	5.0	10	0.4	0.3	3.5	15.0
Coefficient of thermal expansion, D-696; 10^{-5}/°F	2.6	1.7	2.2	2.5	—	0.88
Thermal conductivity, C-177; Btu · in/ (h · in · ft^2 · °F)		6	—	0.3	0.34	0.2

A large use of epoxies is to combine them with glass fabrics to produce laminated printed circuit boards (FR-4); another use of epoxies is to manufacture conformal coatings, varnishes, and adhesives. These uses are discussed in Chap. 8.

Epoxies can be used as compression or transfer molding compounds, where they are reacted with phenolic resins to produce parts with high dimensional stability and low outgassing properties. These materials are described in MIL-P46892 and are used for bobbins, connectors, and chip carriers. Typical properties for epoxy molding compounds are given in Table 1.1.

1.2.1.4 Phenolics.   Phenolics are selected for their low cost, high-temperature capability, and high mechanical strength. They have poor arc resistance. Phenolic resins are the reaction product of phenol and formaldehyde.

They are available as molding compounds, liquid varnishes, foams, laminates, and as a component of adhesives. Phenolic materials can be molded by compression, transfer, or

TABLE 1.1 Properties of Thermosetting Plastics (*Continued*)

	Alkyds		Poly-ester GPO-3	Polyimide	Polyure-thane	Mineral-filled silicone
	MAG	MAI-60				
Dielectric constant, D-150						
60 Hz	6.3	5.6	4.5	3.5	6	3.6
10^6 Hz	4.7	4.6	—	3.4	3	3.7
Dissipation factor, D-150						
60 Hz	0.04	0.10	0.05	0.0025	0.1	0.005
10^6 Hz	0.02	0.02	—	0.01	0.04	0.003
Dielectric strength, D-149; V/mil	400	375	300	6500	500	425
Volume resistivity, D-257; $\Omega \cdot$ cm	10^{14}	10^{13}	—	10^{18}	10^{14}	10^{15}
Arc resistance, D-495; seconds	>180	180	>180	230	120	240
Specific gravity, D-792	2.24	2.07	1.95	1.4	1.1	2.05
Water absorption, D-570; % 24 h	0.08	0.07	0.5	2.9	0.2	0.15
Heat deflection temperature, D-648; at 264 lb/in², °F	350	>400	—	680	190	>500
Tensile strength, D-638; lb/in²	3000	6000	9000	17,000	1000	6500
Impact strength (Izod), D-256; ft · lb/in	0.3	9.5	8.0	1.5	25	0.5
Coefficient of thermal expansion, D-696; 10^{-5}/°F	3	2	2	2.8	25	2.8
Thermal conductivity, C-177; Btu · in/ (h · in · ft² · °F)	7.2	3.6	4	6.8	0.1	3.1

injection or they can be laminated. Typical properties of phenolic moldings are listed in Table 1.1.

Phenolics are used as chip carriers, connectors, bobbins, and with kraft paper in copper-clad laminates for printed wiring boards.

1.2.1.5 Polyesters. Thermosetting polyester resins are useful for their low-pressure processing conditions, low cost, good electrical properties, and high arc resistance. Their chemical resistance is low. They are formed by the reaction of an organic acid with an alcohol and then further reacted with a vinyl monomer such as styrene and an organic peroxide. These resins are usually unsaturated (double valence bonds are present).

Polyesters can be produced in a variety of forms. In Sec. 1.2.6 polyethylene terephthalate film is described. The resins can also be cast, but the heat of the exothermic reaction is high. As moldings, polyesters can be divided into alkyds and unsaturated polyesters.

TABLE 1.2 Curing Agents for Epoxies

Chemical type	Characteristics	Typical curing agents
Aliphatic amine	Room-temperature or elevated-temperature cure; fast; low heat resistance; high exotherm	Diethylene triamine; dicyandiamide (Dicy)
Aromatic amine	Long pot life; elevated-temperature cure; higher heat resistance	Metaphenylene diamine; diamino diphenyl sulfone
Anhydride	Safer; very long pot life; elevated-temperature cure; very high heat resistance	Hexahydrophthalic anhydride
Catalytic	Good pot life; high exotherm; good physical properties	Piperidine; boron trifluoride ethylamine (Lewis acid)

1. *Alkyds.*   These molding compounds are formulated from polyester resins, monomers (often DAP), catalysts, mineral fillers, pigments, and fibers. They are usually supplied as free-flowing powders which can be compression, injection, or transfer molded. They retain excellent electrical properties up to 350°F. Alkyds are specified in MIL-M-14G as the following types:

Type MAG	Mineral filled, arc resistance
Type MAI-60	Glass-fiber filled, electrical properties, and impact strength
Type MAT-30	Mineral and glass-fiber filled, heat, track, and arc resistant, high impact strength
Type MAI-30	Similar to MAT-30, but with better mechanical properties and easier molding

Table 1.1 presents the properties of alkyds.

2. *Unsaturated polyesters.*   These materials can be used as bulk molding compounds (BMC), sheet molding compounds (SMC), hand lay-up, spray-up, resin transfer molding, laminates, filament windings, and pultrusion.

The unsaturated polyesters of most interest to the electronic designer are the laminate grades defined by NEMA.[2] These laminates are fabricated from random glass-fiber mat, fillers (usually aluminum oxide trihydrate), and polyester resins. Six grades are given:

Grade GPO-1	General purpose
Grade GPO-2	Flame resistant
Grade GPO-3	Flame, arc, and track resistant
Grade GPO-1P	Punching grade of GPO-1
Grade GPO-2P	Punching grade of GPO-2
Grade GPO-3P	Punching grade of GPO-3

Properties of grade GPO-3 are listed in Table 1.1. Polyester sheets are also described in ASTM D-1532[3] and MIL-P-24364.[4]

Polyester moldings are used in bobbins, terminal boards, housings, and connectors. Films are used as wire insulation, coil insulation, and protective covers. Pultrusions are used as bus supports and spaces. Radomes are often filament wound or compression molded.

1.2.1.6 Polyimides.

Polyimide polymers are excellent in heat resistance, at cryogenic temperatures, and in radiation resistance. Their electrical properties and low outgassing make them favorites for use in extreme environments of space and high temperatures. They are sensitive to alkalies, organic acids, and moisture.

The polymers are often copolymerized with other molecules and are considered either thermosetting or thermoplastic, but from a rigorous definition most lack cross-linking and are truly thermoplastics with a melting point close to the decomposition point. Typical copolymers are amideimide, imide phenolics bismaleimide, epoxy imides, and polyester imides.

The polyimides are sold as powders or as solutions in highly polar solvents. In finished form polyimides are available as wire insulation, film, moldings, laminates, coatings, sleeving, tapes, fibers, and fabrics. The properties of a pure polyimide are given in Table 1.1. The dielectric properties are little affected by increasing either temperature or frequency, as shown in Fig. 1.3. The response of volume resistivity with temperature is given in Fig. 1.4.

Polyimides are used extensively in the electronics industry. They are found as multilayer circuit board coatings, chip carriers, laminates, film insulation, flexible cable, flexible circuits, tape, tape wire wrap, sleeving, moldings, coil bobbins, wire enamel, and fiber reinforcements.

1.2.1.7 Polyurethanes.

Polyurethanes are elastomeric-like materials with unusual toughness, tear resistance, and abrasion resistance.

The polymer is formed by reacting a diisocyanate with a glycol. The availability of many starting ingredients makes many polymers possible. The polyol can be based on an ether or a polyester. The polyether-based polymers are better in moisture resistance. Linear polymers, which are thermoplastic, can also be manufactured.

Electronic designers use polyurethanes as embedding compounds or conformal coatings. They can also be molded by reaction injection molding (RIM) for automotive parts, produced as paints or wire enamels, and foamed.

Urethanes are sensitive to chlorinated solvents, ketones, and strong acids and bases. They should not be used above 250°F. Table 1.1 lists the properties of a flexible polyurethane.

1.2.1.8 Silicones.

Silicones have very high temperature capabilities. They can be used from −86 to 480°F, and they have excellent arc resistance. Silicones are different from organic polymers based on carbon and hydrogen. Instead of a carbon molecular chain, silicones have silicon and oxygen.

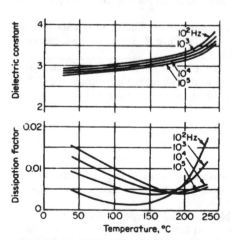

FIGURE 1.3 Effect of temperature on electrical loss properties of polyimide.

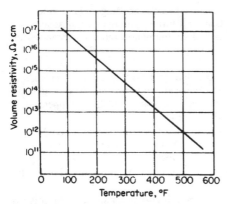

FIGURE 1.4 Volume resistivity versus temperature for unfilled polyimide.

Silicones are produced as liquids, resins, and elastomers. In electronics, they are compression molded to form shaped parts, used as wire enamels, laminated in sheets, cast to embed parts, fabricated as elastomers in tapes, sleeving, and tubing, combined with solvents to form varnishes, and combined with elastomers to make adhesives.

Molding compounds are made by combining a resin with mineral fillers and glass fibers. Table 1.1 lists properties of a mineral-filled silicone. The materials are little affected by temperature up to 500°F. Figure 1.5 shows how constant the dissipation factor of silicones is with temperature. Silicone laminates are standardized by NEMA[2] as grade G-7.

Silicones are prepared as gums for conversion into elastomers. Vulcanization is by heat and a catalyst—usually peroxides. Gums are compression or transfer molded and extruded. In electronics they are usually found as wire insulation or sleeving. Sleeving can be made heat shrinkable for tight fit on conductors and connections. Elastomers are defined in federal specifications MIL-W-8777C and MIL-W16878. Silicone elastomers are highly arc- and track-resistant. They are classified as nonburning and form a nonconducting ash of silica.

Silicones can be cast from solvent-free liquids and used to embed components. Conformal coatings can be silicone-based and are described in Chap. 8. Similarly, the same resins can be combined with a solvent and used as a varnish.

Silicone adhesives can be compounded to achieve a wide range of properties, including pressure-sensitive varieties. A popular type is the room-temperature-vulcanizing (RTV) adhesive, which polymerizes with water at ambient conditions, releasing acetic acid or ethyl alcohol.

1.2.2 Thermoplastic Plastics

Full descriptions of all thermoplastic plastics and their properties are given in the *Modern Plastics Encyclopedia.*[1] The following discussions describe the unique features for each thermoplastic plastic.

Thermoplastic plastics are fabricated by blow molding, extrusion, foaming, injection molding, rotational molding, stamping, and vacuum forming. Thermoplastics are used as pure polymers or enhanced with fillers and reinforcements. These materials are often combined to form alloys.

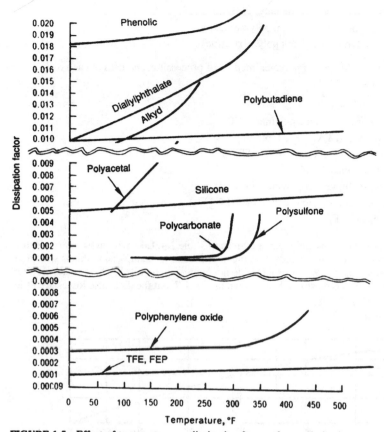

FIGURE 1.5 Effect of temperature on dissipation factor of several plastics.

1.2.2.1 Acrylics. Acrylics are optically clear—approaching glass—with superior weather resistance. They burn as readily as wood. They are made by polymerizing methyl methacrylate and are used as conformal coatings.

1.2.2.2 Fluoropolymers. These plastics have a wide variety of properties that make them indispensable to electronic circuit designers. They have high heat resistance, are inert to almost all chemicals, do not burn, and have high arc resistance, low dielectric losses, zero water absorption, and a low coefficient of friction. They are limited in being relatively soft, hard to process, and expensive.

A wide variety of copolymers is available. In general, these modifications result in thermoplasticity, improved mechanical properties, narrower temperature range of performance, less attractive electrical properties, some chemical sensitivity, and better creep resistance.

The pure polymer, polytetrafluoroethylene (PTFE), is intractable and must be formed by powder metallurgy methods, cold extrusion (like lead), or latex processing methods. The polymerization methods result in three forms of the polymer:

1. Granular polymer (suspension polymerization)
2. Fine powder polymer (emulsion polymerization)
3. Dispersion polymer (emulsion polymerization)

Table 1.3 shows the type of polymer, method of processing, and uses of each resin. The fluoroplastic copolymers are

Polytetrafluoroethylene (PTFE)

Fluorinated ethylene-propylene polymer (FEP)

Perfluoroalkoxy resin (PFA)

Ethylene-tetrafluoroethylene copolymer (ETFE)

Polyvinylidene fluoride (PVDF)

Polychlorotrifluoroethylene copolymer (PCTFE)

Ethylene-chlorotrifluoroethylene copolymer (ECTFE)

Polyvinyl fluoride (PVF)

Property values for these polymers are given in Table 1.4. Like with most materials, the electric strength of PTFE is dependent on the thickness of the sample, and this is shown in Fig. 1.6. The electrical properties of PTFE do not change up to 400°F. There is some effect of frequency on the electric strength, as shown in Fig. 1.7, but the dielectric losses are little changed.

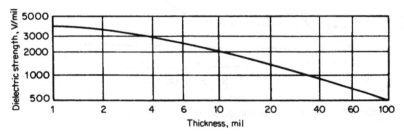

FIGURE 1.6 Dielectric strength of PTFE as a function of thickness.

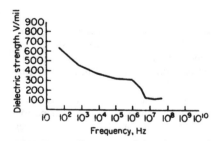

FIGURE 1.7 Short-term dielectric strength, in oil, of PTFE fluorocarbon versus frequency. Breakdown path 1/16 in; electrode diameter 3/4 in; radius 1/8 in; 40-s test.

TABLE 1.3 Polyetetrafluoroethylene Resins and Their Uses

Type	Resin grade	Method of processing	Description	Main uses
Granular	Agglomerates	Molding, preforming sintering, ram extrusion	Free-flowing powder	Gaskets, packing, seals, electronic components, bearings, sheet, rod, heavy-wall tubing; tape and molded shapes for chemical, electrical, mechanical, and nonadhesive (adherent) applications
	Coarse	Molding, preforming sintering	Granulated powder	Tape, molded shapes for chemical, mechanical, electrical, and nonadhesive applications
	Finely divided	Molding, preforming sintering	Powder for use where highest quality, void-free moldings are required	Molded sheets, tape, wire wrapping, tubing, gasketing
	Presintered	Ram extrusion	Granular, free-flowing powder	Rods and tubes
Fine powder	High-reduction-ratio resin	Paste extrusion	Agglomerated powder	Wire coating, thin-walled tubing
	Medium-reduction-ratio resin	Paste extrusion	Agglomerated powder	Tubing, pipe, overbraided hose, spaghetti tubing
	Low-reduction-ratio resin	Paste extrusion	Agglomerated powder	Thread-sealant, tape, pipe liners, tubing, porous structures
Dispersion	General purpose	Dip coating	Aqueous dispersion	Impregnation, coating, packing
	Coating	Dip coating	Aqueous dispersion	Film, coating
	Stabilized	Coagulation	Aqueous dispersion	Bearings

Source: From Kirk and Othmer.[5]

TABLE 1.4 Properties of Fluoropolymers

	PTFE	FEP	PFA	ETFE	PVDF	PCTFE	ECTFE	PVF
% fluorine	76.0	76.0	76.0	59.4	59.4	48.9	39.4	41.3
Melting point, °C	327	265	310	270	160	218	245	200
Upper use temperature, °C	260	200	260	180	150	204	170	110
Density, g/cm^3	2.13	2.15	2.12	1.7	1.78	2.13	1.68	1.38
Oxygen index, %	>95	95	95	28–32	44	—	48–64	—
Arc resistance, seconds	>240	>300	>300	72	60	2.4	18	—
Dielectric constant	2.1	2.1	2.1	2.6	9–10	2.5	2.5	9
Dissipation factor	0.0002	0.0002	0.0002	0.0008	0.02–0.02	0.02	0.003	0.002
Tensile strength, lb/in^2	5000	3100	4300	7000	6200	6000	—	—
Specific gravity	2.2	2.17	2.17	1.7	1.78	2.2	—	—
Water absorption, % 24 h	0	0	0.03	0.03	0.06	0	—	—
Electrical strength, V/mil	480	600	500	400	280	600	—	—

Source: From *Modern Plastics Encyclopedia.*[1]

DuPont has another material, Teflon AF,* which unlike other fluoropolymers, is totally amorphous. It is totally clear, useful to 300°C, and has a high coefficient of friction and a dielectric constant of 1.89 at 10 GHz.

Because the fluoropolymers are so useful in electronics, there are many applications in the military sector, and a large number of federal specifications for fluoronated plastics exist. These are well described in the *Handbook of Electrical and Electronic Insulating Materials.*[4]

Fluoropolymers are used in electrical, mechanical, and chemical applications. They are produced as films, moldings, coatings, wire and cable insulators, release films, tubing, microfibers, sleeving, yarn, flexible circuits, and tapes.

1.2.2.3 Liquid Crystal Polymers.

Liquid crystal polymers (LCPs) are new polymers exhibiting high-temperature resistance, resistance to chemicals, low thermal expansion, high physical strength, and an unusual ability to be molded into thin sections. The name is derived from the ability of the polymer to retain a crystalline structure both in melt and in solid form. The molecules are in long rigid fibrous form and orient in flow to produce parts that are in effect self-reinforced. This gives parts high tensile strength, excellent creep resistance, and high radiation resistance. Typical properties for glass-filled LCPs are shown in Table 1.5. The coefficient of thermal expansion is very low and close to that of glass, as shown in Table 1.6. Modern surface mount technology (SMT) utilizes vapor-phase or infrared soldering techniques, which put an excessive thermal burden on components. Figure 1.8 compares LCPs with other materials relative to cost, performance, and soldering method. LCPs are considered nonburning with an oxygen index of 50 and a UL-94 rating of V-0.

The low melt viscosity permits them to be molded into intricate shapes with cross sections of as little as 10 mil. Warpage control and low mold shrinkage are exceptional. LCPs are used as chip carriers, sockets, connectors, bobbins, and relay cases.

1.2.2.4 Nylons.

Nylon (polyamide) polymers have excellent mechanical strength and are tough and abrasion-resistant. All nylons absorb moisture to equilibrium with the surrounding humidity. This moisture absorption plasticizes the polymer; it increases the flexural strength and impact strength while decreasing the resistivity, dielectric strength, and dielectric properties. Figure 1.9 shows the increase in the dielectric constant with humidity. Typical properties for nylon 6,6 are given in Table 1.5.

Nylons are processed with all the typical thermoplastic processing methods. Much nylon is extruded into fibers for textiles. Nylon based on lactams can be cast from the monomer into large shapes. Nylons are available as monopolymers, copolymers, and reinforced compounds. Molded nylons are used as coil bobbins, wire ties, connectors, and wire jackets.

Applicable military standards are:

LP-395	Plastic molding material, nylon, glass-film reinforced
MIL-P-46180	Plastic molding material, polyamide, glass-film reinforced
MIL-P-46101	Plastic molding material, polyamide

1.2.2.5 Polycarbonates.

Polycarbonate plastics are flame-retardant, high in impact strength, clear, and temperature-resistant.

---

*Teflon AF is a registered trademark of E. I. du Pont de Nemours & Company, Inc.

TABLE 1.5 Properties of Thermoplastic Plastics

	Glass-filled liquid crystal	Nylon 6,6	Polycarbonate	Polyesters			Imides		
				Polyarylate	Polybutylene terephthalate	Polyethylene terephthalate	Polyamide-imide	Polyetherimide	Polyimide
Dielectric constant, D-150									
60 Hz	4.3	4.0	3.17	3.08	3.3	3.8	3.5	3.0	3.43
10^6 Hz	3.9	3.6	3.05	3.13	3.1	3.6	3.5	3.15	3.42
Dissipation factor, D-150									
60 Hz	0.010	0.14	·0.009	0.002	0.002	0.0059	—	—	0.003
10^6 Hz	0.006	0.04	0.01	0.02	0.03	0.0016	0.02	0.0013	0.0116
Dielectric strength, D-149; V/mil	870	385	425	610	420	650	600	831	560
Volume resistivity, D-257; $\Omega \cdot$ cm	10^{15}	10^{14}	8.2×10^{16}	2×10^{14}	1.4×10^{15}	10^{15}	1.2×10^{17}	6.7×10^{17}	10^{16}
Arc resistance, D-495; seconds	180	120	120	125	190	123	125	126	230
Specific gravity, D-792	1.72	1.14	1.22	1.20	1.31	1.39	1.4	1.29	1.43
Water absorption, D-570; % 24 h	0.01	1.2	0.15	0.2	0.09	0.2	0.28	0.25	0.47
Heat deflection temperature, D-648; at 264 lb/ft², °F	510	194	288	320	130	106	525	420	680
Tensile strength, D-638; lb/in²	29,000	11,800	9000	10,000	8500	10,500	26,900	15,200	18,000
Impact strength (Izod), D-256; ft · lb/in	2.5	2	18	5.5	1.0	0.65	2.5	1.0	1.0
Coefficient of thermal expansion, D-696; 10^{-5}/°F	0.3	4.5	3.75	4.0	4.0	3.8	2.0	3.1	2.8
Thermal conductivity, C-477; Btu · in/(h · ft² · °F)	—	1.7	1.41	1.47	1.1	3.6	1.7	0.125	6.8

	Glass-filled PEEK	Polyphenylene oxide	Polyphenylene sulfide	Poly-sulfone	Polyaryl-sulfone	Polyether-sulfone	Polyphenylene-sulfone
Dielectric constant, D-150							
60 Hz	—	2.6	3.1	3.06	3.51	3.5	3.44
10^6 Hz	—	2.6	3.2	3.03	3.54	3.5	3.45
Dissipation factor, D-150							
60 Hz	—	0.0004	0.0004	0.0008	0.0017	0.001	0.00058
10^6 Hz	—	0.0009	0.0007	0.0034	0.0056	0.006	0.00764
Dielectric strength, D-149; V/mil	400	700	520	425	383	400	371
Volume resistivity, D-257; $\Omega \cdot$ cm	10^{16}	10^{17}	3×10^{15}	5×10^{16}	7.7×10^{16}	10^{18}	3.5×10^{15}
Arc resistance, D-495; seconds	—	75	200	122	81	116	41
Specific gravity, D-792	1.46	1.06	1.8	1.24	1.37	1.37	1.29
Water absorption, D-570; % 24 h	0.12	0.07	0.03	0.22	0.10	0.43	—
Heat deflection temperature, D-648; at 264 lb/ft², °F	550	265	500	345	400	400	400
Tensile strength, D-638; lb/in²	23,000	9600	13,000	10,200	12,000	12,200	10,400
Impact strength (Izod), D-256; ft · lb/in	1.6	7.0	1.0	1.3	1.6	1.6	1.2
Coefficient of thermal expansion, D-696; 10^{-5}/°F	1.45	4.4	2.0	3.1	2.7	3.1	3.1
Thermal conductivity, C-477; Btu · in/(h · ft² · °F)	—	1.5	—	1.8	—	1.8	2.0

TABLE 1.6 Coefficients of Thermal Expansion

Material	10^{-5} (in/in)/°F	10^{-5} (cm/cm)/°C
Liquid crystal (GR)*	0.3	0.6
Glass	0.4	0.7
Steel	0.6	1.1
Concrete	0.8	1.4
Copper	0.9	1.6
Bronze	1.0	1.8
Brass	1.0	1.8
Aluminum	1.2	2.2
Polycarbonate (GR)	1.2	2.2
Nylon (GR)	1.3	2.3
TP polyester (GR)	1.4	2.5
Magnesium	1.4	2.5
Zinc	1.7	3.1
ABS (GR)	1.7	3.1
Polypropylene	1.8	3.2
Epoxy (GR)	2.0	3.6
Polyphenylene sulfide (GR)	2.0	3.6
Acetal (GR)	2.2	4.0
Epoxy	3.0	5.4
Polycarbonate	3.6	6.5
Acrylic	3.8	6.8
ABS	4.0	7.2
Nylon	4.5	8.1
Acetal	4.8	8.5
Polypropylene	4.8	8.6
IP polyester	6.9	12.4
Polyethylene	7.2	13.0

*GR—Glass-fiber reinforced.
Source: From Tortolano.[6]

All thermoplastic fabrication methods are used with polycarbonates. Materials are available with glass fiber, mineral filler, and lubricating compounds. Typical properties for a general-purpose grade of polycarbonate are given in Table 1.5. The impact strength of polycarbonates is exceptional, ranking among the highest of all plastics. Figure 1.10 shows the drop-weight impact resistance of polycarbonate and other plastics (from ASTM D-3029[8]), and Fig. 1.11 shows the creep resistance. These polymers do absorb water, and highly stressed parts are sensitive to many solvents and chemicals. Military purchase specification LP-393 applies to the material. Electronics designers use them as bobbins, connectors, and terminal boards.

1.2.2.6 Polyesters and Polyarylates. Thermoplastic polyesters include polyarylates (PA), polybutylene terephthalates (PBT), polyethylene terephthalates (PET), and copolymers of these.

The polymers are characterized by stable electrical properties over the range of temperatures and frequencies, clarity, high arc resistance, and low dielectric constant. Typical properties of these compounds are given in Table 1.5. The dielectric strength of PBT with temperature is compared in Fig. 1.12 to that of other plastics. They are manufactured with

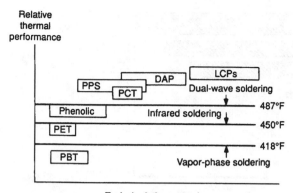

FIGURE 1.8 Minimum heat requirments for soldering methods. (*From Rose.*[7])

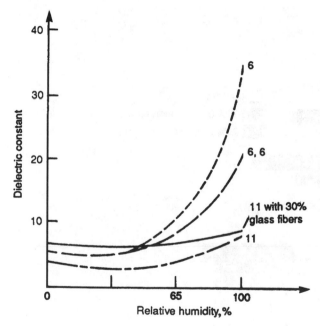

FIGURE 1.9 Dielectric constant of nylons as a function of humidity at 30 kV. (*From Shugg.*[4])

all normal thermoplastic molding processes. The melt viscosity is very low, permitting thin-walled parts.

Polyarylates are used to mold three-dimensional circuit boards in a unique two-shot molding process incorporating plated-up conductor paths.[10] Typical electronic applications include insulating film, connectors, terminal blocks, and fuse holders.

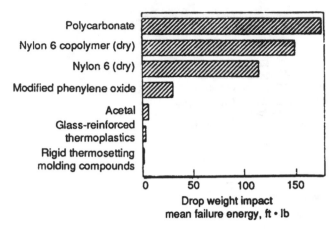

FIGURE 1.10   Impact resistance of polycarbonate and other plastics. (*From ASTM D-3029.*[8])

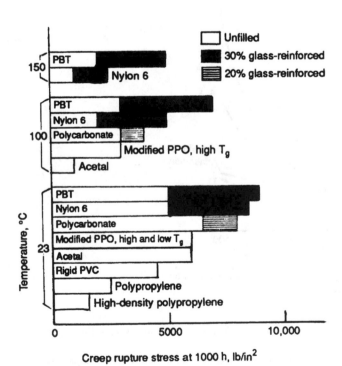

FIGURE 1.11   Creep strength of polycarbonate. (*From Modern Plastics Encyclopedia.*[1])

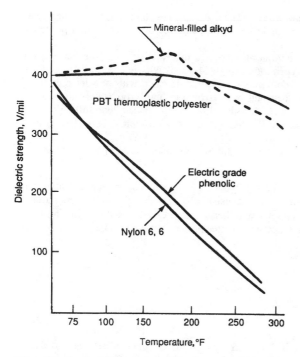

FIGURE 1.12   Dielectric strength of PBT at 50% relative humidity, 60 Hz. (*From ASTM D-149.*[9])

1.2.2.7 Polyetherimides.

This polymer has low dielectric losses over a wide temperature range, good mechanical strength, and is easily molded into complex shapes.

The aromatic structure of polyetherimide gives it high radiation resistance with less than 6 percent loss in tensile strength after 500-Mrad exposure to cobalt 60 at 1 Mrad/h. It withstands boiling water well. The resistivity of polyetherimide is very high. Table 1.5 gives the properties of various imide-based polymers.

The polymer is available as plain polymer or with fiberglass reinforcement. It is processed in normal thermoplastic manufacturing equipment.

Electronic uses include circuit-breaker housings, chip carriers, pin connectors, bobbins, fuse blocks, and circuit board insulators which can be vapor-phase soldered. The heat resistance makes it useful in automotive engine components. It can be electroless plated.

1.2.2.8 Polyethylene and Polypropylene.

Known as olefins, the polymers polyethylene and polypropylene and their copolymers are inert and abrasion resistant, with excellent electrical properties.

Olefins are the lightest of all plastics. The specific gravity of polypropylene is 0.91 and for polyethylene it varies from 0.92 to 0.96. The dielectric constant for each is 2.3, and the dissipation factor is 0.0002 at 60 Hz. Water absorption is extremely low.

The chief electronic use for these olefins is for wire and cable, both as primary insulation and as a jacket. Polypropylene film is used as a capacitor dielectric. Molded polypropylene is used for fuse blocks and battery cases. Both polymers burn readily.

1.2.2.9 Polyimide and Polyamide-Imide.   These plastics are characterized by their resistance to elevated temperatures, low dielectric losses, and wear resistance. Their properties are shown in Table 1.5.

Polyimides have very high radiation resistance at fluxes to 10^9 rad. They have a low coefficient of expansion. The oxygen index is 44. The resistivity is extremely high. Polyimides are hard to mold and are compression molded or made with powder metallurgy techniques. They are chemically resistant but dissolve in highly polar solvents and are attacked by hot caustic. They are available as an alloy with polyphenylene sulfide.

Polyamide-imides can be injection molded but require a postcure. Their oxygen index is 43. They are very sensitive to steam.

These plastics are used for connectors, circuit boards, radomes, and films.

1.2.2.10 Polyetheretherketones.   These polymers have good properties at elevated temperatures, excellent electrical properties, and resist burning with little smoke generation.

The most common polymer is polyetheretherketone (PEEK), but the following polymers, which relate to the ether and ketone linkages present, are also available:

Polyketone (PK)

Polyetherketone (PEK)

Polyetherketoneketone (PEKK)

Polyetheretherketoneketone (PEEKK)

Polyaryletherketone (PAEK)

Properties of PEEK are given in Table 1.5 for a 20 percent glass-filled material. These polymers can be used continuously at 500°F. They are stronger, stiffer, and tougher than fluoroplastics and have very low creep and a flexural modulus of 2×10^6. They withstand most chemicals and are highly steam resistant. They are manufactured on standard thermoplastic machines and are used as wire and cable insulation and connectors.

1.2.2.11 Polyphenylene Oxide.   Polyphenylene oxide (PPO) resins have high heat resistance and low water absorption. Many grades exist, with heat distortion temperatures ranging from 180 to 317°F. Table 1.5 gives properties of the PPO polymer. Federal purchases are made under MIL-P-46115.

All conventional thermoplastic equipment can be used to process PPO. It is used in computers, connectors, fuse blocks, relays, and as bus bar insulation.

1.2.2.12 Polyphenylene Sulfide.   This inherently flame-retardant plastic has good resistance to chemicals and high tolerance for heat. The fluid resin easily wets glass fibers, permitting unusually high loading of reinforcements. No solvents exist for polyphenylene sulfide (PPS), and very few materials will react with it even at high temperatures. Low dielectric losses are attractive. Table 1.5 gives properties of a glass-filled compound.

It is molded like all thermoplastics. One special grade is available for encapsulating integrated circuits. PPS is used in connectors, sockets, and microwave components.

1.2.2.13 Polystyrene.   This polymer has very low dielectric losses. It is attacked by many chemicals and solvents and has a low resistance to temperature. Its chief use in electronics is in striplines, where its dielectric constant of 2.45 and dissipation factor of 0.0001 make it unique.

1.2.2.14 Polysulfone, Polyarylsulfone, Polyethersulfone, and Polyphenylenesulfone.
Sulfones are highly temperature-resistant with excellent electrical properties. These polymers have high heat distortion temperatures. All are highly creep resistant. Flexural properties are little affected from 30 to 350°F. They are very hydrolytically stable and among the best of all polymers in radiation resistance. Properties are shown in Table 1.5. They conform to MIL-P-461204 and MIL-P-46185.

These polymers are easily extruded and injection molded. Electroless plating grades are available. They are used as circuit board substrates, television components, and multipin connectors.

1.2.2.15 Polyvinyl Chloride. Polyvinyl chloride (PVC) is a flexible polymer with resistance to most chemicals. It must be compounded with plasticizers to make the properties useful. It can be used only at modest temperatures. The chief applications in electronics are as low-temperature wire insulation, cable jackets, tubing, and sleeving.

1.2.3　Elastomers

The original elastomer is natural rubber (chemically 95 percent *cis*-1,4polyisoprene), obtained as the sap from rubber trees. As polymer chemists evolved more and more polymers resembling natural rubber in properties, the term elastomer has grown to represent these materials, rubber being reserved for its original use.

ASTM D-1566[11] defines an elastomer as a "macromolecular material that returns rapidly to approximately the initial dimensions and shape after substantial deformation by a weak stress and release of the stress." Such elongations typically exceed 100 percent. Elastomers behave differently than other polymers, and their elasticity makes certain properties of critical importance when compared to the more rigid polymers.

1.2.3.1 Aging. Elastomers are affected by the environment more than other polymers. Thermal aging usually increases cross-linking and the chain length of the elastomer, thus increasing stiffness and hardness and decreasing elongation. Radiation has a similar effect.

Elastomers are sensitive to oxidation and in particular to the effects of ozone. Ultraviolet radiation acts similarly to ionizing radiation, so some elastomers do not weather well. Environmental effects are especially noted on highly stressed parts. Some elastomers are particularly affected by hydrolysis. Complete chemical reversion has been experienced where the polymer depolymerizes back to a liquid state.

1.2.3.2 Creep. Creep of elastomers refers to a change in strain when stress is held constant. Special terms are used for elastomers. Compression set (ASTM D-395'2) is creep that occurs when the elastomer has been held at either constant strain or constant stress in compression. Constant strain is most common and is recorded as a percent of permanent creep divided by original strain. Strain of 25 percent is common. Permanent set is deformation remaining after a stress is released.

1.2.3.3 Hardness. The hardness of elastomers is a measure of the resistance to deformation measured by pressing an instrument into the elastomer surface. Special instruments have been developed, the most common being the Shore durometer. Figure 1.13 shows the hardness of elastomers and plastics. On the Shore A scale, 0 is soft and 100 is hard. On this scale, a rubber band is 27, and an auto tire is 70.

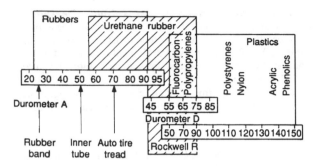

FIGURE 1.13 Hardness of rubber and plastics.

1.2.3.4 Hysteresis. Hysteresis is energy loss per loading cycle. This mechanical loss of energy is converted into heat in elastomers and is caused by internal friction of the molecule chains moving against each other. The effect causes a heat buildup in the elastomer, increasing its temperature, changing its properties, and aging it. A similar electrical effect can occur at high frequencies when the dissipation factor of the elastomer is high.

1.2.3.5 Low-Temperature Properties. As temperatures are decreased, elastomers tend to become stiffer and harder. Each material exhibits a stiffening range and a brittle point at the glass transition temperature. These effects are usually time dependent.

1.2.3.6 Tear Resistance. This is a measure of the stress needed to continue rupturing a sheet elastomer after an initiating cut or notch. Elastomers vary widely in their ability to withstand tearing.

1.2.3.7 Tensile Strength, Elongation, and Modulus. In tension metals behave in accordance with Hooke's law, and in strain they react linearly to the yield point. Polymers and plastics (unreinforced) deviate somewhat from linearity (logarithmically). Special tensile tests are used for elastomers per ASTM D-412.[13] Elastomers are not generally designed for tensile service, but many other physical properties of the elastomers correlate with tensile strength.

1.2.3.8 Types of Elastomers. A large number of chemically different elastomers exist. ASTM D-1418[14] describes many of these. Table 1.7 lists elastomers and their properties.

   Natural Rubber (NR). NR is still used in many applications. It is not one uniform product but varies with the nature of the plant producing the sap, the weather, the locale, the care in producing the elastomer, and many other factors. A variant of NR, gutta-pereha, was used in most of the early electrical products, especially cable. It has excellent electrical properties, as shown in Table 1.8, low creep, and high tear strength. On aging it reverts to the gum.

   Isoprene Rubber (IR). IR is similar in chemistry to NR, but it is produced synthetically. Polyisoprene constitutes 97 percent of its composition. It is more consistent and much easier to process than NR.

   Acrylic Elastomer (ABR). The heat resistance of ABR is almost as good as that of fluorinated compounds and silicones. It also ages well but is sensitive to water. Its chief use is in contact with oils.

TABLE 1.7 Chemical Description of Elastomers

ASTM D-1418	Chemical type	Properties
NR	Natural rubber polyisoprene	Excellent physical properties
IR	Polyisoprene synthetic	Same as NR, but more consistency and better water resistance
ABR	Acrylate butadiene	Mechanical elastomer; excellent heat and ozone resistance
BR	Polybutadiene	Copolymerizes with NR and SBR; abrasion resistance
CO	Epichlorohydrin	Chemical resistance
COX	Butadiene-acrylonitrite	Used with NBR to improve low-temperature performance
CR	Chloroprene, neoprene	Withstands weathering; flame-retardant; chemical resistance
CSM	Chlorosulfonated polyethylene	Colors available; weathering and chemical resistance; poor electrical properties
EPDM	Ethylene-propylene terpolymer	Similar to EPM; good electrical properties; resists water and steam
EPM	Ethylene-propylene copolymer	Similar to EPDM: good heat resistance; wire insulation
FPM	Fluorinated copolymers	Outstanding heat and chemical resistance
IIR	Isobutylene-isoprene, butyl	Outstanding weather resistance; low physical properties, track resistance
NBR	Butadiene-acrylonitrite, nitrile, Buna N	General-purpose elastomer; poor electrical properties
PVC/NBR	Polyvinyl chloride and NBR	Colors available; weather, chemical, and ozone resistance
SBR	Styrene: butadiene, GRS, Buna S	General-purpose elastomer; good physical properties; poor oil and weather resistance
SI (FS1, PS1, VS1, PVS1)	Silicone copolymers	Outstanding at high and low temperatures; arc- and track-resistant; resist weather and ozone; excellent electrical properties, poor physical properties
T	Polysulfide	Excellent weather resistance and solvent resistance
U	Polyurethane	High physical and electrical properties

TABLE 1.8 Electrical Properties of Elastomers

ASTM elastomer	Dielectric strength, V/mil	Dissipation factor tan δ	Dielectric constant	Volume resistivity, $\Omega \cdot$ cm
COX	500	0.05	10	10^{15}
CR	700	0.03	8	10^{11}
CSM	700	0.07	8	10^{14}
EPDM	800	0.007	3.5	10^{16}
FPM	700	0.04	18	10^{13}
IIR	600	0.003	2.4	10^{17}
NR	800	0.0025	3	10^{16}
SBR	800	0.003	3.5	10^{15}
SI	700	0.001	3.6	10^{15}
T	700	0.005	9.5	10^{12}
U	500	0.03	5	10^{12}

Butadiene Elastomer (BR).   This elastomer is used to copolymerize with SBR and NR in tire stocks.

Epichlorohydrin Elastomer (CO, ECO).   These elastomers are flame-retardant because of the presence of chlorine. Their electrical properties are modest, but they age well and resist most chemicals. Dissipation factors are high.

Carboxylic Elastomer (COX).   These elastomers have good low-temperature performance, excellent weather resistance, and extremely good wear resistance. Electrical properties are average.

Neoprene (CR).   Neoprene (chloroprene) was the first synthetic elastomer and is widely used in industry. It is nonflammable and resists ozone, weather, chemicals, and radiation. However, it is highly polar and has a high dissipation factor and dielectric constant.

Chlorosulfonated Polyethylene (CSM).   This elastomer is similar to CR, with some improvement in electrical properties and with better heat resistance. It is available in colors and often used in high-voltage applications.

Ethylene-Propylene Terpolymer (EPDM).   This elastomer is synthesized from ethylene, propylene, and a third monomer, a diene. The diene permits conventional sulfur polymerization. The elastomer is exceptionally resistant to radiation and heat. The glass transition temperature is –60°C. Electrical properties are good.

Ethylene-Propylene Copolymer (EPM).   Often used as a wire insulation, it is being replaced by EPDM because processing qualities are somewhat inferior to those of EPDM.

Fluorinated Elastomers (FPM).   There are several fluorinated polymers—fluorocarbons, fluorosilicones, and fluoroalkoxy phosphazenes. The elastomers can be used to 600°F, do not burn, are unaffected by most chemicals, and have excellent electrical properties. In thermal stability and aging, only the silicones are better. Physical properties are high, but so is the cost.

Butyl Rubber (IIR).   This elastomer is highly impermeable to water vapor. Its nonpolar nature gives it good electrical properties. Compounded with aluminum oxide trihydrate, it has exceptional arc and track resistance. Butyl has good aging characteristics and good flexibility at low temperatures.

Nitrile Rubber (NBR).   This elastomer is resistant to most chemicals, but its polarity gives it poor electrical properties, so its major use is in mechanical applications.

Polyvinyl Chloride Copolymers (PVC/NBR).   Similar to NBR, it can be colored and is used in wire and cable jackets.

GRS (SBR).   GRS stands for government rubber, styrene, a nomenclature derived during World War II when natural rubber was not available in the West. It is used in mechanical applications.

Silicone Elastomers (SI).   Composed of silicon and oxygen atom backbones, the silicones have the highest temperature ability (600°F), a large temperature range (−100 to 600°F), and excellent electrical properties. They do not burn and are arc resistant. Physical properties are modest.

Polysulfides (T).   This elastomer weathers best of all and is highly chemical resistant. Dissipation factors are excellent (as low as 0.001). Physical properties are modest.

Polyurethanes (U).   These elastomers are either ester- or ether-based. Ester-based elastomers are poor in water resistance. They are excellent in electrical applications with outstanding physical properties. Abrasion resistance is particularly high. They become stiff at low temperatures. They can be compounded like regular elastomers, used as cast elastomers, or injection molded like thermoplastics.

1.2.4   Plastics Processing

There are many plastics manufacturing processes, and often the designer has some freedom in selecting how an electronic component may be made. Some processes, such as embedding, may be done at ambient conditions, but in general both high temperatures and high pressures are a part of the processes. This is because most processes consist of melting the polymer, causing it to flow, and then freezing the polymer in its desired form. Generally the viscosity of the liquid polymer is so high that considerable pressure is required to force it into its final configuration. Thermosetting polymers "freeze" by a chemical reaction converting the liquid to a solid, while thermoplastic polymers truly freeze by a phase change from liquid to solid. These fundamental differences in polymers dictate the plastic processing method used.

Most of the plastics processing and manufacturing methods are only of passing interest to the electronics designer, because they are performed by special manufacturers with the proper facilities. Only the largest of the electronic manufacturing plants can afford the substantial investment in equipment needed. However, the processes of embedding and conformal coating are usually done at the circuit level, and more emphasis on these processes is given in this handbook.

1.2.4.1   Compression and Transfer Molding.   These are two separate processes which can often be accomplished in the same molding press but with different molds and processing parameters.

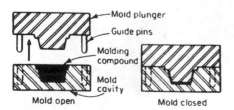

Mold open   Mold closed

FIGURE 1.14   Compression molding.

In compression molding, thermoset materials are placed into a hot mold. The materials may be free-flowing granules, puttylike, or preshaped into pieces (preforms). The mold is closed by a hydraulic press, softening, liquefying, and flowing the material throughout the mold. The process is illustrated in Fig. 1.14. After the material polymerizes (cures), the mold is opened, the part is pushed from the mold with ejector pins, and the cycle repeats. Typical conditions are mold temperature 350°F, mold pressure 100 lb/in², and cure time 3 min. Since the plastic is chemically converted, scrap cannot be reused.

Compression molded parts are low cost, and high volumes are possible. Molds must withstand high pressures and temperatures and are expensive. Therefore production volumes must be large enough to amortize the mold investment.

Transfer molding is also used for thermosetting materials and is similar to compression molding. A transfer press is shown in Fig. 1.15. In this process the same materials as those used in compression are placed into a heated cavity (pot), usually on top of the mold. The hot mold is closed, and the transfer ram pushes the liquefying material down into the mold, along channels called runners, through a constriction to provide back pressure called a gate, and into the mold, which shapes the part, as shown in Fig. 1.16.

The great advantage of the transfer process is that the dimensions of the part are better controlled, and inserts in the mold can be positioned exactly in place. The process is often used to embed electronic components and is fully described in Chap. 6. Transfer molded parts can be low cost, and high volumes are possible. Transfer molds are even more expensive than compression molds because of their increased complexity, and high throughput is needed to amortize the mold cost.

1.2.4.2 Injection Molding.   Injection molding was developed for thermoplastic materials. Plastic, in the form of small pellets, falls from a feed hopper into a screw which forces the material forward through several heating zones, as shown in Fig. 1.17. As the plastic is transported, it is heated to a highly viscous liquid state, arriving at the front zone fully melted, de-aired, dried, and compressed to up to 20,000 lb/in². The flights of the screw vary in each zone. When enough melt has accumulated, the screw is forced forward hydraulically as a piston, pushing the melt through a nozzle, down runners to the mold, much as in transfer molding. This mold is cooled—usually with chilled water—to freeze

FIGURE 1.15   Compression-transfer molding press. *(Courtesy of Hull Corp., Hatbro, PA.)*

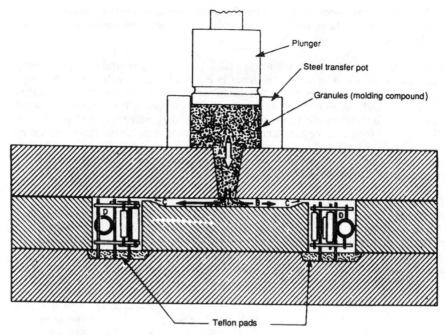

FIGURE 1.16 Transfer or plunger molding. A—sprue; B—runner; C—gate; D—cavity with module to be embedded.

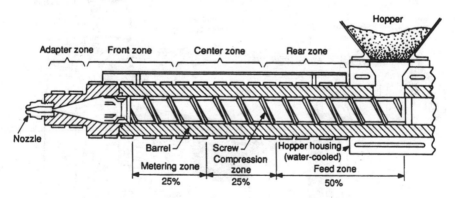

FIGURE 1.17 Screw section, injection molding machine.

the plastic as quickly as possible. The removal of heat from the plastic is the time-limiting factor, which determines the overall cycle time, from seconds to minutes. The thermoplastic molding materials are not altered substantially by this process, so sprues, runners, and scrap parts can be reground and blended with virgin material for reuse. Injection molded parts are of extremely low cost because of the automated nature of the process and the very low cycle time. However, the molds are extremely complex, must withstand very high pressures, and are very costly. Production must be considerable to justify the mold cost and setup expense.

If the screws are modified and the molds heated instead of cooled, certain thermoset materials can also be fabricated this way at reduced cost and with improved dimensions, but this scrap is not reusable.

1.2.4.3 Extrusion. Extrusion is a thermoplastic process very similar to injection molding. A cutaway view of an extruder is shown in Fig. 1.18. Plastic pellets from the hopper are pushed forward by a screw through several heating zones where they melt, dehumidify, are de-aired, and are compressed. An extrusion die is placed at the extruder end, which forms the resin into the correct cross section. The melted extrudate is cooled quickly, usually in a water bath, forming a continuous strand, which is removed on a conveyor and cut into the desired lengths. As in all thermoplastic processes, scrap is reused. When the extruder is equipped with the appropriate die heads, the process can be used to extrude jackets and insulation onto wires and cables.

1.2.4.4 Thermoforming. Thermoforming is also called *vacuum forming* or *drape forming*. It makes use of the fact that most thermoplastic resins do not have a well defined melting point but rather a broad range of temperatures wherein the polymer softens to a rubberlike consistency before true melt occurs. In thermoforming (Fig. 1.19) a plastic extruded sheet is clamped into a form and held above a room-temperature mold. The mold has air passages drilled through it. A movable heater is positioned above the sheet and

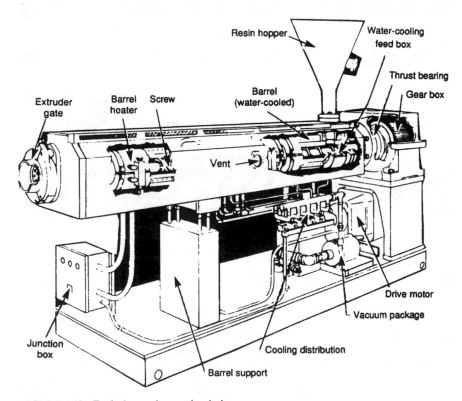

FIGURE 1.18 Typical extruder, sectional view.

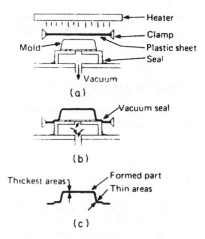

(a)

(b)

(c)

FIGURE 1.19　Thermoforming.

radiant heat warms and softens the sheet (Fig. 1.19a). At the proper moment the sheet is forced downward onto the mold and a vacuum is applied to the underside of the mold while the heater is removed (Fig. 1.19b). The sheet is held conformed to the surface of the mold until it cools. The formed part is then removed and the clamped edges are trimmed (Fig. 1.19c). Scrap can be returned to the extruder.

Molds are very inexpensive since they must withstand only 1 atm of pressure and in fact are often made of wood. The process is rapid, but tolerances are poor and dimensions highly variable, particularly at corners. The process lacks the throughput ability of injection molding, except for automated lines for small parts.

1.2.4.5 Laminating. To laminate is to press into a thin sheet. This process is used mostly to manufacture reinforced thermoset plastics, but some thermoplastics can also be laminated. Laminates in the electronics industry generally refer to sheets made from several layers of paper, fabric, or fiber bonded together with a resin. Specific laminates used for printed circuit board substrates are discussed in Chap. 8. Laminating is a special case of compression molding. Laminates are made by impregnating a web (such as paper or fabric) with a liquid resin in a solvent, heating this in an oven to remove the solvent and partially polymerize the resin, and cutting the sheets to size. The sheets are then stacked together and molded between pressing plates in compression presses with many heated shelves or openings. Pressure is applied by closing the press, and the sheets are cured at conditions such as 500 lb/in^2, at 350°F for 1 h. The process is inexpensive, with high throughput, and the product is made to close tolerances. While the molds (pressing plates) are not expensive, the entire line of equipment is capital intensive.

1.2.4.6 Pultrusion. Pultrusion is similar to extrusion, except that the product is pulled through the process rather than pushed. It is used to produce thermoset fiber-reinforced products. Continuous glass fibers (often combined with mat or fabric) are pulled from the roving feed packages and through a resin impregnating tank where the liquid resin saturates the fiber bundles. The wet fibers pass through a preformer which orients the fibers and from there into a heated die with the desired cross section. Here the resin cures. The product is removed by the puller, which looks like opposing tractor treads, and a cutoff wheel cuts the sections into the desired lengths. The process operates continuously with little supervision and is of low cost.

Fiber orientation is primarily in the direction of travel, and crosswise strengths are reduced. Shapes are limited and large sizes may not be possible. Polyester resins are the easiest to use in pultrusion, but some manufacturers offer epoxies and silicones.

1.2.4.7 Reaction Injection Molding (RIM). RIM is a type of injection molding used to produce thermoset shapes, usually of large sizes. Polyurethane and polyester resins are used. Reactive liquid components are prepared separately and then pumped into a mixing head where they are thoroughly combined. The mixed liquid resins are then forced into a heated mold which may also contain fiber-mat reinforcement. The part cures like a compression molded part. Advantages are low cost and large parts. Disadvantages are the limited choice of materials, the rather low-performance physical properties, and the difficulty of controlling the reaction process. Electronics designers find enclosures and cabinets can be RIM processed.

1.2.5 Casting, Embedding, Encapsulation, Impregnation, and Potting

Electronic designers must often protect and strengthen the components of an electronic circuit and usually choose one of these methods. The terms tend to be used interchangeably but there are differences.

1.2.5.1 Definitions

Casting.   Forming a plastic part by pouring a liquid resin system into a mold, curing the part, and removing it for use.

Embedding.   Surrounding a part or component positioned in a mold with a liquid resin system, curing the resin, and removing it for use.

Encapsulation.   Completely enclosing a component in a coating of a viscous resin system by dipping, spraying, or embedding with or without a mold. The coated component is usually irregular in shape.

Impregnation.   Filling the interstices of an electrical component (coil, fabric, winding) with a low-viscosity resin system to consolidate the component.

Potting.   Similar to embedding, except the mold (can or case) remains a part of the system.

Transfer Molding.   A special type of embedding, which is fully covered in Chap. 6.

These processes are useful to the electronics designer for several reasons.

1. *Environmental protection.*   Critical components are isolated from water and chemicals. They can better withstand gases and humidity. Sunlight is shielded, and thermal cycling becomes easier.

2. *Electrical integrity.*   The materials used in these processes are excellent insulators. They provide breakdown resistance and decrease losses, and they add arc and track resistance. They increase surface and volume resistivity, and spacing can be reduced.

3. *Mechanical integrity.*   These processes ensure that components do not move. Contaminants are isolated and immobilized; abrasion resistance, fatigue, and creep are enhanced. The package becomes more rugged, stronger, abrasion-resistant, and better resists impact and handling.

1.2.5.2 Important Properties of Casting Materials.   Because of the nature of the processes, there are certain parameters and properties of materials that are unique to these systems. They must be understood and controlled for successful electronic applications. This is of greater importance than with other plastic processing techniques.

Viscosity.   Viscosity is the property of a fluid which makes it resistant to flow. It is measured in centipoise (cP), and the viscosity of water is 1 cP. Table 1.9 lists the viscosities of common materials. Common viscosities for casting compounds are 30,000 cP. Low viscosities make it easier to impregnate coils of fine interstices and to remove volatiles. If the viscosity is too low, the materials may run out of the mold at joints. A viscosity greater than about 50,000 cP requires all manufacturing processes be done in vacuum and a pressure cycle is needed. Viscosity is very temperature-dependent; the addition of fillers increases viscosity.

Shrinkage.   As resins polymerize or cure, the molecules are confined and positioned closer together. This causes a volumetric decrease called *shrinkage*. Shrinkage varies with

TABLE 1.9 Viscosity Comparison Chart

Material	Viscosity	
	SSU*	cP*
Acetone	—	0.3
Benzene	—	0.5
MEK	—	0.4
Toluol	—	0.6
Xylol	—	0.6
Water	—	1
Turpentine	—	1
Milk	—	3
Ethylene glycol	80	16
Linseed oil (raw)	155	28
Oil (core)	157	29
SAE 10 motor oil	345	65
Corn oil	375	72
SAE 20 motor oil	600	125
Maple syrup	700	144
SAE 30 motor oil	1,000	200
Castor oil	1,200	240
SAE 90 transmission oil	2,860	590
SAE 140 transmission oil	10,300	2,200
Syrup (Karo brand, light)	11,600	2,500
Ketchup	230,000	50,000†
Vaseline petroleum jelly	290,000	64,000
Mustard	322,000	70,000†
Tomato paste	875,000	190,000†
Shortening	5,500,000	1,200,000†

*All viscosities, both Saybolt Universal (SSU) and centipoise (cP), are approximate.

†Materials are thixotropic, that is, material will be lower in viscosity with agitation, stirring, etc.

each polymer but ranges from 0.1 percent for some epoxies to more than 7 percent for some polyesters. Shrinkage imposes rather severe stresses on embedded components and may even crack the casting. It is related to the curing conditions and can also be affected by the use of fillers. It is possible to coat an embedded component with a flexible "barrier coating" to distribute the shrinkage stresses more evenly.

Exotherm.   As these liquid systems polymerize, the heat of reaction causes a rise in temperature called *exothermic reaction.* Very high temperatures can result. The low thermal conductivity of these resins makes it difficult for this heat to dissipate. High temperatures during cure cause differences in thermal expansion between molds, resins, and embedded components, and internal stresses can become quite high, causing cracking. Many embedded components are affected by temperatures above 220°F. Excessive exotherm can cause a runaway thermal condition and even ignite the resin. Exotherm can be controlled by the resin chemistry, cure rate, use of fillers, and mass of the resin.

Exotherms for resin systems can be measured accurately. Gel time measures the time for a resin to increase in a controlled bath to 150°F. Peak exotherm is the maximum temperature reached, and time to peak is the time to reach that temperature.

Cure Time.   This is the total time for a part to be fully cured or cured well enough to be removed from a mold. It depends on the chemistry of the system, the heating facilities, and the mass of the system. Long cure times are costly and tie up production equipment.

Moisture.   Most of these casting processes involve the use of vacuums and the additional of mineral fillers. Therefore careful control of the moisture present in the equipment, the resin, and the fillers is important to obtain void-free castings. If these resins are exposed to high voltages, such voids result in partial discharges (corona), which are highly destructive to organic systems and result in short life cycles.

Thermal Shock.   Systems with embedded components are very sensitive to thermal shock. Temperature changes cause movement or stresses between the resin and the embedment and may result in cracking or crazing. The coefficient of thermal expansion of plastics is much higher than that of metals and ceramics, and this enhances that movement (see Table 1.10). The more flexible the system, the better the resistance to thermal shock. In general, thermal shock resistance is rated polyurethanes > silicones > epoxies > polyesters. Thermal shock testing is defined in military specification MIL-1-16923 and calls for cycling castings from −55 to 125°C ten times without failure.

Thermal Conductivity.   The thermal conductivity of a resin system may be 1000 times less than that of copper. The addition of fillers helps improve the thermal conductivity modestly.

1.2.5.3 Resins for Casting

Epoxy.   By far, these resins are the most commonly used for casting and embedding. They are easily handled, can be formulated with a wide range of properties from rigid to flexible, withstand chemicals and common environmental effects, and have very low shrinkage. Their adhesion to embedded components is exceptional, and low viscosities make impregnation easy. Exotherms are mild and costs are attractive. However, they can cause dermatological problems, and some people are easily sensitized—particularly to the curing agents. While formulations can be very flexible, those flexible materials tend to

TABLE 1.10 Coefficients of Thermal Expansion for Several Packaging Materials

Material	10^{-6} (cm/cm)/°C
Silicone elastomers	275–300
Unfilled epoxies	100–200
Filled epoxies	50–125
Epoxy, glass laminate (z axis)	100–200
Epoxy, glass laminate (xy axis)	12–16
Aluminum	20–25
Copper	15–20
95% alumina ceramic	6.3
Type 400 steels	6.3–5.6
Glass fabric	5.1
Borosilicate glass	5.0
Silicon	2.4
Nickel-iron alloy (30 Ni–61 Fe)	1.22
Quartz	0.3
Kevlar fabric (axial)	−2.0
Kevlar fabric (transverse)	60.0

have poor physical properties. Tear resistance is poor. The properties of an epoxy casting resin are shown in Table 1.11.

Polyester. Polyester resins are the lowest-cost embedding resins and have the lowest viscosity. They are usually reacted with styrene, but higher-temperature performance is obtained by using triallyl cyanurate in place of styrene at some increase in viscosity. With accelerators, polyesters can cure at room temperature. Curing is very rapid—much faster than for all other systems. Exotherms can be very high, and castings are often highly stressed. Fillers help, in particular calcium carbonate. A range of flexibilities can be formulated. Polyesters have high shrinkage, poor thermal shock resistance, less humidity resistance, and poorer adhesion to metals than the other materials. Electrical properties are very good.

Silicone. Silicone embedding compounds can be formulated over an extreme range of flexibilities. Silicones can be prepared as gels. These provide high stress relief on embedded components before they are embedded in a more rigid composition. Gels can be penetrated with a probe to inspect circuits, and the gel heals itself when the probe is removed. The most common use of these silicones is in elastomeric formulations. Clear, transparent formulations are available, but most have some filler or pigmentation. In general, heat must be applied to cure the material, but special RTV silicones exist. The RTVs

TABLE 1.11 Properties of Typical Embedding Resins

	Epoxy	Polyester	Silicone	Urethane
Dielectric constant, D-150				
60 Hz	3.9	4.7	2.7	5.7
10^6 Hz	3.2	—	2.7	3.4
Dissipation factor, D-150				
60 Hz	0.04	0.017	0.001	0.123
10^6 Hz	0.03	—	0.001	0.03
Dielectric strength, D-149; V/mil	450	325	550	400
Volume resistivity, D-257; $\Omega \cdot cm$	10^{15}	10^{14}	10^{15}	10^{13}
Arc resistance, D-495; seconds	150	135	120	180
Specific gravity, D-792	1.15	1.2	1.05	1.0
Water absorption, D-570; % 24 h	0.15	0.3	0.12	0.4
Heat deflection temperature, D-648; at 264 lb/in², °F	380	260	<70	<70
Tensile strength, D-638; lb/in²	9000	10,000	1000	2000
Impact strength (Izod), D-256; ft · lb/in	0.5	0.3	No break	No break
Coefficient of thermal expansion, D-696; $10^{-5}/°F$	5.5	7.5	4.0	15
Thermal conductivity, C-177; Btu · in/(h · ft² · °F)	1.7	1.7	1.5	1.5
Linear shrinkage; %	0.3	3.0	0.4	2.0
Elongation, D-638; %	3	3	175	300

react with atmospheric moisture, producing acetic acid or ethyl alcohol as a reaction product. Applications must therefore permit a supply of moisture and a means of dispersing the reaction products. RTV silicones are highly viscous pastes, with viscosities of about 70,000 cP, so they are usually used as encapsulants. Rigid solventless silicones are not as widely used as other types because their resistance to thermal shock is inferior, and they are difficult to process, but they have superior thermal stability.

Silicones have very low exotherms and low residual stresses. Their viscosities are only slightly affected by temperature. They can be used from -100 to $600°F$, and their dielectric properties are the best of the embedding resins. Their high cost is often a problem, and their mechanical strength is not exceptional. Adhesion to other plastics is poor, to metals fair, and to ceramics good.

Urethanes. Urethanes are very tough, abrasion-resistant, and tear-resistant. They are used for their elastomeric properties without fillers. Their impact and thermal shock resistance is exceptional. Easy to fabricate, they can be cured at room temperature or with elevated temperatures. Electrical properties are good and adhesion to other components is very high, approaching that of epoxy. They are not high-temperature materials and must be used below $250°F$. Urethanes can be reacted from either polyester or polyether constituents. Polyester-based urethanes are very poor in moist environments, but some formulations based on butadiene are very good. Moisture interferes with the urethane polymerization process, so embedding systems and materials must be kept dry. Some of the isocyanates and the urethane coreactants are toxic, and one should have full health and safety information available on all urethanes.

Foams. Most of the embedding resins can be made into low-density foams by adding blowing agents. Polyesters are not foamed because their cure is affected, while silicone and epoxy pour-in-place foams are often used in embedding applications. However, it is the urethanes which lend themselves to low-density foams. They can be made to foam using by-products of polymerization or by adding foaming ingredients. They can foam at room temperature or at elevated temperatures. The electrical losses of foams are excellent, but they have poorer dielectric strength, mechanical properties, and thermal conductivity.

1.2.5.4 Fillers for Plastics. Fillers are used extensively in embedding formulations, but also in most other types of plastic materials such as laminates, molding compounds, elastomers, sheets, and pultrusions. They are of particular use to electronics engineers who are designing, casting, and embedding formulations, because they lend so many attractive properties to the embedding compositions.

Properties of Fillers

1. *Viscosity.* Fillers increase viscosity and make the embedding process easier to control. Some fillers (silicas) make the formulations thixotropic, and this thickness keeps settling from occurring and permits high filler loading.

2. *Cost.* The fillers usually cost less than the resins on a volume basis. However, they increase the density.

3. *Exotherm.* Fillers reduce exotherm better, match coefficients of expansion, and help prevent cracking.

4. *Shrinkage.* Shrinkage is reduced. The effect is mostly a bulk effect, and hence the higher the filler volume, the less the shrinkage.

5. *Thermal conductivity.* Fillers increase thermal conductivity and help in thermal management. The more filler used, the better, but since the resin is still in the continuous phase, the effect is not as great as wished.

6. *Reinforcement.* Many fillers in rod or fiber form improve the physical properties.

7. *Shelf life.*   Many fillers help stabilize the embedding compound.

8. *Hardness.*   Fillers usually increase the hardness of the system.

9. *Electrical properties.*   The electrical properties may be either improved or harmed. The dielectric constant and the dielectric strength are particularly sensitive.

Types of Fillers. The shape, size, amount, wettability, and chemical nature of fillers greatly affect their use in plastic materials and decide the embedding formulation properties. Fillers are used in quantities of from 5 to 80 percent by weight, depending on the needs of the designers. The following fillers are used with plastics.

1. *Mineral fillers*

Aluminum oxide trihydrate (aluminum oxide with water of hydration). This filler has an unusual effect. When heated over 200°C, it loses the hydrated water. This lends flame resistance to the formulations. The loss of the water also cools electric arcs, making many polymers more arc- and track-resistant.

Beryl (beryllia, beryllium oxide). Beryl has very high resistivity and greatly improved thermal conductivity. Powders are toxic.

Calcium carbonate ($CaCO_3$). This filler is of low cost and can be used at high loadings. It is particularly useful with polyesters. It can have a thixotropic effect and eases machining.

Calcium silicate ($CaSiO_3$; calcium metasilicate, woolastonite). A long fibrous rod structure, it gives reinforcing properties but reduces impact; it helps thermal conductivity and thermal shock resistance.

Clay (china clay, kaolin, aluminum silicate; $Al_2O_3 \cdot 2SiO_2 \cdot 2H_2O$). A very common and useful filler, it decreases shrinkage and cost and has good machinability and good electrical properties.

Mica (muscovite, phlogopite). Reinforcing fillers with platelike particles, they are often coated with a primer to better bind to resins and improve impact. They lend excellent electrical properties, in particular dielectric strength, have lower cost, and prevent cracking.

Sand (SiO_2; silica, crystalline silica, diatomaceous earth, precipitated silicas, quartz). These are available from particles of less than 0.1 μm to coarse beach sand. Some embedments are formed by surrounding the article in dry coarse sand and vacuum filling the mold. Fine silicas act as thixotropic agents in viscosity control. Diatomaceous earths give plastic a smooth surface. The silicas add thermal conductivity, low cost, good electrical properties, and improved compressive strength. They are hard to machine.

Talc (talcum, soapstone, steatite, magnesium silicate $Mg_3Si_4O_{10}(OH)_2$). Talc controls cracking, is of low cost, and greatly cases machining.

2. *Metal fillers.*   Powders can be made of aluminum, copper, bronze, gold, silver, platinum, and other materials. They are added to resins to improve fllermal conductivity and to make the polymers more electrically conductive. Conducting inks, adhesives, and coatings are often metal-filled.

3. *Special fillers*

Antimony oxide (antimony trioxide; Sb_2O_3). This filler has a synergistic effect with bromine and chlorine present in flame-retardant polymers and greatly improves this property. It causes machining difficulties.

Carbon black (graphite). This is often used as a reinforcing filler in elastomers.

Glass spheres. Glass as well as some plastics can be made into small (10 μm) solid or hollow spheres. In electronics these are used to make syntactic foams useful for their light weight. They do not help thermal conductivity but, when filled with air, they decrease the dielectric constant substantially.

4. *Fibers.*   Many fibrous fillers are used in electronic insulations. Glass fiber is by far the most common and is discussed in Chap. 8.

5. *Organic fillers.*   Examples are wood flour, starches, cellulose, and nut shells. These organic fillers are used in molding materials, particularly phenolics, but not generally in embedding applications.

1.2.5.5 Embedding Specifications.   Embedding and casting compounds are described in the following specifications.

MIL-1-16923	Insulating compound, electrical, embedding
MIL-M-24041	Molding and potting compound, chemically cured, polyurethane
MIL-S-23586	Sealing compound, electrical, silicone rubber, accelerator required
ASTM D-1674	Polymerizable embedding compounds used for electrical insulation

1.2.6 Film, Sheet, and Tape

A film is a polymer thinner than 15 mil (ASTM D-2305[15]). Above that thickness it is called sheet. Tape is slit film or fabric often coated with an adhesive. Virtually all of the thermoplastic resins described previously can be made into films. Films are made by four processes.

1. *Extruded.*   Films are made by attaching a suitable die to an extruder. Some films are blown. In this case the extruder extrudes a tube vertically and air is blown into the tube, expanding the dimensions, thinning the walls, and orienting the film. The tube is then cut open to form a film. Extruded film can be oriented in the longitudinal direction by using rolls of differing speeds. Orientation in the transverse direction is done with tenter frames used for textiles. Oriented film has improved physical properties.

2. *Cast film.*   A polymer is dissolved in a solvent and spread out as a wet film on a continuous highly polished metal belt. The thickness of the wet film is controlled with a die, metering rolls, or doctor blades. The film is transported through an oven where the solvent is removed, the polymer melted, and flow obtained. The film is then stripped from the belt. Polycarbonates and polyimides are formed this way.

3. *Calendering.*   Films can be formed by passing a polymer sheet between heated rolls. This method is seldom used now.

4. *Skiving.*   Films can be cut mechanically on a lathe from round billets of plastic, much like plywood is prepared. Some fluoropolymers are skived.

Films tend to have better dielectric strength, flexibility, tear resistance, and fatigue life than sheet plastics. Tensile strength and elongation are greatly improved, particularly in oriented films. Electrical properties are those of the sheet materials. The most useful films in electronic design are polyester (polyethylene terephthalate), polyimide, fluorocarbon, polyethylene, polycarbonate, and polysulfone.

1.2.7 Adhesives

Adhesives are used in a variety of electronic designs. Adhesives represent a specific use of polymers which emphasizes their melting ability and their ability to bond to many different surfaces. Adhesives are specialty chemicals. They are carefully formulated by suppliers to enhance the natural ability of polymers to stick to surfaces.

There are a host of adhesives, and the technology of adhesion covers chemistry, physics, mechanical design, and materials science. The entire subject is too broad for a packaging handbook, and reference should be made to the *Handbook of Plastics and Elastomers*[16] for full information. The following adhesives are used in electronics manufacturing:

Flat cable.	Films are bonded to substrates.
Cable assemblies.	Conductors are bonded to jackets.
CRT.	RTV systems are used to attach components.
Connectors.	Rigid adhesives are used to bond to circuit boards.
Rigidizing components.	Large components are bonded to boards to stiffen and remove stress from solder joints.
Board repair.	Repair of circuits requires bonding of new circuit elements.
Wire hold down.	Leads and wires are immobilized.
Die attach.	Dies are attached to substrates or LCC.
Hybrid attach.	Hybrid components are bonded to substrates.
Heat sinks.	Heat sinks are bonded to circuit boards.
Waveguides.	Waveguides are bonded in place.
Enclosures.	Components are bonded to enclosures, and enclosures are fabricated by bonding.
Surface mount.	Thermally conductive materials are used for attaching components. Adhesives for surface mount (SM) are classified as follows:

Type I	All SM; adhesives not needed unless reflow is used.
Type II	Mixture of SM and DIP; adhesive used if chips are on back side.
Type III	DIP on one side and SM on other; adhesives always used.

1.2.7.1 Thermosetting Adhesives

Epoxies. These are the most commonly used adhesives. They can consist of either one component or two components. Two-component adhesives are usually dispensed with twin dispensing cylinders, each emitting an equal volume and one containing the hardener while the other has the resin. Epoxies are often filled with silver to increase thermal conductivity. Conductivities of 20 Btu • in/(h • ft • °F) are common. Most adhesives used to adhere surface-mount components contain silver to help in thermal management.

Some one-component epoxies are supplied frozen to increase their shelf life. When they are to be used, they must have time to come to thermal equilibrium and be protected from condensation. Cure conditions for epoxies vary from room temperature to low temperatures (80°C) and high temperatures (>150°C). Epoxy adhesives are also supplied as sheets, often supported on a fabric or film. This form is advantageous when bonding large flat surfaces such as heat sinks to laminates. Suppliers will die cut these sheets to shape. Sheet adhesives may be plain, filled with silver, or filled with other mineral fillers.

Polyimide. Polyimides can be silver-filled for thermal conductivity. The adhesives have low viscosity and can be cured at 180°C. Shrinkage is low and parts have low stress.

Silicone. Silicone adhesives have a wide range of viscosities, from waterlike to thixotropic. They act as stress relievers from −65 to 200°C. Water-clean, ultrapure silicones are used as adhesives in electrooptical devices. RTV silicones cure at room temperature, emitting acetic acid or ethyl alcohol. They are usually one-part systems, but two-part systems are available. Silicones do not off-gas appreciably.

1.2.7.2 Thermoplastic Adhesives

Hot Melt Adhesives. These adhesives are supplied in stick or rod form to be inserted in special heat guns. The guns melt the resin and extrude it through a nozzle for application, where it freezes. The adhesives are acrylic, nylon, phenoxy, or olefins. The advantages of hot melts are (1) fast set time; (2) no fixtures needed; (3) low cost; and (4) they can bond polyethylene and polypropylene.

Sheet Adhesives. Similar to hot melts, they are ethylene-acrylic acid copolymers. They are placed on the substrate, heated to 150°C, and then cooled to freeze.

Pressure-Sensitive Adhesives. These adhesives are used mostly as tapes. They are based on elastomers, silicones, or acrylics. These may also be thermoset polymers.

1.2.7.3 Elastomeric Adhesives.

Adhesive formulations are available based on many of the elastomers described earlier in this chapter. As adhesives, they are generally dissolved in a solvent. In use the adhesive is applied by spray or brush to the surfaces to be bonded, and the solvent is permitted to evaporate, usually about 20 min. Then the surfaces are joined. One must be careful at this point that alignment is correct since adhesion occurs as soon as the surfaces touch. These adhesives are useful when one of the adherends is flexible or when great flexibility or stress relief is desired at the bond line.

1.2.7.4 Ceramic Adhesives.

These adhesives are usually based on a low-melting glass formulation. The adhesives are thixotropic pastes with silver filler and an organic polymer base. The adhesives are fired in a programmed temperature cycle, which evaporates and sublimes the polymer, melts the glass, and wets the substrate. Cool down is slow to control crazing. The adhesive loses one-half its volume during firing, which reaches at least 390°C.

1.2.7.5 Cyanoacrylate Adhesives.

Most adhesives cure by heat or solvent removal. Cyanoacrylate adhesives cure by the exclusion of air. As two surfaces coated with a cyanoacrylate are brought together, the resin displaces air. In the absence of air the resin polymerizes. This gives very rapid set times, and adhesion is possible to a wide variety of substrates. Some substrates require the presence of a primer. Bond lines must be thin, and the adhesives cannot be used to fill gaps.

1.2.7.6 Bonding Process.

Satisfactory adhesions require four steps: preparing the surfaces, preparing the adhesives, bonding, and curing.

1. *Preparing the surfaces.* This is a critical step in adhesive bonding. Adhesion is dependent on the adhesive thoroughly and intimately wetting and covering the adherend. The presence of any contaminant (oil, oxide, dirt, chemicals) will prevent good wetting. Cleaning methods include abrading, sanding, solvent wash, vapor degrease, chemical etch, and ultrasonic.

2. *Preparing the adhesive.* Two-part adhesives must be mixed with care to ensure that mixing is complete and not contaminated. Poor mixing is often the cause of poor joints. Two-part adhesives can be supplied in side-by-side hypodermic cylinders so that a push on a common plunger dispenses each component evenly. Thorough mixing is still required.

3. *Bonding.* Liquid and paste adhesives can be applied by spray, brush, hypodermic needle, or spatula. An array of sophisticated automated dispensing equipment is available.[17] Surface-mount shops use screen printing, pin transfer, and needle dispensing. Solid adhesives are supplied as sheets, die-cut forms, preforms, and "glob tops." The adhesives must cover the adherends uniformly and be free from contamination. Adherends must be brought together in correct alignment and restrained there by suitable jigs or fixtures until cure is complete.

4. *Cure.*   The adherends must not be moved, vibrated, or shaken until cure or setting of the adhesive has progressed far enough that the bond line is reasonably solid. Adhesive manufacturers usually state a "handling time." High-temperature cures require ovens with good heat distribution or conveyor ovens with programmed temperature ramps.

1.2.8 Polymer Thick Films

In the past, thick-film technology has been based on ceramic binders for resistors, insulators, conductors, and thermistors. These thick films are made from a polymer binder, solvent, glass frit, or ceramic powder and a conducting or semiconducting filler. The films are pastelike and screened onto a substrate. The solvent is removed and the polymer cured. This assembly is then fired to burn off the polymer, melt the conductor, and fuse the glass. The process is described in detail in Chap. 7.

In polymer thick films (PTFs), the inorganic ceramic or glass is replaced with a polymer matrix. This eliminates the firing step at high temperatures and makes low-temperature cures possible. PTF ink consists of three components: a conductive or semiconductive filler, a solvent, and a polymer. The polymer is usually thermosetting, but thermoplastics are also used. The polymer acts as the binder for the filler. The solvent controls the ink viscosity and flow characteristics. The following circuit components can be obtained:

Conductor | The fillers used are silver, copper, and silver alloys of platinum or palladium. Conductivities can reach 20 mΩ/□. Higher conductivities require copper plating. Silver-based conductors should be protected with a conformal coating to prevent moisture-induced silver migration.

Resistor | The fillers used are carbon particles or carbon mixed with copper or silver. The resistance is controlled by the volume content of the filler. Organic solvents control the viscosity.

Insulator | Conductive fillers are not used in this mixture of polymer and solvent. Solventless mixes are also used and in fact many use regular solder masks.

PTF can be used in multilayer applications[18] where its light weight can be of great advantage. The chief advantages of PTF are the same as for regular cermet thick films, except that PTF eliminates the high firing temperatures (850°C), which makes the processing less costly and permits the use of a variety of substrates. FR-4 epoxy-glass substrates are common, but CEM-1 is also used, as are ceramics and thermoplastics.

There are several disadvantages to PTF: (1) It is difficult to solder to PTF conductors and plating may be required. (2) Films do not level well, which results in problems with multilayers. (3) Substrates may warp when the inks are cured. (4) Long conductor traces have high resistivity. (5) Stability of properties is often not as good as for cermet thick films.

PTF is being used successfully in telecommunications, the automotive industry, and small computers. One instance where PTF is used is in molded circuit or three-dimensional circuit devices,[19] which are described in Chap. 8.

1.3 METALS

The field of metals (see also Recommended Reading no. 8) is extremely broad, and many publications exist. This section covers metals in three areas: (1) steels and nonferrous metals, including the distinguishing properties or design considerations; (2) corrosion considerations; (3) plated, metallic, and inorganic finishes.

Metals can be classified as ferrous or nonferrous. Ferrous metals are those that have iron as their primary element, and nonferrous metals have other elements. Of the ferrous metals, steels are the most commonly used in the electronics industry, while many of the nonferrous metals find special uses in electronics.

Steels.   Steels can be categorized in many ways, each with its advantages for a particular situation. One breakdown of general and specialty steels is given in Table 1.12, which also indicates the distinguishing property and general uses for each steel type.

Nonferrous Metals.   These metals are elementary or alloyed aluminum, beryllium, cobalt, copper, lead, magnesium, nickel, zinc, and two groups of precious metals (gold, palladium, platinum, rhodium, silver) and refractory metals (tungsten, tantalum, titanium, molybdenum, niobium or columbium, zirconium). The important considerations and distinguishing properties for these metals are given in Table 1.13.

1.3.1   Corrosion of Metals

Broadly speaking, corrosion can be defined as materials deterioration which is caused by chemical or electrochemical attack, in the presence of electrically conductive solutions (electrolytes). Electrolytes are common in nearly all environments. When, for example, a steel part is in contact with another metal in the presence of an electrolyte such as salt in water, an electric current is generated that induces corrosion.

In a desert climate, absence of moisture will prevent corrosion, even though contaminants are present; distilled water is an insulator and is not corrosive, but water found in nature contains enough contamination to make it conductive, and therefore corrosive, in the presence of a dissimilar metal junction.

The degree of attack is also influenced by the position of the dissimilar metal in relation to steel in the galvanic series shown in Table 1.14, with the least noble metal at the top. This order is for usual types of electrolytes met in outdoor environments; with other conditions, such as specific types of chemical exposure, the order may vary.

A less noble metal (higher in the list than steel) will be consumed when in contact with steel in the presence of an electrolyte. A magnesium rod, for example, will corrode and protect the steel lining in a hot-water tank. Zinc, cadmium, and aluminum as plated coatings offer sacrificial protection to steel and also serve as barriers to moisture. Even when these coatings are porous, the steel underneath is not attacked.

More noble metals such as tin, nickel, copper, and chromium will cause steel to corrode. When used as a plated coating, these metals must offer a complete barrier to moisture; a pinhole or crack in the coating will cause formation of rust in the presence of moisture.

Corrosion between dissimilar metals can be prevented by one of three means: (1) excluding moisture with an impervious barrier, (2) using a less noble or sacrificial metal, or (3) electrically insulating the dissimilar metal.

1.3.2   Electroplated and Chemically Deposited Coatings

Chemically deposited coatings constitute another one of the groups of materials vitally important to the electronic packaging engineer. While used primarily with metals, chemically deposited coatings are also used with plastic materials. The reasons for such coatings on metals are decorative and protective requirements, and reflective and thermal surface requirements for aerospace applications. When used with plastics, chemically deposited metallic coatings often provide shielding for critical electrical circuits.

TABLE 1.12 Distinguishing Properties and Uses of Some General and Specialty Steels

Designation	Distinguishing property	Uses
Carbon steels, AISI C1030–C1080	Hardenable	Crankshafts, linkages, screw fastenings, cutting blades, springs
Carbon steels, AISI C1015–C1022, C1117–C1118	Carburizable	Structural shapes, wear-resistant machine parts, cams, bearing journals, gears, piston rings
Carbon steels, 135, N, EZ	Nitridable	Shafts, cylinder liners, cams, rollers, gears
Carbon steels, classes 60,000–100,000	Inexpensive, generally applicable, hardenable	Structural members, welded frames, electromagnet cores, transformer cores
High-strength steels, classes 45–60, D-6A, MX-2, 300 M	Workable, machinable, and weldable	Structural shapes, rivets, bolts, engine frames, turbine casings, pipes, and high-pressure tubing
Free-cutting steels, AISI B1111–B1113, C1211–C1213	Easy machining and case-hardenable	Fastening devices, washers, spacers, shafts
High-speed steel, high-molybdenum, M1–M44	Highly hardenable	Cutting tools, surfaces useful at moderate cutting speeds but can handle high cutting pressures
High-speed steel, high-tungsten, T1–T15	Very hard	Cutting surfaces, instrument-bearing journals
Tool steels, chrome H10–H19, tungsten H20–H26, molybdenum H41–H43	Hardenable, tough	Forging dies, striking platens, plastic molds
Metallurgical powders, ferrous F series, stainless SS series	Magnetic, corrosion-resistant	Coil cores, armatures, magnetic linkages, light-duty gears, other machine parts
Austenitic stainless steels, AISI 200 and 300 series	Weldable, corrosion-resistant	Forged shapes, weldments, tubing, plumbing fittings
Martensitic stainless steels, AISI 400 series, except ferritic types	Hardenable, corrosion-resistant	Valves, impellers, ball bearings, springs, cutting devices, gears
Ferritic stainless steels, AISI 406, 430, 446	Free-machining, corrosion-resistant even at elevated temperatures	Chemical containers, tubing, flame tubes, nozzles
Age-hardenable stainless steels, wide variety of types, such as W, 17-7PH, and AM350 series	High strength, corrosion-resistant	Structural members, pressure vessels, chemical processing equipment

Source: From Katz.[20]

TABLE 1.13 Distinguishing Properties of Some Nonferrous Metals

Metal	Distinguishing properties
Aluminum and aluminum alloys	Aluminum is a versatile metal with combinations of light weight and tensile strength up to 125,000 lb/in^2. Its excellent resistance to corrosion in many environments is made possible by the protective, highly adherent oxide film which develops in air, oxygen, or oxidizing media. High-strength aluminum alloys generally are not as corrosion-resistant as the high-purity or moderate-strength aluminum alloys. Aluminum is available in a great number of alloys and tempers. The metal can be fabricated economically by all common processes and can be cast by any method known to foundry workers.
Beryllium	Beryllium provides high stiffness-to-density ratios, high strength-to-density ratios, and excellent dimensional stability. Modulus of elasticity of 44×10^6 lb/in^2 and density of 0.066 lb/in^3 are common to all forms of beryllium. High thermal conductivity and low thermal expansion coefficient contribute to dimensional control with temperature variations.
Beryllium copper	Beryllium copper combines strength, wear resistance, electrical and thermal conductivity, and ease of fabrication properties. Age-hardened beryllium copper alloys can provide tensile strengths up to 215,000 lb/in^2 with endurance strengths (at 10^8 cycles) of approximately 40,000 lb/in^2. A Rockwell hardness up to C-45 broadens the alloy's usefulness in applications requiring resistance to wear. Electrical conductivities reach 50% IACS, with a thermal conductivity range of 775–1600 Btu/ (h · ft^2 · °F). Beryllium-copper alloys can first be fabricated or machined in the unhardened state. Desired properties can then be imparted to the finished product by a simple low-temperature thermal treatment.
Beryllium nickel	Beryllium nickel is the highest-strength nickel alloy at temperatures from room to 900°F. After precipitation hardening, tensile values over 270,000 lb/in^2 with 230,000 yields can be obtained in wrought alloys. Like wrought forms of beryllium copper, beryllium nickel is normally supplied to a fabricator in relatively soft condition. The alloy can be formed using conventional methods. The parts can then age by a simple cycle in an ordinary furnace to achieve outstanding mechanical properties. The alloy exhibits the general corrosion resistance of nickel.
Cobalt and cobalt alloys	The retention of hardness and strength at high temperature is a distinctive feature not only of cobalt-base alloys but of other cobalt-containing alloys. Materials are available for use under stress up to 2000°F and, at no load, up to 2400°F. Cobalt is the element with the highest Curie temperature (2050°F). The metal is used by itself or in alloys (cobalt-nickel, cobalt-phosphor, cobalt-nickel-phosphor) for memory and other magnetic devices. It is an important alloying element in a series of permanent magnets as well as soft magnetic materials. Cobalt and its alloys are remarkable for their low coefficient of friction and nongalling characteristics. They can often be used as high-temperature bearings without lubrication.

TABLE 1.13 Distinguishing Properties of Some Nonferrous Metals (*Continued*)

Metal	Distinguishing properties
Copper and copper alloys	Copper has the highest electrical conductivity of any metal except pure silver. Copper alloys are easily fabricated and set the standard for nonferrous alloys in most fabrication operations. The brasses are ideally suited to cold-forming operations such as deep drawing, bending, spinning, and stamping. The ready solderability of copper is vital in electrical conductivity. The generally excellent corrosion resistance in natural environments accounts for the wide use of copper and copper alloys. These metals tarnish superficially in moist air, but the pleasing colors of the surface films developed after further exposure often are a plus factor. Copper and its alloys are resistant to corrosion by most natural waters—fresh as well as brackish and salt.
Lead and lead alloys	Lead is one of the most easily formed metals. At room temperature it is approaching the plastic state (melting point about 620°F) and is easily rolled, extruded, cast, or often shaped by hand. Lead is resistant to most active chemicals. It is quite stable as a metal or in compounds, because of its chemical family and high atomic weight. Lead, with a density of 0.41 lb/in^3, is the lowest-cost high-density material. The element has the rare, balanced combination of neutrons and protons that makes it an excellent shield against gamma rays and X-rays.
Magnesium and magnesium alloys	Magnesium, with a specific gravity of 1.74, has long been recognized as the lightest structural metal. Magnesium is well suited to modern die-casting processes. The metal flows easily and readily fills complex dies and thin sections. The most common magnesium alloys in regular use are in the magnesium-aluminum-zinc system for all cast and wrought forms. Magnesium-zinc-zirconium alloys, both cast and wrought, provide improved properties. Magnesium-thorium-zirconium, magnesium-thorium-zinc, and magnesium-thorium-manganese are the alloy systems used for retention of usable properties at elevated temperatures. Magnesium-lithium alloys are the lightest commercial magnesium alloys. A magnesium-lithium alloy, with specific gravity only 1.35, can be obtained in various mill forms. Magnesium-lithium alloys have improved cold-workability. They are fusion-weldable by the inert-gas shielded arc and also can be electrical-resistance spot-welded. They are receptive to the same chemical treatment as other magnesium alloys. Although magnesium is sometimes used unpainted, protective and decorative finishes are applied in most cases. Application of electrolytic coatings is sensitive and requires good controls.
Nickel and nickel alloys	Nickel alloys, in general, are stronger, tougher, and harder than most nonferrous alloys and many steels. Their most important commercial mechanical property, however, is their ability to retain strength and toughness at elevated temperatures. The range of alloys based on nickel as a major constituent is wide enough to combat a great variety of corrosive environments. Generally, reducing conditions retard, while oxidizing conditions accelerate, corrosion of nickel. Nickel-base alloys often have the excellent corrosion resistance of elemental nickel, as well as that of other elements they contain.

TABLE 1.13 Distinguishing Properties of Some Nonferrous Metals (*Continued*)

Metal	Distinguishing properties
Zinc and zinc alloys	Resistance to corrosion of wrought and cast zinc products is generally retained throughout the product's life. As coating, zinc's protection continues when the underlying material (generally steel) eventually becomes exposed. Corrosive attack is then directed to the zinc rather than the base metal. Zinc's low melting point facilitates low-cost production in cast, wrought, or coating form. While the melting point is in the soldering temperature range, zinc alloys have physical and mechanical properties suitable for many structural applications. Most of the finishes applied to other metal products can also be applied to zinc die castings. These include (1) mechanical: buffing, polishing, brushing, and tumbling; (2) electrodeposited: copper, nickel, chromium, brass, silver, and black nickel; (3) chemical: chromate, phosphate, molybdate, and black nickel; (4) organic: enamel, lacquer, and varnish; (5) plastic.
Precious metals	Precious metals are platinum, gold, palladium, iridium, rhodium, osmium, ruthenium, and silver. The precious metals are highly resistant to many corrosive environments either in their pure form or in alloys. Platinum is the most generally applicable but the others of the group are widely used. Silver is tarnished by sulfide environments. Gold and silver do not oxidize to form a scale at elevated temperatures, but silver does dissolve considerable oxygen and this must be considered in its use. The basic precious-metal alloy system used for structural purposes at elevated temperatures is platinum-rhodium. Pure platinum may be used in cases where the hot-strength requirement is not high. Where the strength requirements are higher than can be attained with platinum-rhodium alloys alone, platinum-rhodium alloys are used to sheath higher-strength materials such as molybdenum, which does not have adequate oxidation resistance. The stability and wide range of high electrical properties make the precious metals useful in a number of areas. Among the useful properties are stable thermoelectric behavior, high resistance to spark erosion, tarnish resistance, and broad ranges of resistivities and temperature coefficients of electrical resistance.
Tungsten	Tungsten is stronger than any other common metal at temperatures over 3500°F. Tungsten's melting point of 6170°F is higher than that of any other metal. The electrical conductivity of tungsten is approximately one-third that of copper, much better than the conductivity of nickel, platinum, or iron-base alloys. Conversely, the resistivity in fine wire form has been exploited in many lamp and electronic applications in which tungsten wire serves as a light-emitting or electron-emitting filament.
Molybdenum	On a strength basis, pure molybdenum is generally considered the most suitable of all refractory metals at temperatures between 1600 and 3000°F. Small amounts of other refractory metals with molybdenum form alloys that have much greater strength-to-weight ratios at higher temperatures. Thermal conductivity of molybdenum is more than three times that of iron and almost half that of copper. Abrasion resistance of molybdenum is generally outstanding at high temperatures.

TABLE 1.13   Distinguishing Properties of Some Nonferrous Metals (*Continued*)

Metal	Distinguishing properties
Tantalum	Tantalum is the most corrosion-resistant of the major refractory metals. It closely duplicates the corrosion resistance of glass. Tantalum is an easy-to-fabricate metal. Spinning, deep drawing, and severe bending can be performed without tears, cracks, or excessive peeling. Ductile, nonporous welds can be made easily by using TIG, resistance, or electron beam welding. Tantalum finds important use in electrolytic capacitors because it forms tantalum oxide (Ta_2O_5). This provides a high dielectric constant and good dielectric strength.
Columbium	At temperatures from 2000 to 3000°F (in vacuum or inert atmosphere), columbium alloys give the best performance on a strength-to-weight basis among metals that exhibit ductile welds. Columbium has excellent corrosion resistance, including resistance to liquid alkali metals. Columbium is the major component in high-field, superconducting alloys for use at cryogenic temperatures. Columbium has good nuclear properties, important in such applications as fuel-element cladding for nuclear propulsion reactors. The material is quite ductile, but is susceptible to air and hydrogen embrittlement at elevated temperatures. While columbium is easily worked when pure, it becomes difficult to work when highly alloyed. At high temperatures, columbium absorbs oxygen, nitrogen, and hydrogen and becomes brittle. It must be alloyed for high-temperature applications. Columbium is a reactive metal and accordingly reacts with coatings at high temperature.
Titanium	Titanium, with a density of 0.161 lb/in³, is classed as a light metal, being 60% heavier than aluminum (0.10 lb/in³), but 45% lighter than alloy steel (0.28 lb/in³). Titanium-base alloys are extremely strong, with ultimates starting at about 30,000 lb/in² and reaching in the neighborhood of 200,000 lb/in². Strengths of 160,000 lb/in² are attainable in the general-purpose (Ti6A1-4V) grade. The strength of titanium alloys is accompanied by excellent ductility. Titanium is virtually immune to corrosion from the atmosphere. Its corrosion resistance is excellent in most oxidizing environments and many mildly reducing environments. Titanium-base alloys provide excellent fatigue properties. Titanium is the only known structural metal with a corrosion-fatigue behavior in salt water practically identical to that in air.
Zirconium	Zirconium displays excellent resistance to many corrosive media. The metal resists both acids and alkalies, in particular alkali-and-chloride solutions, some inorganic acids, and chlorine-saturated water. Resistance covers a wide range of temperature and concentration. Zirconium offers a low-absorption cross section for the slow thermal neutrons necessary to sustain a chain reaction. The material is therefore used for reactor components. Zirconium is lighter than steel and heavier than titanium. When impure, it is very hard and brittle. When pure, it is soft, malleable, and ductile. Its mechanical properties resemble those of mild steel, although they change rather rapidly as temperature is elevated. For example, Zircaloy-2 has a yield strength of 65,000 lb/in² at room temperature, but drops to 35,000 lb/in² at 700°F.

Source: From Wigatsky.[21]

TABLE 1.14 Galvanic Series of Metals under Atmospheric Conditions

Corroded end	Brass
(anodic, or least noble)	Copper
Magnesium	Bronze
Magnesium alloys	Copper-nickel alloys
Zinc	Silver solder
Aluminum 1100	Nickel (passive)
Chromium	Inconel (passive)
Cadmium	Titanium
Aluminum 2017	Stainless, ferritic (passive)
Steel or iron	Stainless, austenitic (passive)
Stainless, ferritic (active)	Silver
Stainless, austenitic (active)	Graphite
Lead-tin solder	Gold
Lead	Platinum
Tin	Monel
Nickel (active)	Protected end
Inconel (active)	(cathodic, or more noble)

Source: From Bronson.[22]

1.3.2.1 Electroplated Metallic Coatings. An electroplating solution is an electric field in a conductive liquid solution (electrolyte) with current entering the solution at the anode and emerging at the cathode. The cathode consists of the parts being plated. The anode and cathode can be considered uniform potential electrodes in an electrostatic field, with the anode positive and the cathode negative. If the field is plotted between the electrodes for different configurations, high-potential gradients can be seen at projections and low-potential gradients at recesses. Since the current density is proportional to the potential gradient, projections will receive high current density and recesses will receive low current density. The rate of deposition is proportional to the current density. Therefore, on irregularly shaped parts, the plating thickness may vary considerably. To have adequate thickness in recesses, the average thickness may have to be two to three times the minimum specification thickness. A discussion of individual electroplates follows.

Cadmium Plating. Electrodeposits of cadmium are used extensively to protect steel and cast iron against corrosion. Because cadmium is anodic to iron, the underlying ferrous metal is protected at the expense of the cadmium plate, even if the cadmium becomes scratched or nicked, exposing the substrate.

Cadmium is applied usually as a thin coating (less than 1 mil thick), intended to withstand atmospheric corrosion. It is seldom used as an undercoating for other deposited metals, and its resistance to corrosion by most chemicals is low. It is used to coat parts that are made of dissimilar metals because of its ability to minimize galvanic corrosion. Its excellent solderability is an advantage.

Zinc Plating. Zinc is anodic to iron and steel and therefore offers more complete protection when applied in thin films (0.3 to 0.5 mil) than do similar thicknesses of nickel and other cathodic coatings. Because it is relatively cheap and readily applied in barrel, tank, or continuous plating facilities, zinc is often preferred for coating ferrous parts when protection from atmospheric and outdoor corrosion is the primary objective. Normal electroplated zinc without subsequent treatment becomes dull gray in appearance after exposure to air. Bright zinc that subsequently has been given a bleached chromate conversion coating or a coating of clean lacquer is used as a decorative finish. Such a finish, although less durable than heavy nickel-chromium, in many instances offers better corrosion protection than thin coatings of nickel-chromium at lower cost. Plating of zinc on gray iron and malleable iron presents serious operational difficulties; so cadmium is preferred.

Copper Plating. Copper electrodeposits are used as underplates in multiple-plate systems, as stop-offs, for heat transfer, and in electroforming. Although copper is relatively corrosion resistant, it is subject to rapid tarnishing and staining when exposed to the atmosphere. Consequently, electrodeposited copper is rarely used alone in applications where a durable and attractive surface is required.

Nickel Plating. Nickel plate, with or without an underlying copper strike, is one of the oldest protective-decorative electrodeposited metallic coatings for steel and brass. Nickel plate will tarnish, taking on a yellow color during long exposure to mildly corrosive atmospheres, or turn green on severe exposure. The introduction of chromium plate overcame the tarnishing problem and led to a great increase in the use of nickel as a component of protective-decorative coatings in various combinations with copper and chromium.

Chromium Plating (Decorative). Decorative chromium plating is a protective-decorative coating system in which the outermost layer is chromium. The layer is applied over combinations of plated coatings of copper and nickel. The function of this system is twofold: (1) to provide the basis metal with protection against corrosive environments, and (2) to maintain, in service, an appearance conforming to an agreed-on standard.

Chromium Plating (Hard). Hard chromium plating, commonly known as industrial or engineering chromium plating, differs from decorative chromium plating in the following ways.

Hard chromium deposits are intended primarily to restore dimensions of undersized parts or to improve resistance to wear, abrasion, heat, or corrosion, rather than to enhance appearance.

Hard chromium normally is deposited to thicknesses ranging from 0.1 to 20 mil (and for certain applications to considerably greater thicknesses), whereas decorative coatings seldom exceed 0.1 mil.

With certain exceptions, hard chromium is applied directly to the base metal and is usually ground to a final dimension; decorative chromium is applied over undercoats of nickel or of copper and nickel, and is either buffed or used in the as-plated condition. Whereas decorative coatings generally contain pores, the heavier hard coatings are more likely to be smoother, but they may contain microcracks.

Tin and Tin-Lead Plating. Electrodeposits of tin are corrosion-resistant and nontoxic, posses excellent solderability, and are noted for their softness and ductility. Although tin plate is quite tarnish-resistant, it does not offer sacrificial protection in the air when plated on steel. To provide corrosion protection of a substantial nature, tin deposits must be thick enough to be virtually nonporous. Tin-lead electrodeposits are widely used in the electronics industry for solderability purposes, particularly on printed-circuit boards. A tin-lead electrodeposit is, of course, essentially a solder coating.

Silver Plating. Silver electrolytes are the most widely used commercially both for decorative and for engineering plating. With proper control the electrolyte composition and plating procedures, smooth, dense, fine-grained, strongly bonded silver plate ranging from less than 1 mil to more than 1/4 in thick may be deposited from cyanide baths on most metals and alloys. The smoothness and brightness of cyanide silver deposits can be controlled by adding to the bath small quantities of sulfur-bearing compounds such as ammonium thiosulfate and carbon disulfide. These additives also help to suppress burning and treeing in areas of high current density.

Gold Plating. Gold is ideal because it (1) resists tarnish and corrosion of various chemicals, (2) maintains a low electrical contact resistance, (3) resists high-temperature oxidation, and (4) is solderable after extended periods of storage.

1.3.2.2 Immersion and Electroless Coatings. Immersion and electroless coatings applied without the aid of an electrical current in the plating bath are important in the electronics industry. Nickel is commonly deposited by one or both of these techniques. Perhaps most commonly applied are copper, nickel, gold, tin, and silver.

These coatings have two advantages. (1) Application is made from a nonelectrolytic solution. Variable potential gradient points do not occur on the part being plated and hence much more uniformity of deposit is obtained than from an electrolytic bath. (2) Nonconductors, such as plastics, can be given a basic nonelectrolytic metallic coating using electroless techniques, which can in turn be followed by electrolytic deposition of other metals onto the nonmetallic base material. The basics of immersion plating and electroless or catalytic reduction can perhaps best be described by outlining their application in nonelectrolytic nickel plating.

The aqueous immersion plating bath is capable of depositing a very thin (about 0.025-mil) and uniform coating of nickel on steel in periods of up to 30 min. The coating is porous and possesses only moderate adhesion, but these conditions can be improved by heating the coated part at 1200°F for 45 min in a nonoxidizing atmosphere. (Higher temperatures will promote diffusion of the coating.)

The electroless nickel-plating process uses a chemical reducing agent (sodium hypophosphite) to reduce a nickel salt (such as nickel chloride) in hot aqueous solution and to deposit various thicknesses of nickel on a catalytic surface. The deposit obtained from an electroless nickel solution is an alloy containing from 4 to 12 percent phosphorus and is quite hard. The hardness of the as-plated deposit can be increased by heat treatment. Because the deposit is not dependent on current distribution, it is uniform in thickness, regardless of the shape or size of the plated surface.

1.3.2.3 Conversion Coatings. A conversion coating can be either a nonelectrolytic chemical treatment or an electrolytic chemical treatment that modifies a reactive metallic surface to a relatively inert inorganic film. Nonelectrolytic chemical treatments include phosphate, chromate, and oxide coatings. Typical of electrolytic chemical coatings are anodic treatments for aluminum- and magnesium-based alloys and other treatments such as anodizing for aluminum- and magnesium-based alloys. These conversion coatings then are primarily used to convert the surface of active metals such as magnesium and aluminum to a less chemically active or inert surface.

While chromate and oxide coatings are used predominantly on aluminum and magnesium, phosphate coatings are also used on steel and zinc metals.

1.3.2.4 Aluminum Coatings. Aluminum may be coated either by electroplating or by conversion coatings. Electroplating of aluminum requires particular precautions due to the high surface activity of the raw aluminum surface. For this reason the common practice for electroplating on aluminum is to apply first a zinc coating onto the surface of the aluminum, a process known as zincating. Subsequent electrodeposited coatings can then be applied on the zinc coating. A listing of the coatings which can be applied to aluminum and aluminum alloys is shown in Table 1.15. Because of the problems in electroplating aluminum, conversion coatings and anodized coatings are widely used on aluminum.

Conversion Coatings on Aluminum. There are proprietary compounds on the market for coating aluminum (such as Alodine,[*] Iridite,[†] and Kenvert[‡]). Their chief advantages are corrosion resistance, paint base, and ease of application. The corrosion resistance

---

[*]Alodine is a registered trademark of Amchem Products, Inc.
[†]Iridite is a registered trademark of Allied Research Products, Inc.
[‡]Kenvert is a registered trademark of Conversion Chemical Corporation.

TABLE 1.15 Chemically Deposited Coatings for Aluminum and Aluminum Alloys

Treatment	Purpose	Application	Operation	Finish and thickness
Zinc phosphate coating	Paint base	Wrought alloys	Power spray or dip. For light to medium coats, 1–3 min at 130–135°F	Crystalline, 100–200 mg/ft²
Chromium phosphate coating	Paint base or corrosion protection	Wrought or cast alloys	Power spray, dip, brush, or spray. For light to medium coats, 20 s to 2 min at 100–120°F	Crystalline, 100–250 mg/ft²
Sulfuric acid anodizing	Corrosion and abrasion resistance, paint base	All alloys; uses limited on assemblies with other metals	15–60 min, 12–14 A/ft², 18–20 V, 68–74°F; tank lining of plastic, rubber, lead, or brick	Very hard, dense, clear. 0.0002–0.0008 in thick. Withstands 250–1000-h salt spray
Chromic acid anodizing	Corrosion resistance, paint base; also as inspection technique with dyed coatings	All alloys except those with more than 5% Cu	30–40 min, 1–3 A/ft², 40 V dc, 95°F; steel tanks and cathode, aluminum racks	0.00002–0.00006 in thick, 250-h minimum salt spray
Chromate conversion coating	Corrosion resistance, paint adhesion, and decorative effect	All alloys	10 s to 6 min depending on thickness, by immersion, spray, or brush, 70°F; in tanks of stainless, plastic, acid-resistant brick or chemical stoneware	Electrically conductive, clear to yellow and brown in color, 0.00002 in or less thick, 150–2000-h salt spray, depending on alloy composition and coating thickness
Chemical oxidizing	Corrosion resistance, paint base	All alloys, less satisfactory on copper-bearing alloys	Basket or barrel immersion, 15–20 min, 150–212°F	May be dyed, 250-h minimum salt spray
Electropolishing	Increases smoothness and brilliance of paint or plating base	Most wrought alloys, some sandcast and diecast alloys	15 min, 30–50 A/ft², 50–100 V, less than 120°F, aluminum cathode	35–85 μm depending on treatment
Zinc immersion	Preplate for subsequent deposition of most plating metals, improves solderability	Many alloys, modifications for others particularly regarding silicon, copper, and magnesium content	30–60 s, 60–80°F; agitated steel or rubber-lined tank	Thin film

TABLE 1.15 Chemically Deposited Coatings for Aluminum and Aluminum Alloys (*Continued*)

Treatment	Purpose	Application	Operation	Finish and thickness
Electroplating	Decorative appeal and/or function	Most alloys after proper preplating	—	Same as on steel
Chromium	—	—	Directly over zinc immersion coat, 65–70°F, 6–8 V, 200–225 A/ft^2; transfer to bath at 120–125°F if copper has or copper and nickel have been applied	
Copper	—	—	Directly over zinc, or follow with copper strike, then plate in conventional copper bath	
Brass	—	—	Directly over zinc, 80–90°F, 2–3 V, 3–5 A/ft^2	
Nickel	—	—	Directly over zinc, or follow with copper strike, then plate in conventional nickel bath	
Cadmium	—	—	Directly over zinc, or follow with copper or nickel strike, or preferably cadmium strike, then plate in conventional cadmium bath	
Silver	—	—	Copper strike over zinc using copper cyanide bath, low pH, low temperature, 24 A/ft^2 for 2 min, drop to 12 A/ft^2 for 3–5 min; plate in silver cyanide bath, 75–80°F, 1 V, 5–15 A/ft^2	
Zinc	—	—	Directly over zinc immersion coating	
Tin	—	—	Directly over zinc immersion coating	
Gold	—	—	Copper strike over zinc as for silver, then plate in conventional bath	

Source: From Lisman.[23]

imparted to aluminum by using these dips is excellent, and in some cases better than chromic acid anodizing. As a paint base these materials are superior, and it is recommended that one of the proprietary compounds be used in order to obtain better adhesion. The ease of application of these coatings is a very important factor in replacing anodizing, and because of this the cost is one-third to one-half that of anodized coatings.

Anodizing. Anodizing is an electrochemical process by which the oxide film present on aluminum can be increased in thickness. This oxide is an inherently corrosion-resistant film, which normally has a thickness in the untreated form of 0.00000052 in. Subsequently, after an anodizing treatment, the thickness of this oxide (Al_2O_3) will be increased by 500 to 2000 times the original thickness.

The procedure for the formation of this heavy oxide consists of making the aluminum part to be anodized the anode in a direct-current circuit. The current passes from the cathode through a solution of sulfuric acid, chromic acid, or other electrolyte to the anode. This procedure results in the liberation of nascent oxygen at the anode, which reacts with the metal to form a film of aluminum oxide at the surface of the metal.

Two general types of anodizing are normally used, the sulfuric acid process and chromic acid anodizing. Sulfuric acid anodizing for aluminum is the most widely used anodizing process and the most economical. The coating varies from a clear, transparent film to one that is opaque or translucent; it usually allows some of the metallic sheen of the base metal to remain. The differences in color depend on the particular alloy, the length of treatment, and variations in the composition of the bath. The film thickness ranges from 0.0001 to 0.001 in. When unsealed, the coating is porous, absorbent, and very hard. In order to give maximum corrosion protection, it must be made nonabsorbent by sealing. It provides very good corrosion and abrasion resistance and is a very good paint base.

Chromic acid coatings are opaque and gray in color, varying from light gray for purer aluminum to greenish gray on certain alloys. The greenish tinge, when present, is said to be due to reduced chromium compounds. The thickness of the film ranges from 0.00003 to 0.001 in, and the average is about 0.00005 in. The thickness can be varied with bath conditions and length of treatment. The coating formed by the chromic acid process is not as porous or as absorptive as that formed by the sulfuric acid process. This makes it less desirable for dyeing purposes, but its protective qualities are equally as good and it is preferable in some instances, such as assemblies. Unlike the results of sealing the sulfuric acid anodic coating, sealing a chromic acid process coating does not increase, but sometimes may even decrease, its corrosion resistance. The film produced with chromic acid is not so porous. This is because the higher temperature and longer treatment time of the process favor the formation of the more pore-free, hydrate crystal aluminum oxide coating. Second, hot-water sealing leaches out some of the protective chromate absorbed from the bath, resulting in reduced protection. A coating that is dyed, however, is still sealed in nickel acetate or hot water to protect the color.

Hard Anodic Coatings. The thickest anodic coatings formed on aluminum are produced by the hard anodic coating techniques. Proprietary processes include Alumilite,[*] Duranodic,[*] Martin,[*] Sanford,[†] and Hardas.[‡] Because a thickness of 1 mil or more is one of the primary requirements of a hard anodic coating, an electrolyte that enables the growth process to continue is essential. The other primary requirement is hardness. This term is somewhat ambiguous in describing this coating, however, because the aluminum oxide of the so-called hard anodic coating is actually no harder than that of conventional anodic coatings. The electrolyte and other conditions are such, however, that the thick aluminum oxide coating produced has a dense structure and resists abrasion and erosion better than the conventional anodic coatings. This high density results from anodic oxidation at high current density in an electrolyte, which has a minimum solvent effect on the aluminum oxide formed. The electrolyte is sulfuric acid (low solvent action) with or without additions of oxalic acid. Oxalic acid additions result in even less solvent action; consequently, denser coatings are produced.

---

[*]Alumilite, Duranodic, and Martin are registered trademarks of Aluminum Company of America.
[†]Sanford is a registered trademark of the Sanford Process Company, Inc.
[‡]Hardas is a registered trademark of Hardas Ltd.

Conventional anodic coatings produced in sulfuric acid (15 percent by weight) electrolytes at 70°F use current densities of about 12 A/ft^2. Hard anodic coatings are formed in electrolytes of 15 percent or less concentration, in a temperature range of 25 to 50°F, at current densities of 24 to 36 A/ft^2. All of these conditions favor high coating ratios, even on alloys that normally produce porous anodic coatings.

Parts with thin sections or sharp edges must be treated with extreme care to produce satisfactory hard anodic coatings. Such areas, particularly in the copper-bearing alloys, promote "burning" or dissolving of the base metal. Sharp corners and edges produce high-current-density areas in conventional anodic-coating methods, and this condition is accentuated by the higher current densities used in hard anodic processing.

1.3.3 Magnesium Coatings

Magnesium, even more than aluminum, has a very reactive surface which necessitates close control of plating and finishing operations.

The Dow electroplating process for magnesium consists of the initial application of an immersion zinc coating followed by copper striking and electroplating in standard plating baths. With this process any metal that can be electrodeposited can be applied to magnesium. The success of the process depends almost entirely on the adhesion and uniformity of the zinc immersion coating. This coating is approximately 0.0001 in thick. Electron diffraction analysis shows the coating to be pure metallic zinc. Copper- and silver-plated surfaces have been joined by soft soldering without disrupting the adhesion. Deposits have been hammered and bent without failure. Heavy chromium plates have been applied to articles for wear resistance without a tendency for peeling of deposits. General corrosion tests (salt spray, high humidity, interior and exterior exposure) over a period of years indicate good performance. Service conditions usually determine the entire plating cycle required in plating on any base metal. Thus, for exterior use, service testing to determine the proper plating thickness on magnesium parts is recommended. In many current electronic applications, hot flowed tin, gold, or silver electroplates are used over either a copper or a copper-nickel plate. Standard tin, gold, or silver electroplating baths are used. Such coatings are used primarily for RF grounding or hermetic sealing.

The Dow electroless nickel process for magnesium permits the chemical deposition of a nickel coating directly on magnesium with no corrosive attack on the base metal surface. The nickel plate applied by this process can be used as a final coat or as a strike coating (0.0001 to 0.0002 in) over which standard plates (such as bright nickel, chromium, tin, cadmium, or zinc) can be deposited. No electric current is required. The coating can be built up to any desired thickness. The process coats the surface wherever the solution touches it and covers all surfaces evenly. This uniform deposition rate gives closer dimensional tolerances and better maintenance of detail than electroplating. The auxiliary anode and special racking commonly required to "throw" the plate inside blind holes and recessed areas are not needed with electroless nickel plating. If maximum adhesion and corrosion resistance are needed, the zinc immersion plus copper strike should be used prior to the electroless nickel plating.

1.3.4 Selective Plating

This process is a form of localized plating, sometimes known as brush plating or Selectron.* It is electrolytic solution plating using very small quantities of plating solution

---

*Selectron is a registered trademark of Selectron Ltd.

applied by means of a small brush or stylus which carries an electric current. Many metals can be electrolytically deposited by this process. Its use allows either the plating of small assemblies or the repair of plated parts which are not practical to replate in a tank electroplating bath.

1.3.5 Electroforming

Electroforming essentially consists of electroplating onto a master form or mandrel, which is removed after plating, thus leaving an electroplated shell or form whose contour takes the shape of the mandrel or form on which it was plated. Low-melting metals which subsequently can be melted out of the electroformed part are commonly used for mandrels.

Electroformed objects are normally somewhat thicker than electroplated parts. Electroforming is, of course, most widely used for the fabrication of metal parts, where this form of fabrication offers unique advantages. Electroformed parts can be made from many metals using a variety of electroplating solutions. Since electroformed parts usually have a considerable thickness buildup (compared with plating) electroforming involves more plating time than simple electroplating.

Some of the key advantages of electroforming include controllability of the metallurgical properties of an electrodeposited metal over a wide range of selection of metals, adjustment of plating bath composition, and variation of conditions of deposition. In addition, with the proper choice of matrix material, parts can be produced in quantity with a very high order of dimensional accuracy; reproduction of fine details is of a high order; there is almost no limit to the size of the end product; and shapes can be made that are not possible using any other fabrication method.

Among the disadvantages, on the other hand, is the cost, which is often relatively high. Also, the production rate may be comparatively slow, sometimes measured in days. There are also limitations in design: the electroform cannot have great or sudden changes in cross section or wall thickness unless these can be obtained by machining after electroforming. In addition, because of the exactness of reproducibility, scratches and imperfections in the master will also appear in all pieces processed from it.

1.4 CERAMICS AND GLASSES

Ceramics and glasses (see also Recommended Reading no. 8) serve as electrical insulators in electronic packaging applications. Selection is based on mechanical, thermal, and chemical properties, assembly characteristics, and availability. Because ceramics and glasses differ in assembly procedures, they are discussed separately.

1.4.1 Ceramics

The great majority of traditional ceramic materials are electrical insulators. This handbook is limited to those ceramics which are commonly used, or will be used, in electronic packaging applications. A broader, less focused discussion of ceramic insulators is found in Buchanan.[24]

Ceramic insulators for electronic manufacturing, packaging, and assembly are listed in Table 1.16 along with materials properties. An indication of the degree of commercialization is also provided. The most important one of all the listed materials is aluminum oxide.

TABLE 1.16 Electronic Packaging Ceramics

Ceramic	Dielectric constant	Dissipation factor tan δ, %	Thermal conductivity, W/m · K	Flex strain, MPa	Thermal expansion, ppm/°C	Density, kg/m³
AlN*	10	0.100	150	300	2.7	3.2
Alumina	9.7	0.060	40	385	7.2	4.0
BeO	6.8	0.100	300	170	7	2.9
BN (hex)	4.1	0.100	60	110	3.8	2.2
Borosilicate glass	3.7	2.600	2	50	3.2	2.1
Cordierite	4.5	0.400	2.5	70	2.5	2.7
Forsterite	6.2	0.050	3.3	170	9.8	2.9
Glass (slide)	7.2	6.000	2	50	9.2	2.9
Low fire tape*	8.0		3	200	7.9	2.9
Mullite	6.8	0.500	6.7	140	4	3.1
Quartz	4.1		2	140	13	2.6
SiC	40		270	450	4.3	3.2
Steatite	5.7	0.100	2.5	170	4.2	2.7
Vitreous SiO_2	3.8	0.004	1.6	50	0.5	2.2

*Emerging technology.

Aluminas. The aluminum-oxide-based ceramic insulators are a natural extension of traditional, triaxial, porcelain (clay, feldspar, flint) ceramics. Starting from the compositions of fine grades of dinnerware, electrical porcelains wore formulated with higher levels of aluminum oxide because of improved strength (Table 1.17). The trend toward higher alumina contents brought the additional benefit of high thermal conductivity, which made high-alumina ceramics superior to the alternatives of the day. Aluminum oxide is 20 times higher in thermal conductivity than most oxides. The flexure strength of commercial high-alumina ceramics is 2 to 4 times greater than that of most oxide ceramics.

The drawbacks of alumina ceramics, which are increasingly drawing attention, are their relatively high expansion ($\approx$7 ppm/°C) compared to the chip material silicon ($\approx$4 ppm/°C) and the moderately high dielectric constant ($\approx$10). The thermal expansion is in fact an intermediate value between the extremes of the common electronic packaging and device materials silicon and copper. Acceptability tends to reduce to the relative determination of the materials compatibility problem that is easier to design around.

Aluminas are fabricated from aluminum oxide powders with various percentages of sintering promoters,[25] which produce the so-called glassy phase. The latter additives reduce the densification temperatures to between 1500 and 1600°C. Based on the final application, the powders may be pressed, extruded, or prepared in slurries for slip casting or tape casting. In the latter technique a slurry of powders is prepared, usually incorporating organic-based binders, dispersants deflocculants, and solvents.[26] By drawing the "milk-shake"-like slurry under a knife-edge, called a doctor blade, onto a moving easy-release coated polymer or glass surface, a thin coating is formed which becomes a flexible sheet after drying. This "green" (unfired) tape may be peeled from the supporting surface and cut or punched to shape and subjected to further processing. The resulting sheets of material are densified at an elevated temperature in an atmosphere reflective of the subsequent processing of the green tape. If the alumina is not metallized, air firing is used. Without any further processing, the resulting product is the common alumina substrate.

TABLE 1.17 Typical Properties of High-Alumina Ceramics

Property	Aluminum oxide content, %							
	85	90	90*	94	96	99.5	99.9	99.9
General								
Specific gravity	3.41	3.60	3.69	3.62	3.72	3.89	3.96	3.99
Color	White	White	Black	White	White	Ivory	Ivory	Transparent white
Surface finish,† μin	63, 39, 8	63, 20, 3.9	63, 39, 3.9	63, 51, 12	63, 51, 12	35, 20, 3.9	20, 35, <1.2	24, 35, <1.2
Water absorption	None	None	None	None	None	None	None	None
Gas permeability‡	None	None	None	None	None	None	None	None
Mechanical								
Hardness, R45N	73	79	75	78	78	83	90	85
Tensile strength, lb/in²								
25°C	22,000	32,000	33,000	28,000	28,000	38,000	45,000	30,000
1000°C	—	15,000	—	15,000	14,000	—	32,000	15,000
Compressive strain, lb/in²								
25°C	280,000	360,000	350,000	305,000	300,000	380,000	550,000	370,000
1000°C	—	75,000	—	50,000	—	—	280,000	70,000
Flexural strength, lb/in²								
25°C	43,000	49,000	53,000	51,000	52,000	55,000	80,000	41,000
1000°C	25,000	—	—	20,000	25,000	—	60,000	25,000
Modulus of elasticity, lb/in²	32×10^6	40×10^6	45×10^6	41×10^6	44×10^6	54×10^6	56×10^6	57×10^6
Shear modulus, lb/in²	14×10^6	17×10^6	18×10^6	17×10^6	18×10^6	22×10^6	23×10^6	23.5×10^6
Thermal								
Specific heat, cal/(g · °C)	0.22	0.22	0.25	0.21	0.21	0.21	0.21	0.21
Coefficient of expansion, 10^{-6} (in/in)/°C								
25–200°C	5.3	6.1	7.2	6.3	6.0	7.1	6.5	6.5
25–800°C	6.9	7.7	8.1	7.6	8.0	8.0	7.8	7.8
25–1200°C	7.5	8.4	—	8.1	8.4	—	8.3	8.3
Conductivity, cgs								
20°C	0.035	0.040	0.030	0.043	0.059	0.085	0.093	0.095
400°C	0.016	0.017	0.018	0.017	0.024	0.028	0.032	0.032
800°C	0.010	0.010	—	0.010	0.013	0.015	0.015	0.015
Maximum service temperature,¶ °C	1400	1500	1500	1700	1700	1750	1900	1900

Electrical

	1	2	3	4	5	6	7	8
Dielectric strain, V/mil								
0.25 in	230	240	220	210	220	135	235	240
0.050 in	510	460	430	370	425	415	450	440
0.01 in	—	800	840	580	720	720	760	720
Dielectric constant								
1 kHz	10.1	9.9	9.8	9.0	8.9	22.0	8.8	8.2
1 MHz	10.1	9.8	9.7	9.0	8.9	9.8	8.8	8.2
100 MHz	10.1	—	—	9.0	8.9	—	8.8	8.2
Dissipation factor								
1 kHz	0.0005	0.0020	0.0002	0.0011	0.0002	0.3000	0.0006	0.0014
1 MHz	0.00004	0.0002	0.0003	0.0001	0.0001	0.0200	0.0004	0.0009
100 MHz	0.00006	—	—	0.0002	0.0005	—	0.0004	0.0009
Loss index								
1 kHz	0.0050	0.020	0.002	0.010	0.002	6.6	0.005	0.011
1 MHz	0.0004	0.002	0.003	0.001	0.001	0.200	0.004	0.007
100 MHz	0.0006	—	—	0.002	0.004	—	0.004	0.007
Volume resistivity, $\Omega \cdot$ cm								
25°C	—	$>10^{15}$	$>10^{14}$	$>10^{14}$	$>10^{14}$	$>10^{14}$	$>10^{14}$	$>10^{14}$
1000°C	—	1.1×10^{7}	—	1.0×10^{6}	5.0×10^{5}	4.0×10^{4}	8.6×10^{5}	—
T_e value, § °C	—	1170	—	1000	950	—	960	850

*Opaque.

†Values pertain to as-fired, ground, and polished conditions, respectively.

‡No helium leak through plate 0.25 mm thick by 25.4 mm diameter at 3×10^{7} torr vacuum versus about 1 atm helium pressure for 15 s at room temperature.

¶No load.

§Temperature at which resistivity is 1 M$\Omega \cdot$ cm.

Source: Coors Porcelain Company.

The surface finish of ceramics used to be quite limited by the method of manufacture. Considerable refinement in processing has made excellent surface finishes possible without further processing. Generally aluminas may have surface finishes of 3 to 25 µm/in as a result of normal processing. For very smooth finishes (2 µm/in) the surfaces have to be lapped or polished. Glazing offered a cheap improvement over processed surfaces. Today, there are processing techniques that can make further improvements in surface finish.[27] Only time will tell whether these processes can be price-competitive with lapping and polishing.

Alumina ceramics, like most ceramic materials, are rarely used apart from being bonded to metals. The means by which this is accomplished frequently dictates the processing technique. Metallization of aluminas is usually accomplished by either high-temperature firing or low-temperature thick-film processing. Fired, shaped aluminas are usually postmetallized by refiring the formed article after coating with a slurry of molybdenum and manganese powder or tungsten metal powder.[28] Based on the purity of the alumina, glass powder may be added to the metal powder. The mechanism of metallization requires an internal glassy phase or the added glassy phase to penetrate the interstices of the refractory metal powder as well as the intergranular phase of the aluminum oxide. This is accomplished by firing in slightly reducing and moist atmospheres at temperatures above 1500°C. The resulting metallization is usually plated electrochemically with nickel or copper, or both.

Alumina substrates may be metallized by thick-film processing.[29] Thick-film processing has traditionally been performed in oxidizing atmospheres at moderate temperatures, 800 to 1000°C. The metals are the precious metals—gold, silver, platinum, palladium, and their combinations. A glassy phase is usually incorporated in the thick-film paste.

Each metallization system has advantages and disadvantages. The advantage of the high-temperature metallization schemes with molybdenum or tungsten is their moderate cost. The disadvantage is the high resistivity of the resultant films. These relative merits are reversed for thick-film materials. Another advantage of thick-film metallization is that the metallization is often applied by the device fabricator and not the ceramic vendor. This can allow for greater efficiency in design modification.

For many reasons, the use and amount of glassy phase in high-alumina ceramics vary considerably, in particular in substrates. Normally these ceramics are identified by the percent of aluminum oxide they contain. The quoted figures generally range between 90 and 100 percent.[30] This designation is misleading in the sense that it does not refer to the percentage of crystalline aluminum oxide, but to the weight percent of chemically analyzable Al_2O_3. Often one-third of the glassy phase is Al_2O_3 and is included in the total analysis. This means that the proportion of crystalline Al_2O_3 is usually less than the reference percent. This becomes important when selecting the grade of high-alumina ceramic because of thermal conductivity. In a high-alumina substrate, for example, if the volume percent glass is above 9 percent, the crystalline Al_2O_3 grains lose their connectivity[31] and thermal conductivity deteriorates rapidly. Although the composition of the glassy phase and its volume fraction are usually considered proprietary, the effect of the glassy phase on conductivity in high-alumina ceramics is very pronounced (Fig. 1.20). The effect on strength and the dielectric constant is much smaller.

A final comment on high-alumina manufacturing relates to the manufacture of cofired multilayer ceramics.[32] Because of areal and resolution limitations in two-dimensional circuitry, high-density packages, chips, and discrete devices are frequently attached to the planar lead termini of three-dimensional circuitry formed by a layering and interconnecting strategy called multilayer. In the cofired version, green tapes have feed-through vias. They then have conductor lead traces carefully applied by silk screening. The various circuit layers are applied to separate green sheets. The resulting sheets are stacked, the via

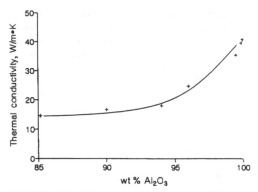

FIGURE 1.20 Effect of weight percent Al₂O₃ on thermal conductivity.

holes are made conductive, and the layers are registered. The assembly may have over 40 layers which, after stacking, are initially consolidated in a laminating press using low temperatures ($\approx 100°C$). The laminated stack is then fired in the manner previously described. When the green tape is Al_2O_3-based, the metallization must be a refractory metal such as tungsten or molybdenum because of the high firing temperature. This results in higher lead resistivities than thick-film metallization.

Until recently, only Al_2O_3-based cofired multilayer substrates were available. An alternative to cofired circuitry was produced by building up multiple thick-film applications and firing to alumina substrates. Recently, however, a number of low-fire tape systems[33] based on extensions of thick-film compositions have been commercialized, which permit cofired multilayer processing in thick-film equipment. These have targeted Al_2O_3-matching expansions and lower-expansion, lower-dielectric-constant systems. The lower dielectric constant and expansion, contrasted to Al_2O_3-based cofired multilayer materials, come at the expense of poor thermal conductivity.

Efforts to find alternatives to Al_2O_3 arise from a number of concerns. Logic applications find speed limited by the high dielectric constant of the substrate and the high resistivity of the lead traces. Power applications find the thermal conductivity of Al_2O_3 inadequate for thermal management. Increases in chip size find the thermal expansion mismatch between substrate and die no longer manageable. Of the alternatives that are emerging it is clear that no single material on the realistic horizon will answer all of Al_2O_3's deficiencies. Discussion of these alternatives will be focused on understanding these trade-off issues.

Beryllia. Beryllium oxide (BeO)-based ceramics are in many ways superior to alumina-based ceramics.[34] The major drawback is the toxicity of BeO. Beryllium and its compounds are a group of materials which are hazardous and must be handled in such a way that proper precautions are taken to prevent potential accidents. Using proper care, the material has been used in electronic applications from automotive ignitions to defense systems. To understand the safe handling of this material, the reader is referred to Powers.[35]

Beryllium oxide materials are particularly attractive for electronic application because of their electrical, physical, and chemical properties, particularly their high thermal conductivity, which is roughly 10 times higher than that of the commonly used alumina-based materials. Figure 1.21 compares the thermal conductivity of BeO to that of alumina and some emerging alternative materials. As can be seen, BeO is of lower dielectric constant and thermal expansion than alumina, but slightly lower in strength.

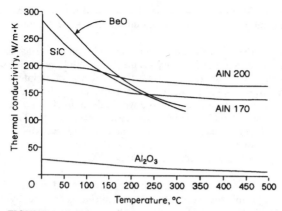

FIGURE 1.21 Thermal conductivity of ceramic substrate materials.

Beryllia materials are fabricated in much the same way as alumina materials, although the toxic properties of the powders mandate that they be processed in laboratories equipped to handle them safely. This results in a limited number of suppliers. Simple cutting, drilling, and postmetallization are also handled by especially equipped vendors. Reliable thin-film systems may be applied to BeO substrates. Such coatings may necessitate less elaborate safety precautions and are often applied by the device fabricators. The general safety requirements for the source substrate and package materials impact design and prototype scheduling, but not unreasonably. Prices are considerably higher than for alumina-based materials. Thus use is generally restricted to applications where thermal management considerations cannot be served in any other way.

Aluminum Nitride. Aluminum nitride (AlN)-based substrates[36] and packages began to appear in the late 1980s as an alternative to the toxicity concerns of BeO. As Fig. 1.21 shows, the thermal conductivity of AlN is comparable to that of BeO but deteriorates less with temperature. The dielectric constant of AlN is comparable to that of alumina and thus a liability in high-speed device applications, but its thermal expansion is low (4 ppm/°C) and comparable to that of silicon. Because of its expansion, it is being evaluated in conjunction with large silicon die or silicon substrate attachments, commonly called multichip designs.

The fabrication of AlN-based materials is very different from that of alumina- and beryllia-based materials and must be factored into application considerations. The high thermal conductivity is attributable to the presence of AlN and prevention of its oxidation to Al_2O_3. Thus AlN powders are sintered at very high temperatures in N_2 atmospheres. Because the powders are costly and sintering is difficult, the product costs are comparable to those of beryllia materials. As Fig. 1.21 indicates, thermal conductivity occurs in three ranges, the highest being the most difficult and costly to fabricate. The lowest conductivity level is associated with cofire metallization. Because tungsten undergoes unfavorable reaction with nitrogen, cofired AlN is fired in vacuum. This inhibits sintering, which results in lower thermal conductivity ($\approx 2 \times$ alumina).

Because AlN is a relatively new material and because it is a nitride and not an oxide, glass bonding characteristics are not well defined. Therefore thick-film coating systems are in the early stages of development and reliability is not firmly established. Nevertheless, the major thick-film manufacturers are active in the development of compatible systems.[37] AlN reacts with the alkaline environments frequently used in cleaning/degreasing operations, and they must be avoided. To summarize, while alumina- and beryllia-based materials and their thick-film companions are technically mature materials, AIN is only emerging at the beginning of the 1990s.

Silicon Carbide. Silicon carbide (SiC)[38] has found some special application as a substrate material because of its high thermal conductivity (Table 1.16). While normally an electrical conductor, SiC has been able to function as an insulator because of an inter-granular insulating phase which interrupts continuity between the individual grains of the polycrystalline sintered material. Conductivity can be observed within the individual grain feature size of 40 Am. A major drawback of SiC is its high dielectric constant ($\approx$40). The material is more costly than alumina-based materials, but comparable to BeO and AIN at this time. There are no known development efforts of thick-film systems comparable to that known for AIN. It would thus appear that this material should be viewed as a selective alternative to BeO and AIN.

Boron Nitride. Boron nitride ceramics[39] are used occasionally when thermal conductivity or machinability concerns are prominent. Hexagonal boron nitride is formed by pyrolytic deposition or by hot processing boron nitride powders in the presence of a borate glass phase. The resulting material is soft and easily machined. Of chief interest is the thermal conductivity of the material, which is about twice that of the better aluminas, that is, 60 W/ m • K. It cannot be metallized nor sealed to; so its use is very limited.

1.4.2 Traditional Ceramics

There are a number of ceramics that are used occasionally in electronic applications.

Steatites   Lower firing ceramics offering price advantages over alumina ceramics.

Forsterite   A low-loss insulator with moderately high thermal expansion, which makes it more compatible with some metals.

Cordierites   Low-expansion magnesium silicates which are useful for applications experiencing thermal shock.

Lava   The chief attractiveness of this natural alumino or magnesium silicate is its softness and machinability. Once machined, it may be fired at moderate temperatures into a dense ceramic article. This gives "do-it-yourself" capabilities to many laboratory and development activities.

This potential is also met with a glass ceramic available from a major glass supplier.[40] The advantage of the latter is that it does not have to be fired and machining dimensions are perfectly retained.

1.4.3 Glasses

Glasses are important materials in electronic applications, but serve in more varied roles than the ceramic materials just described.[41] Glasses may be used for insulating purposes, for their surface smoothness, for their controllable thermal deformability, or for their ability to serve as a bonding phase. They are used as substrates, delay lines, passivation layers, capacitors, resistor and conductor bonding phases, package sealants, and insulating bushings.

Glasses are noncrystalline solids. They have the random atomic structure of a liquid that gets frozen in as the melted glass is cooled. The feature that permits glasses to remain amorphous rather than crystallize is the very high viscosity of the melted material. This phenomenon is frequently used to advantage when glasses are employed. At suitable temperatures, glasses will slowly deform and fill cavities for insulator feedthroughs. Due to its high surface tension, glass will also fill and be retentive in the cavities of a filler powder, which may be insulator (capacitors), semiconductor (resistors), or conductor in thick-film applications.

Since glasses are not definite crystals but are closer to chemical mixtures, they may have an infinite range of compositions. In spite of this reality, almost all glasses are based on four glass-forming oxides: SiO_2, B_2O_3, P_2O_5, and occasionally GeO_2. The first three may be intermixed. To these basic oxides are added a number of softeners, or "fluxes," the oxides of Li, Na, K, Rb, Cs, Pb and extenders, the oxides of Mg, Ca, Sr, and Ba. Another common group of nonfunctional additives are colorants, the oxides of Co, Mn, Fe, and Cr.

Table 1.18 lists some common glasses of electronic packaging along with some properties. It is best to think of glass as a liquid that has been cooled to a temperature where it has become so stiff that for all practical purposes it is rigid. This viewpoint helps in understanding the various viscosity points that are frequently reported for glasses (Fig. 1.22). The different points refer to temperatures where the glass has become less stiff than at room temperature. These points are defined in some cases by how long it takes for the strain to be relieved in hours (the strain point) or in minutes (the annealing point) by viscous flow at the temperature, or how long it takes for a fiber to deform under its own weight (the softening point). The working point is where the glass may be "pushed around." These temperatures are used to determine where particular glasses may be sealed, bonded, and so on.

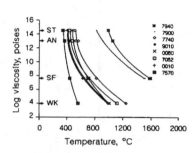

FIGURE 1.22 Glass viscosity as a function of temperature. ST—strain point; AN—annealing point; SF—softening point; WK—working point.

Table 1.19 shows the compositions of some commercial glasses. The additives in glass have a great effect on the viscosity points, thermal expansions, and electrical properties, as Figs. 1.22 and 1.23 show. These are the major design parameters used in glass selection. The thermal conductivity for most glasses is very similar and very low. The flexure strength of glasses is very similar between compositions, also.

The thermal expansion of glasses varies from ≈0 ppm/°C for pure SiO_2 glass to ≈12 ppm/°C for some compositions. There is one feature of thermal expansion that is peculiar to glass and is important for designers to understand. In addition to the absence of structure, glasses are distinguished by their nonlinear thermal expansion. Glasses undergo an abrupt increase in thermal expansion within 20 to 50°C of their softening points (Fig. 1.23). The temperature at which the effect begins is called the transformation point T_g. The commonly quoted thermal expansion coefficients for glass describe the expansion below T_g. Bonding to metals and ceramics occurs at temperatures beyond T_g, so the selection of matching thermal expansions must take this into account. Data this complete are normally only obtained by special request from the manufacturer. Compatibility selection must consider this detail.[42]

Electrical resistivity is inversely controlled by the alkali present. In order to reduce alkali levels, PbO is frequently used to reduce processing temperatures without elevating conductivity. A brief discussion of the common glass systems follows.

TABLE 1.18 Properties of Common Electronic Packaging Glasses

Glass* code	Type	Thermal expansion 0–300°C, ppm/°C	Viscosity data, °C				Density, g/cm³	Young's modulus, 10^6 lb/in²	Poisson's ratio	$\log_{10}$ of volume resistivity			Dielectric properties at 1 MHZ and 20°C		
			Strain point	Annealing point	Softening point	Working point				25°C	250°C	350°C	Power factor, %	Dielectric constant	Loss factor, %
0010	Potash soda lead	93	395	435	625	985	2.86	8	0.21	17.+	8.9	7.0	0.16	6.7	1
0080	Soda lime	92	470	510	695	1005	2.47	10.0	0.24	12.4	6.4	5.1	0.9	7.2	6.5
7052	Borosilicate	46	435	480	710	1115	2.28	8.2	0.22	17.0	9.2	7.4	0.26	4.9	1.3
7570	High lead	84	340	365	440	560	5.42	8.0	0.28	—	10.6	8.7	0.22	15.0	3.3
7740	Borosilicate	33	515	565	820	1245	2.23	9.1	0.20	15.0	8.1	6.6	0.50	4.6	2.6
7900	96% silica	8	820	910	1500	—	2.18	10.0	0.19	17.0	9.7	8.1	0.05	3.8	0.19
7940	Fused silica	5.5	990	1050	1580	—	2.20	10.5	0.16	—	11.8	10.2	0.001	3.8	0.0038
9010	Potash soda barium	89	405	445	650	1010	2.64	9.8	0.21	—	8.9	7.0	0.17	6.3	1.1

*Corning Glass Works.

TABLE 1.19 Estimated Commercial Glass Compositions

Glass code*	Type	SiO_2	Na_2O	K_2O	CaO	MgO	Li_2O	PbO	B_2O_3	Al_2O_3
0010	Electrical	63	8	6	0.3	—	—	22	0.2	0.5
0080	Lamp	74	16	1	5	3	—	—	—	1.0
0120	Lamp	56	5	9	—	—	—	30	—	—
7050	Sealing	67	5	1.0	—	0.3	—	—	25	1.7
7070	Low loss	70	—	0.5	0.1	0.2	1–2	—	28	1.1
7570	Solder seal	3	—	—	—	—	—	75	11	11
7720	Electrical	73	2	2	—	—	—	6	17	—
7740	Labware	80	3.5	0.5	—	—	—	—	14	2
7900	96% SiO_2	96	0.2	0.2	—	—	—	—	3	0.6
7910	Fused SiO_2	99.5	—	—	—	—	—	—	—	—
8870	Sealing	35	—	7	—	—	—	58	—	—

*Corning Glass Works.

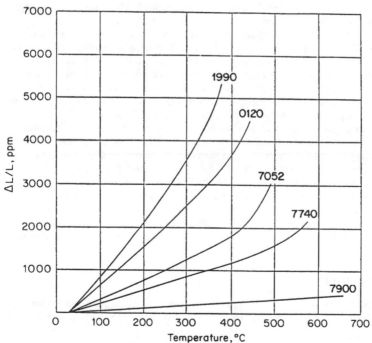

FIGURE 1.23 Thermal expansion of commercial glasses.

Pure SiO₂ Glass. Vitreous silica or fused silica, often inaccurately called fused quartz, is used only rarely and often where thermal shock considerations are primary. Coated and uncoated windows are typical applications. Pure SiO_2 glass is costly because of its exceptional melting temperature (>1700°C). Pure silica glass has the lowest dielectric constant of any ceramic material, 3.8.

96%SiO₂ Glass. Where the properties of pure SiO_2 are wanted at lower cost, a novel synthesis technique[43] permits the fabrication of 96%SiO_2 glass with very similar properties to 100%SiO_2. This glass is easily shaped and is thus appealing where complicated shapes, impossible to form from pure SiO_2 glasses, are wanted. The glass is formed by melting a lower

temperature glass, which after forming and heat treatment separates into two intertwined glass phases. One of the phases is easily digested in acid, leaving a skeleton of 96%SiO_2. This skeleton sinters densely at temperatures below 900°C, leaving a 96% silica glass.

Soda-Lime Glass. These compositions are based on Na_2O, CaO, SiO_2, and lesser amounts of such oxides as K_2O, MgO, or Al_2O_3. They define the bulk of common glasses, windows, tumblers, and so on. They are of moderately high thermal expansion and are not good in thermal shock applications. The alkali varies between 12 and 15 wt %, typically resulting in dielectric constants of 6 to 9. These glasses, with and without small levels of PbO, are frequently used as the compositions of feed-through bushing insulators, that is, hermetic metal packages. Their proven chemical durability is attested by their application in food processing.

Borosilicate. These glasses are the common laboratory and cookware glass compositions. They are of moderately low thermal expansion and show excellent chemical durability. They have high electrical resistivities and low dielectric constants. By themselves they are not a common electronics material, but they are the glassy binder phase in some of the newer low-temperature firing substrate systems.

Lead Alkali Borosilicate (Solder/Sealing Frits). These glasses are often based on PbO, B_2O_3, Al_2O_3, BiO, ZnO, and SiO_2 and are widely used in sealing and adhesive applications because of their low melting and softening temperatures.[44] They are used as powder additives by the vendors of thick-film pastes. Their compositions are considered proprietary but are compounded with a view to the ranges of materials compatibility requirements in semiconductor processing. PbO and BiO are additives that exhibit various compatibility problems and are included or not as the material requires.

The low-melting glasses are also used as hermetic "glues" for applying ceramic or metal lids to packages. In some cases these glasses are formulated to crystallize (devitrify). The advantage is that the crystallized products are frequently of much different properties than the glasses. This is a means of getting low or very high expansion.

Glass Ceramics. Glass ceramics[45] are glasses that have been designed to be crystallizable after forming, at temperatures below the glass melting and deformation temperatures. The resulting material is predominantly crystalline and offers special properties. Corning-ware* is the most familiar example. These materials may be twice as strong as common glass, or they may be machinable. They are used in electronics in specialty applications.

1.4.4 Glass Joining

For high-reliability devices, electrical connection is made by metal penetrations through insulating glass bushings.The glass is directly bonded by fusion to the feed-through pin

---

*Corningware is a registered trademark of Corning Glass Works.

and package wall, or bonded ring. Because glasses are so much weaker than ceramics, expansion matching is more critical. Remembering that glasses undergo a sharp nonlinearity in their thermal expansions, matching is not accomplished by matching published low-temperature coefficients. To join a metal and a glass, the total contractions are matched from a temperature referred to as the set point. This temperature is defined as the temperature where glass and metal joined at elevated temperature experience mutual stress upon further cooling. Above this temperature the glass dissipates stress by viscoelastic relaxation. Below this point the stress cannot be dissipated.

A matching contraction seal normally requires the following relation between published thermal expansion: $\alpha_g = \alpha_p = \alpha_r$ (Fig. 1.24). A limited series of alloys have been formulated which undergo an anomalous thermal expansion nonlinearity reminiscent of glass. These are based on 40–50%Ni-Fe alloys. These metals, combined with appropriate glasses, form the only truly matching seals and are used in the most severe thermal shock applications. They are frequently used in feed-through applications as depicted in Fig. 1.24.

Another approach to glass-to-metal compatibility often used in glass-to-metal feedthroughs is known as the compression seal. In this technique, a true expansion-matching pin and glass insulator combination is selected. The ring is selected from a higher-expansion alloy, typically 1010 steel. The expansion relationships are shown in Fig. 1.24. Upon cooling from the sealing temperature, the metal attempts to contract more than the pin and glass, which results in the glass being thrust in compression. Ceramics are extremely strong in compression and the seal is rendered much stronger by this strategy. Full potential is not realized because of tendencies for the sealing glass to form menisci with the pin and ring, which exposes a region of tension. Properly traded off, a more reliable seal is achieved. The feedthroughs described here are brazed or soldered to metal packages.

Housekeeper Seals. Housekeeper seals, which are named for their inventor,[46] are glass-to-metal seals that make no allowance for contraction matching. Instead, the metal is shaped to a feather edge and embedded in the softened glass. The malleability of the metal is used to absorb the mutual strain between glass and metal. Such seals are used with copper or Kovar, for example. This is the only reliable metal joining scheme for the zero-expansion 96% + glasses.

These seals are found in high-power vacuum tubes, arc tubes, and lamps. This is a seal of last resort, and because of the required feathering, probably is ruled out of most feedthrough applications because of the high electrical resistance.

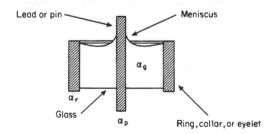

Matching expansion seal: $\alpha_g = \alpha_p = \alpha_r$

Compression seal: $\alpha_g < \alpha_p < \alpha_r$

FIGURE 1.24 Cross section of glass-to-metal feed-through seal.

1.4.5 Glazes

Glasses are often applied to ceramics to improve their smoothness and reduce the physical absorption of contaminants. The glass coatings are very similar to the glass compositions already described and may be applied from powder slurries at the same time as the original firing or in postfiring.

1.4.6 Diamonds

Gemstone diamonds have been revered for their brilliance, hardness, abrasion resistance, and beauty for centuries. During the decade starting in 1940, scientists described the crystalline forms of graphite and diamonds and prepared reliable phase diagrams. From these, General Electric and ASEA of Sweden manufactured the first synthetic stones using very high pressures and temperatures.[47] These diamonds were small and irregular and found immediate use as abrasives in grinding and cutting applications. Gemstone-quality stones up to 4 carats have been produced by General Electric, Sumitomo Electric, and DeBeers.

Recently scientists have found it possible to make diamond films using established chemical vapor deposition (CVD) processes. Such a system uses a mixture of 95 percent hydrogen and 5 percent of some hydrocarbon, such as methane, as reactants. Exposed to microwave energy, the reactants form a plasma, and the carbon deposits on a nearby substrate form a diamond film.[48]

Diamond films are of use in electronic packaging as heat sinks. Such diamond-coated substrates are now being marketed.[*] Type 2 diamonds have a thermal conductivity of 20 W/cm • K, almost 5 times that of copper and silver. Solid diamond pieces are also embedded in copper forms. Such heat sinks are available commercially for use with laser diodes.[†] Note that this thermal conductivity of diamonds is the highest of any material, and it is one of the few materials known that is both an electrical insulator and a thermal conductor.

Diamonds may be a future semiconductor. Boron and phosphorous doped into diamond film create n and p junctions and produce a diamond semiconductor. Electrons move faster in diamonds than even in gallium arsenide, and since they have a higher dielectric strength, they can deliver higher power levels. Diamonds also have greatly improved radiation resistance. They are ideal for space and defense electronics. Additional information on diamonds is found in Chap. 2.

REFERENCES

1. R. Juran (ed.), *Modern Plastics Encyclopedia,* McGraw-Hill, New York, 1985.

2. NEMA Ll-1, "Industrial Laminated Thermosetting Products," National Electrical Manufacturers Assoc., Washington, D.C., 1983, sec. 11.

3. ASTM D-1532, "Specification for Polyester Glass-Mat Sheet Laminate," Am. Soc. for Testing and Materials, Philadelphia, Pa., 1988.

4. W. T. Shugg, *Handbook of Electrical and Electronic Insulating Materials,* 3d ed., Van Nostrand Reinhold, New York, 1986, pp. 53, 267.

*Air Products and Chemicals, Inc., Allentown. Pa.
†Dubbeldee Diamond Company, Mt. Arlington, N.J.

5. R. E. Kirk and D. F. Othmer, *Concise Encyclopedia of Chemical Technology,* Wiley, New York, 1985, p. 514.

6. F. W. Tortolano, "Rising Star in the Plastics Lineup," *Des. News,* p. 56, Nov. 20, 1989.

7. J. Rose, "Engineering Resins for Connection Devices," *Connection Technol.,* p. 12, May 1989.

8. ASTM D-3029, "Test Method for Impact Resistance for Flat Rigid Plastic Specimens by Means of a Tap (Falling Height)," Am. Soc. for Testing and Materials, Philadelphia, Pa., 1990.

9. ASTM D-149, "Test Method for Dielectric Breakdown Voltage and Dielectric Strength of Solid Electrical Insulating Materials at Commercial Power Frequencies," Am. Soc. for Testing and Materials, Philadelphia, Pa., 1990.

10. M. S. Reisch, "Electronic Uses Spur Growth of High Performance Plastic," *Chem. Eng. News,* p. 46, Sept. 4, 1989.

11. ASTM D-1566, "Terminology Relating to Rubber," Am. Soc. for Testing and Materials, Philadelphia, Pa., 1990.

12. ASTM D-395, "Test Method for Rubber Property—Compression Set," Am. Soc. for Testing and Materials, Philadelphia, Pa., 1989.

13. ASTM D-412, "Test Method for Rubber Properties in Tension," Am. Soc. for Testing and Materials, Philadelphia, Pa., 1987.

14. ASTM D-1418, "Practices for Rubber and Rubber Lattices—Nomenclature," Am Soc. for Testing and Materials, Philadelphia, Pa., 1990.

15. ASTM D-2305, "Method of Testing Polymeric Films Used for Electrical Insulation," Am. Soc. for Testing and Materials, Philadelphia, Pa., 1987.

16. C. A. Harper, *Handbook of Plastics, Elastomers and Composites,* 3d ed., McGraw-Hill, New York, 1996.

17. A. J. Babiarz, "Adhesive Dispensing for Surface Mount Assembly," *Printed Circuit Assembly,* p. 8, July 1989.

18. R. Keiler, "Polymer Thick Film Multilayers: Poised for Takeoff," *Electron. Packag. Prod.,* pp. 35–38, Aug. 1987.

19. F. Wood and J. Templin, "Molded Electronic Packaging," *Printed Circuit Assembly,* pp. 31–34, Apr. 1990.

20. 1. Katz, "A Reconsideration of Metals," *Electrochem. Des.,* June 1966.

21. V. W. Wigatsky, "Nonferrous Metals," *Des. News,* Sept. 1965.

22. H. Bronson, "Dissimilar Metals," *Prod. Eng.,* June 1963.

23. T. Lisman, *Metals Handbook—Heat Treating, Cleaning and Finishing,* vol. 2, 8th ed., Am. Soc. for Metals, Materials Park, Ohio, 1964.

24. R. C. Buchanan, *Ceramic Materials for Electronics,* Marcel Dekker, New York, 1986.

25. J. R. Floyd, "How to Tailor High-Alumina Ceramics for Electrical Applications," *Ceramic Ind.,* pp. 44–47, Feb. 1969; pp. 46–49, Mar. 1969.

26. R. E. Mistler, D. J. Shanefield, and R. B. Runk, "Tape Casting of Ceramics," in G. Y. Onoda, Jr., and L. L. Hench (eds.), *Ceramic Processing before Firing,* Wiley, New York, 1978, pp. 411–448.

27. R. Block, "CPS Microengineers New Breed of Materials," *Ceramic Ind.,* pp. 51–53, Apr. 1988.

28. D. M. Mattox and H. D. Smith, "The Role of Manganese in the Metallization of High Alumina Ceramics," *J. Am. Ceram. Soc.,* vol. 64, pp. 1363–1369, 1985.

29. R. E. Cote and R. J. Bouchard, "Thick Film Technology," in L. M Levinson (ed.), *Electronic Ceramics,* Marcel Dekker, New York, 1988, pp. 307–370.

30. "Coors Ceramics—Materials for Tough Jobs," Coors Data Sheet K.P.G.-2500-2/87 6429.

31. W. D. Kingery, H. K. Bowen, and D. R. Uhlmann, *Introduction to Ceramics,* Wiley, New York, 1976, 637 pp.

32. B. Schwartz, "Ceramic Packaging of Integrated Circuits," in L. M. Levinson (ed.), *Electronic Ceramics,* Marcel Dekker, New York, 1988, pp. 1–44.

33. H. T. Sawhill, A. L. Eustice, S. J. Horowitz, J. Gar-EI, and A. R. Travis, "Low Temperature Co-Fireable Ceramics with Co-Fired Resistors," in *Proc. Int. Symp. on Microelectronics,* pp. 173–180, 1986.

34. "Ceramic Products," Brush Wellman, Cleveland, Ohio, undated.

35. M. B. Powers, "Potential Beryllium Exposure while Processing Beryllia Ceramics for Electronics Applications," Brush Wellman, Cleveland, Ohio, undated.

36. N. Iwase and K. Shinozaki, "Aluminum Nitride Substrates Having High Thermal Conductivity," *Solid State Technol.,* pp. 135–137, Oct. 1986.

37. E. S. Dettmer, H. K. Charles, Jr., S. J. Mobley, and B. M. Romenesko, "Hybrid design and processing using aluminum nitride substrates," ISHM 88 Proc., pp. 545–553, 1988.

38. E. S. Dettmer and H. K. Charles, Jr., "AIN and SiC Substrate Properties and Processing Characteristics," *Advances in Ceramics,* vol. 31, Am. Ceramic Soc., Columbus, Ohio, 1989.

39. "Combat Boron Nitride, Solids, Powders, Coatings," Carborundum Product Literature, form A-14, 011, effective Sept. 1984.

40. *Machineable Glass Ceramic,* Corning Glass Works, Corning, N.Y., undated.

41. G. W. McLellan and E. B. Shand, *Glass Engineering Handbook,* McGraw-Hill, New York, 1984.

42. W. H. Kohl, *Materials and Techniques for Electron Tubes,* Reinhold, New York, 1960, pp. 394–469.

43. W. P. Hood, U.S. Patent 2,106,744, Feb. 1, 1938.

44. Ceramic Source, 90, vol. 5, *Am. Ceramic Soc.,* Columbus, Ohio, 1990, 356 pp.

45. P. W. McMillan, *Glass-Ceramics,* Academic Press, London, 1979.

46. W. G. Housekeeper, U.S. Patent 1,294,466, Feb. 18, 1919.

47. R. K. Bachmann and R. Messler, "Diamond Thick Films," *Chem. Eng. News,* pp. 2440, May 15, 1989.

48. A. S. Brown, "Diamonds Shine Brightly in Aerospace's Future," *Aerospace America,* pp. 12–37, Nov. 1987.

RECOMMENDED READINGS

1. *Plastics for Electronics,* William M. Alvino, McGraw-Hill, 1995.

2. *Multichip Module Design, Fabrication and Testing,* James J. Licari, McGraw-Hill, 1995.

3. *Design Guidelines for Surface Mount and Fine-Pitch Technology,* 2d ed., Vern Solberg, McGraw-Hill, 1996.

4. *Flip Chip Technologies,* John H. Lau, McGraw-Hill, 1996.

5. *Ball Grid Array Technology,* John H. Lau, McGraw-Hill, 1996.

6. *Flexible Printed Circuitry,* Thomas H. Stearns, McGraw-Hill, 1996.

7. *Handbook of Plastics, Elastomers, and Composites,* 3d ed., McGraw-Hill, 1996.

8. *Electronic Materials and Processes Handbook,* 2d ed., McGraw-Hill, 1994.

CHAPTER 2

THERMAL MANAGEMENT

Frank E. Altoz
Consultant
Baltimore, Maryland

2.1 INTRODUCTION

Thermal management of electronic packaging is an all-inclusive method involving the selection, analysis, testing, and verification of a cooling design for the purpose of producing a reliable end product. The increasing emphasis on thermal management in electronic design stems from the large number of microelectronic devices in systems, their high heat densities, and the exponential nature of the component failure rates with temperature. Since the invention of the silicon integrated circuit (IC) in 1958 by Kilby,[1] there has been in excess of a sixfold increase in the number of circuits per chip, even though chip sizes have only increased from 3 by 5 mm in the early wafers to approximately 10 by 15 mm today. In addition, the feature size of circuit chips is now less than 1 μm, compared to initial sizes of 100 μm, thus permitting closer spacing, which in turn has led to higher heat densities. The classical means of cooling components through natural and forced convection air cooling is no longer a satisfactory solution to many problems. Cold plates and other enhancement techniques are being applied in order to lower temperatures so that the new reliability goals may be achieved. Moreover, there are no standard cooling techniques that are applicable to electronic packages in today's environment. The proliferation of various types of computers, from lap-type and hand-held models to mainframe supercomputers, leads to almost as many cooling approaches as there are types of machines. In addition, since the packaging techniques for the various components, such as power supplies, digital processors, and RF circuitry, differ, the same is true for cooling, which in turn necessitates a tailoring of the cooling concept within the total packaging approach. In most cases, however, the application of basic heat transfer principles will prove adequate in managing an electronic thermal design. This chapter will further that goal by providing a broad overview of various aspects of thermal design and analysis. The emphasis is on the fundamentals and the tools necessary to solve most thermal problems. There is also a brief introduction to less conventional cooling methods, such as heat pipes, immersion, microchannel, and jet cooling. The information in this chapter can be used to conduct trade-off studies involving the selection of an optimum packaging scheme.

The initial sections review basic heat flow theory as applied to electronic packaging. Next the external environment and reliability are treated as each affects the thermal design. This is followed by an explanation of air and liquid cooling systems, coolant selection, cold-plate design parameters, cabinet cooling, and a brief section on fan selection.

Section 2.7 outlines the steps necessary to obtain junction temperatures in flow-through and inline-conduction cooling systems. These are compared to illustrate the advantages of each. The remaining sections of this chapter deal with immersion cooling as it relates to high heat density, the more recent techniques of jet and microchannel cooling, and an overview of computer technology as a means of solving complicated thermal networks.

There exist a number of reference books that deal specifically with thermal design of electronic equipment.[2–4] In addition, numerous reports and conference papers are referenced within this text.

2.1.1 Units

The heat transfer units most widely used by practicing engineers in the United States in the domain of electronic cooling are a hybrid set that includes watts, British thermal units (Btu) per hour, inches, seconds, degrees Celsius, degrees Fahrenheit, and so on. This blending of U.S. customary and metric units has evolved due to a combination of factors, not the least of which have been the mechanical engineer's traditional use of U.S. customary units and the electrical engineering experience based on metrication. Gradually the trend appears to be moving toward a more common system of SI units, the International System of Units, which includes such units as watts, meters, degrees Celsius, kilograms, and seconds, but it will be difficult to arrive at a uniform, consistent set of SI units in the near future. While at present the papers published in the heat transfer journals are in SI units, the body of experience represented by the industrial technical community far outweighs in number the users and proponents of SI in academia. It should also be noted that because manufacturing drawings contain for the most part U.S. customary units (inches), this bias strongly affects the carryover of data into the computer software programs used in CAD/CAM, where the merging and interfacing is easier to accomplish in terms of hybrid units. Finally, in some cases we will display alternate sets of units; the appearance of two sets indicates an option, with a preference given to the set listed first. The following nomenclature and set of units will be used in this chapter unless otherwise specified. Table 2.1 lists the conversion factors.

Letter Symbols

$a =$ free convection parameter, $1/(\text{ft}^3 \cdot {}^\circ\text{F})$

$A =$ constant; conduction area, in^2

$A_f =$ fin area, ft^2

$A_s =$ surface area, ft^2

$b =$ plate spacing, fin height, ft

$C_p =$ specific heat, $\text{Btu}/(\text{lb}_m \cdot {}^\circ\text{F})$

$C =$ free convection constant; conductance, $\text{Btu}/(\text{h} \cdot {}^\circ\text{F})$

$C_{sf} =$ boiling constant

$\text{CHF} =$ critical heat flux, W/in^2, W/cm^2

$d =$ diameter, ft

$d_h =$ hydraulic diameter, ft

$f =$ Fanning friction factor, dimensionless

$g =$ gravitational constant $= 32.17 \text{ ft/s}^2$

$G =$ mass flow per area, $(\text{lb/lh} \cdot \text{ft}^2)$

$\text{gpm} =$ flow, gal/min

$h_c =$ convective heat transfer coefficient, $\text{Btu}/(\text{h} \cdot \text{ft}^2 \cdot {}^\circ\text{F})$

TABLE 2.1 Unit Conversion Factors

Length
  12 in = 1 ft
  2.54 cm = 1 in
  1 μm = 10^{-6} m = 10^{-4} cm

Mass
  1 kg = 2.205 lb_m
  1 slug = 32.16 lb_m
  453.6 g = 1 lb_m

Energy
  1 ft $\cdot$ lb_f = 1.356 J
  1 kWh = 3413 Btu
  1 hp $\cdot$ h = 2545 Btu
  1 Btu = 252 cal
       = 777.9 ft $\cdot$ lb_f

Pressure
  1 atm = 14.696 lb/in^2 = 2116 lb_f/ft^2
       = 1.01325 $\times$ 10^5 N/m^2
       = 407 in of water
  1 in Hg = 70.73 lb_f/ft^2

Viscosity
  1 centipoise (cP) = 2.24 lb_m/(h $\cdot$ ft)
  1 (lb_f $\cdot$ s)/ft^2 = 32.16 lb_m/(s $\cdot$ ft)

Thermal conductivity
  1 cal/(s $\cdot$ cm $\cdot$ °C) = 242 Btu/(h $\cdot$ ft $\cdot$ °F)
  1 W/(m $\cdot$ °C) = 0.5779 Btu/(h $\cdot$ ft $\cdot$ °F)

Force
  1 dyn = 2.248 $\times$ 10^{-6} lb_f
  1 lb_f = 4.448 N
  10^5 dyn = 1 N

Useful conversions to SI units

Length
  1 in = 0.0254 m
  1 ft = 0.3048 m
  1 mi = 1.60934 km

Area
  1 in^2 = 0.0006452 m^2
  1 ft^2 = 0.092903 m^2

Volume
  1 in^3 = 1.63871 $\times$ 10^{-5} m^3
  1 ft^3 = 0.02832 m^3
  1 gal = 231 in^3 = 0.004546 m^3

Mass
  1 lb_m = 0.45359 kg

Density
  1 lb_m/in^3 = 2.768 $\times$ 10^4 kg/m^3
  1 lb_m/ft^3 = 16.0185 kg/m^3

Force
  1 dyn = 10^{-5} N
  1 lb_f = 4.448 N

Pressure
  1 atm = 1.01325 $\times$ 10^5 N/m^2
  1 lb_f/in^2 = 6894.76 N/m^2

Energy
  1 Btu = 1055 J
  1 ft $\cdot$ lb_f = 1.356 J
  1 cal (15°C) = 4.1855 J

Power
  1 hp = 745.7 W
  1 Btu/h = 0.293 W

Heat flux
  1 Btu/(h $\cdot$ ft^2) = 3.154 W/m^2
  1 Btu/(h $\cdot$ ft) = 0.9613 W/m

Thermal conductivity
  1 Btu/(h $\cdot$ ft $\cdot$ °F) = 1.7303 W/(m $\cdot$ °C)

Heat transfer coefficient
  1 Btu/(h $\cdot$ ft^2 $\cdot$ °F) = 5.677 W/(m^2 $\cdot$ °C)

Specific heat
  1 Btu/(lb_m $\cdot$ °F) = 4186.7 J/(kg $\cdot$ °C)

Thermal resistance
  1 (h $\cdot$ °F)/Btu = 1.896 °C/W

Dynamic viscosity
  1 lb_m/(ft $\cdot$ s) = 1.488 (N $\cdot$ s)/m^2

---

h_{fg} = latent heat of evaporation, Btu/lb
h_r = radiation equivalent heat transfer coefficient, Btu/(h $\bullet$ ft^2 $\bullet$ °F)
H = height, ft; heat spreading factor, dimensionless
j = Colburn heat transfer factor, dimensionless
k = thermal conductivity, Btu/(h $\bullet$ ft $\bullet$ °F)
K = forced convection constant

$\ell =$ length, ft

$m =$ mass, lb_m; fin constant

$\dot{m} =$ weight flow, lb/min

$mb =$ fin parameter, dimensionless

MTBF $=$ mean time between failures, h

$n, m =$ constants

$N =$ streamwise component position, numbered from leading edge of PCB; $N = 1$, 2, 3... dimensionless; speed, r/min

$p =$ pressure, inches of water, lb/in^2

$P =$ perimeter, in; power, W

$\Delta p =$ pressure drop, inches of water, lb/in^2

$q =$ heat flow, dissipation, W, Btu/h

$q' =$ heat flow per unit length, W/in

$\ddot{q} =$ heat flux per unit area, W/in^2, W/cm^2

$Q =$ heat, Btu

$\dot{Q} =$ volumetric flow rate (CFM), ft^3/min

$r =$ radius, ft, in

$R =$ thermal resistance, (°F • h)/Btu, °C/W; reliability, percent failure-free operation; gas constant, $= 53.3$ ft/°R for air

$S =$ pitch of component in flow direction, ft

SG $=$ specific gravity

$t =$ time, h; temperature, °C, °F; fin thickness, in

$t_a =$ ambient temperature, °C, °F

$t_f =$ film temperature, °C, °F

$t_b =$ bulk liquid temperature, °C, °F

$T =$ absolute temperature, °R

$\Delta T =$ temperature difference, °C, °F

$u =$ viscosity, $lb_m/(ft • h)$

$V =$ velocity, ft/min; volume, ft^3

$w =$ width, in; fin opening, in

$x =$ distance, ft

Greek Letters

$\alpha =$ absorptivity of surface to solar radiation; fin aspect ratio, dimensionless

$\beta =$ coefficient of thermal expansion, 1/°F; heat exchanger parameter (surface area/volume), ft^2/ft^3

$\gamma =$ earth albedo (fraction of solar radiation incident on the earth which is reflected)

$\delta =$ gap distance, in; thickness, in

$\Delta =$ prefix indicating finite increment

$\varepsilon =$ emissivity of surface, dimensionless

$\rho =$ density, lb_m/ft^3; reflectivity of surface to radiation, dimensionless

$\sigma =$ surface tension, lb_f/ft; density correction factor, dimensionless

$\sigma r =$ Stephan-Boltzmann radiation constant, $= 0.1713 \times 10^{-8}$

$\tau =$ time constant, h; transmittance for radiation

ζ = fin efficiency

Dimensionless Numbers

F_a = radiation view factor
F_e = emissivity factor for radiation
G_r = Grashof number
N_s = specific speed
N_u = Nusselt number
P_r = Prandtl number
R_a = Raleigh number, = $G_r P_r$
R'_a = channel Raleigh number
R_e, R_f = Reynolds number

Subscripts

a = ambient, air
a–f = air to fin
b = bulk, base
c = case, chip, channel
c–a = case to air
c–hs = case to heat sink
e = earth
f = flotation, film, fin
i = specific, internal
hs = heat sink
in = inlet
int = interface
j = junction, jet
j–a = junction to air
j–c = junction to case
e = liquid, length
r = radiation
s = surface, solar
t = thermal
v = vapor
w = wall
1, 2 = designated locations

2.2 BASIC HEAT FLOW THEORY AS APPLIED TO ELECTRONIC PACKAGING

The process of heat transfer occurs between two points as a result of a temperature difference between them. Thermal energy may be transferred by three basic modes—conduction, convection, or radiation. Of equal importance is the convection process associated with the change of phase of a fluid; examples are condensation and boiling. Commonly used heat pipes incorporate both of these processes.

This chapter is for the most part concerned with uniform or steady-state heat flow, which exists when the temperature at each point in the heat flow path does not vary with time. The only exception to this is Sec. 2.2.6, which gives a brief review of transient heat flow, where the temperature is time-dependent prior to reaching a steady-state equilibrium.

2.2.1 Conduction

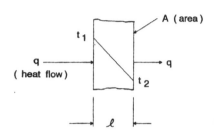

IGURE 2.1 One-dimensional heat flow cross plane surface.

Heat transfer by conduction in solids occurs whenever a hotter region with more rapidly vibrating molecules transfers its energy to a cooler region with less rapidly vibrating molecules. Examples abound of conduction heat transfer in electronic equipment. The heat flow from the junction of a chip in an IC package to its case is a common example; other examples are a metallic core or heat from a printed circuit board (PCB) to the board edge, the conduction at the heat sink end to a chassis wall across a dry interface, the heat spreading within a chassis wall, and the conduction from a component mounted on a metal bracket to a cold plate.

Conduction heat transfer represents the most common form of heat transfer in electronic equipment.

In the case of a single plane wall, as shown in Fig. 2.1, with parallel surfaces l ft apart maintained at temperatures t_1 and t_2 °F, the heat flow q, in Btu/h, across a constant area A, in ft^2, is given by the one-dimensional Fourier equation

$$q_{1-2} = \frac{kA}{\ell}(t_1 - t_2) \quad \text{(Btu/h)} \quad (2.1)$$

The thermal conductivity k determines the quantity of heat transferred per unit cross-sectional area for a given temperature difference across the wall of thickness ℓ. Thus an aluminum wall transfers approximately 700 times more heat than an elastomer made of silicone for the same geometry and temperatures difference. The values of thermal conductivity for solid materials used in electronic equipment range over four orders of magnitude. Tables 2.2 and 2.3 list these values for commonly used materials. In general, there are three classifications of material to be dealt with—metals, insulators, and semi-insulators. The metals are seen as the best conductors of heat (Table 2.2), whereas the insulators, such as epoxy, polyimide, Teflon,[*] or nylon, are the poorest (Table 2.3). Materials between metals and insulators include the dielectric materials used to isolate components from electrical potential as well as the doped silicon and gallium arsenide die materials for use in microelectronic devices. Although the thermal conductivity k varies with temperature, it is generally acceptable to use the values at 80°F listed in Tables 2.2 and 2.3 for metals and insulators since their conductivity varies only 10 percent for most metals and less than 5 percent for insulators in the range of 0 to 200°F. However, in the case of semiconductors, their thermal conductivity varies significantly over temperature, as illustrated in Fig. 2.2, so that specific manufacturers' data sheets must be relied on for more accurate property

---

*Teflon is a registered trademark of E.I. du Pont de Nemours & Company, Inc.

TABLE 2.2 Thermal Conductivity of Electronic Packaging Materials at 80°F

Material	Btu/(h · ft · °F)	W/(m · °C)
Diamond (type 2A)	1155	2000
Silver	242	419
Copper	228	395
Gold	172	298
Beryllia	140	242
Aluminum 1100H18	126	218
Aluminum 6063 T6	116	201
Aluminum 6061 T0	100	173
Aluminum 6061 T6	90	156
Aluminum 5052	80	139
Aluminum 2024 and 7075	70–75	121–130
Silicon	68	118
Eutectic bond	39.4	68
Iron	36	62
Phosphor bronze	30	52
Solder 80-20 Au-Sn	30	52
Gallium arsenide	29	50
Monel	15	26
Alumina (99% Al_2O_3)	14.5	25
Solder (Pb-In)	12.7	22
Alumina (96% Al_2O_3)	12.0	21
Kovar	9.6	16.6
Stainless steel	8.6	15
Metal frame type "42"	6.2	10.7
Titanium	5.8	10.0
Ferrite	3.6	6.2

TABLE 2.3 Thermal Conductivity of Electronic Insulating Materials at 80°F

Material	Btu/(h · ft · °F)	W/(m · °C)
Borosilicate glass	0.67	1.67
Ablefilm 550K	0.45	0.78
Mica	0.41	0.71
Epoxy (conductive)	0.2–0.5	0.35–0.87
Polyimide	0.19	0.33
Epoxy glass fiberglass	0.14	0.24
Ablefilm 550 dielectric	0.14	0.24
Nylon	0.14	0.24
Polytetrafluoroethylene	0.14	0.24
Epoxy (dielectric)	0.13	0.23
Solder mask	0.12	0.21
Mylar	0.11	0.19
Silicone rubber	0.11	0.19
Neoprene rubber	0.11	0.19
Styrofoam	0.02	0.034
Air	0.015	0.026

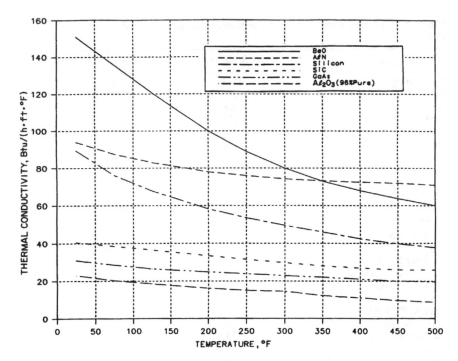

FIGURE 2.2   Thermal conductivity of dielectric electronic materials.

data. When surfaces having different thicknesses and conductivities are joined together, as for example within a microelectronic device, the concept of thermal resistance is useful in expressing the heat flow parameters. Equation (2.1) may then be rewritten for the case of a series of resistances within a group of materials, where the same heat q passes through each material,

$$q_{1-n} = \frac{t_1 - t_n}{R_t} = (t_1 - t_n)\left[\frac{1}{\dfrac{\ell_1}{k_1 A_1} + \dfrac{\ell_2}{k_2 A_2} + \ldots + \dfrac{\ell_n}{k_n A_n}}\right] \tag{2.2}$$

Thus the ability to conduct heat in a solid is represented as an inverse function of resistance R_t or as the summation of ℓ/kA terms having units of (°F • h)/Btu or °C/W. Alternatively, an equally useful concept is that of conductance C_t or the inverse of resistance $1/R_t$ which represents the total resistance as expressed in Eq. (2.2). Problem 2.1 illustrates the use of the resistance concept in solving for the temperature gradient within solid materials.

Problem 2.1:   In Fig. 2.3 a silicon chip measuring 0.07 by 0.07 in, 0.025 in thick, is mounted to an alumina case of 0.025-in thickness with conductive epoxy 0.003 in thick. Determine the internal resistance chip to case and its temperature rise for a heat dissipation of 0.25 W.

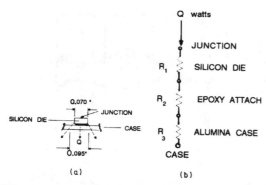

FIGURE 2.3 Die resistance path, junction to case.

Solution: The total resistance from chip junction to case R_{j-c} is the sum of the individual resistances R_1, R_2, and R_3, as illustrated in Fig. 2.3b. The temperature rise from chip to case is obtained from Eq. (2.2) by solving for R_t as follows:

$$R_t = \frac{\ell_1}{k_1 A_1} + \frac{\ell_2}{k_2 A_2} + \frac{\ell_3}{k_3 A_3}$$

$$= \frac{0.025 \times 12}{68 \times 0.07 \times 0.07} + \frac{0.003 \times 12}{0.4 \times 0.07 \times 0.07} + \frac{0.025 \times 12}{15 \times 0.082 \times 0.082}$$

$$= 22.2 (°F \cdot h)/Btu \ (42.1 °C/W)$$

Note that in the calculation of R_t, the area A_3 is increased slightly over the chip area to account for heat spreading in the case. The assumption of a constant spreading angle of 26.6° is used to approximate this simplified condition; however, it is only to be used in cases where the thickness-to-length dimension $(0.025/0.070 = 0.357)$ is less than 2.0. For a treatment of the subject of heat spreading, the reader is referred to David[5] (see also Kennedy[33] and Sec. 2.7.2). The temperature rise from chip to case in Fig. 2.3 equals the resistance R_t multiplied by the dissipation of 0.025 W, or 19.3 °F (10.7°C).

In this example, it is assumed that the surface of the chip is heated uniformly which yields a constant temperature on the die surface, whereas in many multichip applications, such as large-scale and very-large-scale ICs (LSICs and VLSICs), there are significant variations in local heat dissipation on the chip surface, leading to large surface temperature differences. The latter requires two- and three-dimensional heat flow nodal models, which are best solved by computer programs, as explained in Sec. 2.11.

2.2.1.1 Contact Interface Resistance. There are many electronic applications that involve the conduction of heat across interface surfaces. These vary from small individual components bolted to cold plates, such as diodes or power transistors, to much larger surfaces covering many square inches. Typical applications include computers, automobile electronics, power supplies, and a broad range of business equipment. A typical interface of this type is depicted in Fig. 2.4a, which shows constricting heat flow lines crossing

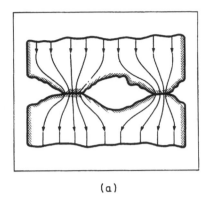   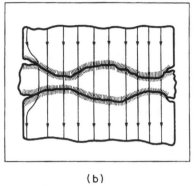

(a)   (b)

FIGURE 2.4 Heat flow across contacting surfaces. (a) Without interstitial material. (b) With interface material. (*Reprinted with permission of Chomeric Corp., Woburn, Mass.*)

from one material to the other. The air between the two materials, which have a high impedance to heat flow, is shown schematically with no lines crossing. This discontinuity at the engaging surface presents an opportunity for large temperature gradients, unless extreme care is exercised in the design of the joint. Both the air gap and the points of contact contribute to the total heat flow across the interface. In general, for two surfaces forced together under pressure, the contact is not complete and the actual heat transfer area of the joint is only a small fraction (< 5 percent) of the apparent contact area. Sometimes a filler is used to improve the conductance of the joint, as shown in Fig. 2.4b, by flowing into the interstices created by the engaging hills and valleys. This material must be of high conductivity and also quite thin, so as not to add to the joint impedance. Fillers usually result in greater improvement in contact resistance when used between rough surfaces, as opposed to surfaces that are initially smooth. Typically fillers are silicone-type greases, elastomeric materials, both filled with conductive metals and unfilled, and soft metals such as indium, aluminum, and copper. Conductive ceramics, such as boron nitride, aluminum oxide, and others, are sometimes used as fillers in elastomers to improve conductivity and to provide dielectric insulation when voltage gradients are present.

The interface resistance in commonly used engineering units is expressed in $(°C \cdot in^2)/W$ and is obtained empirically from tests on the various interface models. The variables affecting the resistance, and hence the temperature drop across an interface, are numerous. They consist of a combination of the pressure and hardness of the materials in contact, the surface finish, and the flatness on the contacting surfaces, which affect the number and area of the high-point engagements and the type of surface coatings used. Anodized aluminum (aluminum oxide), for example, exhibits higher thermal resistance than bare aluminum when mated with bare aluminum. The increase ranges from 20 to 50 percent at pressures of 25 to 150 lb/in^2 in one reported set of data.[6] Smaller surfaces of below 1 in^2 are more easily controllable as to surface finish and flatness, and therefore have lower impedance than large surfaces, which are not as flat and which vary in the quality of the surface finish. Design engineers often resort to soft metal foils (<0.005 in thick) and conductive elastomers to fill the voids covering large expanses. The other common method is to use grease or a paste filler to fill the voids. Thin grease films have the lowest interface resistance, but are not favored by all designers because of the associated handling problems.

Over the past 30 years, there have been a great deal of published data on interface resistance, with a considerable number of mathematical models proposed. Only recently, however, has there been any significant progress toward predicting the contact interface impedance,[7] and thus far only under the most simple controllable conditions. Therefore, due to the complexity of the problem, it is necessary that tests be performed to obtain the joint resistance values, which depend not only on the materials in contact, but also on the installation conditions present in specific applications. The important parameters to be recorded in the testing program must include the following:

1. Flatness, in/in, of each surface.
2. Surface finish or roughness, in micrometers, for each surface.
3. Contact pressure, in lb/in^2, of the engaging surfaces. It is well to test under pressure changes beyond at least 50 lb/in to obtain the characteristic curve of
4. interest, that is, impedance versus pressure.
5. Simulation of the interstitial material, such as air, liquid, and metal foil.
6. A Brinell hardness reading of each of the engaging surfaces.
7. Average temperature of the assembly under test.

The limited test data in Fig. 2.5 show the contact resistance versus pressure for four general material combinations: (1) bare and dry 6061T6 aluminum surfaces, each having 32-μin finish, (2) an elastomeric 0.015-in-thick sheet of conductive material, (3) a conductive elastomeric of higher durometer than (2), 0.009 in thick, and (4) a silicon paste material* commonly used in electronic systems. The data shown can be used as a general guide in noncritical applications where the interface resistance is to be approximated to a level of ±20 percent. In these experiments, the sample parts were approximately 2 by 4 in in area and held to a flatness of 0.001 in/in. It should be noted that data were also published by manufacturers of interface materials, which quote resistance values of 0.3 to as low as 0.2 (°C in^2)/W and lower for material between flat plates under pressures exceeding 200 to 300 lb/in^2. These are generally used in power devices where electrical isolation may also be a requirement and where surface areas are on the order of only 1 in^2.

When used in space applications, the lack of air reduces the joint conductance for materials such as dry aluminum on aluminum significantly—a factor of 2 is sometimes used as a guide, although this may vary considerably with each application. It should also be noted that although lower resistance values have been reported for dry aluminum interfaces, those given in Fig. 2.5 are considered to be conservative.

In the case of airborne applications from sea level to 70,000 ft, the contact resistance increases with altitude, with the increase being more pronounced on rougher rather than on smooth surfaces. A fairly complete set of data is reported in Evans and Lannon[6] for 5052H34 aluminum having anodized surfaces and for pressures to 150 lb/in^2. The sea-level data in Evans and Lannon[6] show an average of 35 percent lower resistance as compared to the Fig. 2.5 data for the same 32, μin-finish material combinations.

2.2.2 Free or Natural Convection

Heat transfer by natural convection occurs as a result of a change in the density of the fluid, which causes fluid motion. Convective heat transfer between a heated surface and the surrounding fluid is always accompanied by a mixing or intermingling of fluid adjacent to the surface. In many low-dissipation systems, natural convection heat transfer is simple to implement and is relatively reliable. In contrast to this, a mechanically induced

---

*Manufactured by Wakefield Engineering Company as #120 Wakefield paste.

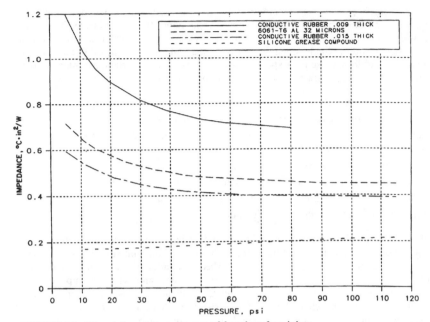

FIGURE 2.5 Thermal contact resistance of four interface joints.

motion by means of a pump or fan results in a forced convection process. The latter is capable of supporting potentially greater heat fluxes with higher film coefficients h_c, in Btu/(h • ft^2 • °F), at the surface. Natural convection film coefficients in air typically range from 5 to 20 percent of those for forced convection, and in low-dissipation packages at moderate ambient temperatures (50 to 100°F and sea level), these may be sufficient to obtain adequate cooling. The general equation defining the convective heat transfer either free or forced from a surface at t_s into a gas or fluid at a surrounding temperature t_0 is given by

$$q_c = h_c A_s (t_s - t_0) \text{ (Btu/h; W)} \tag{2.3}$$

where the variable of particular interest is h_c, the heat transfer film coefficient at the surface. Experiments have shown that the film coefficient h_c, expressed within the dimensionless Nusselt number N_u, is directly related to the product of the dimensionless ratio involving the Grashof and Prandtl numbers G_r and P_r or the Raleigh number R_a, $(G_r \times P_r)$. Equations (2.4) through (2.6) express these relationships:

$$N_u = C_1 (G_r \times P_r^n) = C_1 R_a^n \tag{2.4}$$

and

$$N_u = h_c \frac{\ell}{k} = C_1 (a \ell^3 \Delta T)^n \tag{2.5}$$

where the parameter a is a function of the fluid properties, or

$$a = \frac{g\beta C_p \rho^2}{uk} \qquad (2.6)$$

Figure 2.6 illustrates the temperature dependence of the parameter a and hence, the film coefficient for air, sulfur hexafluoride SF_6, and the fluorinert liquid FC-77.[*] These fluids

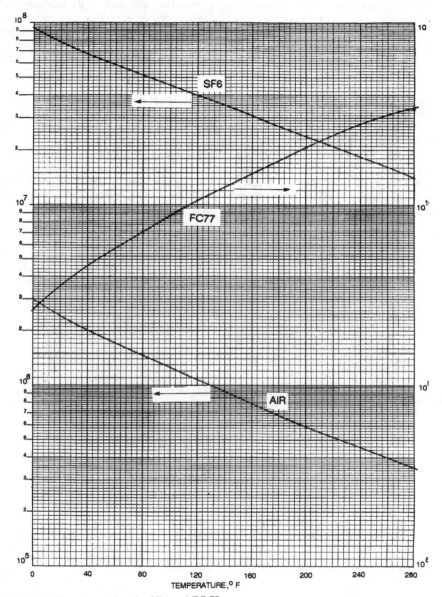

IGURE 2.6 Parameter a for air, SF_6, and FC-77.

are representative of typical electronic coolants in general use, as in the case of air and, more specifically, in power supplies (SF_6) and in liquid-cooled radar transmitters and test equipment (FC-77). The much higher a values for liquid FC-77 as opposed to the two gases illustrate the superiority of liquids over gases in natural convection cooling. In all free convection applications, the fluid properties used in determining a are evaluated at the mean temperature of the film or the average of the surface and bulk fluid temperatures. The constants C_1 and n in Eq. (2.5) have been reported on extensively in the literature, and a compilation of these appears in Table 2.4. Examples of natural convection cooling include the exterior walls of electronic packages, PCBs mounted within electronic enclosures, and components mounted on brackets or support structures having low dissipation. In Table 2.4, the constants C_1 and n are seen as strongly influenced by the geometry, physical size, and spatial orientation, as well as whether laminar-free convection, $n = 0.25$ ($10^3 < a t^3 \Delta T < 10^9$), or turbulent conditions, $n = 0.33$ ($a t^3 \Delta T > 10^9$), prevail. The equations are recommended for use whenever the orientation and size of the surface are known and the surface temperature or rise above the ambient fluid is known or can be estimated. An estimated surface temperature used for determining a must always be checked in arriving at a proper heat balance for a given application.

Typical h_c values in air at sea-level range from 0.6 to 1.2 Btu/(h • ft^2 • °F) [0.106 to 0.211 W/(m^2 • °C)] in equipment packages. At altitudes other than sea level, the sea-level values for h_c are reduced by the square root of the pressure ratio to obtain the coefficients at higher altitude. Problem 2.2 illustrates a method for determining the heat loss from the exterior of an electronic package in air at sea level.

Problem 2.2:   The processor shown in Fig. 2.7 dissipates 400 W and is internally cooled with liquid. Determine the heat loss due to natural convection from the exterior surfaces of the box when exposed to an ambient temperature of 77°F (25°C). The measured surface temperatures are 100°F for the vertical surfaces, 113°F for the top surface, and I00tF for the bottom surface. Assume all six surfaces convect heat to the ambient air.

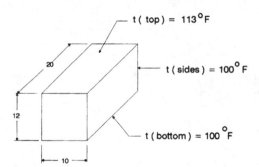

FIGURE 2.7   Electronic processor package.

Solution: There are three calculations required for h_c namely, the four side surfaces, the top surface, and the bottom surface. Table 2.4 is used to obtain constants C_1 and n for use in Eq. (2.5), and a is taken from Fig. 2.6. The heat loss is determined from Eq. (2.3). Note that t, the characteristic dimension for the top and bottom surfaces, is the geometric mean, namely, $2 \ell_2 \ell_2 / (\ell_1 + \ell_2)$, or 1.11 ft.

---

*FC-77is a registered trademark of 3M Company.

TABLE 2.4 Constants C_1 for Natural Convection in Air and Two Values of n

$R_a = a\ell^3\Delta T$ (where ℓ is limited to 2 ft)	Laminar, 10^3–10^9 $n = 0.25$	Turbulent, $>10^9$ $n = 0.33$
Vertical plate*	0.55	0.13
Horizontal plate facing up*	0.71	0.17
Horizontal plate facing down*	0.35	0.08
Horizontal cylinder*	0.45	0.11
Sphere (ℓ = radius)*	0.63	0.15
Small parts (wires, components, etc.)†	1.45	—
Components on PCBs†	0.96	—
Small components in free air†	1.39	—

*From Brown and Marco,[8] table 7.2, p. 136.
†From Steinberg,[3] table 5.1, p. 108.
Reprinted with permission of the publishers.

The Raleigh number is $R_a = a\beta\Delta T$, and the properties are at the average film temperature for a,

$$R_a(\text{side surfaces}) = 1.37 \times 10^6 \times 1^3(100 - 77) = 31.5 \times 10^6$$

$$R_a(\text{top surfaces}) = 1.30 \times 10^6 \times 1.11^3(113 - 77) = 64.0 \times 10^6$$

$$R_a(\text{bottom surfaces}) = 1.30 \times 10^6 \times 1.11^3(113 - 77) = 43.1 \times 10^6$$

Then

$$h_c(\text{sides}) = C_1 \frac{k}{\ell} R_{a^n} = 0.59 \times \frac{0.015}{1.0}(31.5 \times 10^6)^{0.25} = 0.006 \text{ Btu/(h} \cdot \text{ft}^2 \cdot °\text{F)}$$

$$h_c(\text{top}) = 0.71 \times \frac{0.015}{1.0}(64.0 \times 10^6)^{0.25} = 0.86 \text{ Btu/(h} \cdot \text{ft}^2 \cdot °\text{F)}$$

$$h_c(\text{bottom}) = 0.35 \frac{0.015}{1.0}(43.1 \times 10^6)^{0.25} = 0.38 \text{ Btu/(h} \cdot \text{ft}^2 \cdot °\text{F)}$$

The total heat loss Q_τ is the sum of the losses from all six surfaces,

$$Q_\tau = \sum_{i=1}^{6} Q_i = \sum_{i=1}^{6} h_c A_i (T_i - T_a)(\text{Btu/h})$$

$$= \frac{2[0.66 \times 10 \times 12(100\text{-}77)]}{144} + \frac{2[0.66 \times 12 \times 20(\ 100\text{-}77)]}{144}$$

$$= 25.3 + 50.6 + 43 + 12.1 = 131 \text{ Btu/h (38.4 W)}$$

In electronic cooling, there is a large class of equipment that lends itself to natural convection cooling. This category includes stand-alone packages such as modems and small computers having an array of PCBs mounted within an enclosure. These are commonly encountered natural-convection heat transfer examples deserving special attention. In many of these cases the long vertical channels develop boundary layers that merge near the entrance so that fully developed flow exists along much of the channel and the local heat transfer coefficient h_c remains essentially constant. At the entrance of the PCBs, or in the case of short vertical channels, the heat transfer coefficient h_c is greater and is defined in terms of the flat isolated plate, as in Eq. (2.5). The more typical case of fully developed flow is best expressed in terms of a channel Raleigh number R_a as where b is the spacing between the plates. In a similar manner the channel Nusselt number N_{u0} is defined with respect to the surrounding ambient fluid at to as

$$R_a = \frac{ab^3 \Delta Tb}{\ell} = \frac{ab^4 \Delta T}{\ell} \tag{2.7}$$

where b is the spacing between the plates. In a similar manner the channel Nusselt number N_{u0} is defined with respect to the surrounding ambient fluid at t_0 as

$$N_{\mu 0} = h_c \frac{b}{k} \tag{2.8}$$

Proper cooling of an array of plates or PCBs by natural convection requires that adequate spacing exist between adjacent members. Ordinarily this suggests a large spacing but, in practice, the plates need only be brought together by approximately two boundary layers or at a channel Raleigh number $R_a = 463$. This space b at $R_a = 463$ is from Bar-Cohen and Rohsenow,[9] as is the maximum spacing b_{max}, defined for isothermal plates,

$$b_{max} = \frac{4.64}{p^{0.25}} \tag{2.9}$$

where

$$p = a \frac{\Delta T_0}{\ell} \tag{2.10}$$

For any other closer spacing, the defining equation for the channel Nusselt number, as derived in BarCohen and Rohsenow,[9] is for the common case of laminar flow at low temperature gradients for symmetrical isothermal plates,

$$N_{u0} = h_c \frac{b}{k} = \left(\frac{576}{R_a^2} + \frac{2.873}{R_a^{0.5}} \right)^{-0.5} \tag{2.11}$$

Problem 2.3 illustrates the application of Eqs. (2.7)–(2.11) in solving natural convection problems associated with arrays of PCBs or vertical plates in general.

Problem 2.3:   An array of vertical PCBs 6 in high and 9 in wide are at a uniform temperature of 140°F and are spaced 0.50 in apart. If the surrounding ambient temperature is 80°F, determine the heat transfer coefficient h_c. Determine the theoretical optimum spacing between boards and calculate the heat entering each channel from two adjacent boards.

Solution: Solution: From Fig. 2.6, the heat transfer parameter a at an average film temperature of 110°F is $1.15 \times 106/(°F \cdot ft^3)$. Then the Raleigh number R_a for an isolated plate is

$$R_a = a\ell^3 \Delta T = 1.15 \times 10^6 \times 0.5^3 (140 - 80) = 8.625 \times 10^6$$

and

$$h_c = C_1 \frac{k}{\ell} R_a^n = 0.59 \times \frac{0.0160}{0.5} (8.625 \times 10^6)^{0.25} = 1.02 \text{ Btu/(h} \cdot \text{ft}^2 \cdot °\text{F)}$$

The convective heat transfer coefficient h_c for a channel spacing of 0.5 in is determined from the channel Raleigh number as follows:

$$R_a = ab^4 \frac{\Delta T}{\ell} = 1.15 \times 10^6 \times 0.04167^4 \frac{140-80}{0.5} = 415.9$$

and the Nusselt number related to the inlet temperature t_0 is, from Eq. (2.11),

$$N_{u0} = \left(\frac{576}{R_a^2} + \frac{2.873}{R_a^{0.5}} \right)^{-0.5} = \left(\frac{576}{415.9^2} + \frac{2.873}{415.9^{0.5}} \right)^{-05} = 2.63$$

from which the film coefficient in the channel is determined as

$$h_c = N_{u0} \frac{k}{b} = 2.63 \times \frac{0.0158}{0.04167} = 0.99 \text{ Btu/(h} \cdot \text{ft}^2 \cdot °\text{F)}$$

Note that the value for h_c in the channel is nearly identical to that of the free isolated plate, which suggests that each plate acts independently of the other at the 0.50-in spacing. The maximum spacing b_{max} from Eqs. (2.9) and (2.10) corresponds to the maximum cooling from the plates as follows:

$$b_{max} = \frac{4.64}{p^{0.25}} = \frac{4.64}{\left(\frac{a\Delta T}{\ell} \right)^{0.25}} = \frac{4.64}{1.15 \times 10^6 \left(\frac{60}{0.5} \right)^{0.25}} = 0.0428 \text{ ft } (0.51 \text{ in)}$$

The conductive heat corresponding to the 0.51-in spacing is, from each surface of the plates,

$$q_c = h_c A_s (T_w - T_0) = 0.99 \times \frac{9 \times 6}{144} (140 - 80) = 22.3 \text{ Btu/h } (6.5 \text{ W})$$

In the sample problem, if one reduces the spacing to 0.25 in, the film coefficient becomes 0.63 Btu/(h $\cdot$ ft^2 $\cdot$ °F) and the convective heat transfer from one surface of the PCB is only 14.3 Btu/h or 4.2 W. Therefore, as expected, reducing the spacing below the maximum value causes the boundary layers to intermerge, thus inhibiting heat transfer. Of course, the spacing in practice is often determined by factors other than the optimum heat transfer, such as connector size or the component protrusions from the PCB. In any event, the procedure outlined may be used to approximate the heat transfer and to evaluate the need for forced convection cooling.

2.2.3 Forced Convection

The process of heat transfer resulting from fluid flow across a heated or cooled surface is called forced convection. In this case, the temperature gradient is confined to a very thin layer of fluid adjacent to the surface so that the heat flows through a relatively thin boundary layer. In the main stream outside this layer it is assumed that isothermal conditions exist. Therefore the analysis of the convection process becomes one of defining the heat transfer within this boundary layer, which may be either laminar or turbulent, depending on the Reynolds number. Turbulent flow is characterized by a random velocity of particles which enhances heat transfer, whereas in laminar flow the particles near the surface travel parallel to the axis in a uniform manner.

There are many applications of forced convection cooling in electronic packaging. Airflow across PCBs in computers is a prime example. Others include flow through finned heat exchanger cold plates, across discrete components within an enclosure, or in tubes used to cool a chassis. In all cases it is well to distinguish whether the flow is fully developed laminar or turbulent, as opposed to flow that takes place at the entrance or leading edge of a surface. The equations defining the heat transfer coefficient h_c under these various conditions are examined in this section in terms of the dimensionless parameter groups consisting of Reynolds (R_e) Prandtl (P_r) and Nusselt (N_u) numbers. Both the flow inside tubes and the flow along flat plates is examined.

When fully developed laminar flow $(R_e < 2100)$ occurs inside a tube, the Nusselt number N_u in the limiting case is a function of the tube aspect ratio. For circular and square tubes it ranges between 2.9 and 4.4. Other shapes and aspect ratios yield values ranging from 2.4 to 6.5.[10] A more common condition encountered in electronic cooling is at the entrance to tubes, where the heat transfer coefficient h_c is higher than in the region downstream. The generally accepted equation is based on the work of Seider and Tate,[11] who proposed the following empirical equation:

$$N_u = h_c \frac{d_c}{k} = 1.86 R_e^{0.33} P_r^{0.33} \left(\frac{d_h}{l}\right)^{0.33} \left(\frac{u}{u_w}\right)^{0.14} \tag{2.12}$$

The fluid properties in Eq. (2.12) are evaluated at the bulk fluid temperature, except that the dynamic viscosity u_w is at the wall temperature and $(u/u_w)^{0.14}$ accounts for the temperature dependence of viscosity at the surface boundary layer. In most applications the variation in temperature between wall and fluid is small and the $(u/u_w)^{0.14}$ term can be assumed to be 1.0 in Eq. (2.12). When the flow is fully developed and turbulent within the tube $(R_e > 8000)$, Dittus and Boelter[12] recommend an equation in the form

$$N_u = h_c \frac{d_h}{k} = 0.023 R_e^{0.8} P_r^{0.4} \left(\frac{u}{u_w}\right)^{0.014} \tag{2.13}$$

where the fluid properties are evaluated as in Eq. (2.12) at the bulk fluid temperature except for u_w, the viscosity at the wall temperature. If the flow is not fully developed, as in the entrance region of a tube, Holman[13] recommends the use of the following equation:

$$N_u = h_c \frac{d_h}{k} = 0.036 R_e^{0.8} P_r^{0.4} \left(\frac{d_h}{\ell}\right)^{0.055}, \quad 10 < \frac{\ell}{d_h} < 400 \tag{2.14}$$

In electronic cooling applications the forced convection equations are used in calculating film coefficients h_c in the air gap between PCBs, in antenna tubing where high heat fluxes require liquid coolant, in heat exchanger passages, and in the routing of liquid within an electronic chassis. The case of airflow between PCBs is one that is based on experimental data since tubular flow does not describe this condition adequately.

For laminar flow parallel to a surface ℓ ft long, the average film coefficient $\bar{h}_c$ over the length ℓ is, for $R_{e\ell} < 50 \times 10^4$,

$$\bar{N}_u = \bar{h}_c \frac{\ell}{k} = 0.664 P_r^{0.33} R_{e\ell}^{0.5} \tag{2.15}$$

where $\ell \leq$ 2ft, and for turbulent flow, where $R_{e\ell} > 50 \times 10^4$ (see Holman[13]),

$$\bar{N}_u = \bar{h}_c \frac{\ell}{k} = P_r^{0.33} 0.036(R_{e\ell}^{0.8} - 836) \tag{2.16}$$

In the use of Eqs. (2.15) and (2.16) the film temperature of the fluid is estimated and can be confirmed through an iterative process once h_c and the final temperatures are determined. The forced convection coefficients vary widely in electronic equipment, generally falling in the range of 3 to 20 Btu/(h • ft^2 • °F) [17 to 113 W/(in^2 • °C)] for air flowing in tubes. For flat surfaces, air traveling between 500 and 2000 ft/min results in average surface coefficients of 3 to 6 Btu/ (h • ft^2 • °F) [17 to 34 W/(in^2 • °C)]. On the exterior of pod surfaces mounted on aircraft, the skin acting as a heat dissipater is capable of developing much higher coefficients in accordance with Eq. (2.16). For example, at $R_{e\ell} = 13 \times 10^6$, corresponding to a station 3 ft from the leading edge, an aircraft traveling at Mach number 0.7 at sea level will develop an average h_c of 79 Btu/(h • ft^2 • °F) [448 W/(in^2 • °C)]. Unfortunately, from a heat transfer standpoint the skin may also be quite elevated in temperature, depending on free stream conditions, so that the transfer of heat, although efficient, likely will occur at high temperature, causing high internal electronic temperatures. Therefore the pod surface cannot be relied upon in that case and other means must be devised for dissipating or storing heat during high-speed missions.

For the case of fluid traveling across small discreet components such as resistors or diodes, the experimental correlations of data by Knudsen and Katz[14] expressed in the equation

$$\bar{N}_u = \bar{h}_c \frac{d_h}{k} = C_2 R_e^m P_r^{0.33} \tag{2.17}$$

and Table 2.5 can be used for determining the average gas or liquid heat transfer coefficient h_c. In Eq. (2.17) the fluid properties are determined at the average film temperature. The application of Eq. (2.17) is best illustrated by an example.

TABLE 2.5 Constants for Use with Equation (2.17)

R_e	C_2	m
0.4–4	0.989	0.33
4–40	0.911	0.385
40–4000	0.683	0.466
4000–40,000	0.193	0.618
40,000–400,000	0.0266	0.805

Source: From Holman,[13] table 6.1. Reprinted with permission of McGraw-Hill, Inc.

Problem 2.4: Air flows across a dual-in-line package (DIP) with a length of 0.2 in on a PCB at 400 ft/min. Calculate the surface impedance from case to air at an average film temperature of 100°F and compare this value to published data.

Solution:

$$R_e = \frac{\rho V \ell}{u} = \frac{0.070 \times 400 \times 60 \times 0.2}{0.046 \times 12} = 608.7$$

Then

$$\bar{N}_u = C_2 R_e^m P_r^{0.33} = 0.683 \times 608.7^{0.466} 0.7^{0.33} = 12.1$$

and

$$\bar{h}_c = \bar{N}_u \frac{K}{\ell} = 12.1 \times \frac{0.0158}{0.0167} = 11.4 \ \text{Btu/(h} \cdot \text{ft}^2 \cdot °\text{F)}$$

Then

$$R_{t-a} = \frac{1}{\bar{h}_c A} = \frac{1}{11.4 \times 0.00416} = 21.1 \ (°\text{F} \cdot \text{h)/Btu} \ (39.9°\text{C/W)}$$

From the Motorola publication[15] for a 16-lead ceramic DIP, R_{j-a} = 54.0°C/W at 400 ft/min and R_{j-c} = 15QC/W; therefore R_{c-a} = 54 − 15 = 39°C/W. The variation between calculated and published values is unusually close for heat transfer calculations, where 10 to 20 percent variation is the norm.

Forced convection cooling of an array of components on a PCB represents one of the most common applications of heat transfer cooling of electronics.

Equations (2.12) through (2.17) do not adequately describe conditions where components project into the air stream within confined channels. Instead, test data are often relied upon to better define the heat transfer coefficient h_c, as in the following expression:[16]

$$h_c = 0.93 \frac{\ell}{\ell'} + 0.225 \frac{(\rho V)^{0.8}}{(N\ell)^{0.36}} \left(\frac{\ell'}{\ell} - 1 \right)^{0.13} \tag{2.18}$$

where ℓ is the component length in the flow direction, ℓ' is the component pitch, and N represents the station or location of the component. The component DIP in Problem 2.4, when placed in a PCB assembly, will exhibit a lower average h_c due to the confined boundary-layer flow. At 400 ft/min at the entrance location (N = 1), Eq. (2.18) yields an h_c of 14.1 Btu/(h $\cdot$ ft $\cdot$ °F), whereas at station 8 (N = 8), near the exit of the PCB, h_c is only 7.0, or approximately 50 percent of the entrance value. The pitch ℓ' in Eq. (2.18) was assumed to be 0.35 in (0.0292 ft) and length ℓ was 0.2 in (0.0167 ft), as in the sample problem.

2.2.3.1 *Pressure Drop.*

The flow of air in a duct or channel results in a pressure drop along the length of passage. In forced convection systems, high heat transfer coefficients and high pressure drop are concomitants, but the design challenge is to extract the greatest film coefficient h_c in a system having a specified or allowable friction loss. The latter is defined for flow in tubes and pipes where the friction factor f is a function of the Reynolds number R_c. Assuming fully developed streamline or turbulent airflow, the duct loss expressed in inches of water pressure drop is given by

$$\Delta p = 2.305 \times 10^{-10} \left(4f \frac{\ell}{d_h} + K \right) \frac{G^2}{\rho} \qquad (2.19)$$

where f represents the Fanning friction coefficient and may be estimated from Fig. 2.8 and K is the loss factor associated with bends, turns, expansions, and contractions. The term G is mass flow per unit area [lb/(h • ft^2)] and ρ is the average density of the air within the flow passage. For other than round tubes, refer to Kays and London[10] for f values, or calculate an equivalent hydraulic diameter from the equation

$$d_h = 4 \frac{A_c}{P} \qquad (2.20)$$

and determine the Reynolds number and f value from Fig. 2.8. In Eq. (2.20), A_c is the flow cross-sectional area and P is the perimeter of the flow passage. Losses due to bends as well as exits and entrances may be minimized by good design practice. A Crane report[17] may be used for determining the K factors for these losses. The pressure drop for liquid coolants is treated in Sec. 2.5.

2.2.4 Radiation

Radiation refers to the transfer of energy by electromagnetic wave propagation. The wavelengths between 0.1 and 100 μm are referred to as thermal radiation wavelengths. The ability of a body to radiate thermal energy at any particular wavelength is a function of the

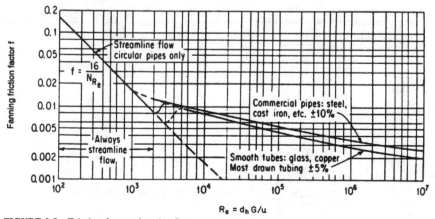

FIGURE 2.8 Friction factors for pipe flow.

body temperature and the material characteristics of the radiating surface. Figure 2.9 shows the ability to radiate energy for an ideal radiator, a blackbody, which by definition radiates the maximum amount of energy at any wavelength. The total energy radiated, in Btu/(h • ft^2), by a blackbody at any particular temperature is given by the area under that specific temperature curve in Fig. 2.9 and is equal to $\sigma_r T^4$ where σ_r is the Stephan-Boltzmann radiation constant.

The incident radiation on any surface is partially absorbed, reflected, and in some cases transmitted through the surface. This situation may be expressed by $\alpha + \rho + \tau = 1$. Bodies such as metals and most plastics which do not transmit radiation, are opaque, and the equation reduces to $\alpha + \rho = 1$. Glass and quartz are examples of materials that will transmit a portion of the incident radiation. Most materials encountered in electronic equipment packaging are opaque. However, there is also a class of optical materials used in infrared sensor equipment which is tailored to operate over specific wavelengths of energy. The lenses and mirrors for this type of equipment transmit and reflect efficiently over wavelengths corresponding to narrow bands of energy.

Materials or objects that act as perfect radiators are rare. Most materials radiate energy at a fraction of the maximum possible value. The ratio of the energy radiated by a nonblackbody to that emitted by a blackbody at the same temperature is called the emissivity ε. The emitted energy from a surface is given by $\varepsilon \sigma T^4$. The ability of a surface to absorb the total incident radiation on it is called the absorptivity α. As in the case of emissivity, the absorptivity is also dependent on the material and temperature of the emitter, that is, the wavelengths of the incident radiation. Some materials, such as the dark-color paints, exhibit the property of having $\alpha \approx \varepsilon$ when absorber and emitter are at temperatures of $\pm 100°F$ of each other. These materials are defined as gray bodies. Table 2.6 lists some gray-body emissivity values.

The basic equation for radiation heat transfer between two gray surfaces is given by

$$q_{(1-2)} = 0.171 \times 10^{-8} F_e F_a A_s (T_1^4 - T_2^4) \tag{2.21}$$

where F_e is a function of the emissivity values of the two surfaces between which the radiation occurs and their relative geometries, and F_a is the "view" factor, or the amount of radiation interchange area seen by the radiating surface. Table 2.7 illustrates the procedure for calculating F_e and F_a. For other shapes and areas see Brown and Marco.[8]

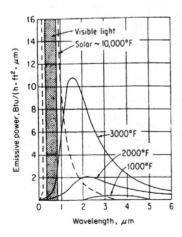

FIGURE 2.9 Energy distribution of blackbody.

TABLE 2.6 Emissivities of Surfaces at 80°F

Surface type	Finish	Emissivity ε
Paint	Black (flat lacquer)	0.96–0.98
Paint	Gray	0.84–0.91
Paint	White	0.80–0.95
Paint	White epoxy	0.91–0.95
Paint	Aluminum silicone	0.020
Metal	Nickel	0.21
Metal	Aluminum	0.14
Metal	Silver	0.10
Metal	Gold	0.04–0.23
Special surfaces		
Metal	Aluminum (sandblasted)	0.41
Metal	Aluminum (black anodized)	0.86
Metal	Nickel (electroless)	0.06–0.17
Metal	Aluminum (machine-polished)	0.03–0.06
Metal	Gold (electrodeposited or polished)	0.02
Ceramic	Cermet (ceramic containing sintered metal)	0.58
Metal	Brass (highly polished)	0.030
Metal	Copper (polished)	0.018
Metal	Nickel (polished)	0.070
Metal	Silver (polished)	0.02–0.03
Glass	Smooth	0.9–0.95
Metal	Alodine* on aluminum	0.15

*Alodine is a registered trademark of Amchem Products, Inc.

TABLE 2.7 Radiation View and Emissivity Factors

Configuration	Area on which heat transfer is based	F_a	F_e
Infinite parallel planes	A_1 or A_2	1	$\dfrac{1}{1/\varepsilon_1 + 1/\varepsilon_2 - 1}$
Completely enclosed body, small compared with enclosing body (subscript 1 is enclosed body)	A_1	1	ε_1
Completely enclosed body, large compared with enclosing body (subscript 1 is enclosed body)	A_1	1	$\dfrac{1}{1/\varepsilon_1 + 1/\varepsilon_2 - 1}$
Concentric spheres or infinite concentric cylinders (A_1 is inside A_2)	A_1	1	$\dfrac{1}{1/\varepsilon_1 + (A_1/A_2)(1/\varepsilon_1 - 1)}$

The concept of a radiation heat transfer coefficient is useful when comparing radiation with other modes of heat transfer. The term is given approximately by

$$h_r = 6.85 \times 10^{-9} F_e F_a \bar{T}^3 \; [\text{Btu/(h} \cdot \text{ft}^2 \cdot \text{°F)}] \qquad (2.22)$$

Typically, the h_r values for external surfaces of electronic equipment vary between 0.7 and 1.76 Btu/(h $\cdot$ ft^2 $\cdot$ °F) for temperatures of 100 to 200°F. The heat loss or gain in terms of the radiation coefficient is expressed as

$$q_r = h_r A_s (T_1 - T_2) \text{ Btu/h} \tag{2.23}$$

In Problem 2.2, using gray external painted surfaces at an average temperature of 104°F (564°R) and an ambient T_2 of 77°F (537°R), yields a radiation coefficient h_r of 0.97 Btu/(h $\cdot$ ft2 $\cdot$ °F). The resulting heat dissipation from radiation alone is 203.7 Btu/h or 49.7 W. Substituting an iridite or chemical finish on the external surface in place of the paint reduces the radiation loss to only 12 W.

2.2.4.1 Space Cooling. Radiation heat transfer plays a major role in the cooling of electronics in the hard vacuum of space. Likewise, conduction heat transfer is an equally important mode, particularly with respect to the interfaces which normally have increased resistance due to the lack of air at the interface (Sec. 2.2.1.1). On the exterior of a space vehicle or surface, the input heat sources consist of the following:

1. .Direct solar radiation q_s, = 443 Btu/(h $\cdot$ ft^2)
2. Solar radiation reflected or the albedo from a planet, such as the earth, q_a
3. Direct thermal radiation from the earth q_e

Other parts of the vehicle also couple radioactively with the surface of interest, that is, the radiating surface and their view factors and surface radiative characteristics must be specified for each individual ease. Neglecting the reflected solar radiation and the thermal radiation from adjacent surfaces of the vehicle, the heat balance for steady-state conditions for a surface in the space environment is in accordance with the equation

$$q_i + 443\alpha F_{a_s} A_s + 443\gamma\alpha F_{a_a} A_s = \sigma\Sigma F_a A T^4 \tag{2.24}$$

where q_i is the internal heat generation, α represents the absorptivity of the surface to solar radiation, γ the albedo or fraction of the solar radiation incident on the earth which is reflected (0.35 under normal conditions), and F_{a_s} and F_a are the view factors associated with the direct solar radiation and reflected solar radiation or albedo from the earth; the other terms are as defined previously. Table 2.8 presents absorptivity values for some common materials. It is not unusual to encounter a wide range of values associated with materials such as paints or the various surface finishes for the metals listed in Table 2.8. Therefore, as in many thermal applications, accurate absorptivity values will necessarily require empirical test data for best results. Table 2.8 is nevertheless useful as a guideline for choosing general categories of materials and finishes for solar irradiation applications. A high α/ε ratio suggests high solar absorption relative to the surface emission, hence higher surface temperatures. Good thermal surfaces for space radiation are those having $\alpha/\varepsilon = 0.25$ or lower and consist of special paints or, in some applications, multilayer insulation, which is alternate layers of aluminum and Mylar with nylon netting in between and with second surface aluminum/Teflon outer layers. These surfaces have α/ε values of 0.15 or less and are designed especially to have high solar reflectivity properties combined with high emissivity at normal operating temperatures. The multilayers are used as outer blankets to provide thermal insulation for the inner electronic packages. The view factors in Eq. (2.24) are complicated functions of the space vehicle in orbit, which vary at each position with time. These are geometric dependent functions involving the radiating surface,

TABLE 2.8 Solar Absorptivity

Surface type	Finish	Absorptivity α_s
Paint	Black	0.79–0.94
Paint	Yellow	0.54
Paint	Gray	0.49–0.57
Paint	Blue	0.48
Paint	White epoxy	0.3–0.35
Paint	White	0.19–0.33
Paint	Aluminum silicone	0.23
Metal	Alodine* on aluminum	0.4
Metal	Nickel	0.33–0.45
Metal	Gold	0.30
Metal	Silver	0.25
Metal	Black anodized	0.78
Metal	Aluminum sheet (commercial sheet)	0.39
Metal	Aluminum (sandblasted)	0.60
Metal	Aluminum (machined-polished)	15.25
Metal	Gold (polished or vacuum deposited)	0.14
Ceramic	Cermet (ceramic containing sintered metal)	0.65
Multilayer insulation	0.001-in double-sided aluminized Kapton	0.10
Multilayer insulation	Aluminum, Mylar layers, outer surface of aluminized Teflon	0.12

*Alodine is a registered trademark of Amchem Products, Inc.

the sun position, and the earth. Look-up tables or arrays listing the values of F_{a_s} and F_{a_a} are contained in computer programs such as TRASYS* tailored for space orbital missions. As an example, to solve the problem of determining the surface temperature of a space vehicle, the steady-state equation (2.24) is replaced with a finite-element analysis system represented by a series of lumped heat capacity nodes and resistances defining the physical system. The heat transfer within each of these nodes is neglected with respect to the adjacent nodes so that the smaller the node, the more accurate will be the computer solution. In some cases the thermal network involves the coupling of thousands of nodes, requiring the use of a mainframe computer to obtain rapid convergence to a solution in the iterative process. Because of the possibility of relatively sudden changes in heat input during an orbital mission as the vehicle moves into and out of the earth's shadow, it is not unusual to encounter large temperature excursions depending on the thermal mass. An α/ε ratio of 0.4, for example, yields a surface temperature at steady state of 106°F (41.1°C) from Eq. (2.24) with full sun and with no internal and no earth heat input. Under these conditions, without solar heat the surface temperature will cool down at a rate limited only by the thermal mass of the radiating surface and its thermally coupled system. Thermal control is often achieved in this case by utilizing space shutters or venetian-blind-type devices, which open and close as the surface temperature varies. Thus in the hypothetical case referred to, a closure of the shutter during the earth's shadow portion of the orbit serves to conserve lost energy. In some cases, heat is also applied in conjunction with or without shutter control to maintain proper temperature control.

---

*TRASYS (thermal radiation analyzer system) is a registered trademark of Martin Marietta Corporation. It is a computer program for up to 1000-node capacity used for specular, diffuse, or diffuse/specular radiant interchange surfaces.

The electronic packages within the spacecraft are designed thermally in the usual fashion, as in most conductive systems, except that particular attention to interface joints is critical. Without air, interface resistance values increase and the use of solid interstitial fillers is often mandatory. These require testing in a hard vacuum to obtain resistance values and to assure credibility of the design. In some cases the interface material must also be tested for outgassing because of potential deleterious effects in the space environment, particularly in conjunction with the design of optical sensor equipment.

2.2.5 Selecting the Cooling Mode

In the early stages of design it is often necessary to choose a cooling approach to satisfy the thermal requirements. As an aid to selection, Fig. 2.10 depicts the range of surface flux and temperature rise made possible for different cooling modes, types of surfaces, and flow conditions. The heat transfer coefficients derived from the figure are listed in Table 2.9 for convenience. Each surface is identified along with the restrictions associated with each case. The heat-sink temperature used in Fig. 2.10 is 80°F and the fluid properties are evaluated at the film temperature, except in the case of tube flow (curves K and L), where the bulk fluid temperature properties at 80°F are used. Care should be exercised in the use of Fig. 2.10 and Table 2.9 since these data apply only to the specific cases analyzed in accordance with the applicable equations listed. Nevertheless, the information can be useful in determining modes of heat transfer likely to result in acceptable temperatures for equipment operation. Alternatively, other more efficient convective modes, such as boiling and microchannel heat transfer, not included in the figure, are addressed later in this chapter. A further aid in the selection of a suitable cooling mode is summarized in Fig. 2.11, which expresses a range of heat transfer coefficients corresponding to several types of coolants. Natural and forced convection are the coolant modes represented, along with

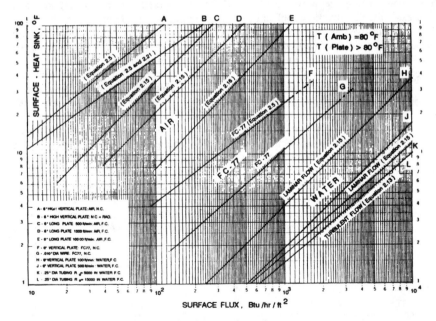

FIGURE 2.10 Surface heat flux for various heat transfer modes.

TABLE 2.9 Film Coefficient h_c for Radiation and Convection Heat Transfer

Flux, Btu/(h · ft²)	h_c, Btu/(h · ft² · °F)	Reference curve in Fig. 2.10
10–120	0.7–1.2	A
10–240	1.6–2.4	B
10–280	2.8	C
10–490	5.0	D
100–1140	11.0	E
100–1300	22–66	F
400–2000	83–105	G
1000–7500	250	H
10^3–10^4	550	J
10^3–10^4	715	K
10^3–10^4	970	L

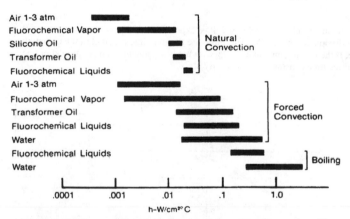

FIGURE 2.11 Comparison of coolants and heat transfer modes. (*From Chou and Simons.[18] Reprinted with permission of the publisher.*)

boiling water and fluorochemical liquids, which are an order of magnitude better than the same coolants when used in the forced convection mode.

2.2.6 Transient Heating and Cooling

Electronic systems undergo alternate heating and cooling cycles during their lifetimes. When a system is first turned on, the temperature of the components rises until a balance is reached between the heat dissipation and the rate at which heat is removed from the system. A perfectly insulated equipment would continue to rise in temperature at a rate inverse to its thermal capacitance mC_p, but in actuality the existence of a heat path from source to sink limits the temperature increase. This gives rise to a parameter referred to as the thermal time constant, which determines the rise time of temperature with respect to steady-state equilibrium. In simple terms, the thermal time constant λ expressed in the equation

$$\lambda = R_{tm}C_p \tag{2.25}$$

is the product of thermal capacitance and thermal resistance. It represents a time corresponding to a temperature difference of 63.2 percent of the steadystate value over the starting, or initial, temperature. Figure 2.12 shows the rise time of a component or element of a system in nondimensional form, that is, the number of time constants versus rise in temperature as a percentage of the steady-state rise. At three time constants, the temperature difference in Fig. 2.12 is 95.0 percent of its steady-state value, and in many applications this is considered as steady state. The temperature T_t for a linear system during heating at time t starting from an initial temperature T_i is expressed in the equation

$$T_t = T_i + \Delta T(1 - e^{-t/\lambda}) \tag{2.26}$$

for the transient case, where ΔT_0 is the steady-state temperature differential.

Problem 2.5 illustrates the use of Eqs. (2.25) and (2.26) as applied to Problem 2.1. In this example, the thermal response of the chip is determined with respect to the chip case. Values of specific heat for some common electronic materials appear in Table 2.10.

Problem 2.5: The silicon chip in Problem 2.1 is exposed to a step voltage at time $t = 0$ resulting in a heat dissipation of 0.25 W. Determine the thermal time constant of the chip with respect to its case. If the initial temperature is 25°C, what is the junction temperature after one time constant, at three time constants, and at steady state?

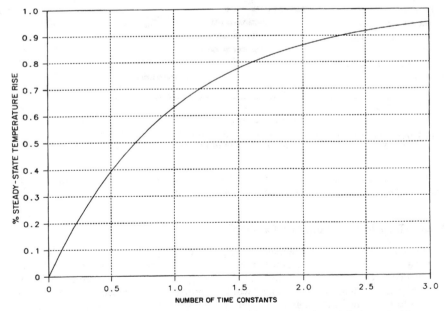

FIGURE 2.12 Percent steady-state temperature rise with respect to thermal time constant λ.

TABLE 2.10 Specific Heat C_p of Common Electronic Packaging Materials at 80°F

Metals	Btu/(lb · °F)	Nonmetals	Btu/(lb · °F)
Aluminum	0.21–0.23	Epoxy, cast	0.3–0.35
Copper	0.09	Alumina	0.20
Iron	0.11	Silica	0.19
Silver	0.06	Nylon	0.40
Gold	0.03	Rubber	0.45
Nickel	0.10	Glass, pyrex	0.20
Steel	0.11	Beryllia	0.24
Inconel	0.13	Mica	0.21
Magnesium	0.23	Epoxy fiberglass	0.30
Phosphor bronze	0.09	Air	0.24
Lead	0.03	Silicon	0.18

Solution: For silicon, density $\rho = 0.084$ lb/in^3, specific heat C_p (Table 2.10) = 0.18 Btu/(lb · °F), and resistance R_{j-c} (Problem 2.1) = 22.2°F · h/Btu. Then

$$\gamma = mC_pR_t = (0.084)(0.07 \times 0.07 \times 0.025)0.18 \times 22.2 = 4.11 \times 10^{-5} \text{ h}(0.15 \text{ s})$$

The steady-state temperature rise of the junction is based on the resistance from junction to case, as calculated in Problem 2.1,

$$\Delta T_0 = R_t q = 42.1°C/W \times 0.25 \text{ W} = 10.5°C$$

Then, after one time constant,

$$T_j = t_i + \Delta T_0(0.632) = 25 + 10.5 \times 0.632 = 31.7°C$$

After three time constants, $t = 3 \times 4.11 \times 10^{-5} = 1.23 \times 10^{-4}$ h, and the temperature is

$$T_{t=3\lambda} = T_i + \Delta T_0(1 - e^{-t/\lambda}) = 25 + 10.5(1 - e^{-3}) = 35.0°C$$

At steady state, the junction temperature is determined as follows:

$$T_j = T_i + \Delta T_0 = 25 + 10.5 = 35.5°C$$

In a similar manner, the transient cooling may be expressed in terms of the time constant λ and the steadystate temperature drop ΔT_0,

$$T_t = T_i - \Delta T_0 e^{-t/\lambda} \qquad (2.27)$$

The silicon chip in Problem 2.5 cools down at a rate that is equivalent to its heating rate so that if the cool down is assumed to begin at steady state, say 35.5°C, the temperature after the power is turned off at one time constant, or 0.15 s, will be 28.9°C, that is, 35.5 – 10.5 × 0.632 = 28.9°C.

2.3 *THERMAL ENVIRONMENTS AND COOLANT SPECIFICATIONS FOR ELECTRONIC SYSTEMS*

Sophisticated thermal models and computer analysis programs depend on an accurate representation of the external environment and the flow and temperature of the internal coolant within the units in order to achieve valid temperature predictions. These constraints therefore become an important part of the design process. For most electronic equipment some type of active cooling, either air or liquid, is essential. This may simply consist of mounting the equipment in an air-conditioned room and utilizing a blower to draw air into the unit for cooling. The many midsize computers are typical of these. More elaborate cooling installations, such as the mainframe computers supplied by IBM Corporation and others, draw conditioned air from a cooling source at the floor or depend on a direct feed of water flow for cooling. These commercial-type systems operate over more narrow temperature bands relative to the much wider temperature excursions of military and space systems aboard aircraft, ships, satellites, and so on. Moreover, not only is the surrounding ambient environment for these military type equipments much broader, but so also is the coolant supply temperature. As an example, Fig. 2.13 illustrates the amount of air [in lb/(min • kW)]

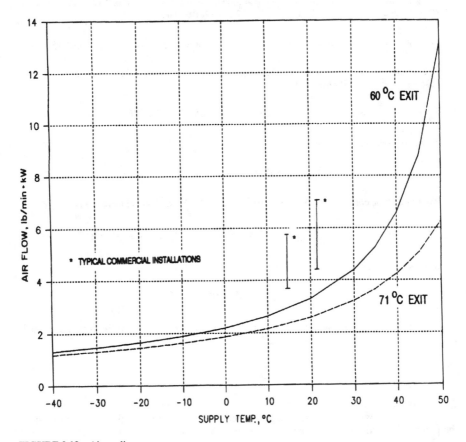

FIGURE 2.13   Air cooling system curves.

delivered as a function of supply temperature (in °C) for electronic equipment installations mounted aboard typical fighter aircraft such as the F-16 and F-18. Also shown in the figure are some operating ranges for several typical commercial computer installations. For such installations, exit temperatures for the air leaving the equipment are always lower than the 60 and 71°C shown in Fig. 2.13 and commonly specified for military installations. So also are the resulting maximum junction temperatures, which in commercial equipment may be 85°C as compared to 100 to 110°C for military hardware (see Fig. 2.17).

In the case of liquid cooling, the delivered flow is expressed in gal/(min • kW) at specific supply temperatures, and the quantity is always less than air for the same inlet and exit temperatures because the specific heats are greater for liquid. Three coolant curves are shown in Fig. 2.14, ethylene-glycol and water (62 percent glycol by weight), polyalpha-olefin, and Coolanol 25R. Since the exit temperatures are 60°C in all three cases shown, the variation in the quantity of coolant is attributed to the differences in the specific heats of the coolants. The two curves in Fig. 2.13 are based on the equipment heat q (in W) being absorbed by the coolant supply flow $\dot{m}$ (in lb/min) for air in accordance with the equation

$$T_{out} = T_{in} + \frac{0.1316q}{\dot{m}} \; (°C) \tag{2.28}$$

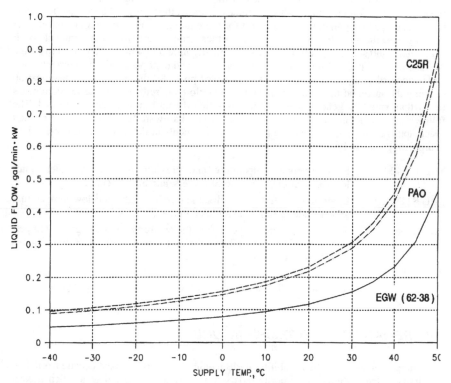

FIGURE 2.14 Liquid cooling system curves for 60°C exit temperature. EGW—ethylene-glycol and water; PAO—poly-alphaolefin; C25R—Coolanol 25R.

The curves in Fig. 2.14 are for liquid coolant and are based on the equation

$$T_{\text{out}} = T_{\text{in}} + \frac{0.00379q}{\text{gpm } C_p \text{ SG}} \ (^\circ\text{C}) \qquad (2.29)$$

where the flow is expressed in gal/(min • kW), C_p is the coolant specific heat in u Btu/(lb • °F), and SG is specific gravity. In addition to specifying the system flow rates, most hardware specifications also contain the operating pressure and the allowable pressure drop for the equipment when coolant is supplied from an external source. Typically, air pressure drops range from 1 to 10 in of water and for liquid coolants from 5 to 50 lb/in^2. Knowing the flow rate and the allowable pressure drop for a system, the designer must then allocate the resources to achieve an efficient, reliable thermal management system.

The surrounding thermal environments are often of secondary consideration and are dependent on the equipment application, whether it be airborne or ground-based. In general, when equipment is being supplied a source of coolant, the surrounding ambient does not affect the internal temperature of the electronics significantly. Only in cases where the units are cooled by natural means or by ram air or skin cooling in aircraft will the environment play a major role. There are many factors which define the equipment environment. For example, in aircraft applications, the aircraft speed, altitude, and mission profile as well as the equipment location, whether in a pressurized or a nonpressurized region, are important considerations. A careful study of these factors leads to an in-service environmental specification data sheet which includes, among other things, the storage, transportation, and operating ambient temperature for the equipment. For space applications, the defining specification includes mounting surface temperature for heat removal, orbit parameters, and so on; for missiles, the skin and structure temperatures are used to define the heat removal paths. Table 2.11 lists the operating and nonoperating environments for a wide range of equipment installations. Whenever possible, the equipment specification defining the surrounding environment should be tailored to reflect actual usage conditions rather than simply a general classification to which the unit may never be exposed. The following government specifications refer to military hardware for ground, airborne, and space applications; these define the thermal environments in detail for various classes of equipment purchased.

MIL-E-5400	Environmental temperature and pressure for airborne electronics located in aircraft, missiles, boosters, and allied vehicles
MIL-STD-810	Environmental test method standards for high- and low-temperature and altitude testing of avionic equipment
AR-70-38	An Army specification covering the environmental design for temperatures, humidity, and solar radiation for four world operating regions (hot, basic, cold, and severe cold)
MIL-STD-210	A standard defining the environmental conditions (temperature, humidity, sand, dust, etc.) for worst-case worldwide operation

2.4 RELIABILITY AND TEMPERATURE

The dramatic increase in the reliability of electronic systems of the past decade from double digits to triple digits for airborne radars,[19] for example, has led to a much greater emphasis being placed on reliability. Concurrently, since heat density and packaging density have also increased, the burden on increasing reliability continues to fall on quality

TABLE 2.11 Electronic Equipment Environments and Heat Sinks

Application	Nonoperating ambient temperature range, °C	Operating ambient temperature range, °C	Comments
Ground			
Inside without temperature control	< −40 to 71	10 to 40	General noncritical commercial installations; air-cooled
Inside with temperature control	< −40 to 71	15 to 25	Air-conditioned rooms; computer rooms air-cooled
World extremes	−62 to 95	−54 to 71	Military gear exposed to the elements
Shipboard	−54 to 71	0 to 50	Internal mounted equipment; for external equipment refer to specification
Aircraft			
Radome	−62 to 95	−54 to >100*	Antenna can be free-convection or liquid-cooled
Equipment bay	−62 to 95	−54 to 71	Coolant supplied in accordance with Fig. 2.13
Cockpit	−54 to 71	0 to 32	Cockpit units self-air-cooled
Missiles			Forced-convection skin cooling during flight heat storage
Captive flight	−54 to 95	54 to >100*	
Free flight	—	−54 to >100	
Spacecraft			Conduction to mounting structure is primary heat sink; external equipment subject to wide range of cold to hot, depending on orbit
Launch pad	−30 to 71	—	
In orbit	0 to 40	0 to 40	

*Short duration periods.

improvements of the components and in particular on cooling. Component failure rates λ_i are strongly dependent on temperature and hence on the cooling efficiency. This dependence is best illustrated by reference to Figs. 2.15 and 2.16, which show the failure rates of an electronic mix of components, digital, analog, and RF, as a function of the junction or case temperature. Similar curves exist for other components, such as diodes, capacitors, and resistors. In Fig. 2.15 the mix of chips shown is normalized as to size, complexity, and quality of the environment, and the linear analog circuit is seen to be much more sensitive to temperature, both absolute and in rate of change, than either the digital CMOS or the bipolar, which suggests an apportionment of cooling resources for optimum reliability. This is in fact being done in some present designs; see, for example, Minor[19] and Altoz et al.,[20] which illustrate a mathematical approach to flow apportionment in electronic systems. In Fig. 2.16 there is no normalization of components, the programmable array logic (PAL) circuit being physically larger and more complex than either of the other digital-type circuits, the microprocessor and the 256K DRAM. The approximately 2:1 failure rate variation again suggests some type of cooling allocation, particularly at higher temperatures. The bases for the curves in Figs. 2.15 and 2.16 are the predicted failure rates from MIL-HDBK-217, and the assumption is made that each part operates at the same stress ratio and temperature in a specific environment. Environments do differ, so that failure rates also vary; ground installations, for example, are usually less severe than equipment mounted in aircraft or aboard naval vessels. However, these figures illustrate the dependence of the failure rate on temperature and are only of qualitative value and not absolute.

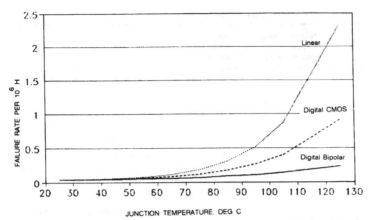

FIGURE 2.15 Component failure rates over temperature; digital and analog components. (*From MIL-HDBK-217.*)

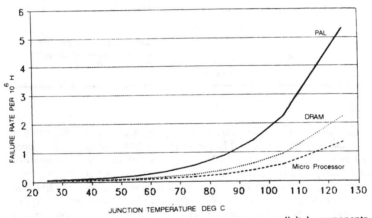

FIGURE 2.16 Component failure rates over temperature; digital components. (*From MIL-HDBK-217.*)

Recent studies of electronic equipment have shown that the field reliability of equipment is temperature-related, with respect to both free- or natural-convection-cooled and forced-cooled units. Moreover, for avionics equipment, Air Force studies have concluded that 20 percent of reported field failures are temperature-related and, in addition, temperature cycling has been associated with system reliability and induced failures caused by a combination of temperature and vibration. There is a significant body of knowledge, therefore, that argues for lower-temperature operation as a means of improving electronic equipment reliability.

The reliability of an electronic system comprising a group of components is most simply stated as the probability, expressed in percent, of operating continuously over a specified period of time with no failures. For most solid-state electronic devices, the reliability handbooks which establish a common basis for comparing competitive designs utilize an Arrhenius-type failure-rate model of the form

$$\lambda_i = B_i e^{-Ai/\lambda j(T)} + Ei \tag{2.30}$$

where the coefficients A_i, B_i, and E_i are independent of temperature. Considering that each part has a particular failure rate λ_i expressed in number of failures per million hours, the mean time between failures MTBF for a group of components constituting a system is expressed as

$$\text{MTBF} = 1/\sum_{i=1}^{n} \lambda_i(T) \tag{2.31}$$

The reliability R, which is the probability of no failures over the operating time t, is, in terms of MTBF,

$$R = e^{-t/\text{MTBF}} \tag{2.32}$$

Therefore for a module consisting of 1000 devices, each having an assumed failure rate λ_i of 1.0 ppm, the MTBF is 1000 [from Eq. (2.31)], and the probability of operating with zero failures over only 100 h is 90.5 percent [from Eq. (2.32)]. Furthermore, assuming there are 10 identical assemblies or modules constituting a package, the overall package MTBF is only 100 h and the reliability for a 100-h operating time is reduced to 36.8 percent. The inclusion of nine additional modules effectively lowers the reliability by 59 percent.

In some cases we would like to know the MTBF required to achieve a given reliability. For example, in the case cited, if the desired reliability of the 10-module system is 0.95, that is, 95 percent probability of failure-free operation for 100 h, the resulting system MTBF, from Eq. (2.32), is 1949 h, or 19,490 h per module.

As to what is considered a proper level of operation, Fig. 2.17 illustrates typical junction temperatures for equipment presently operating in a large number of field applications. The acceptable operating range for semiconductor junctions is shown to be generally 40 to 60°C (see Fig. 2.15), although even fewer failures occur, down to as low as 0°C. Below 0°C, reliability is uncertain and some semiconductors cease operating, only to return to operation at higher temperatures with no apparent permanent damage. It must be recognized that the allowable junction temperature for any system required to meet a spec-

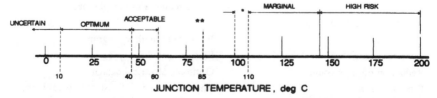

FIGURE 2.17 Temperature spectrum of operating junctions.

ified reliability may vary considerably due to many factors, including parts count, type of components, and dissipation levels. Nevertheless, the upper limit for commercial applications is usually set at 85°C, and for military equipment the acceptable upper limit is 100 to 110°C for *all* semiconductors in power supplies and processors. It should also be recognized that other electronic components, diodes, capacitors, resistors, and so on, are also sensitive with respect to temperature, even though in general these components do not drive systems with regard to reliability. Nevertheless, in determining MTBF, all components become contributors and as such figure into the overall computation. The reliability improvements or system enhancements which are directly temperature-dependent, as well as those that are not temperature-related, are illustrated in Table 2.12.

The net benefit of a successful reliability implementation program is a significant savings in life-cycle costs which accrue from low or maintenance requirements and a need for fewer replacement spares.

A recent development in reliability introduces a means of measuring temperature stress within an electronic system through the use of time stress measurement devices (TSMDs). Future systems of a complex nature will include provisions for the measurement of actual service-life stress levels, particularly in aircraft, as a means of correlating equipment failures in flight that cannot be duplicated on the ground. Actual stress levels during the service life may vary widely, and if predicted stresses are less than the actual stresses, equipment will experience a shorter life than the required, or design, service life. Actual hardware failures are caused by material fatigue failures due to vibration, thermal cycling, elevated temperature, duty cycles, corrosion, shock, and other contributors, which can all lead to an increase in thermal stresses. An increase in thermal stress may result from any of the following.

TABLE 2.12   Reliability Improvement Factors

	Implementation and trades
Temperature-dependent	
Reduction of temperature levels	Lower heat dissipation.
	Improve thermal coupling from component to heat sink.
	Lower heat-sink temperatures.
Optimization of thermal design	Secure proper placement of components on PCBs for maximum MTBF.*
	Allocate cooling resource or greater coolant flow to sensitive units.†
Improvement in temperature analysis	Use conduction-cooled modules as opposed to flowthrough to lower gradients and temperature amplitudes.
Non-temperature-dependent	
Burn-in of parts	Institute reliability program.
Parts screening	Institute reliability program.
Reduction of parts count	Optimize electrical design.
Process control	Improve solder and assembly techniques, inspection, and control.

*From Dancer and Pecht.[21]
†From Minor[19] and Altoz et al.[20]

1. Higher than calculated thermal resistance between a module heat sink and its cold wall, resulting from an incorrect clamping force on the wedge clamp or corrosion at the interface

2. An increase in the number of thermal cycles due to ground power on-off cycles or an increase in ground operating time where cooling is reduced

3. An increase in stress level caused by a greater number of flight thermal cycles

4. More frequent and more stressful missions than originally planned

5. Restricted or reduced cooling air or liquid

6. An increase in vibration level due to more frequent or stressful missions

The TSMD allows an identification of the actual stress levels at module locations deemed sensitive over the entire service life of the equipment. These devices may incorporate any or all of the features that measure power-on time, the number of on-off cycles, the number and magnitude of temperature cycles, an alert when temperatures exceed an upper limit, the number and duration of mode changes, and other stresses such as vibration levels or voltage variations. In practice, device outputs will be fed into computer storage locations for proper logging and interpretation. This type of program will attempt to quantify better the reliability of an equipment with its true service operating condition so that proper corrections can be implemented to reduce the system life-cycle cost.

2.5 COOLANT SELECTION AND HEAT TRANSFER ANALYSIS OF LIQUID SYSTEMS

2.5.1 System Types

There are two types of liquid systems used extensively in electronic cooling—direct and indirect. In the direct cooling system approach, the coolant flows over the component to remove heat from the surface and as such must be capable of sustaining a voltage gradient; alternatively in cold-plate cooling, which is indirect, there is no such requirement. The direct method is an efficient means of heat removal because the coolant is closest to the heat source. Even so, in critical applications where the heat flux rates are high, surface temperatures are best determined through actual measurements of temperature rather than exclusive dependence on analytical predictions. On the other hand, the more conventional applications invoke the use of Eq. (2.17) and Table 2.5 (Sec. 2.2.3) in calculating the heat transfer film coefficient h_c from which the surface temperature is determined using Eq. (2.3). Some examples of direct liquid cooling of electronic equipment and the coolants used in these systems appear in Table 2.13. Common to these applications is the necessity for the maintenance of coolant purity, which is obtained through a means of filtration consistent with the dielectric requirements. Experience has shown that in direct cooling of components, the system must be absolutely airtight in order to prevent air or moisture from degrading the heat transfer performance of the coolant. In addition, moisture absorption can lead to a chemical breakdown in some coolants, which in turn lowers the flash point, hence creating a potential safety hazard. Extreme care is thus required in selecting a dielectric for direct cooling applications.

The category of indirect liquid cooling refers to the use of cold plates for the absorption of heat. Military systems of this type are driven by the criteria of separation of coolant

TABLE 2.13 Common Examples of Direct Liquid Cooling

Coolant	Application
FC-77	Cray-2 supercomputer
FC-104, EGW, deionized water	Laser target illumination for electrooptical systems
C25R	Radar transmitter and TWTs on F-15 and F-16 fighter aircraft and others
FC-77, EGW	Antenna and klystron tubes for E3-AWACS radar system
C25R, PAO	TWTs on F-16 and ATF radars

and components as defined in MIL-STD-454, Requirement 52. Some recent examples of this type of cooling include cold plates for use in antenna array modules having solid-state circuitry and in high-heat flux power supply modules in aircraft applications. Commercial applications include mainframe computers where a refrigerant system provides a heat sink for the liquid used to cool high-heat chip packages.

2.5.2 Coolant Selection

The choice of coolant in these applications, whether direct or indirect, often depends on the forced convection figures of merit derived from the four basic coolant properties, C_p, u, k and ρ. Table 2.14 lists these properties at 1 atm for five coolants at the three temperatures of -40, 77, and $200°F$ (-40, 25, $93.3°C$). It is important to note that absolute viscosity [in lb/(ft h)] varies widely with temperature and as such has significant implications in the coolant selection process. Figures 2.18 and 2.19 illustrate the coolant figures of merit as functions of temperature for straight finned or tubular channel flow operating in the laminar and turbulent regimes. Figure 2.18 derives from Eq. (2.12) for h_c at the entrance to the cold plate (Seider and Tate[11]), whereas Fig. 2.19 is for fully developed turbulent flow for $\ell/d > 10$. These cases define typical flow conditions in liquid systems used in electronic applications. Equations (2.33) and (2.34) express the figure of merit FOM for these two cases of interest. For laminar flow at the entrance to the cold plate,

$$\text{FOM} \approx k^{0.66} \rho^{0.33} C_{p\,0.33}$$ (2.33)

For fully developed turbulent flow in the cold plate,

$$\text{FOM} \approx \frac{k^{0.6} \rho^{0.8} C_p^{0.4}}{u^{0.4}}$$ (2.34)

Good design practice for cold plates operating in the laminar region, which is common in liquid systems, dictates that critical components be located near the entrance of cold plates in order to benefit from the low inlet coolant temperature but also to take advantage of the higher cold-plate film coefficients. As to coolant selection, the heat transfer figures of merit shown in Figs. 2.18 and 2.19 are to be used judiciously along with other criteria deemed important to specific applications. These parameters include the following:

Toxicity
Pour point

TABLE 2.14 Coolant Properties of Five Coolants at Different Temperatures

	2 = cSt PAO*			C25R†			FC-77‡			EWG (62-38)§			Water		
	-40°F	77°F	200°F	-40°F	77°F	200°F	-40°F	77°F	200°F	-40°F	77°F	200°F	-40°F	77°F	200°F
C_p	0.49	0.54	0.58	0.37	0.44	0.52	0.22	0.25	0.28	0.66	0.74	0.82	—	0.99	1.0
u	532	14.9	3.4	184	12.0	3.2	19.2	3.44	1.15	485	12.3	2.30	—	2.1	0.74
k	0.086	0.083	0.079	0.079	0.075	0.073	0.040	0.037	0.033	0.23	0.22	0.21	—	0.35	0.39
ρ	52.4	49.3	46.4	59.3	56.2	52.4	121	111	99.2	70	67.4	64.3	—	62.4	60.1

*Chevron Chemical Company.
†Monsanto Company.
‡3M Company.
§E. I. du Pont de Nemours & Company, Inc.

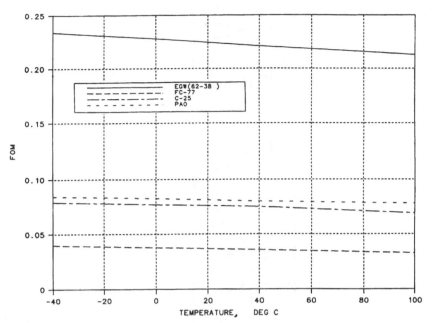

FIGURE 2.18   Figures of merit for liquid coolants in laminar flow.

Maximum wet wall temperature

Flammability

Cost per gallon

Freeze point

Material compatibility

Corrosion

Pressure drop characteristics

Water absorption sensitivity

The importance of these and other criteria in coolant selection is best treated through the assignment of weighting factors.

2.5.2.1 Fin Efficiency. The film heat transfer coefficient h_c for liquid heat exchangers and cold plates, unlike those for air, is generally high [>100 Btu/(h • ft2 • °F)]; therefore fin efficiency plays a major role in the calculation of heat transfer. Within the fin proper and on the surfaces exposed to the coolant, the heat distribution is not uniform from base to tip so that unequal amounts of heat enter the coolant. Furthermore, actual film coefficients are seldom constant at the fin surface, which adds further to the complexity, but for most calculations h_c variations are ignored in the analysis. The geometry for a typical fin, being an element of a heat exchanger, is shown in Fig. 2.20. Fin efficiency ζ is defined as the ratio of the actual heat transferred by the fin to the heat that would be transferred if the fin were at the base temperature. Mathematically, the heat transfer q_f from the fin at temperature T_f to a coolant at temperature T_b (bulk) is in accordance with Eq. (2.35) derived by Hofman,[13]

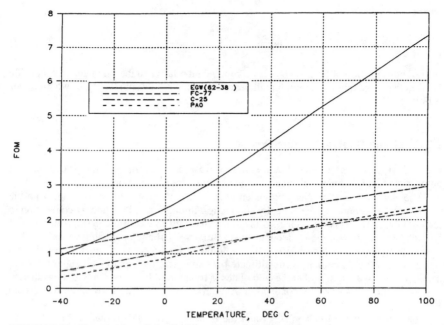

FIGURE 2.19 Figures of merit for liquid coolants in turbulent flow.

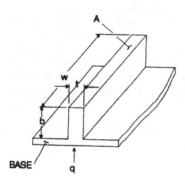

FIGURE 2.20 Fin geometry.

$$q_f = (h_c PkA)^{0.5} \tanh{(mb)} (T_f - T_b) \qquad (2.35)$$

where P is the perimeter of the fin ($= 2w + 2t$) in Fig. 2.20 and the other terms are as defined previously. Assuming that the fin thickness t is small compared to the width w, and that the top of the fin is insulated (no heat transfer), a conservative assumption, the fin efficiency ζ is then

$$\zeta = \frac{\tanh{(mb)}}{mb} \qquad (2.36)$$

where

$$mb = \left(\frac{2h_c}{kt}\right)^{0.5} b \qquad (2.37)$$

The optimization of fin geometry is accomplished relative to the fin material, fin profile bt, and cost. Note that maximum fin efficiency occurs at $b = 0$ (tanh $0 = 1$), or where no fin is present.

2.5.3 Liquid Cooling System Design

Figure 2.21 depicts an indirect liquid system for use in electronic cooling. It comprises a pump, cold plates, a heat exchanger, and an accumulator or reservoir.

The electronic system heat is transferred into the cold plates by conduction from the component heat source to the plate, by conduction through the plate, and by convection to the cold fluid. This type of exchanger is commonly used in electronic systems because of its high heat transfer efficiency [conductance, Btu/ (h • °F)] and packaging versatility. The heat-sink exchanger in Fig. 2.21 is a liquid-liquid or liquid-gas exchanger, depending on whether air or liquid coolant is the ultimate heat sink. In this type of exchanger the electronic heat from the hot fluid in the cold plates is transferred by convection to the wall separating the fluid passages, by conduction through the wall, and by convection to the cold fluid.

The thermal control valve (TCV) and the bypass valve (BV) shown in Fig. 2.21 are optional. Across the heat exchanger they serve to bring the system up to temperature during warm-up and at the pump they effect pressure relief due to possible flow blockage or low-temperature operation. Not shown in this simplified system is any temperature- or flow-control apparatus for maintaining accurate delivery conditions demanded of some systems. In analyzing a system of this type, the variables associated with the system are illustrated in the flowchart of Fig. 2.22. Each block represents a key parameter to be considered in the overall design. To illustrate the procedure, the first step in the process is to determine the overall system requirements, namely, the total and local heat dissipations at the cold plates and the characteristics of the heat sink as to temperature, flow, and whether it is air- or liquid-cooled. The coolant selection is made on the basis of a comparison

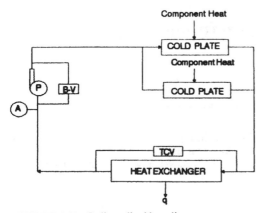

FIGURE 2.21 Indirect liquid cooling system.

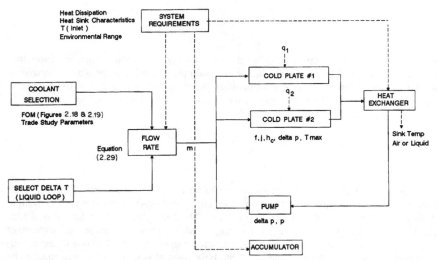

FIGURE 2.22 Flowchart of liquid system design variables.

between the coolant figures of merit and the other parameters listed which best character-ize the application. A priority assigned to these will aid considerably in the evaluation pro-cess. Selection of the allowable temperature rise in the liquid loop is often made on the basis of choosing 18°F (10°C) as a starting point. Experience has shown that reasonable flow rates are achieved for this rise in temperature for conventional systems without imposing an undue penalty on the system pump power or on the heat transfer within the cold plates. The flow rate is next determined from Eq. (2.29) in gal/min or in lb/min, and all the remaining variables associated with the cold plates, the heat exchanger, and the pump are based on this calculated flow.

2.5.4 Liquid Cold Plates

The cold plate heat transfer design is best expressed in terms of a friction factor f and a heat transfer j, both being functions of the Reynolds number. For the viscous pressure drop through the heat exchanger core, neglecting the entrance and exit losses,

$$\Delta p_{core} = 4f\frac{\ell}{d_h}\left(\frac{\rho V^2}{2g}\right) = 4f\frac{\ell}{d_h}\left(\frac{G^2}{2g\rho}\right) \tag{2.38}$$

In Eq. (2.38) the velocity head term $(1/2g)\rho V^2$ is replaced by $G^2/2g\rho$, where G is the mass flow per unit area, or lb/(h • ft^2), a more commonly used heat exchanger parameter.

The total system pressure drop in Fig. 2.21 includes the cold-plate and heat exchanger viscous core drops and the inviscid entrance and exit losses within the exchangers as well as line pressure drops throughout. The Crane paper[17] may be used to obtain loss coeffi-cients through valves, expansion, and contractions in the equation

$$\Delta p = \frac{kG^2}{2g\rho} \tag{2.39}$$

The pump characteristic curve showing flow versus pressure drop is then used together with superposition of the system resistance to obtain the operating point. This procedure is similar to that of determining the fan operating point described in Sec. 2.6. In the case of heat transfer, the thermal j, or Colburn factor, is used in determining the film coefficient h_c from the equation

$$h_c = \frac{jGC_p}{P_r^{0.66}} \tag{2.40}$$

Both the Fanning friction factor f and the Colburn factor j are obtained from Fig. 2.23 corresponding to common lanced and offset fin configurations ranging from 0.050 in high, 20 fins/in, 0.125 in lanced or offset, to 0.125 in high, 20 fins/in, 0.125 in lanced or offset. These curves are approximations of the f and j factors commonly used in heat exchanger and liquid cold-plate design. Considerable variation can exist in the data due to test conditions and, specifically, the temperature of the bulk fluid at which the test was run. In the curves in Fig. 2.23 the upper f and j values correspond to fins closer to 0.1 in high; lower values are for shorter fins. With respect to the heat exchanger design, there are companies which specialize in the design and fabrication of these types of exchangers. They use computer programs extensively, which define basic fin design characteristics of a proprietary nature that become inherent features of these in-house programs. The theory of exchanger design goes beyond the scope of this section, and only basic concepts relative to cold plates are discussed. Problem 2.6 illustrates the procedures to be followed for determining

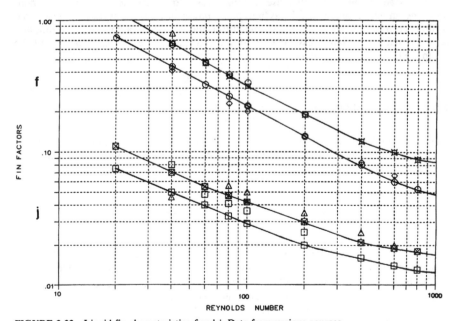

FIGURE 2.23   Liquid fin characteristics f and j. Data from various sources.

both pressure drop and heat transfer characteristics of ordinary electronic cold-plate designs.

Problem 2.6: An electronic cold plate 20 in long and 8 in wide dissipates 1500 W. Determine the local temperature rise in the plate near the coolant exit where the surface flux is double the average value. Assume the coolant is C25R flowing at 2 gall min at an inlet temperature of 21.1°C (70°F).

Solution: The procedure followed uses Fig. 2.23 for the f and j fin characteristics after determining the Reynolds number at the average coolant viscosity.

1. $$\dot{m} = \text{gpm} \times 8.33 \times \text{SG} = 2 \times 8.33 \times 0.9 = 15 \text{ lb/min}$$

2. $$T_{out} - T_{in} = \frac{0.00379q}{\text{gpm } C_p \text{SG}} = \frac{0.00379 \times 1500}{2 \times 0.44 \times 0.9} = 7.2°C$$

3. The average coolant bulk temperature is

$$\bar{T}_b = \frac{T_{out} + T_{in}}{2} = \frac{21.1 + 28.3}{2} = 24.7°C \ (76.5°F)$$

4. The properties of the coolant at 76.5°F from Table 2.14 (using 77°F values) are

$$u = 12 \text{lb/(ft} \cdot \text{h)}$$
$$C_p = 0.44 \text{ Btu/lb}$$
$$k = 0.075 \text{ Btu/(h} \cdot \text{ft}^2 \cdot °F)$$
$$\rho = 56.2 \text{ lb/ft}^3$$

5. Select a 0.10-in-high fin, 18 fins/in, 0.125 lanced and offset. The cold-plate fin stock is 7.5 in wide by 0.1 in high.

$$G = \frac{\dot{m} \times 8640}{A_c} = \frac{15 \times 8640}{7.5 \times 0.1 \times 0.85} = 203,294 \text{ lb/(h} \cdot \text{ft}^2)$$

where the factor 0.85 is the percentage of open flow area based on 0.006-in-thick fin material.

6. Then, $$R_e = \frac{d_h G}{u} = \frac{0.000518 \times 203,294}{12} = 87.8$$

7. From Fig. 2.23 at $R_e = 87.8$, $j = 0.045$ and $f = 0.035$.

8. $$h_c = j\frac{GC_p}{P_r^{0.66}} = 0.045\frac{203,2948 \times 0.44}{\left(\dfrac{0.44 \times 12}{0.075}\right)^{0.66}} = 242.9 \text{ Btu/(h} \cdot \text{ft}^2 \cdot °F)$$

9. Fin efficiency ζ is based on mb for the fin,

$$mb = \left(\frac{2h_c}{kt}\right)^{0.5} b = \left(\frac{2 \times 242.9 \times 12}{100 \times 0.006}\right)^{0.5}\frac{0.1}{12} = 0.82$$

$$\zeta = \frac{\tanh(mb)}{mb} = \frac{0.670}{0.82} = 0.823$$

10. The fin core pressure drop is

$$\Delta p = 4f\frac{\ell}{d_h}\left(\frac{G^2}{2gp}\right)$$

$$= 4 \times 0.35 \times \frac{20}{0.062}\left(\frac{203,294^2}{1.2018E11 \times 56.2}\right) = 276 \text{ lb/in}^2$$

where the constant 1.2018E11 is a conversion factor to yield pressure in lb/in^2.

11. The heat transfer area is

$$A_s = \beta \times V = \frac{670 \times 7.5 \times 0.1 \times 19.5}{1728} = 5.67 \text{ ft}^2 \text{ (total for cold plate)}$$

where β =ft2 heat transfer area/ft^3 volume.

12. The fin cold-plate conductance is

$$C = \zeta h_c A_s = 0.823 \times 242.9 \times 5.67 \times \quad = 1133.6 \text{ Btu/(h} \cdot {}^\circ\text{F)}$$

or the conductance per unit area is

$$C = \frac{1133.6}{8 \times 20} = 7.08 \text{ Btu/(h} \cdot {}^\circ\text{F} \cdot \text{in}^2)$$

13. The temperature rise is

$$\Delta T_{\text{fin C25R}} = \frac{q}{C} = \frac{2 \times 1500 \times 3.41}{1133.6} = 9.0{}^\circ\text{F (5}{}^\circ\text{C)}$$

where temperature rise is at the inlet and flux is twice the average value over the cold plate.

2.6   AIR COOLING SYSTEM CONSIDERATIONS

Fans, blowers, and heat exchangers are essential components of air cooling systems. Small and medium size computers, for example, use internal fans, whereas mainframe computers operate on air or chilled water supplied from a refrigerated source. Examples of forced-air cooling include the skin cooling of missiles and the cooling by ram air of electronic systems in pods mounted aboard aircraft. On the ground, fan cooling in console racks is quite common, as is the natural or free convection cooling of transit cases and shelter equipment common to military installations. Some cabinets include internal heat exchangers, which are watercooled and fan-driven, and still others utilize induced air where the heat dissipation is low. In the design of air-cooled systems it is often necessary to consider the basic flow and heat transfer parameters associated with heat exchangers, blowers, and fans and the factors affecting cabinet cooling. These subjects are treated in the following sections.

2.6.1 Induced or Draft Cooling of Cabinets

The use of natural convection air cooling for cabinets with low dissipation is consistent with long-term reliability and low-maintenance operation. Large racks and cabinets for use in commercial communications equipment and in ground support having less than 100 W/ft^3 are candidates for this type of cooling.

The driving force in induced-air cooling is the pressure difference caused by a change in air density between the lower and upper regions of a cabinet. This so called chimney effect produces an air motion within the enclosure which cools the component as the air moves from the bottom upward. Resisting this motive force is the internal friction and the losses associated with area changes within the enclosure. The operating point for a cabinet is where a balance exists between the system pressure or resistance and the induced draft. The latter is expressed as the flotation pressure Δpf (in inches of water) in the equation

$$\Delta pf = 0.192 H\rho \ln\left(\frac{T_2}{T_l}\right) \tag{2.41}$$

Here T_2 and T_i are outlet and inlet absolute temperatures (in °R or °K), respectively. In the application of Eq. (2.41), the height H (in ft) is related to the manner in which the cabinet heat is dissipated, that is, H is the full height whenever the entire heat is located at the console bottom, whereas for uniform heat spreading top to bottom, H is equal to one-half the cabinet height. For any other distribution of heat the centroid of heat as measured to the exit or top of the cabinet is the desired height. Figure 2.24 shows a typical cabinet, 18 by 19 by 60 in, with four layers or PCB buckets stacked vertically. The operating point for this cabinet is obtained from a plot of the cabinet resistance made up of the sum of friction loss and the entrance and exit losses as expressed in Eq. (2.19), balanced against the induced draft pressure defined by Eq. (2.41). A simple computer program written in Quick Basic,[*] Table 2.15 is the means for obtaining the curves shown in Fig. 2.25. The variables such as

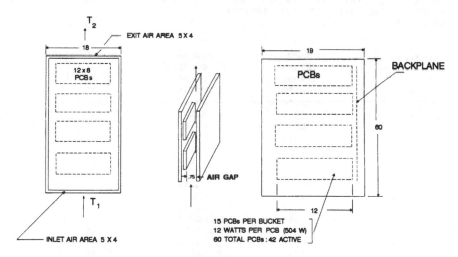

FIGURE 2.24 Typical induced-draft-cooled cabinet.

---

*Quick Basic is a registered trademark of the Microsoft corporation

TABLE 2.15 Computer Program CAB.BAS and Printout of Cabinet Resistance and Induced Draft ($\sigma\Delta p$, in of water, versus flow, lb/min)

```
'PROGRAM; CAB.BAS
'INDUCED DRAFT OF CABINET, SIGIND (IN WATER) AND CABINET RESIS-
TANCE, SIGDELP (IN WATER AT 20 DEG C INLET) AND VARIOUS TE EXITS.
PRINT ''FLOW (LB/MIN) SIGDELP SIGIND TE (DEG C)''
FOR M = 1 TO 20
TE = 25 + M * 2
TAVG = (20 + TE) / 2
RHOAVG = 39.7148 / (492 + TAVG * 1.8)
VISC = (3.378E-07 + 6E-10 * (32 + 1.8 * TAVG) - 2.42E-13 *
(32 + 1.8 * TAVG) ^2) * 115812
SIGMA = RHOAVG / .0765
N = 131.6 * .504 / (TE - 20)
'FLOW PARAMETERS FOR PCB
DHPCB = 4 * 12 *.75 / 25.5
ACPCB = .75 * 12 * 42 / 4
GPCB = 8640 * N / ACPCB
REYPCB = GPCB * DHPCB / VISC
FPCB = 22.5 / REYPCB
'FLOW PARAMETERS FOR ENTRANCE/EXIT OPENINGS
AC = 5 * 4
GE = N * 8640 / AC
VHE = GE ^2 * 2.305E-10 / RHOAVG
DELP = 1.5 * VHE + 2.305E-10 * 4 * FPCB * 32 * GPCB ^2 / DHPCB /
RHOAVG
SIGDELP = SIGMA * DELP
'INDUCED DRAFT PARAMETERS
HAVG = 2.5
DELIND = .192 * HAVG * RHOAVG * LOG ((TE + 273) / (273 + 20))
SIGIND = SIGMA * DELIND
PRINT INT(N * 10) / 10, INT(SIGDELP * 1000001) / 1000001, INT
(SIGIND * 1000001) / 1000001, INT(TE)
NEXT M
END
```

FLOW(LB/MIN)	SIGDELP	SIGIND	TE(DEG C)
9.47	.07589	.00081	27
7.36	.04593	.00104	29
6.02	.03077	.00126	31
5.1	.02204	.00147	33
4.42	.01656	.00168	35
3.9	.0129	.00189	37
3.49	.01034	.00209	39
3.15	.00847	.00229	41
2.88	.00706	.00248	43
2.65	.00598	.00267	45
2.45	.00513	.00285	47
2.28	.00445	.00304	49
2.13	.00389	.00322	51
2	.00344	.00339	53
1.89	.00306	.00356	55
1.79	.00274	.00373	57
1.7	.00247	.0039	59
1.61	.00223	.00406	61
1.54	.00203	.00421	63
1.47	.00185	.00437	65

hydraulic diameter of the channel formed by the PCBs (DHPCB) and flow area (ACPCB) arc purposely written with the numerical component terms to better illustrate the procedure. Note also that the air viscosity (VISC) is in units of lb/(ft • h) as a function of the average air temperature (TAVG) within the cabinet. It is a simple matter to extend the program by introducing a DO loop for determining the crossover of the two curves, or the operating point. The cabinet itself includes 42 active PCBs out of a total of 60 slots, having an average dissipation of 12 W each. At an inlet temperature of 20°C (68°F), the cabinet resistance line in Fig. 2.25 represents the solution to Eq. (2.41) obtained by assuming various exit temperatures (TE), hence flow rates N (in lb/min) in the computer program. The flow resistance line is based on the channel friction loss calculated from Eq. (2.19) along with an assumed 1.5 velocity head loss at the entrance and exit. Note that the pressure drop in Fig. 2.25 is expressed as $\sigma \Delta p$, where a is the density ratio in accordance with the equation

$$\sigma = \frac{\bar{\rho}_a}{0.0765} \tag{2.42}$$

and $\bar{\rho}_a$ is the average density.

The performance of this cooling system as measured by the flow at the operating point is markedly improved by the addition of a fan (see Problem 2.7, Sec. 2.6.2).

2.6.2 Fans and Blowers

In the selection of fans and blowers, the flow and pressure head are the major parameters of interest, but also important are the noise level, life, ac (60 Hz or 400 Hz) or dc operation, size, and, in some cases, operation at other than sea level.

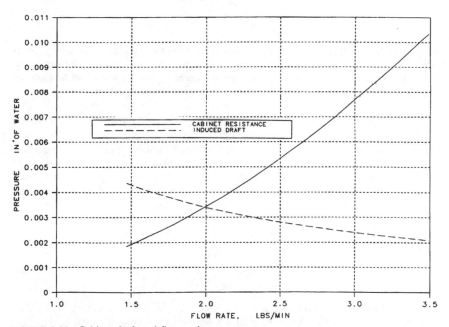

FIGURE 2.25 Cabinet draft and flow resistance curves.

Fan laws are sometimes used to compare the operation of a specific fan under varying assumptions of speed, flow, density, and size:

speed change

$$\dot{Q}_2 = \frac{\dot{Q}_1 N_2}{N_1} \tag{2.43}$$

$$P_2 = P_1 \left(\frac{N_2}{N_1}\right)^2 \tag{2.44}$$

density change

$$\dot{Q}_2 = \dot{Q}_1 \tag{2.45}$$

$$P_2 = P_1 \left(\frac{\rho_2}{\rho_1}\right) \tag{2.46}$$

size change

$$\dot{Q}_2 = \dot{Q}_1 \left(\frac{d_2}{d_1}\right)^3 \tag{2.47}$$

$$P_2 = P_1 \left(\frac{d_2}{d_1}\right)^2 \tag{2.48}$$

The relationships in Eqs. (2.43) through (2.48) express the fan laws under the stated conditions. The assumption of constant efficiency is inherent in these equations, whereas practical fan designs will alter this assumption. The air power P_a (in W) at a given flow (in ft^3/min) and pressure (in inches of water) is in accordance with the equation

$$P_a = 0.1175 \Delta p \dot{Q} \tag{2.49}$$

Fans are sometimes used to simply purge air from a cabinet in order to prevent heat buildup, or they can provide air circulation within enclosures, as in ground equipment racks, or in many cases they provide a high-velocity air flow over components to increase the convection coefficient h_c. These many uses cover a broad spectrum of pressure and flow conditions, leading to the development of different wheel designs tailored to match a wide range of applications. The general classification of blowers is best defined in terms of "specific speed." Specific speed N_s is expressed by the equation

$$N_s = \text{rpm} \frac{\dot{Q}^{0.5}}{\Delta p^{0.75}} \tag{2.50}$$

in terms of the flow $\dot{Q}$ (in ft^3/min) and the pressure Δp (in inches of water). Figure 2.26 shows the range of specific speeds for several wheel designs commonly used in electronic cooling. These devices are illustrated in Fig. 2.27. The propeller fan is a high-pressure

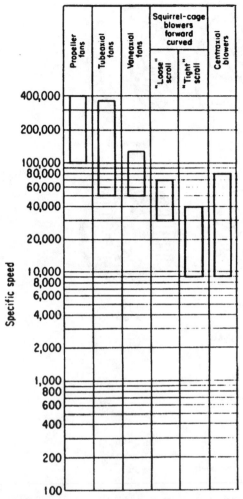

FIGURE 2.26   Specific speed of electronic cooling fans and blowers. (*Reprinted with permission of Rotron Corporation, Woodstock, N.Y.*)

device used mostly as a circulating fan. The tubeaxial fan provides higher pressure versus flow than a propeller fan and represents a logical extension of fan design. Vaneaxial fans are compact high-frequency (400-Hz) units whose airflow is parallel to the motor shaft. The impeller and correctional vanes of these units are airfoil designs installed to develop maximum efficiency. The other centrifugal impellers in Fig. 2.27 vary in blade configuration, which is either radial, forward, or backward curved. In small sizes, the forward-curved blades often provide the best overall performance.

The performance of fans and blowers is represented by a constant-speed plot of pressure head, $\sigma\Delta$ verses flow $\dot{Q}$. A typical example of a constant speed fan is shown in Fig. 2.28 along with a system impedance curve. The intersection of the system impedance

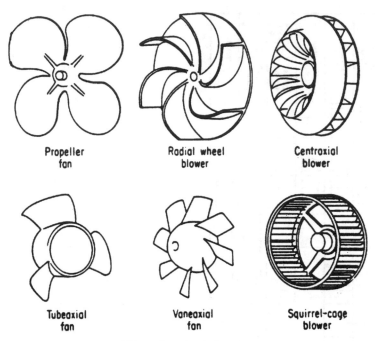

Propeller
fan

Radial wheel
blower

Centraxial
blower

Tubeaxial
fan

Vaneaxial
fan

Squirrel-cage
blower

FIGURE 2.27   Fan and blower impeller designs.

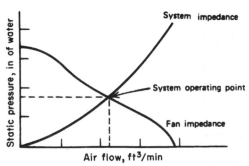

FIGURE 2.28   Typical fan and system operating point.

curve and the fan curve defines an operating point. In each case the designer must either calculate the system resistance, as in Sec. 2.6.1, or run flow-pressure tests on the equipment to obtain the resistance curve. Superposition of the latter on the fan curve establishes an operating point for the system. The system resistance measured in inches of water is expressed in the equation

$$\sigma \Delta p = A_m^{.n} \qquad\qquad (2.51)$$

where $\dot{m}$ is the mass flow (in lb/min) and the constant n varies between 1.3 and for most electronic systems. Constants A and n can be measured precisely through a laboratory test on a breadboard unit. Care must be exercised in recording ambient conditions during any test since the pressure involves the density correction σ. Problem 2.7 illustrates the steps requires to obtain the operating point for a fan.

Problem 2.7: The cabinet depicted in Fig. 2.24 is to be converted into a fan-cooled cabinet in order to lower the component temperatures. A further 10 PCBs are also added, each dissipating 15 W, for a total cabinet dissipation of 654 W. Select a fan based on a desired temperature rise of only 18°F (10°C) inlet to outlet having an operating speed of 1750 or 3450 r/min (60 Hz). The cabinet inlet air temperature is 70°F (21.1°C).

Solution:

1. From Eq. (2.28),

$$T_{out} = T_{in} + \frac{0.1316q}{\dot{m}}$$

or, rearranging terms,

$$\dot{m} = \frac{0.1316 \times 654}{10} = 8.6 \, \text{lb/min}$$

2. The average density is

$$\bar{\rho} = \frac{p}{RT} = \frac{14.7 \times 144}{53.3(460 + 79)} = 0.0737 \, \text{lb/ft}^3$$

3. The air flow is

$$\dot{Q} = \frac{\dot{m}}{\bar{\rho}} = \frac{8.6}{0.0737} = 116.7 \, \text{ft}^3/\text{min}$$

4. The cabinet pressure resistance line is based, as previously explained (Sec. 2.6.1), on the assumption of a total velocity head loss of 1.5 at the entrance and exit to the cabinet plus a friction loss within the cabinet based on a rectangular flow path between the PCB components. A slight modification to the CAB.BAS computer program (Table 2.15) to account for an increase in PCB flow area (52 versus 42 active boards) yields the following equation for cabinet resistance:

$$\sigma \Delta p = 8.5 \times 10^{-4} \dot{m}^{1.95}$$

5. At a flow of 8.6 lb/min (116.7 ft³/min), the pressure drop Δp is, from Eq. (2.51),

$$\frac{0.0737}{0.0765} \Delta p = 8.5 \times 10^{-4} \times 8.6^{1.95} \quad \text{and} \quad \Delta p = 0.059 \, \text{in of water}$$

6. The specific speed is

$$N_s = \text{rpm}\frac{\dot{Q}^{0.5}}{\Delta p^{0.75}} = 1750\frac{116.8^{0.5}}{0.059^{0.75}} = 157,98$$

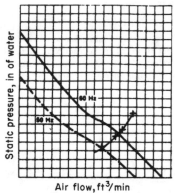

Static pressure, in of water

60 Hz

60 Hz

Air flow, ft^3/min

FIGURE 2.29 Vaneaxial fan curve. *(Reprinted with permission of Rotron Corporation, Woodstock, N.Y.)*

7. From Fig. 2.26, the chart indicates a propeller fan as a likely candidate, and further examination of catalog data reveals that a type BS-6501 satisfies the requirements.[22] A plot of this fan's curve along with the cabinet resistance line defines an operating point at the intersection of 120 ft^3/min and 0.062 in of water pressure drop (Fig. 2.29). The fan operates at 1750 r/min, 60 Hz, and draws only 10 W at full load.

2.6.2.1 Fan Operation over Altitude.

The lower density with increasing altitude often poses a difficult cooling problem with fans because mass flow is proportional to density and fan speed often remains constant. Unless there is no significant drop in the inlet or ambient temperature to the fan with increasing altitude to compensate for a decrease in mass flow, an oversized fan will be required at sea level. Consider the previous example of the cabinet having an inlet density of 0.0737 lb/ft^3 at sea level and 8.6-lb/min flow. At an altitude of 40,000 ft, the air density at 70°F is only 0.014 lb/ft^3, or 19 percent of the sea-level density. Since the fan operates essentially at constant speed over altitude, at 40,000 ft it will deliver only at most 1.63-lb/min airflow, and thus will raise the exit temperature from the sea-level value of 31.1°C to 73.7°C, a significant increase over that originally planned for at sea level. Obviously altitude operation presents a unique problem in fan selection, which differs from the selection problem to meet the sea-level operating conditions. A possible approach in this case is to size the fan for operation at the highest altitude and then allow for an overcooling condition at sea level. This is not the most efficient approach, but it is a necessary means for obtaining the proper cooling over the entire altitude regime. In the example cited, the flow, in ft^3/min, necessary at altitude for the same 10°C rise specified at sea level is 1/0.19, or 5.26 times the flow at sea level (614 ft3/min). But the fan laws [Eq. (2.38)] indicate that pressure is proportional to flow or speed squared and/or fan diameter squared so that, assuming the fan size does not change,

$$\Delta p_2 = \Delta p_1\left(\frac{N_2}{N_1}\right)^2$$

$$= 0.059\left(\frac{614}{117}\right)^2 = 1.62 \text{ in of water}$$

The new fan must have a sea-level operating point of 614 ft^3/min (45.25 lb/min) at 1.62 in of water pressure. The actual performance will be greater because the sealevel system impedance curve will intersect the fan curve at a higher flow. Assuming a fan were available at the higher speed, this represents close to a 15:1 air power P_a increase as determined from Eq. (2.49).

A better solution to this problem is to relax the exit air temperature requirement at altitude from 10°C to a more reasonable compromise that will not deter from reliable operation. Each application differs in this respect and must be dealt with on an individual basis. Another possibility is to use a fan driven by an induction motor which changes speed with load. These devices, marketed by specialists in the manufacture of air-moving devices, are capable of speed adjustments of up to 2:1 and more in some cases, depending on load conditions. It is then obvious that the 5.26 mass flow ratio (sea level to 40,000 ft) needed to maintain a 10°C air rise over altitude can be reduced to 2.6:1 provided a fan is available having a variable-speed slip motor that doubles in speed from sea level to 40,000 ft. This also reduces the power requirements since both pressure and flow at sea level are reduced significantly.

2.6.3 Air Cold-Plate Heat Exchangers

Cold-plate exchangers are common heat transfer devices used for heat removal in electronic systems. The application of these ranges from the flow through-type modules which utilize a sandwich-type fin structure with PCBs mounted on either side of the fins to the more conventional wall-type heat exchangers common to air-cooled box assemblies. Still others are used in conductively coupled assemblies mounted directly on the exchanger surface. There are also special devices, such as transmitter traveling-wave tubes (TWTs), where the air passes through the fin stock surrounding the highest heat flux region of the tube, namely, the collector and body. The design parameters in each of these applications includes airflow, inlet air temperature, heat flux, and pressure drop. The heat exchanger types comprising a number of core plate assemblies, each thermally coupled to the other, as in air-to-air or air-to-liquid exchangers, are not explicitly treated in this section, although the heat transfer principles apply equally to these designs.

There are many companies specializing in the manufacture of air heat exchangers and cold plates, some of whom also maintain in-house design capability. It should be noted that unlike liquid fin structures, which are vacuum or dip brazed, the air cold plates are most often adhesive-bonded since the pressure requirements are minimal, usually under 10 in of water (0.36 lb/in²). The type of fins used in air heat exchangers are similar to those used in liquid systems, except that air fins can be thinner and of greater height because the film coefficients are lower and the efficiencies therefore are higher.

The design of air fin exchangers follows that of liquid fins, except that in the case of air fins a greater source of empirical data is available in the open literature.[7,23] The parameters of interest include the fin geometry, that is, number of fins per inch, fin height, thickness, and β, the ratio of heat transfer area to volume, as well as the usual system parameters of heat flux, inlet temperature, flow rate, and allowable pressure drop. In the vast majority of electronic cooling applications, the film coefficients for the cold plates, with the exception of ram air cooling applications, range from 8 to 20 Btu/(h • ft² • °F) and fin efficiency is exceptionally high (>90 percent). Therefore fin thicknesses range between 0.004 and 0.008 in for 3003 aluminum material, and 0.006 in is typical for most applications. The cold-plate design requires the determination of the friction factor f and the heat transfer parameter j, or Colburn factor. Each of these is used to calculate the pressure drop from Eq. (2.19) and the film heat transfer coefficient as expressed in the equation

$$h_c = j \frac{GC_p}{P_r^{0.66}} \tag{2.52}$$

The fin-to-air temperature rise of the cold plate ΔT_{f-a} is determined from the fin conductance $\zeta h_c A_f$ and heat q in the equation

$$\Delta T_{f-a} = \frac{q}{\zeta h_c A_f} \tag{2.53}$$

The characteristic f and j factors in Eqs. (2.19) and (2.52) are functions of the Reynolds number R_e and can be found in Kays and London[10] for a number of fin configurations. As an example, for commonly used rectangular "straight" or plain fins having fully developed laminar flow and an aspect ratio of 3.0 to 4.0, the f and j factors are expressed in the following equations. For $R_e \leq 2000$ and $\ell/d_h = 100$,

$$f = \frac{18}{R_e} \tag{2.54}$$

$$j = \frac{5.0}{R_e} \tag{2.55}$$

For fully developed turbulent flow ($R_e > 8000$), the f and j factors are in accordance with the equations

$$f = \frac{0.048}{R_e^{0.2}} \tag{2.56}$$

$$f = \frac{0.025}{R_e^{0.22}} \tag{2.57}$$

Note that the aspect ratio α is defined as w_c/b, where b is the fin height and w_c the fin opening or the spacing between fins. In the case of lanced and offset, or "strip," fins, Weiting,[23] based on the data of Kays and London,[10] summarized the f and j factors over the Reynolds number as follows. For $R_e \leq 1000$,

$$f = 7.661\left(\frac{x}{d_h}\right)^{-0.384} \alpha^{-0.092} R_e^{-0.712} \tag{2.58}$$

$$j = 0.483\left(\frac{x}{d_h}\right)^{-0.162} \alpha^{-0.184} R_e^{-0.536} \tag{2.59}$$

For $R_e \geq 2000$,

$$f = 1.136\left(\frac{x}{d_h}\right)^{-0.781} \left(\frac{t}{d_h}\right)^{0.534} R_e^{-0.198} \tag{2.60}$$

$$j = 0.242\left(\frac{x}{d_h}\right)^{-0.322} \left(\frac{t}{d_h}\right)^{0.089} R_e^{-0.368} \tag{2.61}$$

In these equations t is the fin thickness, in inches, and x is the fin length in the flow direction, with the hydraulic diameter d_h and the aspect ratio α as defined previously.

In most cases, an interactive procedure is necessary to make a proper fin design selection. Therefore, whenever repeated calculations are required, it is best to computerize the fin data, that is, the f and j parameters as expressed in Kays and London[10] or by Eqs. (2.54) through (2.61) as a function of the Reynolds number, and proceed with the selection based on the design requirements. The input data consist of heat dissipation, inlet temperature, flow rate, and cold-plate area, and the output is heat transfer coefficient h_c, fin efficiency ζ, air-to-fin temperature rise ΔT_{a-f}, and pressure drop Δp. A printout of fin parameter data permits the selection of the best fin for the given application, and in most cases it will then be possible to obtain the actual fin hardware from a choice of several suppliers.

2.6.3.1 Cold-Plate Operation over Altitude. The operation of cold plates in electronic systems at altitudes other than sea level requires verification as to performance adequacy of the design when exposed to higher-altitude operation. It is often necessary to maintain a certain mass flow in order to properly cool an electronics unit, and because the air density is reduced at increasing altitude, a higher velocity results in greater pressure drop. For air cold plates it is common practice to test the units and plot a curve of $\sigma \Delta p$ versus mass flow m on log-log paper. The result is a constant slope which usually varies from 1.2 to 1.9, depending on whether laminar or turbulent flow conditions prevail. This conclusion is readily apparent when one examines Eqs. (2.58) and (2.60) defining the friction factor f for lanced and offset fins in conjunction with Eq. (2.19). As an example of the pressure drop variation experienced over altitude, consider a cold plate having a slope of 1.5 and a pressure drop of 2 in of water at sea level when operating at an average temperature of 40°C. The following calculations illustrate the method used in determining the pressure drop with the air at 25°C, the pressure corresponding to 40,000-ft altitude, and with the same mass flow at altitude.

For the cold plate,

$$\sigma \Delta p = km^{1.5}$$

For a perfect gas at sea level,

$$\rho = \frac{p}{RT} = \frac{14.7 \times 144}{53.3(77 + 460)} = 0.074 \text{ lb/ft}^3$$

$$\sigma_{\text{sea level}} = \frac{\rho}{0.0765} = \frac{0.074}{0.0765} = 0.967$$

For a perfect gas at 40,000-ft altitude,

$$\rho = \frac{2.73 \times 144}{53.3(77 + 460)} = 0.074 \text{ lb/ft}^3$$

$$\sigma_{40,000} = \frac{0.0137}{0.0765} = 0.1795$$

Therefore, the pressure drop at 40,000 ft for the same mass flow is calculated as follows:

$$\Delta p_{40,000} = \frac{\Delta p_{\text{sea level}} \sigma_{\text{sea level}}}{\sigma_{40,000}}$$

$$= \frac{2.0 \times 0.967}{0.1795} = 10.8 \text{ in of water}$$

The example illustrates that an over fivefold increase in pressure drop for the same mass flow occurs when operating at a pressure corresponding to 40,000-ft altitude. It also points out the obvious difficulty in having a fan-cooled system based on a constant speed. In order to maintain the same mass flow from sea level to 40,000 ft, the fan at sea level will of necessity be required to overcool the system in order to supply the required mass flow at altitude (see Sec. 2.6.2.1).

2.7 JUNCTION-TO-HEAT-SINK TEMPERATURE GRADIENTS

This section examines the series of temperature rises that occur from junction to heat sink for various levels of packaging and component types. The end result is to make possible a reliability prediction as well as to determine where excessive junction temperatures exist throughout the system. There are, in general, three classifications of cooling, which describe the passage of heat from the component chip to its respective heat sink—in-line conduction, flow-through, and flow-over components. The in-line-conduction method is further subdivided into two categories—surface-mount technology (SMT) type components and plated-throughhole (PTH) type components. For SMT designs, as shown in Fig. 2.30, vias or PTHs located beneath the component serve to transfer the heat from the package to the PCB heat sink and from there to the edges of the module and on into the wall heat exchangers. The size and number of holes vary depending on dissipation, but typically the hole sizes range from 0.010 to 0.025 in in diameter with 0.0014 in of copper plating (1 oz.) on the inside. A component measuring 1 by 1 in on a side and mounted on a G-10 epoxy board 0.060 in thick has a thermal resistance of 9.7°C/W in the

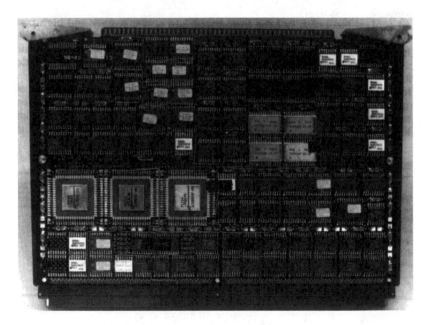

FIGURE 3.30   In-line-conduction flat-pack PCB.

z direction, or normal to the fibers without vias. With 100 vias spaced on 0.10-in centers and 0.020 in in diameter the same component has an equivalent resistance of only 0.48°C/W through the vias. The vias, therefore, lower the temperature gradient by a factor of 20 from 19.4°C to under 1°C at 2-W dissipation. When the cooling method is in-line conduction having DIP-type components, aluminum or copper rails located under the components form a laddertype heat sink which transfers heat to the board edges as in the SMT designs.

An example of a flow-through assembly for SMT components appears in Fig. 2.31. The components, which are not shown in the figure, are bonded to the multilayered board and the cooling air enters and leaves the assembly through the end caps, which include a tapered seal. The ultimate sink, in this case air, is closer to the junctions than in either of the in-line-conduction techniques; therefore junction temperatures are lower on the average. The same technique using liquid coolant has recently been introduced into power supply and high-powered RF modules, wherein the components are cold-plate mounted in SMT fashion on both sides and where the unit includes blind-mated liquid-coupled quick disconnects. The use of this type of design is expected to increase as the module heat continues to increase.

In some mainframe computers the cooling technique in common use consists of multichip modules which interface with either air or liquid supplied from an external source and placed in close proximity to the chips. These designs are quite efficient thermally, as evidenced by the data in Table 2.16, which lists the module parameters, including the thermal resistance for seven production computers. Table 2.16 also lists both internal (RINT) and external (REXT) resistance values, which refer to chip-to-heat-sink and heat-sink-to-fluid resistance, respectively.

In still another category of cooling, the flow-over-component technique, a liquid such as FC-77[*] passes directly through a module to cool the modules, as in the CRAY super-

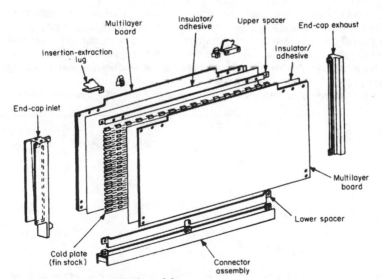

FIGURE 2.31 Flow-through module.

---

[*]FC-77 is a registered trademark of 3M Company.

TABLE 2.16 Multichip Module Parameters

Technology	Total dissipation, W	Maximum chip Q, W	Heat flux, W/cm^2	Heat density, W/cm^3	RINT,* K/W	REXT,* K/W
Mitsubishi HTCM	36	4	0.83	0.4	3.0	4.3
Hitachi RAM	6	1	0.8	0.5	10.1	24.6
Honeywell SLIC	60	>0.5	0.9	0.2	—	—
NEC SX LCM	250	>5.4	1.6	0.3	—	—
IBM 4381 IMPNG	90	3.8	2.2	0.5	9.0	8.0
IBM 3090 TCM	500	7.0	2.2	0.4	7.2	1.5
NTT	377	15.1	4.2	8.4	2.8	0.5

*Based on the chip heat dissipation. RINT—internal thermal resistance; REXT—external thermal resistance.
Source: From Bar-Cohen.[24] © 1987 IEEE. Reprinted with permission.

computer. This approach permits close physical spacing of circuitry, resulting in extremely fast computer speeds. In the more traditional flowover-component case, air is used as the coolant, in which case its quality must be maintained and controlled through the use of in-line filters. It is instructive to note that military specifications restrict the use of airflow over components except in special cases, in accordance with MIL-STD-454, Requirement 52.

2.7.1 Package Characterization

In determining component junction temperatures on PCB assemblies, the approach to be followed is to sum the individual parts consisting of the series of temperature rises from package junction to heat sink. This simplification for the most part assumes the paths to be one-dimensional, whereas in actuality there arc two- and three-dimensional heat transfer paths, which can only be solved analytically through the use of computer codes utilizing finite-difference modeling techniques. These are often either proprietary in nature or too complex to be treated in this section. For details on the multidirectional modeling approach, see Lee et al.,[25] Furkay,[26] and Sec. 2.1.1.

The thermal resistance of a chip package is defined in terms of its junction-to case resistance,

$$\theta_{j-c} = \frac{T_j - T_c}{q_c}(°C/W) \tag{2.62}$$

or its junction-to-air resistance,

$$\theta_{j-a} = \frac{T_j - T_a}{q_c}(°C/W) \tag{2.63}$$

where T_j is the average surface temperature of the chip, T_c that of the case of the package, and T_a the ambient or local air temperature adjacent to the package. In-line-conduction and flow-through cooling methods consisting of cold plates sandwiched between PCBs as described previously require a knowledge of θ_{j-c} for determining the junction temperature,

TABLE 2.17 Package-Type Thermal Impedance $\theta_{j\text{-}c}$, in °C/W*

Package	Small-scale integration (≤ 12 gates)		Medium-scale integration ($12 > 100$ gates)		Large-scale integration (> 100 gates)		Standard nonhermetic plastic package	
	DIP†	FP†	DIP	FP†	DIP	FP†	DIP	FP
14 leads	28	22	28	22	21–26	22	65	—
16 leads	28	22	28	22	19–22	22	53	—
18 leads	28	22	27	22	18–20	22	53	—
20 leads	28	22	27	22	18	22	53	—
22 leads	28	22	26	22	16–17	22	53	—
24 leads	28	22	26	22	14–17	22	53	—
28 leads	28	22	25	22	16	22	53	—
40 leads	28	22	24	22	8–15	22	50	—
48 leads	28	—	23	—	7–14	—	50	—
64 leads	28	—	23	—	6–13	—	50	—

*Gate count is specified in MIL-HDBK-217. Random logic devices only.
†From MIL-M-38510G, appendix G.

whereas for flow-over components in either the natural or the forced convection mode, the necessary parameter is $\theta_{j\text{-}a}$. The manufacturer's data sheet listing the $\theta_{j\text{-}c}$ and $\theta_{j\text{-}a}$ resistance values forms the basis for determining PCB component junction temperatures. Table 2.17 is a compilation of typical $\theta_{j\text{-}c}$ values taken from military specification MIL-M38510G for small-, medium-, and large-scale IC packages having 14 to 64 leads per package. The resistance value $\theta_{j\text{-}c}$ may be thought of as a figure of merit for the internal package design consisting of a series of individual resistances for the silicon chip, solder, adhesive, base material, leads, and so on. Despite the limitations inherent in Table 2.17 and in the vendor data for $\theta_{j\text{-}c}$ and $\theta_{j\text{-}a}$ it is common practice at the present time to use these values in the thermal analysis of PCBs. However, it is well to note that possible errors can accrue from indiscriminate use of vendor data for $\theta_{j\text{-}c}$ and $\theta_{j\text{-}a}$. In critical applications, the requisite $\theta_{j\text{-}c}$ and $\theta_{j\text{-}a}$ should always be obtained experimentally or by detailed computer modeling of the chip package. The reason for this lies in the dependence of $\theta_{j\text{-}c}$ on its environment, mounting attachment, and power dissipation.[27] Bar-Cohen et al.[28] propose the use of a modified $\theta_{j\text{-}c}$ to account for the thermal coupling of all the individual external package surfaces (top, bottom, and sides) and leads in a specific configuration. Andrews[27] argues that the package thermal resistance is a function of up to eight variables, which need to be defined in every equipment application.

Furkay[26] examines $\theta_{j\text{-}c}$ and $\theta_{j\text{-}a}$ for SMT devices having 68, 84, 100, and 124 leads and shows variations in natural and forced convection cooling for different operating conditions and internal package construction. Siegal[29] concludes that the thermal resistance $\theta_{j\text{-}c}$ is not constant and in fact varies with power dissipation and the ambient environment. Other investigators have reached similar conclusions; see, for example, Pasqualoni et al.[30] For data on specific devices, see Wesling[31] for $\theta_{j\text{-}a}$ on a 149-lead VLSI pin grid array package cooled with flowover air and having extended fins. Measurements by Hopkins et al.[32] show that high pulsed short-duration transistors having high capacitance within plastic IC packages are not adequately defined in terms of $\theta_{j\text{-}c}$ when operating under pulsed load conditions. These references are listed to caution the reader in the application of $\theta_{j\text{-}c}$ and $\theta_{j\text{-}c}$ values, which for now remain lacking with respect to total accuracy and completeness. The use of Table 2.17 is recommended for initial design trade studies where relative rather than absolute values are significant.

2.7.2 Heat Spreading

The heat spreading within semiconductor packages is characterized by heat flow from a relatively small die surface to a case of larger physical dimensions. Therefore, heat spreading takes place in three dimensions, which is often modeled in a thermal network using the finite-difference approach. Kennedy[33] presented an analysis in graphic form from which an equivalent resistance is calculated without the use of a thermal network program. The results are for a cylindrical source with uniform flux into a cylinder element having the appropriate boundary conditions assigned. In the case of chips mounted within packages, the conversion of noncircular configurations to equivalent circular form with small error can be related to Fig. 2.32, which depicts the spreading resistance factor H^2 as a function of the die and case geometry. The spreading resistance R is then determined from the equation

$$R = \frac{H_2}{k\pi a} \tag{2.64}$$

where a is the heat-source equivalent radius, k is thermal conductivity, and H^2 is found from Fig. 2.32. The boundary condition in this case assumes no leakage from the heat source (wire bonds, convection, etc.) and all heat flow occurs in the component case hav-

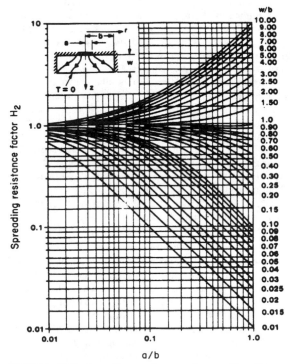

FIGURE 2.32 Spreading resistance factor H_2. (*From Kennedy.[33] Reprinted with permission of International Business Machines Corporation.*)

ing a constant temperature along its heat-sink surface. A sample problem illustrates the procedure.

Problem 2.8: Determine the overall thermal resistance of the silicon ship in Problem 2.1 assuming the alumina case is 0.3 by 0.3 in and rests on an isothermal heat sink. The chip dimensions are 0.07 by 0.07 in, Q.025 in thick, and the die attachment is 0.003-in-thick conductive epoxy.

Solution: The total resistance is composed of the sum of three resistances, R_1, R_2, and R_3, the chip, the attachment interface material, and the case resistance, respectively.

$$R_t = R_1 + R_2 + R_3$$

$$= \frac{\ell_1}{k_1 A_1} + \frac{\ell_2}{k_2 A_2} + R_3$$

$$= \frac{22.733 \times 0.025}{68 \times 0.07 \times 0.07} + \frac{22.733 \times 0.003}{0.4 \times 0.07 \times 0.07} + R_3 = 1.71 + 34.8 + R_3$$

$$= 36.5°C/W + R_3$$

To determine R_3, the resistance through the case, proceed as follows:

$$a = \left(\frac{0.07 \times 0.07}{3.14}\right)^{0.5} = 0.0395 \text{ in (equivalent chip radius)}$$

$$b = \left(2\frac{0.3 \times 0.3}{3.14}\right)^{0.5} = 0.169 \text{ in (equivalent chip radius)}$$

$$\frac{w}{b} = \frac{0.025}{0.169} = 0.148$$

and

$$\frac{a}{b} = \frac{0.0395}{0.169} = 0.234$$

From Fig. 2.32, $H_2 = 0.50$, and from Eq. (2.64),

$$R_3 = \frac{H_2}{k\pi a} = \frac{0.50 \times 1.8944}{15 \times 3.14 \times (0.0395/12)} = 6.1°C/W$$

Note that compared to the more approximate resistance calculated in Problem 2.1, the case resistance is within 2 percent when accounting for heat spreading. Then,

$$R_t = 36.5 + R_3 = 36.5 + 6.1 = 42.6°C/W$$

2.7.3 Flow-Over Component Channel Cooling

The best examples of this type of cooling can be found in cabinet installations where the airflow occurs over arrays of components mounted on PCBs. Because it is a commonly encountered problem, much effort has been devoted recently toward obtaining a better the-oretical understanding of the heat transfer and pressure drop for this type of configuration. Most studies treat the IC components as idealized rectangular blocks with uniform con-ductivity and heat generation. The hydrodynamic nature of the flow differs from that of flow between a pair of smooth parallel plates (Sec. 2.2.2) in that the components cause an obstruction to the flow, creating recirculating zones between components and downstream from the last component. These effects in some studies have been carefully modeled using solid modeling representation and discretization of the IC package and PCB along with computer flow simulation programs. In most design applications it is sufficient to describe the condition at the critical and average component locations by determining the heat transfer coefficient h_c at these points.

Fig. 2.33 shows an array of components which in practice is deployed along a surface of a PCB and in turn is part of a cluster of PCBs within a cabinet or other type of structure. In the simplest case, the components (flat packs or DIPs) are all of equal height B and completely fill the array, whereas in the more general case, the IC components are of vary-ing heights and usually a finite number of them are missing from the PCB at random loca-tions. Some reported results, as in Sparrow et al.,[34] present the data in ratio form, which compares the heat transfer of the specific component with that for the basic array without odd-size components. They found that the deployment of odd-size components within an array of constant-height components enhanced the heat transfer for those neighboring components depending on the relative heights in the airstream. Missing components also enhanced the heat transfer as much as 40 percent. For a tall module, the enhancement is due to the direct impingement of air on the protruded element, whereas for short compo-nents, enhancement may be due to the recirculation behind the taller component. These enhancements occur at the expense of increased pressure drop. The significant problem in array thermal design lies in the determination of a suitable heat transfer coefficient h_c at the IC's location or station along the array. The geometric variables associated with chan-

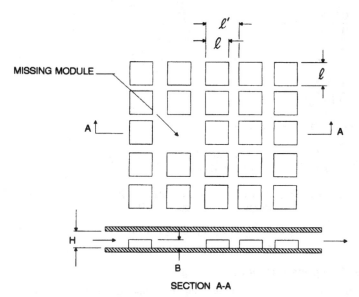

FIGURE 2.33 Array of IC components on PCB.

nel array flow are ℓ, B, H, and ℓ' as depicted in Fig. 2.33. Test data by Moffat et al.[35] demonstrated the importance of H/B in determining the film coefficient h_c for fully developed flow. However, no clear correlation has been established for a fixed H/B in terms of the other variables and in terms of the channel Reynolds number R_a. The safest approach therefore in dealing with array cooling is to either conduct experiments on the actual PCB or search the literature for similar configurations having ratios of H/B, ℓ/B and ℓ'/ℓ close to that under consideration. The problem example fits the latter and is based on the results of extensive testing conducted at the Cranfield Institute of Technology by Wills.[16] Here the DIPs and flat-pack PCBs were on 0.40- to 0.80-in centers with H/B close to 3. Equation (2.65) expresses the Nusselt number for fully developed channel flow in an array,

$$N_u = 0.19 R_e^{0.7} \left(\frac{\ell'}{\ell} - 1 \right)^{0.13} P_r^{0.33} \tag{2.65}$$

Fully developed flow occurs downstream from the leading edge by the fifth or sixth row ($N = 5$ or 6) or sooner, depending on the component-to-pitch distance ℓ/ℓ', as shown by Moffat et al.[35] (see Fig. 3.2 in BarCohen and Krauss[2]). For the film coefficient h_c [in Btu/(h • ft^2 • °F)] in the entrance or developing flow region, Wills' equation, expressed in U.S. customary units, is as follows:

$$h_{c_N} = 0.93 \frac{\ell}{\ell'} + \frac{0.225(V\rho)^{0.8}}{(N\ell')^{0.36}} \left(\frac{\ell'}{\ell} - 1 \right)^{0.13} \tag{2.66}$$

where V is in ft/min at station position N, ℓ and ℓ' are in feet or inches, ℓ' being the lateral (cross-stream) component dimension in ft, and ρ is in lb/ft^3.

The film coefficient h_{cN} is also applicable to heat transfer from the PCB in those cases where the component heat is well-coupled through a conductive interface. The copper layers in the board then act as fins in transferring heat from the PCB to the lower air temperature flowing by. In most cases the $h_c(N = 1)$ value at the first component ($N = 1$), where the boundary layer is thin, is approximately twice that of the fully developed value downstream. Most investigators reporting data on arrays support the concept of a decreasing film coefficient from the leading edge based on an accumulation of test data. In Problem 2.9 the film coefficient is calculated for a DIP array of components.

Problem 2.9: A 7.2- by 5.0-in PCB is populated with 96 ceramic DIPs on a 12× 8 pattern and is cooled with air passing over the surface. The DIPs (0.2 by 0.7 by 0.16 in high) are mounted with the 0.7-in side in the flow direction in 8 columns, pitched at 0.9 in, and 12 rows at 0.35-in pitch and protruding 0.20 in from the PCB. The boards, which are 0.07 in thick, are mounted on 0 5-in centers. Determine the film coefficient of the first and last rows of components. Assuming a uniform heat of 0.20 W per component, what is the junction temperature of these components assuming an airflow of 0.25 lb/min through the channel, a θ_{j-c} value of 25 for all DIPs, and an inlet temperature of 25°C?

Solution:

1. PCB power dissipation is

$$q = 0.20 \times 96 = 19.2 \text{ W}$$

2. From Eq. (2.28),

$$T_{out} = T_{in} + \frac{0.131q}{\dot{m}}$$

$$= 25 + \frac{0.131 \times 19.2}{0.25} = 35°C$$

3. The average air temperature is

$$\bar{T} = \frac{25 + 35}{2} = 30°C(86°F)$$

and the average density is

$$\rho = \frac{39.715}{86 + 460} = 0.073 \text{ lb/ft}^3$$

4. The DIP geometry parameters are

$$\frac{H}{B} = \frac{0.5 - 0.07}{0.25} = 1.73$$

$$\frac{\ell}{\ell'} = \frac{0.7}{0.9} = 0.778$$

$$\frac{\ell'}{\ell} = 1.286$$

The flow area is

$$A_c = 0.44 \times 5 - 12 \times 0.2 = 1.72 \text{ in}^2(0.0119 \text{ ft}^2)$$

5. The velocity of air through the PCB channel is

$$V = \frac{\dot{m}}{A_c \bar{\rho}} = \frac{0.25}{0.0119 \times 0.073} = 288 \text{ ft/min}$$

6. From Eq. (2.66) at station 1,

$$h_c(N = 1) = 0.93 \times 0.778 + \frac{0.225(288 \times 0.073)^{0.8}}{(1 \times 0.9/12)^{0.36}}(1.286 - 1)^{0.13}$$

$$= 6.3 \text{ Btu/(h} \bullet \text{ft}^2 \bullet °F)$$

7. At station 8, or at the exit of the PCB, the film coefficient is calculated from Eq. (2.661 using N = 8,

$$h_c(N = 8) = 0.93 \times 0.778 + \frac{0.225(288 \times 0.073)^{0.8}}{(8 \times 0.9/12)^{0.36}}(1.286 - 1)^{0.13}$$

$$= 3.4 \text{ Btu/(h} \bullet \text{ft}^2 \bullet °F)$$

8. Using five sides for the heat transfer area for the DIP,

$$A_s(\text{DIP}) = \frac{0.2 \times 0.7 + (0.7 + 0.2)2 \times 0.16}{144} = 0.000297 \text{ ft}^2$$

9. The junction temperature at the exit is

$$T_j = T_{\text{air}} + \theta_{j-c}q + \frac{q}{h_c A_s} = 35 + 25 \times 0.2 + \frac{0.2 \times 1.8944}{3.4 \times 0.000297} = 77.7 \,^\circ\text{C}$$

10. The fully developed Nusselt number from Eq. (2.65) at the exit of the PCB is

$$N_u = 0.19 R_{e\,0.07}\left(\frac{\ell'}{\ell} - 1\right)^{0.13} P_{r\,0.33}$$

$$= 0.19\left(\frac{0.9 \times 8 \times 288 \times 0.073 \times 60}{12 \times 0.045}\right)^{0.7} (1.286 - 1)^{0.13} 0.7^{0.33} = 130.2$$

$$\text{and } h_c = N_u \frac{k}{\ell} = \frac{130.2 \times 0.0153}{7.2/12} = 3.3 \text{ Btu/(h} \cdot \text{ft}^2 \cdot \,^\circ\text{F)}$$

Therefore, the fully developed film coefficient calculation agrees well with the local coefficient as determined in step 7.

2.7.4 In-Line-Conduction Cooling of PCBs

Recent trends in military packaging stress the need for standardization of module designs. Typical of this movement is the standard electronic module SEM "E" heat sink shown in Fig. 2.34. This type of module consists of a 0.125-in central metal core, which conducts the heat from the PCB components mounted on one or both sides of the heat sink. The footprint component mounting area for the SEM "E" is 27 in^2 of mounting surface for each PCB side. Basically these modules, which feature the in-line-conduction cooling technique, are distinguished by a more uniform temperature throughout and by a better heat storage and transient response as compared to the flow-through type of module.

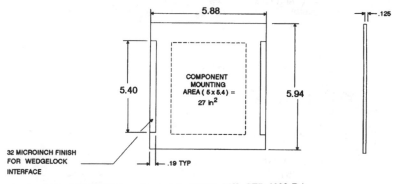

FIGURE 2.34 SEM "E" type heat sink. (*From MIL-STD-1389-D.*)

Figure 2.35 shows an in-line-conduction cooled electronic package with modules thermally coupled to the upper and lower heat exchangers. These exchangers ordinarily consist of 3003-type aluminum-plate fin material, 0.006 in thick, and having 10 to 18 or, in some cases, as many as 24 fins per inch. The fin stock is either dip- or, more likely, vacuum-brazed to the chassis in the case of liquid cooling, or it is bonded to the chassis when air is the coolant. The coolant temperature rise associated with the coolant flow and module heat is determined from Eq. (2.23) or (2.24) (Sec. 2.3). The heat path in these applications is from the component package through the PCB, the metal core, the tab interface, and finally the wall exchangers located at the ends. Figure 2.36 illustrates the heat path and is a more detailed representation of the tab interface, commonly a typical wedge-clamp configuration. The edge guides, which are commercially available devices, vary somewhat in all designs but must provide an interface pressure between the wall tab and the heat-sink core so as to minimize the thermal resistance. The wedge clamp shown in

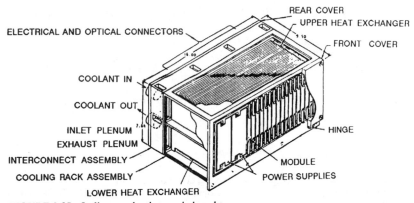

FIGURE 2.35   In-line-conduction cooled package.

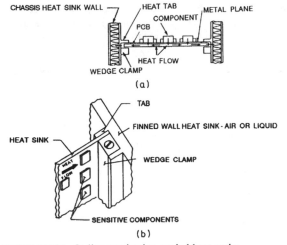

FIGURE 2.36   In-line-conduction cooled heat paths.

Fig. 2.36*b* typically consists of three or as many as five or more segmented inclined surfaces, which force the heat sink against the wall tab whenever torque is applied to the screw at the top. The interface resistance, expressed in (°C • in²)/W, is a function of contact pressure, surface finish, flatness, and operating altitude or ambient pressure. It is best determined experimentally through testing; however, a representative value is 0.50°C • in²/W for clean aluminum surfaces having a minimum of 30 lb/in² pressure at sea level and where each surface finish is at least 64 μin rms (see Sec. 2.2.1.1).

The resistance model for the module from the chip junction to its ultimate heat sink is shown in Fig. 2.37. Here the assumption of one-dimensional heat flow leads to the equation

$$T_j = q_c \theta_{j-c} + \Delta T_{c-hs} + \Delta T_{hs} + \Delta T_{int} + \Delta T_{f-a} T_c \qquad (2.67)$$

where q_c is the chip power dissipation, in W, and the ΔT, in °C, are the individual gradients, case to heat sink (*c-hs*), heat sink (*hs*), interface (int), and fin to air (*f-a*). T_c is the coolant temperature at the specific location. The terms in Eq. (2.67) are shown in Fig. 2.37 as a series of thermal resistances. As an example, the resistance from the case R_{c-hs} is composed of the package interface bond adhesive Locktite[*] or some equivalent, which replaces the high-impedance air gap, the vias near the heat sink, which vary in number and

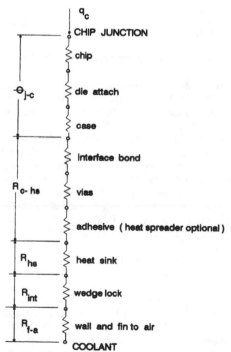

FIGURE 2.37 Resistance path of in-line-conduction system.

---

size (see Sec. 2.7), and the PCB adhesive between the board and the metal heat sink. The latter often consists of thin layers of prepreg used to bond the epoxy PCB laminate at elevated temperature and pressure to the metal heat sink. In some cases a thin aluminum heat spreader, 0.003 to 0.005 in in thickness, is placed under the vias to spread heat evenly throughout the fin stock and board area. A basic ℓ/ka conduction term is used to calculate the individual resistances for all but the fin to-air impedance (see Sec. 2.3). In Fig. 2.37 for the PCB this includes the interface bond, the vias, and the adhesive, which together constitute R_{c-hs} Not shown, but in parallel to these three resistances, is the epoxy board itself which, in the case of 0.060in-thick G-10 material, is an order of magnitude greater than the sum of those resistances shown in the figure and, as such, is neglected. The heat-sink resistance R_{c-hs} represents the heat-path resistance from the component location to the nearest heat exchanger edge. By assuming a uniform heat distribution within the PCB, it is possible to determine an average heat-sink resistance for each side of the PCB from the centerline to one edge of the heat sink using the equation

$$R_{hs} = \frac{7.58\ell}{kA_c} \quad (°C/W)$$
(2.68)

where ℓ is one-half the core width, in inches, k is the heat-sink conductivity, in Btu/(h • ft • °F), and A_c in in^2, is the cross-sectional area for heat transfer. From the equation

$$\overline{\Delta T}_{hs} = 7.58\frac{q'\ell^2}{kA_c}$$
(2.69)

the average temperature drop in the heat sink, in °C, for a uniform heat q', in W/in length, is found.

Next, at the interface between the heat sink and the wall, the usual practice is to assume an interface resistance ΔT_{int} of 0.50°C • in^2/W, which is an empirical value verified by tests on aluminum surfaces having a minimum pressure of 30 to 50 lb/in^2, engaging surfaces flat to 0.001 in/in, and surface finishes of better than 64,μin rms (Sec. 2.2.1.1). This interface area can potentially be a source of high impedance, depending on the type of engagement between the module and chassis. For example, where a rigid or nonfloating connector is used, a finite space or gap at the wall interface must not only be closed but also pressurized to at least 30 lb/in^2. A floating connector type of engagement is a solution to this problem, but in some installations this is not always possible. Hence the wedge clamp must then be called upon to bend the heat sink, not an altogether satisfactory condition considering the multiplicity of solder joints present on the PCB. It is always best, where possible, to run experiments to verify the integrity of the interface configuration; otherwise refer to Fig. 2.5 for interface values.

The fin-to-air impedance ΔT_{f-a} in Eq. (2.67) is dependent on the choice of fin stock, as explained in Secs. 2.6.3 and 2.5.4. These fin selections for air or liquid cases must always be consistent with the system requirements, as explained in Sec. 2.3.

In the other form of in-line-conduction cooling, as characterized by the PTH type component, the temperature gradients are similar to those described. The heat sink in the case of a DIP-populated PCB, for example, consists of a solid metal sheet 0.060 in thick, having irregular cutouts to accommodate the component leads through the metal heat sink. The rails underneath the DIPs are 0.170 in wide and 0.060 in high and serve to conduct the heat to each of the edges.

The temperature gradients ΔT for a SEM "E" conduction-type PCB module (Fig. 2.34) are shown in Fig. 2.38. Average junction temperatures T_j may be closely approximated with the aid of Fig. 2.38 and through the use of Eq. (2.67) modified as

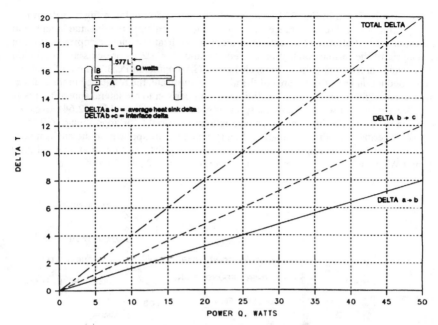

FIGURE 2.38 SEM "E" temperature gradients ΔT.

$$\overline{T}_j = q_c \theta_{j-c} + \Delta T_{c-hs} + \Delta T_{hs-\text{wall}} + T_w \qquad (2.70)$$

At 30-W dissipation, a PCB module with 69 flat packs having an average wall temperature T_w of 40°C will operate at an average junction temperature of 64.5°C. In Eq. (2.70) the heat-sink-to-wall gradient from Fig. 2.38 is 12.0°C; θ_{j-c} equals 22°C/W from Table 2.17, and the average chip dissipation of 0.43 W is based on the 30 W of module heat. The case-to-heat-sink ΔT is generally low and can be assumed to be 3°C so that 24.5°C represents the overall chip-to-wall ΔT for 30 W in a SEM "E" type module. This simple example demonstrates the advantage of using a simplified analysis technique, which is often required in performing trade studies. This approach, however, is not recommended as a substitute for a more rigorous computer analysis, which determines the temperature of all components.

2.7.5 Flow-Through Cooling of PCBs

Modules having PCB heat dissipations in excess of 40 to 50 W are generally cooled with flow-through cooling, as opposed to the in-line-conduction method. There are two main reasons for this: (1) the fin stock heat transfer area in flow-through modules is several multiples of that of an associated inline-conduction wall exchanger, which leads to lower coolant-to-fin ΔT, and (2) the coolant in the flow-through module can be brought closer to the package chips, which results in a lower overall thermal resistance of chip to sink. The last point is best illustrated by an example depicting the heat path from chip to sink for a flow through module populated by configurable gate arrays (CGAs). The local module cross section is shown in Fig. 2.39a along with a resistance model in Fig. 2.39b. In this case, θ_{j-c} consists of three resistances in series—the silicon chip, the die attachment, and the alumina case. Following the same one-dimensional heat conduction approach used in the analysis of the in-line-conduction cooled module, Table 2.18 presents a compilation of resistance values from chip to fin stock for a CGA-populated PCB. A note of caution: for the conduction heat transfer area of the alumina substrate, the resistance in the z direction or the straight through path is negligible in comparison to the lateral spreading resistance. Figure 2.32 is used to obtain the spreading coefficient H_2, and Eq. (2.64) gives the resistance in °C/W. The fin to-air resistance is again a variable of approximately 0.15 to 0.3°C/W in terms of module dissipation. In a typical aircraft application the environmental control system (ECS) can be assumed to be delivering 0°C cooling air at a rate of 2.19 lb/(min kW), which results in a 60°C exit temperature [Eq. (2.23)]. Therefore the average

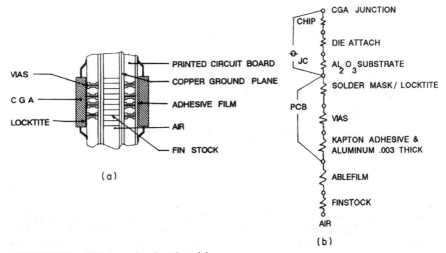

FIGURE 2.39 CGA thermal path and model.

TABLE 2.18 CGA Resistances, Chip to Heat Sink

	ℓ, in	k, Btu/ (h · ft · °F)	A, in^2	R, °C/W
Silicon chip	0.025	68	0.0512	0.16
Eutectic bond Au-Pb die attachment	0.003	39.4	0.0512	0.03
Substrate Al$_2$O$_3$	—	—	—	1.10*
Solder mask and Locktite	0.005	0.12	0.123	7.70
100 vias 0.013 in in diameter and 0.001-in Cu-plated	0.062	225	0.00571	1.10
Kapton adhesive	0.0015	0.12	0.107	2.66
Ablefilm adhesive	0.003	0.45	0.245	0.70
Fin stock	Variable with airflow, etc.		$\theta_{j\text{-}hs}$ =	13.5 °C/ W

*See Fig. 2.32 for spreading resistance factor H_2.

junction temperature for a flow-through module of 60W dissipation (0.4 W/chip) can be calculated using the equation

$$\bar{T}_j = \bar{T}_{\text{fin}} + \theta_{j-hs} \tag{2.71}$$

For a 50°C fin temperature it is found to be 55.4°C.

2.7.6 Comparing the Cooling of In-Line-Conduction and Flow-Through Modules

In the present state of the art, both types of modules are finding application in electronic cooling. Both digital and RF-type modules are being packaged using in-lineconduction and flow-through designs. The choice of cooling mode depends on system considerations, such as reliability goals, the drive and necessity for standardization, the coolant available, and also the usual packaging constraints, involving circuit partitioning, size, and weight factors. Table 2.19 is offered as a guide in performing trade studies for inline-conduction versus flow-through cooling. Regarding the in-line modules, these are characterized as being more uniform in temperature due to a near constant sink temperature at each specific location (wall edge) whereas in the flow-through design the total coolant temperature rise takes place from inlet to outlet within each module. Therefore each module must be balanced as regards flow to achieve the same exit temperature, whereas in the in-line package it is a relatively easy matter to maintain identical flow in the two wall heat exchangers. Furthermore, in the flow-through electronic package the inlet plenum distribution duct requires a more careful design to assure that each module receives its allotted flow of coolant—tapered plenums are often used, for example, in the air distribution to each module. For an insight into the design of a tapered plenum see the *Fluid Flow Data Book.*[36] A significant advantage of the flow-through module, however, lies in its ability to handle greater module dissipations or, alternatively, to operate the chip junctions at a lower average temperature due to a reduced thermal resistance over the in-line cooled module. Nevertheless, the final choice must be based on the relative importance of those factors listed in Table 2.19 along with other unlisted additions and constraints of the application.

TABLE 2.19 Comparison of In-Line versus Flow-Through Modules

In-line	Flow-through
More uniform module temperature	Lower average junction temperature (higher power capability)
Liquid or air as coolant	Air only as coolant*
Standard module design	No set standard anticipated
Higher module weight	Lower weight per cubic inch
Series flow with modules operating at different temperature levels	Parallel flow, all modules subject to same thermal levels
Components surface-mounted	Components surface-mounted
Excellent transient response and heat storage capacity	Low heat storage capacity (5 to 10°C/ min component temperature rise)
More area for high dissipators	LRU coolant distribution to modules critical

*Some modules use liquid coolant.

2.8 IMMERSION COOLING OF HIGH-HEAT-DENSITY PACKAGES

This section presents an overview and guidelines for the packaging engineer on the subject of liquid immersion heat transfer of electronics. Power supplies, both low- and high-voltage, fall into this category, as do microcircuits having high heat density. The efficient heat transfer provided by nucleate and pool boiling has long been recognized as a superior cooling technique, but only recently have applications developed as a result of increased packaging flux densities. One of the early papers by Kobayashi et al.[37] describes the cooling of submerged semiconductor rectifiers in Freon R-113[*] having flux densities of 100 W/in². More recent papers on microelectronic immersion cooling include Park and Bergles[38] on comparing Freon R-113 and the fluorinert liquids FC-72 and FG87,[†] and Yokouchi et al.[39] on bubble formation in boiling and temperature overshoot in high-density packages. An in-depth review of the saturated boiling process and the characteristics of common commercially available liquids appears in Danielson et al.[40] and a list of additional references on the subject appears in Armstrong[41] and Rohsenow.[42]

Figure 2.40 illustrates the general types of liquid-filled packages, those having vapor space above the components as in Fig. 2.40a and b, and those that are completely filled, as in Fig. 2.40d and e. In some high voltage power supplies, as in Fig. 2.40c, the space above the component is filled with a gas such as SF_6, which provides increased arc suppression at low temperature where the vapor dielectric properties are much reduced. Packages that are completely filled require an expansion device to account for volumetric changes of the fluid over temperature, whereas units with vapor space usually are designed to withstand

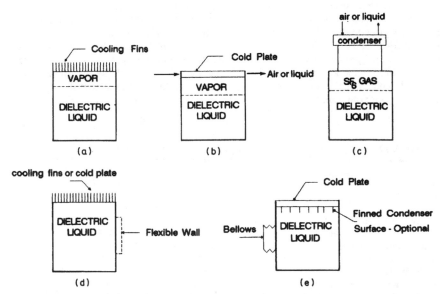

FIGURE 2.40 Types of liquid-filled enclosures. (*a*) Ambient cooled. (*b*) Forced convection cooled. (*c*) Forced convection cooled, external condenser. (*d*), (*e*) Filled enclosure.

---

*Freon R-113 Is a registered trademark of E.I. du Pont de Nemours & Company, Inc.
†FC-87 is a registered trademark of 3M Corporation.

both positive and negative pressure differentials caused by temperature and altitude changes in airborne applications. A completely filled unit is best for heat transfer and voltage arc suppression, but it is heavy and requires a bellows; partially filled units on the other hand eliminate the need for bellows and may have poorer heat transfer and voltage suppression, especially when the components arc in the vapor space. For all of the units depicted in Fig. 2.40, the existence of some noncondensable gas will degrade the thermal performance by impeding the heat transfer at the interfaces. It is good practice, therefore, to pump the air from the package prior to filling and to degas the coolant since in many cases the liquid can absorb as much as 50 percent of its volume as gas at room temperature. Note that the condensers shown in Fig. 2.40 are located either internal or external to the package. In all cases the operating temperature of the liquid is related to the heat-sink or condenser temperature under steady-state conditions.

The heat transfer process for immersion-cooled components is explained by means of Fig. 2.41, which illustrates the classical boiling curve with only slight modification. A heated microelectronic component immersed in liquid exhibits a superheat or temperature rise ΔT_{w-sat}, from the wall to the liquid saturation temperature to a degree depending on the surface heat flux. At low flux levels, A to B, the heat transfer is by free convection without phase change, and the natural convection film coefficient equations in Sec. 2.2.2 are applicable. Between B and C, an increase in flux level results in the formation of vapor bubbles and boiling occurs. Along this portion of the curve, higher flux rates produce activation of more nucleation sites with only small increases in temperature difference. The nucleate boiling regime from B to C in Fig. 2.41 is characterized by a continuous stream of bubbles, which travel from the component surface to the vapor interface in a partially filled enclosure or to the condenser surface in a filled unit. At C, the critical heat flux, or "burnout," is reached, and the component is completely blanketed with vapor. The bubbles form so rapidly at the surface that a large increase in temperature results from C to D and beyond to the unstable region. Upon a reduction in flux rate, the operating point moves along to F and G and then along the nucleate boiling line to B. Although the general shape of the curve in Fig. 2.41 is common for the coolants used in immersion applications, the relative positions of the curves are dependent on the coolant and the surfaces in contact. Surface treatments such as selective etchants and the deposition of alumina on silicon surfaces as well as the development of "dendridic" heat sinks are recent innovations intro-

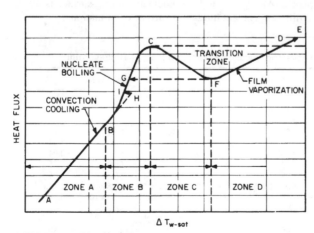

FIGURE 2.41 Liquid boiling curve.

duced to increase the number of nucleation sites for improved heat transfer.[43] Another phenomenon to be dealt with in boiling heat transfer is the hysteresis curve in Fig. 2.41, identified as line BHI. This portion of the curve is less defined and shows a delay in nucleate boiling and an extension of the free-convection curve *AB*. Repeated test runs often result in lowering the incipience of boiling (point *H*), as does surface treatment, and other means are used to promote nucleation at the surface. For additional information on incipient boiling the reader is referred to Park and Bergles.[38]

2.8.1 Coolant Selection for Nucleate Boiling Applications

The coolant candidates for direct immersion cooling are the class of fluorinated fluorocarbons such as FC43, FC-88, and FC-77, and the less expensive and lighter chlorofluorocarbons such as Freon R-113. Mixtures of the fluorocarbon compounds may also be formulated to obtain desired boiling points for specific applications. In addition, there are other electrical considerations, such as high dielectric strength, high volume resistivity, and low dielectric constant. In contact with components the fluids must exhibit compatibility and thermal stability over a wide range of operating temperatures. Also important are low toxicity and low flammability, particularly in airborne applications. These are inert fluids having low surface tension and viscosity, but higher densities than most liquids. Another characteristic that marks this group of liquids is their large gas absorption capacity, which can be as high as 50 percent by volume. Table 2.20 lists the relevant characteristics of five common candidates having a wide range of boiling points for comparison and for selection in specific cooling applications. The heat transfer variables to be dealt with in applications are the predictions of the wall-to-liquid temperature differences $\Delta T_{\text{w-sat}}$, at specific flux levels, which depend on the fluid and surface in contact and on the incipience of boiling. The references cited in Sec. 2.8 offer some test data in this regard, but for specific applications it is best to perform laboratory testing.

2.8.2 Nucleate Boiling Equations

The expression for wall superheat $\Delta T_{\text{w-sat}}$, in °F, for a nucleate boiling surface is due to Rohsenow[42] and is stated for different coolant fluids and surface combinations,

TABLE 2.20 Coolant Properties for Nucleate Boiling Systems at Boiling Point

	FC-43	FC-72	FC-77	FC-84	R-113
Boiling point at 1 atm, °F	345	133	207	181	118
Pour point, °F	−58	−130	−166		88
Liquid density, lb/ft³	90	92	92	91	88
Vapor density, lb/ft³	1.07	0.74	0.93	0.77	0.43
Dynamic viscosity, lb/(ft · h)	0.88	1.01	1.01	0.94	1.13
Conductivity, Btu/(h · ft · °F)	0.038	0.031	0.033	0.031	0.041
Specific heat, Btu/(lb/°F)	0.32	0.28	0.30	0.29	0.25
Heat of vaporization, Btu/lb	32.9	40.6	38.7	36.8	67.8
Surface tension, lb$_f$/ft	3.08×10^{-4}	5.82×10^{-4}	5.48×10^{-4}	5.28×10^{-4}	1×10^{-3}
Coefficient of expansion β, 1/°F	6.67×10^{-4}	8.88×10^{-4}	7.77×10^{-4}	8.33×10^{-4}	9.44×10^{-4}
Dielectric strength, kV	42	38	40	48	—
Resistivity, Ω · cm	3.4×10^{15}	1×10^{15}	1.9×10^{15}	5.6×10^{15}	—
Dielectric constant	1.68	1.72	1.75	1.71	2.40

$$\Delta T_{w-\text{sat}} = \frac{C_{sf} h_{fg} P_r^{1.7}}{Cp} \left[\frac{\ddot{q}}{u h_{fg}} \sqrt{\frac{\sigma}{g(\rho_\ell - \rho_v)}} \right]^{0.33} \tag{2.72}$$

In many cases, Eq. (2.72) can be simplified and reduced to the general form

$$\ddot{q} = C(\Delta T)^n \tag{2.73}$$

where the constant C represents the combined effect of the fluid properties at a specific temperature or range and the surface in contact, $\ddot{q}$ is in W/in, and ΔT in °C. Table 2.21 is a compilation of empirical published data from which approximations of wall superheat are obtainable assuming relatively invariant properties over the nucleate boiling range of temperatures.

Equation (2.74) in Zuber et al.[45] expresses the critical heat flux in nucleate boiling, commonly referred to as CHF, the point at which a vapor blankets the surface and shown in Fig. 2.41 as point C,

$$\ddot{q}_{\text{crit}} = \frac{\pi}{24} h_{fg} \rho_v \left[\frac{\sigma g^2 (\rho\ell - \rho_v)}{\rho_v^2} \right]^{0.25} \left(1 + \rho \frac{v}{\rho\ell} \right) \tag{2.74}$$

where p is mass density, in lb_m/ft^3, and σ is surface tension, in lb_f/length; other units are h_{fg} in Btu/lb and $\ddot{q}_{\text{crit}}$ in Btu/(h • ft^2).

Design heat flux values in liquid-immersion applications should be below this calculated value by a suitable margin of 15 to 20 percent. Table 2.22 lists the CHF values from Eq. (2.74) for several candidate immersion coolants at 1 atm pressure; water is listed as a benchmark indicator, and the other fluids are common types used as dielectric coolants.

TABLE 2.21 Nucleate Boiling Constants C and n for Use in Equation. (2.73)

Coolant	Reference	n	C_1	Comment on test conditions
Water	Jakob and Hawkins[44]	4.0	3.59×10^{-3}	Horizontal surface
Freons and fluorocarbons	Armstrong[41]	3.41	3.6×10^{-3}	Data reported on various surfaces for freons and fluorocarbons; limit values from regression analysis where 95% of all data fall within constants
FC-77	Danielson et al.[40]	3.54	4.83×10^{-3}	0.010-in-diameter wire samples in fluorinert FC-77

TABLE 2.22 Critical Heat Flux Values for Coolants Using Equation (2.74)

Coolant	Boiling point at 1 atm, °F	Critical heat flux	
		W/in^2	W/cm^2
FC-43	345	76	12
FC-77	207	104	16
R-113	118	307	48
FC-84	181	107	17
Water	212	715	111

2.8.3 Design Guidelines

Free-convection, single-phase immersion cooling using a dielectric coolant such as Freon R-113, or a fluorocarbon such as FC-77, generally produces acceptable temperature rises for flux rates of under 5 W/cm^2 (32 W/in^2). Small surfaces characteristic of microelectronic devices yield film coefficients up to 50 percent greater than those predicted from Eq. (2.5)[46] in Sec. 2.2.2. Therefore it is best to use short interruptive surfaces if possible in liquid-immersion systems for maximum heat transfer. Higher flux rates result in nucleate boiling with much smaller rates of temperature rise, that is, surface-to-liquid ΔT values, as the flux is increased compared to free-convection temperature rises.

The design of immersion cooling systems must include an examination of all of the internal heat producing surfaces for maximum flux rates in order to maintain an adequate margin of safety with respect to the critical heat flux. For Freon-12 and Freon-22 the CHF value as reported by Goodling et al.[47] is 40 W/cm^2 for upward-facing surfaces, whereas Ma and Bergles[48] report only 10 W/cm^2 as the critical heat flux for Freon R-113; Eq. (2.74) yields 48 W/cm^2 for Freon R-113. By varying the spacing between heated surfaces, Yokouchi et al.[39] showed that the critical heat flux increases approximately 50 percent, from a 0.30- to a 0.40-in gap. In the case of PCBs, these should be mounted vertically, with high dissipaters located at the bottom of the board. Cooling fins can be attached to extra high power elements or to the PCBs themselves. Dielectric coolants suitable for immersion are typically those that exhibit large superheats of 10 to 30°C; hence one must take account of this in designing liquid systems. Tests by Yokouchi et al. show reductions in overshoot of only 5°C in fluorinert coolants by varying the composition of FX-3250 when combined with FX-3300, confirming previously reported cases of various degrees of overshoot with different coolants and in some cases even with the same coolant through retests. The use of jet impingement boiling described in Sec. 2.10.1 reduces the overshoot significantly in some reported cases.

Fins used on component surfaces to enhance cooling efficiency as in pin grid array packages are best oriented in the horizontal or vertical planes for unrestricted heat transfer, as opposed to downward-facing surfaces, in which case bubble entrapment may lead to serious overheating, particularly in the case of highflux surfaces. Bubble escape is a necessary requirement in nucleate boiling in order to achieve free flow and liquid replacement near the boiling surface. The testing of enclosures using Plexiglas or glass windows can often reveal these trouble spots, which are correctable in the design phase. Surface treatment of cooled surfaces such as silicon to enhance nucleation can produce up to 50 percent increases in film coefficients as compared to smooth surfaces. Some common treatment methods include the following:

Mechanical surface treatment, such as "vapor blasting'

Etching techniques to roughen the surface

Dendritic surface treatment to increase area- forming a brush of nickel powder with a magnet and plating the needles in place

Laser treatment

Machining microchannels or grooves into the surface

Future immersion systems will likely contain chip-treated surfaces to enhance the heat removal capacity. In addition, major improvements are possible by using forced movement of liquid past the heat-dissipating surface or through jet impingement, as described in Sec. 2.10.

2.9 COOLING WITH HEAT PIPES

Heat pipes are used in electronic systems to transport heat efficiently from one point to another. Unlike conduction systems, which rely on the thermal conductivity of solid materials, heat pipes operate with fluids in the boiling/ evaporative and condensing modes for heat transfer. Their advantage lies in that relatively large amounts of heat transport take place within the pipe, with only small internal resistance from source to sink, hence correspondingly small temperature gradients. Heat pipes have for many years occupied a special place in the electronic industry, mostly in applications deemed too difficult to solve by conventional means. Thus they appeared early in space applications, where the need for spacecraft temperature equalization and for transferring heat to radiative surfaces without power-driven pumps proved advantageous. The reader is referred to Chi[49] for a description of these systems. At present the use of heat pipes is expanding into areas such as PCBs, where high dissipation levels create excessive temperature gradients in the module heat sinks and where weight is critical. Potential applications are also appearing in high-power hybrid or multichip packages, where novel arrangements are being devised to overcome conventional high-resistance heat paths. There is no doubt that as chip heat densities move from the present 30-W/cm^2 (193-W/ in^2) level to flux densities on the order of 100 W/cm^2 (645 W/in^2) or higher, there will be further opportunities to explore the potential of heat pipes. As with all relatively recent innovations, there is a certain reluctance on the part of designers to implement heat pipes in many cases due to suspicions regarding reliability, cost, producibility, and so on. However, as more applications evolve, this concern may well disappear. In any case, it is always important to perform tradeoff studies to determine whether a heat pipe is the correct design choice for any proposed application.

2.9.1 Principles of Operation

A simplified description of heat-pipe operation can be made by reference to Fig. 2.42. Here a round or rectangular tube contains an internal wick, usually made of a fine-screen mesh material and attached to the inner wall of the tube through diffusion bonding or brazing. The central core region of the enclosure is open, thus permitting the free flow of vapor to pass whenever there is heat applied to any surface of the pipe. This causes liquid in the evaporator or heated region to boil from within the mesh, and in doing so, the vapor then seeks a lower temperature region in the pipe, which in the figure is represented as the condenser located at the opposite end of the evaporator. The inner surface of the pipe lined with the mesh or wick material is saturated at the time the pipe is filled. Then during operation, the evaporation of liquid depletes the mesh, causing the liquid-vapor interface at the evaporator to enter into the wick surface, which causes a capillary pressure to develop. A pumping action then takes place, with the liquid within the wick flowing from condenser

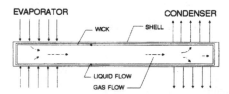

FIGURE 2.42 Heat-pipe component designations.

to evaporator, ready for reevaporation. The process continues with constant replenishment of liquid at the evaporator, provided a sufficient capillary pressure is maintained within the system. There is, however, an upper limit to pipe operation based on capillary pumping alone, which effectively restricts the heat capacity of a given pipe. Because capillary pressure is quite low, heat-pipe sensitivity extends to orientation with respect to gravity as well as to other design constraints. These include, in addition to capillary pumping limits, sonic or choke flow of vapor similar to the compressible flow of gas in a nozzle, entrainment of liquid caused by high velocity gas flow, and dryout due to excessive boiling flux rates. Despite these limitations, properly designed heat pipes are capable of high heat transport rates with low temperature gradients in comparison to conventional techniques, and as such offer an alternative solution to some difficult heat problems. The conductance values (in W/°C) of heat pipes are often orders of magnitude better than those of solid materials, particularly when the heat transfer occurs over large distances. One application area of recent interest is in the cooling of PCBs, where heat levels exceed 40 to 50 W. In these cases there are two basic heat-pipe configurations of interest—vapor chamber and embedded. These two concepts are shown in Fig. 2.43. In the vapor-chamber approach, there is a frame, wick, and cover sheet joined to form a single vacuum tight enclosure. The embedded pipe concept consists of several flat heat pipes sandwiched between cover sheets and bonded into a frame to form the completed assembly. It is not the intent here to explore the variables relating to these designs or to any other heat pipes, but the interested reader is referred to Chi[49] and Bienert and Skrabek[51] for detailed information on the subject. There are at present a sufficient number of qualified suppliers ready to design and deliver heat pipes for specific applications; therefore it is not necessary or even recommended that one undertake heat-pipe production unless serious consideration is first given to the many material selection issues and the manufacturing procedures and processes associated with heat pipes.

2.9.2 Embedded Heat Pipes for Modular PCBs

There are many special applications that warrant the use of heat pipes in electronic cooling. For the most part, however, there are few if any that are known to be in large-scale production, on the order of several thousand. The possibility exists that PCBs will be the first to reach that goal if (1) costs can be further reduced and (2) dissipation levels go

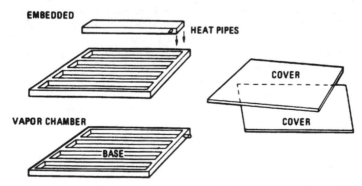

FIGURE 2.43 Basic embedded and vapor chamber concepts for cooling PWBs. (*From Basilius and Minning.*[50] © *1985 IEEE. Reprinted with permission.*)

beyond 50 and, more likely, beyond 100 W. Recent designs in this area have tended to favor the embedded concept, possibly because of its lower sensitivity to gravity as compared to the vapor-chamber heat pipe or possibly due to the flexibility of the embedded design with respect to the PCBs. The conventional embedded design consists of a number of embedded pipes on the order of 0.090 by 0.50 in in cross section and extending over the entire width of the module. The number of embedded pipes depends on the dissipation level, size of the module, and temperature objectives of a specific design. Figure 2.34, which depicts a standard SEM "E" type module, is one of the more likely candidates for embedded heat pipes. Before examining the thermal performance of this assembly, it is well to note that in addition to the previously mentioned design limits, namely, capillary pumping, entrainment, sonic flow, and critical heat flux in boiling, there exists an added constraint in the SEM "E" module application due to the physical layout of the assembly. The clamping edge distance, for example in the standard module has been established as 0.19 to 0.22 in for military applications, and therefore the thermal impedance resulting from the 2.5in^2 interface area is 0.20°C/W. At 50-W dissipation, there is a 10°C gradient and at 100 W it is 20°C if one assumes a 0.5°C • in^2/W resistance (see Fig. 2.5). In addition, at higher heats, the condenser area of the pipe extends beyond the edge distance and toward the center of the pipe assembly, and this definitely affects thermal performance. Furthermore, data showing surface temperature variations of the heat pipe from the center to the edge must be carefully interpreted as to edge location because the largest gradient takes place at the very end due to condenser heat transfer limitations. This effect is demonstrated in Fig. 2.44, which shows the edge ΔT as a function of dissipation in an embedded heat-pipe module. Figure 2.44 clearly illustrates the advantage of doubling and even tripling the edge distance from 0.22 in to larger values at heat levels beyond 25 W. Commercial designs are less restricted in this respect than are the military designs. Similarly, a vertical or off-axis tilt lowers the performance, but in most cases a proper orientation can be accounted for in the installation. For applications where aircraft maneuvers take place and the heat pipe is in an unfavorable attitude, usually the system mass is sufficient to allow for operation during these transient periods with a minimum of temperature increase.

Test data for a vertically oriented SEM "E" module with embedded heat pipes appear in Fig. 2.45 along with calculated and measured ΔT values for a solid aluminum 6061-T6

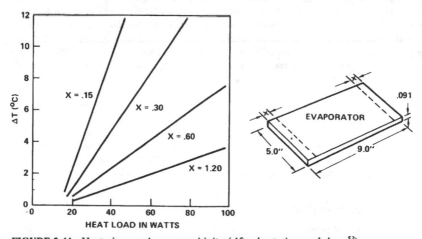

FIGURE 2.44 Heat-pipe condenser sensitivity (*After heat-pipe workshop.*[52])

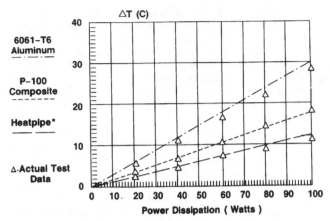

FIGURE 2.45 Comparison of solid metal, composite material, and embedded heat pipe of SEM "E" dimensions. *Includes condenser temperature drop. (*Tests conducted at Westinghouse Electric Company, 1987.*)

material, 0.10 in thick. Also shown are test data for a P-100 composite heat sink having graphite fibers embedded within an epoxy base. The thermal conductivity of the graphite composite is 180 Btu/(h • ft • °F) and the density is 0.065 lb/in^3. In this test the row of thermocouples sensing the edge temperature was 0.38 in from the end of the heat sink, or as close as possible to the interface area. Data taken further in-board from the edge indicate less of a temperature gradient from center to edge, thus making the pipe appear more efficient in heat transfer. Nevertheless, at 100 W, a nearly 3:1 improvement can be noted over the solid aluminum assembly (29 versus 11°C), having a comparable weight within 5 percent of the heat-pipe assembly. An assembly wider than 5.88 in when compared with a solid conductor will favor the embedded pipes even more because the solid heat-sink center-to-edge gradient is directly proportional to the distance squared from the center [Eq. (2.68), Sec. 2.7.4].

As regards composite-material heat sinks, these appear as favorable substitutes for aluminum, being 35 percent lighter and in some cases two and three times better in conductivity. However, the interface can pose a problem in that the conductivity normal to the fibers is dependent on the embedment material, which is generally of low conductivity. There also may be a material compatibility problem at the interface between the composite heat sink and metal engaging surface. As of this writing, development work is proceeding and the somewhat favorable comparison in Fig. 2.45 must be tempered by the lack of clear solutions to the use of composites. There are at the present time many composites under investigation, some of which include thermal coefficient of expansion (TCE) matching with leadless chip carrier substrate materials for use in module assemblies. This development is driven by the greater selection of leadless parts compared to the other microelectronic devices and by cost advantages in military systems, which operate over wide temperatures. Although Fig. 2.45 represents a typical PCB application, similar comparisons favoring heat pipes at higher power are even more beneficial.

2.9.3 Heat-Pipe Specifications

The application of heat pipes requires proper documentation through a requirements specification due to their unique features and due to their interfacing with heat sinks and sources. Some of the most important elements covering the external heat-pipe requirements are the following.

1. *Thermal transport* q_{max}. The maximum heat transport capacity of the heat pipe in its nominal operating mode. As an example, in the case of a PCB assembly, q_{max} represents the total component dissipations or the evaporator heat load in watts.

2. *Heat flux* q. A maximum flux level, in W/in^2, at a specific location in the evaporator. The characteristic design features of a heat pipe to meet a flux condition involve proper coolant selection in conjunction with adequate heat spreading within the shell. As in any boiling process, there is a burnout limit (critical heat flux) for any heat-pipe coolant material.

3. *Thermal resistance* R_{int}. The total resistance of the heat pipe, in °C/W, from the evaporator to the condenser determines the temperature drop for a given heat load. Care must be exercised to take account of the two external interface resistances at the evaporator and the condenser in addition to the internal pipe resistance from evaporator to condenser.

4. *Orientation*. Heat pipes are attitude-sensitive; therefore the normal orientation relative to the gravitational force vector is important to specify. Significant differences may occur due to orientation, particularly in vapor-chamber designs, so that worst-case conditions must be treated accordingly.

5. *Cold start*. There are heat-pipe applications that demand operation over a wide temperature band. In those cases the time to reach steady state may be critical and must be specified.

6. *Temperature variation*. The heat-pipe advantage extends to temperature uniformity as well as to low thermal impedance; therefore total allowable variation in temperature across the evaporator is an important consideration in some applications.

7. *Interfaces*. The attachment of the heat pipe to the structure not only involves the dimensional tolerances of flatness, finish, and so on, but must also take account of the thermal impedance as mentioned.

8. *Environmental*. The temperature variation, vibration, shock, and so on, to which the heat pipe will be exposed, must be specified in each application. Recent three-axis vibration and shock tests on SEM "E" embedded-type heat pipes indicate that the thermal performance during and after exposure to the levels specified in MIL-STD-810D, that is, 6 to 3000 Hz, 0.08g^2/Hz vibration, and 15g shock impulses, 11-ms duration,[53] are sustainable.

The manufacture of heat pipes requires a high level of quality control, which can easily lead to problems if not properly addressed and implemented. The process includes gas evacuation prior to charging the system with fluid and, as such, demands the use of ultra-clean parts free of contaminants, either solid, liquid, or gaseous. Any contaminant present within the enclosure will affect heat-pipe performance and life. Material selection is also critical and must not lead to undesirable corrosion products. This applies equally to the parts as well as to the fluid, which must be of high purity and free of noncondensables. Any foreign substances within the enclosure will degrade the thermal performance over time. These factors all add to the overall cost of production, which can present a limitation to the use of heat pipes in some commercial applications.

2.10 COOLING TECHNIQUES FOR HIGH HEAT DENSITY

2.10.1 Jet Impingement Cooling

As the scale of circuit integration continues to increase and the need to place IC chips closer together leads to chip fluxes beyond the present 30 W/cm^2 (193 W/ in^2) toward 100 W/cm^2 (645 W/in^2), further advances in cooling will be necessary to match these conditions. Jet impingement and microchannel cooling are two methods presently under investigation. Jet impingement cooling of microelectronic chips is accomplished by passing a coolant through a capillary tube or orifice aimed at the surface to be cooled. A typical arrangement is shown in Fig. 2.46. A liquid under pressure in the chamber is allowed to pass through an orifice plate and directly onto the microchip component. Shown in the figure is a leadless chip carrier attached to a substrate through a series of solder bump joints. The dielectric coolant strikes the chip and absorbs its heat dissipation. The development model shown is not representative of a system in production. However, in addition to microchip cooling, there are applications in electronics which utilize spray cooling or direct impingement cooling of component surfaces. One such application is a variable-speed constant-frequency (VSCF) generator for use aboard commercial and military aircraft. The dielectric coolant in this case passes through small-orifice jets aimed directly at high heat dissipators, mostly diodes, and power transistor assemblies located within a partially filled liquid reservoir. Since these units are exposed to wide temperature extremes, the coolant flow at low temperature through the nozzles is much reduced, and this reduction must be taken into account in the thermal design. The oil that cools the electronic components also cools the generator and stator. There are two modes of operation possible with liquid jet impingement—single-phase and two-phase cooling. In addition, the jet can be free or submerged. In the submerged case the cavity is filled with the coolant while in the free mode, as shown in Fig. 2.46, the liquid jet is exposed to a gaseous environment. The most efficient heat transfer takes place in the free liquid impingement case when the cooling mode is a combination of boiling and free convection. Ma and Bergles[48] have reported experiments with Freon R-113 where up to 100 W/cm^2 over a small chip surface

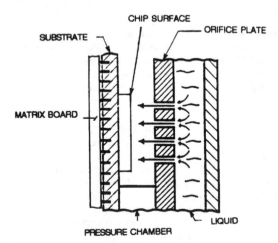

FIGURE 2.46   Jet impingement configuration.

(0.2 by 0.2 in) was cooled using jet impingement boiling. These experiments and others[*] have found that significant improvements in heat transfer are possible with jet impingement above that normally encountered in the boiling mode alone for immersed components. The combination of local boiling and forced convection leads to an extension of the critical heat flux for Freon R-113 beyond the 20- to 40-W/ cm^2 free zone to 70 W/cm^2 with jet impingement under saturated liquid conditions, to as high as 100 W/ cm^2 when the liquid is subcooled. As in boiling alone, the surface condition and coolant in contact are important parameters in determining the heat transfer. The advancement of wafer scale integration (WSI) with its packaging and electrical advantages may well result in the initial development of novel jet impingement techniques for IC cooling implementation into hardware.[54] The design of this type of system must take into account the ultimate heatsink or boiling-point temperature of the jet coolant and the heat-path resistance from chip to sink. Goodling et al.[47] report on the use of Freon-12 and Freon-22 operating below room temperature with jet impingement on the backside of a 4.0-in water. Their tests reveal flux densities of 100 W/cm^2 and higher as the critical heat flux. However, despite this and other reported progress, future implementation into a production system using ICs will require compatibility tests to verify the coolant stability in contact with the electronic enclosure and the chip surface over long periods. These same experimenters found that the temperature overshoot in boiling could be much reduced using jet impingement as compared to the normal free saturation boiling mode. In the jet impingement experiments reported to date, small high-velocity jets of under 50 ft/s are directed to the backside of chip surfaces for local heat removal; present experimentation has been done mostly with dielectric liquid fluorocarbons and with freons and water. The options in this type of cooling include free jet versus submerged jet, type of coolant, spacing of the jets, and the distribution of liquid to large array surfaces.

2.10.2 Impingement Cooling Analysis

Single-phase freejet impingement cooling is influenced by many variables, such as jet diameter d, velocity V, number of jets n per source, jet-to-source distance x, jet configuration, size of heat source area $\ell \times \ell$ and coolant properties. Based on the single-phase freejet tests of Jigi and Dagan,[54] the correlating data for FC-77 and water were found to agree well with Eq. (2.75) for jet distances to the heat source falling within the limits of $3 < x/d < 15$,

$$\bar{N}_u = 3.84 R_e^{0.5} P_r^{0.33} \left(0.008 \frac{\ell}{d_h} + 1 \right) \tag{2.75}$$

Their tests were for nozzle diameters d_h of 0.020 to 0.040 in and jet velocities of under 50 ft/s. Equation (2.75) holds true for small surface dimensions of $\ell < 0.5$ in, which in most cases is typical of microelectronic devices. Larger surfaces with multijet impingement can also be analyzed provided the number of jets per unit heat source area is comparable with the test conditions specified. Application of the equations is illustrated by the following example.

Problem 2.10: A single-phase freejet impingement nozzle is placed in the center of a 0.5- by 0.5-in heated surface a distance of 0.125 in from the backside. FC-77 is made to pass

---

*See references listed in Ma and Bergles.[48]

through a 0.040-in tube at a rate of 0.125 gal/min to cool the plate. Determine the average film coefficient h_c and the surface temperature of the heat source at steady state if the supply coolant is at 25°C and the heat load is 20 w or 80 W/in ^{2} (12.4 W/cm^2).

Solution:

1. Since the Prandtl and Reynolds numbers are based on the fluid properties of the film at the heat source surface, we will assume the rise at the surface to be 20°C as a first estimate. Thus the film temperature of the FC-77 is calculated as follows:

$$\bar{T}_f = \frac{T_c + T_s}{2} = \frac{25 + 45}{2} = 35°C$$

Then the fluid properties at 35°C film temperature are

$$\rho = 109 \text{ lb/ft}^3$$
$$u = 2.9 \text{ lb/(ft • h)}$$
$$k = 0.036 \text{ Btu/(h • ft • °F)}$$
$$C_p = 0.252 \text{ Btu/(lb • °F)}$$
$$P_r = 20.3$$

2.
$$\frac{x}{d} = \frac{0.125}{0.04} = 3.125$$

$$A_s = \frac{0.5 \times 0.5}{144} = 1.736 \times 10^{-3}\text{ft}^2$$

3.
$$\dot{m} = \frac{0.125 \text{ gal}}{\text{min}} \times \frac{14.55 \text{ lb}}{\text{gal}} = 1.82\text{lb}/\text{min}$$

4.
$$G = \frac{\dot{m}\ 8640}{A} = \frac{1.82 \times 8640}{0.785 \times 0.04^2} = 12.52 \times 10^6 \text{ lb/(h • ft}^2_j$$

5. The velocity head is

$$\frac{G^2}{\rho k} = \frac{(12.52 \times 10^6)^2}{109 \times 1.2018 \times 10^{11}} = 12.0 \text{ lb/in}^2$$

6.
$$R_e = \frac{dG}{u} = \frac{(0.04/12)12.52 \times 10^6}{2.9} = 14,390$$

7.
$$\bar{N}_{u\ell} = 3.84 R_e^{0.5} P_r^{0.33}\left(0.008\frac{\ell}{d}n + 1\right)$$

$$= 3.84 \times 14,390^{0.5} \times 20.3^{0.33}\left(0.008 \times \frac{0.5}{0.040} + 1\right) = 1368.4$$

8. $$\bar{h}_{c\ell} = \bar{N}_u \frac{k}{\ell} = 1368.4 \times \frac{0.036}{0.5/12} = 1182 \text{ Btu/(h} \cdot {}^{\circ}\text{F)}$$

9. $$\Delta \bar{T}_s = \frac{q}{h_c A_s} = \frac{1.8944 \times 20}{1182 \times 1.736 \times 10^{-3}} = 18.5 {}^{\circ}\text{C}$$

10. The assumed temperature rise of 20°C is within 10 percent of the calculated value; therefore no further iterations are required.

Note: Performance may be improved by increasing the number of jets and/or decreasing the jet size. Even greater improvements occur with boiling freejet impingement.

A large increase in heat transfer is possible with jet impingement boiling where the dependence on surface condition and fluid in contact is most important. In the case of submerged jet cooling on small surfaces, Ma and Bergles[48] report significant increases in the critical heat flux or burnout as compared to the same surfaces when exposed to immersion boiling alone. They found that the pool boiling curve is coincident with the jet boiling curve and can in fact be extrapolated to values of critical heat flux for jet impingement boiling on the order of five times the pool boiling values. Not all experimenters report similar results. At the present time there are no generalized equations defining jet boiling for either submerged or free conditions. In the case of nonboiling jet impingement in a submerged liquid, the stagnation-point Nusselt number within the potential core correlatable to within 10 percent is in accordance with Eq. (2.76),[48]

$$N_u = 1.29 R_e^{0.5} P_r^{0.4} \qquad (2.76)$$

A comparison of the submerged Nusselt number at stagnation [Eq. (2.76)] with the freejet Nusselt number [Eq. (2.75)] for average conditions on small areas indicates a 3:1 improvement in these single-phase heat transfer cases.

2.10.3 Microchannel Cooling

The advent of high-density chips as in very-large-scale integration (VLSI) and wafer-scale integration (WSI) has led to investigations involving convective heat transfer designs which are located in close proximity to the heat source on the chip surface. In microchannel cooling, small rectangular or round openings are provided, often within the silicon substrate, for the passage of liquid coolant. The channel dimensions used in the early experiments by Tuckerman and Pease[55] were typically 0.002 in wide and 0.012 in deep with 0.002 in of solid material between fins. These experiments carried out with water resulted in a low thermal impedance of 0.09°C/W, corresponding to 1000-W/cm^2 heat flux and surface temperatures of under 120°C. Other experimenters reported comparable results, thus verifying the low thermal impedance inherent in microchannel cooling. Figure 2.47 illustrates the typical microchannel configuration. In the figure, the cover plate shown is sealed by bonding to the heat sink with an adhesive material. The liquid coolant passes from the inlet plenum through the microchannels and out the exit manifold. There are two types of plenums shown—side-fed and end-fed. Heat from the IC elements flows by conduction through the substrate and into the coolant by convection.

The small rectangular passages are fabricated in silicon by precision sawing or by orientation etching. In another technique, the channels are first machined into a separate heat sink and then attached to the silicon. Recently, much smaller size passages have been fab-

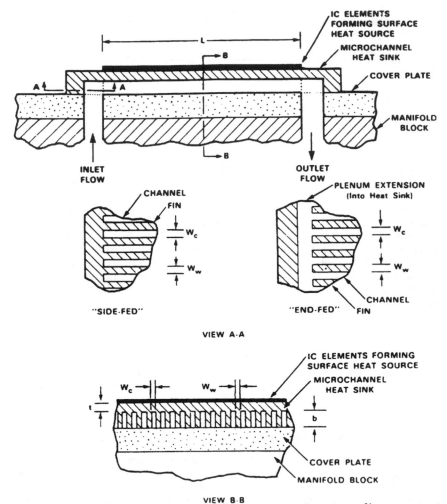

FIGURE 2.47 Schematic view of a microchannel heat sink. (*From Phillips.*[56] *Reprinted with permission of Lincoln Laboratory, Massachusetts Institute of Technology, Lexington, Mass.*)

ricated than those used in the earlier experiments.[57] Microfabrication techniques developed in the electronics industry permitted the manufacture of flow channels ranging from several hundred micrometers (100 μm = 0.003937 in) to a few hundred angstroms (100 Å = 0.00003937 in).[57]

The reason for striving to develop small openings is that as the hydraulic diameter decreases, the film coefficient h_c increases inversely. Moreover, since large decreases in film resistance ($1/h_c A$) are limited mostly to area improvement, the drive toward increasing the film coefficient h_c becomes apparent. In fully developed laminar flow for a channel aspect ratio b/Wc of 5 in Fig. 2.47, the Nusselt number from Kays and London[9] (p. 104) is equal to 5.7. Therefore the film coefficient for water corresponding to channel dimensions of 0.010 by 0.002 in is 7545 Btu/(h • ft^2 • °F) at an average bulk temperature of 100°F. Film coefficients of this order makes possible the support of substantial heat flux rates at nominal wall temperatures.

Although this method of cooling appears to have great potential, there may be significant difficulties in implementation from an engineering viewpoint, not the least of which is coolant plugging within the small passages. Performance degradation due to even slight contamination over time is a concern that must be addressed in the near future. Also, the flow behavior of fluids in small passages as regards pressure drop and heat transfer presents another valid concern. Recently investigators at the University of Pennsylvania[57] have reported on their studies of pressure drop and found that a significant departure from the classical Navier Stokes equations, which describe fluid flow, takes place as the channel size decreases, indicating that at some point the conventional equations are not applicable. These results are preliminary to date and further experimentation on pressure drop and heat flow is anticipated.

2.11 COMPUTER PROGRAMS IN THERMAL ANALYSIS

The one-dimensional approach for solving heat transfer problems illustrated thus far is satisfactory in a variety of cases. However, in many electronic cooling systems a detailed component temperature distribution is usually required to determine the overall system reliability and in determining the adequacy of the thermal design. In those cases the analyst must resort to more sophisticated means. Since there are few thermal analysis applications that lend themselves to closed analytic treatment, the implementation of numerical finite differences or finite element analysis with the aid of computer software programs is commonly employed. This has been the basis of significant advances in the art of analytical thermal design over the past 20 years. At present a wide choice of programs is available based on the lumped-parameter system approach. In general, most numerical techniques of this type involve the subdivision of a configuration into a grid network consisting of nodal elements having a specific volume and interconnected to adjacent nodes through a network of conductances. The thermal properties and temperature of the nodes are assumed to be uniform throughout each segment. In addition it is assumed that the heat transfer through each face of the node is expressed in terms of the basic Fourier conduction equation or by simplified convection and radiation terms. These equations for conduction, convection, and radiation may be written in conductance terms for the three basic heat transfer modes as follows:

$$q_{\text{cond}} = \frac{kA}{\Delta X} \qquad\qquad = \frac{kA}{\Delta X} \qquad q_{\text{cond}} = G_{\text{cond}}\Delta T$$

$$q_{\text{conv}} = h_c A \Delta T \qquad\qquad G_{\text{conv}} = h_c A \qquad q_{\text{conv}} = G_{\text{conv}}\Delta T$$

$$q_{\text{rad}} = \sigma F_e F_a A (T_1^4 - T_2^4) \qquad G_{\text{rad}} = \sigma F_e F_a A \qquad q_{\text{rad}} = G_{\text{rad}}(T_1^4 - T_2^4)$$

In the case of fluid flow, one-way conductors are defined as

$$G_f = \dot{m} C_p \qquad q = G_f(T_1 - T_2)$$

where $\dot{m}$ is the mass flow rate and C_p represents specific heat. The computer program will solve the network of nodes as defined by the three types of heat transfer modes G_{cond}, G_{conv}, G_{rad} and the fluid conductors G_f for both steady-state and transient cases. Three types of nodes emerge in these problems— diffusion, arithmetic, and boundary-layer types. The diffusion nodes are the lumped masses consisting of a temperature T_i, specific capacitance $(\dot{m} C_p)_i$, or C_i, and in some cases a heat load q_i. Arithmetic nodes are the types that have temperature, no thermal mass, and possibly heat loads, whereas boundary-layer nodes have only fixed temperature, independent of the thermal network. Steady-state problems include only arithmetic and boundary nodes, whereas transient problems must also include diffusion-type nodes. In transient solutions, the computations are done using finite increments of time, or time steps, and the parameter change is directly related to the size of the node and the time step. It is important to select a suitable time step because too small a step leads to excessive computations (long computer times), and too large a step causes inaccuracies and possible instability. An acceptable time step can be determined through the use of Eq. (2.77),

$$\Delta T = \frac{C_i}{\sum G_{ij}} \tag{2.77}$$

applied to each diffusion node, and then selecting the lowest value of Δt for each iteration.[58] Since the speed and storage capacity of computers vary widely, networks having large numbers of nodes will require various running times on different machines. Also, the method of handling large numbers of equations through successive iteration or matrix inversion affects running times. At present computer software continues to be upgraded, with some programs designed to run on personal computers, where machine operating speeds are also undergoing rapid changes.

One of the common software programs used in thermal analysis is the CINDA family. Originally developed as the Chrysler improved numerical differencing analyzer, several derivatives are now available for use in thermal analysis. The systems improved numerical differencing analyzer (SINDA) and the Martin interactive thermal analysis system (MITAS) are two such versions. These are flexible programs, which users often modify to adapt to their specific types of problems. The inputs required for running CINDA include network conductances, heat dissipation, material thermal properties, boundary conditions, and so on. A large system containing many nodes is sometimes broken down into separate groupings, such as modules within a package, and each module contains its own thermal network subject to specific boundary conditions and treated as a separate problem. Then in the system model, the module can be characterized by a limited number of nodes, or even a single node having the proper time constant and conductance which accurately defines the unit's thermal properties. This approach i can also be applied to single-chip and multi-chip packages, where the chip temperatures are accurately obtained through the use in some cases of thousands of nodal elements. There are existing procurement specifications which require the manufacturer to perform this type of analysis as a means of verifying the thermal performance and overall junction-to-case impedance, in °C/W, of a module.

Software programs are proliferating in areas closely related to thermal temperature calculations. The complex flow of air across PCBs and in manifold distribution systems is modeled and solved with programs such as PHOENICS and FLOTRAN. PHOENICS[*] is a general-purpose computational fluid dynamics program used in the simulation of fluid

---

*Distributed and licensed by CHAM of North America, Huntsville, Ala.

flow, heat transfer, mass transfer, and even chemical reactions. The FLOTRAN* program is a finite-element program dealing with convective heat transfer and flow distribution in the coolant and around the components. It solves for flow patterns, pressure distribution, and film coefficients. Many other software programs are marketed that relate to thermal management. Selecting the most appropriate system depends on the type of problems to be solved, the flexibility of the program, and the cost involved.

Finally there are graphics computer programs which are capable of producing an isothermal map, often in color gradations of the temperature of a solid body such as a cold plate. A widely used interactive graphics program of this type is PATRAN,[†] a finite-element analysis software program that produces displays of various functions such as temperature. It is often used in conjunction with other software or in-house developed programs to generate the nodal network and the temperatures.

2.12 THERMAL MEASUREMENTS, TESTING, AND FLOW BALANCING

Achieving acceptable thermal performance of electronic equipment requires a realistic test plan that defines the goals and conditions under which the tests are to be conducted. A successful test plan must reflect the use of the hardware in its true environment subject to conditions of actual field usage with respect to cooling air or liquid and duty cycle (heat dissipation). The parameters that affect component temperature levels are those to be tested and include heat dissipation, coolant flow, temperature and pressure conditions, the environment surrounding the equipment, and the proper simulation of the internal system resistances. In an overall program, the tested equipment will vary from single components to complete systems and may include preliminary mock-ups as well as the final hardware, depending on the stage of testing. At the heart of the testing is the measurement of temperature, which is commonly done with thermocouples mounted on components and parts buried deep within electronic enclosures. Accurate readings are obtained with copper-constantan #36 gauge type thermocouples properly installed on the surfaces to be measured. To measure components, the thermocouples are attached to the outside surface of the part at a location that least interferes with the cooling process. The thermocouple leads are often flattened and spring-loaded for good contact to the surface in order to obtain the best accuracy. An adhesive bond, such as epoxy or an equivalent, is the preferred method of attachment to assure consistent readings. Infrared thermography and laser beams are other techniques used for measuring surface temperature, in this case without contact as, for example, on PCBs to detect hot spots. Both infrared and laser techniques require visibility of the surface, which limits their practicality with respect to system hardware. Other temperature-sensing methods make use of paints, lacquers, and crayons, which change irreversibly in appearance when reaching a set temperature.

These devices measure the maximum temperature level reached and are useful during flight testing of hardware where the proper instrumentation is not always available for recording temperature data. Temperature measurements of coolant inlet and exit together with flow rate can be used along with current and voltage readings to verify the heat dissipation level of assemblies [see Eqs. (2.28) and (2.29), Sec. 2.3, for calculating heat dissipation]. The technique used for measuring the junction temperature varies and can be

---

*Distributed and licensed by Compuflow, Inc., Charlottesville, Va., and Swanson Analysis systems, Houston, Pa.

†PATRAN is a registered trademark of PDA Engineering Company, Costa Mesa, Calif.

applied directly or indirectly. In the indirect method, a calibrated curve is used, which is based on the forward voltage drop at low current for a diode diffused directly on the chip surface near the hot spot to be measured. This electrical technique is preferred to other methods such as liquid crystals, laser, or infrared techniques which require delidding of the component to perform the measurement. Although the electrical method can be performed on a lidded part and is noncontacting, its disadvantage lies in not always being near the chip hot spot or in not being in an acceptable location on the chip for an electrical measurement.

The measurement of parameters for system testing of hardware is often done in conjunction with a computer acquisition and control system which enables the simultaneous storage and recording of many quantities. This also provides the means to conduct temperature surveys on equipment used to determine actual time constants associated with different parts of systems.

2.12.1 Thermal Testing

The thermal testing of electronic systems takes place at the overall systems, unit, and module level, as well as at the component level. The purpose of these tests is to uncover thermal problems, evaluate designs, verify the assumptions used in the design, and in general confirm the predictions that the system can operate satisfactorily under the assumed environmental conditions.

A thermal evaluation test is a system test performed once on the equipment to determine its adequacy and need not be repeated unless major design changes are made on the equipment. The qualification or environmental test performed on military hardware usually includes limited thermocouple instrumentation used for this purpose. In some cases a more comprehensive approach, consisting of a totally separate evaluation test, often is performed on critical thermal equipments. Regardless of the specificity, any degree of thermal evaluation can be most useful in verifying temperature predictions as well as the dissipation levels and flowbalancing parameters. Even the most rudimentary of tests, consisting of a limited number of thermocouple placements, is certainly beneficial and necessary as a means of measuring thermal performance. At the opposite extreme of testing, a system qualification may well include a complete thermocouple installation with closely controlled environmental features duplicating the system exposure. The latter approach is a routine occurrence with respect to the testing of spacequalified hardware, which must be fully tested and proven prior to launch. These tests are often performed in environmental space chambers held at low pressure ($<10^{-6}$ torr) using solar radiators and cryogenic panels to simulate the space environment. The acceptance test procedure for these systems includes temperature surveys and the correlation of test data with temperature predictions based on the extensive use of thermal modeling.

There are also other tests of a thermal nature performed on equipment in support of attaining a total quality product. These include environmental stress screening (ESS) and burn-in testing. The purpose of these tests is to use vibration and temperature cycling as a means of inducing early failures on equipment in the factory rather than in the field. Optimum screening to be most effective will ferret out the highest fraction of deficient items by causing their failure in the shortest period of time without adversely affecting or degrading the test article. There are no universally acceptable procedures regarding the levels or rates of change or the temperature limits or number of cycles used in screening tests. These are evolutionary; they arc tailored for specific equipment based on past experience and not intended to overstress equipment beyond its operating limits. Typical temperature-screening cycles are performed on modules between room temperature and both high and low extremes; for military units these range from $-54°C$ to 95 or $125°C$ nonoperating, with narrower limits on operating units of between -40 and $71°C$. Nonpowered units are cycled at 10 to $20°C/min$ and powered units at lower rates of 5 to $10°C/min$ for up to 50 cycles in typical test regimes.

Quality testing also includes burn-in screening, which essentially is intended to weed out bad components through elevated temperature tests run over an extended period of time. The flaws detected during these tests include insulation breakdown, cracks and fracture of parts, material degradation, and corrosion. Thermal cycling tests combined with vibration during environmental stress screening testing typically reveal open-circuit solder joints, cracks, hermetic seal contamination, IC bond failures, open PCB connections, and so on.

2.12.2 Flow Balancing

Electronic systems must be flow-balanced in order to assure that each unit receives the proper allocation of air or liquid coolant. The pressure drop or flow impedance across individual units is the common denominator in the setting of flow rates. Separate bench tests are used to establish the flow-pressure characteristics of each unit, so that in a system test one need only measure the pressure at each unit to determine the flow conditions. Therefore, for a system comprising a number of separate packages cooled from a common source, the pressure drop serves as the calibration for flow in each unit. In ground-based commercial equipment, the cooling air or liquid undergoes only nominal changes so that the system balance remains essentially constant. In the case of flight hardware or in military equipment exposed to a wide range of flow conditions, balancing is performed at the most frequent operating point and operation at other conditions usually results in a certain degree of unbalance, to an extent depending on the design of the individual units. For example, in a system comprising a number of units using different fin stock in the cold plates, the flow resistances as expressed in Eq. (2.51) are not matched. This can stem from the variation of the exponent n from unit to unit, which not only is due to fin stock variations but also may be the result of operating the cold plates in different regimes—laminar, turbulent, or transition.

Ideally, a reasonable objective in design is to strive for uniform exponents for all the cold plates, in which case the off-operating points will remain in proper balance. In practice, however, this is seldom, if ever, achieved. Alternatively, there is another design objective, which is to utilize to the fullest all the pressure drop made available from the source rather than across orifice plates used for balancing. Heat load variations from unit to unit as well as dissimilar configurations often preclude satisfying all these objectives.

The control of unit impedance characteristics is documented in the cold-plate specification, which typically allows for a 10-percent variation from the design curve. Replacing units in the field or elsewhere is then made possible without significantly altering the system flow balance, provided that units adhere to the manufactured specification limits. Out-of-tolerance cold plates can be caused by manufacturing variations in fin stock, that is, number of fins per inch, height, and thickness, and by bonding or brazing differences due to material thickness variations. A quality cold-plate product includes strict adherence to the pressure-flow curve defined in the specification.

Flow bench tests are performed early in the design to establish the flow impedance curves of cold plates. Successful designs are the product of fully utilizing the available pressure drop as heat transfer conversion within fin stock. This requirement demands a minimum of inviscid pressure loss, which is not always possible due to the distribution of coolant throughout a system. Operation over wide temperature extremes of the coolant is an important design consideration with respect to heat transfer since conductance values may vary considerably. As an example, operation at nominal conditions in the turbulent region may prove satisfactory, whereas a lower inlet temperature and flow could possibly place the operation in the laminar region, resulting in possibly higher cold-plate temperatures. Hence the importance of flow bench tests is indicated as a means of establishing the flow regime during system design. Unfortunately the most direct method in determining design adequacy, namely, heat transfer measurements under heat load conditions, is seldom if ever performed as a separate test because of cost and time constraints. It is only in the most critical of applications, that complete testing of both flow impedance and heat transfer are conducted.

REFERENCES

1. J. S. Kilby, "Invention of the Integrated Circuit," *IEEE Trans. Electron Dev.,* vol. ED-23, pp. 648–654, 1976.

2. A. Bar-Cohen and A. D. Krauss, *Advances in Thermal Modeling of Electronic Components and Systems,* vol. 1, Hemisphere Publ., Washington, D.C., 1988.

3. D. S. Steinberg, *Cooling Techniques for Electronic Equipment,* Wiley, New York, 1980.

4. A. W. Scott, *Cooling of Electronic Equipment,* Wiley, New York, 1974.

5. R. F. David, "Computerized Thermal Analysis," *IEEE Trans. Parts, Hybrids, Packag.,* vol. PHP-13, Sept. 1977.

6. R. Evans and R. B. Lannon, "The Effects of Reduced Atmospheric Pressure on Thermal Contact Resistance and Electronic Component Forced Air Film Coefficients," Rep. TR/6042/C80/04, U.S. Dept. of the Navy, 1980.

7. M. M. Yovanovich and V. W. Antonetti, "Application of Thermal Contact Resistance Theory to Electronic Packages," in A. Bar-Cohen and A. D. Krauss (Eds.), *Advances in Thermal Modeling of Electronic Components and Systems,* vol. 1, Hemisphere Publ., Washington, D.C., 1988, chap. 2.

8. A. I. Brown and S. M. Marco, *Introduction to Heat Transfer,* 2d ed., McGraw-Hill, New York, 1951.

9. A. Bar-Cohen and W. M. Rohsenow, "Thermally Optimum Spacing of Vertical, Natural Convection Cooled, Parallel Plates," *J. Heat Transfer,* Feb. 1984.

10. W. M. Kays and A. L. London, *Compact Heat Exchangers,* McGraw-Hill, New York, 1984.

11. E. N. Seider and C. E. Tate, "Heat Transfer and Pressure Drop of Liquids in Tubes," *Ind. Eng. Chem.,* vol. 28, p. 1429, Dec. 1936.

12. F. W. Dittus and L. M. K. Boelter, *Univ. of California (Berkeley) Pub. Eng.,* vol. 2, p. 443, 1930.

13. J. P. Holman, *Heat Transfer,* 3d ed., McGraw-Hill, New York, 1972.

14. J. D. Knudsen and D. L. Katz, *Fluid Dynamics and Heat Transfer,* McGraw-Hill, New York, 1958.

15. *Motorola MECE Integrated Circuits,* ser. D, Tech. Inform. Ctr., Motorola Semiconductor Products, 1985.

16. M. Wills, "Thermal Analysis of Air-Cooled PCBs, pt. 1," *Electron. Prod.,* pp. 11-E8, May 1983.

17. "Flow of Fluids through Valves, Fittings and Pipe," *Crane Tech. Paper 410,* 18th printing, Crane Co., 1979.

18. R. C. Chu and R. E. Simons, "Thermal Management of Large Scale Digital Computers," in R. T. Howard, S. S. Furkay, R. F. Kilburn, and G. Monti, Jr. (Eds.), *Thermal Management Concepts in Microelectronic Packaging from Component to System,* ISHM, Silver Spring, Md., 1984, p. 197.

19. S. M. Miner, "Maximizing Electronic System Reliability through Optimized Distribution of System Coolant," presented at the Winter Annual Meeting, San Francisco, Calif., Dec. 10–15, 1989, ASME paper 89-WA/EEP-10.

20. F. Altoz, P. Brach, and D. Rosen, "Reliability Optimization—A Method for Thermal Designs," in *Proc. 1982 Reliability and Maintainability Symp.* (Los Angeles, Calif., Jan. 26–28, 1982).

21. D. Dancer and M. Pecht, "Component Placement Optimization for Convectively Cooled Electronics," *IEEE Trans. Reliability* vol. 38, June 1989.*

22. *Catalog Data Reference Book—Rotron Custom Airmovers,* 5th ed., Rotron, Woodstock, N.Y.

23. A. R. Weiting, "Empirical Correlations for Heat Transfer and Flow Friction Characteristics of Rectangular Offset-Fin Plate-Fin Heat Exchangers," *ASME J. Heat Transfer,* pp. 488–490, Aug. 1975.

24. A. Bar-Cohen, "Thermal Management of Air and Liquid-Cooled Multichip Modules," *IEEE Trans. Components Hybrids Manuf. Technol.,* vol. CHMT-10, June 1987.

25. Y. C. Lee, H. T. Ghaffari, and J. M. Segelken, "Internal Thermal Resistance of a Multi-Chip Packaging Design for VLSI-Based Systems," in *Proc. IEEE 38th Electronics Components Conf.,* pp. 293–301, 1988.

26. S. S. Furkay, "Thermal Characterization of Plastic and Ceramic Surface-Mount Components," presented at the IEEE 4th Annual Semi-Therm Conf., San Diego, Calif., Feb. 1988.

27. J. A. Andrews, "Package Thermal Resistance Model: Dependency on Equipment Design," presented at the IEEE 4th Annual Semi-Therm Conf., San Diego, Calif., Feb. 1988.

28. A. Bar-Cohen, R. Eliasi, and T. Elperin, "θ_{jc} Characterization of Chip Packages—Justifications, Limitations and Future," presented at the IEEE 5th Annual Semi-Therm Conf., 1989.

29. 29. B. S. Siegal, "Factors Affecting Semiconductor Device Thermal Resistance Measurements," in *Proc. IEEE 4th Annual Semi-Therm Conf.* (San Diego, Calif., 1988), pp. 12–18.

30. A. Pasqualoni, J. Crane, and M. Diorio, "Thermal Dissipation Characteristics of a Non-Hermetic Dual-in-Line Package," in *Proc. IEEE 4th Annual Semi-Therm Conf.* (San Diego, Calif., 1988), pp. 67–77.

31. P. B. Wesling, "Thermal Characterization of a 149-Lead VLSI Package with Heat Sink," in *Proc. IEEE 4th Annual Semi-Therm Conf.* (San Diego, Calif., 1988), pp. 62–65.

32. T. Hopkins, C. Cognetti, and R. Tiziana, "Design with Thermal Impedance," in *Proc. IEEE 4th Annual Semi-Therm Conf.* (San Diego, Calif., 1988), pp. 55–61.

33. D. P. Kennedy, "Heat Conduction in a Homogeneous Solid Circular Cylinder of Isotropic Media," Tech. Rep. TR00.15072.699, IBM Corp., Dec. 1959.

34. E. M. Sparrow, A. Yanezinoreno, and D. R. Otis, "Convective Heat Transfer Response to Height Differences in an Array of Block-Like Electronic Components," *Int. J. Heat Transfer,* vol. 27, no. 3, 1984.

35. R. J. Moffat, D. E. Arvizu, and A. Ortega, "Cooling Electronic Components: Forced Convection Experiments with an Air-Cooled Array, presented at the 23rd AICHE-ASHE National Heat Transfer Conf., Denver, Colo., 1985."

36. *Fluid Flow Data Book,* Genium Publ., Schenectady, N.Y., 1984.

37. G. Kobayashi, M. Fukushima, O. Tanaka., and S. Mitsumoto, "Boiling and Condensing Heat Transfer for Cooling of High Power Semiconductor Rectification Equipment," *Mitsubishi Elec. Eng.,* pp. 24–33, 1974.

38. K. A. Park and A. E. Bergles, "Boiling Heat Transfer Characteristics of Simulated Microelectronic Chips with Fluorinert Liquids," Tech. Rep. HTL-40, College of Eng., Iowa State University, Ames, Aug. 1986.

39. K. Yokouchi, N. Kamehara, and K. Nuva, "Immersion Cooling for High-Density Packaging," presented at the IEEE 37th Electronic Component Conf., Boston, Mass., May 11–13, 1987.

40. R. D. Danielson, L. Tousignant, and A. Bar-Cohen, "Saturated Pool Boiling Characteristics of Commercially Available Perfluorinated Inert Liquids," in *Proc. 1987 ASME/JSME Thermal Engineering Joint Conf.,* vol. 3, pp. 419–430.

41. R. Armstrong, "Evaporative Cooling with Freon Dielectric Liquids," Bull. EL-11, E. I. du Pont de Nemours & Co., 1966.

42. W. M. Rohsenow, "Pool Boiling," in G. Hetsroni (Ed.), *Handbook of Multiphase Systems,* McGraw Hill–Hemisphere, New York, 1982.

43. S. Oktray and A. F. Schmeckenbecher, "Preparation and Performance of Dendritic Heat Sinks," presented at the Annual Meeting of the American Electrochemical Society, San Francisco, Calif., May 12–17, 1974, paper 81.

44. M. Jakob and G. Hawkins, *Elements of Heat Transfer,* 3d ed., Wiley, New York, 1957.

45. N. Zuber, M. Tribius, and J. W. Westwater, "The Hydrodynamic Crisis in Pool Boiling of Saturated and Subcooled Liquids," in *International Development in Heat Transfer,* no. 27, ASME, 1963.

46. K. A. Park and A. E. Bergles, "Heat Transfer Characteristics of Simulated Microelectronic Chips under Normal and Enhanced Conditions," Heat Transfer Lab. Rep. HTL35, ISU-ERI-Ames-86211, Iowa State University, Ames, 1985.

47. J. S. Goodling, N. Williamson, R. C. Jaeger, C. D. Ellis, and T. D. Slagh, "Wafer Scale Cooling Using Jet Impingement Boiling Heat Transfer," presented at the ASME Winter Meeting, Boston, Mass., Dec. 13–18, 1987.

48. C. F. Ma and A. E. Bergles, "Boiling Iet Impingement Cooling of Simulated Microelectronic Chips," in *Heat Transfer in Electronic Equipment,* vol. 28, 1983.

49. S. W. Chi, *Heat Pipe Theory and Practice—A Sourcebook,* McGraw-Hill–Hemisphere, New York, 1976.

50. A. Basilius and C. P. Minning, "Improved Reliability of Electronic Circuits through the Use of Heat Pipes," presented at the 37th National Aerospace and Electronics Conf., Dayton, Ohio, May 20–24, 1985.

51. W. B. Bienert and E. A. Skrabek, *Heat Pipe Design Handbook,* NASA Rep., Contract NAS9-11927, Dynatherm Corp., Aug. 1972.

52. *Proc. Printed Wiring Board (PWB) Workshop* (Mar. 25–26, 1986), Hughes Aircraft Co., Torrance, Calif.

53. H. J. Tanzer, K. D. Gier, and C. T. Numa, "PWB Heat Pipe Substrates Subjected to Dynamic Environments," Test Rep., Hughes Aircraft Co., Electronic Dynamics Div., Mar. 1989.

54. L. M. Jigi and Z. Dagan, "Experimental Investigation of Single Phase Multi-Jet Impingement Cooling of an Array of Microelectronic Heat Sources," in W. Aung (Ed.), *Proc. Int. Symp. on Cooling Technology for Electronic Equipment* (Honolulu, Hawaii March 1987), pp. 265–283.

55. D. B. Tuckerman and R. F. W. Pease, "Ultra-high Thermal Conductance Microstructures for Cooling Integrated Circuits," *in Proc. IEEE 32nd Electronics Components Conf.,* pp. 145–149, 1982.

56. R. J. Phillips, "Forced Convection, Liquid-Cooled, Microchannel Heat Sinks," Tech. Rep. 787, AF Contract F19628-85-C-W2, Lincoln Laboratory, Massachusetts Inst. of Technol., Cambridge, Mass., 1988.

57. J. Harley, J. Phahler, and H. Baw, "Transport Processes in Micron and Submicron Channels," presented at the ASME Winter Meeting, San Francisco, Calif., Dec. 10–15, 1989.

58. L. E. Frank, "Cinda Enhancements at Hughes for High Power Density Electronic Packaging," presented at the 23rd National Heat Transfer Conf., Denver, Colo., Aug. 4–7, 1985.

RECOMMENDED READINGS

1. *Multichip Module Design, Fabrication, and Testing,* James J. Licari, McGraw-Hill, 1995.

2. *Ball Grid Array Technology,* by John H. Lau, McGraw-Hill, 1996.

CHAPTER 3

CONNECTOR AND INTERCONNECTION TECHNOLOGY

Robert S. Mroczkowski

AMP Incorporated
Harrisburg, PA

3.1 CONNECTOR OVERVIEW

This chapter provides an introduction to electrical/electronic connectors. Both the function and structure of connectors will be discussed. Due to space limitations the discussion will be limited. References are provided for those interested in more detailed discussions of individual topics.

3.1.1 Connector Function

A functional definition of a connector is:

> An electromechanical device which provides a separable interface between two electronic subsystems without an unacceptable effect on signal integrity or loss in power.

The key elements in this definition are "electromechanical," "separable," and "unacceptable." A connector is an electromechanical device because it uses mechanical means, spring deflections, to create an electrical interface. These separate functions also tie into the other two key words. Providing "separability" is basically a mechanical function, while ensuring that the connector does not introduce "unacceptable" electrical effects on the system is an electrical function which is dependent on the mechanical stability of the connector contact interfaces.

Separability is the major reason for using a connector, and it may be required for a number of reasons, among them being ease in manufacturing of subassemblies, repair or upgrading or, increasingly, portability of electronic equipment. Separability requirements are generally stated in terms of the mating force and number of mating cycles a connector must support without degradation. Mating forces become increasingly important as the number of positions in the connector, or pin count, increases. Connectors are now available with over a thousand positions. Depending on the function of the connector, the required number of mating cycles can vary from a few to several thousand.

In this chapter, the major electrical parameter to be discussed will be connector resistance. The allowed change in resistance over the application life of the connector will be used as the measure of "unacceptable." The meaning of "unacceptable" is strongly application dependent. In particular, the allowed resistance change depends strongly on whether the connector is used in a signal or power application. Signal applications are, in general, more tolerant of increases in resistance than power applications.

3.1.2 Connector Applications: Levels of Interconnection

Levels of interconnection is a concept that allows for the description of connectors in terms of their application. While a number of levels of interconnection approaches exist, this chapter will follow that of Granitz.[1] In this approach the levels of interconnection are defined by the two points within the electronic system which are being connected and not the connector type or function. The six levels of interconnection are illustrated and described in Fig. 3.1.

The levels of interconnection concept highlights several different connector issues and requirements. In general, as the level of interconnection increases the number of mating cycles a connector will experience increases. The exposure of the connector to untrained users increases in the same manner. The standards which a connector must meet also vary with the level of interconnection. Currently standards requirements are more detailed at levels 2 and 5 and 6 than at the other levels. Standards on levels 3 and 4 are, however, increasing. Levels 3 through 6 are the levels which include more "traditional" connectors. Pin count requirements peak at level 3 and, in general, decrease from levels 4 through 6. It should also be noted that some connector types, in particular cable connectors, see applications at more than one level appearing at levels 4 and 5. The discussion in this chapter

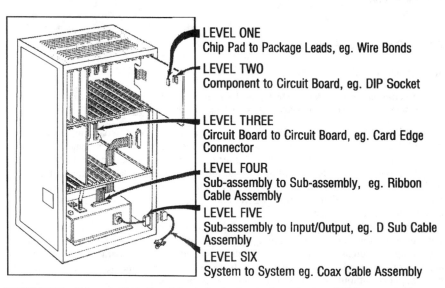

LEVEL ONE
Chip Pad to Package Leads, eg. Wire Bonds
LEVEL TWO
Component to Circuit Board, eg. DIP Socket

LEVEL THREE
Circuit Board to Circuit Board, eg. Card Edge Connector
LEVEL FOUR
Sub-assembly to Sub-assembly, eg. Ribbon Cable Assembly
LEVEL FIVE
Sub-assembly to Input/Output, eg. D Sub Cable Assembly
LEVEL SIX
System to System eg. Coax Cable Assembly

FIGURE 3.31   Schematic illustration of the six levels of packaging. *(Courtesy AMP Incorporated.)*

will be limited to levels 3 through 6, the levels in which traditional connectors are used. The interconnection devices in level 2 are more properly described as *sockets*. Socket requirements, however, are becoming increasingly similar to those of connectors as socket pin counts increase.

3.1.3   Connector Types

One method of characterizing connector types follows the levels of interconnection approach in that the connector type is defined by the permanent connection. According to this scheme there are three major connector types:

- board-to-board, levels 3 and 4
- wire-to-board, levels 4 and 5
- wire-to-wire, levels 5 and 6

In this discussion, the term *wire* also includes cables of various constructions. A brief description of each type follows, with additional discussion to follow in Section 3.9.

3.1.3.1 Board-to-Board Connectors   Board-to-board connectors have experienced the greatest impact of the advances in microprocessor technology. Tremendous increases in chip functionality and, therefore, input/output (I/O) requirements require similar increases in pin count in level 3 connectors. Level 3 connectors have also experienced more demanding requirements in impedance control because they are closest to the chip and face the greatest demands on maintaining signal integrity. These two requirements, especially pin count, have led to a transition from card-edge connectors to multi-row, two-piece connectors to satisfy high-density connector requirements.

3.1.3.2 Wire-to-board Connectors   Wire-to-board connectors, particularly at level 4, may also face controlled impedance requirements. These connectors are increasingly cable assemblies rather than discrete wire to more effectively address the wire handling requirements for high-pin-count connectors.

3.1.3.3 Wire-to-wire Connectors   Wire-to-wire connectors are generally farther away from printed wiring boards and chips and, with the exception of coaxial connectors, face less demanding impedance requirements. Wire-to-wire connectors are often external to equipment and, therefore, ruggedness and grounding/shielding requirements may become important considerations.

3.1.4   Connector Applications: Signal and Power

The two basic connector applications are *signal* and *power*. Signal applications are characterized by low current and voltage requirements. Power contacts/connectors, in contrast, generally address higher current and, often, higher voltage applications.

3.1.4.1 Signal Applications   Signal applications span the range of currents from microamps to hundreds of milliamps, with this range expanding in both directions as electronic applications increase in complexity. Driving voltages for signal applications are generally a few volts.

An additional requirement facing signal connectors is impedance control, as mentioned in the previous section. Connector design for impedance control is a major area of development in connector technology. Approaches using ground pins, open pin field connectors, and ground planes are being developed (Section 3.8)

3.1.4.2 Power Applications   Power contacts/connectors face an additional requirement: thermal management. Managing the Joule, or I^2R, heating which accompanies current flow becomes an important design/selection criterion.

Two approaches to power distribution are dedicated power contacts and using multiple "signal" contacts in parallel. The application considerations for these two approaches differ significantly and will be discussed in Section 3.8.

This completes the overview of connector function. Attention turns to connector structure—how the design and materials of manufacture of a connector affect connector performance.

3.1.5   Connector Structure

A connector can be described as consisting of four components as outlined by Mroczkowski (2) and schematically illustrated in Fig. 3.2. At this point, each component will be considered in an overview fashion with a more detailed discussion to follow.

3.1.5.1 Contact Interface   There are two types of contact interfaces that must be considered: separable and permanent. The separable interface, created when the plug and receptacle halves of the connector are mated, has already been mentioned as the primary reason for using a connector. Permanent interfaces must also be created between the connector and the functional elements that are being connected. Both metallurgical (soldered and brazed/welded) and mechanical (crimped, insulation displacement and compliant pin) permanent connections are widely used. Figure 3.2 illustrates a crimped connection and a compliant press-in connection. Separable connections will be discussed in Section 3.6 and permanent connections in Section 3.7. A stable, low-resistance contact interface requires that a metallic contact interface be established. Control and elimination of surface films is a major factor in creating such an interface. It is the contact finish which has the strongest impact on the formation and disruption of surface films.

3.1.5.2 Contact Finishes   The contact finish has two major functions, to provide corrosion protection for the contact spring and "optimizing" film management to allow the creation of a metallic interface. The two basic classes of contact finishes—noble (gold and other precious metals) and non-noble (tin, silver and nickel)—differ significantly in the design considerations for, and mechanisms of, film management. Contact finishes will be discussed in Section 3.3.

3.1.5.3 Contact Spring   The contact spring also has two functions. Electrically, the contact spring provides continuity between the permanent and separable connections. To achieve acceptable performance the contact spring must be metallic. Mechanically, the contact spring must be capable of creating the contact forces which establish and maintain the separable and permanent contact interfaces. The contact spring must also be capable of forming the desired permanent connection, mechanical or metallurgical. Mechanical permanent connections may require severe forming, such as in crimped connections, or spring characteristics, such as in IDC connections. Metallurgical connections, in particular soldered connections, usually rely on a plating, generally tin-lead, on the spring to maintain solderability.

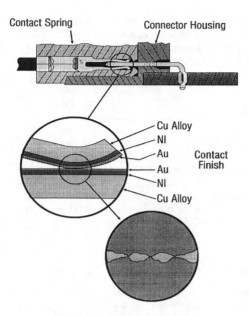

Cu Alloy
NI
Au
Au
NI
Cu Alloy

Contact
Finish

FIGURE 3.32 Schematic cross section of a connector. Four components of the connector are illustrated and described in the text. *(Courtesy AMP Incorporated.)*

Copper alloys are the dominant contact spring materials due to a good combination of mechanical and electrical characteristics as will be discussed in Section 3.4.

3.1.5.4 Connector Housing The connector housing also performs electrical and mechanical functions. Electrically the housing insulates the contacts from one another. Mechanically, the housing captures and supports the contact springs and maintains the contact spacings of both the separable and permanent connections. Dimensional control of the separable interface must be maintained to ensure appropriate mating of the connector. Dimensional control of the permanent interface is necessary to ensure proper registration of the permanent connections with the wire/cable or *printed wiring board* (PWB). A variety of thermoplastic polymers with acceptable electrical and mechanical properties are used in connectors. Assembly requirements—in particular, surface mounting technology—strongly influence polymer selection for connector housings. The application temperature is also an important consideration in material selection. Connector housings will be discussed in Section 3.5.

3.1.5.5 Connector Design The performance and reliability of connectors depends on the material selection and design of these four connector components. For separable connections, the contact finish, the normal force and mating interface geometries determine the connector contact resistance as well as the mating force and mating durability of the connector. For permanent connections minimizing the magnitude of connection resistance and maximizing its stability are major considerations. Ensuring formation of a metallic interface and a suitable residual force distribution through controlled deformation of the conductors and connectors are dominant considerations for mechanical permanent connections. The connector housing must maintain its insulating characteristics and dimensional integrity throughout the manufacturing/assembly process and over the desired operating life of the connector. Selection of an appropriate connector for a given application requires consideration of the operating environment, mechanical, chemical and thermal, as well as the electrical requirements the connector must meet.

3.2   THE CONTACT INTERFACE

It can be argued that the structure of the contact interfaces determine the electrical and mechanical performance of a connector. In fact, a not completely facetious description of a connector has been given as "contact interfaces held together by supporting structures" [Whitley, private communication]. An understanding of the basic structure of contact interfaces is necessary to understand their importance in connector performance. The following discussion will be expressed in terms of separable interfaces, but similar considerations apply to the interfaces of permanent mechanical connections.

3.2.1   Contact Interface Morphology and Contact Resistance

As mentioned previously, the contact interface is created when the plug and receptacle contacts are mated. Williamson[3] provides an informative analogy for the process of creating a contact interface in terms of bringing two mountainous regions in contact, e.g., Vermont on top of New Hampshire. This analogy illustrates the importance of surface roughness on contact interface formation. Figure 3.3 schematically illustrates the creation of a contact interface. Only the high spots on the surface, called *asperities* or *a-spots,* actually come in contact. Due to their small size and radii the asperities deform plastically even at low applied loads.[4] The asperities deform to create a contact area sufficient to support the applied load. For typical connector interfaces only a small portion of the surfaces, a percent or less, are in contact. The distribution of the contact spots is determined by the geometries of the mating surfaces. For example, as illustrated in Fig. 3.3, spherical surfaces in contact will result in a circular a-spot distribution.

This interface morphology gives rise to an electrical resistance, termed constriction resistance, at the contact interface. The source of constriction resistance, as described by Holm[5] is a fundamental effect and can be illustrated by consideration of Fig. 3.4. The asperity microstructure of contact interfaces causes current flow to be "constricted" to flow only through the asperity contact spots, hence the terminology "constriction resistance." According to Holm,[5] for a single spot the constriction resistance is given by

$$R_c = \rho/2a \qquad (3.78)$$

where ρ is the resistivity of the material and a the radius of the asperity contact spot. For the purposes of this discussion, it can be stated that the asperity contact distribution affects contact resistance as illustrated in Fig. 3.5. The equations inset in Fig. 3.5 indicate the constriction resistance for a single asperity and for multiple asperities distributed over a geometric area depending on the contact geometry. The multispot equation

$$R_c = \rho/2na + \rho/D \qquad (3.79)$$

where n is the number of asperity contacts and D the diameter of the area over which the contacts are distributed is more relevant to typical contact interfaces. The first term indi-

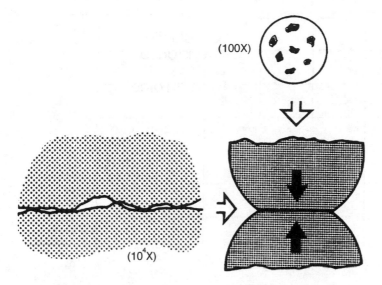

FIGURE 3.33 Schematic illustration of the morphology of a typical contact interface. *(Courtesy AMP Incorporated.)*

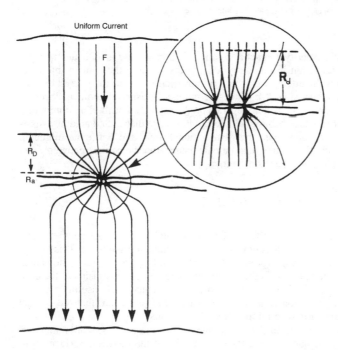

FIGURE 3.34 Schematic illustration of the current distribution at a contact interface which results in constriction resistance.

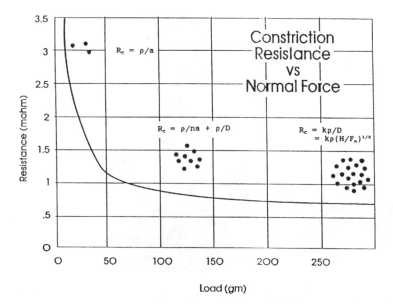

FIGURE 3.35 Typical relationship between contact resistance, R_c and contact force, F_n resulting from the variation in asperity number and distribution as contact force increases. Equations relating contact resistance to interface parameters are shown. *(Courtesy AMP Incorporated.)*

cates the effect of multiple asperity contact resistances in parallel and is indicated as R_a in Fig. 3.4. The current constriction in this case occurs very close to the contact interface. The second term indicates the effect of the current constriction due to the overall distributed contact area and is indicated as R_D. When the number of asperity contacts becomes large, several tens, the second term, which depends on the asperity distribution, dominates. In such cases the contact interface behaves as though the entire distributed area of a-spots is conducting. Under these conditions, the distributional area, and hence its diameter, assuming a circular asperity distribution, can be calculated from the contact geometry and hardness of the materials in contact according to:

$$R_c = k\rho(H/F_n)^{1/2} \tag{3.80}$$

where k is a coefficient including the effects of surface roughness, contact geometry and elastic/plastic deformation, H is the hardness and F_n the contact normal force. This simple equation, however, has a complex interpretation in that, in addition to the interactions in k, in real cases the appropriate hardness is the composite hardness of the finish and the contact spring. Despite these limitations the basic form of the equation is useful in understanding material/design effects on constriction resistance, in particular, that the contact normal force is a major factor in determining contact resistance.

It should also be noted that the calculation assumes metallic contact; that is, any surface films are completely displaced due to the deformation of the surfaces. Surface films, whether present initially or occurring during connector application, are a major degradation mechanism for contacts interfaces. The contact finish is a major factor in film management because it determines the types of films which will form and how readily they can be disrupted. This topic will be discussed in more detail in Section 3.3.

If surface films are not displaced, they result in a resistance in series with the constriction resistance. Film resistances can be very high and are highly variable. The variability arises from a combination of several factors. Film resistance depends on the thickness, composition, and structure of film. Each of these factors, in turn, depends on the conditions under which the film was formed. The composition of the environment, temperature and humidity are particularly influential in determining film characteristics. For these reasons, avoiding film effects by disruption and displacement is an important aspect of connector design. The contact normal force and geometry are two design parameters that strongly affect the effectiveness of film displacement.

3.2.2 Contact Interface Morphology and Mechanical Performance

The same asperity contact distribution responsible for constriction resistance determines the mechanical characteristics of friction and wear at the contact interface. Friction, in turn, influences the connector mating force and the mechanical stability of the contact interface. Wear processes impact on the number of mating cycles a connector can experience before showing any effect on performance. Both friction and wear processes depend on asperity contact interfaces as illustrated in Fig. 3.6.

As discussed in Bowden and Tabor,[4] friction and wear processes depend on the location at which the interfaces separate during relative motion. Plastic deformation of asperity contacts can lead to cold welding at the asperity interface. In fact, the strength of the cold welded interface may be higher than the cohesive strength of the base metal due to the work hardening that occurs during deformation. Two junctions are illustrated in Fig. 3.6. Junction *a* is "stronger" than junction *b* having experienced a larger amount of plastic deformation, therefore, cold welding. Under a force tending to shear the contact interface, junction *a* may break away from the original interface resulting in a wear particle and

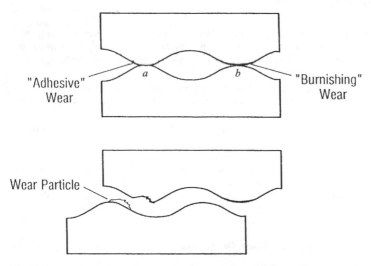

FIGURE 3.36 Schematic illustration of the asperity structure of contact interfaces as it relates to friction and wear mechanics. *(Courtesy AMP Incorporated.)*

metal transfer, as shown in the lower part of Fig. 3.6. Junction *b*, with lower deformation, and, therefore, less work hardening and cold welding, may separate at or near the original interface so that little wear or metal transfer results.

The wear process at junction *a* is referred to as adhesive or galling wear, while that at junction *b* is burnishing wear. Wear tracks in the two cases are rough or smooth, respectively. It should also be noted that if the transferred wear particle from junction a breaks loose it may act as an abrasive at the interface and lead to a third wear mechanism, abrasive wear as discussed in Antler.[6] The same process, separation of cold welded junctions under a shear stress, is the source of friction. The friction force, in fact, is the force necessary to shear the multiple cold welds which occur at contact interfaces. Friction forces at the separable interface are a major factor in determining connector mating force, as will be discussed in Section 3.6.

3.2.3 Summary

The asperity model of contact interfaces provides insight into both the electrical and mechanical characteristics of contact interfaces. Contact interface morphology depends, simply speaking, on the surface roughness, the force on the interface and the geometry of the surfaces in contact. Surface roughness strongly influences the number of asperity contacts created. The force on the interface, the contact normal force, and the hardness of the surface determine the total contact area. As mentioned, it is a composite hardness of the contact finish and the spring material that determines the effective hardness. The geometries of the contact springs determines the area over which the asperities are distributed. This explains why the contact finish, contact normal force, and geometry are major design parameters, and each will be discussed in more detail—contact finishes in Section 3.3 and contact force and geometry in Section 3.6.

3.3 THE CONTACT FINISH

As mentioned earlier, the two major reasons for using a contact finish are: corrosion protection for the base metal of the contact spring and optimization of the properties of the contact surface with respect to establishing and maintaining the contact interface, in particular, a metallic contact interface.

Section 3.2 discussed the structure of the contact interface and its relationship to constriction resistance and mechanical durability. Surface films and film resistance were briefly mentioned. But film management, during mating as well as during use, is arguably the most critical factor in connector performance. The selection of the contact finish is one of the basic means of controlling film effects as will be discussed.

In the great majority of connectors the contact finish is applied by electroplating. The electroplating process will not be discussed here. Basic references relevant to connector electroplating include Durney[7] and Reid and Goldie.[8] Clad or inlay finishes see limited use in connectors. An overview of cladding processes relevant to connectors is available in Harlan.[9]

3.3.1 Contact Finish and Corrosion Protection

In most cases, connector contact springs are made from copper alloys that are subject to corrosion in typical connector operating environments—through oxidation or sulfidation, for example. Application of a contact finish, in effect, seals off the contact spring from the environment and prevents corrosion. For this function, the contact finish itself must be corrosion resistant, either intrinsically (noble metal finishes) or by forming passivating surface films (tin and nickel).

3.3.2   Contact Finish and Interface Optimization

Optimization of the contact interface in this context is film management. There are two approaches to film management, first, avoiding film formation and, second, disruption of the films which may exist during the mating of the connector. These two approaches define the difference between noble, or precious metal, and non-noble finishes.

Noble finishes—gold, palladium and some alloys of these metals—are intrinsically free, or relatively free, of surface films. Under such conditions, metallic contact is straightforward, at least initially. The potential for film formation at other points in the connector system, in particular exposed contact spring surfaces, must be given consideration, however. Factors that must be considered include maintaining the nobility of the contact surface against extrinsic factors such as contamination, base metal diffusion, and contact wear. As will be discussed, nickel underplates are important in providing such protection.

Non-noble finishes, in particular tin, are covered with a surface film, usually an oxide. The utility of tin contact finishes derives from the fact that the oxide film is easily disrupted on mating and metallic contact areas are readily established. The application concern with tin finishes is reoxidation of the tin, in the form of fretting corrosion, the major degradation mechanism for tin contact finishes.

Silver is best considered as non-noble since it is subject to sulfide and chloride corrosion. Nickel, too, is non-noble because it readily forms a surface oxide.

The difference between noble and non-noble finishes is reflected in a number of connector design considerations, as will be discussed.

3.3.3   Noble Finish Overview

Noble metal contact finishes are systems in which each component performs multiple functions. A noble metal contact finish system consists of a noble metal surface, usually gold, over an underplate, usually nickel, over the contact spring material, generally a copper alloy. Typical thicknesses of the platings are 0.4 to 0.8 μm for the gold, or in some cases gold over palladium alloy, and 1.25 to 2.5 μm for the nickel underplate.

The function of the noble metal surface is to provide a film free metallic contact surface. But the presence of a precious metal surface does not in itself guarantee a film free surface. The finish must be continuous. Discontinuities in the finish, such as porosity and scratches, can result in corrosion at sites where base metal is exposed. Wear through of the noble surface metal also serves to expose base metal. Wear through of the finish can occur through mating of the connector or mechancially or thermally induced relative motions of the contacts during application. Such small scale motions, of the order of a few to a few tens of micrometers, are referred to as fretting. The importance of fretting action is that it can lead to contact resistance degradation under a variety of application conditions as will be discussed.

Diffusion of the base metal consituents of the contact spring to the surface can also result in surface films. Each of these potential degradation mechanisms is moderated by the nickel underplate.

3.3.3.1 Noble Finishes and the Nickel Underplate   For a detailed discussion of the functions of a nickel underplate see Mroczkowski.[10] At this point, a summary of the benefits of such an underplate will suffice.

1. Nickel, through the formation of a passive oxide surface, seals off the base of pore sites and surface scratches reducing the potential for corrosion of the underlying copper alloy contact spring.
2. Nickel provides an effective barrier against the diffusion of base metal consititents to the contact surface where they could result in films.[11]
3. Nickel provides a hard supporting layer under the noble surface which improves contact durability.[6]
4. Nickel provides a barrier against the migration of base metal corrosion products reducing the potential of their contaminating the contact interface.[10]

The first three benefits allow for equivalent or improved performance at reduced noble metal thicknesses. The effects of discontinuities have been moderated, the noble metal is not used as a diffusion barrier and the durability is improved. The fourth benefit allows a reduction in the size of the contact area which must be covered with the noble metal finish. All these functions serve to maintain the nobility of the contact surface finish at a reduced cost.

The most common noble metal contact finishes are gold, palladium and alloys of these metals. These finishes vary in their degree of "nobility" and will be considered separately.

3.3.3.2 Gold   Gold provides an ideal contact finish due to excellent electrical and thermal characteristics and corrosion resistance in virtually all environments. Pure gold, however, is relatively soft and gold contact platings are usually alloyed, with cobalt being the most common, at levels of a few tenths of a percent to increase the hardness and wear resistance. A typical hardness for soft gold is in the range of 90 $Knoop_{25}$ while cobalt hardened gold contact finishes show hardnesses in the range of 130 to 200 $Knoop_{25}$. Because of this combination of characteristics gold is the dominant contact finish for connectors which must provide high reliability in demanding applications.

Cost reduction objectives have led to the development of selective plating practices (to reduce the required area which must be plated) and reductions in thickness (permitted by the use of nickel underplates) to reduce gold finish costs. Further cost reductions can be realized by the use of alternative noble metals, most commonly gold flashed palladium or palladium alloy.

3.3.3.3 Palladium   Palladium is also a noble metal, but does not equal gold in corrosion resistance or electrical/thermal conductivity. It is, however, significantly harder than gold (200 to 400 $Knoop_{25}$), even without alloying, which improves durability performance. Palladium can catalyze the polymerization of organic deposits and result in contact resistance degradation under fretting motions. Palladium is, therefore, not as noble as gold although the effect of these factors on connector performance depends on the operating environment as discussed by Antler and Sproles.[12] Whitley, Wei, and Krumbein[13] state that, in most applications, palladium is used with a gold flash (of the order of 0.1 μm) to provide a gold contact interface.

3.3.3.4 Noble Metal Alloys   The performance of gold alloys in connector applications has been mixed. The major problems have occurred due to the loss of corrosion resistance which often accompanies alloying using base metals. When this factor is accounted for, or avoided by selection of appropriate alloying agents, satisfactory performance can be realized in some connector applications. The major gold alloy used in connectors is WE1, a gold/silver/platinum (69/25/6) alloy. WE1 is available only in inlay form and sees limited use.

There are two palladium alloys in use, palladium(80)-nickel(20) and palladium(60)-silver(40) as reviewed by Whitelaw[14] and Antler, Drozdovich, and Haque,[15] respectively. The palladium-nickel alloy is electroplated and the palladium-silver finish is primarily an inlay. In general these finishes include a gold flash to counter the lower corrosion resistance of these alloys compared to gold.

3.3.4 Non-noble Finishes

Non-noble contact finishes differ from noble finishes in that they always have a surface film. The utility of non-noble finishes depends on the ease with which these films can be disrupted during connector mating and the potential for recurrence of the films during the application lifetime of the connector.

Three non-noble contact finishes, tin (in this discussion "tin" includes tin-lead alloys, most commonly 93/7 tin/lead for mechanical connections and 60/40 tin/lead for soldered connections), silver and nickel are used in connectors. Tin is the most commonly used non-noble finish; silver offers advantages for high-current contacts, and nickel is used in high-temperature applications.

3.3.4.1 Tin   The utility of tin as a contact finish is a result of the ease which the tin oxide surface is disrupted and displaced. The mechanism of disruption is schematically illustrated in Fig. 3.7.

The oxide disruption is facilitated by the large disparity in mechanical properties between the tin and tin oxide. The tin oxide is thin (a few tens of nanometers), hard and brittle. In contrast the underlying tin, generally in the range of 2 to 5 μm in thickness is soft and ductile. This combination of properties results in cracks in the oxide under an applied load. The load transfers to the soft ductile tin which flows easily, opening the cracks in the oxide, and tin then extrudes through the cracks to form the desired metallic contact regions.

Generation of a tin contact interface for tin finishes is, therefore, relatively simple. Normal force alone may be sufficient, but the wiping action that occurs on mating of connec-

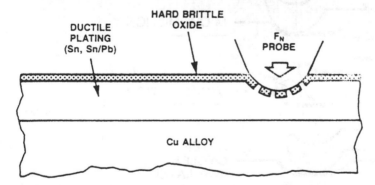

FIGURE 3.37   Schematic illustration of the mechanics of tin oxide disruption under an applied force. *(Courtesy AMP Incorporated.)*

tors virtually ensures oxide disruption and creation of a metallic interface. The potential problem with tin, however, is the tendency for reoxidation of the tin at the contact interface if it is disturbed. This process is referred to as fretting corrosion as discussed by Bock and Whitley.[16] As mentioned previously, *fretting* refers to repetitive small scale motions, of the order of a few to a few tens of micrometers. The "corrosion" aspect of fretting corrosion is the reoxidation of the surface as the tin surface is repeatedly exposed, as schematically illustrated in Fig. 3.8. The end result is a buildup of oxide debris at the contact interface leading to an increase in contact resistance.

Under severe conditions, fretting corrosion can lead to open circuits in a few hundred fretting cycles. Fretting corrosion is the major degradation mechanism for tin contact finishes. Driving forces for fretting motions include mechanical (vibration and disturbances/shock) and thermal expansion mismatch stresses.

Two approaches to mitigating fretting corrosion in tin and tin alloy finishes are high normal force (to reduce the potential for motion) and contact lubricants (to prevent oxidation). Each has been used successfully, and each has its limitations. High normal forces limit the durability capability of the finish, which is already low due to tin being very soft, and result in increased mating forces, which limit the realizable pin count in tin-finished connectors. Contact lubricants also have limitations. They require secondary operations for application, either during connector manufacture or in assembly, have limited temperature capability, and may also result in dust retention.

Tin contact finishes, however, can provide acceptable electrical performance at both low (millivolts and milliamps) and high (volts and amps) circuit conditions if fretting corrosion can be avoided. Unfortunately, the susceptibility of a connector to fretting is application dependent and often difficult to assess.

3.3.4.2 Silver   Silver is considered a non-noble contact finish due to its reactivity with sulfur and chlorine. Silver films differ in their effects on connector performance. Silver

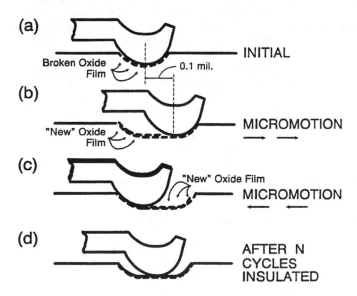

FIGURE 3.38   Schematic illustration of the mechanism of fretting corrosion. *(Courtesy AMP Incorporated.)*

sulfide films tend to be soft and readily disrupted. In addition, silver sulfide does not result in fretting corrosion. Silver chloride films, however, are harder, more adherent, and more likely to have detrimental effects on contact performance. In addition, silver is susceptible to electromigration, which can be a problem in some applications. The end result is that silver sees limited use in connectors. Despite these limitations, silver is a candidate for high-current contacts due to its high electrical and thermal conductivity and resistance to welding. Silver finish thicknesses are generally in the range of 2 to 4 μm.

3.3.4.3 Nickel In the discussion of nickel underplates, it was noted that nickel forms a passivating oxide film that reduces its susceptibility to further corrosion. This passive film also has a significant effect on the contact resistance of nickel. In this case, the favorable disparity in properties that gives tin its utility is not present. Both the nickel oxide and the base nickel are hard. Under these conditions, cracking and separation of the oxide and extrusion of the nickel to the surface are more difficult to realize. For this reason, nickel finishes require higher contact normal forces to ensure film disruption. The self-limiting oxide on nickel, however, makes it a candidate for high-temperature applications. It should be noted that nickel contact finishes are susceptible to fretting corrosion by the same mechanism as tin finishes. This potential should be taken into account in connectors using thin noble metal finishes. Wear-through of the finish due to mating cycles or fretting action, which exposes the nickel underplate, can result in fretting corrosion.

3.3.5 Selection of a Contact Finish

Selection of an appropriate contact finish depends on the application requirements the connector must meet. Of particular concern are the required number of mating cycles, the operating environment, and the electrical requirements.

3.3.5.1 Mating cycles The number of mating cycles a connector must support varies from a few to several thousand. Level 3 connectors generally require a low number of mating cycles, while level 6 requirements may run into the thousands.

The mating durability of a connector depends on the contact finish, the normal force, and the contact geometry. With respect to contact finishes, the dominant factor is the hardness of the finish. From this viewpoint, the *relative* durability performance of typical finishes is summarized in Table 3.1. It should be noted that the durability performance of a contact finish depends very strongly on the contact geometry, normal force, surface roughness, and state of lubrication of the interface. The statements in Table 3.1 are based on results from flat test coupons traversed by a spherical rider. Gold-flashed palladium or palladium alloy finishes provide the highest durability capability. Silver and tin finishes are severely limited in the number of mating cycles they can support. Contact lubricants are often used to improve the mating cycle capability of noble finishes. An appropriate contact lubricant can improve durability capability by an order of magnitude under favorable conditions.

In addition, durability depends on the finish thickness. The relationship between mating cycles and thickness is roughly linear for a given contact force/geometry configuration.

The effect of normal force on mating life has two aspects. In general, wear rates increase with mating force. In addition, the wear mechanism may change from burnishing to adhesive wear as normal force increases as discussed in Antler.[17,18]

The effect of contact geometry on mating cycles results from localization of the contact areas. High curvatures result in narrow contact wear tracks and increased wear for a given value of normal force.

TABLE 3.1   Selected Mechanical Characteristics of Typical Contact Finishes.

	Hardness (Knoop@25 g)	Durability	Coefficient of Friction
Gold (soft)	90	fair	0.4/0.7
Gold (hard)	130/200	good	0.3/0.5
Palladium	200/300	very good[*]	0.3/0.5
Palladium(80)-nickel(20)	300/450	very good[*]	0.2/0.5
Tin	9/20	poor	0.5/1.0
Silver	80/120	fair	0.5/0.8
Nickel	300/500	very good	0.3/0.6

[*] with gold flash

3.3.5.2 Operating Environment   Operating temperature and corrosion severity will obviously impact finish selection. Corrosion of contact surfaces, in particular for noble metal contact systems, has received a great deal of attention in recent years in terms of both the composition of the environments and the corrosion mechanisms in those environments. For example, see Bader, Sharma and Feder,[19] Abbott,[20,21] Mroczkowski[22] and Geckle and Mroczkowski.[23] These workers have shown that chlorine and sulfur are the major contributors to corrosion mechanisms in noble metal plated connectors. Under test conditions intended to simulate typical connector operating environments,[21] the corrosion resistance of noble metal finishes decreases in the following order: gold, palladium, and palladium(80)-nickel(20) alloy.[22] The importance of these differences will depend on the severity of the operating environment. As mentioned, the nickel underplate used in noble contact finishes reduces the corrosion susceptibility of the connector. In addition, shielding of the contact interface by the connector housing reduces corrosion by limiting access of the environment to the contact interface.

Somewhat surprisingly, tin finishes show good stability with respect to corrosion. This is due to the passivating surface oxide, mentioned previously, which protects the surface from further corrosion in most operating evironments. Fretting corrosion, however, must always be taken into consideration for tin finishes.

Temperature limitations for noble metal finishes show a similar pattern to corrosion resistance. In both cases, the order is determined by the presence of the non-noble constituent in the palladium alloys, because it is the alloying agents that are susceptible to corrosion. Hard gold finishes, even though the alloy content is of the order of tenths of a percent, are also subject to oxidation as operating temperatures increase. In general, soft, or pure, gold finishes are recommended for temperatures above 125°C.

Tin has a temperature limitation due to an increasing rate of intermetallic compound formation, a reduction in the already low mechanical strength and an enhanced oxidation rate. The interaction of these factors results in a recommendation that tin not be used above 100°C in conventional connector designs.

Thermal cycling is another important environmental consideration and is arguably the major driving force for fretting corrosion of tin systems. In addition, thermal cycling accelerates the effects of humidity on connector degradation.

3.3.5.3 Electrical Requirements   In general operating voltages for electronic connectors are decreasing, but the opposite is true for currents. While metallic contact interface performance is not dependent on circuit voltage or current, corrosion/film effects on contact resistance do show such a dependence. Unfortunately current/voltage effects on interfaces containing films are highly unstable. The voltage necessary for film breakdown is variable, being dependent on the film structure and thickness as discussed by Wagar.[24] For this reason relying on voltage, especially low voltages, to effect film breakdown is suspect. In addition, the resistance realized on dielectric breakdown may depend on the current flowing in the circuit at the time. Such variability is not acceptable in electronic connectors, so mechanical disruption of surface films and avoiding corrosion during applications are critical to connector performance.

In low-current applications, non-noble contact systems are more susceptible to noise due to interface films and variations in contact resistance as shown by Abbott and Schrieber.[25] In high-current, high-power applications where low interface resistance must be maintained, film resistance susceptibility must be minimized to avoid thermal runaway due to Joule heating as resistance increases.

In principle, because both noble and non-noble interfaces can be metallic, both systems are capable of equivalent performance. In practice, however, the greater susceptibility of the non-noble systems to film formation result in a reliability risk factor.

3.3.6 Summary

Selection of a contact finish depends on a wide range of application requirements, primarily durability and resistance stability. Precious metal finishes provide greater inherent stability, but tin finishes may be acceptable if the conditions necessary to reduce fretting susceptibility can be satisfied. Table 3.1 provides a brief compilation of contact finish characteristics of relevance to connector performance. Antler[26] provides a recent summary of contact finish practices and trends.

3.4 CONTACT SPRINGS

The contact spring performs two separate functions in a connector:

1. As a mechanical element, it provides the contact force that produces and maintains the separable contact interface and allows for the creation of the permanent connection interface.
2. As an electrical element, it carries the signal, or power, from the separable interface to the permanent interface and then to the circuit element to which it is connected.

This discussion will be directed primarily to the first function, but a few words on the second are in order.

3.4.1 Contact Spring and Electrical Requirements

The resistance of a connector consists of three components as schematically illustrated in Fig. 3.9.

R_{CONN}, is the resistance introduced at the permanent connection, whether it is soldered or mechanical. The bulk resistance, R_B is that introduced by the contact spring and depends on the spring material resistivity and the geometry of the spring. The third component, the resistance of the separable interface, is commonly referred to as the contact resistance, R_C. These three components are present in both halves of the connector, the separable interface resistance being shared in common. For a typical connector the values of these resistances are of the order of tens/hundreds of microohms, several milliohms and a milliohm, respectively.

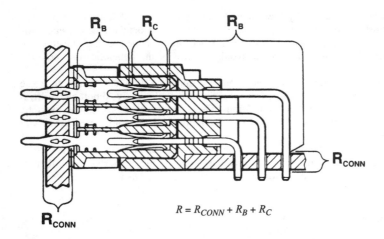

$$R = R_{CONN} + R_B + R_C$$

FIGURE 3.39 Schematic connector cross section illustrating the sources of connector resistance. Permanent connection, R_{CONN}, bulk, R_B and contact, R_C, resistance contributions are indicated. *(Courtesy AMP Incorporated.)*

The effect of the contact spring is largest in the bulk resistance, the largest component of connector resistance. The resistivity of the contact spring material, therefore, is an important property. This is particularly relevant for power applications. In such cases, millivolt drop budgets in power distribution can place severe limitations on bulk resistance. In addition, Joule heating, which depends primarily on bulk resistance, and the associated temperature rise can limit the current rating and application temperature of connectors. In signal applications the bulk resistance is of secondary importance in most cases. In such applications stability of the resistance is more important than the initial value.

It should be noted that the range of electrical resistivities of the copper alloys most commonly used in connectors is in the range of four to five times the resistivity of copper. For power contacts higher conductivity materials such as C19500, C19500 and C17410 are often selected to minimize bulk resistance contributions. Table 3.2 contains a compilation of electrical and mechanical properties of copper alloys relevant to connectors.

TABLE 3.2 Selected Properties of Typical Contact Spring Materials

	Young's modulus (E) (10^6 kg/mm^2)	Electrical conductivity (%IACS)	0.2% offset yield strength (10^3 kg/mm^2)	Stress relaxation
Brass (C26000)	11.2	28	40/60	poor
Phosphor bronze (C51000)	11.2	20	50/70	good
Beryllium-copper (C17200)	13.3	20/26	55/95	excellent

3.4.2 Contact Springs and Mechanical Requirements

In contrast to bulk resistance, which is variable only with respect to material selection for a given spring geometry, mechanical design considerations are much more complicated and involve a number of trade-offs. The important factors and trade-offs are somewhat different for separable and permanent interfaces.

3.4.2.1 Separable Interfaces   For separable interfaces, the main function of a contact spring is to provide the required contact normal force, the force between the two mating surfaces. The material properties which are important in this context are the elastic or Young's Modulus and yield strength as these properties strongly influence the deflection characteristics of the spring and the amount of deflection which can be supported while remaining elastic. Stress relaxation resistance is also important because it can reduce the contact normal force over time. A brief treatment of these variables and their effect on contact normal force is in order.

A simple cantilever beam can be used to indicate how variations in materials properties affect the contact normal force. For a cantilever beam, Fig. 3.10, the force versus deflection equation takes the form

$$F = (D/4)EW(T/L)^3 \qquad (3.81)$$

where F is the force resulting from a beam deflection D, E is the elastic modulus of the spring material, and $W, L,$ and T represent the width, length, and thickness of the beam, respectively. Many contact spring configurations approximate this cantilever geometry.

The only materials property in Equation 3.4 is the elastic modulus. The contact force for a given deflection of the beam increases as E increases. For this reason, a high value of E is generally desirable because it provides a higher normal force for a given deflection. Once again, however, for the copper alloys most often used in connectors, E falls in a relatively narrow range of 11 to 14 gigapascals, as indicated in Table 3.2.

A second equation which is useful in illustrating the effects of material properties relates the contact force to the stress in the beam:

$$F = 1/6\sigma T^2(W/L) \qquad (3.82)$$

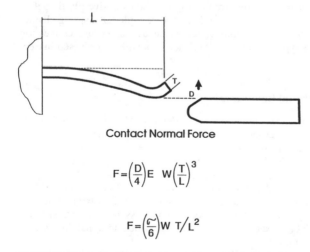

Contact Normal Force

$$F = \left(\frac{D}{4}\right)E \; W\left(\frac{T}{L}\right)^3$$

$$F = \left(\frac{\sigma}{6}\right)W \; T/L^2$$

FIGURE 3.40   Schematic illustration of a cantilever beam. Deflection of the cantilever beam provides the contact force which creates and maintains the contact interface. *(Courtesy AMP Incorporated.)*

This equation shows that the contact force, F, depends on the stress, σ, in the beam. This relationship is particularly important because the maximum normal force which can be realized by a contact spring, while remaining elastic, is determined by the maximum stress the spring can support which is the yield strength of the spring material.

Equations 3.4 and 3.5 show that the material properties having the greatest impact on contact force performance are the elastic modulus, which, in combination with the contact dimensions, determines the spring rate, and the yield strength, which determines the maximum normal force which can be achieved under elastic loading of the beam. The variation in yield strength capability among the copper alloys commonly used in connectors is significant as indicated in Table 3.2. It is also important to note that the formability of copper alloys in the higher strength tempers used in connectors also varies significantly among the alloys. Equations 3.4 and 3.5 also indicate the importance of the contact geometry as all three dimensions of the spring enter into the equations for calculating contact force. Loewenthal et al.[27] provide a detailed analysis of a contact spring design in terms of materials properties. The range of contact spring designs in connectors, for both separable and permanent connections is very large. A few examples will be discussed in Section 3.6.

Another important materials property relevant to contact force is stress relaxation. The effect of stress relaxation is apparent in Equation 3.5. Stress relaxation causes a reduction of the stress in the beam under load as a function of time and temperature due to the conversion of elastic strain to plastic strain. The reduction in stress results in a reduction in normal force. The effects of stress relaxation are materials and temperature dependent as discussed by Horn and Zarlingo.[28]

Thus, from a contact spring/normal force viewpoint, the most important materials properties are elastic modulus, yield strength, and stress relaxation resistance.

The spring properties required for the separable interface must be counterbalanced by the formability required to form the separable contact geometry. These geometries can be quite complex, as discussed in Section 3.6.

3.4.2.2 Permanent Connections Creation of mechanical permanent connections involves the same materials properties, but in a different context. In some cases, such as crimped connections, good formability is required to support the extensive plastic deformations which occur in the crimping process. For others, for example IDC, spring characteristics are important to create the high contact forces needed to ensure connection resistance stability. The various types of permanent connections will be discussed in Section 3.7.

3.4.3 Contact Spring Material Selection

Selection of a material for a contact spring requires consideration of both ends of the contact, the permanent connection and the separable interface. The requirements on each are related, but different as discussed. A listing of materials properties of three basic families of contact spring materials is provided in Table 3.2. In each case, a number of alloys in the family may be used, the data given are for the most commonly used composition. Copper alloy C26000, cartridge brass, is used for low-cost and commercial applications. Alloy C51000, a typical phosphor bronze, is a general purpose alloy used in a large variety of contact springs. C172000, beryllium copper, is used when stress relaxation resistance becomes very important. Bersett (29) and Spiegelberg (30) provide additional discussions and properties of contact spring materials.

In Table 3.2, the tensile strength ranges listed are those typically used in connectors and represent a trade-off between spring performance and formability in manufacturing. A few brief comments on each alloy are in order.

3.4.3.1 Brass Cartridge brass (70% copper-30% zinc) is a low-cost material accounting for its wide usage in consumer applications. Brass has adequate spring properties for room temperature applications and more robust contact designs. While the electrical conductivity of brass is quite good, it has poor stress relaxation resistance, which limits its application temperature to of the order of 75°C in most cases.

3.4.3.2 Phosphor Bronze A number of phosphor bronze alloys are used in connectors with C51000 (4% tin balance copper) being the most common. Phosphor bronzes are used in a wide range of contact spring designs due to a very good combination of formability, electrical conductivity, and spring performance, along with reasonably good stress relaxation resistance at a moderate cost.

3.4.3.3 Beryllium copper The use of beryllium copper is limited by its cost, although lower cost alloys have recently reached the market. Where high strength (miniaturized contacts) and high stress relaxation resistance (elevated temperature applications) are necessary, however, beryllium copper is often the material of choice.

3.4.3.4 Other Alloys There are, of course, many other alloys used in connectors. One connector application that may drive selection of alloys other than those already discussed is power contacts/connectors. In such applications, the electrical conductivity becomes an important consideration and high-conductivity alloys such as C19400 (copper-2.4% iron) and C19500 (copper-1.5% iron-0.6% tin) are more suitable. These alloys have conductivities of 65 and 50 percent of the conductivity of copper respectively and result in lower bulk resistance and Joule heating in high-current applications. C70250 (copper-3.0 nickel%-0.75% silicon) is another stress relaxation resistant alloy seeing increased use in higher temperature applications such as automotive connectors.

3.4.4 Summary

Selection of an appropriate contact spring material for a given connector application requires particular attention to the operating temperature and power requirements which the connector must satisfy. Ensuring the generation and stability of an adequate contact force and sufficient formability to form the separable and permanent contact interfaces are the primary trade-offs in material selection.

3.5 CONNECTOR HOUSINGS

A connector housing has obvious electrical and mechanical functions, but it also can provide environmental benefits. The electrical and mechanical performance of a connector housing depends on the polymer from which the housing is molded, in most cases thermoplastic polymers. The environmental benefit depends primarily on the design of the housing. Consider each separately.

3.5.1 Electrical

The electrical function of a connector housing is to insulate the individual contact springs from one another. The surface and volume resistivities of the polymer, as well as its dielectric breakdown voltage, are the major properties of interest for this function. Most engineering thermoplastics have electrical properties that readily meet typical connector requirements. Polymer materials do differ, however, in the stability of these properties with exposure to temperature, humidity, and chemical exposures. Electrical stability is one selection parameter for polymer materials.

3.5.2 Mechanical

There are two types of mechanical requirements that a connector housing must satisfy. The first relates to support of the contact spring within the housing and the second to dimensional control of the housing.

3.5.2.1 Mechanical Support   In most cases, the contact spring is latched into the housing which must therefore support the spring deflection to varying degrees depending on the housing and spring designs. Polymers differ significantly in their mechanical properties and, in particular, in the variation of those properties with temperature. While stress relaxation is an important consideration in contact springs, the corresponding degradation mechanism in polymers is creep. Creep also results from the conversion of elastic to plastic strain. During creep the material flows away from the point of stress as a function of time and temperature. Creep can result in a loss in contact force or a displacement of the contact from its intended position in the housing.

Another aspect of mechanical stability relates to the handling of the connector during assembly and application. In many cases, the connector housing contains guiding features, to reduce the potential for abusive mating, and latches, to ensure proper engagement of the two halves of the connector and eliminate undesired connector separation in high-vibration environments.

3.5.2.2 Dimensional Control   Dimensional control is important in maintaining the proper spacing of the contact springs at both the separable interface and permanent connection. Variations in dimensions at the separable interface can have a significant impact on the connector mating force. Variations at the permanent connection end can result in an inability to make the required permanent connection, in particular in soldered connections to PWB or IDC connections to small centerline ribbon cable. Dimensional stability depends on the molding characteristics of the polymer, which affect shrinkage and bow/warp of the housing, and the thermal expansion and moisture absorption characteristics of the polymer. The assembly process which the housing must support also affects dimensional stability. Surface mount (SM) soldering requirements are particularly demanding with respect to temperature stability.

3.5.3 Environmental Shielding

While materials properties are critical to ensuring the required electrical and mechanical performance, housing design determines the effectiveness of environmental shielding. Figure 3.11 demonstrates the shielding provided by the housing against a corrosive environment. The coupon shown is silver plated and was exposed, while mated to the card-edge connector shown, to an operating environment containing sulfur, which is highly reactive with silver. The corrosion of the silver is apparent on the exposed section of the

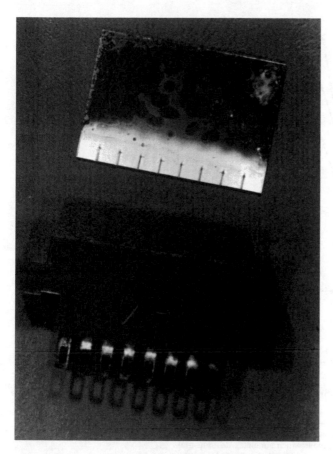

FIGURE 3.41 Photograph of a silver-plated coupon exposed to an environment containing sulfur and chlorine while mated to the card-edge connector shown. *(Courtesy AMP Incorporated.)*

coupon as is the shielding provided where the coupon is within the housing. Figure 3.12 demonstrates the effect of shielding on connector resistance of connectors exposed to a mixed flowing gas environment intended to simulate an industrial environment. The test environment contains *parts per billion* (ppb) levels of chlorine, hydrogen sulfide and nitrogen oxide. The connector was a two-piece post-receptacle design and was exposed to the test environment in both mated and unmated conditions. In addition durability cycling of the connectors was performed before exposure. Data for both gold and tin finished connectors are shown. Note the significant increases in contact resistance for the gold finished connectors exposed unmated and the stability of the resistance of those exposed in the mated condition. The effectiveness of the housing in providing environmental shielding is apparent. Note also the stability of tin finished connectors in both mated and unmated conditions. Tin is non-reactive in this environment due to the protective nature of the tin oxide. It must be noted, however, that this exposure does not stimulate the dominant tin degradation mechanism, fretting corrosion.

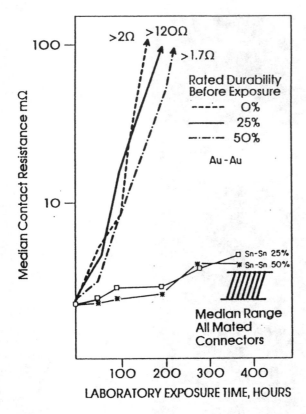

FIGURE 3.42 Connector resistance versus exposure time for tin and gold finished 25 square post connector system exposed mated and unmated to a test environment which simulates an industrial environment. The effect of housing shielding in attenuating the environment is apparent. *(Courtesy AMP Incorporated.)*

3.5.4 Application Requirements

Two types of application requirements must be considered: those during assembly of the product and the functional requirements during use. Assembly processes, in particular soldering, may dictate the selection of the polymer. This is especially true for connectors that must be surface mounted. Temperature stability against both warping and degradation is required. In use requirements must also be considered but, in most cases, are less demanding than assembly requirements. The application temperature may, however, limit the materials options due to polymer creep and its associated effects on mechanical support and dimensional changes. Moisture absorption and sensitivity to humidity and solvents also may be important.

Additional mechanical requirements that may arise involve latches/locks, hinges, and snap-fit features. These may be necessary in assembly of the connector or attaching it to its intended circuitry.

3.5.5 Material Selection

Selection of an appropriate connector housing material, as for a contact spring material, depends on a number of property/performance trade-offs. Before specific materials and trade-offs are considered, a brief discussion of polymer structure as it affects connector performance is in order. Alvino[31] provides a general discussion of plastics for electronic applications.

3.5.5.1 Polymer Structure The word "polymer" means "many mers," where a "mer" is the basic structural unit(s) from which the polymer is synthesized, and is descriptive of polymer structure. In general, polymers consist of long chains of carbon-carbon bonds with a variety of side groups distributed along the carbon-carbon backbone. The polymer type determines the side groups and the length of the carbon backbone chains, the molecular weight, determines the processing and mechanical properties of the polymer.

There are two classes of polymer materials: thermosets and thermoplastics. As the names imply, thermosetting polymers take a permanent set during molding (due to cross-linking or bonding between the chains) and, therefore, cannot be reprocessed. Thermoplastic polymers, which exhibit little if any cross-linking, become plastic (can flow under pressure) at increasing temperatures, which allows them to be readily injection molded and reprocessed. It should be noted, however, that the material for reprocessing (the regrind, sprues, and runners from the injection molding, for example) has lower properties, due to a reduction in molecular weight, than the virgin material. This effect may limit the amount of regrind that can be used, depending on any reprocessing of the regrind and the application requirements. Thermoplastic polymers dominate in connector applications due to their processing advantages and cost effectiveness.

There are two basic classes of thermoplastic polymers: amorphous and crystalline. Once again, the names are descriptive. Amorphous polymers show little internal ordering of the carbon chains, while in crystalline polymers the long chain molecules are aligned to a significant degree. These structural differences give rise to chemical, mechanical, and thermal differences between the two types. In general, amorphous polymers are isotropic, which provides advantages in molding and dimensional stability. These positive traits are counterbalanced by greater thermal sensitivity, lower strength and, generally, lower chemical stability than crystalline polymers.

The alignment of the polymer chains provides enhanced mechanical strength, thermal stability, and solvent resistance to crystalline polymers. The alignment also leads to anisotropy, which is reflected in molding stresses that may result in dimensional instability.

The performance of both amorphous and crystalline polymers is generally enhanced by a variety of additives. Such additives include flame retardants, reinforcing agents (primarily glass fibers) for mechanical strength, lubricants for molding ease, and other additives to enhance thermal stability.

When all the trade-offs are taken into account, crystalline thermoplastics dominate the materials used in connector housings. With this limited background, a few of the more commonly used connector housing materials will be briefly discussed.

Table 3.3 provides a brief compilation of properties of interest in connectors for a small sampling of polymers commonly used in connector housings. A brief discussion of these materials providing general comments on their suitability for connector applications follows. Temperature capabilities, both application temperature and processing, in particular surface mounting technology (SMT) capability, are particularly important in material selection.

3.5.5.2 Polyamides (Nylon) While there are many types of nylon, at one time 6/6 nylon was a dominant connector housing material. 6/6 nylon possesses a good balance of

TABLE 3.3 Selected Properties of Typical Connector Housing Materials

	Flexural modulus (10^6 kg/mm^2)	Heat deflection temperature @264 psi (°F)	UL temperature index (°C)	Dielectric strength (V/M)
Polyamide (6/6)	0.7	466	130	17.4
Polybutylene terephthalate	0.8	400	130/140	24.4
Polyethylene terephthalate	1.0	435	150	26.0
Polycyclohexane terephthalate	0.9	480	130	25.4
Polyphenylene sulfide	1.2	500	200/230	18.0
Liquid crystal polymers	1.5	650	220/240	38.0

mechanical, electrical, and thermal characteristics, good chemical resistance, and good processibility. It has, however, one severe limitation, a tendency to absorb moisture. Moisture absorption has a negative effect on the mechanical properties of nylon and a significantly negative effect on its dimensional and electrical stability. This limitation, in combination with the development of a wide variety of alternative materials, including other nylons such as 4/6, has diminished the range of applications for 6/6 nylon as a connector material. 4/6 nylon has a lower sensitivity to moisture and higher temperature capability—a heat deflection temperature (HDT) of 543°F—which makes it SMT compatible.

3.5.5.3 Polybutylene Terepthalate (PBT) PBT is arguably the most commonly used connector housing material due to its combination of processibility and functional/performance characteristics. PBT is a crystalline material and possesses good electrical and mechanical properties, dimensional and chemical stability, good solvent resistance, and generally acceptable temperature capability. Temperature capability, however, does limit the suitability of PBT for SMT applications (an HDT of 400°F). PBT is generally used in glass reinforced grades, (20 to 40%) for improved mechanical properties.

3.5.5.4 Polyethylene Terepthalate (PET) PET shares the structural and performance characteristics of PBT with an improvement in temperature capability. On the negative side, the creep and warpage of PET are greater than PBT. Creep under load is a concern in connectors in which the contact spring is preloaded against the housing. Once again, glass reinforced grades are used to improve the mechanical properties. PET resins are generally reinforced with 30 to 50 percent glass. The HDT of PET (435°F) is marginal for SMT.

3.5.5.5 Polycyclohexylenedimethylene Terepthalate (PCT) PCT shares the generally good electrical and mechanical characteristics of PBT and PET and has higher temperature capability (HDT 480°F), which makes it suitable for SMT applications.

3.5.5.6 Polyphenylene Sulfide (PPS) PPS is also crystalline. It has high-temperature capabilities (HDT 500°F and a continuous-use temperature of 210°C). The mechanical properties of PPS are also very good, particularly with respect to stiffness. The negative side of this strength, however, is a tendency toward brittleness.

3.5.5.7 Liquid Crystal Polymers (LCPs) LCPs are relatively new materials, available in a number of compositions, which are characterized by a highly aligned rod-like structure which is maintained even in the melt state (the "liquid" for a polymer). The highly crystalline structure results in high stiffness, mechanical strength, and electrical properties that are maintained at elevated temperatures, with UL temperature index values in excess of 220°C. LCPs are SMT compatible with HDT values above 600°F.

3.5.6 Summary

The process of selection of a polymer for a connector housing is complicated by the variety of materials available and the rapid rate of introduction of new materials. The electrical characteristics of engineering polymers are far in excess of typical connector requirements in most cases. Mechanical strength and dimensional stability as well as application factors, including assembly processes such as soldering, are the considerations most likely to influence material selection. In general, more than one material will be acceptable, and selection is often influenced by experience and familiarity as well as technical advantages in a particular process, such as SMT soldering. Walczak, McNamara, and Podesta,[32] Gupta[33] and Hawley[34] provide overviews of polymer materials from different connector application perspectives.

3.6 SEPARABLE CONNECTIONS

This section will provide a brief overview of separable contact interfaces in terms of spring design parameters such as contact force and contact geometry. The interaction of these variables on important performance characteristics such as mating force and durability will be reviewed. Permanent connections will be discussed in Section 3.7 but will be referenced in this section as well because all contacts have both separable and permanent connection requirements.

3.6.1 Contact Designs

A wide variety of separable interface contact designs are used in connectors. The separable connection always has two sides. In general, one side (usually the receptacle) is a spring member, while the other (the plug) is a solid contact, a PWB, post, or pin. Posts and pins differ primarily in geometry, posts being square or rectangular and pins being round. Generically, these systems are referred to as card-edge, post/receptacle or pin/socket connectors.

3.6.1.1 PWBs, Posts, and Pins. Examples of typical plug contacts are provided in Fig. 3.13. A PWB, Fig. 3.13a, is part of a connector system in many level 3 and some level 4 applications. In most cases, the contact area for separable connections to the PWB is a gold/nickel plated pad or land. It is important that the platings on the PWB and the receptacle contact be compatible in performance characteristics.

The 25 square post (0.025 in on a side), Fig. 3.13b, is currently the most common post geometry, although applications using smaller posts, 1 mm and 15 square, are increasing. 25 square technology is used in connectors from levels 3 through 5 and in noble metal and tin finished versions. Noble metal finishes are used in high-performance applications, while tin sees use in both electronic and commercial products. Many post contacts are duplex plated with a tin or tin-lead finish for the permanent connection, in particular for soldered applications, and a noble finish at the separable interface.

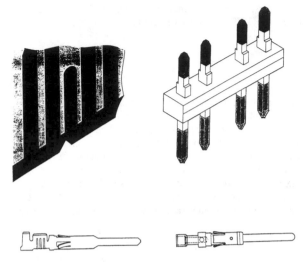

FIGURE 3.43 Schematic illustration of plug contacts, including (*a*) a printed wiring board (PWB), (*b*) a post, and (*c*) stamped and formed and (*d*) and screw machined pin contacts. *(Courtesy AMP Incorporated.)*

In levels 3 and 4, the posts may be inserted directly into the board or included in headers, shrouded or unshrouded, depending on application requirements. The posts are connected to the board by soldering or press-in connections.

While pins (Figs. 3.13*c* and *d*) see limited use in level 4, their prime areas of application are in levels 5 and 6, often as wire-to-wire connections. The majority of pin contacts are designed for crimped permanent connections. They are available in a number of sizes (diameters) depending on the application. Two general types of pins are used: screw machined (Fig. 3.13*c*) and stamped and formed, (Fig. 3.13*d*). Screw machined contacts are generally gold plated and used in military and high-performance systems, while the stamped and formed pins are gold or tin plated and used in electronic and commercial applications. The major differences between screw machined and stamped and formed pins are the presence of a seam in the stamped and formed pin and differences in dimensional tolerances.

3.6.1.2 Receptacle Contacts The majority of receptacle contact designs are cantilever beam geometries, although very complex compound beams are also seen. A few of the more common geometries are shown in Figs. 3.14 through 3.16. The simplest receptacle contact design is a cantilever beam, as exemplified by a card-edge contact, Fig. 3.14*a*, although some card-edge contacts take on more complex geometries such as the bellows design shown in Fig. 3.14*b*.

For post/receptacle systems dual contact beams are generally used, in open or box geometries, as shown in Figs. 3.15*a* through 3.15*d*. In these examples, the dual cantilever beams make contact with opposing sides of the post. Open twin beam contacts predominate over box contacts in commercial applications due to lower cost. Four styles of twin beam contacts are common, but with many variations. The flat stamped contact in Fig. 3.15*a* is commonly referred to as tuning fork. The contact in Fig. 3.15*b* is commonly used in Eurocard connectors. In Fig. 3.15*c,* an open twin cantilever beam contact is illustrated.

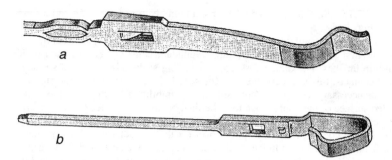

FIGURE 3.44 Schematic illustration of card-edge contacts: (*a*) simple and (*b*) compound cantilever contacts are shown. *(Courtesy AMP Incorporated.)*

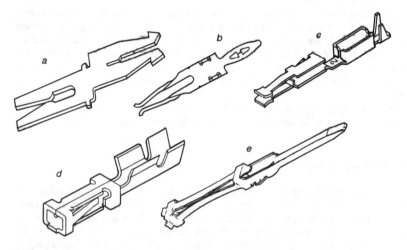

FIGURE 3.45 Schematic illustrations of a variety of receptacle contact geometries that mate to 25 square posts. *(Courtesy AMP Incorporated.)*

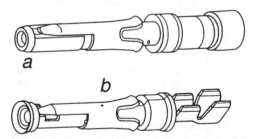

FIGURE 3.46 Schematic illustration of (*a*) stamped and formed and (*b*) screw machined socket contacts. *(Courtesy AMP Incorporated.)*

Figures 3.15*d* and 3.15*e* illustrate box type designs, a twin beam in Fig. 3.15*d* and a four beam version in Fig. 3.15*e*.

The contacts differ in manufacturing practice (and therefore in cost) and cantilever beam design. The first four designs provide dual redundant contacts and are used with both noble and tin finishes. The tuning fork contact mating surface is a sheared surface. For tin-finished contacts, this is acceptable but, for gold finishes, attention to the surface roughness may be necessary to ensure adequate mating durability. All other contacts mate to the mill rolled surface of the contact beams. The dual beam box contact provides additional advantages in lead in protection to minimize the effects of misalignment on mating and, in the design shown, anti-overstress protection for the contact spring. All of these designs are available in tin and gold finishes, including duplex finishes, to satisfy a variety of commercial and electronic applications. The four beam box contact is primarily used in high-performance and military applications and is generally gold plated. The full range of permanent connection technologies is available for receptacles, although not all designs lend themselves to each of the technologies. For example tuning fork contacts are generally intended for IDC applications.

Socket contacts are also available in screw machined and stamped and formed versions for the same markets as discussed for the pins. Examples of screw machined and stamped and formed sockets are shown in Fig. 3.16*a* and 3.16*b*, respectively. Screw machined contacts are generally gold plated, whereas stamped and formed versions are available with noble and tin finishes. Socket contacts, as mentioned, are generally intended for wire-to-wire applications, and crimped connections are the most common permanent connection technology.

There are many other receptacle contact designs in use intended to address specific application requirements. These are not discussed herein, due to space limitations.

3.6.2 Application Issues

Selection of an appropriate connector for a given application requires consideration of a number of design\performance interactions and trade-offs. Among those important from the separable connection perspective are the connector mating force and the contact durability. These performance characteristics, in turn, depend on design parameters such as the contact finish, the normal force, and the contact geometry. Criteria for contact finish selection were discussed in Section 3.3. The following discussion will address the other issues.

3.6.2.1 Connector Mating Mechanics Mating mechanics, in particular the peak mating force, has a significant impact on two important performance factors, the capability of mating high pin count connectors, and contact durability.

Normal force, contact geometry, and engagement length, and their role in the connector mating process, will be briefly described using a post/receptacle contact system as an example. A schematic illustration of the connector mating process is provided in Fig. 3.17. The position of the post in the receptacle and the mating force versus insertion depth relationship are shown simultaneously. There are two phases to the mating process, insertion of the post, phase 1, and sliding, to the final contact location, after the receptacle beams are fully deflected, phase 2.

Phase I: The post enters the receptacle and begins deflection of the contact beams. The insertion force increases with a slope dependent on the coefficient of friction, which depends on the contact finish, the deflection characteristics, or spring rate, of the beam system and the contact interface geometry. During this phase normal force and contact geometry are the key factors which influence the dynamic frictional forces (the state of

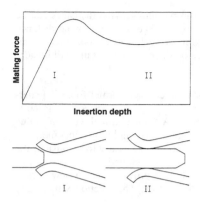

FIGURE 3.47 Schematic illustration of the mating mechanics of a separable interface. Two phases of mating, insertion, and sliding are indicated. *(Courtesy AMP Incorporated.)*

lubrication of the interface is also important because it has a significant effect on the coefficient of friction). For a given state of lubrication, the force/geometry relationship determines the maximum insertion force of the contact system. The peak force depends on the normal force, the angle between the mating surfaces and the coefficient of friction.

It is important to note that this discussion concerns the peak *contact* mating force. The mating force for the *connector* will be the sum of the contact mating forces *plus* any additional force necessary to overcome misalignment of the mating halves of the connector and dimensional variances in the housing. Such effects can add considerably, and may even dominate, connector mating force.

It should also be noted that, under certain conditions of coefficient of friction and geometry, stubbing of the contacts can occur, making mating impossible without damage to the contacts. This "friction lock" condition is addressed by control of the mating geometry.

Phase II: The contact spring has been fully deflected, and the insertion force is that produced by sliding friction, which depends on normal force and the dynamic coefficient of friction. The withdrawal, or retention, force is also determined by these same parameters with the peak force being dependent on the static coefficient of friction, which is higher than the dynamic value. Contact geometry, however, can also influence the magnitude of the retention force.

3.6.2.2 Connector Mating Mechanics and Contact Durability

The relationship between mating forces and contact durability is different in the two phases. In phase I, surface roughness and improper initial mating geometry can lead to catastrophic wear, including skiving of the finish. In phase II, high friction forces can result in adhesive wear and brittle fracture due to high shear forces at the interface. More typically, in particular for lubricated contacts, the sliding wear process, which is dependent on normal force, coefficient of friction and contact geometry results in gradual burnishing wear of the contact surface. Under these conditions the durability will also depend on the length of engagement of the connector, a distance which is typically of the order of two millimeters.

Mating of the connector involves relative motion of the contact surfaces against one another. This motion is referred to as *wipe* and has an important impact on film and contaminant displacement. The effectiveness of wipe on film disruption and contaminant displacement are discussed by Williamson[3] and Brockman, Sieber, and Mroczkowski,[35,36] respectively. Here, it is sufficient to note that wiping action is necessary for reliable connector performance. In most applications, the engagement length of the connector provides wipe far in excess of what is necessary for film and contaminant displacement. In some designs, zero insertion force (ZIF) connectors in particular, wipe distances can be much smaller, and a zero wipe condition must be avoided. The minimum wipe distance depends on normal force and geometry, as discussed in Mroczkowski.[37] For typical connector normal forces and geometries, a wipe distance on the order of 10 mils will be adequate in most applications.

3.6.2.3 Contact Normal Force   In most connectors, the contact normal force is generated by deflection of the receptacle contact beams by the plug. Contact normal force is arguably the most important connector design parameter. A "conventional wisdom" value for minimum normal force is 100 g.[38] As connectors miniaturize and pin counts increase, new limits on normal force come into play so that it becomes necessary to reconsider the context in which the 100 g minimum "requirement" was developed. The case for this statement is argued by Whitley and Mroczkowski[39] who also provide an overview of normal force requirements for precious metal plated connectors. This section will briefly summarize that reference.

The normal force performs two key functions. First, it provides the force that establishes the contact interface as the connector is mated. Second, it maintains the stability of the interface against mechanical disturbances in the application environment. The normal force required for each of these functions is different. To establish a contact interface with an "acceptable" value of contact resistance (a few milliohms) requires only a few grams, as indicated in Fig. 3.18. Maintaining the mechanical stability of the interface is where the remainder of the normal force requirement arises. The value required depends on the contact finish, the design of the connector and the application environment. Fifty grams provides a benchmark for precious metal contacts if the connector design addresses all the relevant application considerations. One hundred grams is a lower limit for tin connectors[40]

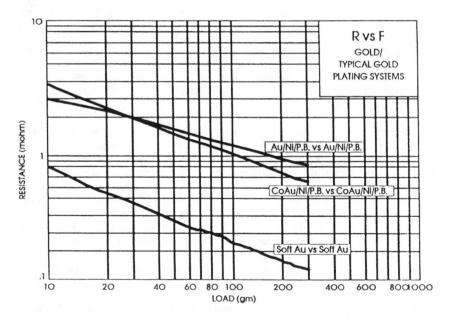

FIGURE 3.48   Contact resistance, R_c, versus contact force, F_n, for a gold-plated contact interface. *(Courtesy AMP Incorporated.)*

to provide the mechanical stability necessary to avoid the important degradation mechanism of fretting corrosion.

3.6.2.4 Contact Geometry　　Another important design parameter which interacts strongly with normal force is contact geometry. Wiping effectiveness, mating force, durability, and contact resistance all vary with the normal force-geometry combination of the separable interface.[41] It has been proposed by Kantner and Hobgood[42] that a parameter identified as "Hertz stress," which includes normal force, contact geometry, and finish and elastic modulus in a simple model, provides a good design guidelines for predicting connector performance. This proposal has been discussed by Fluss[43] and Mroczkowski.[41] Eammons et al.[44] provide another useful study relating connector design to performance. In the authors opinion, these references indicate that a Hertz stress requirement does not provide unambiguous indications of the performance capability of contact interfaces.

3.6.3　Summary of Separable Interfaces

The basic requirements on the separable interface are described in Section 3.1. The connector must provide a low and stable value of resistance while meeting mating requirements, in particular mating force and the number of mating cycles it must withstand without affecting performance. Separable interfaces include a wide variety of contact geometries and forces to satisfy these requirements under an increasing diversity of application conditions.

3.7　PERMANENT CONNECTIONS

Discussion now turns from the separable interface to permanent connections. Such connections occur between the connector and the electronic component or system to which it provides a connection. There are two basic classes of permanent connections: mechanical and metallurgical. Mechanical connections include crimped, insulation displacement, press-in, and wrapped connections. Soldered connections are the predominant metallurgical connections, but brazed and welded connections also are used.

Permanent connections are sometimes described as an extension of the conductor to which they are connected. In that context, the resistance introduced by the connection is expected to be of the same order as a section of the terminated conductor of the same length as the connection. As mentioned previously, permanent connection resistances are of the order of tens to hundreds of microohms, compared to the milliohm or so of separable connections. The magnitude of permanent connection resistance depends on the resistance of the conductor being terminated.

In this discussion, permanent mechanical connections will be considered in two classes, connections to wire/cable and to a PWB. Prior to discussion of the permanent connection technologies, a brief discussion of wire/cable and PWB construction is in order.

3.7.1　Overview of Wire and Cable

The terminology "wire" will be used in this discussion, but these remarks also apply to cables. In this context a cable is viewed as a protected wire (jacketed cable) or a number of wires arranged in a "controlled" geometry (ribbon cable). Coaxial cable will also be considered.

A wire consists of a conductor, the current carrying component, and its insulation, if any. Characteristics of a wire include:

- *Conductor material.* In the majority of electrical and electronic applications, the conductors are annealed copper.
- *Conductor finish.* In many applications, the conductors are bare copper. Common finishes include tin, silver, and nickel. Tin finishes enhance solderability and crimped and IDC permanent connections. Silver is used for high-frequency applications, and nickel for high-temperature applications.
- *Conductor geometry.* Round conductors predominate, while flat and foil conductors are used in specialized applications. This discussion will be limited to round conductors.
- *Conductor construction.* There are two basic round conductor constructions, solid and stranded. Wires are usually characterized by their cross sectional area, in millimeters squared or American Wire Gauge (AWG). Stranded wires are made up of multiple strands of conductors to achieve the total cross section desired. For example, a 26AWG wire can be constructed in several ways as illustrated in Fig. 3.19. The solid wire consists of a single 26AWG conductor having a nominal diameter of 0.0159 in. The other two 26AWG wires shown are 7/34 and 19/38 constructions. The 7/34 construction consists of seven 34AWG conductors having a diameter of 0.008 in to make up the nominal 26AWG cross section. Similarly, the 19/38 construction includes nineteen 38 AWG conductors having a diameter of 0.004 in. The geometric arrangement of the conductors can take several forms in terms of the relative wire positions and the twist or lay of the conductors. The constructions illustrated are concentric constructions in which the geometric relationship of the strands to one another are constrained. Stranded conductors predominate in most applications due to their higher flexibility compared to solid conductors, though they have a slightly lower current carrying capacity than the corresponding solid wire. High conductor count wires are frequently bunch stranded, a construction in with the individual conductors are randomly distributed in the bundle. Bunch stranding provides greater wire flexibility and lower cost than concentric constructions.

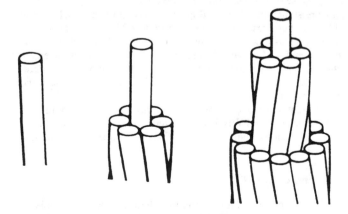

FIGURE 3.49 Schematic illustration of three strand configurations used in stranded conductor wire constructions. *(Courtesy AMP Incorporated.)*

- *Wire insulation.* Insulations are characterized by their material and thickness. Common insulating materials include polyvinyl chloride (PVC), polyethylene (PE), polypropylene (PP), and polytetrafluoroethylene (PTFE). The choice of insulation material and thickness depends on the function the insulation is intended to perform. The two most common functions are electrical insulation and mechanical protection. In some applications the dielectric characteristics of the insulation also become important, for example in coaxial cables intended for high-speed applications. In such cases, foamed insulations are seeing increasing usage to decrease the effective dielectric constand and thereby maximize the propagation velocity of the cable.

A brief discussion of a few common cable constructions is also in order. Schematic illustrations of some common cable constructions are provided in Figs. 3.20 and 3.21.

Jacketed cable, Fig. 3.20a, consists of a collection of discrete wires contained within a single protective jacket. Such cables provide mechanical protection of the wires and reduced wire handling during manufacturing and assembly processes. The wires within the jacket may be of a variety of constructions including discrete, shielded, coaxial, and twisted pair.

Ribbon cable refers to a planar arrangement of multiple wires as shown in Fig. 3.20b. Both round and flat wire ribbon cable are used, but round wire dominates. In addition, the wires may be shielded, coaxial, twisted pair, or transmission line constructions. Ribbon cable also simplifies wire handling and, in addition, reduces the potential for wiring errors by fixing the relative positions of the individual wires. Discrete wire ribbon cable lends itself to mass termination, usually using insulation displacement connection (IDC) technology as will be discussed.

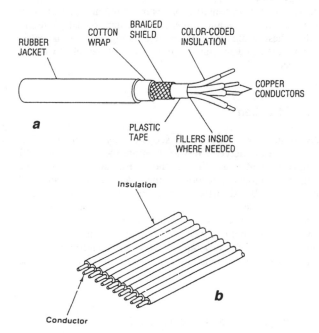

FIGURE 3.50 Schematic illustration of (*a*) jacketed and (*b*) ribbon cable constructions. *(Courtesy AMP Incorporated.)*

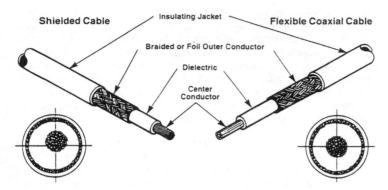

FIGURE 3.51   Schematic illustration of (*a*) shielded and (*b*) coaxial cable constructions. *(Courtesy AMP Incorporated.)*

Twisted pair wires are exactly that: two wires twisted around one another for improved
electrical performance. The details of the improvement are beyond the scope of this discussion, but suffice it to say that the signal is split between the two wires so as to reduce
noise and interference effects on signal integrity.

Transmission line cables consist of alternating signal and ground conductors, which
are generally of different sizes, to improve high-speed digital data transmission performance and provide controlled impedance.

Shielded and coaxial cable are identical in construction, as indicated in Fig. 3.21, with
the essential difference that the concentricity and dimensions of the coaxial cable are
much more tightly specified. It is these geometrical parameters, along with the dielectric
properties of the insulation material, which determine the impedance of the cable. Coaxial
connectors will be discussed in Section 3.9.

3.7.2   Overview of PWB Construction

A detailed discussion of PWB construction is beyond the scope of this presentation. Figure 3.22 contains a cross section of a multilayer PWB to illustrate the features of PWB
construction relevant to making permanent connections to the board. Pads or lands on the
board surface allow the making of separable connections to the board edges, and soldered
connections to lands distributed over the board surface, surface mount connections (SMT).
Plated through holes allow for through hole soldered connections and mechanical press-in
connections. Also shown is a press-in connection with the tail extending through the
board. Such exposed tails provide sites for both connectors or wrapped connections.

With this context, attention now turns to the structure of permanent connections. While
the focus will be on mechanical permanent connections, soldered connections will be
briefly discussed.

3.7.3   Mechanical Permanent Connections to Wire/Cable

The two major wire/cable permanent connections are crimped and insulation displacement
connections.

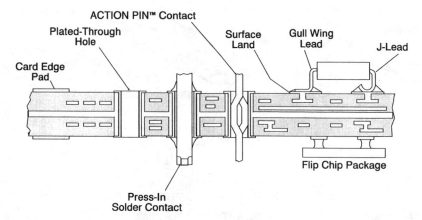

FIGURE 3.52 Schematic cross section of a multilayer printed wiring board. *(Courtesy AMP Incorporated.)*

3.7.3.1 Crimped Connections The principles underlying crimped connections are quite simple. Crimp terminal design and the crimping process, however, require careful attention to the overall crimping system. This system approach cannot be overemphasized, because the controlled deformation of wire and terminal that is responsible for consistent and reliable crimped connection performance derives from ensuring the proper system. The crimping system consists of the proper combination of wire, crimp terminal, and crimp tooling.

While crimped connections can be made to solid wires, stranded wire applications dominate. Copper conductors represent the majority of applications in both bare and tin-plated versions. Annealed copper wire is the most common case, but applications with alloy wire also exist. There are two basic requirements on the wire. First, it must be the correct size; that is, it must fall in the range of wire sizes the crimp terminal is designed to terminate. Second, the insulation must be stripped from the wire to the proper length without damaging the conductors or the stranding excessively. Damage of the conductors will reduce the mechanical strength of the crimped connection and may affect the crimp resistance. Excessive disturbance of the stranding may compromise the insertion of the stripped wire into the terminal, particularly in automatic machinery.

There are two basic styles of crimp terminals: open and closed-crimp barrels, as illustrated in Fig. 3.23. Open-barrel terminals are used in high-volume machine-applied applications because the open-barrel facilitates automatic insertion of the stripped wire into the open barrel which provides a large target area. For this reason, open-barrel crimp terminals are used in the majority of crimped connections. Closed-barrel terminals are often semiautomatically or hand-applied in smaller volume or repair applications. The stripped wire must be inserted axially into the smaller target area of the closed barrel. The crimped connection is made by controlled deformation of the conductors in the crimp barrel.

Figure 3.23 also illustrates a second important element of crimp terminals, the insulation support, or insulation grip. In the insulation support barrel the insulation is gripped by deformation of the tabs to provide a strain relief for the crimped connection. In addition to the strain relief, the insulation support provides protection against the effects of vibration on the crimped connection. The majority of crimp terminals take advantage of the benefits of an insulation support barrel.

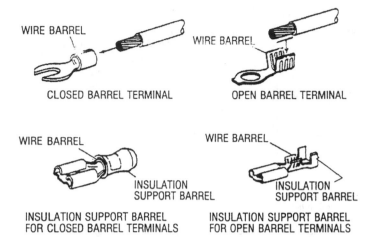

FIGURE 3.53   Schematic illustration of open and closed crimp barrel constructions. *(Courtesy AMP Incorporated.)*

The function of the crimp tooling is to control the deformation of the conductor and crimp terminal because controlled deformation is the source of crimped connection performance as will be discussed. Figure 3.24 schematically illustrates the crimping process in an open barrel terminal. The conductor/crimp terminal cross sectional area is fixed by selection of the appropriate wire/crimp terminal combination. The crimp tooling then controls the deformation by controlling the crimped connection geometry. Two controls on geometry are used. Clearly the shape of the crimp tooling, and the terminal geometry, fix the overall geometry of the crimped connection. Control of the crimp height, illustrated in the inset of Fig. 3.24, is the final control and is accomplished by controlling the closing action of the crimp tooling.

3.7.3.2 The Crimping Process and Crimped Connection Performance   During the crimping process, extensive deformation and relative motion of the conductor(s) against themselves and the crimp terminal body results in a disruption of surface films on the conductors and terminal body producing new film free surfaces. Under the high forces of crimping, contact between these film free surfaces results in the formation of numerous micro cold welds.[45] It is these cold welded joints, between the conductor strands and between the strands and the terminal body, supported by an appropriate residual stress distribution, that provide the electrical and mechanical integrity of crimps. Figure 3.25 shows a cross section of a crimped connection. The deformation of the conductor strands and the terminal are evident.

The mechanical and electrical characteristics of crimps vary with the deformation process and the amount of deformation introduced. The deformation during crimping, therefore, is a parameter which must be controlled to ensure that the properties of the crimp remain in the "optimum" range.

A number of crimp geometries are widely used, examples of which are provided in Fig. 3.26. The F, for folded, crimp (Fig. 3.26*a*) is a commonly used open barrel crimp geometry. W crimps (Fig. 3.26*b*) are used for large wire size closed barrel crimped connections. Military applications often use indent crimping processes as illustrated in Fig. 3.26*c*. The crimp geometry shown is a 4-8 indent crimp.

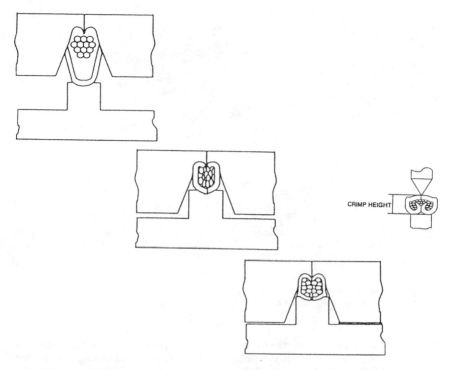

CRIMP HEIGHT

FIGURE 3.54 Schematic illustration of the crimping process for an open-barrel crimp termination. *(Courtesy AMP Incorporated.)*

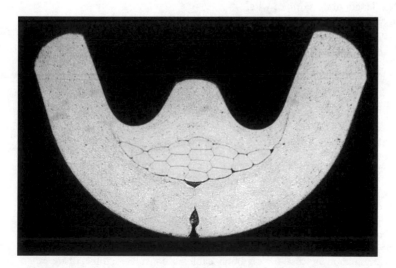

FIGURE 3.55 Photomicrograph of a crimped connection cross section. Note the deformation of each of the individual conductor strands. *(Courtesy AMP Incorporated.)*

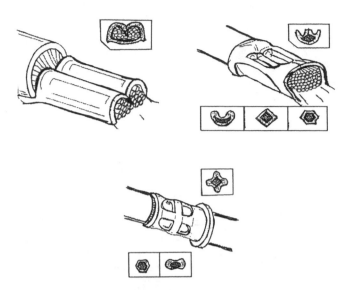

FIGURE 3.56 Schematic illustrations of selected crimped connection cross sections: (*a*) F crimp, (*b*) W crimp, and (*c*) 4-8 indent crimp. *(Courtesy AMP Incorporated.)*

3.7.3.3 Summary The importance of the crimping system—wire/terminal/tooling—is assurance of controlled deformation. If the appropriate wire/terminal/tooling combination is used crimp deformation will be controlled and a repeatable and reliable crimped connection will result.[46] Properly crimped connections provide mechanical strengths approaching that of the wire being terminated and electrical resistances comparable to that of an equivalent length of the crimped wire.

Inspection and process control for crimped connections is simple. Visual inspection of the crimped connection should verify proper stripping of the insulation and insertion of the conductors into the crimp barrel as indicated in the sample inspection document in Fig. 3.27. In addition to the visual examination, the crimp height also serves as a process verification, on crimped connections, and, on some automatic crimping equipment, as a process monitor during the crimping process. A periodic verification of the tensile strength may also be performed.

3.7.4 Insulation Displacement Connections

The second mechanical wire/cable permanent connection is insulation displacement connection (IDC) technology. The principles of IDC technology are schematically illustrated in Fig. 3.28. A wire is simply inserted into a controlled geometry slot in the terminal. As the wire slides down the slot, the wire insulation is displaced and the conductor cross section deformed to create the desired metallic permanent connection interface.

While IDC connections are used on discrete wire, the dominant application is to cable, in particular ribbonized cable (Fig. 3.21*b*). The reason for this dominance becomes apparent when the termination process for a ribbon cable is considered. First, there is no requirement to strip the insulation which reduces wire handling. Second, it is possible to make all the connections at one time using housings in which the contacts are preloaded.

CRIMP INSPECTION

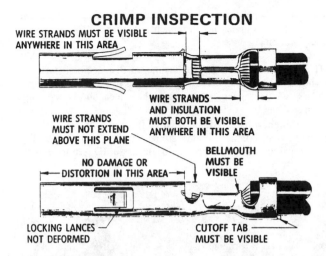

WIRE STRANDS MUST BE VISIBLE
ANYWHERE IN THIS AREA

WIRE STRANDS
AND INSULATION
MUST BOTH BE VISIBLE
ANYWHERE IN THIS AREA

WIRE STRANDS
MUST NOT EXTEND
ABOVE THIS PLANE

BELLMOUTH
MUST BE
VISIBLE

NO DAMAGE OR
DISTORTION IN THIS AREA

LOCKING LANCES
NOT DEFORMED

CUTOFF TAB
MUST BE VISIBLE

FIGURE 3.57 Example of visual crimp inspection criteria. *(Courtesy AMP Incorporated.)*

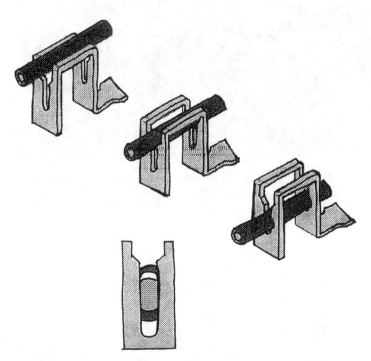

FIGURE 3.58 Schematic illustration of insulation displacement termination technology. *(Courtesy AMP Incorporated.)*

In addition to increasing throughput, wiring errors are eliminated compared to the crimp and insert process necessary for crimped connection approaches. Figure 3.29 shows a variety of IDC cable connector applications.

3.7.4.1 IDC Terminals   There are a wide variety of insulation displacement terminal designs, some of which are shown in Fig. 3.30. The designs differ primarily in the compliance of the beams of the IDC slot which impacts on the residual force capability and range of wire sizes on which the terminal can be used.

The simplest IDC terminal design, which will be used to illustrate the principles of insulation displacement connection mechanics, is the slotted beam, as shown in Fig. 3.30*a*. There are two separate functions which are critical to successful ID terminations:

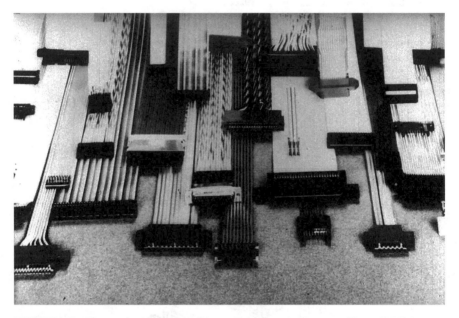

FIGURE 3.59   Photograph of a variety of ribbon cable IDC termination types. *(Courtesy AMP Incorporated.)*

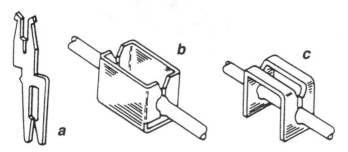

FIGURE 3.60   Schematic illustration of a variety of IDC terminal designs. *(Courtesy AMP Incorporated.)*

1. the displacement of the insulation.

2. the formation and maintenance of the contact area.

Figure 3.31 schematically depicts these two functions. The insulation of the wire musts be displaced at or near the top of the slot to prevent overstressing of the contact beams, Fig. 3.31*a*. The effectiveness of insulation displacement depends on the geometry of the lead-in section of the termination and the thickness and properties of the insulating material on the wire. The terminal lead-in must penetrate the insulation and strip it away from the conductor surface. Lead-in geometry is important, but the hardness of the insulation and the adherence of the insulation to the conductor are equally important. Wire insulations vary significantly in the repeatability with which they can be displaced. PVC insulations dominate in current IDC applications.

After displacement of the insulation, the conductor continues downward in the slot to its final position, Fig. 3.31*b*. It is during the downward motion that conductor deformation takes place producing a wiping action on the beam sides and the conductor(s). The effectiveness of surface film removal during wiping depends on the design of the IDC slot and the surface finish. Deformation of the conductors provides intrinsic cleaning and these surfaces make contact with the sides of the termination beams.

At the final conductor location, the displacement of the beams determines the residual normal force which maintains the integrity of the contact interface. The allowable deformation of the conductor is limited by two factors. The lower limit is dependent on the requirement to establish a contact area, A in Fig. 3.31*b*, which results in an acceptable value of connection resistance. The upper limit of conductor deformation is due to the weakening of the wire and its effect on handling of the connectors during assembly and in their intended application.

3.7.4.2 *Strain Relief* The force required to insert a conductor into an IDC slot is relatively low. It follows that a force applied parallel to the slot may disturb, or even dislodge, the wire. It is for this reason that most IDC connectors include some mechanism for protecting the wire from such forces. The protection may be built into the terminal, housing or separate covers as discussed in Mroczkowski.[47]

3.7.4.3 *Solid and Stranded Wire* IDC technology is suitable for both stranded and solid wire, but more attention to termination design is necessary when stranded wire is used

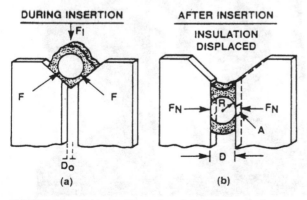

FIGURE 3.61 Schematic illustration of the IDC termination process. Two stages are shown: (*a*) insulation displacement and (*b*) conductor termination. *(Courtesy AMP Incorporated.)*

because control of the conductor deformation is more difficult with stranded conductors due to the fact that the strands can move relative to one another during the insertion process. This effect is illustrated in Fig. 3.32, where examples of solid and stranded wire IDC connections are provided. The rearrangement of the conductor strands can have a detrimental effect on both the magnitude and stability of IDC resistance. The effect on resistance stability is arguably more significant because the strand rearrangement reduces the beam deflection and, therefore, the contact normal force which provides the desired mechanical stability. These concerns can be addressed, however, as indicated by the fact that the majority of IDC application are on stranded wire.

3.7.4.4 Conductor Finish   IDC connections are readily made to both bare and tin-coated conductors. The wiping action on insertion serves to clean the conductor surfaces, enabling the desired metallic interface to be created. Tin-finished conductors, however, show a greater consistency in performance.

3.7.4.5 Summary   IDC technology is primarily used in cable connector applications to take advantage of enhanced productivity through reduced wire handling and reduction in wiring errors. A variety of IDC terminal designs exists to address differing application requirements. For additional discussion of IDC connection technology, see Mitra,[48] Mroczkowski,[47] Wandemacher,[49] and Key.[50]

3.7.5   Mechanical Permanent Connections to Printed Wiring Boards

Two mechanical permanent connection technologies will be discussed, press-in and wrapped connections. Press-in connections are made directly to plated through holes

FIGURE 3.62 Photomicrograph illustrating insulation displacement connections to solid and stranded wire. *(Courtesy AMP Incorporated.)*

(PTH) in the PWB while wrapped connections are made to pins pressed-in or soldered into a PTH.

3.7.5.1 Press-In Connections There are two types of press-in connections: rigid and compliant. In both cases, residual forces maintain the contact normal force. They differ in the relative amounts of elastic and plastic deformation that occurs in the pin, and in the deformation produced in the PTH in the PWB.[51]

Rigid press-in pins undergo little elastic deformation when they are inserted in the PTH and cause plastic deformation and damage to the PTH. Little elastic recovery, or residual force, capability is retained in such connections, which is why compliant press-in pins were developed. A major driving force for compliant pin technologies was to provide an alternative to soldering for through-hole connections. A mechanical permanent connection technology eliminates the potential degradation/contamination effects of the high temperatures and fluxes used in soldering.

All compliant press-in geometries are intended, in one way or another, to produce elastic deformation of the pin, and thereby a residual force to maintain the integrity of the interface between the press-in section of the compliant pin and the PTH. Many designs of compliant press-in pins are in use. Three of the major variants are shown in Fig. 3.33.

There are four basic requirements on the press-in section:

- establishing and maintaining a normal force
- a limit on insertion force
- a minimum retention force
- no unacceptable damage to the PTH.

The normal force requirement is, of course, intended to ensure a low and stable connection resistance. The insertion force limit is necessary to permit the insertion of multiple pin count connectors as well as to limit PTH damage. The minimum retention force requirement depends on the function of the tail of the pin, which extends from the PWB. The requirement on hole damage is to ensure that damage to the PTH and lands of the

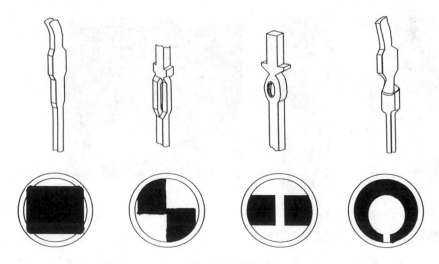

FIGURE 3.63 Schematic illustration of press-in contact geometries. *(Courtesy AMP Incorporated.)*

PWB, including buried layers, does not occur to the point where the performance of the PWB is compromised. Needless to say, some of these requirements are in direct opposition, and a number of compliant press-in section designs have evolved to address these issues and meet these requirements.

Transverse and longitudinal cross sections of the AMP ACTION PIN™ compliant pin geometry in a PTH are shown in Fig. 3.34. The normal force is created by the deflection of the compliant section diagonal of the pin. The original diagonal dimension is shown as the dotted line. The transverse contact occurs at two points. The longitudinal contact extends along the axis of the compliant section.

The 25 square technology is currently dominant, but smaller sizes are seeing increasing use as circuit miniaturization and I/O density requirements continue to increase. Additional discussion on the design and function of compliant press-in pins is provided by Scaminaci,[52] Goel,[51] and Dance.[53]

3.7.5.2 Wrapped Connections Wrapped connections are made to the tails of press-in connections to PTH which extend from the board surface. Figure 3.35 schematically illustrates a wrapped connection. The example shown is a modified wrap in that the first contact of the wire to the board occurs to the insulation. The electrical connections are formed to the stripped wires as the wrapping process continues. The force responsible for creating and maintaining the integrity of the electrical connection is the residual tensile stress in the conductors as they come in contact with the corners of the wrap post. These corners have a controlled radius requirement to facilitate the formation of a metallic contact interface. The tension on the wire is maintained as the conductor is wrapped around the post corners and, therefore, is retained in the conductor sections between the post corners. Wrapped connections exhibit a high redundancy of contact interfaces as readily seen in Fig 3.35.

3.7.6 Summary of Mechanical Permanent Connections

Four mechanical permanent connection technologies have been briefly discussed. *Crimped* and *insulation displacement* connections are wire termination technologies.

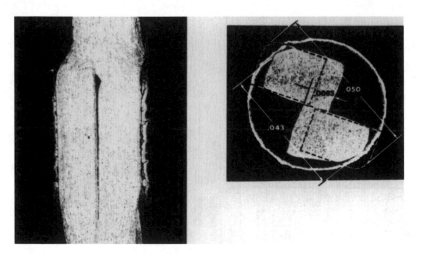

FIGURE 3.64 Photomicrograph of a compliant in termination to a plated through hole. *(Courtesy AMP Incorporated.)*

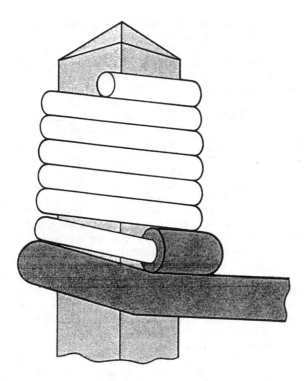

FIGURE 3.65   Schematic illustration of a modified wrapped connection. *(Courtesy AMP Incorporated.)*

Press-in and wrapped connections are connections to PWB, either directly or indirectly. Each technology is widely used in the electrical and electronics marketplace, although they vary in presence across these markets. Each technology derives its utility from the creation and preservation of metallic contact interfaces through controlled deformation of the conductor, wire or PTH, and the terminal. Control of the deformation, through tooling and process controls, is necessary to ensure repeatable permanent connections regardless of the technology used.

3.7.7   Soldered Connections

Numerous texts are available on soldering technology, e.g., Manko.[54] In this discussion, only some of the aspects of soldering relevant to connector technology will be reviewed. There are two major types of connectors for solder connections. They differ in how the solder terminations of the connector interface to the board. The older technology, through-hole technology (THT) relies on soldering the lead into a PTH. Surface mount technology (SMT), as the name implies, relies on soldered connections made to a pad or land on the surface of the PWB. The two technologies differ in several respects.

3.7.7.1 Through-Hole Technology   The THT soldering process is typically accomplished through wave soldering. The solder supply and the soldering heat are provided by the solder wave. This is a significant difference from SMT soldering, as will be discussed. The solder joint in THT is also much larger than that in SMT, extending through the PWB and into a fillet above and below the surface of the board. In addition to this solder joint, however, the soldered post is also mechanically supported by the PWB. These two differences, the solder source and the mechanical support, have a major effect on the design and materials requirements on THT as compared to SMT.

3.7.7.2 Surface Mount Technology For SMT soldering the solder leg of the connector is pressed down into a solder paste which is then reflow soldered, by a variety of techniques, to create the solder joint. The solder source, the amount of solder used, the source of the soldering heat and the structure of the solder joint all differ significantly from that of THT soldering. The solder source is typically a solder paste which is screened or printed onto the PWB in a secondary operation. Control of the amount of solder is critical to SMT performance. Too little solder may affect mechanical strength and too much may lead to solder bridging and shorting of contact pads. The solder heat is generally provided externally to reflow the solder paste, though some wave soldering is used. Reflow technologies which have been used include vapor phase, infrared, hot gas, and convection processes in addition to wave soldering. The solder joint is made only to the surface pad or land with no mechanical support from the PWB itself. The solder joint must support all subsequent mechanical stresses. This fact leads to concerns about solder creep and relaxation during thermal cycling of the PWB during application.

SMT versions of many electronic components, such as chip forms of resistors and capacitors and special IC chips, are available. These components differ significantly from connectors in two respects—size and the stresses they must support. Connectors are much larger than the other components mentioned, which leads to issues in retaining the connectors on the boards prior to reflow and consistent soldering of a larger number of leads than in other components. Connectors are also subjected to larger mechanical loads, particularly due to the mating requirements. These differences create many new and challenging materials, design, and processing requirements. Among the design/material and process issues that must be addressed to ensure the repeatability of SMT soldering processes are:

- registration of the connector leads to the pads
- coplanarity of the leads to ensure contact to the pads
- adequate connector hold down prior to soldering
- solderability of the leads
- the solder supply, sufficiency, and control of bridging
- thermal stability of the housing to withstand the heat of the process
- the ability of the solder joint to withstand mating stresses
- inspection of the solder joints

The economic and performance advantages that can be derived from SMT connectors will lead to continued development of both connectors and processes. Economic advantages include smaller and more cost-effective boards resulting from the elimination of PTH. Performance advantages are also realized by the smaller boards, resulting in reduced signal path lengths and the potential for improved conductor routing on the board to provide enhanced signal transmission quality and speed.

3.7.7.3 Summary Soldering technologies remain widely used in assembly processes with connectors. THT is being replaced, when economics and performance dictate, by SMT, but both will coexist into the foreseeable future.

3.8 CONNECTOR APPLICATIONS

In this section, connector signal and power applications are briefly reviewed. In signal applications, the focus will be on maintaining the integrity of the signal—in particular, the signal rise time and amplitude. For power applications, minimizing the effects of the connector on the distribution of power, primarily current, within the system is the major concern.

3.8.1 Signal Applications

Signal applications can be either analog or digital. In this discussion, the focus will be on digital applications because connector requirements for digital circuits are generally more stringent than those for analog.

The explosive development of microprocessor technology has created two major driving forces for connector technology: miniaturization and "high speed." The effects of miniaturization span virtually all aspects of connector design. With respect to digital applications miniaturization, and the associated reduced spacings between conductors, on boards and in cables and connectors, results in increased crosstalk. Crosstalk, in turn, affects signal integrity. In addition, control of the magnitude and variation of the connector impedance, because it depends on dimensional tolerances, has become a significant connector design consideration.

There are two aspects to "high speed," the signal pulse frequency, or clock frequency, and the rise time of the pulse. In general, the pulse rise time will be the dominant consideration in defining a high-speed application. This is so because the frequency content of a fast rising pulse will include frequencies higher than the clock frequency. The maximum frequency, f_m, in a pulse of rise time t_r can be approximated by

$$f_m = 0.35/t_r \tag{3.83}$$

For a 1 ns pulse rise time, a maximum frequency of 350 MHz is calculated. This frequency exceeds typical clock frequencies, of the order of 100 MHz in 1995 vintage personal computers.

An application must be considered high-speed when transmission line design rules become necessary for any of the components in the system. In general, the transmission line properties of an interconnection become important when the length of the component becomes comparable to the electrical length of the pulse. There are many approaches to selecting an appropriate electrical length. The approach used here relates the component length to the electrical rise length of the pulse which is given by

$$L_r = v_{prop} t_r \tag{3.84}$$

where the signal propagation velocity, v_{prop} is given by

$$v_{prop} = C \varepsilon_{eff}^{1/2} \tag{3.85}$$

where C is the speed of light and ε_{eff} is the effective dielectric constant of the propagation medium. The effective dielectric constant takes into account material composition variances such as the amount of air in a foamed dielectric or the geometry of a polymer/air cavity in a connector housing. An appropriate average of the individual dielectric constant is necessary in such cases.

When the component length is larger than $0.3L_r$, the component should be considered as a transmission line. For connectors and printed circuit boards, an effective dielectric constant of about 4 is appropriate. For 1 and 10 ns rise time pulses, the critical component length becomes 2 and 20 cm, respectively.

From these lengths, it is clear that printed circuit boards and cables, which are typically tens of centimeters in length, should be considered as transmission lines for pulses faster than 10 ns. This is in fact the case, and transmission line design rules have been applied to PWB and cable design for several years. Microstrip geometries on the PWB and controlled impedance cables are commonly used. Connectors are reaching the point where they must be considered as transmission lines as signal rise times break into the subnanosecond regime. In such applications, transmission line considerations such as characteristic impedance and crosstalk replace contact resistance as "key" performance parameters.

For a general discussion of high-speed applications, the reader is referred to Katyl and Simed[55] and Chang.[56] For the purposes of this discussion, a brief review of crosstalk and controlled impedance will suffice.

3.8.1.1 Crosstalk and Controlled Impedance   Crosstalk and characteristic impedance both depend on the materials and geometries of the components so that design changes affect both characteristics, but not necessarily to the same degree or in the same direction. This allows designs to be optimized, depending on whether crosstalk or characteristic impedance is the more important parameter. In cables, crosstalk is arguably the more important due to the length of typical cables while, in connectors, impedance control is dominant due to the inevitable changes in geometry and materials along the length of a connector.

3.8.1.2 Crosstalk   Crosstalk results from the electromagnetic coupling of the field surrounding an active conductor into an adjacent conductor, called the *victim*. The strength of the coupling depends on the proximity and geometry of the active and victim conductors. Crosstalk can be minimized by increasing the separation between conductors and keeping conductor lengths as short as possible. It is this second feature that limits the crosstalk introduced by connectors, because connectors tend to be short in the direction of the signal.

An additional method for reducing crosstalk is to introduce grounds between the signal conductors. Grounds, in effect, attenuate the electric fields and reduce the coupling between signal lines. This approach can be, and is, used in connectors, but is even more important in cables due to their greater length. The use of ground pins in connectors is, in fact, intended more as a means of controlling the impedance of a connector.

3.8.1.3 Controlled Impedance   Impedance control is an important requirement for at least two reasons. Constancy of impedance is necessary to ensure the quality of signal transmission over distance, and matched impedance at interfaces, such as connectors, is critical to minimizing signal reflections. Control of signal integrity is particularly important in digital applications where false triggering or missed triggering of devices can occur due to reflections, positive or negative, caused by impedance mismatches.

The impedance of a component depends on the geometry and materials of the component. Constant impedance requires constant geometries and consistent materials. Impedance control in cables is relatively straightforward in that cables tend to be consistent in both materials and geometry. As cables become miniaturized, however, dimensional tolerances and materials consistency become more difficult to maintain. For connectors, it is a very different situation. Impedance control in a connector is complicated with respect to both dielectrics, a varying mix of air and the polymer of the connector housing, and changes in conductor geometries, both in cross section and curvature. Under these conditions a "constant impedance" connector is intrinsically impossible.

This difficulty leads to two approaches to controlled impedance in connectors which rely on adding grounds to moderate the variations in characteristic impedance. The two approaches are the addition of dedicated ground pins and introducing ground planes within the connector. Examples of these two cases are provided in Fig. 3.36.

3.8.1.4 Open Pin Field Connectors   Varying the number of pins dedicated to grounds in an open pin field connector was the first approach to controlling the impedance and reducing the crosstalk of board-to-board connectors. Many factors, primarily circuit and application related, influence the number of grounds necessary to ensure the required signal integrity. The number of grounds needed is often expressed in terms of the signal/ground ratio. As the signal/ground ratio decreases the electrical performance increases, but the number of pins in the connector which are available to satisfy the I/O requirements decreases. The arrangement of the grounds with respect to the signal lines is also important.

3.8.1.5 Ground Plane Connectors   The introduction of ground planes into the connector is one method for reducing the impact of grounding requirements on the number of available signal pins. Stripline and microstrip geometries have been used in PWB design for a number of years and are now available in connectors. Such connectors are usually designed for 50 Ω impedance and can be used in applications with rise times below a nanosecond. In addition to adding the ground planes, other important design considerations include keeping the conductor lengths as short as possible and minimizing changes in geometry through the connector. Figure 3.37 provides examples of connectors with stripline and microstrip geometries.

Ground plane connectors offer advantages in both performance and density. Performance advantages result from the impedance control and reduced crosstalk due to the better shielding provided by a plane as compared to a pin. Density is improved by the fact that all the pins are available for signal applications. For a 100-position connector, an open pin filed connector with a 1:1 signal/ground ratio has only 50 pins for signals. A ground plane connector would use all 100 pins as signals and can often be built to the same, or only slightly larger, footprint as the open pin field connector. Another potential advantage is that the ground plane can also be used for power distribution providing additional effective density increases.

3.8.1.6 Summary   Signal applications, in particular nanosecond and subnanosecond rise time digital pulse applications, require that connectors be considered as transmission lines. Grounding becomes critical and various grounding schemes must be evaluated with application requirements in mind. In addition, shielding and electromagnetic interference must also be considered. These topics are not considered here, but are discussed by Southard,[57] Sucheski and Glover,[58] and Aujla and Lord.[59]

3.8.2 Power Applications

Power distribution places considerably different requirements on contacts and connectors than those for signal distribution. Power distribution includes both voltage and current considerations. This discussion, however, will focus on current because the Joule heating which accompanies current flow introduces many application related concerns. Power connector design requirements to address voltage related issues involve mainly material selection and dimensions to withstand voltage breakdown effects.

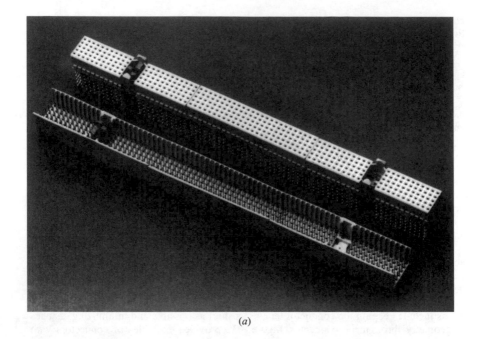

(a)

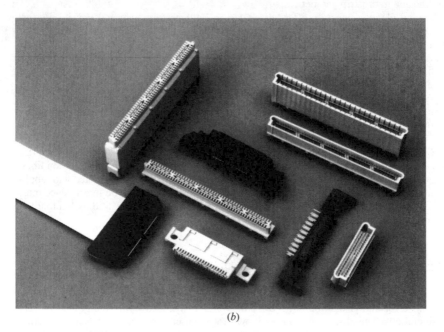

(b)

FIGURE 3.66 Examples of (*a*) open pin field and (*b*) ground plane connector constructions. *(Courtesy AMP Incorporated.)*

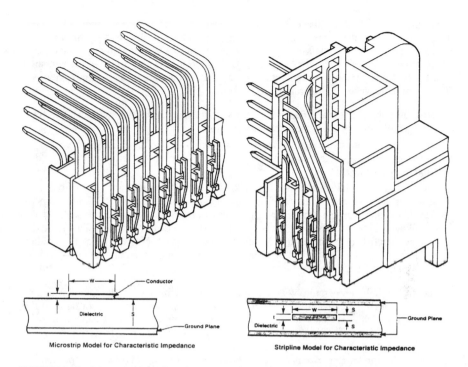

Microstrip Model for Characteristic Impedance

Stripline Model for Characteristic Impedance

FIGURE 3.67   Schematic illustration of (*a*) microstrip and (*b*) stripline ground plane connector constructions. *(Courtesy AMP Incorporated.)*

Power connector requirements differ from those for signals primarily in the magnitude and stability of the connector resistance which must be realized. In addition to the Joule heating, I^2R, mentioned previously, power distribution connectors must minimize the IR drop they introduce into the system. The IR drop includes the bulk resistance of the connector as well as the interface resistance. Higher-conductivity copper alloys are often chosen as spring materials for power contacts in order to minimize bulk resistance as mentioned in Section 3.4. Millivolt drop budgets in power distribution systems are becoming increasingly stringent as PWB systems take on higher functionality and require increasing currents to the board.

There are two fundamentally different approaches to power distribution. They are, first, the use of one or a few dedicated power contacts and, second, using a higher number of signal contacts in parallel. The development of high pin count connectors, especially at level 3, has led to increased use of the second option. However, as board functionality continues to increase both current and I/O counts increase so that the number of pins available for power distribution becomes limited. Hybrid connectors which contain both signal and dedicated power contacts are seeing increasing use to address this concern.

The discussion will begin with some fundamental principles of design for power connectors following Corman and Mroczkowksi.[60]

3.8.2.1 Connector Design and Current Capacity   Important parameters include current (both dc and ac—and the duty cycle), the allowable resistance (both interface and bulk) and temperature (both contact and ambient). The discussion begins with the fundamentals of current flow across a contact interface.

For the purposes of this discussion, the most important aspect of the asperity structure of the contact interface is the size of the asperities because it dictates very high current densities. Joule heating at the asperity contacts can lead to a very high localized temperature referred to as the *supertemperature*, T_s. The supertemperature must be distinguished from the bulk temperature of the contact, T_b, and the temperature of the application, T_a. Three comments on supertemperature are important.

- First, the supertemperature depends on the voltage drop across the interface, therefore, all asperity contacts are at the same temperature.
- Second, the response time of the asperity contacts is of the order of microseconds so supertemperature is determined by the peak current for any currently used power distribution frequency.
- Third, a critical voltage exists at which the asperity contacts will attain a supertemperature sufficient to cause melting of the asperity interface. This voltage limits the peak current the interface can support. Because the voltage is given by IRc, the peak current capability can be increased by lowering the resistance of the interface. This is generally accomplished by increasing the normal force which also increases the contact resistance stability. The critical melting voltages for gold, silver, and tin are 430, 570, and 130 mV respectively.

The contact bulk temperature will be determined by the balance of the Joule heating of the contact, which depends on the current and the bulk resistance of the spring, and the heat dissipation, which depends on the heat sinking provided by the permanent connection to the circuit component. This bulk heating effect is generally referred to as the *contact T-rise* as measured against the ambient or application temperature.

The application temperature is the temperature at which the equipment is operating at the connector location. It may be the ambient temperature, or it may be a higher temperature due to heat generation by the equipment itself.

Supertemperature and T-rise are influenced by two different resistances, supertemperature by the contact interface resistance, and T-rise by the bulk resistance of the contact spring. Interface resistance limits the peak current the contact can carry, as mentioned, while bulk resistance limits the current rating. The T-rise that is generally, and arbitrarily, used to establish the current rating of a contact is 30°C.

For dc, the peak and "average" current are the same. For ac, the "average" current, with respect to T-rise, depends on the shape of the current waveform and the duty cycle. The duty cycle comes into play when it is realized that heat generation depends on power and time, so continuous ac and pulsed ac of the same "average" current will result in different values for T-rise depending on the duty cycle. Wise[61] discusses these issues.

3.8.2.2 Current Rating Current rating of a contact is relatively straightforward. The same is not necessarily true of a connector. Current rating of multiple signal contacts used in parallel to distribute current can be a complex matter because it depends on interaction of the current rating of the individual contacts, the number of contacts carrying current and the distribution of the contacts in the connector housing. This subject is discussed by Corman and Mroczkowski[60] and will only be briefly reviewed here.

The current rating of a contact is generally based on T-rise considerations which depend on the balance of heat generation, generally the Joule heat, and heat dissipation into the medium to which the contact is connected—the PCB or wire. A 30°C T-rise is commonly, and arbitrarily, used as a criterion and is derived from UL practice. The current rating is determined experimentally and, therefore, is highly dependent on the test conditions as discussed by Wise.[62] The current rating also depends on the application temperature since the combination of T-rise and ambient must not exceed the temperature rating of the connector, usually 105°C.

Since connectors contain multiple contacts, current distribution and the thermal effects between the contacts must be considered. The key factors include:

1. The current distribution among the contacts. A true parallel circuit, in which all contacts carry the same current, is preferred. Unequal distribution of current can lead to failures of an overloaded contact which can cascade through the connector.
2. The current carrying contacts should be distributed throughout the connector if possible to minimize thermal coupling between the contacts. The loading of the connector, the percentage of the contacts carrying current, will affect the current rating for the same reason.

3.8.2.3 Power Contact/Connector Summary   Power applications of connectors share all the usual concerns of connector design with a few extra. Special attention must be paid to the magnitude and stability of contact resistance to ensure that voltage drops and temperature increases are minimized. The magnitude of the contact resistance determines the peak current the connector can support, while the overall resistance determines the T-rise. In parallel contact power distribution, attention must also be paid to circuit design to ensure current distributions are uniform, or nonuniformities must be accounted for by allocating additional contacts.

3.9 CONNECTOR TYPES

The discussion of connector types in this section will follow a format similar to that of levels of interconnection. Connector types will be discussed in terms of the permanent connections between the circuit elements being connected. This leads to three different categories: board-to-board, wire-to-board, and wire-to-wire.

3.9.1 Board-to-Board Connectors

The class of board-to-board connectors is often referred to as printed wiring board (PWB) connectors. Level 3 connections are board to board by definition, but many level 4 connections are also board to board, either directly or via cable assemblies. Level 3 connectors must meet both high-performance and high-density requirements—high performance in the sense that they must often meet transmission line requirements, and high density in that the increasing functionality of PWB units has resulted in a significant increase in I/O requirements while miniaturization has maintained or reduced connector size. The end result of these trends in an increase in connector density. Level 3 board-to-board connectors are available in pin counts over 1000 and at contact pitches of 2.54 and 1.27 mm, with 1 mm pitches in development. Level 4 board-to-board connectors see similar high-performance requirements but are generally under 200 positions and most commonly on 1.27 and 1.0 mm pitches.

The typical level 3 packaging structure is a daughter card to mother board connection, with the daughter cards plugged in perpendicular to the mother board and parallel to one another.

In level 3 applications, the mother board is generally a bussing system. In level 4 applications, the mother board may contain functional elements in addition to being a bussing system between the daughter cards. The daughter cards may be different subassemblies of the overall system (level 4).

Such systems may also be configured with parallel boards by using stacking board-to-board connectors, or baby boards plugged in parallel to the daughter boards as illustrated in Fig. 3.38.

Due to their high pin counts, level 3 board-to-board connectors generally include alignment guides, polarizing, and keying features. In addition, mating forces become a major consideration. Level 3 board-to-board connectors are usually used by skilled personnel.

Level 4 connectors, as mentioned, have lower pin counts which reduces mating force concerns.

There are two major classes of board-to-board connectors, card-edge and two-piece, examples of which are shown in Fig. 3.39. The advantages and limitations of each style can be simply summarized.

The major advantage of card-edge connectors is cost, due to the fact that only a receptacle is needed because the PWB provides the plug half of the connector system. The major limitation arises from the tolerances, both thickness and warpage, on printed wiring boards. Variances in board thickness and warpage require that the receptacle contact be able to accommodate a very large deflection range. This, in turn, places severe limitations on the contact spring design to ensure that minimum contact forces and wipe distances are realized. Another packaging related limitation is the fact that such connectors can mate only to the edges of the board.

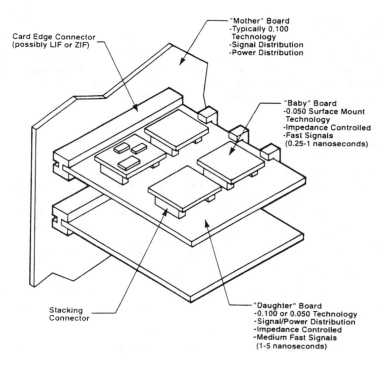

FIGURE 3.68   Schematic illustration of a variety of level 4 board-to-board connection possibilities. *(Courtesy AMP Incorporated.)*

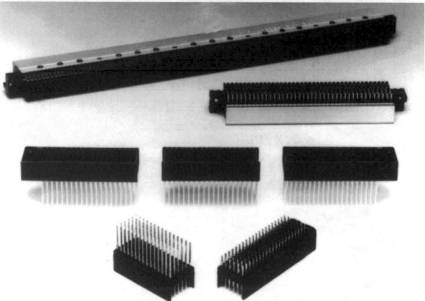

FIGURE 3.69 Examples of (*top*) card-edge and (*bottom*) two-piece board-to-board connectors. *(Courtesy AMP Incorporated.)*

For two-piece connectors the situation is exactly opposite. Cost, relative to card-edge, is the major disadvantage of two-piece connectors. Among the major advantages is tolerance control. Two-piece connectors use stamped/formed posts and receptacles which can meet much tighter tolerances than a PWB. Two-piece connectors are seeing increasing usage for three reasons.

1. shrinking connector centerlines
2. surface mounting requirements
3. increasing performance requirements including controlled impedance and lower signal/ground ratios

Each type of connector will be briefly discussed in more detail.

3.9.1.1 Card-Edge Connectors Through-hole mounting is the most common method of providing the permanent connection of card-edge connectors to mother boards. Both soldering and compliant pin permanent connection technologies are widely used. For the separable interface, card-edge connectors generally consist of two rows of spring contacts that mate with pads on the two sides of the printed circuit board. To meet increasing I/O requirements on the PWB, dual-level contacts are also used. An example of a dual-level connector is shown in Fig. 3.40.

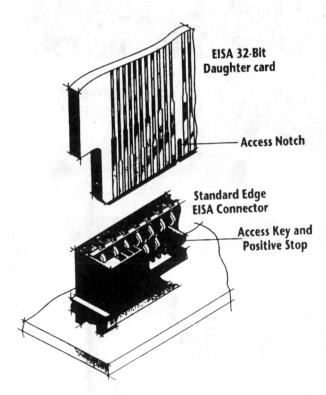

EISA 32-Bit Daughter card

Access Notch

Standard Edge EISA Connector

Access Key and Positive Stop

FIGURE 3.70   Schematic illustration of a dual-level card edge PWB connector construction. *(Courtesy AMP Incorporated.)*

While card-edge connectors are still in wide use, and innovative designs have been employed to enhance their performance, including low insertion force (LIF) and zero insertion force (ZIF) designs, there is an increasing trend toward two-piece connectors as I/O requirements continue to increase.

3.9.1.2 Two-Piece Connectors Two-piece connectors offer advantages with respect to increasing pin count requirements. As mentioned, card-edge connectors are generally two-row connectors, which limits the density of contact pins. Two-piece connectors are not subject to this limitation and are available in up to eight rows and in configurations including ground planes. In addition, two-piece connectors can be placed anywhere on the board, which facilitates increasing packaging density through stacking connectors.

Two-piece connector development is advancing on two fronts. First, surface mountable connectors are now available, and development efforts are continuing. Second, high-performance or high-speed connectors are available which provide improved electrical performance through the use of integral ground planes and controlled impedance as well as improved shielding. In addition, connectors with integrated filters that reduce noise and protect signal integrity are also available.

Two types of two-piece board-to-board connectors are predominant in the marketplace. These are Eurocard and high-density connectors.

Eurocard connectors derive their name from their origin and widespread popularity in Europe. True Eurocard connectors are available in 32, 64, and 96 positions based on one, two, or three rows of the standard 32-contact array on 2.54-mm centerlines. Examples of Eurocard connectors are shown in Fig. 3.41.

The Eurocard contact system consists of a 25 square post and a twin beam receptacle. The standard Eurocard uses a vertical receptacle and a right angle pin connector although an inverse Eurocard is also used. Three performance levels are defined based on both mating cycles, contact resistance and the qualification test programs which are required.

High-density connectors are available with over 1000 positions and in lengths of over 50 cm. Many styles are available using from two to eight rows of two or four beam recep-

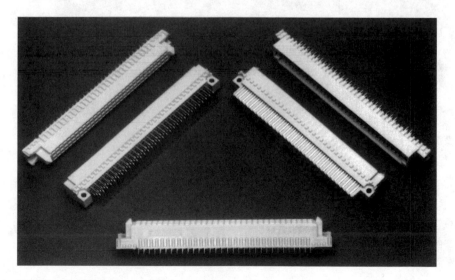

FIGURE 3.71 Examples of Eurocard connectors. *(Courtesy AMP*

tacle contacts. The high pin count connectors are often "sectionalized;" that is, groups of contacts are separated by spaces to incorporate guide pins to assist in connector alignment on mating. Examples of high-density connectors are illustrated in Fig. 3.42.

High-density connectors are also available in hybrids that contain special modules for coaxial, high-current, and fiber optic contacts, as schematically illustrated in Fig. 3.43.

Sequential mating of ground, power, and signal can be accommodated by varying the lengths of the mating pin contacts. Along with density improvements, high-density connectors have also been developed to provide improved electrical performance in two ways. First, the high pin count capability increases the ability to decrease signal-to-ground ratios in open-field connectors. Second, ground planes have been incorporated in a number of configurations to provide microstrip and stripline geometry connectors, as discussed in Section 3.8.

Two-piece connectors are seeing increasing usage due to higher I/O and performance capabilities. I/O capacity is realized by multiple rows and performance by the addition of ground planes to the connector.

3.9.2 Wire-to-Board and Wire-to-Wire Connectors

Wire-to-board and wire-to-wire connectors are combined in this discussion because the focus is on the permanent or wire connection. They differ at the separable interface as defined by the mating element, a board (or board mounted connector) or another wire or cable connector. A dominant technology in wire-to-board connectors is the 25 square technology. Wire-to-wire connectors may also be 25 square, but pin/socket connectors are also common.

FIGURE 3.72   Examples of high-density connectors. *(Courtesy AMP Incorporated.)*

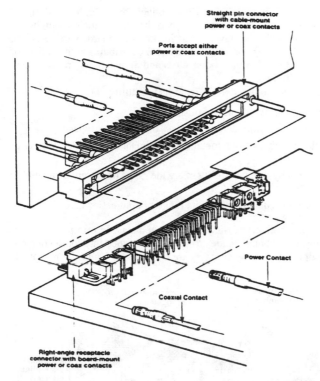

FIGURE 3.73 Schematic illustration of a hybrid connector construction. Signal, coax, and power contacts are shown. *(Courtesy AMP Incorporated.)*

The terminology, "wire," will be used in this discussion, but these remarks also apply to cables. In this context, a cable is viewed as a protected wire (jacketed cable) or a number of wires arranged in a "controlled" geometry (ribbon cable). Wire-to-board connections are used in level 4, while wire-to-wire connectors are predominately levels 5 and 6.

The termination technologies used in wire-to-wire and the wire half of wire-to-board connectors are primarily crimping and IDC, technologies discussed in Section 3.7. Soldering is used to a limited extent but will not be considered here. For wire-to-board connectors, the connection technologies to the board (THT and SMT) have been described in Section 3.7. The discussion of wire-to-wire and wire-to-board types of connectors will be combined because the basic structure of the connectors is similar in both application. And, in fact, some types of connectors are used in both applications.

Wire/cable connector types will be discussed in order of their predominant level of application, from level 4 through level 6.

3.9.2.1 25 Square Technology (Levels 4 and 5) Receptacle connectors mating to a 25 square post represent a significant portion of electronic connectors in both high-performance and commercial applications. At level 4, connectors for both discrete wire and ribbon cable are common. The majority of these applications are based on 2.54 mm centerlines. As the need for miniaturization grows, similar 1.27 and 1.0 mm centerline connector systems based on 15 square posts are emerging.

The variety of level 4 applications ranges from interconnection of high-speed digital electronic subassemblies to connection of processors to peripherals and includes high-current power distribution. 25 square technology is used in board-to-board, wire-to-board, and wire-to-wire applications. This range of functionality also contributes to the variety of connector types. Examples of some 25 square connectors and applications are illustrated in Fig. 3.44. Usage in level 5 is limited.

The popularity of 25 square technology derives from its versatility.

- Modularization of connectors is readily accomplished.
- Common posts provide intermatability across a variety of connector types.
- Connectors are available in both precious metal and tin finishes and with a variety of spring and housing materials.
- Cost effectiveness is realized by the availability of a wide range of automatic assembly equipment.
- Performance capabilities are adequate for a broad range of applications.

Discrete wire connectors use both crimp-snap and IDC contacts. In crimp-snap applications, individual wires, discrete or bundled, are stripped and crimped into loose contacts, generally by automatic equipment, and the contacts are then snapped into the receptacle

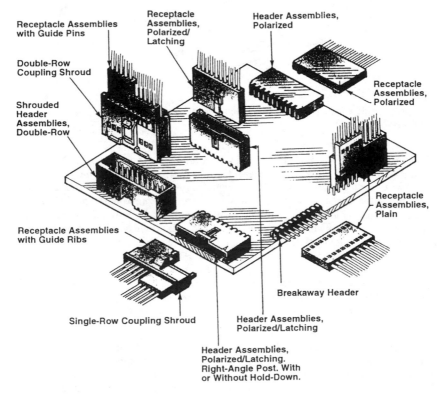

FIGURE 3.74   Schematic illustration of examples of 25 square connector applications. *(Courtesy AMP Incorporated.)*

housing, again automatically. IDC contacts for discrete or bundled wire are generally preloaded into the housing and mass terminated.

Receptacle contact geometries include a variety of designs, some of which were illustrated in Fig. 3.16. Receptacle contacts are generally made from phosphor bronze or beryllium copper alloy.

Both precious metal (gold or palladium alloy based) and tin finishes are used. Duplex plating, gold at the separable interface and tin at the permanent connection, is common on these connectors. Housings are made from a variety of polymers dependent on the application, with PBT as a major housing material.

Board-mounted post contacts may be discrete but most commonly are mounted in headers, shrouded or unshrouded. Brass and phosphor bronze are the most common post materials. Precious metal and tin finishes are both used, including duplex plated versions.

Ribbon cable connectors are characterized by mass termination of multiple conductor cable into preloaded housings via IDC technology. In addition to the economics of mass termination, ribbon cable reduces wiring errors because the individual wires are not handled, and wire locations are fixed by the cable construction. Ribbon cable connectors allow easy daisy chaining, since the connectors can be applied anywhere on a cable, which permits bussing from a single master cable. The majority of ribbon cable used today uses stranded copper conductors, 26 through 30 AWG, on 1.27 mm centerlines. Coaxial and shielded ribbon cable are also used. Cable preparation is more complicated for such applications. Examples of ribbon cable connectors are provided in Fig. 3.45.

Receptacle contact geometries include tuning fork, twisted tulip, and open twin beam contacts. The dominant ribbon cable IDC connection technology is a slotted beam. The contacts are generally manufactured from phosphor bronze or beryllium copper alloys. Precious metal and tin finishes are used.

FIGURE 3.75 Examples of ribbon cable connectors. *(Courtesy AMP Incorporated.)*

The same comments on posts apply as for discrete wire connectors. Some ribbon cable connectors are card-edge connectors and mate to PCBs directly.

3.9.2.2 Pin and Socket Connectors (Levels 4, 5, and 6) Pin and socket connectors are differentiated from post/receptacle designs by geometry. Pins, and the majority of sockets, are round rather than square. In contrast to 25 square posts, pins are seldom directly mounted to a PWB. Instead, connectors containing the pins are board mounted by all the technologies previously discussed. Pin/socket contacts, in various sizes and types are inserted into a wide variety of housing sizes and shapes to meet the diverse needs of level 4 and level 5 application requirements. Among the most common of these connectors are rectangular, circular, soft shell, and subminiature D (also called D sub) connectors.

These connector families have many common components, and a number of hybrid versions with mixed power and signal contacts are used. Coax and fiber optic contacts are less common. The majority of signal contacts are pin/socket designs, both stamped and formed and screw machined. Stamped and formed contacts dominate commercial applications, while screw machined contacts are preferred in military and aerospace applications. The contact materials include phosphor bronze and beryllium copper alloys for the receptacle contacts and phosphor bronze and brass for the pins. Precious metal and tin finishes are used in military/electronic and commercial applications, respectively. Screw machined versions are generally gold plated. Housings are more complex for these families in that many versions contain inserts. Housing materials include PBT and PPS as well as metal shells with inserts varying from polypropylene to thermosets, depending on whether the application is commercial or military.

Rectangular connectors are common in levels 4 through 6. They are characterized by their modularity and diversity with a wide variety of positions and contact combinations, both pin/socket and other geometries available, including signal, power, coax, and fiber optic contacts. These connectors also come in shielded and metal shell versions, primarily for levels 5 and 6. Rectangular connectors are generally discrete wire/cable connectors, regardless of whether they are wire-to-wire or wire-to-board, although ribbon cable use is increasing. Rectangular connectors are used in military, electronic, and commercial applications. Examples of rectangular connectors are provided in Fig. 3.46.

With the exception of the shape of the housing, the same general comments apply to circular connectors. Most circular connectors are used in levels 5 and 6, and include many hybrid connectors to combine power distribution and signal functions. Examples of circular connectors are provided in Fig. 3.47.

Subminiature D is the predominant level 5 connector, seeing wide use as an I/O connector for printers, disk drives, video, and even including low-cost video games. It also sees applications in level 6. The popularity of the line is based on a number of features.

- The line dates back to military connectors of the fifties providing a history of reliable performance.
- They are rugged and easily used, with the D shaped shell providing polarization and the metal shell shielding benefits.
- They are covered in many standards.

In a typical level 5/6 D sub application, one-half of the connector will be bulkhead mounted and the other end connected to an internal PWB by a cable. The level 6 half, which mates to the bulkhead connector, will also be a cable and connects to the desired peripheral through another connector which may be a different style than D sub depending on the peripheral connected.

Subminiature D connectors come in a wide variety of styles, but the standards are based on two pin sizes, 20 (0.04 inch diameter) and 22 (0.03 inch diameter) with five dif-

FIGURE 3.76 Examples of rectangular connectors. *(Courtesy AMP Incorporated.)*

FIGURE 3.77 Examples of circular connectors. *(Courtesy AMP Incorporated.)*

ferent shell sizes to hold a number of contacts ranging from 9 to 50 in size 20 and 15 to 72 in size 22 contacts, with 25 being the most common number.

Both screw machined and stamped and formed pins and sockets are used. Brass pins and phosphor bronze sockets are most common in D sub connectors. D sub permanent connections include solder, crimped, and IDC, with crimping the most common wire connection. Board mounted versions include through-hole and SMT soldered connections with some compliant pin applications.

D sub connectors use an insert within the shell to hold the contacts. Inserts are usually nylon for commercial connectors and diallyl phthallate (DAP) or polyester for military connectors. Gold finishes predominate. Steel shells plated with tin or zinc are most common, but some plastic shells are used in low-cost versions. Examples of D sub connectors are provided in Fig. 3.48.

Miniature ribbon connectors are used in telecommunications, computer peripheral, test equipment, and medical industries. There are numerous standards and some de facto standards covering these connectors.

The miniature ribbon connector family includes board mount and cable connectors in both shielded and unshielded versions. There are also a number of specialty connectors in general use such as gender menders and back-to-back connectors that address particular application requirements. The contact spacing is on 0.085 in centerlines in standard pin counts of 14, 24, 36, 50, and 64 positions. The standard contact finish is gold over nickel, with a 200 mating cycle durability requirement. The mating interface is keystone shaped for polarization.

Cable assemblies may consist of a pair of miniature ribbon connectors or a miniature ribbon and D sub connector, depending on the application. Examples of miniature ribbon connectors are shown in Fig. 3.49. While miniature ribbon connectors are generally used

FIGURE 3.78   Examples of D subminiature connectors. *(Courtesy AMP Incorporated.)*

FIGURE 3.79 Examples of miniature ribbon cable connectors. *(Courtesy AMP Incorporated.)*

in relatively low-speed applications, some shielded versions are suitable for pulse rise times down to 3 ns with proper attention to electrical design.

Soft-shell connectors were developed to fill a need for a small, lightweight, low-cost connectors with current capacities up to 15 A per contact for level 4 and 5 applications in white goods, computers, medical equipment, and automotive markets, among others. To satisfy this diversity of applications, a wide variety of designs and configurations are available, some of which are illustrated in Fig. 3.50.

Soft-shell connectors are generally low pin count, up to about 36 positions, and subject to multiple matings by manufacturers and consumers. The connectors are designed to meet safety and abuse conditions more severe than those of the connectors discussed up to this point. Toward this end, they incorporate a variety of assembly/application aids including polarization systems, lead in chamfers, locking latches, strain reliefs, and contact protection features.

The contacts are generally pin/socket configurations and made from brass or phosphor bronze. Tin finishes predominate in commercial applications, but computer and medical

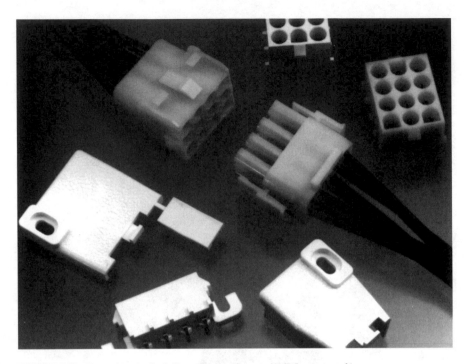

FIGURE 3.80 Examples of soft-shell connectors. *(Courtesy AMP Incorporated.)*

connectors often use selective gold over nickel finishes. Nylon is the most common housing material, with UL 94V-O a common requirement.

3.9.3 Coaxial Connectors

Coaxial connectors differ from the connectors discussed up to this point in that they are specifically intended for high-frequency applications. This difference is reflected in their design and materials of manufacture. A brief discussion of coaxial cable is also included.

3.9.3.1 Coaxial Cable Coaxial cable is the dominant transmission medium for high-frequency level 5 and 6 applications for two reasons. First, it is a controlled impedance transmission medium offering high performance at high frequencies, in the gigahertz range. Second, the coaxial construction provides inherent shielding and electromagnetic interference (EMI) and radio frequency interference (RFI) control. Even at lower frequencies, when the transmission characteristics are not required, EMI/RFI shielding benefits make coaxial cable the medium of choice in demanding applications. The major design challenge for coaxial connectors is to maintain the transmission and EMI/RFI shielding performance of the cable through the connection process and in application.

The basic structure of coaxial cable was illustrated in Fig. 3.21 but is repeated as Fig. 3.51. The four basic parts of a coaxial cable are the center conductor, the dielectric, the outer conductor, and the protective jacket. Using this basic construction, over 1000 different coaxial cables have been developed to meet a broad range of application requirements.

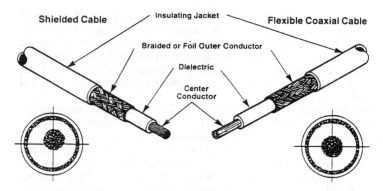

FIGURE 3.81 Schematic illustration of (a) shielded and (b) coaxial cable constructions. *(Courtesy AMP Incorporated.)*

Military applications dominated early developments, and many coaxial cables are described and specified in MIL-C-17. In addition, there are numerous commercial specifications, some highly specialized and some minor variants, to reduce costs of the military specification. Examples of common coaxial cable constructions, which extend beyond the basic geometry to meet increasingly demanding applications, are presented in Fig. 3.52.

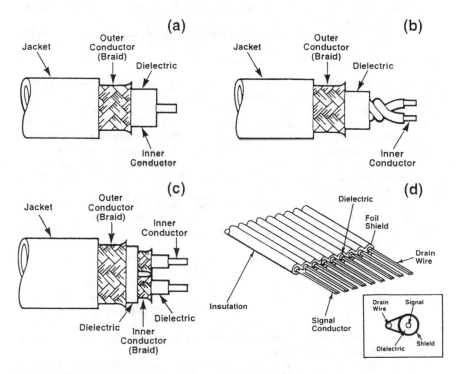

FIGURE 3.82 Schematic illustration of cross sections of selected coaxial cable constructions. *(Courtesy AMP Incorporated.)*

The most common coaxial cable, flexible cable, is illustrated in Fig. 3.52*a*. It uses a braided outer conductor of small gauge wires. While the braided construction provides flexibility, it does not afford complete shielding, due to gaps in the braid. To improve shielding, several braid layers may be used. In some cases, foil shields are also included. A shielding effectiveness of 85 percent is often considered the minimum acceptable. Some flexible cables use only a foil shield, which is lower cost but not as mechanically rugged as the braid construction.

It should also be noted that several technologies have been developed to incorporate air into the dielectrics to reduce the effective dielectric constant and thereby increase the velocity of propagation of coaxial cables in all constructions.

Twinax cable, shown in Fig. 3.52*b*, consists of two inner conductors with a common outer conductor. The inner conductors may be twisted pair or parallel line geometries. The principal application of twinax is balanced mode multiplex transmission in large computer systems. The signal is split between the two conductors and inverted to provide improved noise immunity. Similar to twinax, twin coaxial cable (Fig. 3.52*c*) consists of two coaxial cables surrounded by a common outer conductor.

Coaxial ribbon cable (Fig. 3.52*d*) consists of individual coaxial cables in a ribbon cable construction. It differs from ribbon or transmission cable in the presence of an outer conductor, which is generally a foil and drain wire combination.

A few words on signal transmission in coaxial cable are in order. It is important to note that the electromagnetic field is transmitted through the dielectric, while current flows on the outside of the inner conductor and on the inside of the outer conductor. Performance characteristics therefore depend on the conductor and dielectric materials and the geometric relationships of the cable construction.

Coaxial cable performance parameters of general interest include characteristic impedance, capacitance, *voltage standing wave ratio* (VSWR), velocity of propagation (V_p), and attenuation. These parameters will not be discussed here.

3.9.3.2 Coaxial Connectors

A coaxial connector ideally should appear as an extension of the cable. In other words, the connection process should not have an unacceptable impact on the performance parameters just listed. In practice, however, the connector will always introduce some nonuniformity in impedance, even for controlled impedance connectors. Sources of nonuniformity include variations in the cross section of the contacts, variations in the dielectric medium, and abrupt changes in direction of the contacts. The effect of these nonuniformities will be frequency dependent with the magnitude of the variation increasing with frequency. VSWR is a measure of the connector nonuniformity, and a value of 1.3 is often specified as a limit.

As for cables, a MIL specification, MIL-C-39012, covers many types of coaxial connectors although numerous commercial specifications and proprietary designs are also in common use. Coaxial connector selection is generally determined by the frequency range, cable size, and the coupling method.

The maximum frequency of the application, signal rise time dependent frequency, is a primary consideration. Connector characteristics that are strongly impacted by the application frequency include the coupling mechanism and the connection method.

There are three major coupling mechanisms: bayonet, snap-on, and threaded. Bayonet and snap on coupling, while simple, carry a penalty in overall shielding effectiveness and in noise during vibration. Threaded couplings offer a higher degree of resistance to shock and vibration as well as increased shielding performance due to the continuous shielding provided by this coupling mechanism.

Connection methods are arguably even more important in coaxial connectors than in conventional connectors due to cable preparation requirements and the VSWR penalties that accompany improper connections.

There are three main methods of making connections to coaxial cable:

- solder/clamp
- solder/crimp
- crimp/crimp

The first term describes the connection technique for the center conductor, and the second the connection to the braid. The center contacts are usually pin/socket configurations, as these are the most spatially uniform contact geometries.

In solder/clamp connections the center conductor is soldered to the contact, and the braid is clamped to the connector shell. This method provides easy field repairability but is operator sensitive and does not lend itself well to high-volume production.

Solder/crimp connections are the most common with the center conductor being soldered to the contact and the braid crimped, generally with a ferrule, to the connector shell. This method lends itself to volume production but requires tooling for field repair.

Crimp/crimp connections are especially suited to high-volume manufacturing with dedicated tooling. In addition, crimp/crimp connections provide the most consistent VSWR performance and reliability.

The connector size is determined by two factors. The obvious factor is the cable size. Coaxial connectors are described as subminiature, miniature, small, and large. Cable size, in turn, is affected by both physical constraints and attenuation considerations. Signal attenuation, of course, increases as cable size decreases.

Material selection for coaxial connectors affects both cost and performance. The base materials for the contacts are similar to other connectors with brass, phosphor bronze, and beryllium copper being the usual contact spring materials. Connector shells and ferrules include tin- and nickel-plated brass and stainless steel.

Gold is the most common center contact finish, although silver and tin are also used in lower-cost applications.

The dielectric material is usually either PTFE or polypropylene. PTFE is the superior material due to a wider temperature range and a lower, less lossy, and more stable dielectric constant. Polypropylene is generally used only in commercial applications.

Coaxial connector types are defined in terms of the coupling mechanism, the connection mechanism, and the cable style to which they are connected. Examples of the connector types to be described are presented in Fig. 3.53.

BNC connectors are bayonet coupling connectors, most of which are 50 Ω impedance and suitable for applications up to 4 GHz. Connection styles depend on the application (military or commercial) and include single and dual crimp, single and dual hex crimp, and other proprietary connections. BNC connectors are used primarily as I/O connectors in networks, instrumentation, and computer peripheral applications, level 5.

TNC connectors differ from BNC in that the coupling is threaded, which provides improved performance to 11 GHz and enhanced vibration and shock resistance. TNC connectors are available in essentially the same variety of styles as BNC.

SMA connectors were developed for subminiature coaxial cable and see wide use in avionics, radar, and high-performance test instrumentation applications. They are 50 Ω threaded coupling connectors and meet MIL-C-39012 up to 12.4 GHz. Special crimped SMA connectors are suitable up to 26 GHz. They are available in a similar range of styles to BNC.

SMB connectors are a snap-on coupling version of SMA and, as such, have a lower performance capability, to 4 GHz.

SMC connectors are also 50 Ω threaded coupling connectors, but due to their smaller size they have a lower performance rating than SMA. The reduction in performance comes

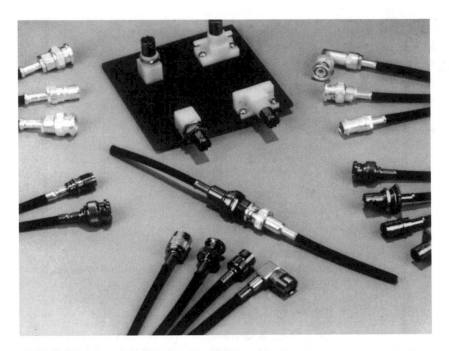

FIGURE 3.83   Examples of coaxial connectors. *(Courtesy AMP Incorporated.)*

from the proportionally larger tolerance variation as connector size is reduced. These variations lead to an increased VSWR.

Coaxial connectors for twinax and other coaxial cable variants are also available. Space precludes discussion of this variety of connector types, many of which are proprietary.

The number and variety of high-performance, high-frequency applications is continually increasing. Coaxial connector types and performance requirements are changing along with this trend. A wide variety of connectors is available for applications including satellite and microwave communications, electronic countermeasures, computer networking, and increasingly sophisticated test equipment.

3.9.4 Summary of Connector Types

This discussion has covered only some basic types of connectors and variants. The variety of connector types exists and keeps increasing to satisfy the ever expanding and increasingly sophisticated functional requirements of electronics technology.

3.10 CONNECTOR TESTING

Connector testing is performed for several reasons, primary among which are development or performance verification and product qualification. In addition, reliability testing is receiving increased attention.

3.10.1 Connector Testing

In general, a test consists of a combination of an exposure or conditioning process, intended to simulate an application condition or environment, followed by a measurement, generally related to some performance characteristic. For example, a heat age test may consist of an exposure to 1000 hours at 125°C to assess the effect of stress relaxation on contact normal force, followed by a contact resistance measurement, a measure of performance.

The conditioning provided by a given exposure may be mechanical (vibration, shock), thermal (steady-state heat age or temperature cycling), or environmental (mixed flowing gas or humidity). Measurements include electrical (contact resistance, dielectric withstanding voltage), mechanical (pull strength, mating force), and visual (dimensions, workmanship).

Connector testing may consist of individual or sequential tests, depending on the purpose. The discussion here will be limited to an overview but with an emphasis on qualification testing. Additional issues of concern for reliability assessment also will be discussed.

Qualification testing is intended to verify that a given product, produced under defined manufacturing processes, is capable of meeting the performance and dimensional requirements of the manufacturers product specification or of an industry derived specification. In general, the specification requirements will have been developed to ensure that the connector will meet the application requirements of the particular market/application for which it was developed.

3.10.1.1 Types of Tests In this discussion, the type of test will be defined by the exposure condition; e.g., vibration followed by contact resistance is a mechanical test, and heat age followed by mating force is an environmental test. A qualification test program will consist of at least the following:

1. sample validation
2. mechanical tests
3. electrical tests
4. environmental tests

Sample validation includes verification of dimensions, color, polarization mechanisms, and other general features of the product family as appropriate, including quality of workmanship. In addition, the materials of manufacture and design parameters such as contact finish thickness and normal force may be verified. The purpose of these evaluations is verify that the samples are representative of the product which is to be qualified.

Mechanical exposures include

- durability cycling
- vibration, sinusoidal, or random
- shock
- acceleration
- cable bending

Basic mechanical measurements include

- contact engaging and separating force
- connector mating and unmating force

- contact retention force in the housing
- connection tensile strength
- latching mechanism strength

Other mechanical tests are used and may be standard or proprietary tests to meet specific product or industry requirements.

Electrical exposures include

- overload current
- current cycling
- dielectric breakdown/withstanding

Electrical measurements cover a broad range, including

- contact resistance
- overall resistance
- termination resistance
- insulation resistance
- T-rise versus current
- capacitance
- VSWR

The same comments apply as for mechanical testing. Contact resistance measures the interface resistance, termination resistance measures the resistance of the permanent connection, and overall resistance measures the resistance across the total connector and insulation resistance the performance of the housing material. T-rise, capacitance, and VSWR are application-dependent tests.

Environmental tests are intended to the verify stability of connector performance under the anticipated conditions of the application environment. The measurements just described may be performed, but only after conditioning such as humidity or temperature exposure. Examples of environmental conditioning include

- humidity (steady-state and/or cyclic)
- temperature/humidity cycling
- mixed flowing gas
- salt spray

3.10.1.2 Test Sequences A qualification testing program generally consists of a number of test sequences which are performed serially or in parallel. The test sequences and test groups are usually specific to an exposure condition, i.e., mechanical, electrical, or environmental. Different sets of samples are used for different sequences because different measurements and fixturing are required. An example of a typical product qualification test program is provided in Figure 3.54.

3.10.1.3 Summary The relevance of qualification testing is dependent on ensuring that the test exposures and measurement methods and procedures are appropriate for the product and the intended application. Testing in addition to product qualification may be directed toward verifying the reliability of the connector in which case application considerations become even more important.

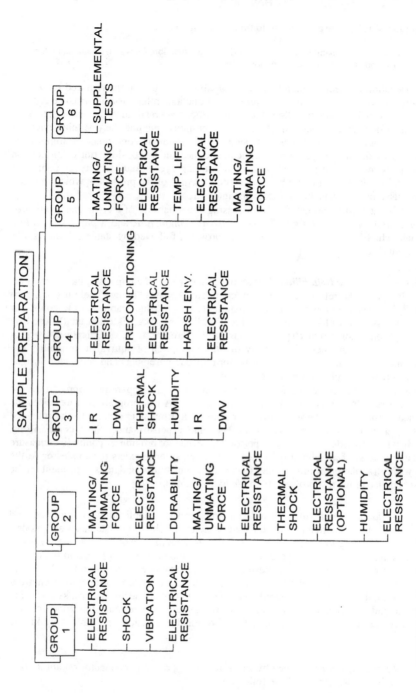

FIGURE 3.84 Generic test plan.

3.10.2 Estimating Connector Reliability

A definition of *reliability* appropriate to this discussion is:

> Reliability is the probability of a product performing a function, for which designed, under specified conditions for a specific period of time.

In evaluating connector reliability, the "probability" would be the required reliability, or failure rate, that the connector must provide. "Function" relates to performance criteria, examples of which include the allowable change in contact resistance, T-rise, or the specified number of mating cycles. The "specified conditions" include the operating environment and application requirements. Operating environment conditions of importance include the application temperature, chemical, and mechanical considerations. Application requirements include the required number of mating cycles and stability of connector resistance. The time factor could be the intended lifetime of the product in which the connector is used or an acceptable maintenance interval.

There are at least three methods for assessing the reliability of a connector: *comparative, prediction based on design and materials of manufacture,* and a *physics of failure* approach, which includes an appropriate test program followed by statistical analysis of the relevant data.

3.10.2.1 Comparative Reliability Reliability assessment via comparison has many limitations. If a particular reference connector, connector A, has a "known" field history, connector B, if it has common materials of manufacture and design parameters to connector A, may be assumed to have a similar reliability. This approach is often tacitly used by comparing connector product specifications, which are based on qualification test programs, against one another. The validity of the assumption of equivalent reliability has become more questionable as connector performance requirements and connector designs have become more sophisticated and demanding.

To address this increased performance sophistication, a comparative approach can be made more relevant by the use of comparative testing. A test program that exercises the reference connector and the alternative connectors in a manner relevant to the intended operating environment can be used to assess relative performance. Such comparisons, while they may provide a hierarchy of performance, cannot provide a quantitative measure of reliability. Even if a "reliability" based on field experience exists for connector A, the validity of extrapolation of this number to the comparative testing data is dependent on the correspondence of the test program to the field conditions.

3.10.2.2 Reliability Prediction In military and aerospace applications, component and connector reliability is often predicted based on generic failure rate data bases and modeling, usually according to MIL-HDBK-217, *Reliability Prediction of Electronic Equipment.* While this approach may be appropriate for some components, its application to connectors is questionable for at least two reasons. First, the data base for failure rate values is, in general, not well documented either with respect to failure criteria or operating environment. Second, the data do not generally distinguish among connector materials and design features intended to enhance connector performance. For these reasons, this approach will not receive further consideration here. For additional discussion see Mroczkowski and Maynard.[63]

3.10.2.3 Physics of Failure The physics of failure approach to connector reliability prediction requires consideration of the following:

1. The active degradation mechanisms must be identified and categorized with respect to their importance in the application of interest.

2. Appropriate tests, acceleration factors, and exposures must be known, defined, or determined for these degradation mechanisms.

3. Failure criteria appropriate to the application of interest must be established.

4. The statistical approach to determining, or calculating, reliability values must be agreed upon.

Each factor will be considered individually.

Degradation or failure of a contact or connector can occur in many ways. It is convenient to divide the mechanisms into two categories, intrinsic and extrinsic. Intrinsic degradation mechanisms refer to those mechanisms, which are related to the design and materials of construction of the contact or connector. Extrinsic mechanisms are those which are related to the application of the contact or connector.

Examples of intrinsic degradation are corrosion, loss of normal force through stress relaxation, and Joule heating leading to temperature related degradation.

Extrinsic degradation includes factors such as contamination and fretting corrosion. Each of these conditions is dependent on the application of the connector, both in manufacturing and use in the final system. Such degradation mechanisms can be qualitatively assessed but, in general, are difficult if not impossible to quantify for use in a determination of connector reliability.

Examples of other degradation mechanisms, which are outside the scope of our definition, include using the connector outside its rated temperature range (both ambient and enclosure related), applying currents in excess of the product specification (in both single and distributed modes), and improper mating practices (mating at excessive angles, pulling on cables, and so on) leading to contact abuse. These mechanisms fall outside our definition in that they violate the "under specified conditions for which designed" clause of our definition of reliability.

Connector manufacturers have control over intrinsic degradation but not over extrinsic factors. Extrinsic degradation can be controlled only by proper specification of product performance by connector manufacturers and proper use of the available information by the user. This joint responsibility ensures that the "under specified conditions for which designed" section of the reliability definition is met. Extrinsic degradation cannot be straightforwardly analyzed, so discussion will be limited to a few aspects of intrinsic degradation to provide a suggested approach to connector reliability evaluation.

Only one failure mode, increases in contact resistance, will be discussed, although many others exist, both mechanical (broken latches, bent pins, and so on) and electrical (crosstalk, leakage between contacts, and so forth).

Consider three degradation mechanisms: corrosion, stress relaxation, and wear. Each of these mechanisms can lead to connector failure, directly or indirectly, through increased contact resistance. Corrosion of the contact interface directly results in increases in contact resistance. Stress relaxation results in loss of contact normal force which, in turn, can lead to contact resistance increases, either directly or through increased susceptibility to mechanical or corrosive degradation. Plating wear can lead to contact resistance degradation if wear through occurs to the contact spring material which may be susceptible to corrosion.

These degradation mechanisms are well established, and the failure cause associated with them—increased contact resistance—is an obvious factor in contact reliability. Thus, information is available relevant to our first concern, degradation mechanisms. Experience with the product, or similar products or applications, will allow us to categorize and rank the degradation mechanisms.

Determination of an acceleration factor for a test exposure requires that a relationship between test methods/data and field experience be developed. Simply put, the objective is to be able to state that X days of test A are equivalent to Y years in application B. In other words, an acceleration factor between the test exposure and field performance can be defined. Such a capability is necessary because one of the components of the definition of reliability is performance "for a specified period of time." Unfortunately, there are few tests for which such a statement can be made. One possible example is the work done at Battelle Columbus Laboratories,[21] which provides an approach and data base from which an acceleration factor for mixed flowing gas (MFG) corrosion testing of noble metal finished connectors can be developed.

Given an acceleration factor, a statistical method to use the data obtained to quantify reliability is needed. There are also a number of additional statistical analysis issues which must be addressed to assess connector reliability. Mroczkowski and Maynard[62] discuss this topic in detail. The type of data, variables or attribute and the acceptance criteria are of particular interest.

In a corrosion test, such as MFG, contact resistance can be measured, and variables data and the distribution of the data can be determined. Alternatively, the contact resistance can be compared to some acceptance criterion and recorded as pass/fail attribute data. The statistical treatment of the data and the sample size needed for a given statistical confidence level varies significantly for these two options.

One parameter which enters into both methods of data handling is the "acceptance" or "failure" criterion applied to the data. For attribute data, it appears directly. In the case of variables data, it will appear in the K factor, a statistical parameter, which reflects the relationship between sample size and the reliability, and confidence level, which can be obtained from a test program. In either case, the choice of "acceptance" or "failure" criterion directly influences the calculated reliability.

As an example of a failure criterion, consider the allowed change in contact resistance after exposure to the test program. What is the appropriate value of contact resistance to use in calculating contact reliability? Consider two possibilities, the product specification and an application related value.

The product specification contact resistance is not the proper choice. This value has a "reliability" aspect to it in the sense that the manufacturer has tested the product design with respect to ensuring that the specified contact resistance will be maintained in its intended range of applications. In effect, the product specification value includes a "safety factor."

In a particular application, a user will have established a value of contact resistance at which the system of interest will cease to function. The value may be 100 mΩ or more in a signal application, or 0.5 mΩ or less for a power contact. The "failure" criterion for contact resistance should be based on this application specific value rather than the product specification contact resistance. The desired requirements on confidence limit and reliability should then be applied with this resistance value as the upper limit of acceptability.

3.10.2.4 From Contact to Connector Reliability   The discussion up to now has been concerned with contact reliability and contact resistance as a degradation mode. The ultimate concern is with connector reliability, which would generally include a number of contacts and degradation modes. A methodology for translating contact reliability, as determined for the desired range of degradation mechanisms, into connector reliability is required. The issue is complicated by two interacting factors. The resistance data is obtained for individual contacts, but the effect of the test on contact resistance is influenced by the fact that the contacts are, generally, in a connector and are part of a system. In other words, the effects of connector housing design and the particular connector system on contact resistance are included in the data. Under these conditions, the reliability being measured is the contact reliability in the specific connector. The contact reliability may not be the same in a different housing geometry or connector system. An important part of developing a reliability testing program is to account for this fact. Aside from this consideration, which is far from trivial, there is another factor that must be considered before contact reliability is used to determine connector reliability, and that is the issue of multiple degradation mechanisms and multiposition connectors.

One approach to this issue is to treat all the degradation mechanisms and individual contact reliabilities as independent. Under this assumption, the contact reliability would be the product of the reliabilities for the individual degradation mechanisms. The connector reliability, in turn, would be the contact reliability to the power of the number of positions in the connector, assuming a failure at any position in the connector results in a system failure.

For example, the reliability for a 16-position connector which is to be qualified for five degradation mechanisms would be given by:

$$R = \{R(t_1)R(t_2)R(t_3)R(t_4)R(t_5)\}^{16}$$

where $R(t_n)$ is the contact reliability as determined for test n, which is intended to simulate degradation mechanism n. If it is assumed that the contact reliability happened to be the same for each test, the connector reliability will be given by the contact reliability raised to the fifth power for the five tests, and to the sixteenth power for the number of positions. For the connector then, the contact reliability will be raised to the eightieth power!

This eightieth power dependence will significantly impact at least two aspects of qualification programs, cost and contact reliability requirements. For a connector reliability of the order of 0.9999 the contact reliability requirement will be of the order of 0.999999. Achieving a reliability of this magnitude is one problem (a significant one indeed), but another is verifying that it has been achieved. The sample size for such a verification could be very large and will affect program cost.

3.10.2.5 Summary of Reliability Assessment

3.10.2.5 Summary of Reliability Assessment   The purpose of this discussion was to present some of the issues and considerations which are pertinent to determining and calculating connector reliability on a statistical basis. In such an approach, a reliability qualification program consists of the following steps:

1. Determine an application specific contact resistance acceptance criterion. A criterion will also be required for any other failure mode which is to be included in the qualification program.

2. Develop a test program to address the expected degradation mechanisms operative in the application. Ranking of failure modes may be considered in this process.

3. Derive acceleration factors, when possible, for the tests to be specified. When this cannot be done, no reliability prediction can be made. In such cases, only comparative performance capabilities can be provided.

4. Decide on the statistical treatment appropriate to the data generated in the qualification program.

5. Calculate the component reliability. It must be emphasized that both the connector manufacturer and the user should agree on the content, approaches, and values to be specified in these steps individually, and in the qualification program in general. In particular, mutually agreed engineering judgements must be made to select appropriate "acceptance/failure" criteria and acceleration factors for the program.

3.10.3 Summary of Connector Testing

The purpose of this section has been to provide an overview of connector testing and reliability evaluation. General practices and procedures have been reviewed, although it must be noted that specific product and application requirements may dictate different approaches. In general, however, the major purpose of connector testing is to verify product qualification so that selection of the product for a given application based on product specification values will result in satisfactory performance in the field.

REFERENCES

1. Granitz, R. F., "Levels of Packaging," *Inst. Control Syst.,* Aug., 1992.

2. Mroczkowski, R. S., "Materials Considerations in Connector Design," in *Proc. 1st Elec. Mat. Conf. of the American Soc. for Materials,* Chicago, IL, 1988.

3. Williamson, J. P. B., "The Microworld of the Contact Spot," in *Proc. 27th Ann. Hole Conf. on Elec. Contacts,* Chicago, IL, 1981.

4. Bowden, F. P. and Tabor, D., *The Friction and Lubrication of Solids,* Clarendon Press, Oxford, U.K., 1986.

5. Holm, R., *Electric Contacts,* Springer-Verlag, New York, NY., 1967.

6. Antler, M., "Wear of Contact Finishes: Mechanisms, Modeling and Recent Studies of the Importance of Topography, Underplate and Lubricants," in *Proc. 11th Ann. Conn. and Interconn. Tech Symposium,* Cherry Hill, NJ, 1978.

7. Durney, L.J., ed. *Electroplating Engineering Handbook,* Van Nostrand Reinhold, New York, 1981.

8. Reid, F. H. and Goldie, W., eds., *Gold Plating Technology,* Electrochemical Publications Ltd., 1974

9. Harlan, C., "Overview: Inlay Clad Metal and Other Electrical Contact Materials," *Metals Prog.,* Nov., 1979.

10. Mroczkowski, R. S., "Connector Contacts: Critical Surfaces." *Adv. Mat. and Processes,* vol. 134, Dec. 1988.

11. Zimmerman, R. H., "Engineering Considerations of Gold Electrodeposits in Connector Applications," presented at 10th Technical Convention on Electronic Components, Milano, Italy, 1973.

12. Antler, M. and Sproles, E.S., "Effect of Fretting on the Contact Resistance of Palladium," *Trans. IEEE CHMT,* 5, 1982.

13. Whitley, J. H., Wei, I. Y., and Krumbein, S. J., "A Cost Effective High Performance Alternative to Conventional Gold Plating on Connector Contacts," in *Proc. 33d Elec. Comp. Conf.,* Orlando, FL, 1983.

14. Whitelaw, K. S., "Gold Flashed Palladium Nickel for Electronic Contacts," *Trans. Inst. Met. Finish.,* 64, 1986.

15. Antler, M., Drozdowicz, M. H., and Haque, C. A., "Connector Contact Materials: Effect of Environment on Clad Palladium, Palladium-Silver Alloys and Gold Electrodeposits," *Trans. IEEE CHMT,* 4, 1981.

16. Bock, E. M. and Whitley, J. H., "Fretting Corrosion in Electrical Contacts," *Proc. 20th Ann. Hole Conf. on Elec. Contacts,* Chicago, IL, 1974.

17. Antler, M., "The Tribology of Contact Finishes for Electronic Connectors: Mechanisms of Friction and Wear," *Plat. and Surf. Fin.,* 75, Oct. 1988.

18. Antler, M., "The Tribology of Contact Finishes for Electronic Connectors: The Effect of Underplate, Topography and Lubrication." *Plat. and Surf. Fin.,* 75, Nov. 1988.

19. Bader, F. E., Sharma, S. P. and Feder, M., "Atmospheric Corrosion Testing of Connectors—A New Accelerated Test Concept," in *Proc. 9th International Conf. on Elec. Contact Phenomena,* 1978.

20. Abbott, W. H., "Field Versus Laboratory Experience in the Evaluation of Electronic Components and Materials," in *NACE Corrosion/83,* Anaheim CA, Paper 234, 1983.

21. Abbott, W. H., "The Development and Performance Characteristics of Mixed Flowing Gas Test Environments," In *Proc 33d IEEE Holm Conf. on Elec. Contacts,* 1987.

22. Mroczkowski, R. S., "Corrosion and Electrical Contact Interfaces," in *NACE Corrosion/85,* Paper 235, 1985.

23. Geckle, R. J. and Mroczkowski, R. S., "Corrosion of Precious Metal Plated Copper Alloys Due to Mixed Flowing Gas Exposure," *Trans. IEEE CHMT,* 14(3), 1991.

24. Wagar, H. N., "Principles of Electronic Contacts," in *Physical Design of Electronic Systems,* vol. III, Chap. 8. Prentice-Hall, Englewood Cliffs, NJ, 1971.

25. Abbott, W. H., and Schreiber, K. L., "Dynamic Contact Resistance of Gold, Tin and Palladium Connector Interfaced During Low Amplitude Motion," in *Proc. 27th Ann. Hold Conf. on Elec. Contacts,* Chicago, IL, 1981.

26. Antler, M., "Contact Materials for Electronic Connectors: A Survey of Current Practices and Technology Trends in the U.S.A.," *Plat. and Surf. Fin.,* 78, June, 1991.

27. Lowenthal, W. S., Harkness, J. C. and Cribb, W. R., "Performance Comparison in Low Deflection Contacts," *Proc. Internepcon/UK84,* 1984.

28. Horn, K. W. and Zarlingo, S. P., "Understanding Stress Relaxation in Copper Alloys," *Proc 15th Ann. Conn. and Interconn. Tech. Symposium,* 1983.

29. Bersett, T. E., "Back to Basics: Properties of Copper Alloy Strip for Contacts and Terminals," *Proc. 14th Ann. Conn. and Interconn. Technology Symposium,* Philadelphia, PA, 1981.

30. Spiegelberg, W. D., "Elastic Resilience and Related Properties in Electronic Connector Alloy Selection," in *Proc. ASM Int'l 3rd Electrical Materials Conf.,* San Francisco, CA.

31. Alvino, W. M., *Plastics for Electronics: Materials, Properties and Design Applications,* McGraw-Hill, New York, NY, 1995.

32. Walczak, R. S., McNamara, P. F., and Podesta, G. P., "High Performance Polymers, Addressing Electronic Needs for the '90's." *Proc. 21st Int. Conn. and Interconn. Tech. Symposium,* Dallas, TX, 1988.

33. Gupta, G. W., "High Performance Resins for VPS/IR Reflow Applications," *Proc. 22d Ann. Conn. and Interconn. Tech. Symposium,* Philadelphia, PA, 1989.

34. Hawley, J., "Design, Tooling and Material Considerations for Connector Housings Made of Glass-filled Crystalline Materials'" *Proc. 25th Ann. Conn. and Interconn. Tech. Symposium,* San Jose, CA, 1992.

35. Brockman, I. H., Sieber, C. S., and Mroczkowski, R. S., "A Limited Study of the Effects of Contact Normal Force, Contact Geometry, and Wipe Distance on the Contact Resistance of Gold Plated Contacts," *Trans. IEEE CHMT,* 11(12), 1988.

36. Brockman, I. H., Sieber, C. S., and Mroczkowski, R. S., "The Effects of the Interaction of Normal Force and Wipe Distance on Contact Resistance in Precious Metal Plated Contacts," in *Proc. 34th IEEE Holm Conf. on Elec. Contacts,* Chicago, IL, 1988.

37. Mroczkowski, R. S., "Contact Wiping Effectiveness: Interactions of Normal Force, Geometry and Wiping Mode," *Proc. Int'l. Conf. on Elec. Contacts and Electromechanical Components,* Beijing, China, 1989.

38. Van Horn, R. H., "The Design of Separable Connectors," *Proc. 20th Elec. Comp. Conf.,* Las Vegas, NV, 1970.

39. Whitley, J. H. and Mroczkowski, R. S., "Concerning Normal Force Requirements for Precious Metal Plated Connectors," *Proc. 20th Ann. Conn. and Interconn. Tech. Symp.,* Philadelphia, PA, 1987.

40. Whitley, J. H., "The Tin Commandments," *Plating and Surface Finishing,* Oct., 1981.

41. Mroczkowski, R. S., "Concerning Hertz Stress as a Connector Design Parameter," *Proc. 24th Ann. Conn. and Interconn. Tech. Symposium,* San Diego, CA, 1991.

42. Kantner, E. A. and Hobgood, L. D., "Hertz Stress as an Indicator of Connector Reliability," *Connection Technology,* Mar., 1989.

43. Fluss, H. S., "Hertzian Stress as a Predictor of Connector Reliability," *Connection Technology,* Dec., 1990.

44. Eammons, W., Chang, J., Stankos, J., Abbott, W., Sharrar, R., Wutka, T., and Stackhouse, A., "Connector Stability Test for Small System Connectors," *Proc. 15th Int'l. Conf. on Elec. Contacts and 36th IEEE Holm Conf. on Elec. Contacts,* Montreal, Canada, 1990.

45. Mroczkowski, R. S. and Geckle, R. J., "Concerning Cold Welding in Crimped Connections," *Proc. 41st IEEE Holm Conf. on Elec. Contacts,* Montreal, Canada, 1995.

46. Whitley, J. H., "The Mechanics of Pressure Connections," presented at EDN Regional Engineers meeting, New York, Dec., 1963.

47. Mroczkowski, R. S., "Conventional versus IDC Crimps," *Connection Technology,* July, 1986.

48. Mitra, N. K., "An Evaluation of the Insulation Displacement Contact," *Proc. 25th Holm Conf. on Elec. Contacts,* 1979.

49. Wandemacher, K. S., "Insulation Displacement Connector Reliability," *Proc. 19th Ann. Conn. and Interconn. Tech. Symposium,* 1986.

50. Key, E., "A New Concept in Wiring Device Termination—Using Zero-Gap Insulation Displacement," *Proc. 21st Ann. Conn. and Interconn. Tech. Symposium,* Dallas, TX, 1988.

51. Goel, R., "Analysis of Press Fit Technology," *Proc. 31st Elec. Comp. Conf.,* Atlanta, GA, 1981.

52. Scaminaci, J., "Solderless Press-Fit Connections—A Mechanical Study of Solid and Compliant Contacts," *Proc. Ninth Annual Connector Symposium,* Cherry Hill, NJ, 1976.

53. Dance, F. J., "The Electrical and Reliability Characteristics of Compliant Press-Fit Pin Connectors," *IPC Tech. Rev.,* Dec., 1987.

54. Manko, H. H., *Soldering Handbook for Printed Circuits and Surface Mounting,* Van Nostrand Reinhold, New York, NY, 1986.

55. Katyl, R. H. and Simed, J. C., "Electrical Design Concepts in Electronic Packaging." Chapter 3 in *Principles of Electronic Packaging,* D. Seraphim, R. Lasky and C. Li, eds. McGraw Hill, New York, 1989.

56. Chang, C. S., "Electrical Design Methodologies," in ASM *Electronic Materials Handbook, Volume 1, Packaging,* ASM, Materials Park, 1989.

57. Southard, R. K., "High Speed Signal Pathways from Board to Board," *Electron. Eng.,* Sept., 1981.

58. Sucheski, M. M. and Glover, D. W., "A High Speed, High Density Board to Board Stripline Connector," *Proc. 40th Elec. Comp. and Tech. Conf.,* Las Vegas, NV, 1990.

59. Aujla, S. and Lord, R., "Application of High Density Backplane Connector for High Signal Speeds," *Proc. 26th Ann. Conn. and Interconn. Symposium and Trade Show,* San Diego, 1993.

60. Corman, N. E. and Mroczkowski, R. S., "Fundamentals of Power Contacts and Connectors," *Proc. 23d Ann. Conn. and Interconn. Tech. Symposium,* Toronto, Canada, 1990.

61. Wise, J. H., "A Method for End of Life Contact Current Rating," *Proc. 21st Ann. Conn. and Interconn. Tech. Symposium,* Dallas, TX, 1988.

62. Wise, J. H., "End of Life Contact Current Rating Method for Non-sinusoidal Wave Forms," *Proc. 23d Ann. Conn. and Interconn. Tech. Symposium,* Toronto, Canada, 1990.

63. Mroczkowski, R. S., and Maynard, J. M., "Estimating the Reliability of Electrical Connectors," *IEEE Trans. on Reliability,* 40 (5), 1991.

RECOMMENDED READING

1. Stearns, T. H., Flexible Printed Circuitry, McGraw-Hill, Nw York, 1995.

CHAPTER 4

WIRING AND CABLING FOR ELECTRONIC PACKAGING

Edward J. Croop
Technology Diversified Services
Pittsburgh, Pennsylvania

4.1 INTRODUCTION

The selection of suitable hookup and interconnecting wires or cables is an important step in the overall design of electronic packaging. Often these wires have been the last areas to be considered, and designs were based on historical or rule-of-thumb ratings.

Hookup wire can be thought of as connecting the components within electronic "black boxes" or chassis wiring, with insulation wall thicknesses below 20 mil, and usually rated at 250 to 600 V. Interconnecting wire usually indicates a cable or harness between black boxes or equipment, more rugged and meeting specifications more rigorous than those for hookup wire.

Recent years have seen many improvements in electronic packaging technology, with a much higher density of components on printed wiring boards, a vastly increased number of input and output leads on integrated-circuit devices, increased speed of chips and devices (VHSIC), an increased number of layers of copper-clad wiring per board, and an entire new technology of surface-mounting devices replacing through-hole mounting and interconnecting, to name a few. As a result, the selection of wiring and cabling can be critical to the success of package and systems designs.

Because of the many advances in this technology, more and more engineers have recognized the importance of optimizing conductors, insulation, shielding, jacketing, and cabling. They are seeking materials that offer the best combinations of size, weight, environmental protection, and ease of shop handling, together with the lowest possible cost, best availability, and least maintenance. If the intended environment is abnormal, proper wire and cable selection should be verified by adequate environmental testing. This should include extended life testing to determine a probable service life extrapolation. In addition, the designer must evaluate the environmental, mechanical, and electrical stresses and consider the compatibility of the materials with all possible encapsulants, fuels, chemicals, explosives, as well as the new and varied types of associated hardware and fabrication techniques.

The purpose of this chapter is to provide the designer and user with sufficient background information and guideposts for the effective selection of hookup and interconnection wire and cable to meet the demands of specific applications.

4.2 WIRE AND CABLE INSULATION

4.2.1 Materials

The major properties of insulating and jacketing materials used with hookup and interconnecting wire and cable can be divided into three main categories: mechanical, electrical, and chemical.

1. The mechanical and physical properties to consider are tensile strength, elongation, specific gravity, abrasion resistance, cut-through resistance, and mechanical temperature resistance (cold bend and deformation under heat).
2. The electrical properties are dielectric strength, dielectric constant, loss factor, and insulation resistance.
3. The chemical properties are fluid resistance, flammability, temperature resistance, and radiation resistance.

Polymers and plastics represent the largest class of wire insulating materials. Thermoplastics have by far the greatest volume usage for hookup and interconnection wire insulation as well as for jacketing and cabling. Table 4.1 lists the more common insulating materials. A more complete description of some of these polymers is given in Chap. 1.

Tables 4.2 and 4.3 list some major properties for the most commonly used wire insulating and jacketing materials. The data are presented only as a guide; variations from the given values result from varying wall thicknesses, processing methods, test methods, and so on. Figures 4.1–4.3 show typical physical design characteristics of hookup and interconnection wire conductors.

TABLE 4.1 Some Common Wire Insulating Materials

Butyl rubber (butyl)
Ethylene propylene diene monomer rubber (EPDM)
Fluorinated ethylene propylene (Teflon* FEP)
Fluorocarbon rubber (Viton*)
Monochlorotrifluoroethylene (Kel-F†)
Neoprene rubber (Neoprene*)
Polyalkene
Polyamide (nylon)
Polyethylene (PE)
Polyethylene terephthalate polyester (Mylar*)
Polyimide (Kapton*)
Polypropylene
Polysulfone
Polytetrafluoroethylene (Teflon* TFE)
Polyurethane elastomers (urethane)
Polyvinyl chloride (PVC)
Polyvinylidene fluoride (Kynar‡)
Silicone rubber (silicone)

*Registered trademarks of E. I. du Pont de Nemours & Company, Inc.
†Registered trademark of 3M Company, Inc.
‡Registered trademark of Pennwalt, Inc.

TABLE 4.2 Mechanical and Physical Properties of Insulating Materials

Insulation	Common designation	Tensile strength, lb/in²	Elongation, %	Specific gravity	Abrasion resistance	Cut-through resistance	Temperature resistance (mechanical)
Polyvinyl chloride	PVC	2400	260	1.2–1.5	Poor	Poor	Fair
Polyethylene	PE	1400	300	0.92	Poor	Poor	Good
Polypropylene		6000	25	1.4	Good	Good	Poor
Cross-linked polyethylene	IMP	3000	120	1.2	Fair	Fair	Good
Polytetrafluoroethylene	TFE	3000	150	2.15	Fair	Fair	Excellent
Fluorinated ethylene propylene	FEP	3000	150	2.15	Poor	Poor	Excellent
Monochlorotrifluoroethylene	Kel-F	5000	120	2.13	Good	Good	Good
Polyvinylidene fluoride	Kynar	7100	300	1.76	Good	Good	Fair
Silicone rubber	Silicone	800–1800	100–800	1.15–1.38	Fair	Poor	Good
Polychloroprene rubber	Neoprene	150–4000	60–700	1.23	Good	Good	Fair
Butyl rubber	Butyl	700–1500	500–700	0.92	Fair	Fair	Fair-good
Fluorocarbon rubber	Viton	2400	350	1.4–1.95	Fair	Fair	Fair-good
Polyurethane	Urethane	5000–8000	100–600	1.24–1.26	Good	Good	Fair-good
Polyamide	Nylon	4000–7000	300–600	1.10	Good	Good	Poor
Polyimide film	Kapton	18,000	707	1.42	Excellent	Excellent	Good
Polyester film	Mylar	13,000	185	1.39	Excellent	Excellent	Good
Polyalkene		2000–7000	200–300	1.76	Good	Good	Fair-good
Polysulfone		10,000	50–100	1.24	Good	Good	Good
Polyimide-coated TFE	TFE/ML	3000	150	2.2	Good	Good	Good
Polyimide-coated FEP	FEP/ML	3000	150	2.2	Good	Good	Good

Source: Martin Marietta Corporation.

TABLE 4.3 Electrical Properties of Insulating Materials

Insulation	Common designation	Dielectric strength, V/mil	Dielectric constant at 10^3 Hz	Loss factor at 10^3 Hz	Volume resistivity, $\Omega \cdot$ cm
Polyvinyl chloride	PVC	400	5–7	0.02	2×10^{14}
Polyethylene	PE	480	2.3	0.005	10^{16}
Polypropylene		750	2.54	0.006	10^{16}
Cross-linked polyethylene	IMP	700	2.3	0.005	10^{16}
Polytetrafluoroethylene	TFE	480	2.1	0.0003	10^{18}
Fluorinated ethylene propylene	FEP	500	2.1	0.0003	10^{18}
Monochlorotrifluoroethylene	Kel-F	431	2.45	0.025	2.5×10^{16}
Polyvinylidene fluoride	Kynar	1280 (8 mil)	7.7	0.02	2×10^{14}
Silicone rubber	Silicone	575–700	3–3.6	0.003	2×10^{15}
Polychloroprene rubber	Neoprene	113	9.0	0.030	10^{11}
Butyl rubber	Butyl	600	2.3	0.003	10^{17}
Fluorocarbon rubber	Viton	500	4.2	0.14	2×10^{13}
Polyurethane	Urethane	450–500	6.7–7.5	0.055	2×10^{11}
Polyamide	Nylon	385	4–10	0.02	4.5×10^{13}
Polyimide film	Kapton	5400 (2 mil)	3.5	0.003	10^{18}
Polyester film	Mylar	2600	3.1	0.15	6×10^{16}
Polyalkene		1870	3.5	0.028	6×10^{13}
Polysulfone		425	3.13	0.0011	5×10^{16}
Polyimide-coated TFE	TFE/ML	480	2.2	0.0003	10^{18}
Polyimide-coated FEP	FEP/ML	480	2.2	0.0003	10^{18}

Source: Martin Marietta Corporation.

4.2.2 Construction and Application Methods

Reference has been made to extrusion, tape wrapping, and coating as methods of applying specific materials to a conductor. Certain insulating materials are available in a tape or solution form only and cannot be obtained in extruded form. Modern engineering thermoplastics such as PEEK and PES have outstanding thermal capabilities, but are difficult to apply to wire because of their high melt temperatures and their insolubility in all but the most powerful (and expensive) solvents. Following is a brief description of methods of application of conventional materials.

Extrusion. Many of the materials listed in Table 4.1 are thermoplastic; that is, they are heated to softness, formed into shape, then cooled to become solid again. Thermoplastics are shaped around a conductor by the extrusion process. This consists in forcing the plastic material under pressure and heat through an orifice concurrently with the conductor.

Tape Wrap. Tape wrapping is the application of insulation in the form of a thin film or tape. The tape is normally applied to the conductor with minimum overlap to provide wire flexibility without baring the conductor. Layers of tape can be built up to achieve the desired insulation wall thickness. Successive layers are wrapped in opposite directions, ensuring coverage.

Coating. Insulation by dip coating is limited almost exclusively to magnet wire and to glass and other insulating fabrics and braided sleeving. For small-diameter Teflon TFE and Teflon FEP applications, a polyimide overcoating of less than 0.001 in enhances the mechanical strength, abrasion protection, and solder resistance of the insulation significantly.

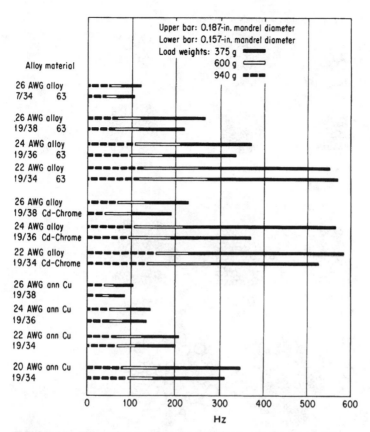

FIGURE 4.1 Flex endurance tests (concentric stranding), annealed copper versus high-strength alloys.[1]

4.3 THERMAL STABILITY AND THERMAL RATINGS OF INSULATION

4.3.1 Thermal Rating Background

As early as 1913, it had become apparent[3] that operating temperatures, hot-spot temperatures, and environmental ambient temperatures have a degrading effect with time on the ability of organic materials to perform their designed electrical, mechanical, and chemical functions. Originally this was believed to be a linear effect, and a "10-degree rule" was later developed[4]—an increase of 10°C in integrated operating temperature would reduce life by 50 percent.

Modern practice treats thermal and other aging phenomena of organic materials as a chemical rate reaction, based on the Arrhenius rate equation

$$L = Ae^{B/T}$$

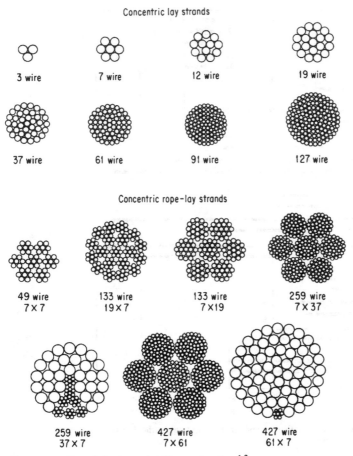

Concentric lay strands

3 wire   7 wire   12 wire   19 wire

37 wire   61 wire   91 wire   127 wire

Concentric rope-lay strands

49 wire
7 × 7

133 wire
19 × 7

133 wire
7 × 19

259 wire
7 × 37

259 wire
37 × 7

427 wire
7 × 61

427 wire
61 × 7

FIGURE 4.2 Stranded wire and cable construction.[1,2]

which plots as a straight line as

$$\ln L = \ln A + B/T$$

where L is the life (usually hours), T is the absolute temperature (kelvins), and A and B are constants determined by the activation energy of the particular reaction.[5] Data are derived by first plotting the change in a property (electrical, mechanical, chemical, weight loss, etc.) of the material on aging at various temperatures to an arbitrarily selected endpoint and then plotting logarithmically the times to reach the endpoint versus the reciprocal of the absolute temperatures of the thermal aging. This usually results in a straight line. Extrapolation of the line can be used to predict life expectancy.

4.3.2 Methods of Determining Thermal Stability

ASTM D-2307 is the most widely used and perhaps one of the simplest methods (twisted pair) of determining the thermal stability of insulated wire and cable. The entire test proce-

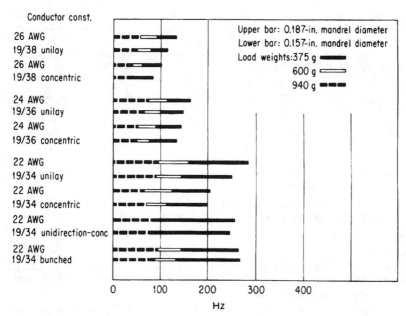

FIGURE 4.3 Flex endurance tests on annealed conductors (copper) in different stranding configurations.[1]

dure is also described in the military (federal) specification J-W-1177. IEEE Std. 1 defines the temperature index and IEEE Std. 98 explains its test procedure.

4.3.3 Thermal Classifications

Table 4.4 summarizes the evolution of insulation temperature classifications.

4.4 SHIELDING

4.4.1 Materials and Construction

Although there is no real substitute for braided copper as a general-purpose shield, certain applications may allow the use of a lighter, less bulky shield construction. The purpose of a shield is (1) to prevent external fields from adversely affecting signals transmitted over the center conductor, (2) to prevent undesirable radiation of a signal into nearby or adjacent conductors, or (3) to act as a second conductor in matched or tuned lines. Table 4.5 presents the shielding effectiveness at tested frequency ranges of the system described. A list of various available shield types follows.

Braided round shield.   Braiding consists in interweaving groups (carriers) of strands (ends) of metal over an insulated conductor. The braid angle at which the carriers are applied with reference to the axis of the core, the number of ends, the number of carriers, and the number of carrier crossovers per inch (picks) determine the percent cover-

TABLE 4.4 Insulation Temperature Classifications

Thermal class (obsolete)	Limiting hot-spot temperature, °C	Old temperature ratings	New temperature ratings	Number range, °C	Temperature index
0	90	Class 90	90	90–104	90
A	105	Class 105	105	105–129	105
B	130	Class 130	130	130–154	130
F	155	Class 155	155	155–179	155
H	180	Class 180	180	180–199	180
>H, H+	Over 180	Class 200	200	200–219	200
		Class 220	220	200+	—

age, which is a measure of the shielding gap (Fig. 4.4). For reference, a solid metal tube is equivalent to 100 percent coverage.

Flat braided shield

Metal tape

Solid shield. Solid or tubular shields may be applied by tube swaging or by forming interlocked, seamwelded, or soldered tape around the dielectric, offering 100 percent shielding as well as a protective armor.

Served shield. A served or spiral shield consists of a number of metal strands wrapped flat as a ribbon over a dielectric in one direction. The weight and size of the shielding are approximately one-half those of braided shields since there is no strand crossover or overlap.

Foil shield

Conductive plastics. PVC and PE compounds have been formulated with conductive additives for the purpose of shielding. Effectiveness has been poor, normally limited to the low, audio-frequency range. As with foil, termination is a problem; a drain wire can be used for terminating.

Conductive yarns. Conductive yarns, such as impregnated glass, provide weight reduction, but low conductivity and difficulty in termination make them suited for special applications only.

Metal shield and conductive yarn. An effective combination of metal strands and conductive yarns has been developed. It provides effective shielding when braided and reduces shield weight. Terminations can be accomplished by normal techniques.

4.4.2 Shield Jackets

It is the basic function of a shield jacket to insulate a shield from ground (structure), other shields, or conductors. Secondarily, a shield jacket may serve as a lubricant in multiconductor cables, enabling free movement during bending; as an insulator to allow varying potentials between adjacent shields; and as a moisture and abrasion barrier. A few of the materials described in Sec. 4.2 are appropriately used as jacketing materials because of either superior mechanical strength or compatibility with the chosen primary insulation system.

Extrusion, braiding, and taping techniques are used in forming shield jackets. Fused tapes and extruded materials are the only reliable methods for assuring a moistureproof jacket. Consistent wall thickness is best maintained with tapes.

TABLE 4.5 Shield Effectivity, Volts Peak to Peak

Sample description	Frequency									
	30 Hz	100 Hz	300 Hz	1 kHz	3 kHz	10 kHz	30 kHz	100 kHz	300 kHz	1 MHz
Control unshielded	<0.600	<1.6	4.0	10.0	12.0	12.5	13.5	12.5	12.5	14.0
Braid, no. 36 AWG, tinned copper, 90% coverage	<0.001	<0.001	0.0025	0.005	0.006	0.00625	0.0075	0.0075	0.007	0.008
Aluminum tape, no. 22 AWG drain wire	<0.001	<0.001	0.001	0.002	0.002	0.002	0.005	0.0085	0.012	0.014
Semiconductive PVC, no. 26 AWG drain wire	<0.001	<0.001	<0.00325	0.020	0.060	0.120	0.240	0.450	0.540	1.85
Scrve, no. 36 AWG tinned copper, 90% coverage, 4 ends reversed	<0.001	<0.001	0.001	0.002	0.0025	0.0025	0.004	0.006	0.0065	0.013
Braid, 8 carriers, no. 36 AWG tinned copper, 8 carriers, conductive glass yarn	<0.001	<0.001	<0.001	0.001	0.00125	0.00125	0.002	0.003	0.005	0.012
Braid, no. 36 AWG flat ribbon silver-plated copper	0.00125	0.003	0.008	0.018	0.024	0.024	0.028	0.028	0.027	0.029
Scrve, no. 36 AWG tinned copper	<0.001	0.003	0.008	0.018	0.026	0.026	0.029	0.029	0.030	0.031
Solid, cadmium-plated copper	<0.001	<0.001	<0.001	<0.001	<0.001	<0.001	<0.001	<0.001	<0.001	<0.001

Source: Martin Marietta Corporation.[6]

4.9

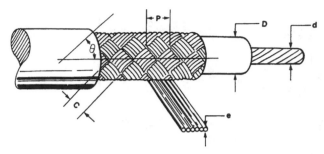

1 *Braid angle*

$$\theta = \tan^{-1}\left[\frac{2\pi(D + e)P}{C}\right], \quad \text{deg}$$

2 *Braid picks per inch*

$$P = \frac{C \tan \theta}{2\pi M}, \quad \text{picks/in}$$

3 *Braid shield weight*

$$w = \frac{nCl}{\cos \theta}, \quad \text{lb/Mft}$$

4 *Braid shield dc resistance*

$$R_{dc} = \frac{r_{dc}}{nC \cos \theta}, \quad \Omega/\text{Mft}$$

where D = diameter under shield, in
d = diameter of center conductor, in
C = number of carriers
e = diameter of end
P = pick (measured in picks per linear inch)
θ = braid angle, deg
w = weight of shield, lb/Mft
n = number of ends in one carrier
l = weight of one end, lb/Mft
M = D + buildup of braid on one shield wall, in
R_{dc} = dc resistance of braid shield, Ω/Mft
r_{dc} = dc resistance of one strand (end) of shield, Ω/Mft

FIGURE 4.4 Braid shield cable design equations. (*Courtesy of Carol Cable Company.*)

However, reliable fusion across interstices of multiple shielded conductors is frequently a problem. Varnished fiber braids allow a large conductor or multiple conductors more flexibility than extruded or taped jackets, but offer limited resistance to moisture and humidity. Figure 4.5 provides abrasion resistance data for shield jackets. In comparing abrasion resistance values, differences in wall thickness should be considered. Some available shield jacket materials follow:

Dacron braid
Fluorinated ethylene propylene (FEP)
Glass braid
Kynar
Polyamide (nylon)
Polyethylene-coated Mylar
Polyvinyl chloride (PVC)
Teflon TFE

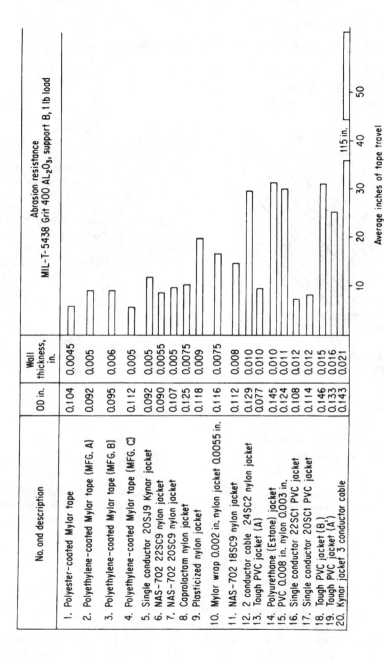

FIGURE 4.5 Shield jacket abrasion resistance. (*Courtesy of Martin Marietta Corp.*)

No. and description	OD in.	Wall thickness, in.	Abrasion resistance MIL-T-5438 Grit 400 AL₂O₃, support B, 1 lb load
1. Polyester-coated Mylar tape	0.104	0.0045	
2. Polyethylene-coated Mylar tape (MFG. A)	0.092	0.005	
3. Polyethylene-coated Mylar tape (MFG. B)	0.095	0.006	
4. Polyethylene-coated Mylar tape (MFG. C)	0.112	0.005	
5. Single conductor 20SJ9 Kynar jacket	0.092	0.005	
6. NAS-702 22SC9 nylon jacket	0.090	0.0055	
7. NAS-702 20SC9 nylon jacket	0.107	0.005	
8. Caprolactam nylon jacket	0.125	0.0075	
9. Plasticized nylon jacket	0.118	0.009	
10. Mylar wrap 0.002 in. nylon jacket 0.0055 in.	0.116	0.0075	
11. NAS-702 18SC9 nylon jacket	0.112	0.008	
12. 2 conductor cable 24SC2 nylon jacket	0.129	0.010	
13. Tough PVC jacket (A)	0.077	0.010	
14. Polyurethane (Estane) jacket	0.145	0.010	
15. PVC 0.008 in. nylon 0.003 in.	0.124	0.011	
16. Single conductor 22SC1 PVC jacket	0.108	0.012	
17. Single conductor 20SC1 PVC jacket	0.114	0.012	
18. Tough PVC jacket (B)	0.146	0.015	
19. Tough PVC jacket (A')	0.133	0.016	
20. Kynar jacket 3 conductor cable	0.143	0.021	

Average inches of tape travel

4.5 DESIGN CONSIDERATIONS

4.5.1 Selection of Wire Gauge

The following factors must be considered for proper wire gauge selection:

1. Voltage drop
2. Current-carrying capacity
3. Circuit-protector characteristics

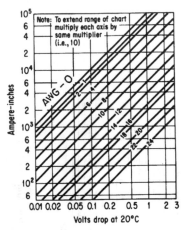

FIGURE 4.6 Direct-current voltage drop of copper wire.[1,7]

Voltage Drop. Voltage-drop calculations should be based on the anticipated load current at nominal system voltage. Voltage drop through an aluminum structural ground return can be considered zero for all practical purposes. Normal voltage drop limits do not apply to starting current of equipment such as motors. The voltage at load during startup should be considered to assure proper operation of the equipment. Figure 4.6 may be used to select the wire gauge. Ampere-inches is the product of the wire length between terminations (inches) and the wire current (amperes).

Current-Carrying Capacity. The following factors must be considered in determining the current-carrying capacity of a wire:

1. Continuous-duty rating
2. Short-time rating
3. Effect of ambient temperature
4. Effect of wire grouping
5. Effect of altitude

Continuous-Duty Rating. This factor applies if a wire is to carly current for 1000 s or longer. The continuous current-carrying capacity of copper and aluminum wire in amperes for aerospace applications is shown in Table 4.6. The following criteria apply to the ratings in the table.

- Rated ambient temperatures:
  57.2°C (135°F) for 105°C rated insulated wire
  92°C (197.6°F) for 135°C rated insulated wire
  107°C (225°F) for 150°C rated insulated wire
  157°C (315°F) for 200°C rated insulated wire
- "Wire bundled" indicates 15 or more wires in a group.
- The sum of all currents in a bundle is not more than 20 percent of the theoretical capacity of the bundle, which is calculated by adding the bundle ratings of the individual wires.

TABLE 4.6 Maximum Current-Carrying Capacity, Amperes

Size, AWG	Copper (MIL-W-5088)		Aluminum (MIL-W-5088)		National Electrical Code	Underwriters Laboratory		American Insurance Association	500 cmil/A
	Single wire	Wire bundled	Single wire	Wire bundled		+60°C	+80°C		
30	—	—	—	—	—	0.2	0.4	—	0.20
28	—	—	—	—	—	0.4	0.6	—	0.32
26	—	—	—	—	—	0.6	1.0	—	0.51
24	—	5	—	—	—	1.0	1.6	—	0.81
22	9	7.5	—	—	—	1.6	2.5	—	1.28
20	11	10	—	—	—	2.5	4.0	3	2.04
18	16	13	—	—	6	4.0	6.0	5	3.24
16	22	17	—	—	10	6.0	10.0	7	5.16
14	32	23	—	—	20	10.0	16.0	15	8.22
12	41	33	—	—	30	16.0	26.0	20	13.05
10	55	46	—	—	35	—	—	25	20.8
8	73	60	58	36	50	—	—	35	33.0
6	101	80	86	51	70	—	—	50	52.6
4	135	100	108	64	90	—	—	70	83.4
2	181	125	149	82	125	—	—	90	132.8
1	211	150	177	105	150	—	—	100	167.5
0	245	175	204	125	200	—	—	125	212.0
00	283	200	237	146	225	—	—	150	266.0
000	328	225	—	—	275	—	—	175	336.0
0000	380	—	—	—	325	—	—	225	424.0

Source: Martin Marietta Corporation.

Short-Time Current Ratings. These ratings apply when a wire is to carry a current for less than 1000 s. The short-time rating is generally applicable to starting loads. Figure 4.7 shows the curves for current ratings for various wire gauges in bundles.

Temperature-Current Relationship. The following equation provides a means of rerating the current carrying capacity of wire and cable at any anticipated ambient temperature:

$$I = I_r \sqrt{\frac{t_c - t}{t_c - t_r}}$$

where I = current rating at ambient temperature t

   I_r = current rating in rated ambient temperature t_r (Table 4.6)

   t = ambient temperature

   t_r = rated ambient temperature (continuous-duty rating)

   t_c = temperature rating of insulated wire or cable (continuous-duty rating)

Figures 4.8 and 4.9 present curves showing the effect of temperature on the current-carrying capacity of copper wire with 10- and 15-mil thick insulation, respectively.

Effects of Wire Grouping. When wires are grouped (bundled, harnessed) together, their current ratings must be reduced because of restricted heat loss. Table 4.6 and Fig. 4.5 take into account reduced ratings based on grouping 15 or more wires.

Effects of Altitude. Air density decreases with increasing altitude. Since lower density reduces the dielectric properties of air, trapped air between conductors and insulators presents a problem. In addition, air is retained in voids that occur primarily in stranded conductors, although voids cannot be entirely eliminated from insulated solid wire

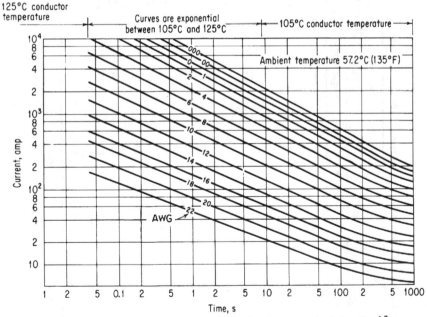

FIGURE 4.7 Short-time working curves for 105°C insulated copper wire in bundles.[1,7]

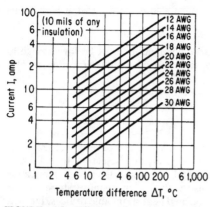

FIGURE 4.8 Effects of temperature on current-carrying capacity of copper conductors (10-mil insulation).[1,8]

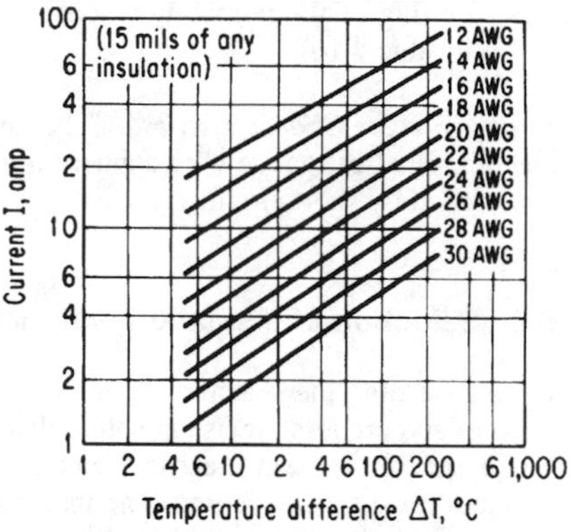

FIGURE 4.9 Effects of temperature on current-carrying capacity of copper conductors (15-mil insulation).[1,8]

construction either. A direct effect of increased altitude on insulated wire is lower corona threshold voltage, resulting in lower peak operating voltage. Wire insulated with organic (polymer) materials should always be operated below the corona extinction voltage, as corona has a degrading effect on all organic materials. Operating below corona threshold but above corona extinction is risky, for a surge may start corona, which will continue until the voltage is lowered to the extinction level.

Figure 4.10 presents curves for the insulation breakdown voltage of air as a function of altitude. In addition to proper derating of the operating voltage on insulated wires and

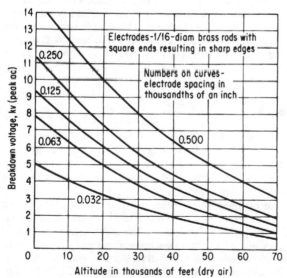

FIGURE 4.10 Average breakdown voltage versus altitude. (*Courtesy of Martin Marietta Corp.*)

cables, derating of the termination spacing in air must also be considered at higher altitudes (see Table 4.16).

Circuit-Protector Characteristics.   The circuit protector rating must be low enough to protect the smallest-gauge wire connected to it against damage from overheating, smoke, or fire caused by short circuits.

4.5.2 Selection of Insulation Material

Factors governing the selection of an optimum insulation material cover a wide range of properties and are less precise in nature than those for wire-gauge selection. In addition to meeting specific system electrical and environmental requirements, related fabrication techniques must be considered. The insulation must withstand mechanical abuse and the heat from soldering or from the application of associated heat shrinkable devices. Furthermore, the insulation must be compatible with encapsulants, potting compounds, conformal coatings, and adhesives. Table 4.7 is a checklist for the selection of insulation.

4.6 INTERCONNECTION AND HOOKUP WIRE

4.6.1 Definitions

For purposes of this discussion, the difference between interconnection and hookup wire is determined by the amount of mechanical stress applied to the wire. Interconnection wire is used to connect electric circuits between pieces of equipment and must withstand rough handling, abrasion when pulling through conduit, and potential accidental damage during installation, which results from the slip impact of hand tools. Hookup wire is used to connect electric circuits within a unit of equipment, or black box. This type of equipment may range from miniaturized airborne elements to a massive ground installation. With today's emphasis on miniaturization, many applications specify the use of hookup wire where interconnection wire was used in the past.

4.6.2 Interconnection Wire

Interconnection wire is by nature bulky; it must withstand severe mechanical abuse. For its major application in the aircraft industry, it must have minimum weight and size to allow greater payload. Insulation suppliers and wire manufacturers are continuing to develop new interconnection wire insulations that will achieve the primary requisite of mechanical resistance consistent with minimum bulk and weight.

   In addition to the development of higher-strength insulation materials, innovations such as the addition of reinforcing mineral fillers to existing resin systems have proved feasible (MS-17411, MS-17412, MIL-W-22759/7B and /8). The addition of filler materials enhances the abrasion resistance of the basic resin system without an increase in wire diameter and with minimum weight penalty. Table 4.8 lists the interconnection wires most widely used in industry. The tabulation is made on the basis of no. 22 AWG insulated conductor so that weight and size can be analyzed and compared for the different types.

TABLE 4.7 Insulation Selection Checklist

Requirement	Considerations
Environment	
Temperature extremes:	
Continuous operating	Refer to Table 4.1.
Short-term operating	May require test that simulates specific application.
Fabrication temperature	Check for soldering-iron resistance in high-density packaging; cure temperatures of encapsulant; compatibility with shrinkable devices, if used.
Storage	Check for embrittlement, long-term storage, low-humidity conditions.
Altitude:	
Outgassing	Weight loss, smoke, condensation.
Corona	Maintain voltage below corona level, especially with insulations susceptible to erosion (Fig. 4.10).
Radiation	
Weather	Moisture resistance, aging, ultraviolet radiation.
Flame	
Fluids	
Electrical	
Capacitance	$C = \dfrac{7.36K}{\log_{10}(D/d)}$ where C = capacitance, pF/ft K = dielectric constant (Table 4.2) D = insulated wire diameter, in d = conductor diameter, in
Dielectric strength	Refer to Table 4.2.
Volume resistivity	Refer to Table 4.2.
Loss factor	Refer to Table 4.2.
Mechanical	
Installation and handling	Check for minimum bend radius, special tooling, clamping, stresses, chafing. Refer to Table 4.1 for abrasion, cut-through, and mechanical resistance.
Operation	Refer to Table 4.1.
Size	Refer to applicable specification for outside dimensions.
Weight	Refer to applicable specification for maximum weight. If not listed, use the following equation for insulation weight: $W = \dfrac{Dd^2}{2}KG$ lb/1000ft where D = diameter over insulation, in d = diameter over conductor, in K = 680 G = specific gravity of insulation (Table 4.1)

Source: From Schuh.[1]

TABLE 4.8 Interconnection Wire Data, Copper Conductors

Basic specification	Class type or MS no.	Size range, AWG	Conductor coating	Primary insulation	Jacket	Voltage rating, V rms	Temperature rating, °C	Diameter rating, maximum	Weight rating, maximum
MIL-W-5086	MS-25190, Ty 1	22–12	Tin	PVC	Nylon	600	−55 to 105	1.0	1.0
	MS-25190, Ty 2	22–4/0	Tin	PVC	Glass nylon	600	−55 to 105	1.1	1.07
	MS-25190, Ty 3	22–4/0	Tin	PVC	Glass PVC nylon	600	−55 to 105	1.25	1.3
	MS-25190, Ty 4	22–16	Tin	PVC	Nylon	3000	−55 to 105	1.46	1.64
MIL-W-8777	MS-25471	22–2/0	Silver	Silicone rubber	Dacron braid	600	−55 to 200	1.25	1.32
	MS-27110	22–4	Silver	Silicone rubber	FEP	600	−55 to 200	1.25	1.61
MIL-W-22759	MS-17411	24–4	Silver	Reinforced TFE	—	600	−55 to 200	1.25	1.82
	MS-17412	24–4	Nickel	Reinforced TFE	—	600	−55 to 260	1.25	1.82
	/7	24–4	Silver	Reinforced TFE	—	600	−55 to 200	1.04	1.36
	/8	24–4	Nickel	Reinforced TFE		600	−55 to 260	1.04	1.36
	/2	22–2/0	Silver	TFE and glass tape	Glass braid FEP	600	−55 to 200	1.07	1.3
MIL-W-81044	/1	24–4	Silver	Cross-linked polyalkene	Cross-linked Kynar	600	−55 to 135	1.0	0.98
	/2	24–4	Tin	Cross-linked polyalkene	Cross-linked Kynar	600	−55 to 135	1.0	0.98
	/5	24–0	Silver	Cross-linked polyalkene	Cross-linked Kynar	600	−55 to 150	1.0	0.925
	/6	24–0	Tin	Cross-linked polyalkene	Cross-linked Kynar	600	−55 to 150	1.0	0.925
	/7	26–20	Silver-high-strength alloy	Cross-linked polyalkene	Cross-linked Kynar	600	−55 to 150	1.0	0.925
	/8	24–0	Silver	Cross-linked polyalkene	Cross-linked Kynar	600	−55 to 150	0.90	
	/9	24–0	Tin	Cross-linked polyalkene	Cross-linked Kynar	600	−55 to 150	0.90	
	/10	26–20	Silver-high-strength alloy	Cross-linked polyalkene	Cross-linked Kynar	600	−55 to 150	0.90	
MIL-W-25038	/1	22–4/0	Nickel clad	TFE and glass tape	Glass braid	600	−55 to 288	1.62	2.26
MIL-W-22759	Class 1	22–4/0	Silver	TFE and glass	Glass braid	600	−55 to 200	1.25	1.82
	Class 2	22–4/0	Nickel	TFE and glass	Glass braid	600	−55 to 260	1.25	1.82
MIL-W-81381	/3	26–2	Silver	Polyimide FEP film	FEP dispersion	600	−55 to 200	0.84	0.89
	/4	26–2	Nickel	Polyimide FEP film	FEP dispersion	600	−55 to 260	0.84	0.89

TABLE 4.8 Interconnection Wire Data, Copper Conductors (*Continued*)

Basic specification	Class type or MS no.	Duty rating*	Cost rating	Availability*	Mechanical properties	Electrical properties	Chemical resistance	Solder	Processing characteristics		
									Bondability	Strippability	Marking
MIL-W-5086	MS-25190 Ty 1	M	1.0	RA	Fair	Fair	Good	Poor	Good	Good	Good
	MS-25190 Ty 2	M	1.5	RA	Fair	Fair	Good	Poor	Good	Fair	Good
	MS-25190 Ty 3	H	1.7	RA	Good	Fair	Good	Poor	Good	Fair	Good
	MS-25190 Ty 4	M	1.3	RA	Fair	Fair	Good	Poor	Good	Good	Good
MIL-W-8777	MS-25471	H	9.6	LS	Good	Good	Fair	Fair	Good	Fair	Poor
	MS-27110	H	9.1	LS	Good	Good	Good	Fair	Poor	Good	Good
MIL-W-22759	MS-17411	H	10.0	RA	Excellent	Good	Excellent	Good	Poor	Good	Poor
	MS-17412	H	11.0	RA	Excellent	Good	Excellent	Good	Poor	Good	Poor
	/7	M	7.8	RA	Good	Good	Excellent	Good	Poor	Good	Poor
	/8	M	8.6	RA	Good	Good	Excellent	Good	Poor	Good	Poor
	/2	M	9.3	LS	Good	Good	Good	Good	Poor	Fair	Good
MIL-W-81044	/1	M	5.6	LS	Good	Good	Good	Fair	Good	Good	Good
	/2	M	4.8	LS	Good	Good	Good	Fair	Good	Good	Good
	/5	M	5.6	LS	Good	Good	Good	Fair	Good	Good	Good
	/6	M	4.8	LS	Good	Good	Good	Fair	Good	Good	Good
	/7	M	7.0	LS	Good	Good	Good	Fair	Good	Good	Good
	/8	L	5.4	LS	Good	Good	Good	Fair	Good	Good	Good
	/9	L	4.6	LS	Good	Good	Good	Fair	Good	Good	Good
	/10	L	6.8	LS	Good	Good	Good	Fair	Good	Good	Good
MIL-W-25038	/1	H	15.0	LS	Good	Fair	Good	Good	Poor	Fair	Poor
MIL-W-22759	Class 1	H	12.2	RA	Excellent	Good	Good	Good	Fair	Fair	Fair
	Class 2	H	12.3	RA	Excellent	Good	Good	Good	Fair	Fair	Fair
MIL-W-81381	/3	M	6.0	RA	Good	Good	Excellent	Good	Fair	Fair	Fair
	/4	M	6.7	RA	Good	Good	Excellent	Good	Fair	Fair	Fair

*L—light; M—medium; H—heavy; RA—readily available; LS—limited sources (fewer than four manufacturers).
Source: Martin Marietta Corporation.

4.6.3 Hookup Wire

Although the term hookup wire is likely to become a catchall for all wire constructions, it represents the area in which by far the highest volume (footage) of military and commercial electronic wire is consumed. Under the demands of miniaturization, many aerospace and missile designs call for hookup wire rather than the stronger, heavier, and bulkier interconnection wire.

The hookup wire field is the most dynamic in the wire and cable industry. Manufacturers must be cognizant of all newly developed resin (polymer) systems, fabrication techniques, and their applicability as a wire insulation. In addition, the manufacturer must be adaptable to the many and varied user requirements, ranging from a conventional back panel application to the complicated environment of spaceflight. Working in a changeable and uncertain environment imposes a substantial burden on both user and manufacturer. Many requirements are necessarily established without the ability to analyze thoroughly either the requirement or the capability of a selected insulation to withstand the required environment. The cost and the availability of test equipment to simulate the many user environments are for all practical purposes prohibitive to a wire manufacturer, and in many cases the manufacturer must depend on user evaluation to determine the suitability of a product.

Table 4.9 presents a summary of hookup wire types, including thin-wall insulations, which have been covered in specifications.

Thin-Wall Insulations.   Certain applications may require the use of ultrathin-wall insulation. Kapton type F (FN) film, used in the construction of MIL-W-81381/1–4 wire, is composed of 1-mil polyimide film coated on one or both sides with 0.5-mil Teflon FEP. The thin-walled wire construction using this Kapton film results in a nominal insulation thickness of 7.5 mil. The use of Kapton type XF (FN) film further reduces the thickness to 5.0 mil. Figure 4.11 compares the cut through resistance of these insulation systems.

Another Kapton thin-film variation has become available: Kapton 120FN616, which utilizes 0.1 mil of Teflon FEP coating on both sides of 1-mil polyimide film in a 5-mil nominal wall construction. This configuration offers greater cut-through and abrasion resistance than type XF (FN) film because of the increased thickness of the stronger polyimide film.

Kapton 120FN616 film also offers the potential of an even thinner wall construction (3-mil nominal wall thickness) with the single-layer wrap construction. Use of the Teflon FEP coating on both sides of the polyimide film makes a single layer construction with a good seal between overlaps feasible. Thus the potential disadvantage of poor strippability caused by placing the Teflon FEP next to the conductor is greatly reduced because of the small amount of Teflon FEP used for sealing.

Automated Termination.   Table 4.10 shows the properties and characteristics of a specialized hookup wire used with automated termination techniques, such as wire wrap and Termi-Point[*] (see Fig. 4.19). These termination techniques require a special set of criteria. The major conductor size is no. 24 AWG with a nominal 10-mil wall of insulation. No. 30 AWG conductor with nominal 5-mil wall of insulation is gaining increasing acceptance.

There are several important considerations for automated termination wire:

1. *Stiffness.*   It is undesirable to have wire that "takes a set" and "pops up" on wiring panels.

---

*Termi-Point (a wire-post termination using metal clips) is a registered trademark of Amp, Inc.

TABLE 4.9 Hookup Wire Data, Copper Conductors

Basic specification	Class type or MS no.	Size range, AWG	Conductor coating	Primary insulation	Jacket material	Voltage rating, V rms	Temperature rating, °C	Diameter rating	Weight rating	Duty rating[a]
MIL-W-16878/1	Type B	32-14	Tin	PVC	—	600	-55 to 105	1.0	1.0	M
MIL-W-16878/1	Type B/N	32-14	Tin	PVC	Nylon	600	-55 to 105	1.15	1.15	H
MIL-W-16878/2	Type C	26-12	Tin	PVC	—	1000	-55 to 105	1.28	1.22	M
MIL-W-16878/3	Type D	24-1/0	Tin	PVC	—	3000	-55 to 105	1.81	1.88	M
MIL-W-16878/4A	Type E	32-10	Silver or nickel	TFE	—	600	-55 to 200 or 260	1.02	1.22	M
MIL-W-16878/5A	Type EE	32-8	Silver or nickel	TFE	—	1000	-55 to 200 or 260	1.21	1.48	M
MIL-W-16878/6A	Type ET	32-20	Silver or nickel	TFE	—	250	-55 to 200 or 260	0.86	0.97	L
MIL-W-16878/7	Type F	24-4/0	Tin, silver, or nickel	Silicone rubber	—	600	-55 to 200	1.21	1.1	M
MIL-W-16878/8	Type FF	24-4/0	Tin, silver, or nickel	Silicone rubber	—	1000	-55 to 200	1.83	1.75	M
MIL-W-16878/10A	Type J	24-4/0	Tin	Polyethylene	—	600	-55 to 75	1.13	0.98	M
MIL-W-16878/11	Type K	32-10	Silver	FEP	—	600	-55 to 200	1.02	1.22	M
MIL-W-16878/12	Type KK	32-8	Silver	FEP	—	1000	-55 to 200	1.21	1.48	M
MIL-W-16878/13	Type KT	32-20	Silver	FEP	—	250	-55 to 200	0.86	0.97	L
MIL-W-22759	MS-21985	28-12	Silver	TFE	—	600	-55 to 200	0.98	1.24	M
MIL-W-22759	MS-21986	28-12	Nickel	TFE	—	600	-55 to 260	0.98	1.24	M
MIL-W-22759	MS-18113	28-8	Silver	TFE	—	1000	-55 to 200	1.17	1.48	M
MIL-W-22759	MS-18114	28-8	Nickel	TFE	—	1000	-55 to 260	1.17	1.48	M
MIL-W-81381	/3	30-12	Silver	Cross-linked polyalkene	Cross-linked Kynar	600	-55 to 135	0.92	0.99	H
MIL-W-81381	/4	30-12	Tin	Cross-linked polyalkene	Cross-linked Kynar	600	-55 to 135	0.92	0.99	H
	/11	30-12	Silver	Cross-linked polyalkene	Cross-linked Kynar	600	-55 to 150	0.92	0.99	H
	/12	30-12	Tin	Cross-linked polyalkene	Cross-linked Kynar	600	-55 to 150	0.92	0.99	H
	/13	30-20	Silver	Cross-linked polyalkene	Cross-linked Kynar	600	-55 to 150	0.92	0.99	H
MIL-W-81381	/1	26-10	Silver	Polyimide/FEP film	FEP dispersion	600	-55 to 200	0.95	1.0	H
MIL-W-81381	/2	26-10	Nickel	Polyimide/FEP film	TFE dispersion	600	-55 to 260	0.95	1.0	H

TABLE 4.9 Hookup Wire Data, Copper Conductors (*Continued*)

Basic specification	Class type or MS no.	Cost rating	Availability†	Mechanical properties	Electrical properties	Chemical properties	Solder iron resistance	Processing characteristics		
								Solderability	Strippability	Marking
MIL-W-16878/1	Type B	1.0	RA	Poor	Fair	Fair	Poor	Good	Good	Good
MIL-W-16878/1	Type B/N	1.2	RA	Good	Fair	Good	Poor	Good	Good	Good
MIL-W-16878/2	Type C	1.1	RA	Fair	Fair	Fair	Poor	Good	Good	Good
MIL-W-16878/3	Type D	1.5	RA	Fair	Fair	Fair	Poor	Good	Good	Good
MIL-W-16878/4A	Type E	6.4	RA	Fair	Excellent	Excellent	Excellent	Poor	Fair	Poor
MIL-W-16878/5A	Type EE	8.8	RA	Fair	Excellent	Excellent	Excellent	Poor	Fair	Poor
MIL-W-16878/6A	Type ET	6.4	RA	Poor	Excellent	Excellent	Excellent	Poor	Fair	Poor
MIL-W-16878/7	Type F	6.4	LS	Fair	Good	Poor	Fair	Fair	Good	Fair
MIL-W-16878/8	Type FF	9.0	LS	Fair	Good	Poor	Fair	Fair	Good	Fair
MIL-W-16878/10A	Type J	1.1	RA	Poor	Excellent	Good	Poor	Poor	Good	Good
MIL-W-16878/11	Type K	5.5	RA	Poor	Excellent	Excellent	Poor	Poor	Good	Fair
MIL-W-16878/12	Type KK	7.8	RA	Poor	Excellent	Excellent	Poor	Poor	Good	Fair
MIL-W-16878/13	Type KT	5.3	RA	Poor	Excellent	Excellent	Poor	Poor	Good	Fair
MIL-W-22759	MS-21985	6.2	RA	Fair	Excellent	Excellent	Excellent	Poor	Fair	Poor
MIL-W-22759	MS-21986	6.2	RA	Fair	Excellent	Excellent	Excellent	Poor	Fair	Poor
MIL-W-22759	MS-18113	8.9	RA	Fair	Excellent	Excellent	Excellent	Poor	Fair	Poor
MIL-W-22759	MS-18114	9.2	RA	Fair	Excellent	Excellent	Excellent	Poor	Fair	Poor
MIL-W-81381	/3	5.1	LS	Good	Good	Good	Fair	Good	Good	Good
MIL-W-81381	/4	4.2	LS	Good	Good	Good	Fair	Good	Good	Good
	/11	5.1	LS	Good	Good	Good	Fair	Good	Good	Good
	/12	4.2	LS	Good	Good	Good	Fair	Good	Good	Good
	/13	5.6	LS	Good	Good	Good	Fair	Good	Good	Good
MIL-W-81381	/1	8.3	RA	Good+	Good	Excellent	Good	Fair	Fair	Fair
MIL-W-81381	/2	9.0	RA	Good+	Good	Excellent	Good	Fair	Fair	Fair

*L—light; M—medium; H—heavy.
†RA—readily available; LS—limited sources (fewer than four manufacturers).
Source: Martin Marietta Corporation.

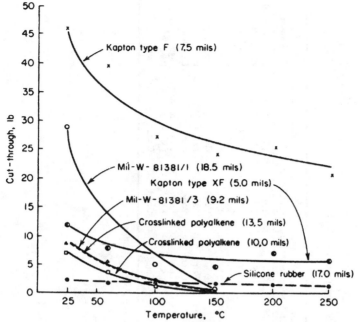

FIGURE 4.11 Cut-through resistance (dynamic).[9,10] Radius 0.005 in; edge 90° bevel; penetration rate 0.05 in/min.

2. *Cut-through.*   To achieve a satisfactory wire wrap, termination pins have very sharp edges.

3. *Long lengths.*   Long uninterrupted wire lengths are desirable for increased efficiency of operation.

4. *Strippability.*   For machine stripping of insulated wire, wall thickness and concentricity must be controlled to close tolerances.

4.6.4 Outer-Space Applications

Light hookup wire configurations lend themselves to outer-space usage because of the prime importance of weight, which directly affects the useful load that can be carried by the space vehicle. The requirements for an outer-space wire are directly related to the performance of the insulation since insulation is the component most susceptible to the environmental extremes of space. Proper selection of insulated wire requires that the environment be well defined for a specific vehicle, mission, and trajectory.

Environment.   The environment discussed here will be limited to those factors peculiar to spaceflight. The ability of an insulated wire to meet previously discussed criteria such as abrasion, rough handling, vibration, and shock are considerations common to all applications and not limited to space usage. The principal space environmental conditions that may have damaging effects on wire insulations are temperature, pressure, and radiation. Figure 4.12 gives a composite picture of the space environment. The extent of environ-

TABLE 4.10 Automated-Termination Wire Data (no. 30 AWG Conductor, Nominal 0.005-in Wall Thickness)

Insulation material	Teflon TFE	TFE/ML	FEP/ML	Vinylidene fluoride	Polyethylene-coated Mylar	Polysulfone	Polyalkene + Kynar	Kapton
Conductor coating	Silver or nickel	Silver or nickel	Silver	Tin	Tin	Tin	Tin or silver	Silver or nickel
Temperature rating, °C	200 or 260	200 or 260	200	135	125	125	135	200 or 260
Cut-through resistance	Poor	Fair	Fair	Good	Good	Excellent	Good	Excellent
Abrasion resistance	Poor	Fair	Fair	Good	Good	Good	Good	Excellent
Dielectric constant	2.1	2.1	2.1	7.7	2.8	3.2	3.4	3.2
Dielectric strength	Good	Good	Good	Good	Good	Good	Good	Good
Flexibility (stiffness)	Good	Good	Good	Fair	Fair	Fair	Fair	Fair
Chemical resistance	Excellent	Excellent	Excellent	Fair	Good	Poor	Fair	Excellent
Cost	Medium	Medium to high	Medium	Low	Low	Low	Low to medium	High
Availability*	RA	LS	LS	RA	LS	LS	LS	DR
Long lengths	Poor	Poor	Fair	Good	Fair	Good	Good	Fair

*RA—readily available; LS—limited sources (fewer than three manufacturers); DR—development required.
Source: Martin Marietta Corporation.

4.24

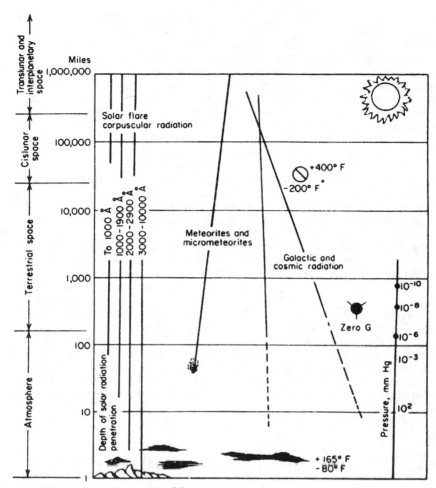

FIGURE 4.12 Space environment.[9,11]

mental extremes to which wiring may be subjected will depend on the specific design of the vehicle and the location of the wire within the vehicle.

Design Considerations

Temperature Environment.   If insulated wire is to be used in non-environmentally controlled areas, Teflon TFE, Teflon FEP, and Kapton insulations offer the broadest range of temperature resistance. These insulations will withstand 180° bending at −184°C without insulation fracture, and maximum continuous temperatures of 260°C for Teflon TFE and 200°C for Kapton and Teflon FEP. Cross-linked polyolefin insulation can be utilized at −184°C; however, it will not take such severe bending and its temperature rating is only 135°C.

Pressure Environment.   The effect of the extremely high vacuum of space on wire insulations manifests itself primarily as an initial weight loss due to a loss of water and absorbed gases. A weight loss rate that does not approach zero rapidly is a serious indication of possible long-term continuous weight loss and slow volatilization of organic materials. This may eventually degrade both the physical and the electrical properties of the wire insulation or seriously affect other equipment in the vehicle. The secondary effect of insulation volatilization of condensable materials is extremely serious for any vehicle with optical systems that may be fogged by these condensables in the spacecraft atmosphere. Table 4.11 presents a comparison of percent weight loss for various insulation materials in ultrahigh vacuum for varying temperatures and times. For comparative purposes this table has been compiled from the results published in Jolley and Reed[12] and Lanza and Halperin,[13] and additional data on Kapton were obtained from du Pont. Novathene is a specially formulated radiation cross-linked polyalkene system designed for space applications. It exhibits somewhat higher weight loss than Teflon or Kapton.

Radiation Environment. All space vehicles are subjected to some form of radiation: solar radiation, Van Allen belt radiation, and radiation from a possible nuclear power source. The effects of solar radiation seem to be negligible within the confines of a spacecraft, and degradation of insulation material is a secondary thermal effect from the absorption of electromagnetic energy.

The radiation resistance of materials varies widely. Figure 4.13 presents a generally broad spectrum of the effects of radiation on humans, electronic components, organics, and inorganics. Of all the organic materials used as hookup and interconnection wire insulation, Teflon TFE shows the lowest resistance to radiation. The threshold dose of Teflon TFE resin in air is approximately 7×10^4 rad. At 1×10^6 rad the tensile strength is about 50 percent of original and the elongation less than 5 percent of original. In the absence of oxygen, Teflon TFE shows less degradation of tensile strength and elongation.

Teflon FEP reacts somewhat differently than Teflon TFE. Cross-linking can occur, and when it is irradiated in the absence of oxygen with doses greater than 2.6×10^6 rad, some initial improvement of physical properties is noted, namely, yield stress and deformation resistance. In terms of elongation, which is an important characteristic for a flexible wire insulation, the radiation tolerance of Teflon FEP is approximately 10 times that of Teflon TFE.[15]

TABLE 4.11 Comparison of Gross Weight Loss in Ultrahigh Vacuum (at 10^{-6} mm Hg)

Sample	Temperature, °F	Time in vacuum, h	Weight loss, %*
Irradiated modified polyolefin	77	240	0.053
Polytetrafluoroethylene	77	240	0.006
Novathene	78	96	0.09
Irradiated modified polyolefin	122	138	0.30
Polytetrafluoroethylene	212	100	0.012
Novathene	212	200	0.66
Novathene	232	42	0.78
Polytetrafluoroethylene	250	120	0.018
Novathene	250	200	0.99
Irradiated modified polyolefin	300	240	1.30
Kapton	392	½	0.036
Kapton	392	30	0.036

*All data based upon insulation weight only.
Source: From Schuh.[9]

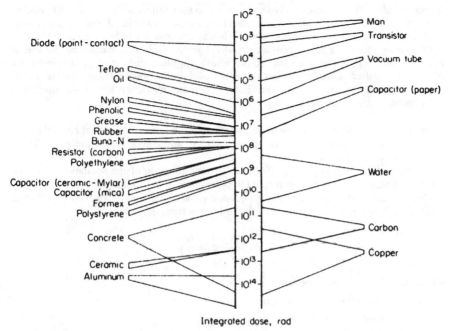

Integrated dose, rad

FIGURE 4.13 Functional radiation-dose thresholds.[9,14]

The effects of radiation on the electrical properties of Teflon TFE and Teflon FEP can be summarized as follows:

1. *Volume resistivity.*   After the threshold dose rate is reached, resistivity decreases rapidly until a dose of 5.5×10^5 rad, where equilibrium occurs. The equilibrium value is greater than $1 \times 10^{12} \; \Omega \cdot$ cm for a 20-mil specimen.[15]

2. *Dielectric strength.*   The dielectric strength of 3-, 5-, and 11-mil specimens showed no change at a dose of 5.7×10^7 rad.[15]

3. *Dielectric constant and dissipation factor.*   The dielectric constant and the dissipation factor of Teflon FEP are unaffected when irradiated to a dose of 8×10^6 rad by x rays, in the absence of oxygen, at measured frequencies of 100 Hz to 100 kHz.[12] These data indicate that Teflon TFE and Teflon FEP insulation may be used satisfactorily in limited-radiation environments and that the limiting level of usage may be increased if the material is kept in an oxygen-free environment, as far as these electrical properties are concerned.

4.7 COAXIAL CABLES

4.7.1 Design Considerations

Coaxial cable consists of a center conductor, an insulation, a shield, and usually an outer jacket. It is essentially a shielded and jacketed insulated wire. The term *coaxial* not only implies construction, but also connotes use at radio frequencies.

Background.   The purpose of a coaxial cable is to transmit radio-frequency (RF) energy from one point to another with minimum loss (attenuation). Loss of RF energy in a coaxial cable can occur (1) in the conductor, where it is a power loss due to heating caused by currents passing through a finite resistance; and (2) in the dielectric, caused by the use of materials with high power factor (dielectric losses). Highfrequency transmission invokes a phenomenon called skin effect, where currents travel on the outer surface (skin) of a conductor and partly through the adjacent insulation material. Hence loss in the insulation itself becomes more significant.

Electrical.   In addition to loss, other important electrical characteristics of coaxial cables are velocity of propagation, impedance, capacitance, and corona extinction point. These are now discussed in detail.

  Velocity of Propagation.   Velocity is an inverse function of the insulation dielectric constant, where $V = 1/K$, and is expressed in percentage of the speed of light.

  Impedance.   The three common impedance values for coaxial cables are 50, 75, and 90 Ω. Impedance can be determined by the following formula:

$$Z_0 = \frac{138}{\sqrt{K}} \log_{10}\left(\frac{D}{d}\right)$$

where Z_0 = characteristic impedance, Ω

  D = diameter over insulation, in

  d = diameter over conductor, in

  K = dielectric constant of insulation

  Capacitance.   It is usually desirable to have minimum capacitance for minimum coupling and crosstalk. Capacitance, like impedance, is a logarithmic function of dimensions and is also dependent on the dielectric constant. The equation for calculating capacitance is

$$C = \frac{7.36K}{\log_{10}(D/d)}$$

where C = capacitance, pF/ft

  D = diameter over insulation, in

  d = diameter over conductor, in

  K = dielectric constant of insulation

  Corona Extinction Point.   This determines the maximum voltage at which a coaxial cable may be operated. The corona extinction point of a cable is determined experimentally by gradually raising the voltage on a sample of cable until corona is detected (Fig. 4.14) and then lowering the voltage until no further ionization is present. If the cable is operated consistently below this level, corona will not occur within the cable. Corona can cause noise at higher frequencies and eventual degradation of organic insulating materials.

Mechanical.   The dependence of significant electrical properties, such as attenuation, capacitance, and impedance, on the relative sizes of conductor and insulation has ramifications with respect to the mechanical strength of coaxial cables. Attenuation can be reduced by increasing the conductor size, which in turn forces an increase in the insulation wall thickness if capacitance and impedance are to be maintained.

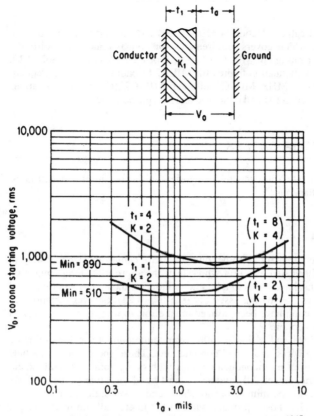

FIGURE 4.14 Corona starting voltage at sea level and 25°C.[16,17]

An additional means of maintaining low capacitance with increased conductor size is by foamed or air dielectrics. The introduction of air into a solid insulation material such as PE or FEP can reduce the dielectric constant to a value as low as 1.4.

Important electrical properties are dependent on the dimensions of conductor and insulation. Any flow or movement of the conductor can affect the electrical properties seriously. For installation, the minimum allowable bend radius should be at least 10 times the cable diameter in order to minimize stresses and preclude any cable deformation.

Environmental. Coaxial cables are normally fabricated with low-loss, low-dielectric-constant insulation materials, as covered in Table 4.3. Teflon TFE, Teflon FEP, PE, and irradiated PE, including foamed versions, are the most common dielectric materials. For coaxial cable application, one additional requirement is imposed on shield or jacket materials: they must be noncontaminating. Shields and jackets must withstand the required environment without allowing any contamination of the dielectric that might affect their loss characteristics. Contamination is usually associated with PVC compounds, which contain plasticizers that can migrate into the dielectric material. In coaxial cable applications the moisture resistance of the cable jacket is important. If water, which has a relatively high dielectric constant, penetrates the core, cable performance can he affected seriously.

4.7.2 Cable Selection

A guide to military coaxial cable selection is MIL-HDBK-216. Specific cable types are documented in MIL-C-17, which covers requirements for approximately 150 different cable configurations. Power ratings of MIL-C-17 cable types are covered in Table 4.12. Table 4.13 presents nominal attenuation figures for MIL-C-17 cables at specific frequencies ranging from 1 to 10,000 MHz. MIL-C23806 and MIL-C22931 are specifications covering semiflexible cables with foamed dielectric and air-spaced dielectric.

4.8   MULTICONDUCTOR CABLES

Multiconductor cables fall into four general categories—airborne, ground electronics, ground support, and miscellaneous.[19]

4.8.1 Airborne Cables

The primary design considerations for airborne multiconductor cables are size and weight. Airborne cables are often fabricated by a user who selects the appropriate interconnection or hookup wires, lays the insulated wires in a bundle or harness, then laces, spot ties, and applies insulating tubing over the wiring assembly (see Fig. 4.23). Several variations of this type of harness construction have been utilized in an effort to reduce size and weight and increase mechanical protection.

Specifications MIL-C-7078 and MIL-C-27500 cover multiconductor cables utilizing interconnection and hookup wires in a round configuration. These specifications include single shielded and up to seven multiconductor cables with or without an overall shield and with or without an overall jacket. All conductors must be of the same gauge; no individually shielded conductors are permitted. Various shield and jacket options are available, offering compatibility with the chosen primary wire. In the interest of minimum weight and size, fillers are not used. Table 4.14 presents recommended MIL-C-27500 options, including construction details and a mechanical usage rating.

4.8.2 Ground-Electronics Cables

The term ground-electronics multiconductor cabling encompasses rack and panel interconnection, equipment cabling installed in conduit (as used with fixed computer and data-processing installations), or cabling placed beneath flooring that is not subjected to extreme mechanical abuse or environment.

Military specification coverage for multiconductor cables in this area of usage is limited; MIL-C-7078, MIL-C-27500, and MIL-C-27072 are used frequently.

The most commonly used heavy-duty construction for this category of cabling contains PVC-insulated primary conductors jacketed with nylon. These are then cabled and jacketed with a PVC sheath.

4.8.3 Ground-Support Cables

Ground-support multiconductor cables for tactical systems should receive early attention from the designer. These cables are tailored to system needs, require considerable lead

TABLE 4.12 Maximum Input Power Ratings* of Coaxial Cable at Different Frequencies, Watts

RG/U cable	Frequency, MHz									
	1.0	10	50	100	200	400	1000	3000	5000	10,000
5, 5A, 5B, 6, 6A, 212	4,000	1,500	800	550	360	250	150	65	50	25
7	4,100	1,550	810	540	370	250	140	70	50	30
8, 8A, 10, 10A, 213, 215	11,000	3,500	1,500	975	685	450	230	115	70	
9, 9B, 214	9,000	2,700	1,120	780	550	360	200	100	65	40
11, 11A, 12, 12A, 13, 13A, 216	8,000	2,500	1,000	690	490	340	200	100	60	
14, 14A, 74, 74A, 217, 224	20,000	6,000	2,400	1,600	1,000	680	380	170	110	40
17, 17A, 18, 18A, 177, 218, 219	50,000	14,000	5,400	3,600	2,300	1,400	780	360	230	
19, 19A, 20, 20A, 220, 221	110,000	28,000	10,500	6,800	4,200	2,600	1,300	620	410	
21, 21A, 222	1,000	340	160	115	83	60	35	15		
22, 22B, 111, 111A	7,000	1,700	650	430	290	190	110	50		
29	3,500	1,150	510	340	230	150	95	50	35	
34, 34A, 34B	19,000	7,200	2,700	1,650	1,100	700	390	140	80	
35, 35A, 35B, 164	40,000	13,500	5,500	3,800	2,500	1,650	925	370	210	
54, 54A	4,400	1,580	675	450	310	210	120	60	40	
55, 55A, 55B, 223	5,600	1,700	2,700	480	320	215	120	60	40	
57, 57A, 130, 131	10,000	3,000	1,250	830	570	370	205	95	20	
58, 58B	3,500	1,000	450	300	200	135	80	40	20	
58A, 58C	3,200	1,000	425	290	190	105	60	25	20	
59, 59A, 59B	3,900	1,200	540	380	270	185	110	50	30	
62, 62A, 71, 71A, 71B	4,500	1,400	630	440	320	230	140	65	40	15
62B	3,800	1,350	600	410	285	195	110	50	31	15
63, 63B, 79, 79B	8,200	3,000	1,300	1,000	685	455	270	130	75	35
87A, 116, 165, 166, 226, 227	42,000	15,000	6,250	4,300	3,000	2,050	1,200	620	480	250
94	62,000	15,500	5,900	4,300	2,900	1,900	1,400	650	480	200
94A, 226	64,000	18,000	9,600	6,800	4,600	3,300	1,750	775	540	250

TABLE 4.12 Maximum Input Power Ratings* of Coaxial Cable at Different Frequencies, Watts (*Continued*)

RG/U cable	Frequency, MHz									
	1.0	10	50	100	200	400	1000	3000	5000	10,000
108, 108A	1,300	360	145	100	70	45	30	15	5	
114, 114A	5,300	1,350	475	345	230	150	85	40	25	15
115, 115A, 235	33,000	9,900	4,200	2,900	2,000	1,380	830	600	450	170
117, 118, 211, 228	200,000	66,000	25,000	19,000	12,800	8,500	4,800	2,200	1,400	490
119, 120	100,000	31,000	13,000	9,000	6,100	4,100	2,400	1,100	770	250
122	1,000	240	100	65	45	30	15	10	5	
125	8,500	2,300	910	620	435	285	165	75	45	
140, 141, 141A	19,000	6,300	2,700	1,700	1,200	830	450	220	140	65
142, 142A, 142B	19,000	5,700	2,600	1,800	1,300	900	530	265	175	100
143, 143A	26,000	8,700	3,750	2,600	1,800	1,250	750	390	275	160
144	51,000	17,000	7,500	5,400	3,700	2,500	1,400	700	440	20
149, 150	7,100	1,900	740	485	315	200	105	45	25	
161, 174	1000	350	160	110	80	60	35	15	10	
178, 178A, 196	1,300	640	330	240	180	120	75	40		
179, 179A, 187	3,000	1,400	750	480	420	320	190	100	73	
180, 180A, 195	4,500	2,000	1,100	800	570	400	240	130	90	50
188, 188A	1,500	770	480	400	325	275	150	80	55	
209	180,000	55,000	22,000	15,000	8,500	6,000	3,400	1,600	1,000	310
281	150,000	47,000	19,000	13,500	8,800	6,000	3,300	1,650	1,150	625

*Power-rating conditions: ambient temperature 104°F; center-conductor temperature 175°F with polyethylene dielectric, 400°F with Teflon dielectric. Altitude: sea level.

Source: Amphenol Corporation.[1R]

4.32

TABLE 4.13 Attenuation Ratings of Coaxial Cables at Different Frequencies, dB/100 ft

RG/U cable	Frequency, MHz									
	1.0	10	50	100	200	400	1000	2000	5000	10,000
5, 5A, 5B, 6, 6A, 212	0.26	0.83	1.9	2.7	4.1	5.9	9.6	23.0	32.0	56.0
7	0.18	0.64	1.6	2.4	3.5	5.2	9.0	18.0	25.0	43.0
8, 8A, 10, 10A, 213, 215	0.15	0.55	1.3	1.9	2.7	4.1	8.0	16.0	27.0	>100.0
9, 9A, 9B, 214	0.21	0.66	1.5	2.3	3.3	5.0	8.8	18.0	27.0	45.0
11, 11A, 12, 12A, 13, 13A, 216	0.19	0.66	1.6	2.3	3.3	4.8	7.8	16.5	26.5	>100.0
14, 14A, 74, 74A, 217, 224	0.12	0.41	1.0	1.4	2.0	3.1	5.5	12.4	19.0	50.0
17, 17A, 18, 18A, 177, 218, 219	0.06	0.24	0.62	0.95	1.5	2.4	4.4	9.5	15.3	>100.0
19, 19A, 20, 20A, 220, 221	0.04	0.17	0.45	0.69	1.12	1.85	3.6	7.7	11.5	>100.0
21, 21A, 222	1.5	4.4	9.3	13.0	18.0	26.0	43.0	85.0	>100.0	>100.0
22, 22B, 111, 111A	0.24	0.80	2.0	3.0	4.5	6.8	12.0	25.0	>100.0	>100.0
29	0.32	1.20	2.95	4.4	6.5	9.6	16.2	30.0	44.0	>100.0
34, 34A, 34B	0.08	0.32	0.85	1.4	2.1	3.3	5.8	16.0	28.0	>100.0
35, 35A, 35B, 164	0.06	0.24	0.58	0.85	1.27	1.95	3.5	8.6	15.5	>100.0
54, 54A	0.33	0.92	2.15	3.2	4.7	6.8	13.0	25.0	37.0	>100.0
55, 55A, 55B, 223	0.30	1.2	3.2	4.8	7.0	10.0	16.5	30.5	46.0	>100.0
57, 57A, 130, 131	0.18	0.65	1.6	2.4	3.5	5.4	9.8	21.0	>100.0	>100.0
58, 58B	0.33	1.25	3.15	4.6	6.9	10.5	17.5	37.5	60.0	>100.0
58A, 58C	0.44	1.4	3.3	4.9	7.4	12.0	24.0	54.0	83.0	>100.0
59, 59A, 59B	0.33	1.1	2.4	3.4	4.9	7.0	12.0	26.5	42.0	>100.0
62, 62A, 71, 71A, 71B	0.25	0.85	1.9	2.7	3.8	5.3	8.7	18.5	30.0	83.0
62B	0.31	0.90	2.0	2.9	4.2	6.2	11.0	24.0	38.0	92.0
63, 63B, 79, 79B	0.19	0.52	1.1	1.5	2.3	3.4	5.8	12.0	20.5	44.0
87A, 116, 165, 166, 225, 227	0.18	0.60	1.4	2.1	3.0	4.5	7.6	15.0	21.5	36.5
94	0.15	0.60	1.6	2.2	3.3	5.0	7.0	16.0	25.0	60.0
94A, 226	0.15	0.55	1.2	1.7	2.5	3.5	6.6	15.0	23.0	50.0

TABLE 4.13 Attenuation Ratings of Coaxial Cables at Different Frequencies, dB/100 ft (*Continued*)

RG/U cable	Frequency, MHz									
	1.0	10	50	100	200	400	1000	2000	5000	10,000
108, 108A	0.70	2.3	5.2	7.5	11.0	16.0	26.0	54.0	86.0	>100.0
114, 114A	0.95	1.3	2.1	2.9	4.4	6.7	11.6	26.0	40.0	65.0
115, 115A, 235	0.17	0.60	1.4	2.0	2.9	4.2	7.0	13.0	20.0	33.0
117, 118, 211, 228	0.09	0.24	0.6	0.9	1.35	2.0	3.5	7.5	12.0	37.0
119, 120	0.12	0.43	1.0	1.5	2.2	3.3	5.5	12.0	17.5	54.0
122	0.40	1.7	4.5	7.0	11.0	16.5	29.0	57.0	87.0	>100.0
125	0.17	0.50	1.1	1.6	2.3	3.5	6.0	13.5	23.0	>100.0
140, 141, 141A	0.30	0.90	2.1	3.3	4.7	6.9	13.0	26.0	40.0	90.0
142, 142A, 142B	0.34	1.1	2.7	3.9	5.6	8.0	13.5	27.0	39.0	70.0
143, 143A	0.25	0.85	1.9	2.8	4.0	5.8	9.5	18.0	25.5	52.0
144	0.19	0.60	1.3	1.8	2.6	3.9	7.0	14.0	22.0	50.0
149, 150	0.24	0.88	2.3	3.5	5.4	8.5	16.0	38.0	65.0	>100.0
161, 174	2.3	3.9	6.6	8.9	12.0	17.5	30.0	64.0	99.0	>100.0
178, 178A, 196	2.6	5.6	10.5	14.0	19.0	28.0	46.0	85.0	>100.0	>100.0
179, 179A, 187	3.0	5.3	8.5	10.0	12.5	16.0	24.0	44.0	64.0	>100.0
180, 180A, 195	2.4	3.3	4.6	5.7	7.6	10.8	17.0	35.0	50.0	88.0
188, 188A	3.1	6.0	9.6	11.4	14.2	16.7	31.0	60.0	82.0	>100.0
209	0.06	0.27	0.68	1.0	1.6	2.5	4.4	9.5	15.0	48.0
281	0.09	0.32	0.78	1.1	1.7	2.5	4.5	9.0	13.0	24.0

Source: Amphenol Corporation.[18]

TABLE 4.14 Multiconductor Cable Options, Recommended by MIL-C-27500

Basic primary wire specifications	Specification symbol	Size range, AWG	Shielded		Jacketed		Shield-jacketed		Voltage rating, V rms	Temperature rating, °C
			Shield style	Jacket style	Shield style	Jacket style	Shield style	Jacket style		
MIL-W-5086:										
MS-25190, Ty 1	A	22–12	T	O	U	1 or 3	T	1, 2, or 3	600	−55 to 105
MS-25190, Ty 2	B	22–4/0	T	O	U	3	T	1 or 3	600	−55 to 105
MS-25190, Ty 3	C	22–4/0	T	O	U	3	T	1 or 3	600	−55 to 105
MS-25190, Ty 4	P	22–16	T	O	U	3	T	1 or 3	3000	−55 to 105
MIL-W-22759/1	D	22–4/0	S	O	U	7	S	6 or 7	600	−55 to 200
MIL-W-22759/2	E	22–4/0	N	O	U	7	N	6 or 7	600	−55 to 200
MIL-C-8777:										
MS-25471	H	22–2/0	S	O	U	4	S	4	600	−55 to 150
MS-27110	F	22–4	S	O	U	5	S	5	600	−55 to 200
MIL-W-22759:										
/5	V	24–4	S	O	U	6 or 7	S	6 or 7	600	−55 to 200
MS-17412	W	24–4	N	O	U	6 or 7	N	6 or 7	600	−55 to 260
/7	S	24–4	S	O	U	6 or 7	S	6 or 7	600	−55 to 200
/8	T	24–4	N	O	U	6 or 7	N	6 or 7	600	−55 to 260
MS-18113	LA	28–8	S	O	U	6	S	6	1000	−55 to 200
/10	LB	28–8	N	O	U	6	N	6	1000	−55 to 200
MS-21985	R	28–12	S	O	U	6	S	6	600	−55 to 200
MS-21986	L	28–12	N	O	U	6	N	6	600	−55 to 260
/2	N	22–2/0	S	O	U	6 or 7	S	6 or 7	600	−55 to 200
MIL-W-25038/1	J	22–4/0	F	O	U	7	F	7	600	−55 to 750
MIL-W-81044/6	M	24–4	S	O	U	4 or 5	S	4 or 5	600	−55 to 135
	MA	24–4	T	O	U	4 or 5	S	4 or 5	600	−55 to 135
	MB	30–12	S	O	U	4 or 5	S	4 or 5	600	−55 to 135
	MC	30–12	T	O	U	4 or 5	T	4 or 5	600	−55 to 135

TABLE 4.14 Multiconductor Cable Options, Recommended by MIL-C-27500 (Continued)

Basic primary wire specifications	Specification symbol	Construction details				
		Mechanical duty rating	Conductor	Primary insulation	Shield	Jacket
MIL-W-5086:						
MS-25190, Ty 1	A	Medium	Tinned copper	PVC/nylon	Tinned copper	(1) PVC
MS-25190, Ty 2	B	Medium; fire-resistant	Tinned copper	PVC/glass, nylon	Tinned copper	(2) Extruded nylon
MS-25190, Ty 3	C	Heavy	Tinned copper	PVC/glass PVC/nylon	Tinned copper	(3) Nylon braid
MS-25190, Ty 4	P	Medium	Tinned copper	PVC/nylon	Tinned copper	(6) Taped Teflon TFE
MIL-W-22759/1	D	Heavy	Silver-coated copper	Teflon TFE	Silver-coated copper	(7) Glass braid
MIL-W-22759/2	E	Heavy	Nickel-coated copper	Tapes and glass braid	Nickel-coated copper	
Mil-C-8777:						
MS-25471	H	Heavy	Silver-coated copper	Silicone rubber	Silver-coated copper	(4) Dacron braid*
MS-27110	F	Medium	—	—	—	(5) Extruded Teflon FEP
MIL-W-22759:						
/5	V	Heavy	Silver-copper	Mineral-filled	Silver-copper	(6) Taped Teflon TFE
MS-17412	W	Heavy	Nickel-copper	Teflon TFE	Nickel-copper	(7) Glass braid
/7	S	Medium	Silver-copper	—	Silver-copper	
/8	T	Medium	Nickel-copper	—	Nickel-copper	
MS-18113	LA	Light	Silver-copper	Extruded Teflon TFE	Silver-copper	(6) Taped Teflon TFE
/10	LB	Light	Nickel-copper	—	Nickel-copper	
MS-21985	R	Light	Silver-copper	—	Silver-copper	
MS-21986	L	Light	Nickel-copper	—	Nickel-copper	
/2	N	Medium	Silver-copper	TFE-glass-FEP	Silver-copper	(6) Taped Teflon TFE (7) Glass braid
MIL-W-25038/1	J	Heavy; fire-resistant	Nickel-clad copper	TFE tapes and glass braid	Stainless steel	(4) Dacron braid*
MIL-W-81044/6	M	Medium	Silver-copper	Polyalkene and Kynar (cross-linked)	Silver-copper	(5) Extruded Teflon FEP
	MA	Medium	Tinned copper	—	Tinned copper	
	MB	Light	Silver-copper	—	Silver-copper	
	MC	Light	Tinned copper	—	Tinned copper	

*Jacket compatible with primary insulation system not available to date. Kynar, polyethylene-coated Mylar, and cross-linked Kynar are proposed additions to specification.
Source: Martin Marietta Corporation.

time for delivery, and can amount to a considerable system cost if no effort at standardization is made. A major manufacturing cost in the production of this type of multiconductor cable is cabling machine setup. Many cable manufacturers require a minimum order of 500 to 1000 ft of cable for a given configuration. After setup, additional cable footage can be produced more economically. Some ground-support multiconductor cable applications follow.

Permanent Installation.   Cables that are buried or placed in conduits, open ducts, troughs, or tunnels are considered permanent cables. These cables are not handled, flexed, reeled, or dereeled except at the time of installation. Either neoprene or PE cable sheaths are preferred.

Portable Installation.   MIL-C-13777 is the basis for the design of heavy-duty portable cable. Thin-wall cable insulation is appropriate for applications of a lighter duty nature, especially where weight is a critical factor.

4.8.4   Flat Flexible Cables

There are two basic types of flat flexible cables, often referred to as tape cable and as flexible printed wiring. The major difference between the two is their manner of construction and general application. Tape cable is used primarily to interconnect individual electronic units; flexible printed wiring is used to provide interconnection within a unit.

Flat-Conductor Cable

Construction.   Flat-conductor cable is constructed by the encapsulation of flat rectangular conductors between layers of dielectric film. Figure 4.15 depicts a manufacturing

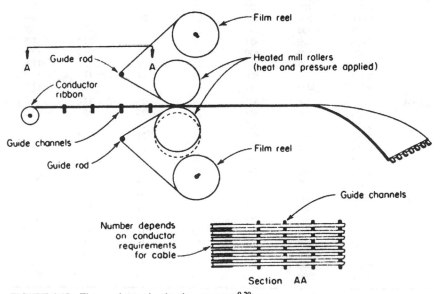

FIGURE 4.15   Flat-conductor laminating process.[9,20]

technique for laminating flatconductor cable. Conductor ribbons are positioned parallel to each other with uniform spacing.

Materials.   Conductors are usually rolled copper, annealed in accordance with federal specification QQ-C576. Other materials and protective coatings may be obtained for conductors to meet unusual application requirements. For selecting the proper conductor size to satisfy current and voltage requirements, refer to Table 4.15 and Figs. 4.16 and 4.17.

Design.   The following important criteria should be considered by the designer and used to formulate optimum cable construction, materials selection, terminations, installation, fabrication, and handling:

1. Electrical requirements
    a. Current-carrying capacity

TABLE 4.15   Copper Conductor Characteristics, Flat Conductor

Flat conductor dimensions		Cross section		Nearest AWG wire size based on equivalent:		Resistance at 20°C, mΩ/ft	Current for 30°C rise, A
Thickness, in	Width, in	mil²	cmil	Cross section	Current rating		
0.0027	0.030	81	102	30	28	100	3.4
	0.045	122	154	28	27	67	3.8
	0.060	162	204	27	25	50	5.1
	0.075	202	254	26	24	40	5.8
	0.090	243	306	25	23	34	6.5
	0.125	338	425	24	22	24	8.2
	0.155	418	527	23	21	19.5	9.2
	0.185	500	630	22	20	16.2	10.7
	0.250	675	850	21	18	12	13.5
0.004	0.030	120	151	28	27	67	4.0
	0.045	180	227	26	25	45	5.2
	0.060	240	302	25	24	34	6.0
	0.075	300	378	24	23	27	7.0
	0.090	360	454	23	22	22.5	7.8
	0.125	500	630	22	20	16.2	10.0
	0.155	620	780	21	19	13	11.8
	0.185	740	930	20	18	11	13.5
	0.250	1000	1260	19	17	8	17.0
0.0055	0.045	248	312	25	24	33	6.0
	0.060	330	415	24	23	25	7.2
	0.075	412	520	23	22	20	8.2
	0.090	495	624	22	21	16.5	9.5
	0.125	687	865	21	20	12	12.2
	0.155	852	1075	20	19	9.5	14.8
	0.185	1020	1285	19	17	8	17
	0.250	1375	1730	18	16	6	21
0.008	0.045	360	454	23	22	23	7.8
	0.060	480	605	22	21	17	9.8
	0.075	600	755	21	20	13.5	11.5
	0.090	720	905	20	19	11.2	13.2
	0.125	1000	1260	19	17	8	17
	0.155	1240	1560	18	16	6.5	20
	0.185	1480	1860	17	14	5.5	23
	0.250	2000	2520	16	13	4.1	26

Source: From *Flexprint Circuit Design Handbook.*[21]

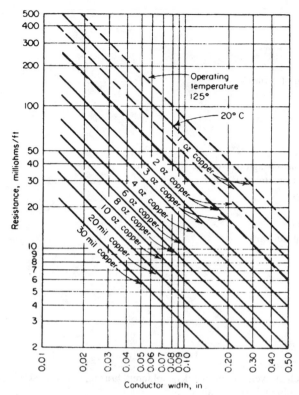

FIGURE 4.16 Etched-conductor resistance.[9,22]

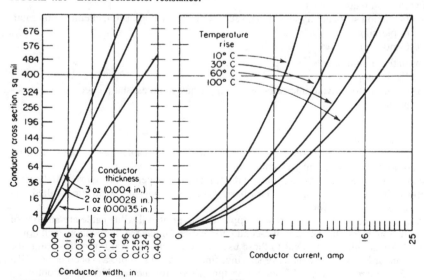

FIGURE 4.17 Etched-conductor current capacity.[9,22]

  b. Voltage drop
  c. Overload rating
  d. Impedance
  e. Capacitance
  f. Shielding requirements
  g. Derating due to stacking of cables

 2. Mechanical requirements
  a. Conductor width
  b. Conductor thickness
  c. Conductor spacing
  d. Insulation tear resistance
  e. Insulation puncture or cut-through resistance
  f. Insulation and conductor flexure resistance

 3. Environmental requirements
  a. Temperature extremes
  b. Flame resistance
  c. Vibration and shock requirements
  d. Temperature cycling
  e. Humidity requirements
  f. Altitude requirements
  g. Materials compatibility
  h. Fungus resistance
  i. Radiation resistance
  j. Aging resistance

 4. Termination requirements
  a. Spacing compatible with connector
  b. Type of termination, such as eyelet, rivet, welded, soldered, brazed, or crimped
  c. Termination type compatible with material

 5. Fabrication requirements
  a. Insulation strippability technique, such as mechanical abrasion, chemical, thermal, or piercing
  b. Conductor support

 6. Insulation requirements
  a. Cable support
  b. Cable routing; avoid interference with high-heat sources, sharp edges, and components

Flexible Printed Wiring

 Construction. The majority of flexible printed-wiring cables are manufactured by standard printed-circuit etching techniques.

 Materials. As was true for flat-conductor cable, copper is the most widely used printed-wiring conductor material. Two basic types of copper are available—rolled and electrodeposited copper foil.

 Design. The design considerations applicable to printed flexible cable are analogs of those previously established for flat-conductor cable. Since the fabrication techniques involved are essentially identical to those used for conventional rigid printed wiring, the design standards apply for conductor spacing, line width, and so on, as specified in MIL-STD-275. Figures 4.16 and 4.17 present design criteria for the selection of the optimum conductor size.

4.8.5 Design Considerations

Design of the cabling installation is an integral part of the mechanical design of any equipment or system requiring electrical interconnection. Design planning for electrical installation should be concurrent with the layout of the mechanical design. Quality installation design requires that each of the following major considerations be thoroughly evaluated and a positive design approach determined.

Environment. Vibration, acceleration, and shock are dynamic environments that are controlling from a design viewpoint. Acceleration places a load on cables, supports, brackets, connectors, and mounting points for black boxes. Wherever possible, these items must be so designed as to be in compression against a structural member. Connectors should be oriented so that the possibility of inadvertent disconnection is minimized. Consideration shall also be given to deceleration forces. Vibration sets up varying stresses in cables and supports, proportional to the mass being supported. Shock, including acceleration and deceleration, also contributes to an overstressed condition.

Other environmental conditions include high dynamic pressure, elevated temperatures, and lowered atmospheric pressure. High temperature has the most severe effect on network installations. Insulating against high temperature and the selection of materials capable of withstanding high temperatures are basic approaches for controlling problems caused by temperature extremes. Low atmospheric pressure is insignificant except in outer-space applications. There the effects can be serious: outgassing and deterioration of plastics and insulation with time.

Several different phenomena are generated by a nuclear burst: radiation, heat pulse, shock wave, and electromagnetic pulses (EMP). The initial radiation may be of such type and intensity over a sufficient time interval that it may dangerously degrade the quality of materials, including metals. The heat pulse and shock wave generate conditions similar to those already described. The EMP can produce voltages and currents through the metallic structure. Thus installation hardware can become electrically energized with electric stresses far higher than those encountered normally.

Ground environmental factors include high and low temperatures, humidity, and dynamic parameters caused by handling and transportation. The levels of these ground environmental factors are usually far less severe than those of flight environment. But despite the lower level, the duration of these stresses is far in excess of normal flight time. The accumulated stress or degradation under these conditions may be quite appreciable.

Routing and Grouping. Interconnecting cables and networks should be designed and installed to minimize the adverse effects of electromagnetic interference and to control crosstalk between circuits. To eliminate the adverse effects, special grouping, separation, and shielding practices should be followed, for which the following general guidelines are recommended:

1. Direct-current supply lines. Use twisted pair; separate from ac power and control lines.
2. Alternating-current power lines. Use twisted lines; separate from susceptible lines; shield ac circuits in which switching transients occur, and ground the shield at both ends.
3. Low-level signals. Use shielded twisted pair and ground the shield at one end.
4. High-level signals. Use shielded twisted pair and ground the shield at both ends.
5. Provide adequate filtering to prevent conducted noise problems.

6. Follow a single-point ground concept where possible. Analyze the flow of parasitic chassis currents, and design the ground conductor for worst case.

7. Plan the separation of signal and power circuits with maximum distance between runs.

8. Keep wire or cable length to a minimum.

9. Locate high-heat-generating wires on the outside.

10. Plan the cable routing in coordination with the structural design effort. Plan cable runs and tie down points in the early design phase for incorporation into the structure. Plan for minimum length; attempt to optimize cable installation; compromise only in the solution of installation and maintenance problems.

11. Hold bend radius of coaxial and multiconductor cables to at least 10 times the cable outside diameter.

4.9 MAGNET WIRE

This field of film-insulated wire is enormous and impacts the insulation materials and practices of electronic packaging in many ways. Magnet wire is used in printed-wiring board jumpers, RF coils, relays, transformers, and inductors, to name a few. Some descriptions of conductors and insulation must be restated in this section in a form applicable to magnet wire.

4.9.1 Conductors

Materials and Construction

Copper. The most common conductor is bare round solid annealed copper wire in accordance with ANSI C7. 1. Square and rectangular copper wire is available as described in ANSI C7.9. Copper strip can be obtained in accordance with federal specification QQ-C-576. Rounded edges should be specified to preclude any roughness or sharp projections.

Aluminum. Although copper is the most widely used conductor material, there is an increase in the application of aluminum conductors. Unfortunately in most applications, a direct substitution of copper is impossible without a significant change in design because aluminum conductors have lower conductivity (62% IACS) and increased brittleness.

Anodized (chemically surface oxidized) aluminum conductors present a unique approach to both mechanical protection and specialized electrical insulation. Figure 4.18 shows the dielectric strength of various anodized aluminum surface thicknesses. Aluminum oxide is inorganic and possesses many desirable electrical insulation

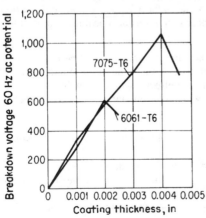

FIGURE 4.18 Dielectric strength of anodized aluminum. (*Courtesy of Martin Marietta Corp.*)

properties, such as resistance to radiation, to aging at high temperatures (melting point 3600°F), and to chemical attack. The film is very brittle and somewhat porous, but organic sealing treatments for protection against moisture are available. Anodized aluminum is a very specialized material. It cannot be a direct substitution for conventional wire enamel, tape, or served insulation.

Conductors for High-Temperature Applications.   The usable temperature range of copper conductors (bare and with protective coatings) is evaluated on the basis of oxidation, melting point, grain growth, and solid-state diffusion.

4.9.2 Insulation Materials

Film Insulation.   A list of magnet wire insulating materials follows:

Acrylic
Ceramic
Ceramic with overcoat
Epoxy
Oleoresinous (plain enamel, black enamel)
Polyamide (nylon)
Polyamide-polyimide (amide-imide)
Polyester
Polyimide (PYRE-ML*)
Polytetrafluoroethylene (Teflon TFE)
Polyurethane
Polyvinyl formal (Formvart†)

PYRE-ML is a more recent, advanced, outstanding magnet wire insulation, the only unvarnished wire enamel that has a 220°C thermal classification. It is more chemically resistant and is compatible with practically all varnishes and encapsulating compounds. As an overcoating, polyimide provides an extremely tough abrasion-resistant film, which exhibits high resistance to nuclear radiation, excellent thermal resistance, and good windability. NEMA MW-16 covers polyimide-coated round wire. Federal specification l-W-1177/l5B, class 220, type M, is also applicable to polyimide-coated round wire.

Textile and Composite Insulation.   One example of this insulation type is glass fiber.

4.10 WIRE AND CABLE TERMINATIONS

4.10.1 Terminating Hardware

Wire and cable terminations are of major importance to design reliability. Careful selection of proper terminating hardware and the reduction of the number of terminations to a minimum should be primary design goals.[23]

---

*PYRE-ML (a polyimide magnet wire enamel) is a registered trademark of E.I. du Pont de Nemours & Company, Inc.
†Formvar (a polyvinyl formal polymer wire enamel) is a registered trademark of Monsanto Inc.

The following conditions must be evaluated in the selection and use of terminations: (1) termination life, (2) connection density, (3) compatibility, (4) environment, (5) preparations, (6) mass production, (7) process control, (8) inspectability (9) current, voltage, and resistance limits, (10) maintenance tools, (11) repairability, including time and skill requirements, and (12) contractual constraints.

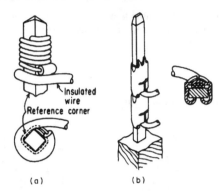

FIGURE 4.19 Wire-post termination systems.[16] (*a*) Gardener Denver's wire wrap. (*b*) Amp's Termi-Point.

A number of wire attachment methods are used in hookup and interconnecting wire terminations, such as (1) crimping, (2) soldering, (3) clamping, (4) welding, (5) wire wrapping,[24] and (6) friction (Fig. 4.19).

Terminating devices normally used electrical installation are (1) studs, (2) lugs (crimp), (3) terminal posts (solder or wire wrap), (4) connectors (with solder or crimp-type contact terminal), (5) splices (crimp or solder), (6) compression screw lugs, (7) screw terminal (usually limited to use on barrier strips), (8) ferrules, (9) taper pins, and (10) pads and eyelets.

Terminals.
Terminal Lugs. Terminal lugs are designed to establish an electrical connection between a wire and a connection point such as a stud.

Terminal Posts. Terminal posts are used on terminal boards in assembly-type wiring and on many components, such as electric connectors (solder type), relays, transformers, lamp holders, and switches.

Design Guides for Terminals. Some rules of terminal and wire termination design follow:

1. Do
   a. Use special prebused connector terminals where required.
   b. Apply supplementary insulation sleeving over axial terminations where continuous insulation is not provided between adjacent terminations.
   c. Ensure that electrical spacings between terminals conform to Table 4.16.
2. Don't
   a. Use solder cap adapters to accommodate additional connectors or larger gauges in connector terminals.
   b. Connect more than three leads to one terminal.
   c. Twist multiple wires or leads to effect terminations.
   d. Terminate more than one wire in a connector terminal.

Terminal Boards. These boards are used for junctions or terminations of wire or cable assemblies as an aid to installation and maintenance.

Stud Terminal Board. The stud terminal board is generally a threaded post with the axial portion of its body firmly anchored into a mounting panel. It requires the use of tools for the attachment of wire lugs.

Barrier Terminal Board. The barrier terminal board is molded of thermoset insulating material. It has integral raised barriers between pairs of screw terminals. Its features are

1. Longer leakage path than the stud type between adjacent terminals

TABLE 4.16 Allowable Voltage between Terminals,* Volts

Minimum air space, in	Creepage distance, in	At sea level			At 50,000 ft			At 70,000 ft		
		Flashover, V rms	Working dc	Working ac	Flashover, V rms	Working dc	Working ac	Flashover, V rms	Working dc	Working ac
†	3/64	800	280	200	300	100	75	200	70	50
1/32†	1/16	1400	490	350	500	190	125	375	125	90
3/64	5/64	2000	700	500	700	210	175	500	175	125
1/16	7/64	2500	840	600	900	315	225	600	210	150
5/64	1/8	3000	1050	750	1050	360	260	675	230	165
3/32	5/32	3600	1260	900	1200	420	300	750	260	185
1/8	3/16	4500	1550	1100	1400	490	350	900	310	225
3/16	1/4	6100	2000	1500	1800	630	450	1100	375	275
1/4	5/16	7300	2500	1800	2000	700	500	1300	455	325
5/16	3/8	8500	2900	2100	2300	810	575	1420	500	355

*The allowable voltage is determined by the actual creepage distance or the minimum air space, whichever provides a lower rating. At 70,000 ft visible corona has been recorded by voltages as low as 350 V rms. Consequently, at these elevations corona may be the limiting factor rather than flashover.

†Continuous insulation should be provided between electrical connections of 1/32 in or less.

Source: Martin Marietta Corporation.

2. More connections in a given length of board

3. Limited current-carrying capacity

4. Poor adaptability to applications with high levels of dynamic stress

Taper Pin Terminal Block.   This terminal block is composed of molded insulating material containing metal inserts designed to hold taper pins.

Specifications.   Terminal boards should be installed in accordance with MILE-7080 for aircraft and MIL-E-25366 for missiles, unless other specific requirements are established.

Splices.   Permanent splices, available for both shielded and unshielded cables, should be used only when absolutely required. Conductor splices in interconnecting wiring should be grouped and located in designated areas selected for ready access. Where leads from electric equipment are spliced into a cable assembly, the splice area should be located as near to the equipment as practical. Nonpermanent splices should be avoided; however, certain special applications may require their use.

Shield Wiring Terminations.   There are two basic shield terminations, (1) terminated to a shield common and (2) floating.

Shield termination merits careful consideration from the design phase through production. A judicious grouping of wires and careful examination of the need for shielding will alleviate termination problems. Some generalized design recommendations follow:

1. Minimize the use of shielded wiring.

2. Avoid shielding of leads less than 4 in long.

3. Provide lead segregation instead of shielding where this is practical.

4. Become familiar with all facets of shield termination techniques (see subsequent paragraphs) to ensure that the techniques fully satisfy the design environment.

Some of the preceding is, of course, not applicable to ac, pulsed, or RF leads and cables with significant EMI radiation potential. The following shield terminations are in general use:

1. Direct shield termination (pigtail)

2. Ferrule termination (crimp attachment)

3. Solder sleeve termination (solder attachment)

Direct Shield Termination.   An established practice is to form the shield braid into a pigtail, as illustrated in Fig. 4.20. No special tooling is required, but the pigtail must be the braid at the breakout point or the shield braid strands. No external pressure should be applied to the breakout point by clamp, tie, or flexure. Supplementary insulation (sleeving) over the breakout of the braid and the braid itself is required.

Shield Ferrule Termination.   Shield termination ferrules are available in two basic types: two-piece preinsulated and single-piece uninsulated. A typical application of the two-piece insulated ferrule is shown in Fig. 4.21.

Solder Sleeve Termination.   Before this termination is selected, compatibility between the solder melting temperature, the shrink temperature of the insulating sleeve, and the temperature resistance of the primary wire insulation as well as that of the jacket over the shield must be determined. The grouping of shield conductors for solder sleeve applications is shown in Fig. 4.22. Solder sleeves permit the use of center strip terminations to

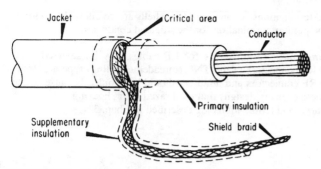

FIGURE 4.20 Shield braid in pigtail. (*Courtesy of Martin Marietta Corp.*)

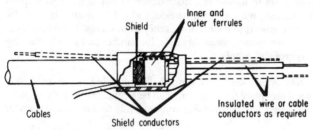

FIGURE 4.21 Insulated ferrule shield termination. (*Courtesy of Martin Marietta Corp.*)

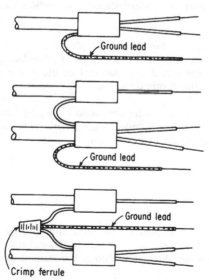

FIGURE 4.22 Typical shield-termination grouping practice. (*Courtesy of Martin Marietta Corp.*)

minimize the bulk of shield terminations at connector backshells and to allow the continuation of shields closer to the point of termination for the shielded conductor.

RF Cable (Coaxial) Termination.   Terminations for RF cables may be selected from MIL-HDBK-216. Straight RF connectors of the TNC (threaded coupling) type are desirable. The right-angle-type RF connectors and adapters should be avoided because of the inherent mechanical weakness of many designs that use brazed metal housings.

Additional interconnections and terminations are described in Chap. 3.

4.10.2   Identification

identification of wiring and cabling includes marking or coding individual wire leads, harness, cables, and termination devices. Wire identification facilitates design control and traceability (such as wiring diagram to hardware), manufacturing efficiency, and maintenance (trouble shooting). Identifying markings on harness and cable assemblies usually provide usage information, interconnection instructions, and part numbers. In addition, the marking may include serial number, source, assembly date, lot number and so on. Marking is also useful for inventory control and supply stock records.

Wire Marking.   MIL-STD-681 is applicable to various wire-marking methods. Color, stripes, bands, and numbers are acceptable. Numbering can be a relatively simple sequential matter, beginning with number 1 and progressing consecutively to the highest number required for an assembly.

Certain military requirements specify a coded marking for individual wires, which includes (1) unit number, (2) equipment identity or circuit function, (3) wire number, (4) wire segment letter, (5) wire gauge, and (6) ground, phase, or thermocouple letter. Hot impression stamping or color banding of the required wire identification is a practical user production marking method.

Harness Marking.   Wire harness assembly marking should be as simple as possible to convey the information required. A simple method is to use a short length of close-fitting insulating sleeving over the harness trunk, adjacent to each termination. Identification can be applied by hot impression stamping of the thermoplastic sleeve.

Cable Marking.   Sheathed cables are identified in a manner similar to wiring harness assemblies. The most significant difference between the two involves the materials used in the actual marking. A cable-marking method for production cables uses a reflective label with a pressure-sensitive adhesive backing.

4.10.3   Associated Hardware and Materials

Insulation Sleeving and Tubing.   Insulation sleeving and tubing serve multiple purposes in electrical assembly and harness fabrication. They are used for insulation, protection from chafing or abrasion, jacketing, strain relief, thermal or chemical protection, and identification.

Materials.   Extruded tubings are made from the plastic and rubber materials listed in Table 4.3.

Braided Insulating Sleeving.   Braided sleeving is made from basic uncoated yarns, lightly treated yarns, or yarns heavily coated with various insulating varnishes or resins. Although practically any yarn can be used for braiding, the material most frequently used in the United States is fiberglass, with some polyethylene terephthalate, acrylic, and nylon yarn also available.

Shrinkable Tubing.   Shrinkable tubing is based on the elastic memory of plastics. Under specific thermal and mechanical conditions, molecules of certain polymers may be overexpanded and then fixed in place in a strained condition. When heated, the material tends to return to its original shape and size as strain is relieved. Solvents may also be used with some materials to relieve strain in order to shrink tubing.

   Materials.   Heat-shrinkable tubing is available in many of the thermoplastics and elastomers listed in Table 4.3. The properties of these heat-shrinkable materials are comparable to those of the basic conventional materials. Heat-shrinkable tubing is available in the following materials:

Butyl rubber
Fluorinated ethylene propylene (Teflon FEP)
Polychloroprene (Neoprene)
Polyolefin (irradiated) (PE, etc.)
Polytetrafluoroethylene (Teflon TFE)
Polyvinyl chloride (PVC)
Polyvinylidene fluoride (Kynar)
Silicone rubber

Table 4.17 presents typical properties of heat-shrinkable tubings.

   Specifications.   Military specification MIL-1-23053 covers heat-shrinkable PVC, polyolefin, and TFE materials.

Wire and Cable Mounting and Spacing Hardware.   Good installation design practice requires adequate space, not only for wiring and cabling (Fig. 4.23), but also for the supporting hardware (clamps, sleeving, grommets, guides, etc., Fig. 4.24). Space is also needed for manipulating tools during the initial installation as well as during maintenance and replacement.[24]

   Hardware.   Cable mounting with the MS-type cable clamp is a proven method. Many variations of this clamp are available from specialty suppliers. The principal advantages of MS clamps are low cost, light weight, high strength, ready adaptability, and ease of installation and servicing. Clamps can be installed on any structure or skin of adequate strength that can be drilled. If the structure or skin cannot be drilled, bonding is recommended. One technique is to bond a cable-supporting device or pad to the supporting area, and then strap, tie, or clamp the cable to it. An alternative technique is to bond the entire cable to the supporting area for a very secure installation. If bonding is the only possible means of attachment, MS nylon, reinforced nylon, or Kynar harness straps, mounting plates, and a compatible bonding material can be used. Specification MIL-S-23190 covers adjustable plastic cable straps for military use.

   Support Spacing.   The spacing of clamps and other cable support tie-down devices (Fig. 4.24) can be determined from experience and developmental mockups. Applicable electrical system specifications generally establish bundle tiedown spacing by stating a maximum distance between supports (MIL-W8160 maximum spacing is 24 in). For adequate design of electric cable installations in missiles and space vehicles, the spacing must be resolved analytically and tested for verification. The spacing will be determined by the

TABLE 4.17 Typical Heat-Shrinkable Tubing Properties

Properties	Irradiated polyolefin					Irradiated PVF$_2$	Flexible PVC	Flexible irradiated PVC	Semirigid irradiated PVC	Neoprene rubber	Silicone rubber	Butyl rubber	PTFE
	Flexible opaque	Flexible clear	Semirigid opaque	Semirigid clear	Dual wall								
Tensile strength, lb/in^2	2500	2500	3000	3000	2000	7000	3000	3000	5000	1900	900	1600	4500
Ultimate elongation, %	400	400	400	400	400	300	300	300	250	220	300	350	250
Brittleness temperature, °C	−60	−85	−60	−90	—	−73	−20	−20	−20	−40	−75	—	−90
Hardness	98A	90A	—	—	—	—	85A	—	—	85A	70A	80A	—
Specific gravity	1.3	0.93	1.3	0.95	0.94	1.76	1.4	1.35	1.4	1.4	1.2	1.2	2.2
Water absorption, %	0.05	0.01	0.05	0.01	0.1	0.1	—	0.6	0.6	0.5	0.5	0.1	0.01
Dielectric strength, V/min	1300	1300	1300	1300	1100	1500	750	750	900	300	300	130	1200
Volume resistance, Ω · cm	10^{15}	10^{17}	10^{15}	10^{17}	10^{16}	—	10^{12}	10^{12}	$>10^{13}$	10^{11}	10^{15}	10^{12}	10^{18}
Dielectric constant	2.7	2.3	2.7	2.4	2.4	—	5.4	—	—	—	3.3	—	2.1
Power factor	0.003	0.0003	0.003	0.0003	0.0005	—	0.12	—	—	—	—	—	0.0002
Fungus resistance	Inert	Inert	Inert	Inert	Inert	Inert	Inert	Inert	Inert	Inert	Inert	Inert	Inert
Fuel and oil resistance	Excellent	Excellent	Excellent	Excellent	Excellent	Excellent	—	Excellent	Excellent	Good	Fair	Fair	Excellent
Hydraulic fluid resistance	Excellent	Excellent	Excellent	Excellent	Excellent	Excellent	—	Excellent	Excellent	Fair	Poor	Good	Excellent
Solvent resistance	Good	Good	Good	Good	Good	Excellent	—	Excellent	Excellent	Fair	Fair	Fair	Excellent
Acid and alkali resistance	Excellent	Excellent	Excellent	Excellent	Excellent	Excellent	—	Excellent	Excellent	Good	Good	Good	Excellent
Flammability	Self-extinguishing	Burns slowly	Self-extinguishing	Burns slowly	—	Nonburning	Self-extinguishing	Self-extinguishing	Self-extinguishing	Self-extinguishing	Self-extinguishing	Burns slowly	Nonburning

Source: From Schuh.[1]

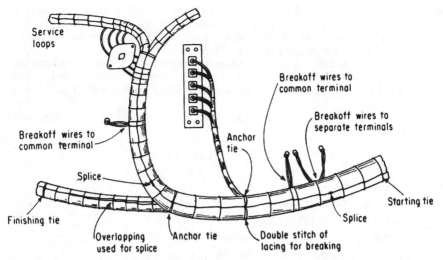

FIGURE 4.23 Cable harness lacing details.[16]

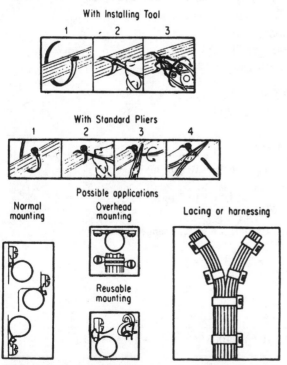

FIGURE 4.24 Cable straps for clamping and harnessing.[16]
Installation: 1—slip strap around wire bundle, rib side inside;
2—thread tip through eye, draw up snug; 3, 4—apply tool,
cinch tight, twist 120°, squeeze to cut off excess.

dynamic environment in which the cabling must perform reliably. Complete design coordination must exist between the structures, dynamics, and electrical installations in order to meet system requirements. Dynamic tests on the development hardware are recommended in a program for verifying the installation.

Drastic changes in cable stiffness or section size caused by the ending or branching of wires may lead to points of dynamic weakness. Firm support is recommended on both sides immediately adjacent to these points, regardless of the spacing of other tie downs or clamps.

The spacing of supporting devices on a high-acceleration missile system can be determined by the formula

$$F = \sum LANG$$

where F = design load of attachment device, lb

L = unsupported length, in

A = unit weight per length for each wire size, lb/in

N = number of each wire size in bundle

G = maximum dynamic environmental load, gs (gravitational)

A sample calculation will illustrate a design example. Given the following harness parameters, find the unsupported length:

10 unshielded wires, no. 20 AWG
10 shielded and jacketed cables, no. 20 AWG
10 twisted, shielded, jacketed pairs, no. 26 AWG

$$G(\text{load}) = 150g$$

$$F = 50\text{lb}$$

$$A_{20u} = 4.02 \times 10^{-4} \text{lb/in}$$

$$A_{20} = 6.36 \times 10^{-4} \text{lb/in}$$

$$A_{26w3} = 4.98 \times 10^{-4} \text{lb/in}$$

F is indicated as 50 lb. However, a safety factor of 2:1 changes this value to

$$F = \frac{50}{2} = 25 \text{ lb}$$

From this,

$$L = \frac{25}{150} \times \frac{1}{10(4.02 \times 10^{-4}) + 10(6.36 \times 10^{-4}) + 10(4.98 \times 10^{-4})}$$

$$= \frac{1}{6} \times \frac{1}{1.536 \times 10^{-2}}$$

$$= 10.8 \text{ in}$$

The sample harness must be clamped or attached every 10.8 in to satisfy the given conditions.

These details indicate a technique evolved under a specific set of requirements and are presented as a guide only.

Support and Clamping of Cables to Connectors. The cable clamp associated with a multipin connector is used primarily to support the wires or the cable terminating at the connector and also to relieve strain from the terminations. Soft telescoping bushings (in accordance with federal specification AN 3420) are available for cables smaller than the cable clamp opening. The bushings permit the cable to be centered and anchored securely without excessive padding. There should be adequate clamping pressure without bottoming the two halves of the clamp. The clamp screw thread engagement should be equal to two-thirds to onehalf times the major nominal screw diameter.

Major differences in size between cable and connector can be corrected with step-up or step-down telescoping extension sleeves instead of bushings.

4.11 SPECIFICATION SOURCES

The following are sources of specifications and standards pertaining to many of the materials discussed in this chapter.

American National Standards Institute (ANSI), 1430 Broadway, New York, NY 10018

American Society for Testing and Materials (ASTM), 1916 Race St., Philadelphia, PA 19103

Institute of Electrical and Electronics Engineers (IEEE), 345 East 47 St., New York, NY 10017

Insulated Cable Engineers Association (ICEA), P.O. Box P, South Yarmouth,MA 02644

Military and federal specifications, DODSS Subscription System and Manual Index, U.S. Government Printing Office, Washington, DC 20401

National Aerospace Standards Association (AIAA), 1250 Eye St. NW, Suite 1100, Washington, DC 20005

National Electrical Manufacturers Association (NEMA), 2101 L St. NW, Washington, DC 20037

Underwriters Laboratories (UL), 207 East Ohio St., Chicago, IL 60611

Since specifications are continually subject to change, it is advisable to consult the latest revisions in meeting technical and contractual requirements. A recent major change involved superseding the entire MIL-W-583C with J-W-1177.

REFERENCES

1. A. G. Schuh, "Wires and Cables," in C. A. Harper (ed.), *Handbook of Electronic Packaging*, McGraw-Hill, New York, 1969, chap. 2.

2. Rome Cable Div., *The Rome Cable Manual of Technical Information,* Rome Cable Corp., Rome, Ga., 1967.

3. C. P. Steinmetz and B. G. Lamme, *Trans. AIEE,* vol. 32, pp. 79–89, 1913.

4. V. M. Montsinger, "Loading Transformers by Temperature," *Trans. AIEE,* vol. 49, pp. 779–792, 1930.

5. T. W. Dakin, "Electrical Insulation Deterioration Treated as a Chemical Rate Phenomenon," *Trans. AIEE,* vol. 67, pt. 1, pp. 113–118, 1948.

6. "Extra Flexible Tactical Cable Report," No. 3, Martin Marietta Corp., Bethesda, Md., Dec. 1964.

7. *Electrical Design,* The Martin Co., Bethesda, Md., 1958.

8. J. C. Reed, "Save Space by Hookup Wire Insulated with Teflon," *J. Teflon,* 1964.

9. A. G. Schuh, "Wires and Cables," in C. A. Harper (ed.), *Handbook of Materials and Processes for Electronics,* McGraw-Hill, New York, 1970, chap. 4.

10. L. L. Lewis, "Ultra Thin-Wall Wire Insulation from Kapton Polyimide Film, Type XF," presented at the Naval Air Systems Command Symp., Oct. 1966.

11. S. Schwartz and D. L. Wells, "Processing of Plastics in Space," *J. Soc. Plastics Eng.,* Aug. 1962.

12. E. E. Jolley and J. C. Reed, "The Effects of Space Environments on Insulation of Teflon TFE and FEP Resins," presented at the Signal Corps Symp., Nov. 1962.

13. V. L. Lanza and R. M. Halperin, "The Design and Development of Wire Insulators for Use in the Environment of Outer Space," presented at the Signal Corps Symp., Dec. 1963.

14. W. J. Prise, "When the Gamma Heat Is On Insulators," *Electron. Des.,* May 23, 1968.

15. J. C. Reed and J. T. Walbert, "Teflon Fluorocarbon Resins in Space Environments," presented at the Signal Corps Symp., Nov. 1962.

16. E. F. Godwin, "Hookup Wires, Multiconductor Cables, and Associated Terminating Devices," in C. A. Harper (ed.), *Handbook of Wiring, Cabling, and Interconnecting for Electronics,* McGraw-Hill, New York, 1972, chap. 3.

17. N. J. Cotter and J. R. Perkins, "Life vs. Voltage Performance of Flat Conductor Cables and Light Weight Round Wire Systems," E. I. du Pont de Nemours & Co., Wilmington, Del., 1968.

18. *Cable Products Catalog,* ACD-5, Amphenol Corp., Oak Brook, Ill., 1970.

19. R. A. Bellino, "How to Select a Multiconductor Cable," American Enka Corp., BrandRex Div., Willimantic, Conn., 1970.

20. K. C. Byram, "Flat Conductor Cabling and Connectors," in *Flexible Flat Cable Handbook,* Inst. of Printed Circuits, Chicago, Ill., 1965.

21. *Flexprint Circuit Design Handbook*, Bull. FT-169, Sanders Assoc., Manchester, N.H., 1965.

22. *Flexible Flat Cable Handbook,* Inst. of Printed Circuits, Chicago, 111., 1965.

23. "Reliable Integrated Wire Termination Devices," Final Rep. ECOM-0394, Contract DAAB07-69-C0394, U.S. Army Electronics Command, Fort Monmouth, N.J., 1969.

24. A. Fox and J. H. Wisher, "Superior Hook-Up Wires for Miniaturized Solderless Wrapped Connections," *J. Inst Metals,* vol. 100, pp. 30–32, 1972.

FURTHER READING

Beitel, J. J., et al.: "Hydrogen Chloride Transport and Decay in a Large Apparatus I. Decomposition of Polyvinyl Chloride Wire Insulation in a Plenum by Current Overload," *J. Fire Sci.,* vol. 4, no. 1, pp. 15–27, 1986.

Cabey, M. A.: "New Silicone Rubber Cable Insulation Promises Circuit Integrity," in *Flaming Environment,* NTIS, Springfield, Va., 1983, 8 pp.

Felsch, C.: "A New Way of Insulating Power Cables with Crosslinked Polyethylene," *Wire World Int.,* vol. 22, p. 294, 1980.

Fischer, T. M.: "Impact of Polyethylene Curing Methods on Wire and Cable Performance," *IEEE Electr. Insul. Mag.,* vol. 5, no. 1, pp. 29–32, 1989.

Frasure, J. W., J. H. Snow, and D. A. Voltz: "Wire and Cable Update," *IEEE Trans. Ind. Appl.,* vol. IA-22, pp. 178–194, 1986.

Gage, C. A., G. Carrillo, E. D. Newell, W. D. Brown, and P. Phelan (eds.): "Wire Harness Automation," in *Proc. Int. SAMPE Symp. and Exhibition,* vol. 33, pp. 787–795, 1988.

Gross, B., J. E. West, H. Von-Seggern, and D. A. Berkley: "Time-Dependent Radiation-Induced Conductivity in Electron Irradiated Teflon Foils," *J. Appl. Phys.,* vol. 51, pp. 4875–4881, 1980.

Hamer, P. S., and B. M. Wood: "Are Cable Shields Being Damaged During Ground Faults?," *IEEE Trans. Ind. Appl,* vol. IA-22, pp. 1149–1155, 1986.

Ishibashi, M., T. Yamamoto, and S. Mogi: "Evaluation of Crosslinked Material for Insulated Electronic Appliance Hook-Up Wires," in *Proc. Int. Wire and Cable Symp.,* vol. 27, Am. Chem. Soc., 1978, pp. 213–219.

Meyer, F. K., and H. Linhart: "New Data on Long Term Stabilization of Polyethylene for Telecommunication Wire Insulation," in *Proc Int. Wire and Cable Symp.* (Cherry Hill, N.J.), NTIS, Springfield, Va., 1983, 10 pp.

Samborsky, A. M.: "Current Rating for Bundled Wires—A Step by Step Procedure," U.S. Naval Electronics Lab., San Diego, Calif., 1967.

Sheppard, A. T., and R. G. Webber: "Polytetrafluoroethylene Insulated Cable for High Temperature Oxygen Aerospace Applications," in *Proc. Int. Wire and Cable Symp.* (Cherry Hill, N.J.), NTIS, Springfield, Va., 1983, 10 pp.

Wolf, C. J., D. L. Fanter, and R. S. Soloman: "Environmental Degradation of Aromatic Polyimide Insulated Electrical Wire," *IEEE Trans. Elect. Insul.,* vol. EI-19, pp. 265–272, 1984.

Zeller, A. F., et al.: "Insulation on Potted Superconducting Coils," *IEEE Trans. Magn.,* vol. 25, pp. 1536–1537, 1989.

CHAPTER 5

SOLDER TECHNOLOGIES FOR ELECTRONIC PACKAGING

Jennie S. Hwang

H-Technologies Group, Inc.
Cleveland, Ohio

5.1 INTRODUCTION

5.1.1 Definition

Soldering has been used as a technique to accomplish the vital function of providing electrical, thermal, and mechanical linkages between two metallic surfaces. Solder paste, by its virtue of deformability and tackiness, is the primary material to make solder connections for surface mount and advanced surface mount industry when it is applied on the mother board (main circuit board) and/or when it is used for IC (integrated circuit) packaging on the module and package level. The deformable form of solder pasted makes it applicable in any selected shape and size and readily adaptable to automation; its tacky characteristics provide the capability of holding parts in position without the need of additional adhesives before the permanent bonds are formed.

5.1.2 Surface-Mount Technology

In this electronic and information age, we witness new technology developments and new product introductions to the marketplace almost on a monthly basis. One of the strongest trends, however, in the electronics interconnection and packaging segment is utilization of the surface-mounting concept to develop superior circuit board assemblies in both performance and cost. This concept has been utilized in hybrid assembly since the 1960s by interconnecting chip resistors, chip capacitors, and bare semiconductor dies on metallized substrates. Nevertheless, the potential of surface mounting was not fully explored and utilized until the early 1980s.

What is surface-mount technology? As the name implies, it is basically the application of science and engineering principles to board-level assembling by placing components and devices on the surface of the printed-circuit board instead of through the board. Although this concept appears to be straightforward, the impact on the production floor is enormous, not only on components and design, but also on materials and equipment. It also narrows the distinction between hybrid circuit assembly and printed-circuit assembly. The specific benefits of surface-mount technology in relation to through-hole technology include:

Increased circuit density

Decreased component size

Decreased board size

Reduced weight

Shorter leads

Shorter interconnection

Improved electrical performance

Facilitated automation

Lower costs in volume production

5.1.3 Industry Trend

Looking at the hierarchy of electronics, semiconductor devices have continuously exhibited improved reliability, reduced feature size, increased wafer size and double complexity every year. As examples, wafer size increased from 3 to 4 inches (75 to 100 mm) in the 1980s to 8 inches (200 mm) today, and likely will increase to 12 inches (300 mm) in the future; circuits have shrunk to submicrometer (0.1–0.5 µm) from several micrometers; IC pin count increased from 40–80 to over 800; IC fabrication techniques and equipment are developing in rapid pace from wet process and microanalysis to x-ray lithography and nanoanalysis.

As more memory-gobbling software are put in use, DRAM has proceed to 64 Mb generation today and is rapidly growing in both demand and capability. Microprocessors now reach 200 MHz using 8-inch wafer and 0.35 micro process. In the meantime, the power supply voltage continues to drop to 3.0 V or below. Marketwise, it is projected that semiconductors alone will exceed $200 billion toward the end of the century.

With the "known good die" being a lingering issue for the board level assembly, the use of packaged surface mount devices continues to be dominating. Across the two decades, the industry has evolved from dual-in-line package (DIP), pin grid array (PGA) to 50 mil surface leadless ceramic chip carrier (LCCC), plastic lead chip carrier (PLCC), small outline IC (SOIC), and to fine pitch quad flat pack (QFP), thin quad flat pack (TQFP) and array packages such as ball grid array (BGA). The development of fine pitch BGAs and chip scale (size) packages are in the works. Scanning over the package evolution, it can be categorized in three generic groups: Through-hole, surface mount, and chip scale/direct die attach.

After over ten years of double-digit growth, the use of surface mount devices has finally exceeded that of through-holes as of 1995, as shown in Fig. 1. It is projected that SMT will continue to enjoy a healthy growth rate at the expense of through-holes, and the introduction of chip-scale (size) packages and direct chip attach will slowly fill the niche areas that require either highest density and speed or smallest size and weight. Various packages differentiate from one another by virtue of functional capabilities and/or physical characteristics. For example, package height of through-hole PGA in 3.5 mm compares with BGA in 2.3 mm; package-to-die size ratio for PQFP is around 8 and TBGA

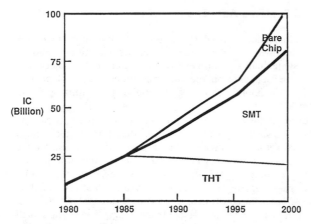

FIGURE 5.1   Market share of surface mount vs. through-hole.

around 5. IC packages having high pin count and new designs are expected to proliferate. Nonetheless, it is interesting to note that, despite the frequent introduction of new packages, SOCI/TSOP IC packages still occupy the largest market share among all at the present time and perhaps in the foreseeable future.

Newer packages such as BGAs, where solder bumps (balls, columns) form array bonding sites on the bottom of carrier substrate without requiring lead frame and wire bonding (on-mother board), have drawn tremendous attention. These array packages would satisfy demands to further shrink down the pitch spacing of peripheral packages when pin count increases above 250. There are pros and cons between these two types of packages in terms of performance/cost ratio and requirements in the key steps of manufacturing.

Adding to the electronic hierarchy another level of packaging as well as another choice of technologies which fits between chip and mother board is multichip module (MCM). MCM where two or more bare chips are mounted and interconnected on a multilayer substrate is packaged on a single functional unit with a defined interconnect density. It can be housed in array or peripheral package using either flip chips or face-up chips.

As a renewed interest, flip chip where a bare chip with solder bumps on its surface is turned upside down and bonded directly onto the substrate without lead frame and wire bonding has served as one of the base technologies to achieve chip size packages, although chip scale packages can also be accommodated by wire bonding.

As IC packages in conventional molded from or in chip size (scale) packages or in flip chip or in other advanced designs are proliferating, the selection among various packages largely depends on the speed, heat dissipation, density (I/O) and cost desired.

As the microelectronics and electronics industries continue to strive for quality and yield in every level of materials, designs, and processes, and as the density of board-level population continues to increase, the demands on the soldering process and solder-paste materials will be increasingly stringent. Furthermore, environmentally friendly production operation will be a required part of future manufacturing. These include CFC-free process and lead-free solders.

In view of these trends, this chapter outlines the fundamentals of soldering and solder paste and the practical techniques and know-how in the key steps of soldering and solder-paste application, as well as new and emerging products and processes. It is hoped that this chapter will provide integrated knowledge in the soldering and solder-paste arena and stimulate much needed innovations in material, design, and process.

5.1.4 Interdisciplinary and Systems Approach

It has been said that the best science and technology are produced by a combination of four elements-an overriding commitment to scientific excellence, vision, intuition, and initiative. Soldering and solder-paste technology is no exception. Therefore, the objective of the researcher is to meet versatile demands on the soldering process and solder-paste material, and to continue to add to the pool of technology by applying and utilizing fundamental sciences and technologies.

From a technology point of view, pastes come from the interplay of several scientific disciplines. Figure 5.2 illustrates the spirit of paste technology.[1] Based on this technology, a number of existing and potential application product lines can be derived. These product lines are composed of organopolymeric vehicles and metallic and nonmetallic particulates, ranging from PM injection molding to EMI shielding composites, cermet thick film, polymer thick film, and solder paste, brazing paste, and adhesives. While each of these product lines has its uniqueness, one common fundamental is paste technology. Sciences and technologies to be utilized in paste technology include metallurgy and particle technology, chemistry and physics, rheology, and formulation technology. In addition, to meet the demands of the ever-changing electronics packaging industry and the accelerating pace of developments, the collaborative effort among user, material supplier, and equipment manufacturer is much needed. They must be involved from the design state onward in order to develop the best suitable product or process system.

5.2 SOLDER MATERIALS

5.2.1 Solder Alloys

The elements commonly used in solder alloys are tin (Sn), lead (Pb), silver (Ag), bismuth (Bi), indium (In), antimony (Sb), and cadmium (Cd). Their melting points are listed in Table 5.1. In addition to tin-lead alloys, binary solder alloys include tin-silver, tin-anti-

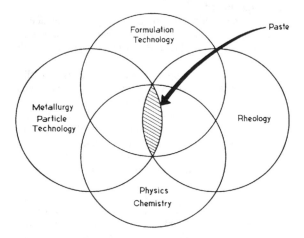

FIGURE 5.2 Market share of surface mount vs. through-hole.

TABLE 5.1   Melting Points of Common Solder Elements

	Sn	Pb	Ag	Bi	In	Sb	Cd
°C	232	328	961	271.5	156.6	630.5	321.2
°F	450	620	1762	520	313	1167	610

mony, tin-indium, tin-bismuth, lead-indium, and lead-bismuth. Ternary alloys include tin-lead-silver, tin-lead bismuth, and tin-lead-indium. With continued development of lead-free solders, new ternary, quartnary, and pentanary systems will proliferate.[2] The solidus and liquidus temperatures of some commonly used compositions are listed in Table 5.2.

Generally, the alloy selection is based on the following criteria:

- alloy melting range in relation to service temperature
- mechanical properties of the alloy in relation to service conditions
- metallurgical compatibility, consideration of leaching phenomenon, and potential formation of intermetallic compounds

TABLE 5.2   Melting Range of Common Solder Alloys

Alloy composition	Melting range, solidus		Melting range, liquidus		Mushy range	
	°C	°F	°C	°F	°C	°F
70Sn/30Pb	183	361	193	380	10	19
63Sn/37Pb	183	361	183	361	0	0
60Sn/40Pb	183	361	190	375	7	14
50Sn/50Pb	183	361	216	420	33	59
40Sn/60Pb	183	361	238	460	55	99
30Sn/70Pb	185	365	255	491	70	126
25Sn/75Pb	183	361	266	511	83	150
10Sn/90Pb	268	514	302	575	34	61
5Sn/95Pb	308	586	312	594	4	8
62Sn/36Pb/2Ag	179	355	179	355	0	0
10Sn/88Pb/2Ag	268	514	290	554	22	40
5Sn/90Pb/5Ag	292	558	292	558	0	0
5Sn/92.5Pb/2.5Ag	287	549	296	564	9	15
5Sn/93.5Pb/1.5Ag	296	564	301	574	5	10
2Sn/95.5Pb/2.5Ag	299	570	304	579	5	9
1Sn/97.5Pb/1.5Ag	309	588	309	588	0	0
96.5Sn/3.5Ag	221	430	221	430	0	0
95Sn/5Sb	235	455	240	464	5	9
42Sn/58Bi	138	281	138	281	0	0
43Sn/43Pb/14Bi	144	291	163	325	19	34
52Sn/48In	118	244	131	268	13	24
70In/30Pb	160	320	174	345	14	25
60In/40Pb	174	345	185	365	11	20
70Sn/18Pb/12In	162	324	162	324	0	0
90Pb/5In/5Ag	290	554	310	590	20	36
92.5Pb/5In/2.5Ag	300	572	310	590	10	18
97.5Pb/2.5Ag	303	578	303	578	0	0

- environment of service compatibility, consideration of silver migration
- wettability on specified substrate
- eutectic versus non-eutectic compositions

5.2.2 Metallurgy

For tin-lead binary alloys, the tin and lead elements have complete liquid miscibility and partial solid miscibility. The phase diagram in Fig. 5.3 shows solid solution regions represented by (Sn) and (Pb), a liquid region represented by L, liquid and solid solution regions represented by L + (Pb) and L + (Sn), and a solid solution mixture region, (Sn) + (Pb). As indicated, the eutectic point is 63 wt percent of Sn at a temperature of 183°C. The solubility of Sn in Pb increase as the temperature drops to the eutectic temperature and then decreases as the temperature continues to drop. The same applies to the solubility of Pb in Sn.

The maximum solubility of Pb is Sn is 2.5 wt percent and of Sn in Pb it is 19 wt percent. During soldering, the molten solder alloy wets the substrate with the aid of fluxes. Since the metallic surfaces involved in soldering tend to get oxidized or tarnished, the wettability depends to a large extent on the chemistry and reactions of fluxes. Nonetheless, the wetting phenomenon follows the basic wetting principle. For a system at a constant temperature T and pressure P,

$$\left(\frac{\partial G}{\partial A}\right)_{PT} = \gamma$$

where G is the free energy, A is the area, and γ is the surface tension. Then, thermodynamic condition for spreading to occur is

$$\Delta G < 0$$

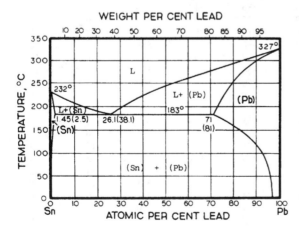

FIGURE 5.3 Phase diagram of Pb/Sn.

The spreading of a liquid with negligible vapor pressure on a solid surface S is as follows:

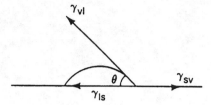

Thus,

$$-\left(\frac{\partial G}{\partial A}\right)_{P,\,T} = \gamma_{sv} - (\gamma_{ls} + \gamma_{vl}\cos\theta)$$

where γ_{ls}, γ_{vl}, and γ_{sv} are liquid-solid, liquid-vapor, and solid-vapor interfacial tension, respectively. Therefore, for spreading to occur,

$$\gamma_{sv} - (\gamma_{ls} + \gamma_{vl}\cos\theta) > 0$$

or

$$\gamma_{sv} > \gamma_{ls} + \gamma_{vl}\cos\theta$$

In general, for a system with liquid to wet the solid substrate, the spreading occurs only if the surface energy of the substrate to be wetted is higher than that of the liquid to be spread.

As the molten solder solidifies during cooling to form solder joints, the cooling process, such as the cooling rate, has a direct bearing on the resulting solder joint as to its microstructure and voids development. Figure 5.4 exhibits the SEM microstructure of 63 Sn/37 Pb melt under slow cooling, and Fig. 5.5 exhibits the microstructure under fast cooling, with other conditions being kept equal.

Because tin-lead solder alloys contain a solvus line and multiple solid phases, as shown in Fig. 5.3, they can be readily affected by heat treatment.

5.2.3 Solder Powder[2]

Alloy powders can be produced by one of the common techniques--reactions and decomposition, electrolytic deposition, mechanical processing of solid particulates, and atomization of liquid alloys.

Alloy powders made from chemical reduction under high temperature are generally spongy and porous. The fine particles of noble metal powders are frequently precipitated by reduction of the salts in aqueous solution with proper pH. The precipitate slurry is then filtered, washed, and dried under highly controlled conditions. A mechanical method is generally used to produce flake-like particles. The metals possessing high malleability, such as gold (Au), silver (Ag), copper (Cu), and aluminum (Al), are most suitable for making flakes.

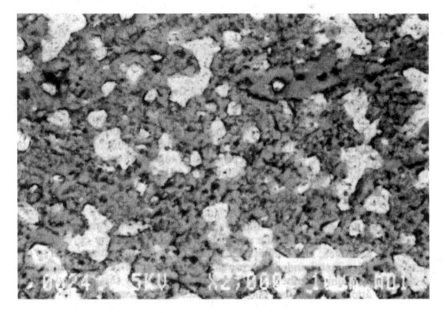

FIGURE 5.4   SEM micrograph of 63Sn/37Pb under slow cooling.

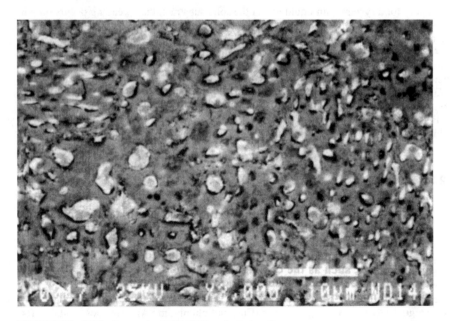

FIGURE 5.5   SEM micrograph of 63Sn/37Pb under fast cooling.

The electrolytic deposition process is characterized by dendrite particles, and it produces high-purity powders. The resulting particle sizes are affected by the type, strength, and addition rate of the reducing agent and by other reaction conditions. The characteristics of the particles are also affected by current density, electrolytes, additives, and temperature. The principle of atomization is to disintegrate the molten metal under high pressure through an orifice into water or into a gaseous or vacuum chamber. The powders produced by this method have relatively high apparent density, good flow rate, and are spherical in shape, as shown in Figs. 5.6 and 5.7. Powders to be used in solder paste are mostly produced by atomization because of its desirable inherent morphology and the shape of the resulting particles. Hence the discussion that follows is concerned with the atomization technique only.

Figure 5.8 is a schematic of an inert gas atomization system with options of a bottom pouring system and a tilting crucible system. The system consists of a control cabinet, vacuum induction furnace, tundish, argon supply line, ring nozzle, atomization tower, cyclone, and powder collection container. The alloy is melted under inert gas at atmospheric pressure to avoid the evaporation of component ingredients. A high melt rate can be achieved. The molten material is then dosed into the atomization tower. The melt is disintegrated into powder at atmospheric pressure by an energy-rich stream of inert gas. The process conducted in a closed system is able to produce high-quality powder.

In addition to inert gas and nitrogen atomization, centrifugal and rotating electrode processes have been studied extensively. The atomization mechanisms and the mean particle diameter are related to the operating parameters (diameter D, melting rate Q, and angular velocity ω of the rotating electrode) and to the material parameters (surface tension at melting point γ, dynamic viscosity η, and density at melting point ρ of the atomized liquid). The relationships among these parameters are presented subsequently.

It has been found that the mean volume-surface diameter d is proportional to the surface tension of the atomized liquid and the melting rate, but inversely proportional to the

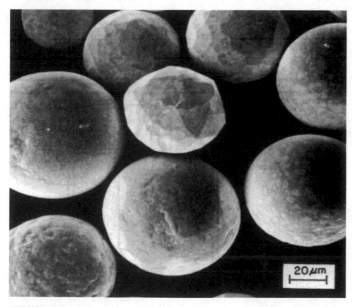

FIGURE 5.6   SEM micrograph of 63Sn/37Pb powder –200/+325 mesh.

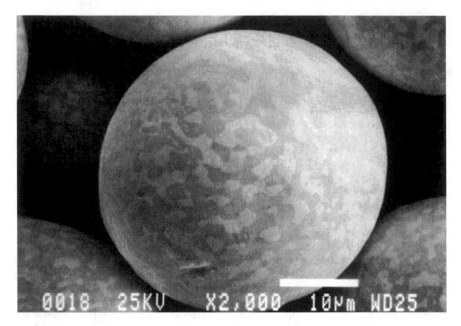

FIGURE 5.7   SEM micrograph of 63Sn/37Pb powder –325/–500 mesh.

angular velocity of the rotating electrode, the diameter of the electrode, and the density of the atomized liquid, expressed by the following relation:

$$d \propto \frac{\gamma^{0.50} Q^{0.02}}{\omega^{1.03} \rho^{0.50} D^{0.58}}$$

where the symbols were defined earlier.[3–5]

The mass proportion of secondary particles P_s is directly related to the angular velocity of the rotating electrode, the density of the atomized liquid, and the melting rate, but inversely proportional to the diameter of the electrode and the surface tension of the atomized liquid, expressed by the following relation:

$$P_s \propto \frac{\omega^{0.33} \rho^{0.56} Q^{1.24}}{D^{0.15} \gamma^{1.05}}$$

Metal powder can also be produced by vacuum atomization, which is believed to yield clean and finer particles. For superfine alloy powder, it is reported that a new atomizing technique is available using pulverizing energy produced by a 50-MPa water pump concentrated at the apex of a conical jet by which the thin stream of molten metal is disintegrated into superfine droplets.[6]

Ultrasonic gas atomization is another technique that produces metal powders successfully.[7] A process of two-stage spinning cup atomization with a liquid quenching is in development to produce fine particles with greater latitude in particle-size control.[8]

5.2.4 Mechanical Properties

The ultimate tensile strength, 0.2 and 0.01 percent yield strength, and uniform elongation of common bulk solder alloys are listed in Table 5.3. In the group of Sn/Pb alloys, the strength decreases with decreasing tin content. This trend is expected with the exception that the eutectic composition does not show the maximum strength. Its origin needs further confirmation.

The 96.5Sn/3.5Ag, 95Sn/5Ag, and 95Sn/5Sb compositions exhibit significantly higher strength and lower elongation. The composition 42Sn/58Bi is particularly strong, yet extremely brittle. In/Sn alloys with high indium content are extremely soft and lack adequate strength. It has been demonstrated that the solder joint strength may not coincide with that of bulk solder alloys due to other external factors, such as solder joint configuration, metallurgical reactions, interfacial wettability, interfacial effect, and the characteristics of other materials incorporated in the assembly.

Figures 5.9 through 5.30 show the creep behavior of solder alloys under a constant load of 920 g (equivalent to 50×10^6 dyn/cm^2 initial stress) at an ambient temperature of 25 + 3°C.

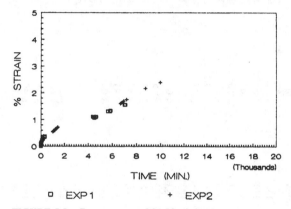

FIGURE 5.9 Creep curve of 96.5Sn/3.5Ag.

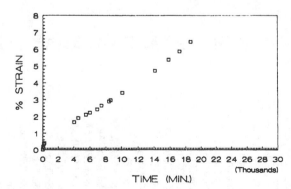

FIGURE 5.10 Creep curve of 95Sn/5Ag.

TABLE 5.3 Tensile Strength and Uniform Elongation of Common Solders

Alloy composition	Liquidus, °C	Solidus, °C	Ultimate tensile strength, 10^3 lb/in²	0.2% yield strength, 10^3 lb/in²	0.01% yield strength, 10^3 lb/in²	Uniform elongation, %
42Sn/58Bi	138	138	9.71	6.03	3.73	1.3
43Sn/43Pb/14Bi	163	144	5.60	3.60	2.77	2.5
30In/70Sn	175	117	4.67	2.54	1.50	2.6
60In/40Sn	122	113	1.10	0.67	0.53	5.5
30In/70Pb	253	240	4.83	3.58	3.08	15.1
60In/40Pb	185	174	4.29	2.89	2.06	10.7
80Sn/20Pb	199	183	6.27	4.30	2.85	0.82
63Sn/37Pb	183	183	5.13	2.34	1.91	1.38
60Sn/40Pb	190	183	4.06	2.06	2.19	5.3
25Sn/75Pb	266	183	3.35	2.06	1.94	8.4
10Sn/90Pb	302	268	3.53	2.02	1.98	18.3
5Sn/95Pb	312	308	3.37	1.93	1.83	26.0
15Sn/82.5Pb/2.5Ag	280	275	3.85	2.40	1.94	12.8
10Sn/88Pb/2Ag	290	268	3.94	2.25	2.02	15.9
5Sn/93.5Pb/1.5Ag	301	296	6.75	3.85	2.40	1.09
1Sn/97.5Pb/1.5Ag	309	309	5.58	4.34	3.36	1.15
96.5Sn/3.5Ag	221	221	8.36	7.08	5.39	0.69
95Sn/5Ag	240	221	8.09	5.86	3.95	0.84
95Sn/5Sb	240	235	8.15	5.53	3.47	1.06
85Sn/10Pb/5Sb	230	188	6.45	3.63	2.62	1.40
5Sn/85Pb/10Sb	255	245	5.57	3.67	2.26	3.50
95Pb/5Sb	295	252	3.72	2.45	1.98	13.70
95Pb/5In	314	292	3.66	2.01	1.79	33.0

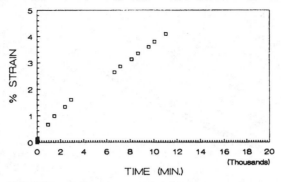

FIGURE 5.11   Creep curve of 95Sn/5Sb.

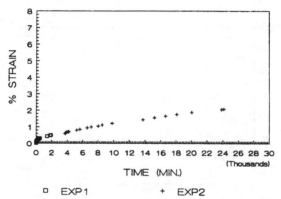

FIGURE 5.12   Creep curve of 5Sn/85Pb/10Sb.

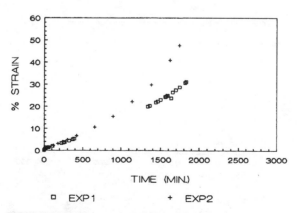

FIGURE 5.13   Creep curve of 85Sn/10Pb/5Sb.

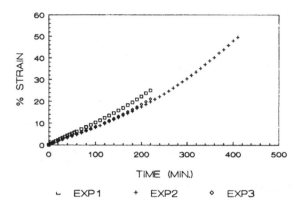

FIGURE 5.14   Creep curve of 63Sn/37Pb.

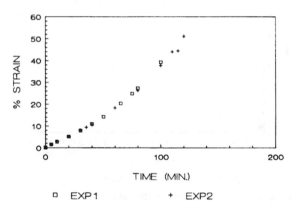

FIGURE 5.15   Creep curve of 60Sn/40Pb.

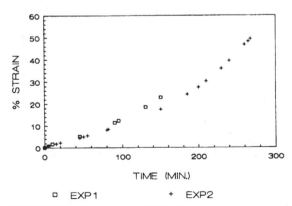

FIGURE 5.16   Creep curve of 80Sn/20Pb.

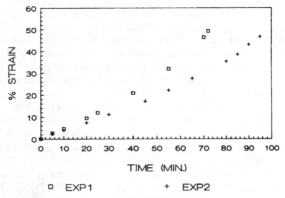

FIGURE 5.17 Creep curve of 25Sn/75Pb.

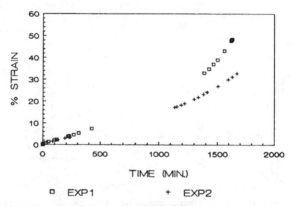

FIGURE 5.18 Creep curve of 10Sn/90Pb.

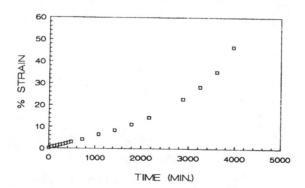

FIGURE 5.19 Creep curve of 5Sn/95Pb.

Alloys of Sn/Ag, Sn/Sb, and 5Sn/85Pb/10Sb impart high creep resistance, as shown in Figs. 5.9 through 5.12. This is primarily attributed to solution hardening as substantiated by their high strength and low elongation. When load is applied, the deformation is hindered by means of either interaction of solute atoms with dislocations or interaction with the formation and movement of vacancies, resulting in the impediment of the dislocation movement. Figure 5.13 is a creep curve for the composition 85Sn/10Pb/5Sb, exhibiting relatively lower creep resistance than 5Sn/85Pb/10Sb. Their melting point may be a factor in creating such a difference.

Figures 5.14 through 5.19 show the creep curves for Sn/Pb compositions. Eutectic 63Sn/37Pb has higher creep resistance than noneutectic compositions—60Sn/40Pb, 80Sn/20Pb, and 25Sn/75Pb. Alloys 10Sn/90Pb and 5Sn/95Pb, however, are benefited by the high melting point of their microstructural continuous phase, resulting in the more sluggish steady-state creep, as shown in Figs. 5.18 and 5.19. This is attributed to lower self-diffusion, although the alloys are ductile and have moderate strengths. The creep curves for Sn/Pb/Ag systems are shown in Fig. 5.20 through 5.24. 62Sn/36Pb/2Ag has the highest creep resistance. Its mechanism, whether through the impediment of grain-boundary

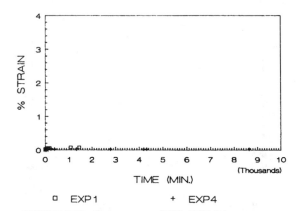

FIGURE 5.20   Creep curve of 62Sn/36Pb/2Ag.

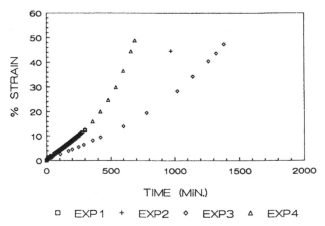

FIGURE 5.21   Creep curve of 15Sn/82.5Pb/2.5Ag.

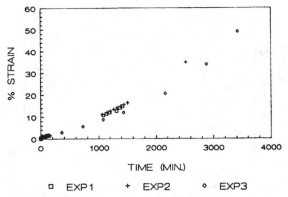

FIGURE 5.22 Creep curve of 10Sn/88Pb/2Ag.

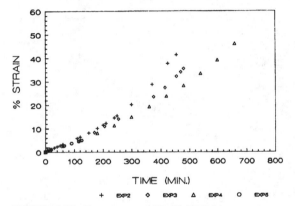

FIGURE 5.23 Creep curve of 5Sn/93.5Pb/1.5Ag.

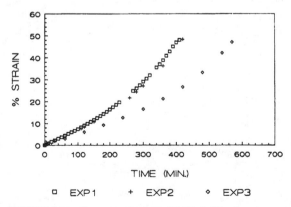

FIGURE 5.24 Creep curve of 1Sn/97.5Pb/1.5Ag.

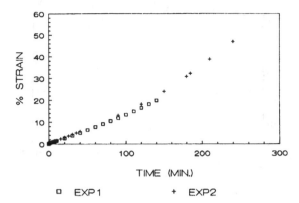

FIGURE 5.25 Creep curve of 42Sn/58Bi.

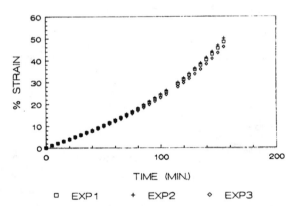

FIGURE 5.26 Creep curve of 43Sn/43Pb/14Bi.

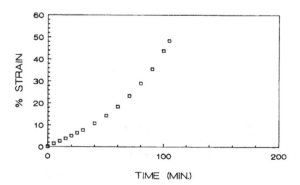

FIGURE 5.27 Creep curve of 30In/70Sn.

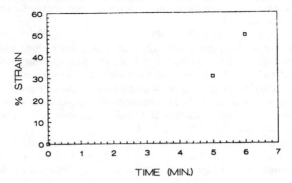

FIGURE 5.28 Creep curve of 60In/40Sn.

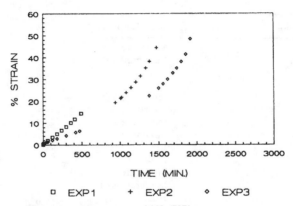

FIGURE 5.29 Creep curve of 30In/70Pb.

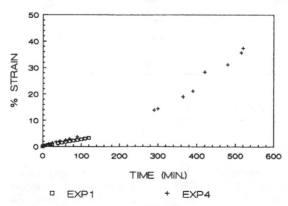

FIGURE 5.30 Creep curve of 60In/40Pb.

sliding due to silver segregation or the result of high activation energy for the dislocation movement, is not substantiated.

Bismuth alloys, 42Sn/58Bi and 43Sn/43Pb/14Bi, though having high tensile strength, are found prone to creep. This may be primarily due to their low melting temperature and the predominance of the diffusion-controlled process. The corresponding creep curves are shown in Figs. 5.25 and 5.26. The In/Sn system has very poor creep resistance, as reflected in Figs. 5.27 and 5.28. The low melting point of their microstructural continuous phase is considered to be a main factor. While In/Pb compositions are ductile, as shown in the elongation data, the single-phase microstructure and moderate melting point provide moderate creep resistance, as shown in Figs. 5.29 and 5.30.

The creep resistance of the various solder alloys, as shown in Table 5.4, is ranked in five groups—low, low-moderate, moderate, moderate-high, and high. As the testing temperature or the applied load changes, a change in the creep behavior of the alloys may result.

TABLE 5.4 Relative Creep Resistance of Common Solder Alloys

Alloy composition	Rank
42Sn/58Bi	Moderate
43Sn/43Pb/14Bi	Low-moderate
30In/70Sn	Low
60In/40Sn	Low
30In/70Pb	Moderate
60In/40Pb	Moderate
80Sn/20Pb	Moderate
63Sn/37Pb	Moderate
60Sn/40Pb	Low
25Sn/75Pb	Low
10Sn/90Sb	Moderate
5Sn/95Pb	Moderate-high
62Sn/36Pb/2Ag	High
15Sn/82.5Pb/2.5Ag	Moderate
10Sn/88Pb/2Ag	Moderate-high
5Sn/93.5Pb/1.5Ag	Moderate
1Sn/97.5Pb/1.5Ag	Moderate
96.5Sn/3.5Ag	High
95Sn/5Ag	High
95Sn/5Sb	High
85Sn/10Pb/5Sb	Moderate
5Sn/85Pb/10Sb	High

5.3 SOLDER PASTE

5.3.1 Definition

Solder paste, by one definition, is a homogeneous and kinetically stable mixture of solder alloy powder, flux, and vehicle, which is capable of forming metallurgical bonds at a set of soldering conditions and can be readily adapted to automated production in making reliable and consistent solder joints.

In terms of functionality, a solder paste can be considered as being composed of three major components. These are solder alloy powder, vehicle system, and flux system. The vehicle primarily functions as a carrier for the alloy powder, a compatible matrix for the flux system, and a basis for a desirable rheology. The flux cleans the alloy powder and the substrates to be joined so that high-reliability metallic continuity results and good wetting can be formed. Both vehicle and flux are fugitive or nonfunctional in nature after completion of the soldering. They are nevertheless crucial to the formation of reliable permanent bonds. On a permanent basis, the alloy powder part is the only functional component in forming a metallurgical bond.

5.3.2 Characteristics

The chemical and physical characteristics of solder paste can be represented by the following parameters:

physical appearance
stability and shelf life
viscosity
cold slump
dispensability through fine needles
screen printability
stencil printability
tack time
adhesion
exposure life
quality and consistency
compatibility with surfaces to be joined
flow property before becoming molten
wettability
dewetting phenomenon
solder balling phenomenon
bridging phenomenon
wicking phenomenon
leaching phenomenon
quantity and properties of residue
residue corrosivity
residue cleanability
solder-joint appearance
solder-joint voids

5.3.3 Fluxes and Fluxing[9]

The fundamental key to good solderability lies in ensuring that the surfaces to be joined are "scientifically" clean. Cleanliness must then be maintained during soldering, so that a metallic continuity at the interface can be achieved. This cleaning process is called fluxing, and the material used is the flux.

Customarily the flux is classified based on its activity and chemical nature, namely rosin-based such as RMA, water-soluble and no-clean.

Fluxes are applied to the surface to react with metal oxides or nonmetallic compounds, thus "cleaning" them from the metal surfaces. Common metal surfaces that are soldered include Sn/Pb, Sn, Cu, Au, Ag, Pd, Au/Pd, Ag/Pd, Au/Pt, Au/Ni, Pd/Ni, and Ni. Each has its own associated fluxing chemistry.

The flux activity can be determined by the combined measurements in water extract resistivity, copper mirror test, halide test, and surface insulation test.

To achieve fluxing, several approaches are available. Flux can be incorporated into the solder paste or inside the solder wire, or it can be applied as a separate chemical directly onto the component and solder paste or inside the solder wire, or it can be applied as a separate chemical directly onto the component and solder pad surface, as with liquid flux used in wave soldering. Still another approach is gas-phase fluxing, which supplies a proper atmosphere to the soldering substrates. Other in-situ cleaning process may render the solder fluxless.

5.3.4 Strength of Fluxes

The fluxing strength depends on the intrinsic properties of the flux agent or flux system as well as the external conditions. The factors include:

- functional group and molecular structure of flux agent
- melting point and boiling point of flux chemicals
- thermal stability in relation to soldering conditions
- chemical reactivity in relation to soldering conditions
- surrounding medium of flux agent
- substrates to be fluxed
- environmental stability (temperature, humidity)
- soldering conditions (temperature versus time, atmosphere)

The effects of molecular structure and medium on the strength of acids and bases are classified as inductive, resonance, hydrogen bonding, solvation, hybridization and steric effects. For commonly adopted inductive effect, the electronic-withdrawing groups adjacent to the carboxylic group of molecules enhance the acidity strength of the carboxylic group as a result of anion stabilization. Conversely, electron-releasing groups decrease the acidity.

5.3.5 Water-Cleaning Flux

Water-cleaning flux is designed so that its residue after soldering can be removed by using either pure water or a water medium with the addition of a saponifier or an additive.

Considering performance, process, reliability and cost, the flux chemistry that takes only water to clean (water-soluble) is the preferred choice. Special notes for using water-soluble solder paste are:

- The cleaning process to avoid flux entrapment and incomplete residue removal, ultrasonic cleaning is an effective aid.
- The soldering process to control the temperature profile, particularly in peak temperature, dwell time at peak temperature to avoid over-heating.

5.3.6 Gas-Phase Flux

Soldering under controlled-atmosphere conditions has been studied in recent years. Such controlled atmosphere can generally be classified as either reactive or protective.

The reactive atmosphere can help the fluxing agent clean component leads and solder pads. This approach has strong merits in solderability and leaves minimal residue. Yet it should be cautioned that a reactive atmosphere is nonselective. It can react with all materials being exposed as long as the conditions meet thermodynamic and kinetic criteria. Therefore all materials of an assembly must be compatible with the reactive atmosphere. With the wide variety of materials used in each assembly and the continued incorporation of new components and materials into assemblies, ensuring this compatibility can be quite a task. The protective atmosphere, on the other hand, primarily functions as oxygen and moisture repellent during soldering without providing external chemical activity. Controlled atmosphere soldering is discussed in Sec. 5.4.6.

5.3.7 No-Clean Flux

From the user's point of view, no-clean flux (especially incorporated directly into solder paste) requires the following:

- minimal amount of residue; ideally none
- residue that is translucent and aesthetically acceptable
- residue that will not interfere with bed-of-nails testing
- residue that will not interfere with conformal coating where applicable
- residue that is nontacky
- residue that stays inert under exposure to temperature, humidity, and voltage bias
- ability to flux effectively without solder-ball formation

Due to the large variety of design and performance requirements in board assemblies, the acceptable amount of residue or the physical and chemical properties of the residue will vary from one application to another. An application-specific approach is needed. Success with a no-clean product requires close communication and collaboration between user and supplier to design a best-fit flux (material) and fluxing (process) system.

Common solder paste tests in chemical and physical characteristics continue to apply to no-clean system. The industry's established test parameters and methods can be used to assess the quality and properties of the assemblies. These include ionic contaminant test, and visual examination. However, the tests for no-clean system have one difference. These tests should be conducted after reflow or soldering. The solder paste chemical make-up measurement in terms of ionic mobility must be also taken after exposure to a specified reflow condition, not before exposure. This procedure is designed to target the characteristics of the residue left on the board, not the as-is paste chemistry.

5.3.8 Comparison between Water-Clean and No-Clean

With proper cleaning process and reflow parameters, a water-soluble process can produce clean assemblies in both function and appearance. In addition, the nature of its chemistry imparts wider fluxing latitude, better accommodating the inherent variations in solderability of components and boards. It requires initial equipment capital, added operating costs in energy and water consumption, as well as the cost of consumables for a closed-loop recycle system.

No clean (air) systems eliminate one process step, clearly an economic advantage. It should be noted that the cleaning process has been perceived as a step to remove residues from solder flux or paste, yet it actually has provided the cleaning function for components and boards for many operations without being noticed. It is not unusual for boards before fluxing and soldering to contain higher amounts of ionic contaminants than after soldering and cleaning. The level of as-received contamination may exceed the acceptable level, since most steps in board fabrication and component plating involve highly ionic chemicals.

For the no-clean system that requires soldering under a protective atmosphere such as N_2, the cost of N_2 may offset or exceed the savings from no-clean, depending on N_2 consumption and the unit cost of N_2 which varies with the location. Other factors that may also complicate the assessment of a no-clean system are solder ball effect and the acceptability of residue appearance.

Nonetheless, both water-clean and no-clean routes are viable application systems. A basic understanding of the principle behind each practice and the compliance with application requirements are essential to the success of implementing either manufacturing system. Table 5.5 summarizes the general feature comparison between water-clean and no-clean, and Table 5.6 illustrates viscosity and metal load of despensing and printing pastes.

TABLE 5.5   Comparison of Water-Soluble versus No-Clean (Air)

	Water-soluble	No-clean (air)
Merits	• Clean assembly in function and appearance • Latitude for solderability variation	• One less process step • Lower operating and capital expenses
Drawbacks	• Extra step of process-cleaning • Operating cost—water, energy, and consumables • Initial capital expenditure	• Unable to remove contaminants from board and components • Often demand higher level of process control • Uncertainty in solder ball effect • Appearance issue • Possible limits for high-frequency application and/or uses that demand extraordinary extension of fatigue

TABLE 5.6   Viscosity and Metal Load of Dispensing and Printing Pastes

Solder paste type	Viscosity, cP[*]	Metal load, wt %
Fine dot dispensing	200,00–450,000	To 88
Screen printing	450,000–1,000,000	To 92
Stencil printing	700,000–1,600,000	To 92

* Centipoise, Brookfield RVT viscometer, TF/5 r/min, 3-min mixing/2-min reading.

5.3.9   Rheology

Paste applicability depends on its rheology, that is, its flow and deformation behavior. The primary driving forces underlying the rheology of solder paste include both kinetic and thermodynamic contributions. Therefore the rheology of solder paste may be affected by the following factors:

- composition, shape, and size of suspended particles
- chemical composition of suspending matrix
- relative concentration of effective ingredients in matrix
- structure of ingredients in matrix
- interactions between matrix and suspended particles either physical or chemical in nature, including wetting and solvation
- volume fraction occupied by suspended particles—usually, the higher the amount of particles, the more deviation from viscous flow
- internal structure and its response to external forces
- interactions among particles and resulting aggregates and flocculants
- temperature

The difficulty of predicting the rheology of such a system is apparent, due to a lack of knowledge of the detailed structure and the nature of forces exerted by molecules or particles. However, its behavior can be characterized. It is also apparent that solder paste is not an elastic material, nor a pure viscous material. Viscoelasticity best describes the behavior of solder paste. The characterization of viscoelasticty and fundamental theories are covered in the literature.[1,10]

The common methods to transfer solder paste consistently and accurately onto the intended solder pads include mesh screen printing, metal mask stencil printing, pneumatic dot and line dispensing, and positive displacement dispensing. Figures 5.31 and 5.32 illustrate the flow behavior of a dispensing paste and a printing paste, respectively.[1] The paste possessing a low yield point and very slight plastic behavior is found most suitable for dis-

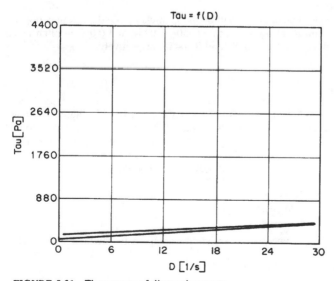

FIGURE 5.31 Flow curve of dispensing paste.

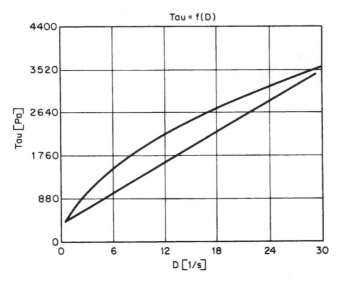

FIGURE 5.32 Flow curve of printing paste.

pensing applications, and a moderate yield point and thixotropy are generally associated with the printing paste. Table 5.6 lists typical viscosity and metal load percentages for dispensing and printing application techniques.

The size distribution of particles suitable for solder paste ranges from 5 to 74 m in diameter, corresponding to –200/+325 mesh, are compatible with the printing standard of 0.050 to 0.25-in pitch land patterns, as well as with dispensing up to 20 gauge. For finer-pitch applications, smaller than 0.025-in pitch or 20 gauge, solder powder smaller than 44 m in diameter is needed.

The printing thickness is another factor. Although the thicker paste deposit may impart "better" joint, the maximum paste thickness for 0.050-, 0.025-, and 0.012-in pitch land patterns are generally limited to 0.015, 0.008, and 0.004 in, respectively.

5.3.10 Formulation

As an example, a typical RMA solder-paste formula may contain 10 to 15 ingredients that provide various intended functions as shown in Table 5.7. The formula may appear to be straightforward. However, designing a viable product capable of delivering all the desired performance characteristics is complex and requires understanding the technologies. The following thinking steps are one route to take in developing a product:

1. Define performance objective.
2. Utilize fundamental technologies.
3. Select raw materials.
4. Understand and anticipate potential synergistic or antagonistic interactions between ingredients.
5. Balance performance parameters.
6. Fine-tune the formula to meet designated specifications.

TABLE 5.7   Ingredients of Typical RMA Solder Paste

Ingredient	Function
Rosin	Rosin system for designated softening point, acid number, thermal stability, fluxing activation, tackiness
Nonhalogen activator	Activator system for accomplishing fluxing action over a wide range of temperatures, rheology
Solvent	Solvent system to accommodate solubility, rheology, temperature compatibility, chemical compatibility
Binder	Providing compatible viscosity, rheology, tackiness
Fluxing modifier	Stabilizing and modifying flux
Rheology modifier	Contributing to targeted rheology

7. Develop production process.

8. Produce consistent product.

A product involves many performance parameters, and some of them are trade-offs. For example, a high metal content is beneficial to solder joints in volume and void and to the residue, yet it makes the paste more prone to drying and difficult to apply. A high-viscosity paste may improve flow control against temperature, yet causes the paste to be difficult to apply. Using highly active fluxing chemicals may improve solderability in some cases, yet their use may leave a more corrosive residue. In such cases, improving the solderability by selecting the proper ingredients without the use of highly active fluxing chemicals is the essence of technology. It should be noted that increasing the flux content does not always improve solderability in terms of wetting or the elimination of solder balling.

After the product is designed, developing a reproducible process for making the paste with consistent characteristics is equally important. It is not an exaggeration, but an indication of the importance of the role of the process, to state that the identical composition formula can produce different products when the process is allowed to vary.

The design of water-soluble and no-clean products follows the same principle, yet specific chemical ingredients differ.

5.3.11   Quality Assurance Tests

Tests to assure the properties and performance of a solder paste can be grouped into five parts: paste, vehicle, powder, reflow, and post reflow. Table 5.8 summarizes the tests in each of the five parts.[1]

5.4   SOLDERING METHODOLOGY

5.4.1   Types

The commercially available reflow methods include conduction, infrared, vapor phase, hot gas, convection, induction, resistance, and laser. Each of these reflow methods has its uniqueness and merits in cost, performance, or operational convenience. For localized and fast heating, laser excels over other methods, with hot air in second place. For uniform temperature, vapor phase ranks first. For versatility, volume, and economy, convection/

TABLE 5.8 Summary of Quality Assurance Tests for Solder Paste

Paste
  Appearance
  Metal content and flux-vehicle percentage
  Density
  Viscosity
  Viscosity versus shear rate
  Cold slump
  Hot slump
  Molten flow
  Tack time
  Dryability
  Dispensability
  Printability
  Shelf stability
  Storage, handling, and safety
Flux/vehicle
  Water extract resistivity
  Copper mirror corrosion
  Chloride and bromide
  Acid Number
  Infrared spectrum fingerprint and other spectroscopies
Solder powder
  Alloy composition
  Particle size, sieve
  Particle size distribution, sedigraph
  Particle shape
  Particle surface condition
  Dross
  Melting range
Reflow
  Solder ball
  Solderability
  Exposure time
  Soldering dynamics
Post reflow
  Cleanliness, resistivity of solvent extract
  Surface insulation resistance, before and after cleaning
  Solder joint appearance
  Solder voids
  Joint strength
  Power cycling
  Temperature cycling
  Vibration test
  Simulated aging
  Thermal shock

infrared are the choice. Conduction heating, however, is a convenience for low-volume and hybrid assembly. For conductive components requiring fast heating and high-temperature soldering, induction heating meets the requirement. Table 5.9 summarizes the strength and limitations of each method.

TABLE 5.9   Outline of Benefits and Limitations of Reflow Methods

Reflow method	Benefits	Limitations
Conduction	Low equipment capital, rapid temperature changeover, visibility during reflow	Planar surface and single-side attachment requirement, limited surface area
Infrared	High throughput, versatile temperature profiling and processing parameters, easier zone separation	Mass, geometry dependence
Vapor phase condensation	Uniform temperature, geometry independence, high throughput, consistent reflow profile	Difficult to change temperature, temperature limitation, relatively high operating cost
Hot gas	Low cost, fast heating rate, localized heating	Temperature control, low throughput
Convection	High throughput, versatility	Slower heating, higher demand for flux activity
Induction	Fast heating rate, high temperature capacity	Applicability to nonmagnetic metal parts only
Laser	Localized heating with high intensity, short reflow time, superior solder joint, package crack prevention	High equipment capital, specialized paste requirement, limit in mass soldering
Focused infrared	Localized heating, suitable for rework and repair	Sequential hating, limit in mass soldering
White beam	Localized heating, suitable for rework and repair	Sequential heating, limit in mass soldering
Vertical reflow	Floor space saving, maintenance of desired throughput	Often more costly

5.4.2   Reactions and Interactions

During soldering, a series of reactions and interactions occur in sequence or in parallel. These can be chemical or physical in nature in conjunction with heat transfer. The mechanism behind fluxing is often viewed at the reduction of metal oxides. Yet in many situations, chemical erosion and dissolution of oxides and other foreign elements acts as the primary fluxing mechanisms. Using a more complex fluxing process in solder paste as an example, the primary steps are represented by the flowchart in Fig. 5.33.

5.4.3   Process Parameters

With the prevalence of infrared and convection reflow, a few more words about furnace profile and furnace operating parameters are pertinent. It should be stressed that the reflow is a dynamic heating process in that the condition of the workpiece is constantly changing

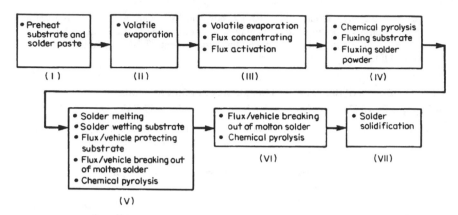

FIGURE 5.33   Flow chart of reflow dynamics.

as it travels through the furnace in a relatively short reflow time. The momentary temperature that the workpiece experiences determines the reflow condition; therefore, the reflow results.

It is ultimately important to establish a correlation between setting the temperature of a given furnace, the measured temperature of the workpiece at each specified belt speed, and the soldering performance. The resulting correlation between soldering performance and temperature setting or profile provides a "workable range" for the assembly.

Under mass reflow operation, both heating and cooling steps are important to the end results. It is generally understood that the heating and cooling rate of reflow or soldering process essentially contributes to the compositional fluctuation of the solder joint. This is particularly true when there are significant levels of metallurgical reactions occurring between the Sn/Pb solder and substrate metals. In the meantime, the cooling rate is expected to be responsible for the evolution of the microstructure.

The key process parameters that affect the production yield as well as the integrity of solder joints include:

- preheating temperature
- preheating time
- peak temperature
- dwell time at peak temperature
- cooling rate

It should be stressed that the reflow in a furnace (infrared, convection) is a dynamic heating process in that the conditions of the workpiece are constantly changing as it travels through the furnace in a relatively short reflow time. The momentary temperature that the workpiece experiences determines the reflow conditions and therefore, the reflow result.

Figure 5.34 illustrates a simulated reflow profile (which is considered to be most desirable temperature versus time), comprising three stages of heating:

1. natural warm-up
2. preheating/soaking
3. spike and reflow

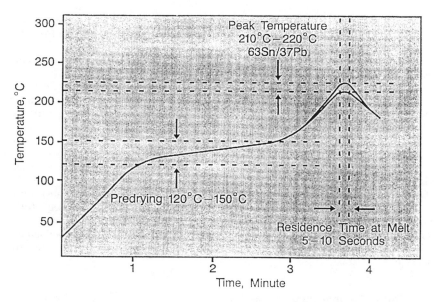

FIGURE 5.34   Reflow profile comprising three stages of heating.

In the natural heating stage, the heating rate of reflow profiles being used in the industry falls in the range of 1–2°C. The parameters of the preheating/soaking stage are important to the reflow results. The desirable preheating rate is less than 1°C/s. During soldering dynamics, the heating process contributes not only to the effectiveness or wetting but also to the extent of metallurgical reactions between solder and the substrates that solder interfaces, particularly the peak temperature and the dwell time at the temperature above the liquidus of the solder.

Several events occur during this stage, as shown in steps (ii) and (iv) of the flow chart (Fig. 5.33). These include temperature set to fit the specific flux activation temperature of the chemical system of the paste, and the time at heat to fit the constitutional make-up of the paste. Inadequate preheating often causes a spattering problem which manifests itself as discrete solder balls. Too high a temperature or too long a time at the elevated temperature result in insufficient fluxing and/or over-decomposition of organic, causing solder balling or hard-to-clean residue (if the no-clean route is adopted). The recommended general conditions for the second stage are 120–150°C for a duration of 60–150 seconds. The third stage is to spike quickly to the peak reflow temperature at a rate of 1.2 to 4.0°C/s. The purpose of temperature spiking is to minimize the exposure time of the organic system to high temperature, thus avoiding charring or overheating. Another important characteristic is the dwell time at the peak temperature. The rule of thumb in setting the peak temperature is 20 to 50°C above the liquidus or melting temperature; e.g., for the eutectic Sn/Pb composition, the range of peak temperatures is 203 to 233°C.

The wetting ability is directly related to the dwell time at the specific temperature in the proper temperature range and to the specific temperature being set. Other conditions being equal, the longer the dwell time, the more wetting is expected but only to a certain extent; the same trend applies at higher temperatures. However, as the peak temperature increases or the dwell time is prolonged, the extent of the formation of intermetallic compounds also increases. An excessive amount of intermetallics can be detrimental to long-term solder joint integrity. Peak temperature and dwell time should be set to reach a balance between good wetting and to expel any non-solder (organics) ingredients from the molten solder before it solidifies, thus minimizing void formation.

For a given system, cooling rate is directly associated with the resulting microstructure which in turn affects the mechanical behavior of solder joints.[2]

It was found that the microstructural variation and corresponding failure mechanisms of solder joints that were made under various reflow temperature profiles are extremely complex. Nonetheless, some correlation between the cooling rate and the basic properties can be obtained.

The copper/solder/copper system is a good example since it is still the most common material combination electronics assemblies. In this system, 63 Sn/37Pb solder joins copper pads (coated or uncoated) on the printed circuit board and the Sn/Pb coated copper leads of IC components.

For the tinned Cu-63Sn/37Pb-tinned Cu assembly, the reflowed solder joints are cooled in five different manners which deliver four cooling rates—0.1°C/sec, 1.0°C/sec, 50°C/sec, and 230°C/sec, respectively, as measured above 100°C. The fifth cooling mode was conducted in a two-step cooling resulting an uneven cooling with an average cooling rate of 12°C/sec. Each of the five cooling modes produced a different development of microstructure of solder joint.[2]

5.4.4 Reflow Temperature Profile

Reflow temperature profile representing the relationship of temperature and time during reflow process depends not only on the parameter settings but also on the capability and flexibility of equipment. Specifically, the instantaneous temperature conditions that a workpiece experiences is determined by:

- temperature settings to all zone controllers
- ambient temperature
- mass per board
- total mass in the heating chamber (load)
- efficiency of heat supply and heat transfer

For furnace-type reflow process, two profiles are taken to illustrate the effect of temperature profile on the reflow results.

Figures 5.35 and 5.36 show the actual temperature profiles of a convection oven, with a relatively lower preheat temperature as shown in Fig. 5.35, and a higher preheat temperature as shown in Fig. 5.36. The importance of the compatibility of solder paste chemistry and the assembly system with the reflow temperature profile can be easily demonstrated. For instance, if the solder paste and the assembly require the temperature profile of Fig. 5.35, performing reflow under the temperature profile of Fig. 5.36 may give rise to the following phenomena:

- deficiency of flux, resulting in solder balls
- overheating of organics, resulting in cleaning difficulty for processes that are designed to include a cleaning step.

On the other hand, if the paste is designed for the higher preheat temperature and/or assembly requires additional heat, using the lower preheat temperature profile can produce the following phenomena:

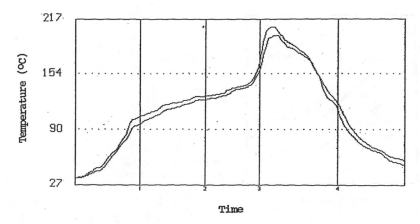

FIGURE 5.35 Convection reflow profile with lower-temperature preheating.

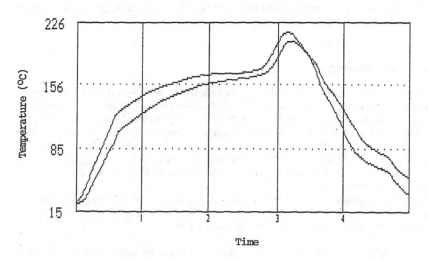

FIGURE 5.36 Convection reflow profile with higher-temperature preheating.

- uneven soldering, resulting in cold solder joints
- an excessive amount of residue remaining or non-dry residue from no-clean paste

The temperature profile with boosted preheating conditions, as shown in Fig. 5.36 is most useful for the assembly that is densely populated with components in large disparity of mass.

Depending on the type of conveyorized furnace, the mass of the assembly and the degree of loading, the major operating parameters to be monitored for effective reflow are the belt speed and the temperature settings of individual zones. The relationship between temperature settings and belt speed; increasing belt speed decreases the resulting peak temperature while other conditions being equal, as shown in Fig. 5.37. Since the required

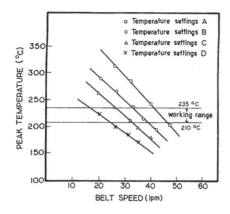

FIGURE 5.37 Reflow oven belt speed vs. peak temperature.

peak temperature is set at 20–50°C above the melting temperature of a solder alloy, the working range of peak temperature is always fixed. For every temperature profile, a relationship between peak temperature and belt speed can be established, and usable range of belt speeds as depicted in Fig. 5.37 can be obtained.

5.4.5 Laser Soldering[1]

Two types of laser have been applied to solder reflow—carbon dioxide (CO_2) and neodymium-doped yttrium-aluminum-garnet (Nd:YAG). Both generate radiation in the infrared region with wavelengths of about 10.6 μm from the CO_2 laser and 1.06 μm for the YAG laser. The wavelength of 1.06 μm is more effectively absorbed by metal than by ceramics and plastics; it is normally reflected by conductive surfaces (metals) and absorbed by organics. In addition to the wavelength difference, the Nd:YAG laser is capable of focusing down to 0.002-in (0.050 mm) diameter in comparison with the CO_2 laser focusing to 0.004-in (0.100-mm) diameter.

The main attributes of laser soldering are short-duration heating and high-intensity radiation, which can be focused onto a spot as small as 0.002 in (0.050 mm) in diameter. With these inherent attributes, laser reflow is expected to

- provide highly localized heat to prevent damage to heat-sensitive components and to prevent cracking of plastic IC packages
- provide highly localized heat to serve as the second or third reflow tool for assemblies demanding multiple-step reflow
- require short reflow time
- minimize intermetallic compound formation
- minimize leaching problems
- generate fine-grain structure of solder
- reduce stress buildup in solder joint
- minimize undesirable voids in solder joint

With these attributes in mind, laser soldering is particularly beneficial to soldering densely packed regions, where local solder joints can be made without affecting the adjacent parts, to soldering surface-mount devices on printed-circuit boards having heat sinks or heat pipes, and to soldering multilayer boards. In addition, it also provides sequential flexibility of soldering different components and it relieves the high-temperature performance requirement for adhesives used for mounting surface-mount devices.

With respect to reflow time, laser soldering can be accomplished in less than a second, normally in the range of 10 to 800 ms. The laser can be applied to point-to-point connections through pulsation, as well as to line-to-line connections via continuous laser beam scan.

The fine-pitch flat-pack devices have been connected to printed-wiring board by using YAG continuous laser beam scans on each side of the package. Both prebumped solder pads and to directly apply solder paste are feasible. In directly reflowing solder paste, although splattering and heat absorption problems have been observed, they are not incurable. To eliminate these problems, the preheating and predrying step is necessary. In addition compatible properties of solder paste have be designed to accommodate fast heating in relation to fluxing and paste consistency, coupled with the proper design in equipment and its settings.

In using the laser, another concern is energy absorption by the printed-circuit boards, which leads to board damage. This is considered to have been corrected by switching from CO_2 laser to YAG laser. Due to the wavelength difference, the energy absorption by polymers can be minimized. Lish[11] has found that sometimes complications may occur. In assembling multilayer polyimide boards by using laser as a second-step reflow, burning was found in the board while it was moving under the laser. The burning was traced to the color pigments contained in the adhesive, which was used for attaching heat sinks. The problem was eliminated by using colorless pigment in the adhesive. This is another clear demonstration that a consideration of all materials in the reflow process is needed.

Regarding the resulting solder joint, a fine-grain microstructure and the formation of significantly reduced intermetallic compounds at the copper and solder (63Sn/37Pb) interface have been observed when laser soldering was used as compared to other reflow methods.[12] Stress buildup in the solder joint due to the difference in thermal expansion coefficients between the materials on both sides of the solder joint is expected when the reflow method requires the whole assembly to be exposed to the soldering temperature. Localized heating and exposure of short duration by using a laser are expected to generate less stress in the joint for the assemblies having materials of different thermal expansion coefficients.

5.4.6 Controlled-Atmosphere Soldering[13]

5.4.6.1 Principle. Based on its function, the atmosphere may be considered as either protective or reactive. Protective atmosphere normally is inert toward a specific soldering material under specific conditions and reactive atmosphere may involve either an oxidizing or a reducing function toward the subject system.

Broadly, chemicals which can provide oxidizing or reducing potential in relation to the specific metal-oxide system and which can generate significant vapor pressure at an operating temperature are expected to contribute to the functional nature of the total atmosphere.

Following is a list of some commonly used atmospheres. Their corresponding nominal compositions are summarized in Table 5.10.

- Dry air
- Nitrogen

TABLE 5.10   Composition of Atmosphere Gases, Percent

Atmosphere	Carbon dioxide (CO_2)	Oxygen (O_2)	Carbon monoxide (CO)	Hydrogen (H_2)	Methane (CH_4)	Nitrogen (N_2)	Trace
Air	—	21.0	—	—	—	78.1	0.9
Nitrogen	—	—	—	—	—	99.8–100	0–0.2
Hydrogen	—	—	—	99.8–100	—	—	0–0.2
Dissociated methanol	—	—	33.3	66.7	—	—	—
Dissociated ammonia	—	—	—	75.0	—	25.0	—
Exothermic gas (air/ gas = 6/1)	5.0	—	10.0	14.0	1.0	70.0	—
Endothermic gas (air/ gas = 2.4/1)	—	—	20.0	38.0	0.5	41.5	—

- Hydrogen
- Nitrogen-hydrogen blends at different ratios
- Dissociated ammonia
- Exothermic gas
- Nitrogen-dopants at different concentrations

The thermal cracking of methanol essentially yields hydrogen and carbon monoxide at high temperatures, as represented by the chemical equation

$$CH_3OH = 2H_2 + CO$$

At low temperatures (below 800°C), side reactions may occur, leading to the formation of H_2, CH_4, CO_2, C.

Each component of the atmosphere gases may function as oxidant or as reducing agent, depending on the temperature and its oxidation-reduction potential relative to that of the materials involved. Among the components of common atmosphere gases, oxygen, water vapor, and carbon dioxide normally serve as oxidants to most metals and metal oxides, and hydrogen and carbon monoxide serve as reducing agents. The ratio of oxidant content to reducing-agent content, in relation to that ratio at equilibrium, indicates whether the resulting atmosphere is oxidizing or reducing.

During soldering, the reactions and interactions of chemicals in the solder paste and between chemicals and the metal surface can be quite complex. In simple terms, the mechanisms may include evaporation, pyrolysis, oxidation, and reduction. The generalized oxidation and reduction reaction can be expressed as follows:

$$xM + yO_2 \Leftrightarrow M_xO_{2y}$$

$$xM_xO_y + yH_2 \Leftrightarrow xM + yH_2O$$

$$xM + yCO_2 \Leftrightarrow M_xO_y + yCO$$

To obtain the thermodynamic equilibrium constant K for each of the preceding reactions,

$$xM(s) + yO_2(g) \Rightarrow M_xO_2(s) \tag{5.1}$$

$$K_1 = \frac{a_{M_xO_{2y}}}{a_M^x a_{O_2}^y}$$

$$M_xO_y(s) + yH_2(g) \Rightarrow xM(s) + yH_2O(g) \qquad (2.2)$$

$$K_2 = \frac{a_M^x a_{H_2O}^y}{a_{M_xO_y} a_{H_2}^y}$$

$$xM(s) + yCO_2(g) \Rightarrow M_xO_y(s) + yCO(g) \qquad (2.3)$$

$$K_3 = \frac{a_{M_xO_{2y}} a_{CO}^y}{a_M^x a_{CO_2}^y}$$

Assuming that the compositions of solids remain constant and the gases behave ideally,

$$K_1 = \frac{1}{P_{O_2}^y} \qquad K_2 = \frac{P_{H_2O}^y}{P_{H_2}^y} \qquad K_3 = \frac{P_{CO}^y}{P_{CO_2}^y}$$

where a represents the activities of the individual reactants as well as the products of reactions (5.1), (5.2), and (5.3), and P_{H_2O}, P_{H_2}, P_{CO}, and P_{CO_2} represent the partial pressure of H_2O, H_2, CO, and CO_2, respectively.

By introducing the relationship between the free energy $\Delta G°$ and the equilibrium constant,

$$\Delta G° = -RT\ln K$$

it is shown that reactions (5.1), (5.2), and (5.3) can proceed in the forward or reverse direction, depending on the temperature and the ratios of P_{CO}/P_{CO_2}, P_{H_2O}/P_{H_2}, and P_{O_2}.

Figure 5.38 shows the standard free energy of formation for some metal-metal oxide systems and CO/CO_2, H_2O/H_2, CO_2/C atmospheres as a function of temperature. Assuming that they are under equilibrium condition and at a soldering temperature of 250°C, lead oxide and copper oxide can be reduced by hydrogen. However, hydrogen is not effective for tin oxides until the temperature reaches 600°C. Equation (5.2) also indicates that the presence of too much water vapor in the furnace atmosphere will cause oxidation to certain metals. The partial pressure of water vapor should therefore be maintained at a constant and defined value.

At a given atmosphere, composition, and dew point, the gas flow rate and the flow pattern of the exhaust systems in the furnace are also important factors to soldering performance. The gas flow rate should be high enough to avoid localized atmosphere buildup as a result of local reactions. In order to achieve the best performance and cost results, the required flow rate is determined by the characteristics of solder paste being used, the furnace belt speed, loading pattern, belt width, and other furnace parameters. The exhaust efficiency and its flow pattern, in combination with the flow rate, dominate the removal of volatile components generated from the pyroloysis and evaporation of chemicals in solder material which, in turn, affects solderability.

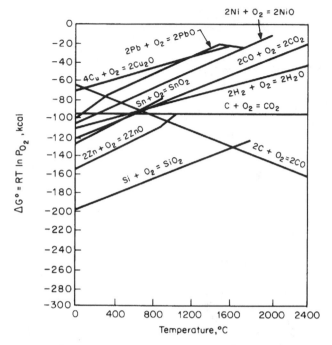

FIGURE 5.38   Standard free energy of formation for metal-metal oxide systems.

It should be noted that a complete burn-out process is normally unfeasible at the eutectic tin-lead or tin-lead-silver soldering temperature when using a solder paste. In order to obtain good solderability and quality solder joints, a metallic continuity at the interfaces between solder and substrate must be formed during soldering. When using a solder paste, a complete coalescence of solder-powder particles has to occur in synchronization with the development of metallic continuity at the interfaces. At the soldering temperature, the atmosphere surrounding the workpiece protects or interacts with the surface of substrates, the solder alloys, and the chemical ingredients in the flux-vehicle system. These interactions determine the chemical and physical phenomena in terms of volatilization, thermal decomposition, and surface-interfacial tension. A controlled atmosphere is expected to deliver a more consistent soldering process.

In addition to consistency, the inert or reactive atmospheres possesses further merits. These include:

- solderability enhancement
- solderability uniformity
- minimal solder balling
- irregular residue charring prevention
- polymer-based board discoloration prevention
- wider process window
- overall quality and yield improvement

The inert and reactive atmospheres are expected to facilitate conventional fluxing efficiency during soldering. It should be noted, however, that performance results rely greatly on the specific atmospheric composition and its compatibility with the solder material, substrate, and chemicals incorporated in the system, which must also be compatible with the soldering temperature profile. Figure 5.39 shows that the solderability under N_2 atmosphere is significantly improved, as solder balls which are formed under ambient air are eliminated.

5.4.6.2 Process Parameters. Gas Flow Rate.

The gas flow rate required to achieve a specific level of oxygen in the dynamic state of the reflow oven is largely controlled by the type of oven, categorically closed-system, semi-closed system and open system. The relationship of the flow rate versus oxygen level within one type of oven and the relationship among the different types of oven are summarized in Fig. 5.40. For a given oven, the required flow rate increases when the allowable oxygen level is lowered. At a given flow rate, when the air tightness in oven construction is reduced, the achievable oxygen level will increase. One practical example is that, at nitrogen flow rate of 800 cfh, the closed system may reach 10 PPM of oxygen and the semi-closed system limit to 500–1,000 PPM.

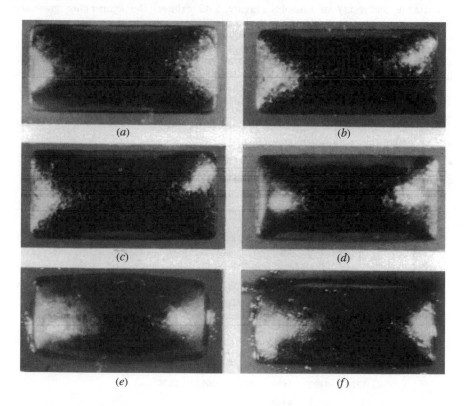

(a) (b)

(c) (d)

(e) (f)

FIGURE 5.39 Solderability performance of solder paste under N_2 atmosphere vs. ambient air: reflowed under (a) N_2, (b) $95N_2/5H_2$, (c) $85N_2/15H_2$, (d) $70N_2/30H_2$, (e) H_2, (f) ambient air.

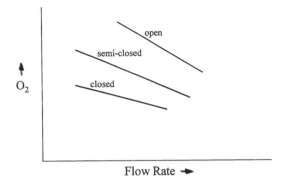

Flow Rate →

FIGURE 5.40   Oxygen level of three types of reflow oven.

As expected, for a given reflow system, the oxygen level is inversely related to gas flow rate as shown in Fig. 5.41. The gas flow rate also affects the temperature distribution and temperature uniformity of assembly. Figure 5.42 exhibits the temperature gradient between the component PLCC-84 and the board surface, indicating that a higher gas flow rate reduces the temperature gradient of an assembly.[14] However, the downside of using high flow rate goes to the higher gas consumption, therefore increasing cost. The cost impact may be mitigated when the design of oven is capable of internal gas recirculation in an efficient fashion.

Humidity and Water Vapor Pressure. Water vapor pressure inside the soldering govern are contributed from:

- The composition and purity of atmosphere
- The reaction product of flux/vehicle chemical system with metal substrates
- The moisture release from the assembly including components and board
- The ambient humidity

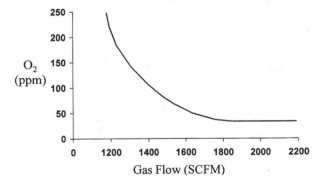

FIGURE 5.41 N_2 flow rate vs. oxygen level.

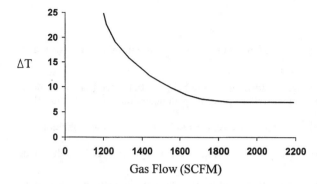

FIGURE 5.42 Temperature gradient vs. flow rate.

Because water vapor is essentially oxidizing to metal substrates that are to be joined by soldering, its partial pressure in the oven affects the overall function of the atmosphere.

The partial pressure of water vapor in an atmosphere gas is conveniently expressed as dew point, that is, the temperature at which condensation of water vapor in air takes place. The dew point can be measured by a hygrometer or dewpointer by means of fog chamber, chilled mirror aluminum oxide technique. The relationship of dew point with the vapor pressure is shown in Figs. 5.43 and 5.44. The relative humidity, RH, is related to the actual vapor pressure of water (or represented by dew point), P_W, and the saturated vapor pressure at the prevailing ambient temperature, Ps, as follows:

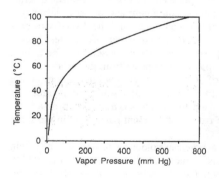

FIGURE 5.43 Dew point vs. vapor pressure (high).

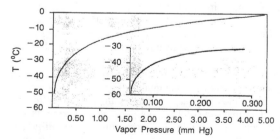

FIGURE 5.44 Dew point vs. vapor pressure (low).

$$RH = \frac{Pw}{Ps}$$

The purity of incoming gas in terms of moisture is normally monitored by measuring the dew point.

Belt Speed. For an evenly spaced loading on the belt, the belt speed not only determines the throughput but also affects other operating parameters that can alter the soldering results.

As examples, the parameters that are affected by the change of belt speed include:

- Peak temperature—at fixed temperature settings, increasing belt speed results in the decrease of peak temperature.
- Atmosphere composition—while other conditions are equal, the change of belt speed may alter the oxygen level (including moisture content).

Temperature. The operating temperature or temperature profile is an integral part of soldering process. It affects the physical activity and chemical reaction of the organic system is solder paste or flux. The operating temperature, particularly peak temperature, changes the wetting ability of molten solder on the metal substrate; wetting ability generally increases with increasing temperature. Chemical reactions and thermal decompositions respond to the rising temperature and the temperature profile.

Oxygen Level. Various studies have focused on the application of no-clean process and on the determination of the maximum oxygen level allowed for using nitrogen-based no-clean soldering process in solder paste reflow and in wave soldering.

Each study was performed with a specific solder paste and flux or with a selected series of paste and flux. Tests were conducted with specific equipment and under a designated process. In view of the continued introduction of new equipment and the diversity of processes coupled with the versatility of solder paste and flux compositions, the test results are expected to represent the specific system (paste, oven, process, assembly) and at best to provide a guideline reference point. For example, a solder paste from the Vendor I to be used with Process A may require a maximum of 20 PPM oxygen level in order to obtain good solderability, grossly solder-ball free and acceptable after-soldering residue. To achieve the similar results, the same paste to be used with Process B may only need a maximum of 300 PPM oxygen. The same could be true for a different paste used in the same process.

The precise oxygen level requirement for a no-clean soldering is impractical to pin down. Instead, the general principle and trends in the relationship between the performance feature and the allowable maximum oxygen level can be derived. Figure 5.45 presents the trend of performance feature merit in relation to oxygen level for a series of solder paste containing various levels of solid contents. For a given performance feature, Figure 5.46 offers the trend of the effect of solid content in no-clean paste on oxygen level requirement during reflow soldering. The performance feature denotes the solderability of the reduction in solder balling.

For convenience, solderability may be monitored by measuring wetting time, wetting force, meniscus rise or wetting angle or by visual wetting quality. The series of the curves represents the generic groupings of no-clean solder paste or flux by the level of solid contents. This is, however, based on the fact that the solid content possesses a good flux system. As shown in Fig. 5.45, performance trend in the increasing solderability and decreasing solder balling is enhanced when the solid content increases, and at a given solid content, beyond a threshold of oxygen level, the performance will significantly drop.

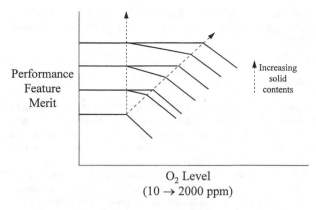

FIGURE 5.45 Performance vs. oxygen level.

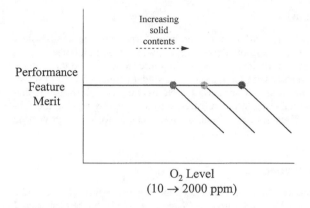

FIGURE 5.46 Solid content in paste vs. performance feature.

It should be noted that increasing solid content delivers increasing amount of after-soldering residue. The required oxygen level may fall in any place within the region, depending on other factors as discussed above. For a given level of performance, the allowable oxygen level will be relieved as the solid content increases, as depicted in Fig. 5.46

5.5 SOLDERABILITY[1]

5.5.1 Definition

Solderability, in a broad sense, is the ability of achieve a clean metallic surface on substrates to be joined during a dynamic heating process so that a good wetting of molten solder on the surface of the substrates can be formed. When using solder paste, solderability requires the additional ability to achieve a clean metallic surface on the solder powder so that a complete coalescence of the solder powder particles can be obtained. Solderability relies both on the fluxing efficiency provided by fluxes or the solder paste and on the quality of the surface of the substrate.

5.5.2 Substrates

Among the common substrates to be soldered, the demand on the flux for good wetting depends on the intrinsic wettability of substrates. The wettability is ranked in the order of Sn, Sn/Pb > Cu > Ag/Pd, Ag/Pt > Ni. Solderability may change due to variation in the quality of the substrate surface. Therefore, using the same flux system may not produce the same results if the condition of the substrate surface varies.

The demand on flux strength also depends on the reflow temperature and techniques. Convection reflow operation under ambient atmosphere requires more fluxes than vapor-phase, hot air, and laser reflow. Inert or reducing atmosphere can modify the reflow performance in wetting as well as in residue characteristics.

5.5.3 Wetting Phenomena

Good wetting is visualized as the formation of a smooth, uniform, and continuous solder coating on the surface of solder pads without dewetting, nonwetting, and pinholes.

Dewetting is the phenomenon of molten solder receding after it has coated the surface, leaving a rough and irregular surface with thick mounds of solder connected by thin solder film. In dewetting, the substrate surface is not exposed. Nonwetting is defined as the phenomenon of molten solder not adhering to the substrate surface, thereby leaving the substrate surface exposed. The molten solder, in such a case, tends to form a high wetting angle ($>90°$).

5.5.4 Solderability of Components

Component leads are commonly made of copper, copper alloys, Alloy-42 (41—42.5 nickel, balance iron), and Kovar (2 percent nickel, 17 percent cobalt, 53 percent iron, 1 percent others).

The leads are normally coated with a coating composition in a range of tin-lead alloys by means of aqueous plating or molten solder dipping. The plating process provides more uniform thickness, which is often porous, and molten solder dip produces a thicker and denser fused coating.

Ideally, device leads are pretinned to assure good solderability. But practically, after tinning, most components undergo operations such as lead forming, encapsulation, or burn-in. There steps can degrade the quality of the lead's surface significantly and, therefore, affect Solderability. In summary, the solderability of tinned leads depends on the following factors:

- composition of base lead materials
- composition of coating
- surface finish and condition of coating
- age of coating
- storage of coating
- thickness of coating

A coating thickness of 0.0003 in (7.6 µm) is most prevalent, and thin coating is often associated with poor solderability. Nonetheless, the ideal coating thickness depends on several practical factors.

The solderability of coated leads, with various shelf times and other conditions, is a concern. Assuming the coating is intact, this concern can be viewed from two aspects: (1) how is the solderability affected by the surface degradation due to oxidation or contamination during shelf time, and (2) how is the solderability affected by the interaction between lead material and tin-lead coating during shelf time or treatment, such as a burn-in test. The surface oxidation of tin-lead alloys normally is not an insolvable problem, since the fluxing is able to take care of it.

The formation of copper-tin intermetallic compounds at the interface of copper-based leads and tin-based coating (Cu_3Sn, Cu_6Sn_5) can readily occur, although it would be extremely sluggish at room temperature, as indicated in the Cu-Sn phase diagram. With this interaction, the tin content at the coating-lead interface will be consumed gradually, resulting in solderability degradation. The consumption rate depends on temperature and time. In this regard, Alloy-42 and Kovar leads are expected to do better than copper leads and high-tin coating. However, high-tin coating, having higher surface tension, normally provides better wettability.

It should be noted that, for leads coated by molten solder dip, copper-tin intermetallic compounds (namely, Cu_6Sn_5) can be formed rapidly during coating. However, due to its intactness, the molten solder dip coating is expected to be relatively more stable during storage than electroplating; leads made of Alloy-42 and Kovar may experience deterioration with age due to moisture permeation through the porous crystalline structure.

In either case, the degradation of the coating is driven by a kinetically controlled process and depends on other practical and environmental factors. Therefore, to assure good quality of the coated surface of leads, the shelf time and storage temperature must be minimized. Using freshly coated leads is ideal.

5.6 CLEANING

5.6.1 Principle and Options

After soldering, the residue surrounding the solder joint can be either removed (cleaned) or left as is. The decision about whether or not to clean up the residue depends on the property and activity of the residue and the reliability desired under specified service conditions.

Residues of water-soluble chemistry which is left on or around solder joints after soldering need to be cleaned. The residues form liquid flux can be a simple system with a small amount of organic acid, high boiling solvents, surfactant and reaction products. Residues from paste may contain a mixture of ingredients, including polar organics, non-polar organics, ionic salts and metal salts.

Key steps in a typical cleaning process are: prerinse – wash – rinse – final rinse – drying. The parameters that affect the cleaning efficiency include:

- temperature of water
- spray pressure
- spray angle
- wash time

- flow rate

- agitation aid

When a compromise among the parameters is needed, higher temperature and higher spray pressure often play more important roles than the flow rate and spray angle. In addition, mechanical agitation aids in dislodging the foreign matters from the board and the clearance between the component and board. Techniques, such as ultrasonic field and centrifugal energy provide effective mechanical force. Centrifugal force which is directly proportional to the square of angular velocity and its parallel direction to the board assist the cleaning operation. Ultrasonic cleaning is to utilize the cavitation defined as the implosion of microscopic vapor cavities within the solution, which is induced by the changing pressure differentials in the ultrasonic field. The pressure differentials occur by the exchange between negative and positive pressure in a liquid region. When the liquid with negative pressure is created, its boiling point drops and many small vapor bubbles are formed. As the pressure changes to positive, the small bubbles implode with great violence. The mechanical soundwave is generated by a high frequency electrical energy wave through a transducer. The cavitation phenomenon provides mechanical agitation and scrubbing effect.

The effectiveness of ultrasonic cleaning depends on the cavitation intensity which in turn is controlled by the magnitude of power or pulse width and by dissolved air in solution. The effect of dissolved air has been illustrated and tested, indicating that the dissolved air can act as an acoustical screen and energy absorber.[15] A deaeration step is needed to remove the air in order to obtain the true vaporous cavitation. It is suggested that the high audible noise level with a pronounced hissing sound and minimal visible bubbles in solution coupled with violent surface activity are the signs of ultrasonic efficiency. It is also suggested that the temperature of the solution is a factor in ultrasonic efficiency. The desirable temperature is approximately in the range of 80–98 percent of the boiling point of the solution.

Due to the concern that ultrasonic may damage wire bonding or other chip components, compatible process parameters are to be identified. Some guidelines are proposed:

Power of ultrasonic cleaner: 30 watts per liter

Ultrasonic range: 30 kHz to 66 kHz

Cleaning time: 3 minutes/cycle of for 5 cycles (not to exceed a total of 15 minutes)

With the narrow gap (clearance) between the components and the board of surface-mount assemblies, the assurance of cleanliness has always been a question. In the dilemma of the efficiency of the cleaning process and the accuracy of cleanliness measurement, a functional test, the surface insulation resistance test, provides an indication of the level of cleanliness for a given assembly. The cleaning efficiency on the production line, however, should rely on the established process and its stringent control, and the tests merely serve as confirmation.

5.7 FINE-PITCH APPLICATION

In addition to the selection of solder paste, stencil thickness versus aperture design, stencil aperture versus land pattern, and stencil selection are major factors contributing to the printing results of solder paste.

5.7.1   Stencil Thickness versus Aperture Design

When printing solder paste, the design of relative dimension of stencil thickness and stencil aperture is to achieve the balance between the printing resolution and the proper
amount of solder deposit in order to avoid starved solder joints or pad bridging. For a
selected stencil thickness, too small a stencil aperture width leads to open joints or starved
joints. Too large aperture width causes pad bridging. Table 5.11 provides the guideline for
designing stencil thickness in relation to aperture.

TABLE 5.11   Guideline of Stencil Thickness vs. Aperture Width

Component lead pitch		Aperture width		Maximum Stencil Thickness	
inch	mm	inch	mm	inch	mm
0.050	1.26	0.023	0.58	0.014	0.35
0.025	0.63	0.012	0.30	0.0075	0.19
0.020	0.50	0.010	0.25	0.0063	0.16
0.015	0.38	0.007	0.18	0.0043	0.11
0.008	0.20	0.004	0.10	0.0025	0.06

5.7.2   Stencil aperture design versus land pattern

To make solder joints by one-pass printing process, the stencil thickness must be selected
for transferring sufficient amount of paste onto the non-fine-pitch solder pads and, while
avoiding the excessive amount of paste deposited onto the fine pitch pads. There are several options to achieve the deposition of a proper amount of solder paste on the land pattern to accommodate a mix of sizes of solder pads. These are:

1. Step-down stencil
   This is commonly achieved by chemically etching the non-fine-pitch pattern area from
   one side of the stencil while etching the step down area for fine pitch pattern to the
   other side during double-sided etch process. Alternatively, step-down area, is etched in
   one foil and non-fine pitch pattern etched in the other foil, then register and glue the
   two foils together.
      The practical step gradient is 0.002 inch (0.05mm), some common combinations
   are:

   > 0.008 inch (0.20mm) for non-fine pitch
   > 0.006 inch (0.10mm) for fine-pitch

   or

   > 0.006 inch (0.15mm) for non-fine pitch
   > 0.004 inch (0.15mm) for fine-pitch

2. Uniform reduction on four sides of apertures
   The dimensions of the fine pitch aperture on stencil are reduced by 10–30 percent in
   relation to those of the land pattern. This reduces the amount of paste deposition on
   fine pitch land pattern, and also provides some room for printing misregistration and
   paste slump, if any.

3. Staggered print
   The opening of the stencil is only one-half length of the solder pad and arranged in an
   alternate manner as shown in Fig. 5.47.

FIGURE 5.47 Staggered print.

For tin-lead coated solder pads, when the paste starts to melt during reflow, the molten solder is expected to flow to the other half of the pad, making the complete coverage. With bare copper or nickel surface, the molten solder may not flow out to cover the area that the paste is not printed on.

4. Length or width reduction
The dimensions of the stencil opening are reduced along the length or along the width by 10–30 percent in relation to that of solder pads, achieving the reduction of the amount of paste deposited.

5. Other shapes
The stencil openings are made with selected shapes such as a triangle or a teardrop in order to achieve the reduced solder paste deposition on fine pitch pattern.

6. Compromised stencil thickness
Instead of using the specific thickness which is considered as the most suitable for a specific land pattern, select a thickness which is practical to both fine pitch and non-fine pitch patterns. For example:

Land pattern pitch combinations		Compromised Stencil Thickness	
mil	mm	inch	mm
50 and 25	1.26 and 0.63	0.007–0.008	0.18–0.20
50 and 20	1.26 and 0.50	0.006–0.007	0.15–0.18

5.7.3 Stencil Selection

The performance of stencils is primarily driven by the foil metal and the process being used to create the printing pattern. Currently five types of stencil materials are commercially available—brass, stainless steel, molybdenum, alloy-42, and electroformed nickel. The processes making the stencils may involve chemical etching, laser-cutting, electropolishing, electroplating, and electroforming. Each type of foil or fabricating process possesses inherent merits and limitations. The key performance of a stencil is assessed by the straight vertical wall, wall smoothness and dimensional precision. In addition, durability, chemical resistance, fine opening capability, and cost are also important factors.

Table 5.12 Compares various stencil materials, and Table 5.13 summarizes the relative performance characteristics of stencil-making techniques.[2]

TABLE 5.12 Comparison of Stencil Materials in Key Performance Areas

Performance	Brass	Stainless steel	Molybdenum	Alloy-42	Ni (electroforming)
Mechanical strength	unfavorable	favorable	favorable	favorable	favorable
Chemical resistance	unfavorable	favorable	unfavorable	favorable	favorable
Etchability	favorable	favorable	favorable	favorable	N/A
Sheet stock availability	favorable	favorable	unfavorable	favorable	N/A
Cost	favorable	less favorable	unfavorable	less favorable	N/A
Fine pitch (openings) Capability	favorable	may need electropolishing	favorable	may need electropolishing	favorable
Unique feature	lowest cost	durable	self-lubricating, smooth wall	—	finest opening

TABLE 5.13 Comparison of Techniques in Building Stencils

Techniques	Characteristics	Capabilities or features
Chemical etching	Most established process, sensitivity of fine pitch capability to process and control, sensitivity of aperture size and vertical wall control	Versatile, economical
Laser cut	Grainy wall surface; sequential cut, not concurrent formation of openings; higher cost; difficulty in making step stencil	Fine pitch capability, no photo tools or resist needed
Electropolishing	Complementary step to produce smooth wall surface	Smooth wall surface

TABLE 5.13 Comparison of Techniques in Building Stencils *(Continued)*

Techniques	Characteristics	Capabilities or features
Ni-plating on aperture wall	Reducing aperture opening, smooth surface	Finer opening
Electroforming	Additive process via electrode position, concern about foil strength, difficulty in making step stencil, suitable for stencil thickness of 0.001 in to 0.012 in	Gasket effect minimizing bleedout, capability of producing very fine opening, no need for electropolishing

5.8 SOLDERING-RELATED ISSUES[2]

5.8.1 Intermetallics vs. Solder Joint Formation

Intermetallic compounds have often been observed at or near the solder/substrate interface as well as in the interior of solder joints. Metallurgically, an intermetallic compound is one type of intermediate phase that is a solid solution with intermediate ranges of composition. Intermetallic compounds form when two metal elements have a limited mutual solubility. These compounds posses new composition of a certain stoichiometric ratio of the two components.

The new compositions have a different crystal structure from those of their elemental components. The properties of the resulting intermetallic compounds also differ from component metals in that they exhibit reduced ductility, density and conductivity. Since tin or tin/lead solder is metallurgically active with most metals that are commonly used in electronics packaging and assembly, the presence of intermetallic compounds is not a surprise during the soldering process. Various intermetallic compositions have been under the equilibrium condition between tin and substrate metals, such as Au, Ag, Cu, Pd, Ni, and Pt. Indium-based solders also interact with these substrate metals, often forming intermetallics. One should note that thermodynamically stable compounds may not always be present, and some intermetallics that do not appear in the equilibrium phase diagram have been identified in soldered systems.

Relating to electronics packaging and assembly, intermetallic compounds may come from one or more of the following processes and sources:

- Intermetallics are formed at the solder/substrate interface during soldering.

- Intermetallics are present in the interior of the solder joint as the inherent metallurgical phases of a given solder composition, such as 95Sn/5Sb and 96Sn/4Ag solder.

- Intermetallics are developed during a service life either along the interface and/or in the interior of the solder joint.

When solder comes in contact with a common metal substrate for a sufficient amount of time at a high enough temperature, intermetallic compounds may form. Below a solder's liquidus temperature, formation is primarily a solid state diffusion process and thus depends highly on temperature and time. While solder is in a molten state, the solubility of the element from substrate into molten solder accelerates the rate of intermetallic formation.

External factors such as the temperature of exposure an the time at the elevated temperature also affect the rate of intermetallic compound formation. Thus, solder reflow conditions such as peak temperature, and the total dwell time at elevated temperatures, influence the rate and extent of intermetallic growth. Also, while in storage or service, the exposure of the assembly is a factor for intermetallic growth in systems.

The thickness of growth between eutectic tin/lead and copper is proportional to the square root of time, coinciding with the diffusion-controlled kinetics. As temperature rises, the rate of formation increases, with the higher tine promoting the process.

The composition of intermetallics at the interface may differ from those of the solder joint interior. Furthermore, the surface condition of the substrate affects the kinetics of intermetallic development. For example, the oxidized surface may show a delayed development of the intermetallic phases, making a thinner layer as compared with a clean surface at a given amount of time. Unlike high tin-containing solder, which tends to form intermetallic compounds with small crystal structures, high lead-containing solder forms high, needle-like crystals.

In brief, the extent of intermetallic formation, the composition of the compounds and their morphology depend on intrinsic factors. These factors include the following:

- the metallurgical reactivity of a solder with a substrate
- soldering (reflow) peak temperature
- dwell time at peak temperature
- the surface condition of a substrate—clean versus oxidized
- the post-soldering storage and service conditions

Intermetallics at the interface can be beneficial or detrimental. Soldering wetting on the substrate followed by the formation of a thin layer of intermetallics is the prevalent mechanism in making permanent solder bonds. However, adverse effects may occur if the intermetallic layer becomes too thick. General acceptable thickness fall in the range of 1 to 5 μm.

The morphology, size and distribution of intermetallic in solder determines their beneficial or detrimental effects on solder joint integrity. In proper properties, the intermetallics in the interior of the solder joint (away form the interface) act as a strengthening phase. In contrast, large and needle-shaped compounds generally weaken the mechanical properties of a solder joint.

The formation of excessive intermetallic compounds has proven to be a frequent source of solder joint failure. Cracks are often initiated around the interfacial area under stressful conditions when an acceptable amount of intermetallics develops along the solder/substrate interface.

The adverse effect of intermetallic compounds on the solder joint integrity is believed to be attributed to the brittle nature and thermal expansion properties of such compounds, which may differ from the interior solder. The difference in thermal expansion contributes to a solder's internal stress development. In addition, excessive amounts of intermetallic compounds impair the solderability of some systems depleting one element of the contact surface. For instance, tin depletion from tin/lead coating on copper leads causes the exposure of Cu_3Sn to oxidation, resulting in inconsistent and/or poor solderability of component leads. In this case, the interface area is composed of gradients with Cu_3Sn phase next to the copper substrate followed by Cu_6Sn_5 phase and lead-rich phase away from the interface line. Also, excessive intermetallics render a dull rough look to solder joints.

A precise bonding process for die attach involving in-situ formation of Cu-Sn intermetallics from vapor-deposited copper-tin multilayer has been introduced. The unique feature of this bonding process is its fluxless nature and its control of intermetallic thickness. The resulting joint is composed of uniformly distributed Sn and Cu_6Sn_5 with a joint thickness of 4.5 μm.

The role of intermetallics, beneficial or detrimental, is determined by the design of an assembly, the service conditions in relation to that design, and the control of the soldering process. Understanding the relationship between them is the key to making reliable solder joints.

5.8.2   Gold-Plated Substrates vs. Solder Joint Formation[2]

Using gold (Au) as a surface coating to resist the oxidation of underlying metals in semiconductor packages and electronics assemblies is a routine practice. Common applications include gold-plating on PCBs, gold-containing thick film circuitry on hybrids, soft gold (24 karat) wire bonding, and hard gold (cobalt or nickel gold) for edge fingers as connectors. However, many in the industry are concerned or uncertain about the full role gold plays in solder.

When a gold-coated substrate is in direct contact with an Sn/Pb solder, the Au combines with the Sn of the solder at a rapid rate due to the metallurgical affinity between Sn and Au, forming Au-Sn intermetallics. Gold-tin intermetallics can affect a solder's physical and mechanical properties, and alter a solder joint's appearance and microstructure.

An Au concentration below 10 percent by weight in Sn/Pb solder slightly increases that solder's initial tensile strength. However, beyond 3 percent, a solder's shear strength slowly drops. Normally, its hardness increases with the addition of gold. This effect is enhanced as the Au content exceeds 7 percent. A solder's ductility is slowly reduced with Au concentrations below 7 percent by weight, and then drops rapidly as the Au content exceeds 7 percent by weight.

Gold can affect a solder's ability to wet and spread. Although a 2 percent Au concentration has no effect on 63Sn/37Pb, concentrations above 2 percent reduce the solder's spreadability and fluidity. For copper plated with Au, a pure Au coating has shown better wetting and spread than alloy Au when soldering with 63Sn/37Pb under identical conditions.

The dissolution rate of Au in solder depends on temperature, time and solder composition. Foreign elements, such as Au, In and Zn, in Sn/Pb solder, retard this dissolution. During a reflow process with a long heating time, the quick dissolution of Au in molten solder causes that solder to wet directly onto the base metal and not the gold coating.

Although one might expect gold's inert nature to provide a base metal with full protection, tests on the aging of gold-electroplated Cu indicate that the solderability as measured by wetting time degrades with aging at a temperature of 170°C. Solderability degrades due to the following causes:

- The diffusion of atmospheric contaminants through the porous Au film results in the oxidation the base material.
- The diffusion of base metal reaches the surface through the coating. The diffusion rate is associated with the Au coating grain size, with smaller grain sizes favoring diffusion.

Gold dissolved in solder alters that solder's microstructure. As the Au content reaches 1 percent by weight, the characteristic needle-shaped phase found in eutectic solder becomes readily detectable in microstructure. The amount of hard phase increases with elevated Au concentrations. At room temperature, the composition of these intermetallics is a mixture of $AuSn_4$ and Sn.

The incorporation of Au may or may not change a solder's physical properties. At concentrations below 10 percent by weight, Au does not significantly affect a solder's electrical or thermal conductivity.

Gold can lower a solder's solidus temperature and increase its liquidus temperature, thereby widening the paste range or creating a pasty range for eutectic solder. This affects a solder's application performance, particularly for solder interconnections. Lowering a solder's "softening" temperature changes it mechanical response to rising temperatures. A eutectic colder is required for applications demanding high solder fluidity, while assemblies with a wide gap to fill find solder with a wide pasty range preferable.

Overall, Au has the most pronounced effect on solder joints in the following areas:

- fluidity
- wettability and spread
- mechanical properties
- phase transition temperature
- microstructure
- appearance

An overly thick Au coating results in a higher Au concentration in solder and an increase in material cost. If the coating is too thin, the surface protection effectiveness may suffer. One should also take into account that surface condition of Au, particularly its porosity, is equally as important to surface protection. An optimum Au application balances surface intactness, concentration in solder after dissolution, and cost.

When a solder's Au content is excessive, the following mechanical and/or metallurgical phenomena may occur:

- premature solder joint fracture due to embrittlement
- void creation
- microstructure coarsening

The upper limit of Au concentration is assessed to be 3 percent by weight. Above 3 percent, deleterious effects can occur in one or more of the aforementioned area. The 3 percent limit cited here is only a guideline. As a rule, one should verify the effect of Au concentration in solder for its performance in a specific electronics package and assembly under a given set of conditions.

To ensure that Au concentrations do not exceed acceptable levels, industry standards call for Au removal immediately prior to soldering. The general guidelines for Au removal are as follows:

- A double tinning process of dynamic solder wave must be used for proper Au removal.
- An Au removal procedure is unnecessary for through-hole components intended for dip or wave soldering attachment, provided that the Au on the leads is less than 0.0025 mm.
- For surface mount parts, Au must be removed from at least 95 percent of the surface to be soldered.

5.8.3 Solder-Joint Voids

Voiding is one of the adverse phenomena in solder-joint integrity and reliability. It is generally expected that a low volume of small, well dispersed voids has little effect on solder-joint integrity; however, high-volume or large-size voids would degrade the joint with respect to its electrical, thermal, or mechanical properties.

For solder joints made from solder paste, the flow characteristics and the thermal and physical properties of the vehicle-flux system as well as the metal load are important factors. To minimize voiding, the processing parameters and the joint design should be optimized. These include the dosage of paste deposit, deposit thickness, joint configuration, reflow time, cooling rate, and wettability. The same paste could generate different voiding in size and concentration if used under different conditions. A quality joint is, therefore, influenced equally by the solder-joint assembling process and by a compatible quality paste. Further discussion can be found in Ref. 2.

5.8.4 Solder Balling

When using solder paste, solder balling in reflow process is a common phenomenon. It has been a continuous effort in soldering process control, in component and board quality, and solder paste design in order to minimize the occurrence of solder balling.

The solder balling phenomenon can be defined as the situation that occurs when small spherical particles with various diameters are formed away from the man solder pool during reflow, and do not coalesce with the solder pool after solidification as shown in Fig. 5.48. Versatile manufacturing environments have revealed two distinct types of solder balling in terms of physical characteristics:

(A) solder balling around any components and over the board

(B) Large size solder balls associated with small and low clearance passive components (e.g. 0805, 1206), being mostly larger than 0.005 inch (0.13 mm).

The type (A) solder balls normally can be removed during cleaning process; type (B) solder balls, however, are difficult to remove using a normal cleaning process. With the implementation of no-clean process, it is obviously desirable to avoid the occurrence of both types of solder balling. With the use of array packages (BGAs), solder balling also becomes more troublesome. In the presence of solder balls, the assembly may encounter the risk of electrical short when any solder balls become loose and mobile during service. Excessive solder balling may also deprive solder from making good solder joint fillet. In general, type A solder balls can be formed for different reasons. The following are the likely sources to be considered:

- Solder paste with inefficient fluxing with respect to solder powder or substrate or reflow profile, resulting in discrete particles which do not coalesce, due either to paste design or to subsequent paste degradation.

- Incompatible heating with respect to paste prior to solder melt (preheating or predry), which degrades the flux activity.

- Paste spattering due to too fast heating, forming discrete solder particles or aggregates outside the main solder pool.

- Solder paste contaminated with moisture or other high "energy" chemicals that promote spattering.

- Solder paste containing extra fine solder particles that are carried away from the main solder by organic portion (flux/vehicle) during heating, resulting in small solder balls.

- Interaction between solder paste and solder mask.

The appearance of solder balls and their distribution often reveal the cause. Solder balls as a result of spattering are usually irregular and relatively large in size (larger than 20 μm

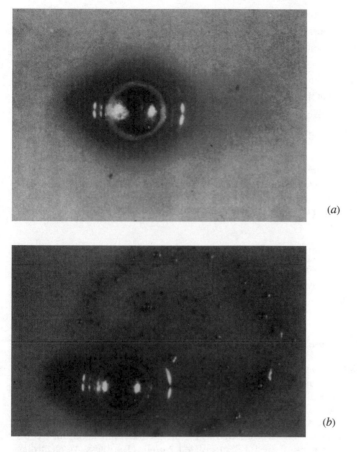

(a)

(b)

FIGURE 5.48   Solder balling.

is not uncommon) and are scattered over a large area of the board; solder balls caused by fine powder in the paste often form a halo around the solder; ineffective or insufficient fluxing results in small solder balls scattered around the joint; and solder mask-related solder balls leave discoloration marks on the board.

Solder paste spattering during reflow can caused by:

- Incompatibility between paste and reflow profile, such as too-fast heating, high volatile content in the paste.
- Hygroscopicity of the paste. When the open time during assembly exceeds the capability of the paste or the paste is exposed to a temperature and/or humidity beyond its tolerance level, the moisture absorbed by the past can cause spattering.

To minimize solder balling during board assembly, several issues need to be addressed:

- selection of a solder paste that is able to deliver performance under the specific production conditions

- understanding of the characteristics of the solder paste selected
- set-up of the reflow process that best fits the solder paste selected
- assurance of consistency and quality of the solder substrate including boards and components
- control of ambient conditions (temperature and humidity)
- control of open time that the paste can accommodate
- assurance of solder mask compatibility with the solder paste
- assurance of complete cure of the solder mask

For large solder balls associated with the small passive components, their formation is largely attributed to the paste slump and flow under the component body between two termination via capillary effect. The slump and flow dynamics can also be affected by the reflow temperature profile, the volume of paste and component placement. In order to reduce the occurrence of these large solder balls, the following parameters are recommended for consideration:

- solder paste rheology—minimizing paste slump
- amount of solder paste deposit—avoiding excess paste
- component placement—avoiding paste spreading during placement
- reflow profile—reducing preheating temperature exposure

5.8.5 PCB Surface Finish

For making sound interconnections, the characteristics and properties of circuit board surface finish are as important as the component leads and termination.

Hot air solder leveled SnPb (HASL) has been successfully used as the surface finish for surface mount and mixed PCBs. As the need of flat surface with uniform thickness becomes increasingly important to forming consistent and reliable fine-pitch solder joints the HASL process often falls short. Alternative to HASL include immersion Sn, electroplated SnPb (reflowed or non-reflowed), electroplated Au/Ni, electroless Au/electroless Ni, immersion Au/electroless Ni, electroplated Sn-Ni alloy and organic coating. When selecting an alternative surface finish for PCB assembly, the key parameters in terms of solderability, ambient stability, high temperature stability, suitability of the use as contact/switch surface, solder joint integrity, wire bondability of those assemblies that involve wire bonding as well as cost, are to be considered.

Basic Process. Three primary techniques to deposit metallic surface finish are electroplating, electroless plating and immersion. Inherently, electroplating utilizing electric current is able to economically deposit thick coating up to 0.000400 in, depending on metal and process parameters. Electroless plating, requiring the presence of a proper reducing agent in the plating bath, converts metal salts into metal and deposits them on the substrate. The immersion plating process, in the absence of electric current and the reducing agent in the bath, deposits a new metal surface by replacing the base metal; plating stops when the surface of base metal is completely covered, thus only a limited coating thickness can be obtained through the immersion process. For both electroless and immersion processes, the intricate chemistry and the control of kinetics are vital to the plating results. The designed process parameters and chemistry including pH and chemical ingredients must be compatible with the solder mask and PCB materials.

Metallic System. Available metallic surface finish on copper traces include Sn, SnPb alloy, SnNi alloy, Au/Ni, Au/Pd, Pd/Ni and Pd. The systems containing noble or semi-noble metals, such as Au/Ni, Au/Pd, Pd/Ni, Pd/Cu are capable of delivering the coating surface with uniform thickness. Those systems imparting a pure and clean surface also provide a wire bondable substrate. In addition, wire bonding generally requires thicker coating, namely, more than 0.000020 in. A unique feature of the Au/Ni system is its stability toward elevated temperature exposure during the assembly process as well as during subsequent service life. When in contact with molten solder of SnPb, SnAg or SnBi, surfaces coated with Sn and SnPb are normally associated with a better spreading and lower wetting angle than others. Of the metallic systems, those containing Ni interlayer are expected to possess a more stable solder joint interface; and in these systems solder is expected to wet on Ni during reflow since noble metals are readily dissolved in solder. The concentration and distribution of noble metals in solder need to be noted in order to prevent any adverse effect in solder joining integrity. For phosphorus-containing plating bath, a balanced concentration of phosphorus in electroless Ni plating is essential. When P content is too high, wettability suffers; and when it is too low, thermal-stress resistance and adhesion strength are sacrificed.

Another characteristic which is important to solderability is the porosity on the surface. A thinner coating is more prone to porosity-related problems although the surface density and texture can be controlled by the chemistry and kinetics.

Organic Coating. Benzotriazole has been well recognized as an effective Cu and anti-tarnish and anti-oxidation agent for decades. Its effectiveness, attributable to the formation of benzotriazole complex, is largely limited to ambient temperature.

As temperature rises, the protective function disintegrates. Azole derivatives, such as Imidazole (m.p. 90°C, b.p. 257°C) and Benzimidzole (m.p. 170°C, b.p. 360°C) have been used to increase the stability under elevated temperature. SMT assembly of mixed boards involves three stages of temperature excursion—reflow, adhesive curing and wave soldering. The reflow step, however, is considered to be potentially most harmful to the intactness of organic coating due to the fact that it is the step with highest temperature and longest exposure time.

Although the performance of organic coating varies with the formula and process, the general behavior of organic coating falls in the following regimen:

- There is a need for compatible flux (generally more active flux).
- For mixed boards, there may be a need for more active flux in wave soldering.
- Thicker coating is more resistant to oxidation and temperature but may also demand more active flux.
- Organic coating needs to be performed as the last step of PCB fabrication.
- At temperatures higher than 70°C, coating may degrade. However, the degradation may or may not reflect on solderability.
- The coating may be sensitive to PCB prebaking process (e.g., 125°, to 1 to 24 hr).
- For no-clean chemistry, may N_2 atmosphere or higher solids content may be required in no-clean paste.
- Steam aging test is not applicable.
- It is not suitable for chip on board where wire bonding is required.

Nonetheless, when fluxing activity and process are compatible, organic coating can be a viable surface finish for PCBs. An additional bonus effect is that the bare copper appearance of the organic coated surface eases the visual inspection of peripheral solder fillets.

Comparison with HASL. Regardless of other deficiencies, HASL however, provides the most solderable surface. When comparing a metallic system with HASL, the HASL process subjects PCBs to high temperature (above 200°C), producing inevitable thermal stress in PCBs. Further, HASL is not suitable for wire bonding.

To make a choice in replacing HASL, many variables are to be assessed. Understanding the fundamentals behind each variable in conjunction with setting proper priority of importance among the variables for a specific application is the way to reach the best balanced solution.

5.9 SOLDER-JOINT APPEARANCE AND MICROSTRUCTURE

5.9.1 Appearance

X-ray techniques, laser thermal techniques, and optical inspection of solder-joint quality have been developed extensively. Nevertheless, visual inspection is being used commonly as an indicator of solder-joint quality, and it is a criterion still being called for by military specifications. Thus a few words are in order about the appearance of solder joints.

The factors affecting solder-joint appearance in terms of luster, texture, and intactness are:

inherent alloy luster

inherent alloy texture

residue characteristics after paste reflow

degree of surface oxidation

completeness of solder powder coalescence

microstructure

mechanical disturbance during solidification

foreign impurities in solder

phase segregation

cooling rate during solidification

subsequent heat excursion, including aging, temperature cycling, power cycling, and high-temperature storage

It is known that the process of solidification from melt is crucial to the microstructure development of an alloy, which, in turn, affects its appearance. Figure 5.49b shows the 63Sn/37Pb off-eutectic microstructure of a two-metal powder mix, while Fig. 5.49a shows the eutectic microstructure of prealloyed powder reflowed under the same condition. The difference is attributed to insufficient reflow in Fig. 5.49b; its microstructure reflects a slightly duller joint.

Microstructure is also related to alloy strength and failure mechanism. It is observed that an ideal Sn/Pb eutectic structure, as shown in Fig. 5.49a, imparts a bright and smooth solder surface. Deviation from the eutectic microstructure is normally visualized as a duller joint.

The heating time at melt is another factor. Figure 5.50 shows the microstructure of two 63Sn/37Pb solder joint surfaces made under a regular infrared heating profile (Fig. 5.50a) and with prolonged exposure at peak temperature (Fig. 5.50b). Figure 5.51 shows the

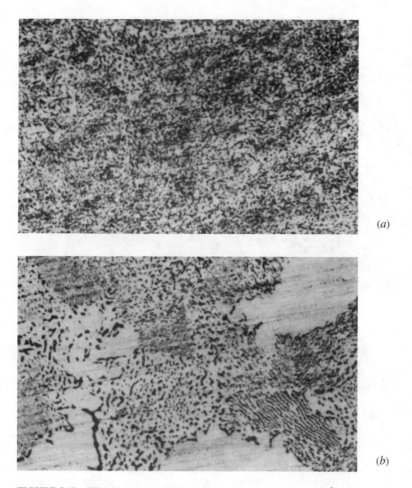

FIGURE 5.49 SEM micrograph of 63Sn/37Pb structures, (*a*) eutectic and (*b*) off-eutectic.

cross sections of these two joints; the excessively heated solder joint in Fig 5.51*b* appears to be rough on the surface.

The residue interference on the molten surface and during its solidification can contribute to a rough texture on the solder surface. Inadequate heating time or temperature will cause incomplete coalescence of solder particles, which also contributes to an unsmooth surface and, possibly, to an inferior solder joint. During cooling, any mechanical agitation that disrupts the solidification process may lead to uneven solder surface. After the completion of reflow and solidification, heat excursion is expected to have a significant impact on solder-joint appearance, whether it is a surface reaction or internal structure change. Heat excursion can come from different sources, such as high-temperature storage, aging, temperature fluctuation during service, and power cycling during functioning.

In many cases, a duller or a rough joint may not be necessarily defective in a functional sense. Test may confirm that the duller and rougher joints have equivalent mechanical

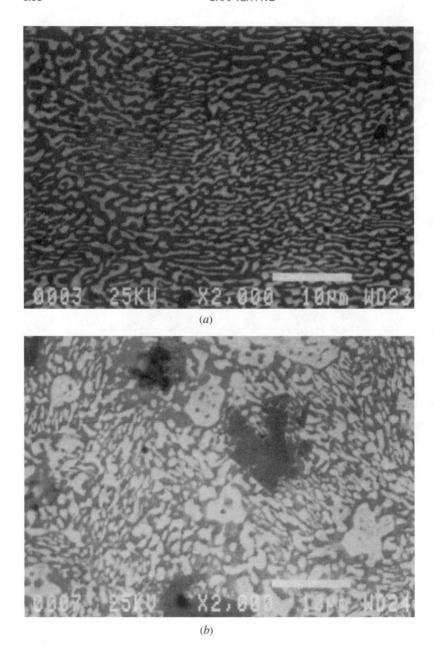

(a)

(b)

FIGURE 5.50   SEM micrograph of 63Sn/37Pb solder joint surface reflowed under
(a) regular and *(b)* prolonged heating.

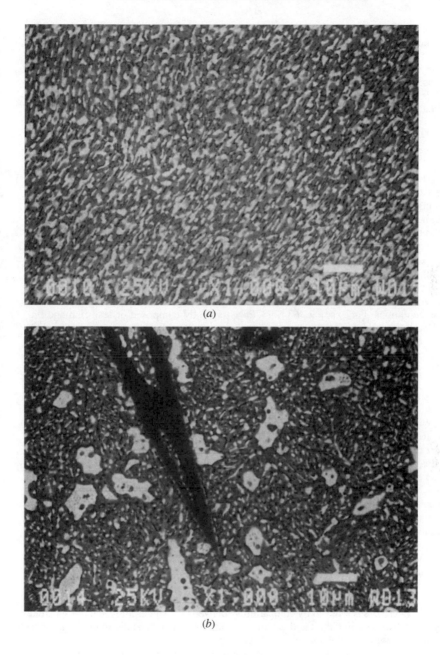

FIGURE 5.51   SEM micrograph of 63Sn/37Pb solder joint surface cross section reflowed under (*a*) regular and (*b*) prolonged heating.

strength. However, one must be prudent in drawing a conclusion regarding joint reliability in relation to joint surface appearance. It should be noted that the surface condition of a metal is considered to be one of the variable that affect its failure mechanism. It should also be noted that the surface appearance may reflect the internal microstructure which, in turn, corresponds to the physical and mechanical properties of the solder joint. In the author's viewpoint, the level of smoothness and brightness that the alloy should show reflect a proper joint having been produced from a proper process.

5.9.2 Microstructure

When dealing with physical objects in the linear dimension larger than 10^2 μm, we work on structure engineering. If we desire to view an object in the scale of 10^{-10} to 10^2 μm, we study material science and nuclear physics. As the scale shrinks to less than 10^{-10} μm, the object becomes intangible and immeasurable. Microstructure essentially falls in the range of 10^{-2} to 10^2 μm. Hence, the understanding of solder joints within this range of dimensions is generally considered adequate in relating material properties to end-use applications.

Solders are normally polycrystallines that consist of an aggregate of many small crystals or grains. Most solder compositions contain multiple metallurgical phases that are physically distinct, and formed and distributed according to given thermodynamic and kinetic conditions. For example, 63 Sn/37Pb is typically composed of lead-rich and tin-rich phases in solid state below eutectic temperature. The finer structure with features smaller than grains and phases is called the submicrostructure. Macrostructure is coarser than microstructure and is discernible to the human eye.

The parameters affecting the formation of a microstructure during the solder joint-making process include heating and cooling. For an assembly prone to the formation of intermetallic compounds at the interface or in the intrinsic solder composition, prolonged heating may produce excessive intermetallic compounds at the interface or in the solder joint. When the solder is liquid, intermetallic compounds at the interface may continue to grow and migrate toward the solder joint interior. In extreme cases, intermetallics may emerge onto the free surface of the solder, causing a change in solder joint appearance. As to the cooling effect the faster its rate, the finer the microstructure becomes. When the cooling rate is slow enough and approaches equilibrium, the microstructure of the eutectic composition normally consists of characteristic lamellar colonies. As the cooling rate increases, the degree of lamellar structure degeneration increases and colonies eventually disappear. Although it is generally accepted that a faster cooling rate creates a finer grain structure in bulk solder, this rule is often complicated by the interfacial boundary and metallurgical reaction at the interface of solder joints. The nature of the substrate and its metallurgical affinity to solder composition can affect the solder joint Microstructure development. It would not be a surprise to see the microstructure of the 63Sn/37Pb joint interfacing with a Ni-plated substrate differ from that of a Cu-plated substrate.

During service life, the integrity of joints made with sound fillet design and good wetting at the interfaces is affected by compatibility between the solder alloy and the substrate metal and subsequent in-circuit and external conditions such as heat dissipation, mechanical load, and environmental temperature fluctuation.

Heat, load, time, and extensive metallurgical interaction between the solder and the substrate metal cause changes in microstructure. Failed solder joints have revealed significant degradation in microstructure otherwise hidden by its as-solidified counterpart. In most cases where the failure is a result of fatigue (fatigue-creep) phenomenon, grain (phase) coarsening has been observed to be a precursor of solder cracks.

If we assess the mechanical properties of a solder joint by using commonly established techniques, then shear strength, creep, isothermal low-cycle fatigue and thermomechanical creep are the top four parameters. For a eutectic solder composition, the shear strength of the solder joint is improved by a very slow cooling rate which results in the formation of a near-equilibrium lamellar eutectic structure. On the other hand, the strength is also enhanced by using a very fast cooling rate as a result of grain size refining. For plastic deformation under creep mode, creep resistance depends on the operating mechanism. When the lattice or vacancy diffusion process is predominating step, creep resistance is often lower with a finer microstructure. This is due to an increased vacancy concentration resulting from a faster cooling rate. Under an isothermal fatigue environment, the relation between microstructure and fatigue resistance is more complex. Nonetheless, microstructure homogeneity is more important to low-cycle fatigue resistance. With thermal cycling fatigue resistance is often associated with decreased grain size.

To examine microstructural features, a 100 to 5000X magnification is needed. The characterization can utilize optical (light) or electron microscopy (desirably, both).

For light microscopy, the solder specimen must be carefully prepared through metallagraphic techniques involving successive grinding and polishing. The technique uses ascending levels of abrasive particle fineness bonded on papers or used as slurry on a cloth-covered wheel. The size of the abrasive particles can range from 23 μm to those of submicrometer size. Then the specimen goes through an etching process. In comparison, scanning electron microscopy (SEM) requires little sample preparation when the sectioning (cutting) of the specimen is well performed. Images from either secondary or backscattered electron signals can be readily obtained. Either provides informative characteristics with distinctive features. By combining information from both images, the microstructure and morphology of a solder joint can be better understood.

For solder joints, the two most information-revealing parameters are elemental composition and microstructure. For a given solder composition, the microstructure in the form of a quality microgram provides "sights" and "insights" into state of solder joint integrity.

5.10 SOLDER-JOINT INTEGRITY

Solder-joint integrity can be affected by the intrinsic nature of the solder alloy, the substrates in relation to the solder alloy, the joint design or structure, the joint-making process, as well as the external environment to which the solder joint is exposed. Therefore, to assure the integrity of a solder joint, a step-by-step evaluation of the following items is warranted:

- suitability of solder alloy for mechanical properties
- suitability of solder alloy for substrate compatibility
- adequacy of solder wetting on substrates
- design of joint configuration in shape, thickness, and fillet area
- optimum reflow method and reflow process in temperature, heating time, and cooling rate
- conditions of storage in relation to the aging effect on the solder joint
- conditions of actual service in terms of upper temperature, lower temperature, temperature cycling, vibration, or other mechanical stress
- performance requirements under the conditions of actual service
- design of viable accelerated testing conditions which correlate with actual service conditions

5.10.1 Basic Failure Processes

Solder-joint failure, in the real world, often occurs in complex mechanisms involving the interaction of more than one basic failure process. Although creep-fatigue is considered to be a prevalent mechanism leading to the eventual solder joint failure, separate test schemes in creep and fatigue are often conducted in order to facilitate the data interpretation and the understanding of the material behavior. The basic processes or factors that are believed to contribute to solder failure during service are (1):

- inferior or inadequate mechanical strengths
- creep
- mechanical fatigue
- thermal fatigue
- intrinsic thermal expansion anisotropy
- corrosion-enhanced fatigue
- intermetallic compound formation
- detrimental microstructure development
- voids
- electromigration
- leaching

5.10.2 Reliability of BGA Solder Interconnections

By virtue of array packages' attributes, BGA interconnections on the mother board generally consist of relatively high number of solder joints per device in comparison with SOICs, PLCCs, or QFPs. The higher number constitutes a higher probability of defect occurrence. This, coupled with less accessibility for inspection, rework and repair, makes the consistency of forming array interconnections and their quality and integrity critically important.

The main concern for the reliability of array interconnections stems from two areas. First, array solder interconnections are less compliant than conventional peripheral-leaded interconnections. The decreased compliance may contribute to reduced performance under a fatigue environment due to the cyclic thermal stress and strain imposed on the system by temperature fluctuation and in-circuit power on/off. The surface mount array interconnection is also relatively new and its applications are still in the infant stage for board-level assembly. Statistically substantiated data are lacking in terms of field performance. A common failure mode of PBGA on PCB interconnection is shown in Figure 5.52. Factors affecting long-term reliability of array solder joints are:

Component Package. The temperature profile to which each solder joint is exposed is a major contributor to the distribution of thermal stress and strain among array solder joints. In addition to the external temperature change, the temperature profile depends on the functional characteristics of the die, the ratio of die to package size, the thermal property of the carrier substrate and power dissipation. A study of IC packages ranging form 81- to 421-pin under power cycling demonstrated that outermost solder joints reached 84.3°C while center joints were 98.9°C (1.2°C below the junction temperature) for an 81-pin package.

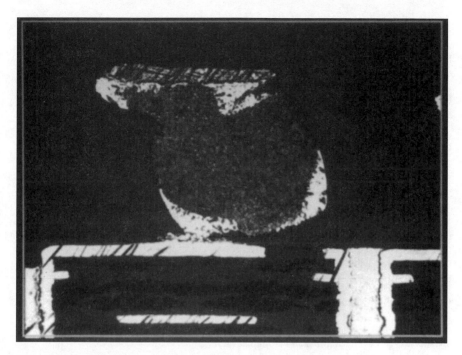

FIGURE 5.52   A common failure mode of PBGA on PCB interconnection.

However, the 421-pin package, having a lower die-to-package size ratio, experienced a large temperature differential between the outer joints (56.2°C) and the center joints (98.5°C).[20] It was found that a 165-pin device that had the largest die size in the components under study had the earliest failure, and its cycle-to-failure was lower than that of the 225-pin device.

It was also found that the solder joints directly underneath the perimeters of the die failure first under temperature cycling. This indicates that solder joint fatigue life depends more on die size than package size, and that the relative location of solder joints to the edge of the die plays an important role in the fatigue performance of solder joints.

Board Materials.   Two characteristics of board material that are most influential to the long-term performance of solder interconnections are planarity and coefficient of thermal expansion (CTE). Poor board planarity adds to the coplanarity problem of the BGA package, contributing to the occurrence of solder joint distortion, which in turn may lead to early failure of the solder joint under cyclic stresses. The CTE of conventional board material (FR-4) is approximately $15 \times 10^{-6}/°C$, while the ceramic carrier substrate of CBGA has nominal CTE of $6–7 \times 10^{-6}/°C$. The CTE of solder material falls in the range of $21–30 \times 10^{-6}/°C$, depending on the alloy. The differential in CTE between the board and carrier substrate results in an additional force to of cyclic plastic deformation in solder joints under temperature-imposed conditions. A closely matched CTE between the board and carrier substrate reduces thermally induced stresses.

Solder Composition.   The solder composition of the BGA carrier solder bumps affects the mechanical behavior of the solder interconnections. In general, the solder that is "stronger" in fatigue and creep resistance is expected to deliver a better service life. The thickness (height) of solder joints between the BGA and board is much larger than that of a fine-pitch QFP. The actual BGA solder joint height depends on the diameter of the bumps and the dimensions of solder pads; for example, the 0.022 in (0.55 mm) BGA solder height compares with a 0.003 in (0.08 mm) height for the QFP. Since the solder height for BGAs is larger, the effect of intrinsic properties of BGA solder material is expected to be more pronounced than for QFPs.

Solder Joint Configuration and Volume . The shape or configuration of solder joints can change the stress distribution and consequently affects failure mode development. Solder joint volume contributes to the kinetics of solder joint crack propagation. In addition, uniformity and consistency in volume and configuration among array solder joints within a package are important.

Other Material: Underfill. Several studies demonstrate that an epoxy that fills the air gap between the solder and the underside of the component is beneficial to the fatigue life of the solder joints. For the plastic BGA, the fatigue life of solder joints for OMPAC (over molded pad array carrier) under temperature cycles of –40 to 125°C improved nearly two-fold with epoxy underfill.[20] The underfill around the chip and the solder of the SLICC (slightly larger than IC carrier) assembly, which is a combination of flip chip and ball array technology, was used to enhance solder joint reliability.[20] The eventual solder joint failure under thermal shock of –55 to 125°C was attributed to the separation of the underfill from the die surface. This loss of adhesion was further related to foreign contamination not thoroughly removed during the cleaning procedure.

A similar enhancement in solder joint performance by means of epoxy underfill was also observed for the assembly of ceramic BGAs.

Manufacturing Process. In addition to the factors that contribute to the long-term performance of solder joints, during their service life. The ability to make high-quality solder joints at the point of production is equally important. Although an existing installed surface mount operation can be directly used to mount operation can be directly used to mount BGAs on a mother board, the setup of process parameters (particularly reflow temperature profile), control of the process and proper ambient conditions are key to making quality solder joints. High humidity and high temperature are generally detrimental to surface mount manufacturing.

Material behavior—in relation to temperature change, component effect and design— is a significant factor in the reliability of interconnections. Understanding each of these areas in conjunction with establishing a quality manufacturing process is the means to full utilization of the merits of BGA packages.

5.10.3 Reliability of Peripheral Solder Joint—Component Lead Effect

Consistency among components in their intended lead dimensions and consistency among the leads of a component are crucial to the quality of solder joints and to the overall yield of manufacturing surface mount assemblies. Although specifications for lead dimensions exist, deviations form these specifications or variations within the specifications in commercial component supplies often contribute to manufacturing problems in terms of quality and yield. This is because physical characteristics affect the long-term performance of solder joints. The effect can come from various sources:

- lead material
- lead length

- lead width
- lead thickness
- lead height
- lead co-planity
- lead material

Common lead materials include copper, Alloy 42 and Kovar. Lead stiffness varies with the design of the component package; however, the intrinsic stiffness follows the general order:

$$copper < Kovar < Alloy\ 42$$

It is believed that less stiff or more compliant lead materials are more favorable to the fatigue life of solder joints, while other conditions being equal.

- *Lead length.*   For QFPs, the lead length is measured from the toe to the contact point with the package body in a horizontal direction. The fatigue life of solder joints was found to increase by 67 percent as the lead length was increased form 0.085 in. (2.13 mm) to 0.1125 in (2.82 mm).[21]

- *Lead width.*   Figure 5.53 shows the effects of lead width on solder joint fatigue life (21); fatigue life decreases as lead width increases. It was found that fatigue life is more sensitive to lead thickness than to lead width. The fatigue life drops rapidly as lead thickness increases, as shown in Figure 5.54.

- *Lead height.*   Figure 5.55 shows the effect of lead height as measured from the contact point of lead and package body to the solder pad in a vertical direction. As can be seen, solder fatigue life increases with increasing lead height.

- *Co-planarity.*   Production defects are often related to the co-planarity of component leads, which includes starved solder joints and open joints. To avoid problems caused

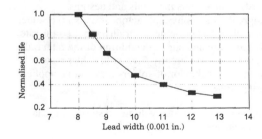

FIGURE 5.53 Solder joint fatigue life vs. QFP lead width.

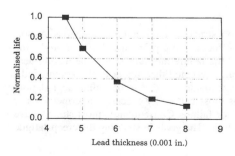

FIGURE 5.54 Solder joint fatigue life vs. QFP lead thickness.

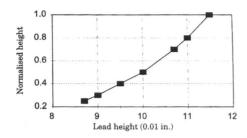

FIGURE 5.55 Solder joint fatigue life vs. QFP lead height.

by poor co-planarity, it is advisable to maintain lead co-planarity in the range of 0.002 in (0.05 mm) for fine pitch components, although 0.004 in (0.1 mm) seems to be an industry accepted value and 0.003 in (0.075 mm) for fine pitch devices. JEDEC 95 specifies co-planarity of individual components.

5.10.4 Challenges in Modeling Solder Joint Life Prediction

It is well recognized that solder joint reliability relies not only on intrinsic material properties, but also on design, the component type, the process that makes solder connections and the long-term service conditions. As electronic IC packages and components continue to change in a rapid pace, it is highly desirable to have a model able to predict the service life and reliably of solder joints under a specific set of conditions. However, to derive such a model is an ever-daunting task. This is primarily due to the complex nature of solder materials in conjunction with the "active" service conditions. Solder materials impart more complex behavior in response to temperature, stress and time than high temperature materials such as steel. Much is to be understood. The challenges are further complicated by the high level versatility in circuit boards with various materials and designs.

For given solder composition and design, the main physical factors affecting the solder material performance are: Temperature, ambient environment, strain range, strain rate, loading wave form, intrinsic microscopic structure and surface condition of the materials. Furthermore, the solder joint is expected to behave differently from bulk solder materials. Hence some established mechanical and thermal behavior of solders may not be applicable. This is presumably due to the following causes:

- the presence of high ratio of substrate surface to solder volume, resulting in a large number of heterogeneous nucleation sites during solidification
- a concentration gradient of elemental or metallurgical composition when the solder joint is formed

Either one of the above conditions may lead to a structure that is not homogeneous. As solder joint thickness decreases, the interfacial effect is more pronounced. Accordingly, the properties of solder joints may be altered and the failure mechanism may be incongruent with that derived for bulk solder. It is generally accepted that, under cyclic strain conditions, the creep-fatigue process essentially accounts for the solder joint degradation, assuming that the interface problems, such as those caused by excessive intermetallics or poor wetting, are not the determining factor for failure. Consequently, most studies have been carried out under creep-fatigue testing mode. The goals in studying the creep-fatigue process are:

- to understand material behavior under cyclic strains which are inevitable encounters during solder joint service life in electronic assembly
- to develop or improve the resistance to degradation under cyclic strains by taking a system approach
- to predict the fatigue life of solder joints so that performance reliability at a given set of service conditions can be designed and assured

Numerous fatigue life prediction methods have been proposed including frequency modified Coffin-Manson (C-M) method, strain range partitioning, fracture mechanics, and finite element analysis (FEA). The methodologies are largely borrowed from the established fatigue and creep phenomena of steels as a result of extensive studies coupled with the field data obtained over a longer period of time. Both the frequency modified C-M method and fracture mechanics-based methods are not capable of handling complex loading wave form, although fracture mechanics can monitor the effect of interfacial crack initiation and propagation on life in a comprehensive manner, and frequency modified C-M method takes frequency effect into consideration. Strain range partitioning is able to deal with the strains in any wave form, yet separating the total inelastic strain range per cycle into creep strain and plastic strain is not easy. FEA also lack the capability of including complex wave form. Increased efforts in tailoring the basic life predication models established for steels are burgeoning in electronics industry. Although the result of efforts may have generated the models that predict solder joint in comparative sense, a true working model has yet to be found.

Service conditions under which solder joints must perform in electronics packages and assemblies often involve random multiaxial stresses and they expose solder joints to creep range in addition to cyclic strains. At this time, sufficient and integrated data of solder joint behavior under such conditions and corresponding damage evolution are much lacking, consequently, some important areas and conditions are grossly ignored in the modeling scheme. Listed below are the areas that either have not been included or not adequately covered. They are in turn considered to be the reasons that contribute to the limitations of the existing models to the real-world applications.

1. effect of initial microstructure
2. effect of grain size
3. effect of microstructure that is not homogeneous
4. change in microstructure versus external conditions
5. multiaxial creep-fatigue
6. identification of presence or absence of crack-free materials at the starting point
7. size of existing cracks, if present
8. effect of interfacial metallurgical interaction
9. joint thickness versus interfacial effect (thinner solder joint imposes increased interfacial effect and decreased conventional fatigue-creep phenomena)
10. damage mechanism-transgranular or intergranular
11. potential damage mechanism shift (from transgranular or intergranular)
12. presence or absence of grain boundary cavitation
13. effect of fillet geometry
14. effect of free surface condition
15. correlation of accelerated testing conditions and the actual service condition
16. testing condition versus damage mechanism

17. service conditioned to include possible variation in chip-power dissipation over time in addition to ambient temperature change and the number of on/off power cycles

18. effect of variation in coplanarity among solder joints

Including the above-listed areas in modeling is not only overwhelmingly time consuming but also extremely difficult. It is a challenge indeed. However, the inclusion of all the above areas in devising a model is necessary to achieve a model's ultimate utility that predicts service life solder joints for a specific application.

REFERENCES

1. J.S. Hwang, *Solder Paste in Electronics Packaging,* Van Nostrand Reinhold, New York, 1989.

2. J.S. Hwang, *Advanced Solder Technology of Comparative Electronics Manufacturing,* McGraw-Hill, New York, 1996.

3. B. Champagne and R. Angers, "Size Distribution of Powders Atomized by the Rotating Electrode Process," *Mod. Devel. In Powder Metall.,* vol. 12, pp. 83–104, 1980.

4. B. Champagne and R. Angers, "Fabrication of Powder by the Rotating Electrode Process," *Int. J. Powder Metall. Powder Technol.,* vol. 16, no. 4, pp. 319–367, 1980.

5. B. Champagne and R. Angers, "Rotating Electrode Atomization Mechanisms," *Powder Metall. Int.,* vol. 16, no. 4, pp. 125, 1984.

6. T. Takeda, Japan Nat. Research Inst. for Metals, Tokyo, Japan.

7. U. Backmark, N. Backstrom, and L. Arnberg, "Production of Metal Powder by Ultrasonic Gas Atomization," *Powder Metall. Int.,* vol. 18, no. 5, 1986.

8. D.L. Erick, Battelle Memorial Inst., Columbus, Ohio.

9. J.S. Hwang "New Developments in Fluxing and Fluxes," *Electron Packag. Prod.,* June 1990.

10. J.D. Ferry, *Viscoelastic Properties of Polymers,* Wiley, New York, 1979.

11. E.F. Lish, Martin Marietta Corp., private communication.

12. E.A. Wright, "Laser versus Vapor Phase Soldering," *Soc. Advancement of Material and Processing Eng.,* 1985.

13. J.S. Hwang, "Controlled Atmosphere Soldering—Principles and Practice," in *Proc. NEPCON West,* pp. 1539–1546, 1990.

14. Norman R. Cox, "The influence of Varying Input Gas Flow on the Performance of a Nitrogen/Convection Oven," *Proceedings,* NEPCON East, 1994. p. 323.

15. James B. Halbert, "Solvent cleaning of SMDs with Boiling and Quiescent Ultrasonics," surface mount compendium, IEPS.

16. J.S. Hwang, "Soldering and Solder Paste Prospects," *Surface Mount Technol.,* Oct. 1989.

17. P. Ford and P.J. Lensch, "Cover Gas Soldering Leaves Nothing to Clean off PCB Assembly," *Electron. Packag. Prod.,* Apr. 1990.

18. E. Small, "No-Clean Fluxes: New Technology Needs New Test," *Circuits Manuf.,* Dec. 1989.

19. J.S. Hwang and N.C. Lee, "A New Development in Solder Paste with Unique Rheology for Surface Mounting," in *Proc. 1985 Int. Symp. on Microelectronics.*

20. J. S. Hwang, *Ball Grid Array and Fine Pitch Peripheral Interconnections,* Electrochemical Publications, Great Britain, 1995.

21. D.B. Barker, I. Sharif A. Dasqupta, and M.G. Pecht, "Effect of SMC Lead Dimensional Variabilities on a lead compliance and solder joint fatigue life," *Journal of Electronic Packaging,* vol. 114, p. 117, June 1192.

P · A · R · T · 2

INTERCONNECTION TECHNOLOGIES

CHAPTER 6

PACKAGING AND INTERCONNECTION OF INTEGRATED CIRCUITS

Charles Cohn
Ming T. Shih
Lucent Technologies, Microelectronics Group
Allentown, Pennsylvania

6.1 INTRODUCTION

One of the most critical levels of electronic packaging is that of packaging and interconnecting integrated circuits (ICs). The ICs are at the heart of electronic system controls, and since they are typically sensitive to electrical, mechanical, physical, and chemical influences, they require special considerations by the packaging engineer. IC packaging is the middle link of the process that produces these systems. Hence, it must respond to demands from both ends, that is, wafer fabrication and device trends upstream (circuit level) and circuit board assembly and system performance trends downstream (system level). This is particularly true in the design, piece part fabrication, and assembly aspects of IC packaging. Today's circuit and system level requirements of high performance, high reliability, and low cost, have placed greater demands on the packaging engineer to have a better understanding of the existing and emerging packaging technologies. Many electronic system performance problems result from a lack of knowledge and understanding of the interaction of the many materials and processes that are a part of this first level of packaging. Packaging engineers must be aware of the latest industry demands and be able to trade-off among the various packaging technologies on what works and what is economically feasible to satisfy these demands. This chapter addresses the current state of IC packaging technology and provides an insight into new IC packaging concepts.

6.1.1 Terms and Definitions

The semiconductor industry has undergone many changes over the past two decades, among them the terminology used in the packaging of ICs. Acronyms have found wide acceptance and usage in the semiconductor industry; many are used in this chapter or will be found in the papers and books referenced. Some of the acronyms that are encountered

by the packaging engineer in daily activities in the office, the laboratory, and on the manufacturing floor are listed in App. A.

An IC package is the housing which assures environmental protection to the IC chip
and provides for complete testing and high-yield assembly to the next level of interconnection. Levels of electronic packaging are defined in Fig. 6.1. This chapter deals with
level 1 packaging, where the IC is assembled into a package such as a quad flatpack
(QFP), pin grid array (PGA) or ball grid array (BGA) using wire bonding (WB), tape automated bonding (TAB) or flip-chip (FC) bumping assembly techniques. The packaged
device, as defined, is then attached either directly to a printed-wiring board (PWB) or to
another type of substrate, which is defined as level 2 packaging. An alternative would be to
assemble the packaged device to an intermediate vehicle, such as a hybrid circuit or a multichip module (MCM). These devices are described in Chapter 7. Hybrid circuits contain
substrates which are typically ceramic with thick or thin film metallization, while MCMs
utilize substrates made of either multilayer PWBs, multilayer ceramic, or silicon with
built-up multilayer dielectric insulators and metal conductors to achieve the required system performance. In this option, a chip may be placed on the substrate naked or bare
(without package) and after interconnection and test, the assembly is sealed to provide
environmental protection to the MCM.

The specific function of the IC package is to protect the chip from mechanical stress,
environmental stress (such as humidity and pollution), and electrostatic discharge during
handling. In addition, the package provides mechanical interfacing for testing, burn-in,
and interconnection to the next-level of packaging. The package must meet all device performance requirements, such as electrical (inductance, capacitance, crosstalk), thermal

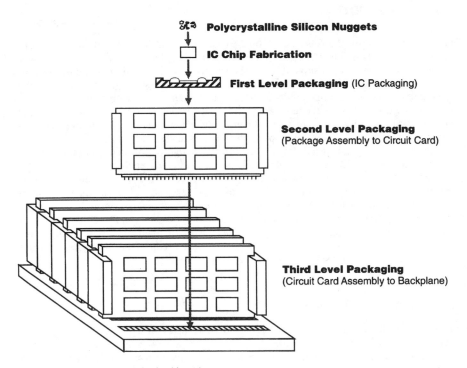

FIGURE 6.1 Electronic packaging hierarchy.

(junction temperature, power dissipation), quality, reliability, as well as fabrication interval, and cost objectives. Until recently, the packaging and interconnecting technologies have not been limiting factors in the performance of most silicon devices in high-volume production. But, as circuit and system demands are increasing, more attention is needed in selecting package technologies, materials and designs that will meet these challenges. Some package designs, may need to provide special design enhancements to accommodate such demands as power dissipation, signal terminations, power and ground distribution, and matched impedance terminations.

IC packages come in a variety of lead arrangements and mounting types that are grouped into families defined by the method of mounting to level 2, through-hole (TH) or surface-mount (SM), and by the physical arrangement of the leads on the package (in-line, perimeter or array). Figures 6.2 and 6.3 illustrate some of the more popular package types used today in the TH and SM families. Section 6.3 describes these and other emerging types in more detail.

6.1.2 Circuit and System Driving Forces

Over the last two decades the electronics industry has grown very rapidly with increases of over an order of magnitude in sales of ICs. In the 1960s, the IC market was dominated by bipolar transistors, but by 1975, digital metal-oxide semiconductors (MOS) emerged as the predominant IC group. Because of MOS's advantage in device miniaturization, low power dissipation, and high yields, its dominance has continued to this day having captured a major market share of IC sales.

Pin - Grid - Array
(PGA)

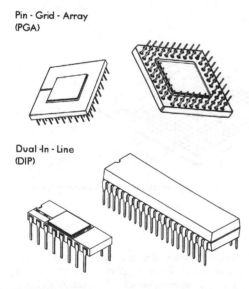

Dual -In - Line
(DIP)

FIGURE 6.2 IC packaging families, TH (through hole) type. Design variations exist within families, such as the way leads are attached to or exit the body, as illustrated in the DIP family, and in the direction the cavity faces with respect to the PWB (up or down), as shown in the PGA family.

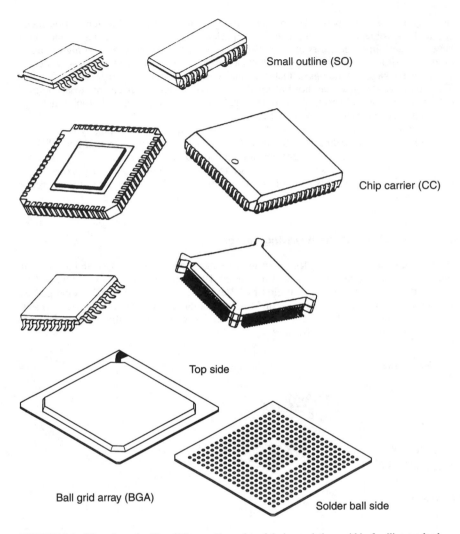

Small outline (SO)

Chip carrier (CC)

Top side

Ball grid array (BGA)

Solder ball side

FIGURE 6.3 IC package families, SM type. Examples of design variations within families are lead geometry (gull wing, J, leadless, or solder balls).

6.1.2.1 Circuit-Level Demands.   IC complexity over the last three decades has advanced from small-scale integration (SSI) in the 1960s to medium-scale integration (MSI) to large-scale integration (LSI), and finally to very-large-scale integration (VLSI), which characterizes devices containing 10^5 or more components per chip. This rate of growth[2] is exponential in nature [curve (*a*) in Fig. 6.4], being very steep up to 1975 and then slowing in recent years because of difficulties in defining, designing, and processing complicated chips. Nevertheless, at the current rate of growth the complexity is expected to reach about 100 million devices per chip by the year 2000.

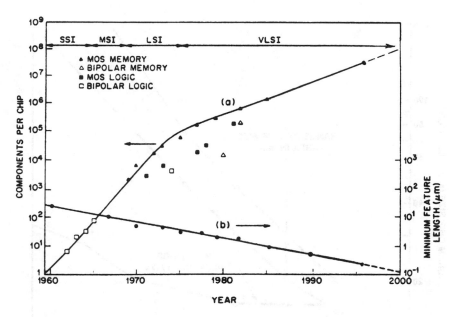

FIGURE 6.4   (a) Exponential growth in number of components per IC chip and (b) exponential decrease of minimum device dimensions *(After Sze[2] and modified),*

Continued reduction of the minimum IC feature dimensions [curve (b) in Fig. 6.4] is a major factor in achieving the complexity levels mentioned. The feature size has been shrinking at an annual rate of 13 percent, thus reaching an expected minimum feature size of 0.18 μm by the year 2000. The packaging and PWB interconnection technologies have not enjoyed an equal rate of advancement over the same period, as the ICs did. The introduction in 1987–88 of plastic bumped quad flat packs (BQFPs) with leads on 0.025-in and 0.020-in (0.64-mm and 0.50 mm) pitch and the projection of a 0.010-in (0.25-mm) pitch by the year 2000 has made it possible for perimeter type packages to stay in contention as an option for high-density packaging. In recent years, the board area necessary to interconnect the chip to the PWB has been significantly reduced by the introduction of ball grid array (BGA) packages with solder balls on a pitch as small as 0.040 in (1.0 mm). In addition, the use of bare chips (no package), with wire bonds, TAB or flip-chip (FC) bonding and chip-scale packaging, have further reduced the interconnection area to where it now approaches IC dimensions.

The rapid increase in the number of components per chip and the performance requirements of these devices create major challenges for packaging designers. Memories have been the driving force in the advancement of leading-edge wafer fabrication technology. However, they do not challenge packaging to the same extent as logic chips, which dissipate more power and require more inputs and outputs (I/Os). Although memory chip complexity and chip size have increased significantly over the last four design generations[4] and this trend will continue in the future (Fig. 6.5), the challenge to packaging in terms of I/O requirements remains essentially constant due to multiplexing techniques on the chip. In contrast, the number of I/O terminals required for logic (such as ASIC) and microprocessor devices continues to increase proportionally to the number of gates on the chip. Figure 6.6 shows the extreme demands[4] placed on packaging by logic and microprocessor

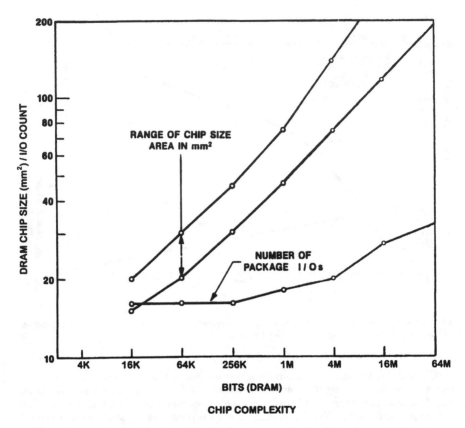

DRAM CHIP SIZE (mm²) / I/O COUNT

RANGE OF CHIP SIZE
AREA IN mm²

NUMBER OF
PACKAGE I / O s

BITS (DRAM)

CHIP COMPLEXITY

FIGURE 6.5   Continuous growth in DRAM complexity and size places little demand on package size and number of I/Os. *(After Striny and modified.[4])*

devices as compared with memories, where the I/O count has remained essentially constant for 4-Kbit to 4-Mbit dynamic random-access memories (DRAMs). Rent's rule is an empirical expression, defined as

$$\text{Number of I/Os} = \alpha \, (\text{gate count})^{\beta}$$

where α and β are constants determined by device design. Rent's rule is used to estimate the number of package I/Os that will be needed as a function of the number of gates in the device design. The example shown in Fig. 6.6 uses these constants and gives a good fit to logic (gate-array) devices.

Further circuit-level demands are due to the improved performance resulting from device miniaturization. One benefit is the reduction of power consumption at the per-gate level. Hannemann[5] reports the trends in circuit power dissipation since 1975 for three major IC groups; complementary MOS (CMOS), gallium arsenide (GaAs), and emitter-coupled logic (ECL). Figure 6.7 illustrates the exponentially decreasing trend in the power per gate for all groups. Figure 6.8, on the other hand, shows that the power dissipation per chip actually increased over the same period of time, including CMOS chips, which have

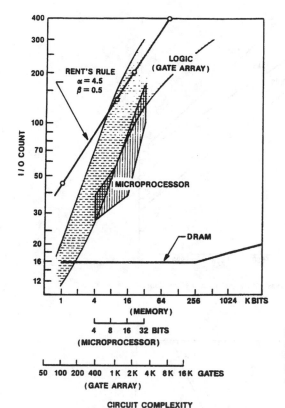

FIGURE 6.6 Comparison of I/O requirements for logic, microprocessor, and memory devices (DRAM) as a function of circuit complexity. *(After Striny[4] and modified.)*

the lowest power dissipating technology of the three. This is explained by the fact that while the power per gate scales linearly with feature size, the power dissipation per chip is largely influenced by the inverse square of the feature-size, as shown below.

$$\text{Power dissipation per chip, } P = f(\text{freq}, C, V^2, \text{gate count})$$

where freq = clock frequency

C = capacitance

V = voltage

gate count = chip area/(feature size)^2

While the clock frequency and gate count have been increasing exponentially over the years (Figs. 6.4 and 6.9), the capacitance and voltage have been decreasing. Therefore, the increase in chip power dissipation is primarily due to the greater number of gates on a chip made possible by the decrease in the feature size.

Device miniaturization has resulted in significant improvement in on-chip switching speeds. These improvements impact the technologies used for interconnection from level 1 to level 2. Off-chip driver rise-time trends[5] for ECL, CMOS, and GaAs are shown in Fig. 6.10. MOS circuits are known to be more sensitive to loading conditions due to their relatively high output impedance. Hence interconnect density is more important in MOS VLSI

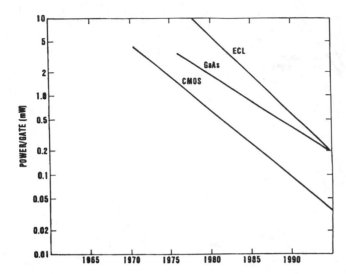

FIGURE 6.7 Trends in circuit power dissipation per gate. *(After Hanne-mann.[5] © 1986, IEEE.)*

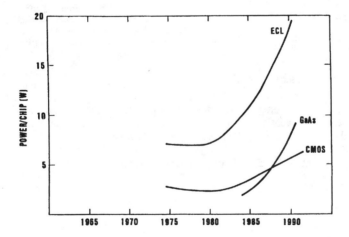

FIGURE 6.8 Trends in circuit power dissipation per chip. *(After Hanne-mann.[5] © 1986, IEEE.)*

systems than for bipolar designs. As the applications for these devices tend toward nano-second and subnanosecond signal rise times, more attention will be directed to the electri-cal design consideration of packages and interconnections.

Reduced unit cost per function is a direct result of miniaturization. The cost per bit of memory chips[2] was cut in half every two years for successive generations of DRAMs. By the year 2000 the cost per bit is projected to be less than 0.003 milicents for a 1-Gbit mem-

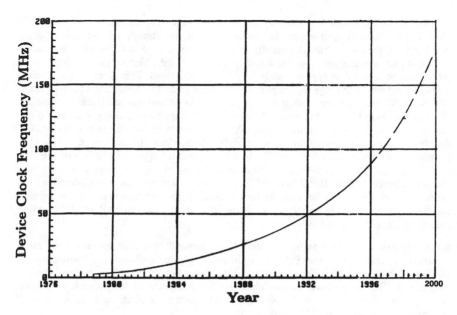

FIGURE 6.9   Average IC performance trends.

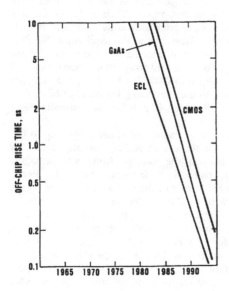

FIGURE 6.10   Off-chip rise times (typical loading). *(After Hannemann.[5] © 1986, IEEE.)*

ory chip. Similar cost reductions are projected for logic ICs. Packaging and interconnection greatly impact the overall cost of the device. The package technology used can be a significant portion of the total device cost,[6] ranging from 10 percent for DRAMs in a low-cost plastic SOJ to 50 percent for complex devices in multilayer ceramic pin grid arrays (CPGAs).

Chip-to-package interconnection, in particular wire bonding, can dictate the cost of the chip. The mechanical limit in gold ball bonding is the minimum pitch between adjacent ball bonds on the chip, which is currently in the order of 0.0035 in (0.09 mm). But, when the bond pads are staggered, as illustrated in Fig. 6.59, the effective pitch (wire to wire pitch) can be reduced to approximately 0.003 in (0.077 mm). This dictates the physical size of the silicon. The active area required is determined by the design rules for the particular wafer fabrication process used in manufacture. As device miniaturization progresses down curve (*b*) in Fig. 6.4, the device designer will be capable of placing more active circuitry, requiring more I/Os, onto each chip. A condition where, for a given number of I/Os, the chip size design is dictated by the effective bond-pad pitch on the perimeter of the chip is known as a *bond-pad-limited* chip. The bond-pad pitch has a profound effect on the chip cost, for example, at the 300-I/O level the chip cost decreases approximately 60 percent as the pitch decreases from 0.007 to 0.004 in (0.18 to 0.10 mm). These considerations are driving the interconnect technology vendors to develop finer-pitch wire bonding capability and forcing the packaging engineer to evaluate alternatives, such as TAB and FC bonding techniques. This is covered in Sec. 6.5.1.

6.1.2.2 System-Level Demands.

6.1.2.2 System-Level Demands.   System-level demands are also driven by the rapid advances in wafer fabrication. Effective utilization of chips with improved performance, that is, faster on-chip switching speeds, requires higher device density at the level 2 interconnection. This is achieved by considering several aspects of IC packaging simultaneously; the physical size of the package, lead spacing or pitch, and the method of attachment to the next level of interconnection.

The DIP family, introduced in the early 1970s, with dual in-line leads on 0.100-in (2.54-mm) centers, becomes prohibitively large[8] (Fig. 6.11) as the number of I/Os increases to levels needed to accommodate the more complex VLSI chips. Pin grid arrays (PGAs), with leads arranged in a 0.100-in (2.54-mm) pitch area array instead of in-line, provide improvement in area savings over the DIP. However, as the requirement for I/Os increases beyond 300 leads, the PGA package sizes again become very large and intolerable. In the late 1980s a different type of package, the ball grid array (BGA), was developed that offers a high-density solder bumped array, down to 0.040-in (1.0-mm) pitch, for surface mounting (SM). The ball grid array (BGA) is thus becoming the package of choice, for 200 + I/Os, because of it's smaller size per I/O count (Fig. 6.11) and better electrical characteristics.

Within a given package there is a distribution of traces that connect the chip to the external leads. Large packages, and in particular those with rectangular shapes such as DIPs, require very long traces to the corner leads of the package. Array-type packages, typically square, also contain a distribution of traces, with the longest trace going to the outermost rows of leads in particular those in the corner of the package. This distribution of trace lengths within the package, in series with similar distributions of trace lengths on the PWB due to the location and placement of large packages, could result in undesirable signal delays, line driver exceeding requirements, or crosstalk between critical signal leads. Hence the demands on packaging are for smaller physical-size packages (see Sec. 6.3.2) that will permit higher-density packaging. This will minimize the variation of trace lengths within the respective distributions and thus reduce their undesirable effects on performance. Figure 6.11 compares the package sizes for some typical surface mount (SM) and through-hole (TH) mount packages.

High-density packaging cannot be achieved with TH-type packages such as DIPs and PGAs. Through-hole (TH) packages require one hole for each pin, typically 0.35-in (0.90-mm) in diameter and are limited to a 0.100-in (2.54-mm) lead pitch by current IC package and PWB design rules and process capabilities. Either perimeter or array-type SM packages with finer lead-pitch capability are needed to satisfy this demand. Figure 6.12 com-

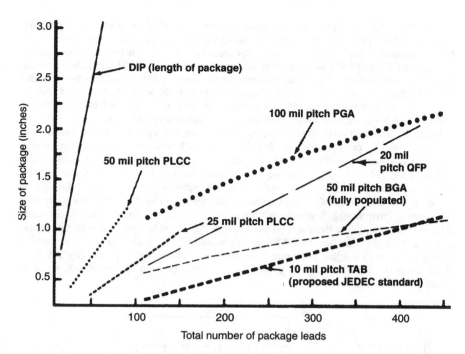

FIGURE 6.11   Sizes of some common TH and SM packages as a function of number of I/O leads. *(After Bregman and Kovac[8] and modified.)*

pares the I/O capability as a function of lead pitch for SM perimeter and array packages of equal size. By the year 2000 some packages will require over 1000 I/Os, an unachievable task for perimeter packages. But, given an array type package like the BGA, with a 0.040-in (1.0-mm) pitch, the feasibility of meeting the over 1000 I/Os capability becomes much more attractive. Perimeter leads for SM packages with a pitch as fine as 0.016 in (0.40 mm) have been achieved, but handling and assembly at these levels has been difficult.

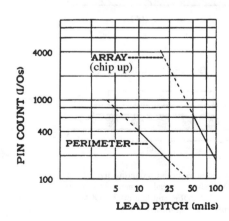

FIGURE 6.12   Comparison of I/O and lead-pitch requirements for perimeter and array type SM packages of identical physical size. At 0.02-in (20-mil, 0.5-mm) lead pitch, an array package can accommodate 20 times as many I/Os as a perimeter type.

The demand for SM packages has placed a great deal of pressure on the IC package technology. New processes such as vapor-phase (VP) and infrared (IR) solder reflow have been developed to implement SM. These processes have placed stringent requirements on the physical design of SM packages that were not required for the TH types. Some of the most important aspects are lead integrity, lead finish, lead geometry, lead true position and thermal effects. In order to ensure a reliable and high-yield (cost-effective) SM attachment, the perimeter leads on a SM package must be coplanar to within 0.002 to 0.004 in (0.05 to 0.10 mm) and the solder balls on a BGA must be coplanar to within 0.006 to 0.008 in (0.15 to 0.20 mm). In contrast, lead lengths are not critical with TH components since the assembly process used usually calls for the leads to be cut off after insertion into the PWB and then crimped to provide mechanical backup. The solder joint is another critical item for SM packages. Since the solder joint provides both the electrical and the mechanical attachment of the packaged device, its integrity must be preserved through all the stress encountered in subsequent assembly and test, and over the life of the system in its operating environment. Solder embrittlement must be avoided; hence, lead-finish materials and the gold plating thickness on BGA solder pads must be specified with care. Variations in package mounting to PWBs are illustrated in Fig. 6.13. The lead geometries shown in Fig. 6.13 *a–g* are in common use today for attachment to practically any substrate material. Each shape has its unique list of advantages and disadvantages, some of which are covered in Sec. 6.3.1.

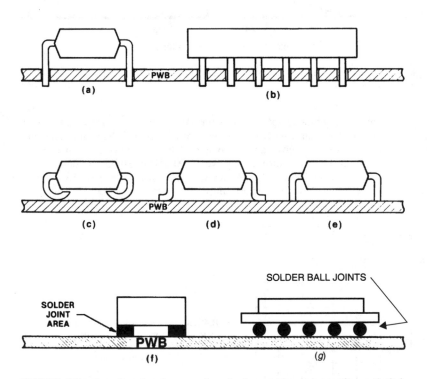

FIGURE 6.13 Examples of packages, package lead geometries, and mounting methods in current use: (*a, b*) typical TH packages, (*c–f*) typical SM packages. Lead geometry options are (*a*) DIP, (*b*) PGA, (*c*) SOJ and PLCC, (*d*) SOIC, SOG, QFP, PQFP, and fine-pitch LDCC, (*e*) butt-leaded package, (*f*) LLCC, and (*g*) BGA. (*After Striny[4] and modified.*)

Reliability of SM packaged devices could be affected directly by the stresses imposed during VP or IR solder reflow. This is particularly true if the packaged device is inadvertently exposed to high levels of moisture during the periods of time between device assembly and board assembly. Immersion of the package into the hot vapor of VP soldering or direct heating by IR occurs so rapidly that even small amounts of moisture, on the order of 0.20 percent by weight, could subject the package to severe internal stresses and initiate cracks from the device surface to the outside. Most of these fine cracks are not readily detected by visual inspection after board assembly because initially they are usually nonfatal to the device and are hidden from view at the package surface. But, if left undetected, the fine cracks may provide a relatively easy path for the ingress of contaminants from fluxes and solvents, used in board assembly, and moisture from subsequent operations such as testing, shipping, storage, or from the ambient. These contaminants may result in conditions at the IC-to-package interface that could lead to premature reliability failures.

The thermal design objective of SM packages is to operate at a junction temperature, T_j, low enough to prevent temperature-activated reliability failures such as chemical corrosion or interfacial diffusion prior to the design end of life. In addition, the electrical performance of the devices, that is, their electrical speed, and the fundamental device physics are affected adversely by temperature. The thermal analysis of the package will need to consider the complete environment to which the package will be exposed over its lifetime. Specific variables such as PWB operating temperature, total PWB power dissipation, local neighbor power dissipation, cooling technique used, thermal properties of the materials used, and any special requirements defined by the system user must be considered. No longer can the packaging engineer specify a generic package for a system and expect it to meet all system requirements without considering all possible conditions that the component will encounter in the field.

Improved quality, reliability, and cost are other system demands receiving much attention today. Systems manufacturers are driving down their costs and consequently are putting great pressure on the quality and reliability of the components that they purchase. Terms such as "zero defect" are used to define quality and reliability objectives and "just-in-time" (JIT) manufacturing to describe the strategy for reducing cost associated with having to carry large in-plant inventories that characterized the assembly process to the middle 1980s. JIT manufacturing requires that the assembly lines be running defect-free, which demands very-high-quality components and assembly processes. Today the quality of the product received by the system user is expected to have incoming defect levels in the parts per million (ppm) or even parts per billion (ppb) range.

Reliability requirements are also becoming more stringent. System users require that packaged devices pass moisture pre-conditioning followed by several reflow simulations, temperature-humidity-bias (THB) test and an accelerated temperature-humidity test such as the pressure-cooker test (PCT). The packaging of more complex high-I/O devices into smaller high-density SM packages will severely challenge the thermomechanical reliability of the package. These new high-density packages will be constructed with a variety of materials exhibiting a wide range of thermal expansion coefficients and strength properties. The packages will be subjected to a variety of temperature stresses such as those encountered in level 1 package assembly, test, burn-in if required, and SM board assembly, with the cumulative effect being an increased potential for failure due to thermally induced stress. The package engineer will be forced to evaluate the thermo-mechanical stresses during the initial stage of the design before committing capital and expense dollars to the project. The analysis is used to gain insight and confidence that the design will have a good chance, at an acceptable risk, of meeting all objectives. Computer-aided design (CAD) tools such as the finite-element analysis can be used to perform these complex two- and three-dimensional analyses.

6.2 FUNDAMENTAL PACKAGE TECHNOLOGIES

IC package technologies fall into three basic categories; plastic molded technology, pressed ceramic technology and laminated ceramic/plastic technology.

In the plastic molded technology, the package is constructed around the IC chip assembled on a metal lead frame in strip form. The IC chip is first mechanically bonded to the lead frame paddle and then electrically interconnected by fine wires from chip bond pads to the corresponding lead frame fingers. If TAB bonding is used, then the IC chip pads are directly connected to the bumped lead frame fingers. The final package configuration is formed by plastic molding around the lead frame sub-assembly. The portion of the lead frame that is external to the package body is subsequently trimmed and formed into specific geometries suitable for either TH or SM attachment.

The pressed ceramic technology package contains a lead frame embedded and glass sealed into a pressed ceramic base. The IC chip is attached and wire bonded to the lead frame and hermetically sealed using a glassed ceramic cap or lid. The leads external to the package body are trimmed and formed to the geometries needed for TH or SM attachment.

Laminated ceramic and plastic packages consist generally of a substrate with an integral metallized or metal conductor fanout, and external terminals which are either leadless, leaded or with solder balls for SM, or leaded for TH attachment.

Leaded types come from the IC package vendor either with the leads ready for TH or SM attachment or as flat-packs (usually with the ends of leads connected by tie bars), where the leads must be trimmed and formed, after chip assembly, to the geometry needed for TH or SM attachment. The substrate body may come with either a cavity to accept the chip or an in-plane configuration where the chip attach paddle and the bond pads are on the same level. The electrical interconnections are done using assembly techniques similar to those used with the plastic molded package types. The substrate material may be either ceramic or plastic, of single or multi-layer construction, depending on the functional requirements. Both the ceramic and plastic substrates can equally provide high-performance, but the ceramic has hermetic sealing capability whereas the plastic substrate is non-hermetic and requires some form of encapsulation for environmental protection of the chip.

A comparison of cost, performance and reliability among the various package technologies show the multilayer ceramic package to be best suited when the requirements call for high performance, hermeticity, and high reliability, but the trade-off is high cost. A multilayer plastic package, with electrical and thermal enhancements (see Sec. 6.2.4), can provide equal or better performance at a lower cost, trading off hermeticity. For the lower end of consumer electronic products the package requirements, in order of importance, are low cost, reliability, and performance. These attributes can be satisfied with plastic molded packages which have recently experienced an improvement in the technology by the proper selection of materials and rigid process controls. Sections 6.2.1 to 6.2.4 discuss packaging technologies that are generally applicable to a variety of package types and families. Section 6.2.5 presents package technology comparisons and tradeoffs.

6.2.1 Molded Plastic Technology

Figure 6.14 illustrates the construction of a QFP in the plastic molded package technology. A metal lead frame, usually copper alloy or Alloy-42 (42 percent Ni, 58 percent Fe) provides the mounting surface for the chip, that is, the chip support paddle. The lead frame also provides the electrical fanout path from the fingers to the outside leads which vary in pitch from 0.040 in (1.0 mm) to 0.016 in (0.40 mm). The lead frame is usually spot plated with gold or silver on the paddle and at the tip of the wire bond fingers to provide reliable

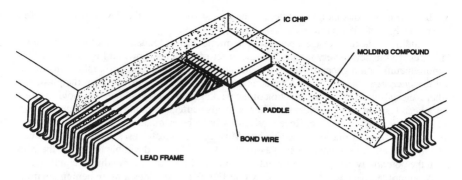

FIGURE 6.14   Sectional view of a typical QFP illustrating molded plastic (postmolded) technology.

chip attachment and wire bonds. Silver spot plating is used where lowest cost is required. Wire bond interconnects are gold wires, typically 0.001 in (0.025 mm) or 0.00125 in (0.032 mm) in diameter, with wire span lengths ranging from 0.050 to 0.180 in (1.27 to 4.57 mm) with the potential of extending the length to 0.200 in (5.08 mm) in the near future. After molding, the external portion of the lead frame must be processed to the final form and lead finish for assembly to the next level. This assembly sequence is shown in Fig. 6.15 and applies to all plastic molded package types. Details such as lead-frame design and format, workholders for chip and wire bonders, molding dies, trim and form tools, and handling hardware will vary according to package types.

The molded package technology described is generally referred to as postmolded plastic technology. Thermosetting (cross-linking) epoxy resin is molded around the lead-frame-chip sub-assembly after the chip has been wire bonded to the lead frame, that is, the package is molded around the chip in situ. The postmolding process is relatively harsh. Major yield and reliability problems, observed in postmolded parts, are wire sweep caused

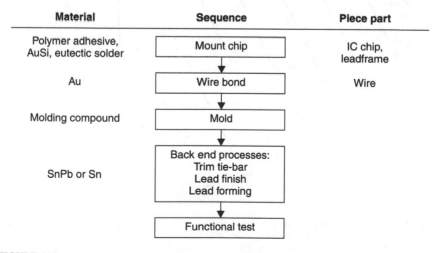

FIGURE 6.15   Assembly sequence for postmolded plastic technology packages.

by the flow of viscous molding materials, and local stresses on the device surface caused by sharp edges of filler particles.

In addition, plastic packages, being non-hermetic, have a tendency of allowing moisture to ingress to the device very rapidly,[9] on the order of days, as shown in Fig. 6.16. The aluminum metallization generally used in IC manufacture is susceptible to rapid corrosion in the presence of moisture, contaminants, and electric fields. Impurities from the plastic or other materials used in the construction of the package can cause voltage threshold shifts or act as catalysts in metal corrosion. Fillers can also affect reliability and thermal performance of the plastic package. Recent advances have been made on the IC devices themselves in the form of improved passivation. Most of the industry uses some type of silicon nitride passivation to achieve a nearly hermetic chip.[10] This is desirable to protect the chip from the potentially hostile environment encountered in the plastic molding technology.

A premolded package[12] concept, shown in Fig. 6.17, provides a more benign environment for the chip. Delicate wire spans and strain-sensitive features on the chip surface are decoupled from the molding process, thus avoiding the stresses associated with post-molded packages. The assembly sequence for a premolded plastic technology package is shown in Fig. 6.18.

The package is fabricated using a preplated lead frame, in strip form with multiple sites per strip, and then either transfer molded, using a thermoset epoxy B molding compound, or injection molded, with a thermoset or a thermoplastic polymer such as polyphenylene sulfide. The lead-frame, usually made of a copper alloy for better thermal performance, is blanket plated with nickel and spot plated with either gold or silver in the paddle area. The external leads may be selectively plated with thin gold to retain solderability for final lead finishing. After molding, the package may need to be deflashed to

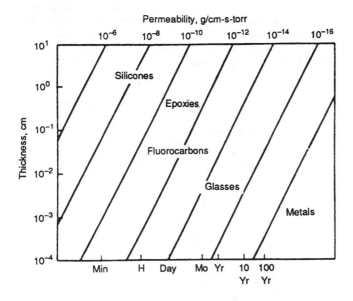

FIGURE 6.16 Calculated time for moisture to permeate various sealant materials (to 50 percent of exterior humidity) in one defined geometry. Organics are orders of magnitude more permeable than materials used for hermetic seals. (*After Traeger,[9] © 1976, IEEE.*)

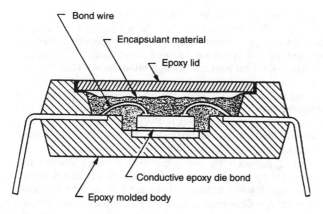

FIGURE 6.17 Cross-sectional view of a completed premolded plastic DIP, including IC chip.

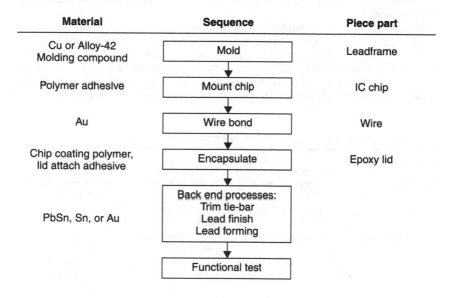

Material	Sequence	Piece part
Cu or Alloy-42 Molding compound	Mold	Leadframe
Polymer adhesive	Mount chip	IC chip
Au	Wire bond	Wire
Chip coating polymer, lid attach adhesive	Encapsulate	Epoxy lid
PbSn, Sn, or Au	Back end processes: Trim tie-bar Lead finish Lead forming	
	Functional test	

FIGURE 6.18 Assembly sequence for premolded plastic technology packages. Package is fabricated using preplated lead frame then either transfer or injection molded using a thermoset epoxy B or a thermoplastic polymer molding compound.

remove any molding compound flash deposited on the chip-bonding areas and then cleaned to remove any residual particulate matter. Details of the fabrication process are described in Parker and Stafford.[12] Chip mounting is done with a polymer adhesive, either conductive (silver-filled) or nonconductive epoxy, or polyimide paste. Wire bonding is either thermosonic or thermocompression ball-wedge bonding using 0.001 to 0.00125-in (0.025- to 0.032-mm) diameter 99.99 percent pure gold wire. This is followed by chip encapsulation with a die coating or preferably a flow coating to fill the entire cavity and lid sealing, thus environmentally protecting the chip and the wedge bond areas in the cavity. Candidate materials for encapsulation include room-temperature vulcanized (RTV) silicone rubbers and combinations of silicon gels and cover coats.[12,13] Back-end processes include trim and form to the final package configuration, be it a DIP,[12,13] or a plastic leaded chip carrier (PLCC).[14] Final lead finish is then applied by solder dipping. Another option is to use the original thin gold plate as the final finish, making sure that solderability is not degraded in the assembly process. A gross leak tight-lid seal will preclude the subsequent ingress of cleaning solvent (used in PWB assembly) into the chip cavity, where it could interact with the flow coat to cause swelling and stress the delicate wire bonds or degrade the interface integrity between chip and encapsulant.

The benign chip environment provided by the premolded package construction sequence is similar to that provided by a ceramic cavity package. Striny and Schelling reported[13] superior temperature-humidity-bias (THB) (85 percent RH and 85°C) reliability performance of 4-Kbit MOS DRAM devices in premolded plastic packages compared with a postmolded package control group containing devices from the same wafer lots. The premolded plastic technology is potentially useful for packaging very sensitive IC devices requiring low-cost assembly. The postmolded package technology, on the other hand, is the more cost effective of the two and is by far the predominant package technology in use today.

The two critical steps in producing a molded product, that meet the device reliability objectives, are selection of the molding material and control of the molding process itself. Overall integrated automation of the assembly process, in particular the automation of chip-bonding, wire bonding, and molding steps, are also important. The polymer most commonly used for plastic IC packages is epoxy. The composition of a typical epoxy molding compound[15], shown in Table 6.1, consists of the epoxy resin portion (approxi-

TABLE 6.1 Composition of Epoxy Molding Compounds

Material family	Example	Purpose	Technology thrust
Epoxy resin	Bisphenol-A epoxy, Novolac epoxy	To cure by a cross-linking reaction	Lower ionic impurities (lower Cl⁻)
Hardener	Amine, anhydride, phenol Novolac	To cure by a cross-linking reaction	—
Reaction accelerator	Tertiary amine	To cure by a cross-linking reaction	—
Filler	Silica powder, alumina powder, glass fiber	Adjust CTE, thermal conductivity, mechanical strength	Low alpha particle (lower uranium)
Flame-retardant agent	Brominated epoxy, Sb_2O_3, hydrited alumina	Flame retardancy	—
Coupling agent	Epoxy silane, amino silane	Better contact between resin and filler	—
Releasing agent	Stearic acid, natural wax, synthetic wax	Release from mold	Balanced releasing contact
Coloring agent	Carbon black, iron black, FeO_3, organic dye	Coloring	Laser marking

Source: From Kakei et al.[15]

mately 25 to 35 percent by weight of the total composition), a hardener, and a curing agent. Today, Novolac epoxies are generally preferred because their higher functionality gives them improved heat resistance. The synthesis of the resins produces sodium chloride as a by-product. Both sodium and chloride ions are deleterious to device reliability. Therefore these by-products must be washed carefully from the resins before they are synthesized into useful molding compounds. Reduction of the chloride content in the epoxy resin is especially effective in decreasing corrosion of the aluminum chip bond pads.

Fillers, such as amorphous or crystalline silica (SiO_2) or glass fibers, and sometimes Al_2O_3 are added to the resin so that the resultant mixture is 65 to 73 percent filler by weight. Amorphous SiO_2 is used when a minimum expansion coefficient is desired, at some sacrifice of thermal conductivity, while crystalline SiO_2 improves thermal conductivity at the expense of the coefficient of thermal expansion. Al_2O_3 has a high thermal conductivity (as a filler material) but is very abrasive to molds. Fillers greatly improve the mechanical strength of the resin and reduce its coefficient of thermal expansion, hence reduce shrinkage after molding. Silica fillers normally contain uranium and thorium, which radiate alpha particles and may cause reliability failures (soft errors) in alpha-sensitive memory devices. The introduction of low alpha radiation molding compounds and the development of devices with structures that are immune to alpha particles has minimized soft errors.

Small amounts of pigments, coupling agents, mold release agents, reaction accelerators, antioxidants, water getters, plasticizers, and flame retarders must also be added to complete and optimize the molding resin. Coupling agents improve the adherence of the resin with the organic fillers, metal lead frames, and silicon chips. This minimizes moisture penetration at the various interfaces and improves device reliability. Mold release agents are used to release the molded part from the mold. Since they can also decrease the adherence of the resin to the lead frame, their behavior and effect on device reliability must be carefully evaluated. Reaction accelerators may decrease the volume resistivity of the molding compound. A flame retardant, normally brominated epoxy and antimony trioxide, is added to meet the industry flammability standard (UL94 V-O). Selection of all additives must be made very carefully to optimize the reliability of the molded device.

It is important to understand the rheological, chemical, and thermophysical properties of molding compounds, both for the molding process and for their interrelation with the reliability of the finished device. The two most important characteristics that must be evaluated for a molding compound are its moldability and its expansion coefficient. The molding process is not cost-effective unless it has high yield. The most common reliability problems can be related to the mechanical quality of the molded part and its ability to withstand thermal stresses. Epoxy molding compound vendors have made significant changes[15] to meet the challenges of VLSI devices.

Thermoset molding materials are usually transfer molded in large multicavity molds (Fig. 6.19). After entering the pot, the preheated molding compound, under pressure and heat, melts and flows to fill the mold cavities containing lead frame strips with their attached ICs. The IC lead frame often has long, fragile lead fingers, particularly in the higher I/O package types. The chip is interconnected to these thin, fragile leads with 0.001 to 0.00125-in (0.025- to 0.032-mm) diameter gold wires at wire spans from 0.050 to 0.180-in (1.27- to 4.57- mm) in length. To avoid damaging this fragile structure, the viscosity and velocity of the molding compound entering the mold must fall within certain ranges. Commercial molding compounds are designed to meet these requirements when molded at approximately 175°C with pressures of about 6 MPA and a mold cycle that ranges from 1 to 5 min. Fig. 6.20 shows an 80-cavity mold used for a 68 I/O PLCC.

To control the velocity of the molten molding compound, each device cavity has a gate, or restriction, to slow the material flow. The fluid mechanics of molding are relatively

CHAPTER SIX

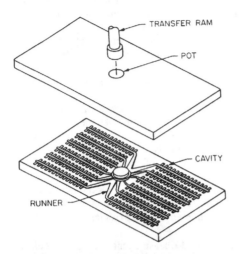

FIGURE 6.19 Schematic of multi-cavity transfer mold showing ram, pot, and runner system. Small packages such as DIPs or SOs use molds containing upward of 168 cavities, while larger packages such as high-I/O PLCCs use molds with fewer cavities, such as the 80-cavity mold of Fig. 6.20. *(After Striny.[4])*

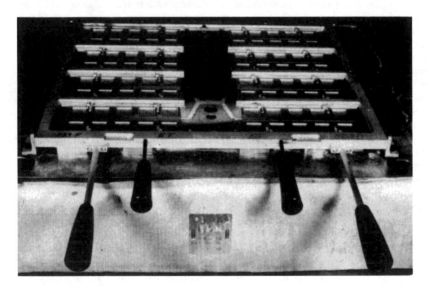

FIGURE 6.20 80-cavity transfer mold for 68-I/O PLCC. Each lead frame accommodates five chips; cavity gate is located in corner of package. *(After Striny.[4])*

complex, because the materials are non-Newtonian. In addition, partial cross-linking can occur during the molding process, affecting the material viscosity. Elaborate mold designs have been proposed to compensate for these variables. An automated multipot approach as shown in Fig. 6.21 can also provide improved control of these molding variables.

Reliability performance criteria for DRAMs or other large complex chips are defined by the level of stresses resulting from expansion mismatches within the package and by

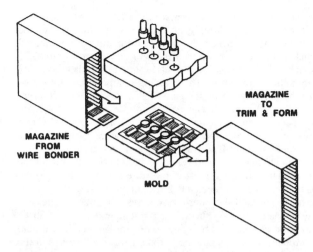

MAGAZINE
TO
TRIM & FORM

MAGAZINE
FROM
WIRE BONDER

MOLD

FIGURE 6.21 Schematic of two-strip multipot molding system. Lead frames are loaded and unloaded directly from and into magazines that are compatible with assembly equipment located upstream (wire bonding) and downstream (trim and form) from molding operation. Schematic illustrates one pot for two cavities and requires fast-curing molding compounds to achieve similar economics as large multichip cavity molds. *(After Striny.[4])*

performance of the packaged device in a humid environment. Stresses are caused mainly by mismatches in the coefficients of thermal expansion of the materials commonly used in packaging and IC chips, as illustrated in Fig. 6.22. Microscale stresses are caused by interactions between the IC active surface and the molding compound filler particles. Macroscale stresses result from interaction due to the relative physical sizes between chips and

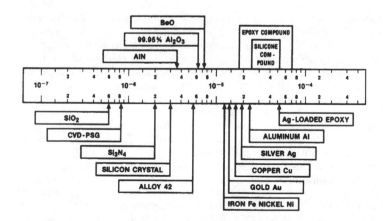

FIGURE 6.22 Coefficients of thermal expansion of materials for semiconductor devices, $°C^{-1}$. CTE of FR-4 and other PWB materials (not shown in figure) fall within 1.2 to 1.5 (10^{-5}) $°C^{-1}$ range. *(After Kakei et al.[15])*

the packages in which they are housed. Humidity performance is affected by the levels of leachable ionic contaminants found in the packaging materials and the integrity of the package and interfaces. Conditions at the interface change[16] as a function of the thermal expansion rates and the thermal sequences that the package will encounter. Figure 6.23 compares the thermal expansion rates of the two leading lead frame materials used today with those of typical epoxy molding compounds. The molding compounds exhibit two distinct expansion rates. Below the glass-transition temperature T_g, the rate $\alpha_1 \approx 20$ ppm/°C and above T_g, the rate $\alpha_2 \approx 60$ to 70 ppm/°C. The expansion rates for the lead-frame materials are linear over the temperature range shown, with Alloy-42 at ≈ 4.3 ppm/°C and copper at ≈ 17 ppm/°C. During cooling from the molding temperature (≈ 175°C), the molding compound contracts faster than either lead-frame alloy and will cause an interference fit between the molding compound and the lead frame. This tight fit will prevent the entry of unwanted contaminants and impede the ingress of moisture into the chip region of the package. Unfortunately, as the package is subjected to back-end processes, such as solder dipping the final lead finish, the stress at the interface reverses from the compressive stress experienced below T_g to a tensile stress at the solder dip temperature of ≈ 250°C. Under this condition a gap may form between the molding compound and the lead frame, depending on the magnitude of the tensile stress and the initial chemical adherence of the molding compound to the lead frame. A similar stress differential exists between the chip surfaces and the molding compound due to the differences in the coefficients of thermal expansion between the silicon and the molding compound, as seen in Fig. 6.22. Gap formation is undesirable because fatal contaminants may find their way to the wire bond pads on the chip and result in reliability failures in the field due to aluminum corrosion. To avoid these problems, much attention must be given in selecting the proper materials, for both lead frame and molding compound, and in specifying the lead-frame finish and design, such as the addition of special features for locking in the leads that will prevent slippage between the critical surfaces.

Fig. 6.23 illustrates how thermal expansion differentials between lead frame and molding compound cause changing conditions at the interfaces. For example, the conditions change from interference fits on cooling to room temperature, to tensile stresses, at the higher temperatures encountered at solder dipping of leads. Similar tensile stresses may be encountered when an SM package is assembled by either VP or IR solder reflow techniques.

Molding compound vendors are constantly upgrading the properties[15] of their materials to achieve improved performance. The macrostress level has been significantly reduced by the use of higher filler levels to lower the coefficients of thermal expansion (matching plastic to the lead frame), by adding silicone modifiers to lower the modulus of elasticity (provide more yielding to stress), and by the addition of proprietary additives that improve adhesion between the chip and the plastic encapsulant, thus improving resistance to crack propagation in the plastic, and providing increased temperature stability under temperature cycling. The use of round-edge filler particles and the use of epoxy fillers to improve adhesion have reduced the microstress on sensitive devices. Automation increases the need for more aggressive molding compound properties which will lead to faster cure cycles and better batch-to-batch control of critical properties.

Molding equipment vendors are being encouraged to provide process flexibility, that is, the ability to mold a wide variety of packages in smaller production runs, with the intent to automate the molding process. Users today have many equipment options, which range from large production molds (Figs. 6.19 and 6.20), handling many packages per shot, to automatic multiplungers with small pots (Fig. 6.21), handling only a few packages per shot. These multiplunger types require lower transfer or injection pressures, resulting in minimum wire sweep, fewer voids, better flash and bleed qualities, and more processing latitude.

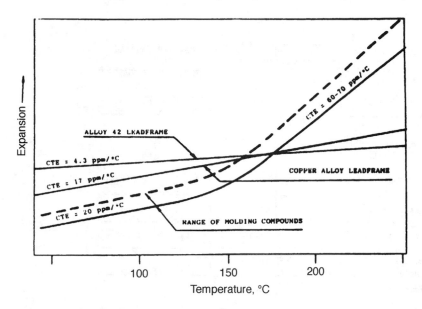

FIGURE 6.23   Thermal expansion rates for molding compounds and lead frames. *(After Murtuza et al.,* [16] *© 1986, IEEE.)*

Today the plastic package technology is widely accepted in the industry as reliable and low-cost. Most of the recent standardization activities in the Electronics Industries Association (EIA), the Joint Electron Device Engineering Council (JEDEC), and the Electronics Industry Association of Japan (EIAJ) are evolving around the registration of new package family outlines. These are covered in Sec. 6.3.

6.2.2   Pressed Ceramic (Glass-Sealed Refractory) Technology

Pressed ceramic technology packages are used mainly for economically encapsulating ICs requiring hermetic sealing. Hermeticity means that the package must pass both gross and fine leak tests and also exclude environmental contaminants and moisture for a long period of time. Further, any contaminants present before sealing must be removed to an acceptable level before or during the sealing process. Glass, as illustrated in Fig. 6.16, is an effective material for achieving a hermetic seal for high-reliability applications. Leak rate, as defined by MIL-STD-883, is that quantity of dry air at 25°C in atmospheric cubic centimeters flowing through a leak or multiple leak paths per second from a high pressure side at 1 atmosphere (760 mm Hg absolute) to the low-pressure side at or less than 1 mm Hg absolute. Standard leak rates are expressed in units of atmosphere cubic centimeters per second (atm • cm^3/s). Procedures for detecting gross and fine leaks in hermetic packages are detailed in MIL-STD-883. Gross-leak tests, such as die penetrants and bubble tests, are capable of detecting leaks in the range of 10^{-1} to 10^{-5} atm • cm^3/s. Fine-leak testing procedures, such as helium and radioactive tracer techniques,[17] will detect leaks in the range from 10^{-4} to 10^{-9} atm • cm^3/s. MIL-STD-883 requires that hermetic packages pass a leak test at 10^{-8} atm • cm^3/s on equipment having a detection sensitivity of at least 10^{-9} atm • cm^3/s.

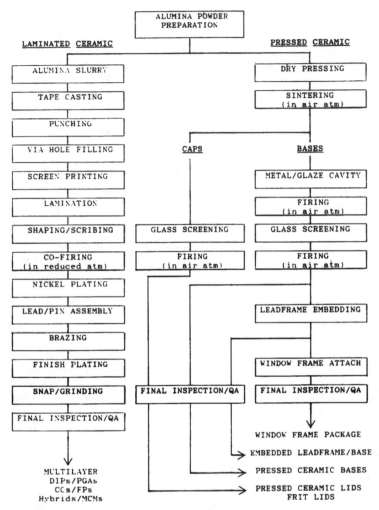

FIGURE 6.24 Process flowchart for piece parts and prefabricated packages for pressed ceramic and laminated ceramic package technologies. *(After Kyocera.[18])*

The process flowchart[18] for fabricating the basic piece parts used in the pressed ceramic packaging technology is shown in Fig. 6.24. Also shown alongside is the process flow sequence for the laminated ceramic package technology, which is described in Sec. 6.2.3. Pressed ceramic piece parts require fewer and simpler process steps, resulting in the lowest-cost piece parts for a hermetic package technology. This technology, when implemented with a lead frame in strip format, that is, multiple sites per strip, has the potential of being automated,[19] and therefore is competitive with the plastic technology for low-cost packaging. The IC package designer has a choice of three pressed ceramic options, as shown in Fig. 6.24, that is, basic piece parts (bases and caps), partially assembled subassemblies (embedded lead frame and base), and a prefabricated package known also as a window-frame package. The IC assembly sequence is described in Fig. 6.25.

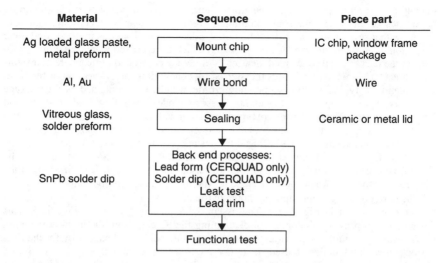

Material	Sequence	Piece part
Ag loaded glass paste, metal preform	Mount chip	IC chip, window frame package
Al, Au	Wire bond	Wire
Vitreous glass, solder preform	Sealing	Ceramic or metal lid
SnPb solder dip	Back end processes: Lead form (CERQUAD only) Solder dip (CERQUAD only) Leak test Lead trim	
	Functional test	

FIGURE 6.25   IC assembly sequence for pressed ceramic package technology.

In the first option, the package is constructed around the IC chip using basic piece parts, for example, ceramic base, lead frame, and ceramic cap. When done in-house, this is the lowest cost option because the IC manufacturer controls all of the assembly process steps. Option 2 is a minor variation of option 1, where the IC manufacturer purchases the lead frame already embedded into the base ceramic and ready for IC chip mounting. This option is attractive to manufacturers who do not have, nor wish to invest in, the equipment needed for embedding the lead frame. These options are known in the industry as the CERDIP or CERQUAD packaging technology. The assembly structure is shown in Fig. 6.26a.

In both options the IC chip is attached to the lead frame using either AuSi eutectic or silver-loaded glass adhesive technologies. Hermetic sealing is done with a glassed ceramic cap or lid. Due to the relatively high temperature used for glass sealing, typically greater

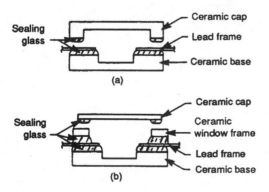

FIGURE 6.26   Structures of (a) CERDIP and (b) window frame. (*Courtesy of Kyocera America, Inc., after Striny.*[4])

than 400°C, an all-aluminum monometallic system is used. This consists of the aluminum bond pads on the chip, ultrasonic aluminum wedge-to-wedge wire bonding, and lead frames with a strip of aluminum deposited or cladded on their lead tips in the second bond area. Gold-aluminum intermetallic compounds, which have caused many reliability problems in the past,[20] are thus avoided. Sealing is done in a controlled-atmosphere conveyor-type furnace,[21] in which the heat is transferred by conduction and convection or by infrared radiation. A typical sealing temperature profile, usually recommended by the package vendor, is shown in Fig. 6.27. Back-end processes include leak testing, lead finishing, and lead trim and form. External leads are typically plated with matte tin, which provides a finish that is compatible with the subsequent lead trim and form operation and final assembly to the next-level packaging. CERDIP lead frames are procured already formed and require only to be finished and trimmed prior to testing. Lead frames for SM types such as the CERQUAD usually come planar and are processed planar through lead finishing, then trimmed and formed into the desired shape for the final package configuration.[22,23]

The third pressed ceramic option, known in the industry as a window-frame package, is the structure shown in Fig. 6.26*b*. The assembly sequence is given in Fig. 6.25. This design concept provides the option of decoupling the IC chip from the high-temperature glass-sealing procedure by providing a low-temperature solder seal capability for the final hermetic seal. This option is used for temperature-sensitive chips that cannot go into a glass-sealed package without degrading their performance or reliability. In addition, this option also allows the use of thermosonic or thermocompression gold ball-wedge wire bonding equipment, which extends this package technology to those manufacturers who have this wire bonding capability but not the ultrasonic aluminum wedge-to-wedge bonders needed to achieve the desired all-aluminum monometallic structure for glass seals. Back-end processes are similar to those used with the other options.

Two types of solder glasses are used today for glass sealing; devitreous or vitreous. Table 6.2 lists the major properties of some of the solder glasses currently in use. Seal glass is an important constituent of the pressed ceramic technology because it influences both electrical performance and reliability of the packaged device. The dielectric constant of the glass used to embed the lead frame can affect the capacitance between the leads, that is, the speed of the I/O signals through the lead frame. Seal glass is exposed to the cavity; hence alpha-particle emission from the glass fillers may affect the soft error reliability of some sensitive devices.

Glass sealing is a complex process that must be understood fully and implemented under rigorous control. Tight temperature control and minimum mechanical and thermal

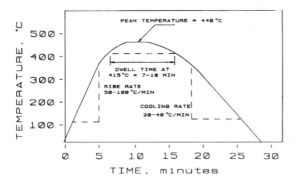

FIGURE 6.27   Typical glass-sealing profile as recommended by package manufacturers. *(After Sharan and Valero.[21])*

TABLE 6.2 Major Properties of Seal Glasses

Filler system	Lead-titanate-zircon	Cordierite	Beta eucryptite	Zircon
Glass type	Vitreous	Vitreous	Vitreous	Devitreous
Sealing temperature, °C	450	430	430	485
Dielectric constant	33	12.2	12.1	17.8
Alpha emission, count/(cm$^2 \cdot$ h)	29	0.8	0.2	41
Mechanical strength, kg/cm^2	550	460	460	650
CTE mother glass, 10^{-7}/°C	110	110	110	—
CTE filler, 10^{-7}/°C	−30/20	8	−80	20
Particle-size distribution	Very fine to coarse	Fine	Fine	Coarse
Filler volume ratio	High	Medium	Medium	Medium
Filler dissolution	Low	High	High	Low

Source: From Loo.[25]

handling are necessary to minimize hermeticity yield loss. The hermeticity of the finished package[25] is influenced by the adhesion of the seal glass to the ceramic, the seal glass to the lead frame, the mechanical strength of the glass, and the dissolution of fillers into the mother glass. Residual stress in the mother glass (due to the mismatch in thermal expansion between ceramic, Alloy-42 lead frame, and glass) and particle-size distribution of the filler influence the mechanical strength, or the ability to provide reliable hermetic seals. Ideally, seal glass should have a low dielectric constant, high mechanical strength, low alpha emission, and low sealing temperature. None of the glasses listed in Table 6.2 come close to the ideal glass definition, each having its advantages and disadvantages. The package designer must know which properties are needed for a particular device and base the selection of the glass on that input. For example, a device not requiring low-alpha-emission sealing glass, low dielectric constant, and low sealing temperature may be sealed using a lead-titanate-zircon system to take advantage of the high mechanical strength. On the other hand, if the device requires low-alpha-emission sealing glass, low dielectric constant, and low sealing temperature, either the cordierite or the beta eucryptite system may be the better choice.

The raw material for the pressed ceramic technology is alumina ceramic (90 percent minimum Al$_2$O$_3$) in black-brown characteristic color. Package vendors[26] are also providing ceramic materials such as aluminum nitride (AlN), silicon carbide (SiC), and low-thermal-expansion sealing glasses (4.7 ppm/°C), for improving hermeticity yields of pressed ceramic. Pressed ceramic parts, made with SiC and low-thermal-expansion glass, were exposed to a sequence of mechanical shock (1500 *g*), vibration variable frequency (20 *g*) and constant acceleration (3000*g*) test conditions. No hermeticity rejections were seen in a sample of 200 packages exposed to this sequence. A second group of 280 parts exposed to a sequence of thermal shock (−65 to +150°C, 15 cycles), temperature cycles (−65 to +150°C, 100 cycles), and moisture resistance tests (10 cycles) also passed with no hermeticity rejections.

6.2.3 Cofired Laminated Ceramic Technology

The cofired laminated ceramic technology is the most reliable IC packaging technology currently available. The technology is constantly responding to the device- and system-level demands for higher performance and higher-density applications. This section addresses the technology for single-chip packages. The cofired laminated ceramic technology is also being used for such applications as hybrids and MCMs, which are covered in Chaps. 7 and 10, respectively.

The process flowchart for the cofired laminated ceramic package technology is shown in Figs. 6.24 and 6.28. A dispersion or slurry of ceramic powder and liquid vehicle (solvent and plasticized resin binder) is first prepared, then cast into thin sheets by passing a leveling or doctor blade over the slurry. After drying, the sheets are in the green-tape stage, ready for cutting to size and punching of via holes and cavities. Custom wiring paths are then screened onto the surface (usually a slurry of tungsten powder), and the via holes are filled with metal.

Several of these sheets are press-laminated together in a precisely aligned fixture, and the entire structure is fired at 1600°C in a reducing atmosphere to form a monolithic sintered body. The cofired laminated refractory ceramic technology is a complex process requiring careful control throughout.

After the laminate has been sintered, it is ready for the finishing operations of lead/pin attachment and metallization plating. Nickel is plated over the tungsten in preparation for lead/pin brazing. The lead/pin material is either Kovar (a Fe-Ni-Co alloy) or Alloy-42 and the brazing material is a Ag-Cu eutectic alloy. After lead/pin brazing all exposed metal surfaces are electroplated or electroless plated (usually gold over nickel) for bondability and environmental protection. Multilayer ceramic packages can be made in lateral dimensions of up to 150 by 150 mm, to a tolerance of ±0.5 percent, and with more than 30 tape layers, in the most advanced processes. A typical single-chip package could contain up to seven tape layers and four screened dielectric layers. The thickness of the taped layer for tape with tungsten-filled vias range from 0.008 to 0.025 in (0.20 to 0.64 mm). Screened dielectric layers are typically 0.001 to 0.003 in (0.025 to 0.076 mm) in thickness.

The cofired laminated ceramic package technology is very effective for constructing complex packages with signal, ground, power, bonding, and sealing layers. Electrical characteristics such as line capacitance, propagation delay, impedance, and inductance can be customized for a particular application by design layout and specification of layer thicknesses. Design guidelines[27] are available from package vendors, and designers are encouraged to get the package vendor involved early in the design stage.

This technology, however, has three technical areas of weakness; hard to control tolerances caused by high shrinkage in processing, a high dielectric constant of 9.5, and a modest thermal conductivity of Al_2O_3. The tolerance problem makes it difficult to use edges as accurate references, and the high dielectric constant affects signal-line capacitive loading. The use of beryllia (BeO), AlN, or SiC, instead of the 90 to 92 percent Al_2O_3, will result in a greatly superior thermal performance and a significantly lower dielectric constant. Properties of advanced ceramics[28] are compared in Table 6.3 with those of Al_2O_3 and silicon. A typical DIP construction is illustrated in Fig. 6.29 and consists of three cofired laminated ceramic layers, with a chip cavity formed by the middle layer and a wire bond ledge formed by the top layer. The construction has two buried metallization layers and one on top. The bottom buried layer provides the metallization for the chip cavity base, while the top buried layer provides the signal fanout traces from the wire bond ledge to the external pins. The different levels can be interconnected by vias, as shown in Fig. 6.30. Top-layer metallization provides the seal ring for hermetic sealing. The seal ring may be tied to a ground plane through a via which may be located near the notch, as illustrated in Fig. 6.29. The same construction applies to a wide variety of package configurations (Fig. 6.31). Packages today come with brazed leads/pins for TH or SM, and leadless perimeter or array pads for SM. The package is delivered to the IC assembly manufacturing line with leads/pins, cavity, and bond sites finish-plated to customer specification, and ready for all subsequent level 1 and 2 assembly steps.

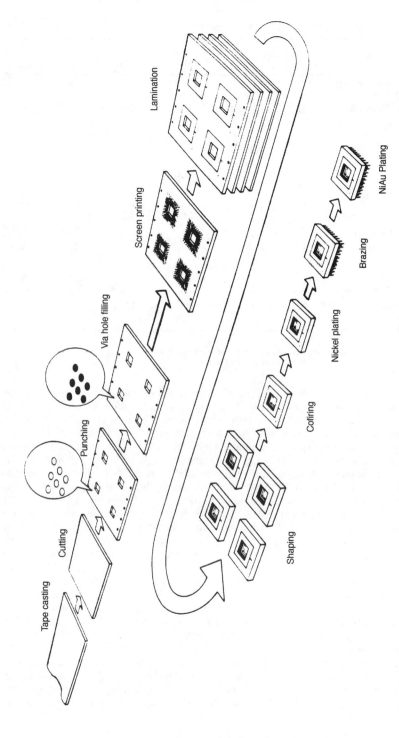

FIGURE 6.28 Process flow chart of laminated ceramic package technology *(After Kyocera.[27])*

6.29

TABLE 6.3 Performance of Advanced Ceramics

	98% Al$_2$O$_3$	AlN	BN	BeO	SiC	Si
Thermal conductivity						
Rank*	1	3	2	3	2	—
Value, W/m · °C	21	170–200	60	250	70	—
Coefficient of thermal expansion						
Rank	1	3	1	1	3	—
Value, 10^{-6}/°C	7.1	4.1	0.0 (75–500°C)	8.0	3.8	4.0
Dielectric constant						
Rank	1	2	3	2	1	—
Value at 10 MHz	10	8.8	4.0	7.0	40	—
Availability, rank	3	(2)†	2	2	(1)	—
Metallizing, rank	3	(2)	1	2	(2)	—
Cost						
Rank	3	(2)	2	1	(2)	—
Value, \$/in^2 (app.)	0.50	2	2–3	3.5–4	2	—
Figure of merit, sum of rankings	12	14	11	11	11	—

*Rank order: best = 3.
†Values in parentheses are predicted to improve over time.
Source: Data from the Carborundum Co. From Spitz.[28]

An assembly sequence flowchart for cofired laminated ceramic IC packages, with wire bond interconnection, is shown in Fig. 6.32. Chip bonding is done with either an AuSi eutectic, AuSn eutectic or polyimide system, to achieve a high yield and highly reliable mechanical and electrical connection. AuSi and AuSn preforms vary from 99.99 percent pure gold, with dopants to provide a reliable electrical back contact, to alloys of a variety of constituents, such as 80Au/20Sn for lower eutectic temperature bonding. Polymer die attachment in hermetic packages have also been used for many years. Filled and unfilled polyimides were found to withstand the hermetic assembly processing temperatures. Electrical interconnections are generally done using thermosonic ball-wedge gold wire bonding with wire diameters typically at 0.001 in (0.025 mm). Wire spans are usually limited to 0.160 in (4.0 mm) to meet shock, vibration, and acceleration requirements. Excessively long wire spans are avoided by going to a two tiered, or in some cases three tiered wire bond construction, as shown in Fig. 6.33. Final hermetic sealing is done with a 80Au/20Sn braze, in a dry nitrogen ambient, either in a belt furnace with a temperature profile peaking at about 350°C or in a controlled-ambient glove box containing a parallel-gap seam sealer. The seam seal brazing operation confines the heat locally to the seal ring area thus the package and IC are not significantly heated by the operation. The environment in the sealed package is predominantly nitrogen and the moisture content is typically at or below 1000 ppm H$_2$O by volume, which meets the military standard requirement for high-reliability applications.

Back-end processes are similar to those for pressed ceramic, that is, leak test, lead trim or lead trim and form,[23] and sometimes lead finish for SM packages. These packages usually come from the supplier with the external leads finished for level 2 assembly to the PWB. The most popular lead finishes available from package vendors range from electro-

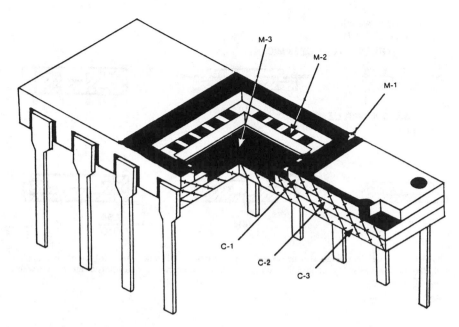

FIGURE 6.29 Isometric sectional view of cofired laminated ceramic technology in DIP configuration. *(After MetCeram.[29])*

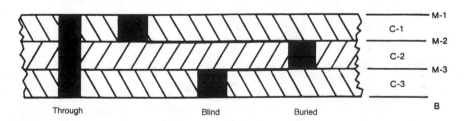

FIGURE 6.30 Isometric sectional view of cofired laminated ceramic technology in DIP configuration. *(After MetCeram.[29])*

plated gold for soldering and socketing to tin plating for soldering applications only. In between, there are several other combinations involving proprietary alloys of Au-Pd-Ni, Au-Ni, and electroless gold and nickel. Some of these finishes are not compatible with one or more of the assembly processes and may degrade solderability to the point where the SM solder reflow quality may be degraded. Leads with thick gold or alloys with significant amounts of gold will cause brittle solder joints that cannot be tolerated in SM. Fine-pitch leaded ceramic chip carriers (LDCCs), may require a solder dip coating to achieve high yields and reliability.

With the advent of high-density, high-pin-count, and high-performance SM package requirements, the ceramic ball grid array (CBGA) was developed to meet these demands. The CBGA is constructed in a similar fashion as other laminated ceramic packages (Fig. 6.31), but instead of perimeter leads and area array pins, the CBGA contains area array

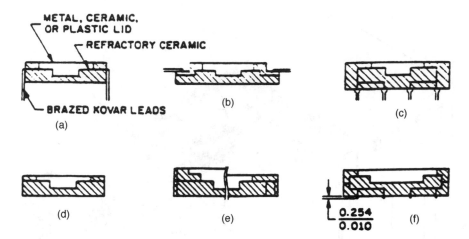

FIGURE 6.31 Cross-sectional sketches of several package types. (*a*) Side brazed, (*b*) top brazed, (*c*) pin grid array, (*d*) leadless with edge metallization, (*e*) leadless with via holes, (*f*) leadless array package with stand-offs. (*After Kyocera.*[27])

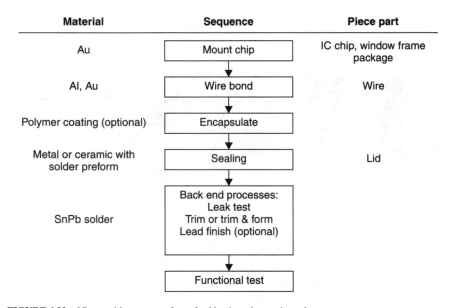

FIGURE 6.32 IC assembly sequence for cofired laminated ceramic packages.

pads for solder ball or column attachment (Fig. 6.34). The substrate consists of a single or several cofired laminated ceramic layers, with molybdenum metallization, and with vias interconnecting the circuitry on the top side to the solder pads on the bottom. The IC chip is interconnected to the alumina substrate by wire bonding or flip-chip techniques, as dis-

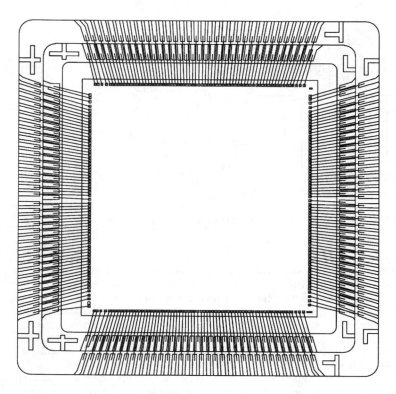

FIGURE 6.33   Tiered wire bond configuration.

cussed in Secs. 6.5.1.2 and 6.5.1.4. The substrate is lidded or capped, to protect the chip, and solder interconnects are attached to the underside to complete the assembly.

The CBGA to board interconnects consist[1] of high-temperature 90Pb/10Sn solder balls, which are attached to the substrate with eutectic 63Sn/37Pb solder. The solder balls are 0.035 in (0.89 mm) in diameter, for grid arrays on 0.050-in (1.27-mm) pitch and 0.025 in (0.64 mm) in diameter for a 0.040-in (1.0-mm) pitch.

To overcome the thermally induced fatigue strain on the solder balls, caused by the CTE mismatch between the alumina substrate and the glass-epoxy PWB, the solder balls are replaced with more compliant solder columns, thus the CBGA becomes a CCGA. The solder columns,[1] which are 0.020 in (0.50 mm) in diameter and 0.050 or 0.087 in (1.27 or 2.2 mm) high, can be either directly attached to the underside of the alumina substrate or a 63Sn/37Pb solder can be used as an interface for attachment. The different types of solder interconnects are shown in Fig. 6.34.

The CBGA and CCGA packages offer high reliability and high performance with pin counts exceeding 1000 and interconnect densities as small as 0.040-in (1.0-mm) pitch.

6.2.4 Laminated Plastic Technology

The laminated plastic packaging technology is a spin-off of the traditional PWB technology where bare chips are mounted directly to the substrate in a technique known as chip

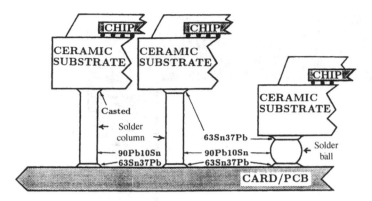

FIGURE 6.34 IBM's ceramic ball grid array (CBGA) showing different types of solder interconnects and attachments. *(After Lau.[1])*

on board (COB).[33] Early vintage laminated plastic packages used FR-4 epoxy substrates, which exhibited high contamination levels due to the flame retardants used. FR-4 epoxy substrates also suffered from a low glass-transition temperature (140°C), making them incompatible with many standard IC package assembly processes. Substrate materials such as high-temperature epoxies, polyimides, and triazines are being used today to overcome these deficiencies. These materials exhibit glass-transition temperatures in the range of 180 to 240°C, thus overcoming a major deficiency of the FR-4 material. In addition, these materials also have a much lower level of contamination than the FR-4 boards.

The substrate construction may contain either single-sided, double-sided, or multilayer circuitry, depending on the application needs, and can be provided with or without a cavity to house the chip. The exposed copper metallization is generally plated with nickel and gold to prevent oxidation and provide wire bondable pads. The most popular package types, using this technology, are the plastic pin grid array (PPGA) for TH mounting and the plastic ball grid array (PBGA) for SM. These packages are described in Sec. 6.2.4.1 and Sec. 6.2.4.2, respectively.

6.2.4.1 Plastic Pin Grid Array (PPGA) Packages. The PPGA, illustrated in Fig. 6.35, is a non-hermetic prefabricated package that houses the IC chip, providing interconnection to the motherboard and protection from hostile environments. The package consists of a square plastic PWB body with round pins press-fitted and solder reflowed to the underside. The pins can be configured in any desired footprint as long as the pin field is in a straight or staggered array of 0.100-in (2.54-mm) pitch and is restricted to the outside of the die attach and wire bond pad areas. The PPGA configuration results in the highest density of pins per package area and highest pin count of any other TH type package. PPGAs come in various forms (Fig. 6.36); chip down facing the motherboard or chip up mounted to the top of the package, double sided or multilayer construction and with or without a chip cavity. In addition, thermally enhanced PPGAs may also contain plated THs thermal vias under the chip or a copper slug to spread the heat dissipated by the chip attached to it.

The PPGA package is constructed[30] by first fabricating the plastic body from a rigid double sided or multilayer PWB by either the additive, semi-additive or subtractive process, using copper conductors. The PWB contains plated through holes, and depending on design complexity, may also have buried and semi-buried vias for electrical interconnection.

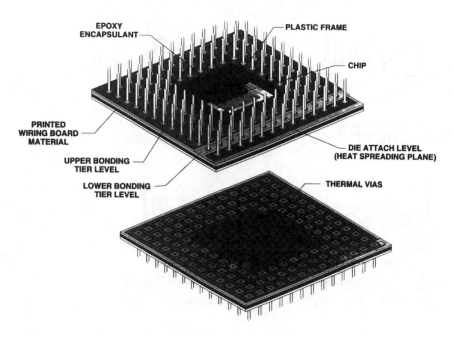

FIGURE 6.35   Multilayer PPGA with chip cavity facing down. *(After Cohn et al.[30] © 1990, IEPS.)*

Pins are made of phosphor bronze,[30] typically 0.018 in (0.46 mm) in diameter, instead of the more expensive Kovar or Alloy-42 used with ceramic packages. The solder plated (60Sn/40Pb or 90Pb/10Sn) pin is inserted through a gold or solder plated through-hole in the substrate and the connection is made by a press fit and solder reflow. A star or knurl is incorporated into the shank of the pin to facilitate a more reliable interconnection.

Solder reflow is accomplished by either dipping the entire length of the pin into a solder bath or laser spot heating the base of the pin. During solder dipping, the solder wicks up into the interface between the plated hole and the shank of the pin, and forms a fillet at the base of the pin. This process provides for a strong reliable contact, but results in a variable solder thickness along the length of the pin which may affect solderability, when tested per MIL-STD-883.

A relatively new solder reflow process uses a laser beam to spot-heat the base of individual pins, causing the solder in the hole-to-pin interface to reflow. Equipment is available that can laser spot-reflow at a rate of 25 pins per second. Sufficient solder must be available at the hole-to-pin interface, that when reflowed, will produce a strong interconnection. Since laser reflow heating only affects the base of the pin, the original solder plated thickness along the length of the pin remains unchanged.

The chip is assembled to the PPGA using most of the same assembly processes and techniques used for the plastic molded technology. The assembly flowchart is shown in Fig. 6.37. The chip is first attached using one of the polymer adhesive systems, such as silver-filled epoxies or polyimides, then wire bonded by either the thermosonic gold ball-wedge or the ultrasonic aluminum wedge-wedge process. The chip is then polymer-coated for environmental protection using a glob-top liquid epoxy, flow-coated silicone gel, or RTV rubber. A resin dam is desirable when encapsulating with flow coats to confine the materials to the IC and wire bond areas. A metal or nonmetal lid, attached with a polymer

**Double Sided,
Chip Down, W/O Cavity**

**4 Metallization Layers
Chip Down W/O Cavity**

**4 Metallization Layers
Chip Down, W/Cavity**

**6 Metallization Layers
Chip Down, W/Cavity**

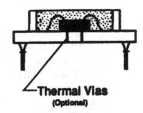

**Double Sided
Chip Up, W/O Cavity**

**4 Metallization Layers
Chip Up, W/O Cavity**

**4 Metallization Layers
Chip Up, W/Cavity**

**6 Metallization Layers
Chip Up, W/Cavity**

FIGURE 6.36   PPGA package configurations. *(After Cohn et al.[30] © 1990, IEPS.)*

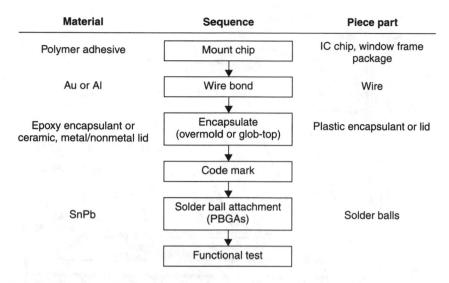

Material	Sequence	Piece part
Polymer adhesive	Mount chip	IC chip, window frame package
Au or Al	Wire bond	Wire
Epoxy encapsulant or ceramic, metal/nonmetal lid	Encapsulate (overmold or glob-top)	Plastic encapsulant or lid
	Code mark	
SnPb	Solder ball attachment (PBGAs)	Solder balls
	Functional test	

FIGURE 6.37   IC assembly sequence for laminated plastic technology packages (PPGAs and PBGAs).

adhesive, is optional and is provided mainly for physical protection when gels and RTV rubber systems are used.

The PPGA comes in either a low-cost or an enhanced, high-performance configuration depending on the application and customer's design objective. The low-cost version, shown in Figure 6.38, consists of a conventional 0.060-in (1.5-mm) thick copper-clad laminate processed as a single or double sided substrate with no cavity. Pins are press fitted and solder reflowed. The chip is encapsulated with either a glob-top liquid epoxy or covered with a silicone gel or RTV and lidded. This low-cost design, of single-layer bonding, is limited to approximately 300 I/Os. The limitations are due to a combination of an inability to fan out a greater number of package bonding pads in the central no-pin area and the constraints of minimum pad pitch design rules. Enlarging the central no-pin area to accommodate an expanded bond-pad fanout, would result in a larger, non-standard PPGA size, for a given pin count, and thus a costlier package. The 300 I/O PPGA limit, for single-layer bonding, is based on a 0.007-in (0.18-mm) package pad pitch and a pad limited chip size with 0.35 µm features.

A performance-driven device,[30] or one having more than 300 I/Os, will require a multilayer substrate with power and ground planes and multiple bonding tiers to yield the appropriate interconnection density within the allowable wire bond design rules. Liquid epoxy or a lid is used to encapsulate the chip. Pins may be inserted to provide either a cavity-up or a cavity-down configuration. A typical cross section through the chip cavity of such a laminated PPGA, showing the multilayers, two-tier bond pads, and thermal vias, is illustrated in Fig. 6.39. The thermal characteristics of a cavity down 223 PPGA, having thermal vias under the chip and heat spreading power and ground planes, are shown in Fig. 6.40. For IC chips dissipating approximately ≥3 W, and depending on the environmental conditions and the maximum T_j allowed, it may be necessary to mount the chip directly to a heat spreading copper slug of a cavity down package configuration, as shown in Fig. 6.41.

The low dielectric constant of PPGA substrate materials (in the range of 4 to 5) and the ability to provide thermal vias under the chip, or a heat spreading copper slug, result in

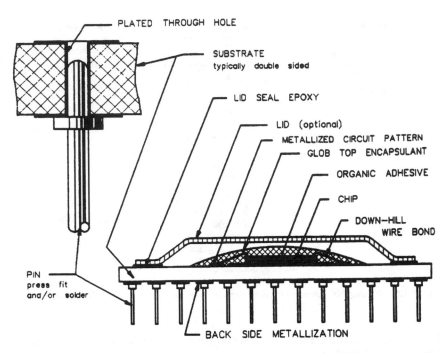

FIGURE 6.38 Cross section of low-cost PPGA in laminated plastic technology. Substrate is either single- or double-sided PWB with no cavity for IC. Chip is environmentally protected by "glob-top" epoxy encapsulant. Metal or plastic lid is optional for physical protection.

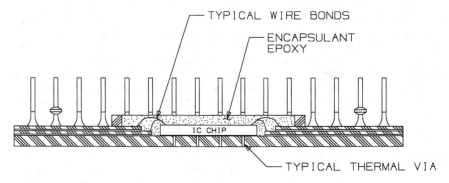

FIGURE 6.39 Cross section of a multilayer PPGA with chip cavity facing down.

equivalent or better electrical and thermal performance when compared to cofired laminated ceramic CPGAs using Al_2O_3.

6.2.4.2 Plastic Ball Grid Array (PBGA) Packages.   The PBGA package (Fig. 6.42) leverages on the PPGA technology, but replaces the pins with solder balls for surface

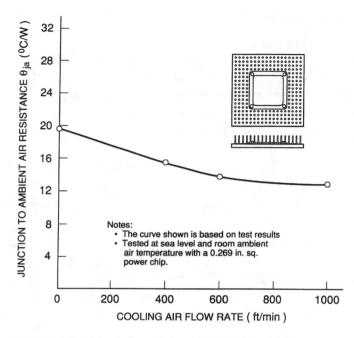

FIGURE 6.40   Thermal characteristics of a 223-I/O PPGA with cavity facing down and thermal vias under the chip.

mounting (SM). The PBGA has emerged as the SM package of choice for high-I/O-count devices. The PBGA offers many attractive features such as:

Increased interconnect density

Multilayer substrate capability

Improved electrical/thermal performance

Bump coplanarity less critical during SM

Self-centering during SM solder reflow

Elimination of fine pitch solder paste printing

Pad-to-pad shorting diminished

Low profile

Less than 50 ppm soldering yield loss

PBGAs were first introduced by Motorola in 1989 for low I/O devices and were then called *over molded pad array carriers* (OMPAC).[3] A PBGA package consists of a chip mounted and interconnected to a square double sided or multilayer PWB substrate with solder balls attached to the underside. Vias interconnect the top surface signal traces to the respective solder pads under the package. The solder pad openings are usually defined by the solder mask and are approximately 0.025 in (0.63 mm) in diameter. The exposed copper metallization is either electroplated or electroless plated with nickel and gold to prevent oxidation and facilitate wire bonding to the bond pads and solder ball attachment to the solder pads. After die-attachment and wire bonding, the assembly is overmolded using

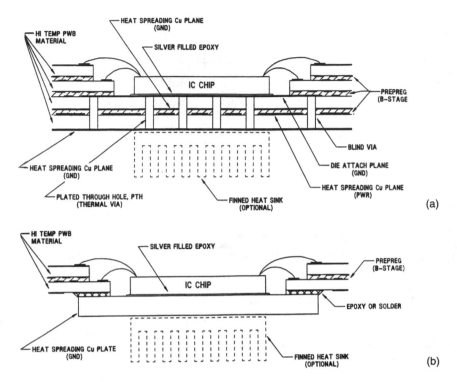

FIGURE 6.41 (a) PPGA package thermal enhancements using heat spreading Cu planes and thermal vias and (b) PPGA package thermal enhancements using a heat spreading Cu plate. *(After Cohn et al.[30] © 1990, IEPS.)*

a transfer molding process. Another encapsulation method is to "glob top" the chip and its bond wires with a liquid dispensed epoxy which is confined within a chip cavity. The 63Sn/37Pb or 62Sn/36Pb/2Ag solder balls are 0.030 in (0.76 mm) in diameter and are arranged in an area array on either 0.060 or 0.050-in (1.50- or 1.27-mm) pitch. PBGAs with solder balls on 0.040-in (1.0-mm) pitch are also being designed but availability is limited because their infrastructure is not well established. The fine pitch PBGAs require smaller diameter solder balls and finer conductor line widths and spaces.

Where device heat dissipation is critical, the PBGA design may contain plated THs "thermal vias" under the chip to provide a direct path for the heat to flow to the solder balls and dissipate into the mother board. PBGAs contain many materials with various coefficients of expansion, thermal conductivities and elastic behavior. Typical PBGA materials are shown in Table 6.4. Maintaining adequate thermal-mechanical performance, mechanical tolerances (e.g., flatness, coplanarity), moisture sensitivity and overall device reliability with these materials is a major challenge to the industry.

The cross section of a typical low-cost PBGA having a double sided substrate with the chip mounted to the top surface, wire bonded, overmolded and solder balls attached is shown in Fig. 6.43. The substrate also contains thermal vias under the chip. In a high-performance PBGA configuration the substrate is multilayered with power and ground planes and typically contains a chip down cavity with single or double bonding tiers and a heat

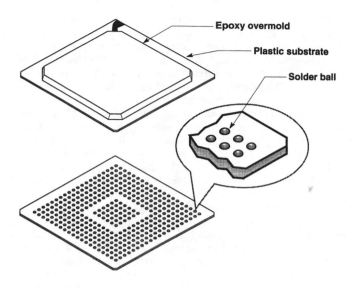

FIGURE 6.42 Plastic ball grid array (PBGA) package (overmolded type).

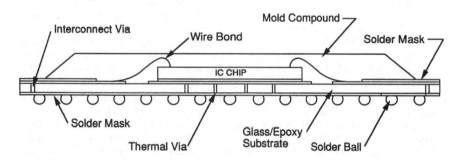

FIGURE 6.43 Cross section through a PBGA with a double-sided metallization substrate. *(After Cohn et al.,[3] © 1995, IEEE.)*

TABLE 6.4 Typical PBGA Materials

Item	Specification
Substrate material	High glass transition temperature (Tg > 160°C)
	Epoxy/glass laminate, e.g. bismaleimide triazine (BT)
Metallization	0.5 oz min. Cu on outside layers and 1 oz min. Cu on internal layers
Plating	20 μinch min. gold over 50 μinch min. nickel
Die attachment	Silver-filled epoxy
Bonding wire	1.5 mils dia. gold wire
Encapsulant	Epoxy overmold or dispensed epoxy
Balls	Eutectic solder, e.g., 62Sn/36Pb/2Ag or 63Sn/37Pb

Source: from Cohn et al.,[3] © 1995, IEEE.

spreading copper slug to which the chip is directly attached, Fig. 6.44. The PBGA substrates are usually fabricated by a PWB vendor, in singulated or strip (multi-site) form to customer's design specifications, and then shipped to subcontract vendors for assembly. Wire bonding is the leading chip interconnection process used today for PBGAs. Flip-chip bonding, although currently not widely used, is expected to become a leading contender in the near future (see Sec. 6.5.1.4). The following is a typical PBGA assembly sequence, processed in strip form and using wire bond interconnection.[3]

Silver-filled epoxy is transfer-printed onto the die attach area, the chip is placed over the epoxy and the assembly is cured. The chip is thermosonically ball-wedge gold wire bonded, and the strip sites are overmolded. After overmolding, the sites are visually inspected, code marked and solder balls are attached to the underside of the individual sites. The individual PBGA devices are separated from the strip by routing or punching and inspected, cleaned, electrically tested, burned-in when required, electrically retested, placed in bakeable shipping trays, baked/bagged, and shipped to customer.

A fully assembled 225 PBGA package, with a double sided BT substrate was qualified by AT&T[3] to the requirements listed in Table 6.5. Prior to qualification testing, the assembled BGAs were first burned-in at 125°C for 24 hours.

The PBGA, being a relatively new package technology, has a number of drawbacks that undoubtedly will be resolved with time and better understanding of the issues. Some of the concerns expressed by the industry are:

1. *Moisture sensitivity.* Many of the materials that make up the present PBGA configurations are susceptible to moisture absorption. As a result, the PBGAs, while being stored in a factory environment awaiting assembly to the PWB, are prone to absorb moisture that may cause internal delamination (popcorning) and cracking during solder reflow. Thus, some semiconductor manufacturers ship the PBGAs pre-baked and in vacuum sealed bags to be opened prior to board assembly. Manufacturers continue to improve the material properties and designs of PBGAs to where it is now possible to achieve a JEDEC Level 3 condition, defined as a floor life of 192 hours in a 30°C/60 percent RH environment before re-baking is required to solder reflow.

2. *Solder joint reliability.* The number of solder joints on a circuit board has increased dramatically over the last few years. Thus, it is critical that the effects on solder joint integrity are understood and solutions are implemented to improve their reliability. This is particularly important when dealing with PBGAs which have solder joints far less compliant than leads. Cycle to failure tests and finite element modeling have been used to predict the life cycle of PBGA solder joints in a given application. Most of the solder joint failures in PBGAs have been attributed to thermal fatigue induced by tem-

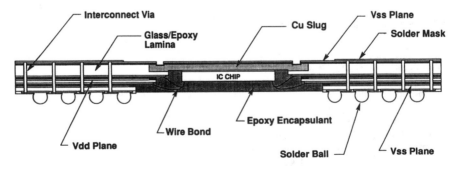

FIGURE 6.44 Cross section through a thermally enhanced PBGA. *(After Cohn et al.,[3] © 1995, IEEE.)*

TABLE 6.5 PBGA Device Qualification

TEST	TEST CONDITIONS	QUAL RESULTS 225BGA-2ML OVERMOLDED (Failures / Sample Size)
HTOB	150°C, 4.1V / 6.2V, 1000 hrs. Mil-Std-883, Mtd. 1005	0 / 105
CLASS (Assembly Simulation Test) + Moisture Sensitivity Test	Bake at 125°C, 24 hrs 30°C / 60% RH, 48 hrs Reflow Water Soluble Flux 2 Reflows EC7, Loncoterge Clean Test followed by THB	0 / 126
THB (Preceeded by CLASS)	85°C / 85% RH 3.3V / 5.5V, 1000 hrs AT&T A90AL0135	0 / 126
Steam Bomb	121°C, 2 Atm, 96 hrs (192 hrs for information) No bias	0 / 102
Temp. Cycling	-65°C to + 150°C 22 min / cy, air to air 300 cy extended to 1000 cy for information Mil-Std-883,Mtd. 1010	0 / 105 (1000 cy)
Thermal Shock	-55°C to +125°C Liquid to Liquid, 100 cy, extended to 300 cy, for information Mil-Std-883, Mtd. 1011, Cond. B	0 / 20 (300 cy)
Moisture Resistance	Mil-Std-883, Mtd. 1004 (w/o bias)	0 / 38
Salt Atmosphere	Mil-Std-883, Mtd. 1009 24 hrs	0 / 15
Solvent Resistance	Mil-Std-883, Mtd. 2015	Not tested (Laser Scribed)
Flammability	UL-94-V-0 + Oxygen Index 28% min.	By Ref. to Vendor's Data
Die Shear Strength	Mil-Std-883, Mtd. 2019	By Ref. to Vendor's Data
Wire Bond Strength	Mil-Std-883, Mtd. 2011	By Ref. to Vendor's Data
Physical Dimensions	Mil-Std-883, Mtd. 2016	0 / 15
X-Ray	Mil-Std-883, Mtd. 2012	0 / 5
ESD (HBM + CDM)	Mil-Std-883, Mtd. 3015.7 + AT&T, X-19435	0 / 9
Latch Up	JEDEC Std. #17	0 / 9

Source: from Cohn et al.,[3] © 1995, IEEE.

TABLE 6.6   Package Attributes and Trade-Offs versus Package Technologies

Attributes and trade-offs	Molded plastic	Pressed ceramic	Laminated ceramic	Laminated plastic
Hermeticity	No	Yes	Yes	No
Power/ground distribution capability	Doable custom design	Not feasible	Standard	Standard
High-density routing capability	Custom TAB design	Not feasible	Standard	Standard
Power dissipation capabilities	Fair	Fair	Excellent	Excellent
Thermal design enhancements	Heat spreaders	New materials	New materials	Thermal vias and heat spreaders
Capability for handling bond-pad-limited chips	Yes	Yes	Yes	Yes
Automated assembly	Excellent	Feasible	Feasible	Feasible
Max. I/O capability	375	196	600+	600+
Cost rank order (lowest = 1)	1	2	3	2+

perature cycling, and caused by the CTE mismatch between the PBGA structure and its PWB attachment surface. Other failure mechanisms may be creep rupture, and solder joint/pad interface cracking due to intermetallics. Solder joint integrity is affected by factors such as: joint location relative to the chip, configuration, intermetallic structure, substrate thickness, number of temperature cycles and range, stress concentration from the sharp edges of the solder mask defining the pads, etc.

The PBGA and its board interconnection are to be designed for reliability at the outset in order to achieve the required service life for a given application.

3. *Solder joint inspection/repairability.*   Other PBGA technology issues of concern to end users are visual inspection of all solder joints and repairability.

Visual inspection of all solder joints is a difficult and sometimes costly process. Inspection techniques have included the use of a stereo microscope with fiber optic lenses and special side mirrors or automated cross-sectional x-ray inspection.

Repairability of BGA assemblies due to defective solder joints is not a simple process. Because BGA solder joints are hidden under the package, the only way to repair a solder joint is to remove the package from the PWB and reball the device. Rework equipment is available that provides component pickup and reflow capabilities.

6.2.5   Comparison of Technologies

A comparison, based on the attributes and tradeoffs of the four contending IC packaging technologies described in Secs. 6.2.1 through 6.2.4, is shown in Table 6.6. Pressed and laminated ceramic technologies have hermetic sealing capability, which provides the best environmental protection for the IC. High-reliability performance is achieved if the sealing operation is done properly, that is, if it provides a leak-tight seal and a cavity free of contaminants, and moisture is controlled to an acceptable level. Nonhermetic molded and laminated plastic technologies can provide equal reliability performance for most applications.

In determining relative electrical performance, the capacitance between adjacent leads is a property that is influenced by the dielectric constant of the material in direct contact with the leads. Table 6.7 lists the materials and dielectric constants for each package technology. Plastic technologies have the potential for best performance due to their lower

TABLE 6.7   Physical Properties versus Package Technologies

Physical properties	Molded plastic	Pressed ceramic	Laminated ceramic	Laminated plastic
Dielectric constant at 1 MHz				
Plastic[38]	4.4			
Seal glass[24]		12–33		
Ceramic[27]		9.5–10	9.5–10	
BT resin				3.8–4.6
Thermal conductivity, W/(cm • °C)				
Plastic[4]	0.007			
Ceramic[4]		0.17	0.17	
BT resin				0.003
CTE,14, ppm/°C				
Chip mount surfaces				
Copper LF	17			
Alloy-42 LF	5.7			
Ceramic		6.2	6.2	
BT resin				14–18
Package body	18–30	6.2	6.2	15
IC chip	2.5	2.5	2.5	2.5

dielectric constants. The high dielectric constant inherent in the laminated alumina ceramic technology can be avoided by the use of thin-film-metal-polymer multilayer structures as used in MCMs (Chapter 10). MCM structures make laminated ceramic competitive with laminated plastic. Electrical performance is further optimized by the use of power and ground planes and high-density routing within the package. Both laminated plastic and ceramic technologies have this as a standard design capability.

Thermal performance is influenced by the thermal conductivity of the basic package materials (Table 6.7) and the use of thermal design enhancements (Table 6.6). The thermal conductivity of alumina ceramic materials is an order of magnitude better than that of plastic materials. Materials such as AlN and SiC, with higher thermal conductivities and heat spreading copper-tungsten slugs have further improved the thermal performance of ceramic packages. Design enhancements, such as thermal vias under the chip and heat spreading copper slugs in laminated plastic, or heat spreaders in molded plastic, have closed the gap and are competing in thermal performance with laminated ceramics.

Mechanical performance is related to internal stresses developed within the package as a result of the mismatches of the coefficients of thermal expansion (CTE) of the different materials used in package construction. CTEs of the silicon IC and of the substrate material to which the chip is mounted, for each technology, are listed in Table 6.7. The mismatches are displayed in Fig. 6.22. The best match is between silicon ICs and ceramic-type packages. Further complications are encountered in technologies where the chip is not decoupled from other materials, such as molding compounds in molded plastic and "glob-top" or molding compounds in laminated plastic technologies. Chips in the pressed ceramic and cofired laminated ceramic technologies are usually decoupled from other materials; however, their assembly processes, which include high-temperature exposure during chip mounting and hermetic sealing, can present problems, particularly for very large chips, greater than 0.500 in (12.7 mm). Other material properties, such as Young's modulus and Poisson's ratio, also affect the magnitude of stresses in packages where the chips are not decoupled. The finite-element analysis is a useful CAD tool for modeling stress conditions in new package designs before committing them to development and manufacturing stages.

The cost of packaging is best addressed by using a rank-order system rather than an absolute cost. There are many factors and assumptions that enter into a cost analysis of packaging, and these are beyond the scope of this chapter. Molded plastic technology packages, such as plastic quad flat paks (QFPs) which are assembled on lead frames, and plastic ball grid arrays (PBGAs) which are assembled on single-layer, multi-site strip substrates, rank low. Whereas, the multilayer laminated technology packages such as CPGAs or PPGAs, which are singulated and need to be assembled in "boats," rank higher.

The ability of a package to accommodate bond-pad-limited chips is an important factor influencing the cost of the chip. For example, the ability to accommodate bond-pad-limited chips in molded plastic and pressed ceramic packages is sometimes limited by the fabrication process for the lead frame. The lead-frame feature size is dictated by its thickness. The use of thinner lead frames and TAB tape, such as in fine pitch high-I/O packages, has extended the capability for the plastic molded technology. The laminated ceramic and plastic technologies solve the bond-pad-limited chip problem for high-I/O devices, by using tiered wedge bond sites, thus bringing the wire bonds closer to the chip and minimizing the span length. The ability to multilayer and to provide high-density routing has made tiered bonding feasible. A tradeoff exists in the cost of packaging bond-pad-limited chips. Cost savings realized by shrinking the chip size are sometimes offset by the higher packaging cost. Each application must be evaluated thoroughly to determine the bottom-line cost for both chip and package.

Table 6.8 compares most of the attributes discussed in this section by rank order, providing a simple format to evaluate trade-offs between the competing package technologies.

6.3 PACKAGE FAMILIES

IC package designs have proliferated over the years into a wide variety of shapes, sizes, number of I/O pins, lead arrangements, and geometries in response to circuit- and system-level demands. This proliferation of package types becomes very uneconomical to end users; for example, original equipment manufacturers (OEMs) using electrically equivalent components from different suppliers must use several PWB designs if the components

TABLE 6.8 Attribute Rank Order versus Package Technologies

Physical properties	Molded plastic	Pressed ceramic	Laminated ceramic	Laminated plastic
Moisture sensitivity	2*	1	1	2
Lead capacitance	1	3	2	1
Power dissipation	3	2	1	‡
CTE mismatch, chip to chip-mount surface	†	1	1	2
Automated assembly capability	1	1	2	**
Capability of accommodating bond-pad-limited chips	1	1	1	1
Cost	1	2	3	2+

* Rank order: best = 1

† Rank order depends on lead-frame material used; 1— —Alloy-42; 3—copper alloy.

‡ Rank order depends on thermal design of substrate; 1+ —design with thermal vias and heat-spreading planes, 2+ —design without thermal vias.

**Rank order depends on package type; 1– —PBGAs (overmolded), 2 —PPGAs.

are in different package types. Assembled packages need to be handled economically and with special precautions, such as electrostatic protection, through a series of back-end processes that include lead trim and form, code marking, electrical testing, electrical burn-in when required, shipping, and subsequent assembly to the next level of packaging. These activities need major industry support in the development of specialized custom equipment and tooling; for example, magazines, sticks, trays, or tape for handling through the back end, test sockets for production testing, sockets and boards for device burn-in, test handlers and electrical contacts for production testing, and automatic placement and insertion equipment for component board assembly. The industry has been very active in the standardization of IC package outlines, which has resulted in footprint and handling compatibility of components from different vendors and the focusing of the above-listed support activities on fewer package types, hence optimizing the overall economics of electronic packaging.

Packages with similar traits are being grouped into families by lead arrangement. These standard outlines then control all dimensions, tolerances, and all other information necessary to define, manufacture, and assemble the package. In the United States, the EIA-JEDEC JC-11 committee is the dominant body developing new package outlines. Over the past several years, JEDEC has been working closely with EIAJ to reduce a further proliferation of outlines and to focus similar outlines from each organization into one worldwide standard outline.

Table 6.9 lists the package families available in each of the package technologies described in Sec. 6.2. They are also broken down by the mounting method to second-level board assembly, that is, TH versus SM. Package families are typically grouped and defined by lead arrangement, such as in-line, perimeter, or area array. The listed packages are EIA-JEDEC, or EIAJ registered or standard outlines. Registration is the first step to having a package outline accepted as an industry standard. To become a standard, a registered package must be generally used in the industry and then approved by a majority of JEDEC JC-11 member companies in an official balloting process. A complete listing of all registered and standard outlines is available in EIA Publication 95.[35] Table 6.10 lists many of these IC packages and also notes the range of typical lead counts that have been standardized in the industry. The next two sections review these package families.

TABLE 6.9 Package Families by Packaging Technology

	Package		Technology	
	Molded Plastic	Laminated Plastic	Laminated Ceramic	Pressed Ceramic
Sealing Method	Molded	Molded/Encapsulated	Metal Lid Seal	Glass Lid Seal
Surface Mount	SOIC	PBGA	CQFP	CERQUAD
	SOJ		LLCC	CQFP
	PLCC		LDCC	
	BQFP		SMPGA	
	MQFP		CBGA	
	TQFP			
	TSOP			
	TAB			
	TapePak*			
Through Hole Mount	DIP	PPGA	DIP	DIP
	SIP	LLCC	SIP	CERDIP
	ZIP		PGA	
	QUIP			

*Registered trademark of National Semiconductor Corporation.

TABLE 6.10 IC Package Families and JEDEC Outlines

Level 2 assembly technology	Packaging technology	Package type	JECED outline	Typical I/Os	Comments
Surface mount	Molded plastic	SOIC	MS-012	8/14/16	0.150" body width
			MS-013	14/16/18/20/24/28	0.300" body width
			MO-046	14/16/20	0.200" body width
		SOJ	MS-023	24/26/28/32	0.300" WB, 0.050" pitch
		MQFP	MS-022		1.0/0.8/0.65 mm pitch, 3.2 mm footprint
				36/44/52	$10 \times 10 \times 2.0$ mm body
				52/64/80	$14 \times 14 \times 2.0$ mm body
				64/80/100	$20 \times 14 \times 2.7$ mm body
				76/88/112	$10 \times 10 \times 3.7$ mm body
				120/128/144/160	$28 \times 28 \times 3.4$ mm body
				184	$32 \times 32 \times 3.4$ mm body
				232	$40 \times 40 \times 3.8$ mm body
			MO-108		1.0/0.8/0.65 mm pitch, 3.2 mm footprint
				44/52	10×10 mm body, 2.0/2.8 mm thick
				52/64/80	14×14 mm body, 2.0/2.8 mm thick
				88	16×16 mm body, 2.0/2.7 mm thick
				64/80/100	$20 \times 14 \times 2.8$ mm body
				112	$20 \times 20 \times 2.7$ mm body
				136	$24 \times 24 \times 3.3$ mm body
				128/144/160	$28 \times 28 \times 3.42$ mm body
				184	$32 \times 32 \times 3.42$ mm body
				232	$40 \times 40 \times 3.8$ mm body
			MO-112		1.0/0.8/0.65 mm pitch, 3.9 mm footprint
				44/52	10×10 mm body, 2.0/2.8 mm thick
				52/64/80	14×14 mm body, 2.0/2.8 mm thick
				64/80/100	$14 \times 20 \times 2.8$ mm body
				120/128/144/160	$28 \times 28 \times 3.42$ mm body
				184	$32 \times 32 \times 3.42$ mm body
				232	$40 \times 40 \times 3.8$ mm body
			MO-143		0.5/0.4 mm pitch, 2.6 mm footprint
				64/80	$10 \times 10 \times 2.0$ mm body
				80/100	12×12 mm body, 2.0/2.7 mm thick
				100	14×14 mm body, 2.0/2.7 mm thick
				128	$14 \times 20 \times 2.7$ mm body
				144/176	20×20 mm body, 2.0/2.7 mm thick
				176/216	$24 \times 24 \times 3.4$ mm body
				208/256	$28 \times 28 \times 3.4$ mm body
				240/296	$32 \times 32 \times 3.4$ mm body
				272/336	$36 \times 36 \times 3.4$ mm body
				304/376	$40 \times 40 \times 3.8$ mm body
		BQFP	MO-069	52/68/84/100/132/164/196/244	0.025" pitch, gull wing
			MO-071	40/52/68/84/132/220	0.020" pitch, gull wing
			MO-086	44/52/68/84/100/132/148/164	0.025" pitch, low profile, gull wing
			MO-089	28/44/52/68/84	0.050" pitch, gull wing
		SOP	MO-155	5	$3.6 \times 4.4 \times 2.5$ mm body, 1.27 mm pitch
		SSOP	MO-152		0.65/0.50/0.40 mm pitch
				8/16/20/24/28/36/48	4.4 mm body width
				24/28/32/36/40/48/56/64/80	6.1 mm body width
				28/32/36/40/48/52/56/64	8.0 mm body width
		TSOP II	MS-025	26/28/44	7.62 mm body family
			MS-024	28/32/36/40/44/50/70	10.16 mm body family
			MO-135	32/36/40/50/54/62/70	12.70 mm body family

TABLE 6.10 IC Package Families and JEDEC Outlines *(Continued)*

Level 2 assembly technology	Packaging technology	Package type	JECED outline	Typical I/Os	Comments
		TSSOP	MO-153		0.65/0.50/0.40 mm pitch
				8/16/20/24/28/36/48	4.4 mm body width
				24/28/32/36/40/48/56/64/80	6.1 mm body width
				28/32/36/40/48/52/56/64	8.0 mm body width
		TQFP	MS-026		1.0 and 1.4 mm thick, 2.0 mm footprint
				32/40	5 × 5 mm body, 0.5/0.4 pitch
				32/40/48/64	7 × 7 mm body, 0.8/0.65/0.5/0.4 mm pitch
				36/44/52/64/80	10 × 10 mm body, 1.0/0.8/0.65/0.5/0.4 mm pitch
				44/52/64/80/100	12 × 14 mm body, 1.0/0.8/0.65/0.5/0.4 mm pitch
				52/64/80/100/120	14 × 14 mm body, 1.0/0.8/0.65/0.5/0.4 mm pitch
				100/128	14 × 20 mm body, /0.65/0.5 mm pitch
				112/144/176	20 × 20 mm body, 0.65/0.5/0.4 mm pitch
				176/216	24 × 24 mm body, 0.5/0.4 mm pitch
				160/208/256	28 × 28 mm body, 0.5/0.4 mm pitch
Surface mount	Molded plastic	PLCC	MS-018	20/28/44/52/68/84	Square, 0.50" pitch
			MS-016	18/22/28/32	Rectangular, 1.27 mm pitch
		TAB	US-001	Pitch/film format/test pads	Body size (mm)
				0.5/35 super/196	14 × 14, 16 × 16, 18 × 18, 20 × 20
				0.5/48 super/260	16 × 16, 20 × 20, 24 × 24, 26 × 26, 28 × 28
				0.5/70 super/436	24 × 24, 28 × 28, 32 × 32, 36 × 36, 40 × 40
				0.4/35 super/244	14 × 14, 16 × 16, 18 × 18, 20 × 20
				0.4/48 super/324	16 × 16, 20 × 20, 24 × 24, 26 × 26, 28 × 28
				0.4/70 super/548	24 × 24, 28 × 28, 32 × 32, 36 × 36, 40 × 40
				0.3/35 super/324	14 × 14, 16 × 16, 18 × 18, 20 × 20
				0.3/48 super/436	16 × 16, 20 × 20, 24 × 24, 26 × 26, 28 × 28
				0.3/70 super/724	24 × 24, 28 × 28, 32 × 32, 36 × 36, 40 × 40
				0.25/35 super/388	14 × 14, 16 × 16, 18 × 18, 20 × 20
				0.25/48 super/532	16 × 16, 20 × 20, 24 × 24, 26 × 26, 28 × 28
				0.25/70 super/876	24 × 24, 28 × 28, 32 × 32, 36 × 36, 40 × 40
				0.5/48 wide/260	16 × 16, 20 × 20, 24 × 24, 26 × 26, 28 × 28
				0.5/70 wide/436	24 × 24, 28 × 28, 32 × 32, 36 × 36, 40 × 40
				0.4/48 wide/324	16 × 16, 20 × 20, 24 × 24, 26 × 26, 28 × 28
				0.4/70 wide/548	24 × 24, 28 × 28, 32 × 32, 36 × 36, 40 × 40
				0.3/48 wide/436	16 × 16, 20 × 20, 24 × 24, 26 × 26, 28 × 28
				0.3/70 wide/724	24 × 24, 28 × 28, 32 × 32, 36 × 36, 40 × 40
				0.25/48 wide/532	16 × 16, 20 × 20, 24 × 24, 26 × 26, 28 × 28
				0.25/70 wide/876	24 × 24, 28 × 28, 32 × 32, 36 × 36, 40 × 40
		TapePAK	MO-094	Pitch (mm)	Ring size (mm)
				0.65/0.5/0.4/0.3	36 × 36, 46 × 46, 56 × 56, 66 × 66, 76 × 76, 86 × 86
			MO-109	0.65/0.5	16 × 16, 26 × 26
	Laminated plastic	PBGA	MO-163	119/153	Rectangular, 1.27 mm pitch
			MO-151	Solder ball matrix (SQ):	1.00 mm ball pitch, body size (mm SQ):
				6/8/10/12/14/16/18	7.0/9.0/11.0/13.0/15.0/17.0/19.0
				20/22/24/26/28/30/32/34	21.0/23.0/25.0/27.0/29.0/31.0/33.0/35.0
				37/39/32/44/47/49	37.5/40.0/42.50/45.0/47.50/50.0
				Solder ball matrix (SQ):	1.27 mm ball pitch, body size (mm SQ):
				5/7/8/9/10/11/13/15	7.0/9.0/11.0/13.0/15.0/17.0/19.0
				16/18/19/21/22/24/26/27	21.0/23.0/25.0/27.0/29.0/31.0/33.0/35.0
				29/31/33/35/37/39	37.5/40.0/42.50/45.0/47.50/50.0
				Solder ball matrix (SQ):	1.27 mm ball pitch, body size (mm SQ):
				4/6/7/8/10/11/12	7.0/9.0/11.0/13.0/15.0/17.0/19.0

TABLE 6.10 IC Package Families and JEDEC Outlines *(Continued)*

Level 2 assembly technology	Packaging technology	Package type	JECED outline	Typical I/Os	Comments
				14/15/16/18/19/20/22/23	21.0/23.0/25.0/27.0/29.0/31.0/33.0/35.0
				25/26/28/30/31/33	37.5/40.0/42.50/45.0/47.50/50.0
		TBGA	MO-149	Solder ball matrix (SQ):	1.00 mm ball pitch, body size (mm SQ):
				14/16/18/20/22/24/26	15.0/17.0/19.0/21.0/23.0/25.0/27.0
				28/30/32/34/37/39/42	29.0/31.0/33.0/35.0/37.0/40.0/42.5
				44/47/49	45.0/47.5/50.0
				Solder ball matrix (SQ):	1.27 mm ball pitch, body size (mm SQ):
				11/13/15/16/18/19/21	15.0/17.0/19.0/21.0/23.0/25.0/27.0
				24/26/27/29/31/33	29.0/31.0/33.0/35.0/37.0/40.0/42.5
				35/37/39	45.0/47.5/50.0
				Solder ball matrix (SQ):	1.50 mm ball pitch, body size (mm SQ):
				10/11/12/14/15/16/18	15.0/17.0/19.0/21.0/23.0/25.0/27.0
				19/20/22/23/25/26/28	29.0/31.0/33.0/35.0/37.0/40.0/42.5
				30/31/33	45.0/47.5/50.0
	Laminated ceramic	CQFP	MO-125	132	0.025" pitch, gull wing
			MO-100	256	0.020" pitch, gull wing
			MO-134	224/256/288/320/351	0.50 mm pitch, ceramic tie bar
		LDCC	MO-044	68/84	0.050" pitch
			MS-007	28/44/52/68	0.050" pitch, J-leaded, type A
			MS-008	28/44/52/68	0.050" pitch, type B
		LLCC	MS-014	16/20/24/32/40/48/64/84/96	Single layer
			MO-041	4/6/18/20/22/24/28/32	0.050" pitch, rectangular
			MS-009	16/20/24/32/40/48/64/84/96	0.040" pitch
			MS-002	28/44/52/68	0.050" pitch, type A
			MS-003	28/44/52/68	0.050" pitch, type B
		CBGA	MO-157	Max. balls	Ceramic, rectangular body
				168/224/336	1.50 mm pitch
				224/304/475	1.27 mm pitch
				340/480/744	1.00 mm pitch
			MO-156	Max. balls	Ceramic square body
				100/144/196/256/289/324/400	1.00 mm pitch
				484/576/676/784/900/1024	
				64/100/121/169/196/225/256	1.27 mm pitch
				324/361/441/484/576/625/676	
				49/64/100/121/144/196/225	1.50 mm pitch
	Pressed ceramic	CERQUAD			
			MO-082	68/84/100/132/164/196	0.025" pitch, gull wing
			MO-084	36/44/52/68/84/100	0.050" pitch, gull wing
			MO-087	28/44/52/68	0.050" pitch, J-leaded
Through hole	Molded plastic	DIP	MS-001	8/14/16/18/20/22/24/28	0.300" row spacing
			MS-010	22/24/28/32	0.400" row spacing
			MS-011	24/28/40/48	0.600" row spacing
			MO-016	36/50/52/64	0.900" row spacing
		SDIP	MS-019	14/16/18/20/22/24	0.070" pitch, 0.300" row spacing
			MS-020	40/42/48/52	0.070" pitch, 0.600" row spacing
			MS-021	64	0.070" pitch, 0.750" row spacing
		SIP	MO-035	11	0.100" pitch
			MO-068	22/24/30/40	0.100" pitch
		ZIP	MO-054	16	0.050" lead center, 0.100" row spacing
			MO-072	19/20/24/28/40	0.050" lead center, 0.100" row spacing
			MO-080	23/26	0.050" lead center, 0.100" row spacing
		QUIP	MO-029	16	5.08/10.16 mm row spacing

TABLE 6.10 IC Package Families and JEDEC Outlines (Continued)

Level 2 assembly technology	Packaging technology	Package type	JECED outline	Typical I/Os	Comments
			MO-030	42	19.05/23.50 mm row spacing
			MO-031	14	5.08/10/16 mm row spacing
			MO-033	42/52/64	17.78/22.86 mm row spacing
			MO-034	42/52/64	0.750/0.925" row spacing
	Laminated plastic	PPGA	MO-083	81/100/121/144/169/196/225 256/289/324/361/400	0.100" pitch
	Laminated ceramic	DIP	MS-015	8/14/16/18/20/22/24/28	0.300" row spacing
				20/22/24/28	0.400" row spacing
				24/28/32/36/40/48/52	0.600" row spacing
				50/64	0.900" row spacing
		SIP	MO-055	16/11	0.100" pitch
		CPGA	MO-066	81/100/121/144/169/196/225 256/289/324/361/400	0.100" pitch, small outline
			MO-067	81/100/121/144/169/196/225 421/481/545/613/685/761 841/1013/1301	0.100" pitch, large outline
			MO-128	145/181/221/265/313/365/421 481/545/613/685/761	0.100 pitch, staggered
	Pressed ceramic	CERDIP	MO-036	8/14/16/18	0.300" row spacing
			MO-058	24/28	0.300" row spacing
			MO-037	22	0.400" row spacing
			MO-103	24/28/32/40	0.600" row spacing
			MO-038	24/28/40	0.600" row spacing
			MO-039	50	0.900" row spacing

6.3.1 Mature Package Families

One of the oldest IC package families still in use today is the DIP family illustrated in Fig. 6.2. A DIP is a through-hole mount package with its leads on a 0.100-in (2.54-mm) pitch and is generally available in three package technologies; molded plastic, pressed ceramic and laminated ceramic. DIPs can accommodate devices with as few as 8 and up to 64 I/Os. The family is defined by width, or lead row spacing, ranging from 0.300 in (7.62 mm) for the most narrow to 0.900 in (22.86 mm) for the widest. A definition of the plastic DIP family by lead row spacing and pin count is given in Table 6.10. The 0.300-in DIP family was the first to be standardized by JEDEC and is identified in Table 6.10 by the JEDEC designation MS-001. Other DIP families that have been standardized are the 0.400-in wide MS-010 and the 0.600-in wide MS-011. MS-001 covers pin counts ranging from 8 to 28, MS-010 covers pin counts from 22 to 32, and MS-011 covers pin counts from 24 to 48. In the MS-001 family, three pin counts at 22, 24 and 28 leads are referred to in the industry as "skinny DIPs" because the bodies were narrowed from the original width of 0.400 in for 22 leads and 0.600 in for 24 and 28 leads (MS-011) to 0.300 in. Another variation, known as "shrink" DIPs (SDIPs), have the same body width as the standard family but with the lead pitch reduced from 0.100 to 0.070 in to provide higher pin-count capability for the same size package as outlined in MS-019, MS-020 and MS-021.

Similar DIP families are defined in the pressed ceramic technology (CERDIP), and in the laminated ceramic technology. Their outlines are registered and available to the user in Publication 95.[35] Pressed ceramic, both CERDIP and window-frame types, come in four body widths, with pin counts and footprints that are identical to those shown in Table 6.10

for the plastic DIP family. Laminated ceramic DIPs, however, come in a wide variety of configurations identified by the method of lead attachment to the ceramic body. Laminated ceramic DIPs can be side-brazed, top-brazed, or bottom-brazed, as shown in Fig. 6.31. Side-brazed DIPs are footprint compatible with the plastic and pressed ceramic DIP families. Top- and bottom-brazed DIPs are generally supplied by a vendor as a flat-pack with leads formed to the specification of the end user. The leads may or may not be footprint compatible with the side-brazed variety.

DIPs are limited in pin count to 64 leads by the physical size of the package, which approaches 3.0 in (76.20 mm) in length at 64 I/Os. At this level the package piece part becomes very expensive, difficult to handle, generally begins to degrade the electrical performance of the device, and uses up a lot of expensive PWB real estate. Area array packages, such as PGAs, were developed to extend the pin-count capability for TH-mount-type packages with a significant saving in PWB area. Figure 6.11 compares the physical size of a PGA package with pins on an area array of 0.100 in (2.54 mm) pitch to the DIP at an in-line 0.100 in (2.54 mm) pitch. The 0.100-in (2.54-mm) pitch PGA family is available in both laminated plastic and ceramic technologies and can provide pin-count capability above 1000 pins. JEDEC registered outline, MO-066, covers a range of CPGA pin arrays from 9 by 9 (81 max. pin count) to 20 by 20 array (400 max. pin count) in both cavity-up and cavity-down configurations. The largest registered CPGA outline, at 1300 pin count, has an overall body size of 2.66 in (67.56 mm) with a staggered pin arrangement.

Other in-line families that are popular today are the single-in-line package (SIP) family, the zigzag-in-line package (ZIP) family, and the quad-in-line package (QUIP) family. All were introduced to meet special product needs, for example, SIPs for memory chips and memory modules. These package families have pins typically on 0.100 in (2.54 mm) pitch and are available to 40 pin counts for ZIPs and to 64 for QUIPs. A QUIP has four in-line rows of leads, with the rows generally spaced 0.100 in (2.54 mm) apart, and the leads in each row staggered with respect to each other to maximize hole-to-hole spacing on the PWB. This package option effectively reduces the package length by one-half over the equivalent DIP of the same pin count. The wide range of IC packages in use worldwide is illustrated in Fig. 6.45.

The first SM packages were introduced in the early 1980s to achieve higher board density and better electrical performance. Initially, industry acceptance of SM did not advance as rapidly as expected. However, the acceptance rate in the 1990s has accelerated and the demand for SM packages have exceeded TH packages. Thus, the availability of new SM packages has expanded greatly into a very formidable list, as seen in Tables 6.9 and 6.10.

Mature SM families today are those introduced in the early 1980s, such as the small-outlined (SO) and PLCC types in molded plastic, CERQUAD in pressed ceramic, and leadless (LLCCs) and leaded (LDCCs) chip carriers in laminated ceramic. These packages have leads on 0.050-in (1.27-mm) pitch instead of the 0.100-in (2.54-mm) used in the TH families. SO is an in-line type defined by body width, or lead row spacing, similar to the way the DIP family is defined. MS-012, and MS-046, both narrow-body SOs, and MS-013, a wide bodied SO, are gull wing (GW) families. The SO package with J-leads (SOJs), a JEDEC registered outline, has the same body width as MS-013, that is, it uses the same plastic molding die, but has a J-formed lead for surface mounting. PLCCs, CERQUADs, LLCCs and LDCCs are perimeter-type packages with the leads on all four sides of the package. The lead geometries vary with the technology used. PLCCs use J-shaped leads designed to provide the highest possible density on the PWB. CERQUAD packages generally come from the vendor as a leaded flat-pack and the device manufacturer or end user does the final lead forming (GW or J) prior to SM. The LLCC is generally restricted to surface mounting on substrates with similar thermal expansion coefficients. This limitation is overcome by using an LDCC. One way to achieve this is by converting an LLCC to

Type	Package	Lead pitch
DIP		• 2.54 mm (100 mil)
Shrink-DIP		• 1.78 mm (70 mil)
Skinny-DIP		• 2.54 mm (100 mil) • Width ½ size
Zigzag in-line		• 2.54 mm (100 mil) • Two rows
Pin grid array		• 2.54 mm (100 mil)
Ball grid array		• 1.5 mm (60 mil) • 1.27 mm (50 mil) • 1.0 mm (40 mil)
Small-outline	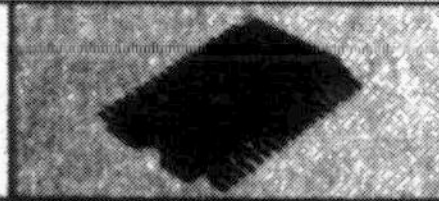	• 1.27 mm (50 mil) • Leads on two sides
Flatpack	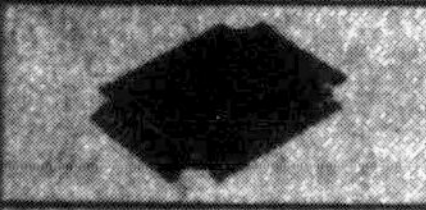	• 1.0 mm • 0.8 mm • 0.5 mm • 0.4 mm • 0.8 mm • Leads on four sides
Leadless chip carrier		• 1.27 mm (50 mil) • 1.00 mm (40 mil)
Leaded chip carrier		• 1.27 mm (50 mil) • J-leaded
Small-outline J-leaded	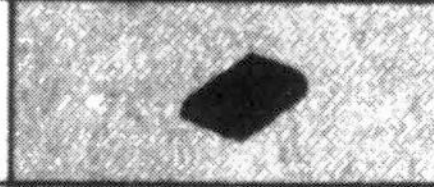	• 1.27 mm (50 mil) • Two rows
Tape automated bonding	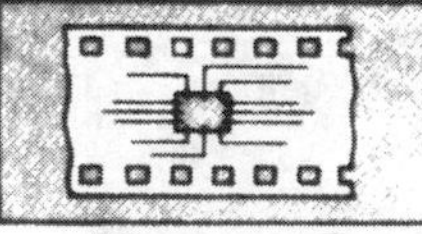	• Thin package

● Plastic   ○ Ceramic

FIGURE 6.45 Commercially available package types. *(After Cohen,[36] and modified.)*

an LDCC by means of one of the commercially available soldered clip-lead techniques. Today LDCCs can be purchased either as leaded flat-packs with the leads formed to the appropriate shape (GW or J) by the device manufacturer or end user, or as finished packages with the leads in final form ready for SM. Examples of SO, PLCC, and CC packages are shown in Fig. 6.3.

The drive for higher pin counts and higher density has been very evident in the development of fine-pitch SM packages. Since the pin count of a perimeter package is a function of package size and lead pitch, as illustrated in Fig. 6.11, to achieve a high pin count with PLCCs, at a 0.050-in (1.27 mm) lead pitch, requires very large packages. Difficulties in molding large plastic packages and maintaining coplanarity of all leads, necessary to achieve high board assembly yields, limited the practical PLCC size to 84 I/Os and 1.190 in (30.23 mm) lead to lead overall.

A new package family has evolved to fill the need for high pin count, namely the quad flat-pack (QFP) with gull wing leads on all four sides, Figs. 6.46 and 6.47. The QFP has a lead pitch that varies from 0.40 to 0.16 in (1.0 to 0.40 mm) and a maximum pin count of 376 leads. Two QFP families have evolved; the EIAJ MQFP version, shown in Fig. 6.46, and the bumpered JEDEC BQFP, shown in Fig. 6.47. The MQFP family is characterized by a fixed body size with variable pitch leads. A variety of pin counts can be manufactured from a single body size using lead frames of various external lead pitches from 0.040 in (1.0 mm) down to 0.016 in (0.40 mm) as shown in Table 6.10 for MS-022, MO-108, MO-112 and MO-143. The BQFP family is characterized by variable pitch leads [0.020 in (0.50 mm), 0.025 in (0.60 mm) and 0.500 in (1.27 mm) pitch] and variable body sizes, with a variety of pin counts. A different molding die is required for each package size, which range from 0.450 in at 52 I/Os to 1.650 in at 244 I/Os, as shown in Table 6.10 for MO-069, MO-071, MO-086 and MO-089. Major advantages of the JEDEC QFP over the EIAJ version are that the bumpers provide for safe handling in tubes, trays, carriers, and tape and reel, and the generous stand-off provides for easy removal of flux from under the package after board assembly. The major advantage of the EIAJ QFP over the JEDEC is the multiple lead counts available per package compared with only one pin count per package size for the JEDEC. EIAJ packages are more difficult to handle without bending leads and the low stand-off, on higher pin counts, causes high stress on solder joints due to

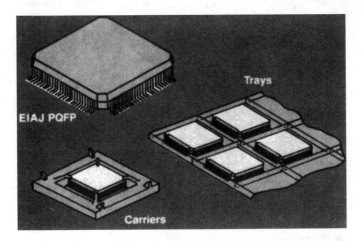

FIGURE 6.46   EIAJ PQFP packaging and handling options. *(After Braden.[37])*

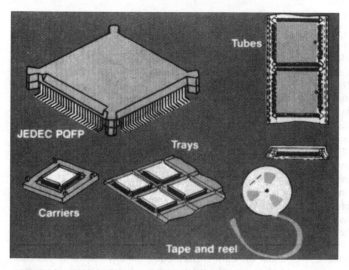

FIGURE 6.47   JEDEC BQFP packaging and handling options. *(After Braden.[37])*

board warp. EIAJ's head start by several years provided valuable experience in high-volume production, which enabled EIAJ to become the industry de-facto standard before the JEDEC version left the ground. Both versions are in use today, but the EIAJ version is gaining in popularity. The EIAJ MQFP family provides pin counts up to 376 I/Os, while the JEDEC BQFP seems to be leveling off at 244 I/Os.

6.3.2 Emerging Package Families

A new package family was developed in response to industry demands for high-pin-count SM packages that are shorter, smaller, thinner and lighter. The new package family consists of thin quad flat-packs (TQFPs) which have been JEDEC registered as MS-026 outlines. Whereas the standard MQFP body thickness varies from 2.0 mm to 3.8 mm, depending on the package body OD, the TQFP body thicknesses are 1.0 mm and 1.4 mm. The TQFP packages vary in lead count from the smallest, 32 I/Os with a body size of 5.0 × 5.0 × 1.0 mm and a 0.80 mm lead pitch, to the largest with a maximum lead count of 256 I/Os in a body size of 28.0 × 28.0 × 1.40 mm, and a 0.40 mm pitch.

As the demand for packages with high I/O count and high performance increases, the industry is turning to BGAs as a favorable alternative to the fine pitch QFPs which are difficult to handle. A great number of BGA outlines have been JEDEC registered (Table 6.10) in both laminated ceramic and plastic technologies. The arrays range from 4 × 4 to 49 × 49 in sizes of 7.0 × 7.0 to 50 × 50 mm. The industry trend for overmolded type PBGAs, is to standardize the PWB substrates in two basic sizes: 27 × 27 × 0.36 mm thick for double sided metallization or 0.56 mm thick for multilayer construction and 35 × 35 × 0.56 mm thick for double-sided and multilayer construction. Each substrate size (27 × 27 mm and 35 × 35 mm) has a given plastic overmold configuration that does not change with different artwork or pin count. This enables the package designer to take advantage of existing mold cavities. The substrates may be either selected from standard open tooled artwork configurations, available from vendor stock, or are custom designed to fit special applications. General purpose substrate designs help reduce assembly turn-around time and inventory logistics problems. In addition, open tool designs eliminate tooling charges and take advantage of possible volume discounts. For PBGAs to ever approach cost parity with PQFPs, the substrates will have to be treated as leadframes.

A new class of IC packages is emerging,[65] with packages that are only slightly larger than the chip itself. The new technology, called chip scale packaging (CSP), comes closest to direct chip attachment, but without the responsibilities usually imposed on the board assembler for handling, attachment, wire bonding and encapsulation of known good die (KGD). Using CSP the above tasks are shifted back to the semiconductor manufacturer, in that the device comes fully tested, burned-in and ready for SM assembly to the PWB by conventional means, like any other IC package. Several chip scale package configurations were developed to provide a lower-cost, high-I/O, high-density compact package. One such package was developed by Tessera Inc, called the micro-BGA (μBGA[TM]).[65] It combines a unique chip interconnection with BGA technology that results in a highly miniaturized package (Fig. 6.48). The μBGA[TM] consists of a copper/polyimide flex-tape substrate that contains single or double sided interconnect circuitry and an array of eutectic solder balls on the back side. Copper/gold ribbon leads extend from the edge of the flex-tape and are thermosonically bonded to the chip's perimeter pads. A silicone elastomer layer between the flex-tape and the chip surface cushions any impact to the chip during socketing or board assembly. The compliant layer also reduces any stresses caused by the CTE mismatch between the silicon and the solder ball joints when surface mounted to a PWB. An elastomeric compound encapsulates the ribbon leads and supports them during package handling.

6.4 PACKAGE DESIGN CONSIDERATIONS

6.4.1 Electrical Design

Electrical performance at the IC package level is of general interest for silicon devices.[38] The ever-increasing speed of today's circuits and their potentially reduced noise margins require great care and consideration in IC package design. Several electrical performance criteria are of interest, namely, low ground resistance (minimum power-supply voltage

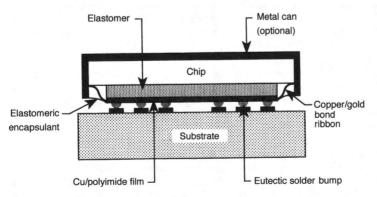

FIGURE 6.48   Typical cross section through a μBGA[TM]. *(After Tessera.[65])*

drop), short signal leads (minimum self-inductance), minimum power-supply spiking due to signal lines switching simultaneously, short-paralleled signal runs (minimum mutual inductance and crosstalk), short-length signal runs near a ground plane (minimum capacitive loading), and the maximum use of matched impedances to avoid signal reflection. These criteria are, of course, not all mutually independent. They may be related through simple geometric variables, such as conductor cross section and length, dielectric thickness, and dielectric constant of the packaging body. These problems are usually handled with transmission-line theory in PWBs, where the lengths of circuit paths are more obviously a cause for concern. A wealth of papers and books cover this topic, and all of the techniques described are generally applicable at the package level.[39–41]

Package inductances,[1] as a function of pin counts, for QFPs and BGAs are shown in Fig. 6.49. The higher inductance of the QFP is attributed to the larger perimeter lead package, and thus longer traces, when compared to the high interconnect density area array of BGAs. Similarly, package propagation delays[14] are also affected by trace lengths, as shown in Fig. 6.50 for different package types.

The most important practical electrical design problem in IC packages is noise reduction. Basically, when a line switches, the voltage induced in the ground line is given by

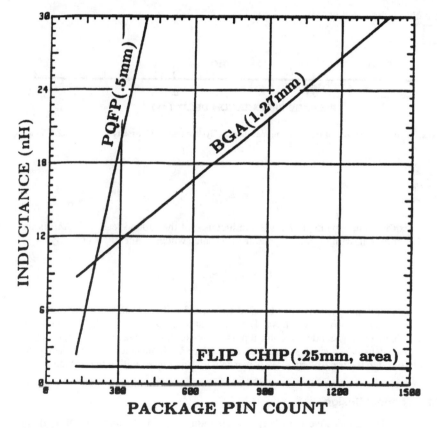

FIGURE 6.49 BGA and PQFP comparison: performance (inductance). *(After Lau.[1])*

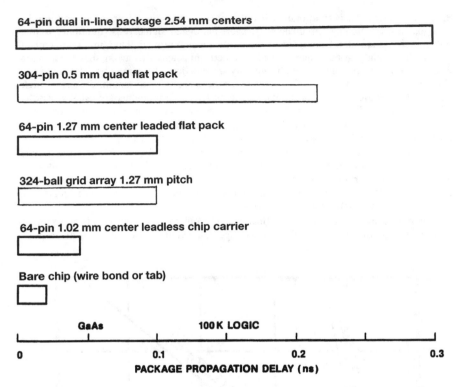

FIGURE 6.50 Comparison of maximum propagation delays for typical IC packages. *(After ICE.[14] and modified.)*

$$V_i = L_g \frac{di}{dt}$$

where V_i is the induced voltage, L_g is the inductance of the ground lead, and di/dt is the derivative of current with respect to time. If j lines are switching, then V_i is given by

$$V_i = L_g \sum_j \frac{di}{dt} j$$

To reduce V_i, multiple grounds must be used to lower L_g. If m ground leads are used, the total inductance is approximately L_g/m. The practical result is that up to 25 percent of the leads may have to be grounded to control noise. This high percentage has a large impact on packaging density and creates an incentive to reduce L_g. Inductance can be reduced significantly through the use of large-area power and ground planes within the package.[30]

6.4.2 Thermal Design

The thermal design objective is to keep the operating junction temperature of a silicon chip low enough to prevent triggering the temperature-activated failure mechanisms for the particular application in which the chip is used. Only in the simplest possible applications can this objective be met, considering the packaged silicon device alone. Usually the packaged device environment must be established for most of the following variables: PWB temperature, total power dissipation on the board, local-neighbor power dissipation, degree of forced-air cooling, space usable by the package both laterally and vertically (between boards), conductivity of the PWB, and ideal performance of the isolated package. The following paragraphs discuss the thermal modeling of a single chip in a package.

In a simplistic heat transfer model of a packaged chip the heat is transferred from the chip to the surface of the package by conduction and from the package surface to the ambient by convection and radiation. Typically, the temperature difference between the case and ambient is small, and hence radiation can be neglected. This model also neglects conduction heat transfer out of the package terminals, which can become significant, particularly with high-I/O VLSI packages. The overall thermal resistance in this model can be considered as the sum of two thermal resistance components θ_{jc} and θ_{ca} defined as

$$\theta_{ja} = \theta_{jc} + \theta_{ca}(°C/W)$$

$$\theta_{jc} = \frac{T_j - T_c}{P}$$

$$\theta_{ca} = \frac{T_c - T_a}{P}$$

where θ_{ja} = junction to ambient thermal resistance

θ_{jc} = junction to case thermal resistance

θ_{ca} = case to a ambient thermal resistance

T_j = average die or junction temperature, °C

T_c = average case temperature, °C

T_a = ambient temperature, °C

P = power, W

θ_{jc}, the conductive thermal resistance for low-power applications, is relatively insensitive to the ambient and is mainly a function of package materials and geometry. With the higher power requirements of most VLSI applications, one must also consider the temperature dependence of materials selected in design. For example, the thermal conductivity of typical materials used in ceramic packaging is temperature-dependent, as shown[42] in Fig. 6.51.

θ_{ca} depends on package geometry, package orientation in the application, and the conditions of the ambient in the operating environment (free or forced convection heat transfer). In actual practice, most designers use experimental methods to determine the thermal resistance of a package configuration. Thermal characterizations are performed with a thermal chip assembled in the given package, which is then mounted into any of the configurations shown in Fig. 6.52. The package is monitored for input power and junction, case and ambient temperatures. The results show that packaged devices with a preferred heat-flow path (most ceramic packages) may be thermally characterized independent of external heat sinking, while those with no preferred heat-flow path (most plastic packages) should be thermally characterized with a specific mounting arrangement and a convection heat-flow environment.[43,44] The use of simple modeling supported by experimentally derived values of thermal resistance is probably adequate for first-order determinations, but should be followed by a more sophisticated program using specially designed test chips and computer modeling.

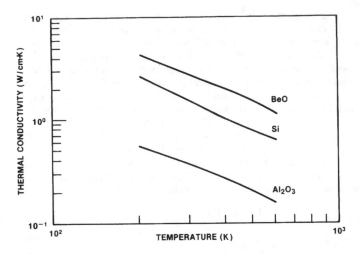

FIGURE 6.51   Influence of temperature on thermal conductivity of IC-related materials. *(After Oettinger.[42])*

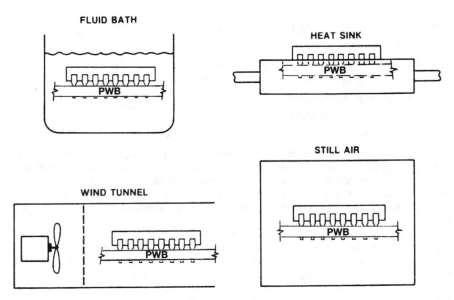

FIGURE 6.52   Typical package-mounting arrangements for thermal characterization. *(After Oettinger[42] and modified.)*

6.4.3 Thermomechanical Design

As the size of an IC chip increases, it presents new challenges to the package designer. Ideally, one prefers to use materials in package construction that are matched in physical properties, in particular the coefficient of thermal expansion. In the real world, however, we see that the materials currently in use have coefficients that vary by orders of magnitude, as shown in Fig. 6.22. The designer could achieve better matches by making engineering tradeoffs that do not degrade the packaged-device yields, performance, or reliability. For example, if a SSI/MSI chip in a plastic molded package is directly attached to an Alloy-42 lead frame, which has a good CTE match to silicon, the thermal stresses will be minimized. But, because of the poor thermal conductivity of Alloy-42, the lead frame may have an impact on the thermal performance of high power dissipating chips. In applications using VLSI, the thermal design requirements are much more demanding and the same Alloy-42 lead frame could bring on higher thermal stresses, resulting in chip cracking or "pop-off" or degradation of performance and reliability due to higher junction temperatures during normal operating conditions.[45] As a result, more than 80 percent of the current plastic molded packages utilize copper lead frames for applications where high thermal performance is required. The thermal stresses caused by the CTE mismatch between the copper lead frame, silicon and molding compound are minimized by using low-stress molding compound and die-attach epoxies.

In the preceding discussion only base materials at the chip-attach interface were considered. In the actual design the chip is attached using a material such as solder, braze, or polymer adhesive and a process step introducing heat to either melt the solder or braze material, or to thermally cure the polymer adhesive. Physical properties of these materials are shown in Table 6.11.[28,46,47] The CTE and stiffness, which is a function of the tensile modulus E, significantly affect the thermal stress and the chip-attach interface as shown in Fig. 6.53. In addition, the thickness of the joint also impacts the thermal stress in the interface as shown in Fig. 6.54. Therefore, thermal stresses increase with increasing E and decreasing thickness. Eutectic chip bonding has a very high E and a thin bond line, hence limiting its usefulness with very large chips to packages where the attachment is made to a substrate with a closely matched coefficient of thermal expansions.

TABLE 6.11   Elastic Moduli and Thermal Conductivities of Materials Used in Packaging*

Material	Tensile elastic modulus E, GPa	Thermal conductivity k, W/(cm • °C)	Application
Bismaleimide triazine (BT) resin	0.21	0.003	Substrate
90–99% Al_2O_3	262	0.17	Substrate
Beryllia (BeO)	345	2.18	Substrate
Common Cu alloys	119	2.64	Lead frame
Ni-Fe alloys (Alloy-42)	147	0.15	Lead frame
Au-20%Sn	59.2	0.57	Die bond adhesive and lid sealant
Au-3%Si	83.0	0.27	Die bond adhesive
63Sn/37Pb	14.5	0.36	Solder balls
Silicon	13.0	0.084	Electronic circuit
Au	78	3.45	Wire metallurgy
Ag-loaded epoxy	3.5	0.016	Die-bond adhesive
Epoxy (fused silica filler)	13.8	0.007	Molding compound

* *Source:* From Refs. 28, 46, 47

Another major mismatch in expansion coefficients of materials used in package construction is between the epoxy molding compound and the chip. This mismatch does not

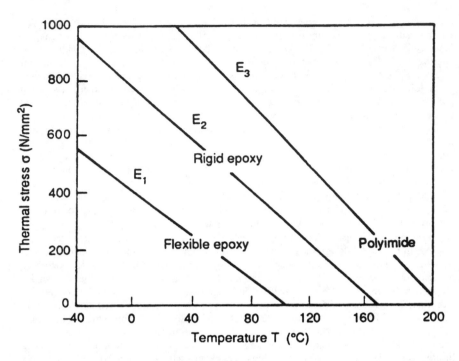

FIGURE 6.53   Thermal stress predicted for flexible epoxy (tensile modulus E_1), rigid epoxy (E_2), and polyimide (E_3) as a function of temperature where $E_3 > E_2 > E_1$. *(After Bolger and Mooney,*[45] *© 1984, IEEE.)*

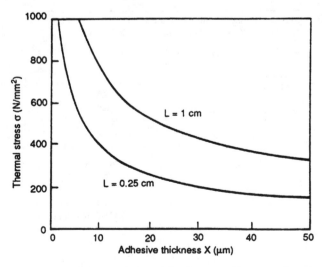

FIGURE 6.54   Effect of chip length L and adhesive thickness on thermal stress. *(After Bolger and Mooney,*[45] *© 1984, IEEE.)*

exist in the pressed and laminated ceramic package technology since no package materials come in direct contact with the IC surface. The chip itself is constructed of materials with over an order of magnitude difference in the CTE (Fig. 6.22), thus compounding the thermal-stress problem in plastic packages.

Analysts have developed sophisticated stress models and introduced computer aids, such as finite-element analysis, to increase their ability to characterize the thermal stresses and deformations throughout the entire plastic package, with the goal to optimize the reliability of the package. The molded plastic IC package is basically a composite structure consisting of a silicon chip, metal lead frame, and plastic molding compound, as shown in Fig. 6.55. A two-dimensional finite-element model of the cross section is shown in Fig. 6.56. The model is divided into small elements, interconnected at a discrete number of nodal points, creating a finite-element mesh. Material properties are assumed to be linearly related to the thermal load. Material interfaces are assumed to have perfect adhesion to each other. The model further assumes an initial isothermal condition of 170°C, corresponding to the molding temperature. The thermal stresses are then calculated at the desired isothermal temperature, which is typically set at –65°C, the lowest temperature expected during service or in transit. Figure 6.57 is a plot of the minimum principal stresses (compression) in both the plastic molded body and the silicon chip. The stress contours represent isobars of equal stress within the package and chip. Severe stress concentration is evident at the lower edge of the chip, which could lead to chip cracking. The use of finite-element analysis has become a powerful tool for quickly predicting the impact of design and process changes with respect to stress-related failure mechanisms. Such an analysis was not possible with classical analytical techniques.

6.4.4 Physical Design—Chip Design Rules

The establishment of, and adherence to good chip design rules is absolutely essential to achieving high yields in IC package assembly. Rules must be generated for the particular package type used and must be compatible with the assembly equipment. These decisions should be made early, preferably before the chip layout is started and definitely before the layout is completed. As the I/O count grows and the active device feature size shrinks, the

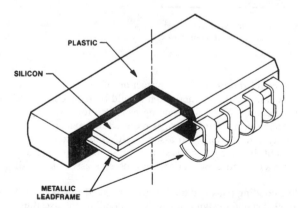

FIGURE 6.55 Plastic package is basically a composite structure consisting of a silicon chip, metal lead frame, and plastic molding compound. *(After Striny.[4])*

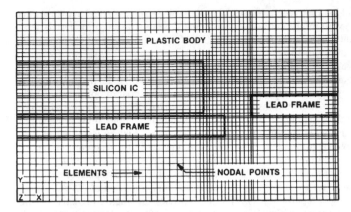

FIGURE 6.56 Plane stress analysis is assumed for two-dimensional model. Model is divided into many small elements interconnected at a discrete number of nodal points. Because of symmetry, model covers only half the IC package. (*After Striny.[4]*)

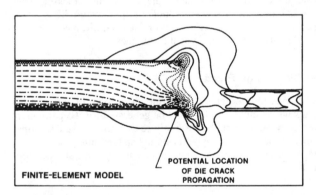

FIGURE 6.57 Finite-element analysis plot showing contours of constant stress (minimum principal stress in compression) in both plastic body and silicon chip. (*After Striny.[4]*)

space required for interconnection could represent a major fraction of the chip area. To avoid this problem, the effective bonding pad sizes and pitch should be reduced. Figure 6.58 shows the consequences of increasing I/O on the bonding pad pitch for several chip sizes.[48] Not only must the pitch decrease, but the tolerances required to produce quality bonds must also decrease. Reliability and the ability to assemble IC chips automatically are affected directly by the chip layout. In the case of VLSI, the chip-to-package interface design must assure high assembly yields and avoid reliability problems associated with poor design rules.

An alternative to smaller in-line bonding pads and pitch is the staggered-row bond pad configuration, as shown in Fig. 6.59. Typical design rules for wire bonding are also illustrated in Fig. 6.59. Design rules recommend wire bond pad sizes and pitch, clearance between pads and the edge of the separated chip, and clearances to internal adjacent metal conductors or other critical design features. Also specified are the maximum allowable

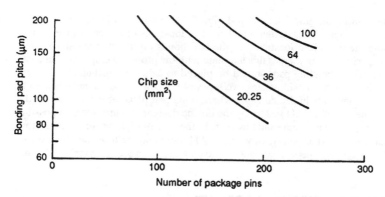

FIGURE 6.58 Bonding pad pitch versus chip led count for several chip sizes. For each number of package pins, pitch is maximum that can be accommodated on that size chip. *(After Otsuka and Usami,[48] © 1981, IEEE.)*

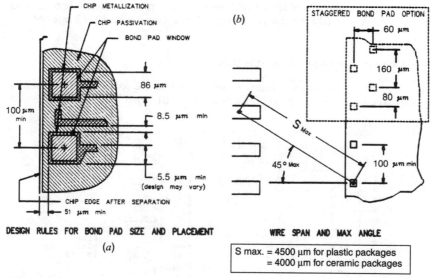

FIGURE 6.59 (*a*) Typical chip design rules for bond pad size and placement. Bond pads should provide 86×86 μm open area and be on at lest 100 μm center-to-center spacing. Adjacent active circuitry should be at least 8.5 μm away from bond pad metallization and fillets should be used at junction of conductors to bond pad. Maximum allowable wire span depends on package type used, with typical values noted. Wire angle, determined by chip-package interface, should not exceed 45° with respect to normal as shown. (*b*) Arrangement of staggered bonding pads resulting in lower pitch than that attainable with single line of pads. Further restrictions on wire angle may be required.

wire bond span length (chip to lead frame or substrate bond site) and the maximum allowable angle of the wire from the normal direction off the chip. This rule attempts to minimize wire sweep in plastic molding, which could lead to potential edge short circuits to silicon and short circuits to adjacent conductors on the lead frames or metallized conductors of the package.

Design rules can be integrated into chip design CAD[4] systems. Figure 6.60 shows a typical package wire bond CAD design template. The locations of all lead frame wedge-bond targets and the optimized locations for the ball bonds on the chip are shown. The chip designer superimposes or merges this template with the proposed chip layout and can immediately see where bonding pads should be placed to achieve a manufacturable and reliable design. Figure 6.60 also illustrates two types of CAD wire-span templates used to verify design rules and to help decide cases where bond pad placement falls outside the optimum zone. The use of a CAD system by the device designer assures the best possible design, particularly for high volume and low cost. In the case of high-performance custom designs, the same design rules apply. However, CAD tools may be too restrictive; therefore, chip and package designers must work as a team to assure a manufacturable design.

6.5 PACKAGE ASSEMBLY PROCESSES

This section covers the level 1 assembly processes used with each of the four package technologies. A detailed listing of all major package families, showing the preferred method of chip attachment and level 1 interconnection, is presented in Table 6.12. Also shown are lead form, lead arrangement, and type of mounting used at level 2 assembly.

6.5.1 Chip-to-Package Interconnection

There are four chip-to-package interconnection options in use today,[49] as illustrated in Fig. 6.61; wire bond (WB), flip-chip (FC), beam lead (BL), and TAB. For wire bonded

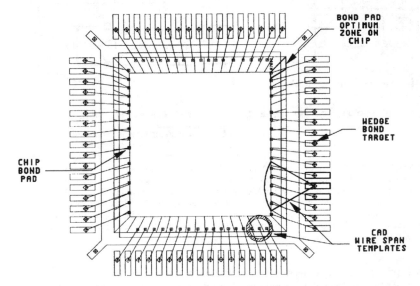

FIGURE 6.60 Example of CAD template for positioning bonding pads on chip in optimum location to achieve high assembly yields and high-reliability performance. Support templates check adherence to wire-span guidelines and provide design flexibility for special cases where optimum zone is too restrictive for electrical layout. *(After Striny.[4])*

TABLE 6.12   Interconnection Levels and Major Process Options

Package	Chip attachment	Level 1 connection	Level 2 connection
Molded plastic	Ag-loaded organic	WB, Au	TH, IL
PDIP	None	TAB	
	Ag-loaded organic	WB, Au	SM, IL, GW/J
SOIC/J	Ag-loaded organic	WB, Au	SM, P, J
PQFP	Ag-loaded organic	WB, Au	SM, P, GW
	None	TAB	
TapePak	Nonconductive organic	TAB	SM, P, GW
Pressed ceramic			
CERDIP	Ag-loaded inorganic	WB, Al	TH, IL
CERQUAD/CQFP	Ag-loaded inorganic	WB, Al	SM, P, GW
Laminated ceramic			
DIP-STB/TB/BB	Metal, AuSi	WB, Au	TH, IL
PGA	Metal, AuSi	WB, Au	TH, AA
LLCC	Metal, AuSi	WB, Au	SM, P
LDCC	Metal, AuSi	WB, Au	SM, P, CL, BZL
FP-LDCC/QFP	Metal, AuSi	WB, Au	SM, P, GW/BZL
MCM-Si on C	Ag-loaded organic	WB, Au	
	None	TAB	
	None	FC	
BGA	Metal, AuSi	WB, Au	SM, AA
	None	FC, SB	
Laminated plastic			
LLCC	Ag-loaded organic	WB, Au	SM, P
PGA	Ag-loaded organic	WB, Au	TH, AA
SMBA/LGA	Ag-loaded organic	WB, Au	SM, AA
BGA	Ag-loaded organic	WB, Au	SM, AA
	None	FC, SB	
Other options			
COB	Ag-loaded organic	None	WB, Au/FC, SB
	None	None	OLB, TAB
TOB	None	ILB, TAB	OLB, TAB
MCM-Si on Si	FC, SB	FC, SB	
TQFP	Ag-loaded organic	WB, Au, WBT	SM, P, GW

* BZL— = brazed lead; J = J-formed lead; P = perimeter lead locus; SB = solder-bumped; SBZ = side-braced; AA
= area array; for all other abbreviations, see App. A.

packages the chip interconnection process generally consists of two steps. In the first step, the back of the chip is mechanically attached to an appropriate mounting surface, for example the lead frame paddle or the substrate die attach area of a laminated ceramic or plastic package. This attachment sometimes enables electrical connections to be made to the backside of the chip. There are three types of chip attachments in use today; metal alloy bonding (AuSi eutectic, AuSn eutectic and soft solders), organic adhesives (epoxies and polyimides), and inorganic adhesives (silver-filled glasses). In the second step, the bond pads on the circuit side of the chip are electrically interconnected to the package by wire bonding. Wire bonding is further split into three options; thermosonic and thermo-

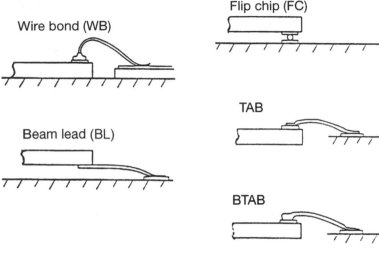

FIGURE 6.61   IC package interconnection options. *(After Khadpe.[49])*

compression ball-wedge using gold wire, and ultrasonic wedge-wedge using either gold or aluminum wire.

The other interconnection options illustrated in Fig. 6.61 (FC, BL, and TAB) generally are done as a one-step process, where the mechanical and electrical connections are provided by the same feature. All four options are used for SSI and MSI packages and all but BL are directly applicable to interconnecting large VLSI chips. The most popular bonding process in general use today for VLSI chips is the thermosonic ball-wedge process.

6.5.1.1 Chip Attachment.   The chip size is growing steadily in proportion to the emergence of improved VLSI technology capabilities. Increased chip size places more stringent demands on the processing and reliability of chip bonding.[50] Chip-bonding technology for large VLSI chips (greater than 12 mm on a side) requires special attention to thermal and stress management, as discussed in Secs. 6.4.2 and 6.4.3. In addition, similar care must be exercised in the selection of a chip-bond process for the particular application, and in the implementation of stringent process controls to assure high-quality bonds. Chip size also impacts wafer fabrication guidelines and assembly automation. These issues influence equipment selection and determine how easily a controllable process can be achieved.

The major choices for a chip attach process include metal alloys (AuSi eutectic, AuSn eutectic or soft solders), organic adhesives (epoxies and polyimides), and inorganic adhesives (silver-filled glasses). Table 6.12 shows the preferred method of chip attachment for the various package types.

The eutectic chip attach process is essentially contamination-free, has excellent shear strength, and assures low-moisture in package cavities. The major disadvantages are that it is difficult to automate, and it places high thermal stresses on the chip due to the high process temperature. Using ceramic substrate materials,[28] such as AlN and SiC, whose coefficient of thermal expansion closely matches that of silicon, will reduce the thermal stresses which can be detrimental to large chips. The eutectic chip attach process is described in Striny.[4]

The organic chip attach process uses an epoxy (with and without silver fillers for electrical and thermal conduction) and has the advantages of being less expensive, more flexible, easy to automate and requires low-temperature curing, which minimizes thermal stresses in large chips. As a result, epoxy chip-bond adhesives are preferred for attaching large chips in both ceramic and plastic packages.

Silver-filled epoxy adhesives are of major interest today as chip-bond materials. The use of silver fillers (typically flakes) makes the epoxy both electrically conductive, to provide low resistance between the chip and the substrate, and thermally conductive to allow a good thermal path between the chip and the rest of the package.

The epoxy chip-attach process can be highly automated and accurate, since the epoxy can be applied, at very high rates, to the die-attach area by transfer-printing, epoxy writing, or syringe dispensing, and the chip placed with high-speed pick and place tools. Accurate chip placement affects automatic wire bonding by yielding greater consistency of wire lengths and improved looping characteristics. In addition, chip placement accuracy also enhances the wire bonder's pattern-recognition performance and efficiency. The chip is removed from its film frame tape using a surface or collect pickup tool and placed over the die attach area contacting the epoxy pattern. Because epoxies are thermosetting polymers (cross-link when heated), they must be cured at elevated temperatures to complete the chip bond. Typical cure temperatures range from 125 to 175°C. In general, epoxy chip bonds are as good or better than their metal counterparts, except in the most demanding applications, that is, applications that require high temperatures, high current through the chip bond, and critical thermal performance requirements.

The inorganic chip attach process is used in the assembly of pressed ceramic CERDIP- and CERQUAD-type packages where the final cure is done at the same time as the high-temperature glass-sealing operation. The chip attach material is a silver-loaded glass system in an organic medium, which is applied as a paste.[53] In the wet form, silver-loaded glass contains 66.4 percent silver, 16.6 percent glass, 1.0 percent resin, and 16.0 percent solvent by weight. After drying and organic burn-out, the system is completely inorganic with 80.0 percent silver and 20.0 percent glass by weight. The diffusion of silicon from the chip backside is important in the development of the bond. Adhesion between the silver glass and the silicon chip is achieved via a glass network structure that starts at the silicon dioxide film on the silicon chip backside, while the bonding mechanism between the silver glass and the ceramic substrate appears to be mechanical in nature[54]. These materials have excellent thermal stability, are lower in cost than the AuSi eutectic, and can provide a void-free bond for high-reliability hermetic parts.

6.5.1.2 Wire Bonding.   Wire bonding to silicon ICs is the most common interconnection process in use today. After attaching the IC to either a lead-frame paddle, the cavity or top surface of a laminated plastic or ceramic substrate, the IC is wire bonded with either gold wire by the thermosonic, thermocompression, or ultrasonic technique, or with aluminum wire by the ultrasonic process. Today gold wires are mostly used for thermosonic wire bonding and aluminum wires for ultrasonic bonding.

Typically, gold wire is ball-wedge bonded, that is, ball-bonded to the chip bond pads (typically aluminum) and wedge-bonded to a lead frame (typically gold or silver plated) or to package bond pads (gold plated), as shown in Fig. 6.62. One advantage of ball-wedge bonding comes from the symmetrical geometry of the capillary tip. The ball is formed by the inner portion of the tip (Fig. 6.62*b*), and then the wedge bond can be placed anywhere on a 360° arc around the ball bond, using the outer portion of the tip (Fig. 6.62*d* and *e*). This capability to dress the wire in any direction from the ball is the key factor that makes this process attractive for high-speed automated bonding, that is, the bonding head or package table does not have to rotate to form the wedge bond. Figure 6.63 shows a typical VLSI device that is ball-and-wedge bonded on today's state of the art automatic wire

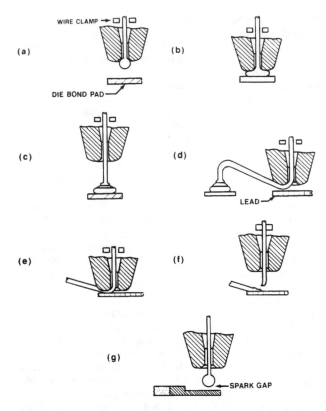

FIGURE 6.62   Tailless ball-and-wedge bonding cycle. (*a*) Capillary is targeted on die's bond pad and positioned above die with ball formed on end of wire and pressed against face of capillary. (*b*) Capillary descents, bringing ball in contact with die. Inside cone, or radius, grips ball in forming bond. In a thermosonic system, ultrasonic vibration is then applied. (*c*) After ball is bonded to die, capillary rises to loop-height position. Clams are open and wire is free to feed out end of capillary. (*d*) Device lead is positioned under capillary, which is then lowered to lead. Wire is fed out end of capillary, forming a loop. (*e*) Capillary deforms wire against lead, producing wedge-shaped bond, which has a gradual transition into wire. In a thermosonic machine, ultrasonic vibration is then applied. (*f*) Capillary rises off lead, leaving stitch bond. At a preset height, clamps are closed while capillary is still rising with bonding head. This prevents wire from feeding out capillary and pulls at bond. Wire will break at thinnest cross section of bond. (*g*). New ball is formed on tail of wire, which extends from end of capillary. A hydrogen flame or an electronic spark may be used to form ball. Cycle is completed and ready for next ball bond. (*After Kulicke and Soffa Industries.*[55])

bonders running in excess of 8 wires per second. The present ball-wedge wire bonding capability is limited to an in-line ball-to-ball pitch of 0.004 in (100 μm). Recent advancements in wire bonders and tool geometries will make it possible to reduce the ball-to-ball pitch to 0.0035 in (90 μm). Further reductions in the wire-to-wire pitch are possible by using a two-row staggered configuration (Fig. 6.59*b*).

FIGURE 6.63   Typical thermosonic ball wire bonds on VLSI chip. *(After Striny.[4])*

Aluminum wire is typically wedge-to-wedge bonded. In a wedge-to-wedge configuration, the orientation of the wedge bond at the chip determines the direction of the wire that terminates in the package wedge bond. The complication of requiring motion in both the bonding head and the work holder results in a slower, more costly, wire bonding process. On the other hand, wedge-to-wedge wire bonding has the capability of bonding to bond pads on tighter centers than ball-wedge bonding. The present in-line wedge bond pad pitch capability on the chip is 0.003 in (76 μm).

The major issue in establishing a quality and reliable wire bonding process is process control. Two types of process tests are in use today; the wire pull test and the ball shear test. A detailed discussion of these tests is presented in Striny.[4]

6.5.1.3 Tape Automated Bonding (TAB).   In the 1960s and 1970s,[49] TAB was used for low-cost, high-volume production of SSI products referred to as "jelly bean" devices. The driving force was strictly low cost, which was achieved by using automated gang bonding of the inner and outer leads, in place of the then very slow manual wire bonding. The process was faster and cheaper than the other options illustrated in Fig. 6.61 and resulted in higher bond strength (reliability) than the other options.

In the 1980s,[49] TAB was pursued because of its many attributes, such as lower inductance, controlled lead geometry, higher bond strength (reliability), and controlled impedance capability, all resulting in better electrical performance than wire bonding. In addition, the characteristic smaller bond pad and pitch significantly reduced the chip size, resulting in better wafer yields and lower chip cost. The physical size of TAB devices also resulted in a significant reduction of real estate at the PWB level.

TAB technology uses a photo-imaging/etching process to produce conductors on a dielectric conductor tape in a "movie-film" format, as illustrated in Fig. 6.64[56] Carrier tape

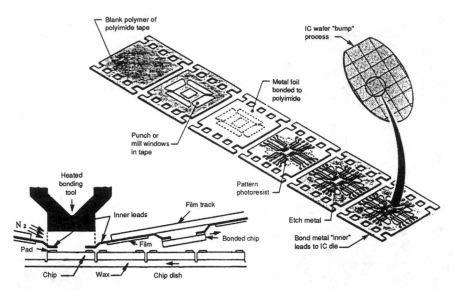

FIGURE 6.64 Overall TAB process diagram with chip-on-tape component manufacture. *(After IPC.[56])*

is stored on reels in widths from 1.4 to 2.8 in (35 to 70 mm). The final tape design has a window with "beam-type" leads that extend over the windows of the tape. Conductors are fanned out from the high-density beams on 0.002 to 0.004-in (0.05- to 0.10-mm) pitch to the external outer bond locations on 0.006 to 0.02-in (0.15- to 0.50-mm) pitch and then beyond to test pads placed on a 0.050 in (1.27 mm) pitch, compatible with conventional test probes. An example[7] of a 300-I/O tape with inner beams on a 0.004 in (0.10 mm) pitch is shown in Fig. 6.65.

In inner-lead bonding (ILB), the beams are bonded to an IC chip that is precisely located under the windows, as shown in Fig. 6.64. Figure 6.66 is a scanning electron microscope (SEM) view of an ILB chip. Once the chips are bonded to the tape, the leads are isolated electrically by a punching operation, and then the chip-in-tape is electrically tested and, if required, burned-in. Finally, the outer lead bonding (OLB) operation transfers the chip, with its leads, to the next-level package. This could be the plastic molding of a QFP with over 164 leads, an MCM, or direct bonding to a PWB or other substrate.

TAB wafer preparation is similar to that for wire bonded wafers. Either a pin-hole-free silicon nitride, silicon dioxide, or a polyimide is deposited at low temperatures. Passivation is selectively removed, leaving a good portion of the aluminum pads exposed. The chip is now in a state where it is ready for wire bonding or sent on for additional TAB fabrication, that is, preparation of the pads for TAB bonding. There are basically two methods of TAB in use: bumped chip, as shown in Figs. 6.67 and 6.68, and bumped tape, as illustrated in Figs. 6.69 and 6.70. Bumped-chip metallurgy is described in IPC.[56]

There are three types of TAB tapes in use today: single-layer, two-layer, and three-layer. Single-layer tape is a conductive foil without any insulating layer. It is the lowest-cost TAB construction and is used in jelly bean products where pre-electrical testing is not required. Two-and three-layer tapes are composed of conductive and insulating layers that are combined by direct deposit of copper on film (two-layer) or by the use of adhesives to hold them together (three-layer). In either tape, the insulator acts as a carrier and isolates the leads electrically. The various types of tape construction are shown in Fig. 6.71. Also

FIGURE 6.65   TAB tape with inner leads on 0.004-in pitch. *(Courtesy of Mesa Technology; after Hoffman.[7])*

FIGURE 6.66   Inner-lead bonded chip. *(Courtesy of Aptos and Mesa Technology; after Hoffman.[7])*

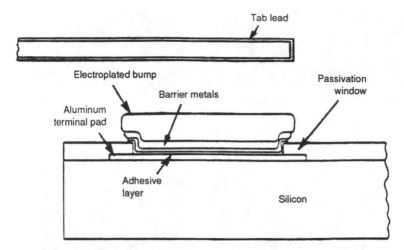

FIGURE 6.67 Interconnection geometry of bumped-chip TAB. TAB lead is either copper foil, ranging in thickness from 2 oz (0.0028 in/0.07 mm) to 1/2 oz, depending on feature sizes required in tape design. Lead is gold plated to facilitate TC ILB. Electroplated bump thickness is typically 0.001 in (0.025 mm). Barrier metals are deposited in three layers, such as 600-Å Ti, 400-Å Pd, and 2000-Å Au. *(After IPC.[56])*

FIGURE 6.68 SEM view of gold bumps on a chip. *(Courtesy of Aptos; after Hoffman.[7])*

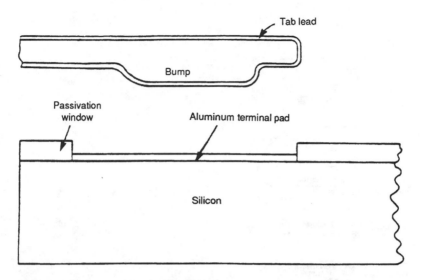

FIGURE 6.69 Interconnection geometry of bumped-tape TAB (BTAB). TAB lead is copper foil with etched bumps and gold plated to facilitate TC ILB. *(After IPC.[56])*

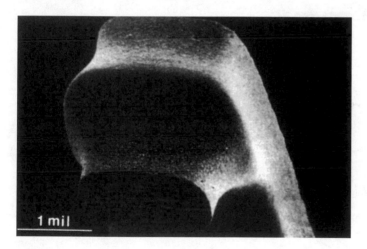

FIGURE 6.70 SEM view of bumped tape lead. *(Courtesy of Mesa Technology; after Hoffman.[7])*

shown is an area array tape, which contacts pads anywhere on the chip surface, and a wire bondable tape, which allows the extension of wire bonding to chips.

There are several advantages of TAB over wire bonding. TAB can accommodate a smaller chip bond pad pitch than wire bonds, as illustrated in Fig. 6.72. An in-line pitch of 0.003 in (0.076 mm) is feasible today and a 0.002-in (0.015-mm) capability is projected for the future. Whereas, wire bonds can accommodate an in-line pitch of 0.0035 in (0.09 mm) today, with future advances it is expected that the pitch will be reduced to 0.003 in

TYPE   CONSTRUCTION

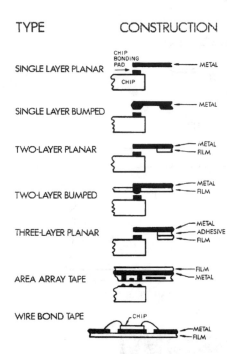

FIGURE 6.71 Various types of tape construction. *(Courtesy of 3M Corporation; after Khadpe.[49])*

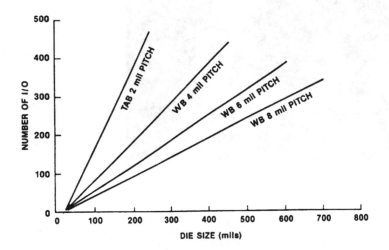

FIGURE 6.72 Number of I/Os and chip size as a function of in-line bond pad pitch on chip. Interconnection technique changes as chip bond pad pitch reduces from 0.008 to 0.002 in (0.2 to 0.05 mm). Above 0.004 in (0.10 mm), both wire bonding and TAB are used, with wire bonding the preferred technique. Below 0.004-in in-line pitch, TAB and FC are the choices.

(0.076 mm). In addition, ICs on TAB tape can be fully tested and burned-in before committing to the next-level of assembly. On the other hand, TAB requires a tape design for each chip size and pad location variation as well as special tooling and expensive bonding equipment for ILB and OLB (second level interconnection). The ICs are typically bumped, even with bumped tape, which adds to the cost. Finally, tight tolerances are required on planarity for gang bonding. If this is not achieved, the bonding yield and device reliability will be degraded. Single-point bonding (generally modified wire bonders) are sometimes used to overcome the lack of planarity.

6.5.1.4 Flip-Chip Bonding.

Flip-chip (FC) bonding is a process that facilitates direct attachment and electrical interconnection of the chip to the substrate circuitry. Whereas, wire bonding requires peripheral pads on the chip, the flip-chip technology utilizes area array terminals that result in the highest I/O interconnection density of any of the other bonding techniques. The terminals on the chip are placed face down (flip-chip) onto matching footprints on the substrate, interconnecting through electrically conductive bumps. The bumps may be either solder or conductive epoxy, depending on the reliability requirements for a given application. Conductive epoxy bumping is a lower cost, lower reliability flip-chip process, that has recently been developed, but its acceptance by the industry has been slow, awaiting further evaluation.

In the higher reliability solder bump interconnection process, the ICs are solder bumped at the wafer level using various bumping techniques, some of which were developed over a twenty year period. One bumping scheme[34] consists of electrolytic tin/lead plating to wettable chip bonding pads, followed by a reflow to form the solder bumps, as illustrated in Fig. 6.73. The solder bumps vary in composition (97Pb/3Sn, 95Pb/5Sn or 60Sn/40Pb) depending on the reflow temperatures that the applicable substrate can withstand. For ceramic substrates, a high melting temperature solder alloy (95Pb/5Sn, melting

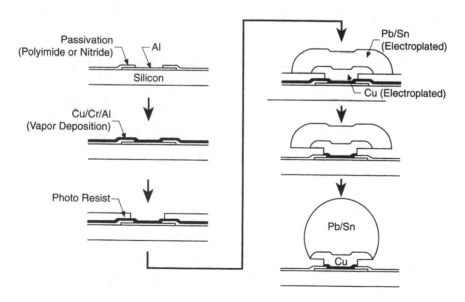

FIGURE 6.73 Typical solder bump-forming process. *(Courtesy of Citizen Watch Co., Ltd.; after Citizen.[34])*

at ≈ 308°C) is preferred in order to avoid remelting of the bumps when the package is reflowed to the PWB. Since organic substrates are assembly processed at lower temperatures, the solder bumps require a lower melting temperature solder such as 60Sn/40Pb which melts at ≈183°C.

Combining a BGA package configuration with flip-chip bonding results in a most effective high-density interconnection that approaches chip scale packaging, as discussed in Sec. 6.3.2. In this configuration, the solder bumped chip is placed face down onto matching wettable bonding pads on a multilayer BGA substrate, and the assembly is reflowed. After flip-chip attachment, the chip is underfilled with an epoxy resin formulated to relieve the stresses induced by the thermal mismatch between the chip and the substrate, and to prevent moisture from getting to the chip surface. The flip-chip can be "glob-topped," capped or a resin dam can be attached to the surface of the substrate and filled with a liquid epoxy to the rim of the chip, leaving it's back surface exposed for heat sink attachment, as shown in Fig. 6.74. After epoxy curing, solder balls are attached to the underside of the substrate to complete the BGA.

The short and stubby solder bump interconnections, between the chip and substrate, provide inductances that are significantly lower than for TAB or wire bonded chips, as shown in Table 6.13. The lower inductances in combination with the higher interconnection density result in improved electrical performance.

TABLE 6.13   Inductance of Chip Attachment Technologies[*]

	Cross section, in	Length, in	Inductance,[†] nH
Wire bond	0.001 (diameter)	0.05	0.9
TAB	0.001 × 0.003	0.05	0.7
Solder bump	0.004 × 0.004	0.002	<0.05

* Based on three-dimensional simulation of 200-I/O chip.
† 25 percent of I/Os are power/ground.
Source: from Bartlett et al.[57]

The major disadvantage of flip-chip bonding is the thermal mismatch between the chip and substrate that may cause the solder bump interconnections to fatigue during thermal

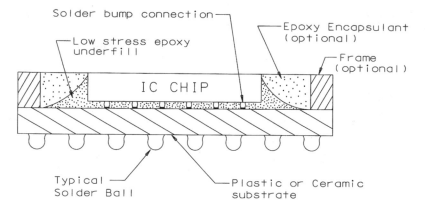

FIGURE 6.74   Plastic ball grid array (PBGA) with solder bumped flip-chip.

cycling. In addition, the high-density interconnection emanating from the chip requires equal high-density circuitry on the substrate. To achieve these circuit densities, the PWB substrates may have to be fabricated by the more expensive additive or build-up process.

Although the infrastructure for flip-chip bonding is currently not industry wide, it is expected that FC will be a leading interconnection technology for the future, particularly for high I/Os and where high electrical performance is required.

6.5.2 Other Assembly Processes

This section looks at other package assembly processes that must be addressed to have a complete package assembly methodology. Among them are cleaning, package lid sealing, encapsulation, alpha particle protection, and assembly environment.

6.5.2.1 Cleaning.   In IC packaging, the most critical cleaning steps are before wire bonding, encapsulation, or lid sealing and before ink code marking. The cleaning process must be chemically compatible with the chip metallurgy. Aluminum has a very narrow range of pH values in which its oxide protects it from corrosion in aqueous solution. Corrosion reactions, with metal dissolution, can occur in both basic and acidic solutions. Two objectives are usually sought in cleaning. One is the removal of organic species that can affect bondability; an organic solvent is usually required for this. The second is the removal of ionic species that can cause corrosion during the life of the device or, in an unusual instance, contribute to surface charge accumulation. Water is a very good solvent for ionic species.

Due to concerns of waste disposal and possible ozone depletion, the use of powerful solvents such as chloroflourocarbon (CFC), and trichloroethylene (TCA) are being curtailed and are expected to be completely phased out by the year 2000. Alternatives to cleaning with CFC and TCA solvents have emerged as possible replacements, such as aqueous, semi-aqueous, flourocarbon inerted organics, oxygenated organics and other cleaning agents. But, the most significant trend has been to shift to a "no-clean" IC assembly operation. The elimination of cleaning during IC assembly was made possible by: (a) requiring that all incoming assembly components be in a certified clean condition and (b) minimizing contamination during assembly by conducting the operation in a Class 10,000 or better clean room environment. In addition, user friendly cleaning processes are being used in assembling BGAs, such as UV-plasma cleaning prior to ink code marking the overmold and the use of aqueous cleaning agents to remove the water soluble flux after BGA solder ball attachment.

The new cleaning trends have had mixed results as manufacturers continue to evaluate the cleaning options and their effect on product reliability.

6.5.2.2 Package Lid Sealing.   The major objective of package sealing is to protect the device from external contaminants during its expected lifetime. Further, any contaminants present before sealing must be removed to an acceptable level before or during sealing. Packages may be hermetically sealed with glass or metals or nonhermetically lidded with polymers. The definition of hermeticity used here includes not only the ability to pass a fine vacuum leak test, but also to exclude environmental contaminants for long periods of time.

Figure 6.16 shows the relative capabilities of several materials to exclude moisture over long time periods. Clearly, organic sealants are not good candidates for hermetic packages. However, in some cases, organic sealants, properly integrated into the package design, do meet the operational definition of a hermetic seal given above.[58] For almost all high-reliability applications the hermetic seal is made with glass or metal. Glass sealing was mentioned previously for pressed ceramic, and the process is essentially the same for lid sealing. Many of the metal alloys used for chip bonding are suitable for lid sealing. A leak-tight metal seal that excludes the external environment can be made without difficulty. The real difficulty has been freeing the package of contaminants, especially water,[52,59] before sealing.

No package should be sealed with greater than 6000 ppm water by volume (dew point 0°C).[59] The technology to measure such moisture levels has been studied thoroughly.[60] The military specification limit is set at 5000 ppm by volume mainly because of the technical difficulties in measuring moisture in small packages. Although 5000 ppm is the specification limit, the package assembler wants to achieve as low a moisture level as possible. A particularly effective method to achieve low moisture levels has been suggested. This method makes use of the ability of atomic silicon to react with water to form SiO.[51] The atomic silicon is formed during the lid seal through the melting of the chip bond material or an added gettering preform. The proposed reaction is:

$$Si + H_2O \rightarrow SiO_2 + 2H_2$$

6.5.2.3 Encapsulation. For applications where hermetic sealing is not a requirement, organic encapsulation is used. Encapsulation technologies may include either molding, potting, glob-top or cavity fill and surface chip coatings when lid sealing. The encapsulant must possess material properties that will enable it to protect the chip from adverse environments, contaminants, package handling, storage and second level assembly. Mechanical strength, adhesion to silicon and substrate, CTE compatible with silicon and substrate, temperature and moisture resistance, electrical insulation, chemical resistance, and flow characteristics are some of the traits to consider when selecting an encapsulant for a given application.

In all cases, the encapsulants are used to cover the chip surface and the fragile wire bonds. It is critical that the dispensing techniques and flow characteristics of the epoxy compound be tailored to the specific configuration in order to prevent air from being trapped or bond wire sweeps that may cause shorts.

Plastic packages, such as QFPs and BGAs, are encapsulated with a novolac based epoxy molding compound, typically consisting of an epoxy resin, accelerator, curing agent, silica fillers, flame retardant and a mold release agent.

Potting compounds, usually consisting of a liquid two-component system, are used for hybrid modules and discrete devices. Glob-top or cavity filled encapsulants are thermoset epoxies or silicone resin filled with inorganic fillers. Glob-top encapsulants are used in low-cost, single-layer PPGAs and on chip-on-board assemblies, whereas cavity filled epoxies are primarily used for encapsulating cavity structured, multilayer substrates for PPGAs and PBGAs.

In cases where polymer lid sealing is utilized, surface chip coatings can be used to protect the aluminum metallurgy from atmospheric contaminants such as moisture, alpha particles and mechanical damage. Silicones, epoxies and polyimides are very effective for this purpose and are used as encapsulants in premolded and laminated PPGAs in combination with a protective lid or cap.

6.5.2.4 Alpha Particle Protection. Soft errors in memory circuits due to alpha particles emanating from packaging materials were first reported [24] in 1978; since then many papers have been published on the subject. Alpha particles are emitted by the decay of uranium and thorium atoms contained as impurities in packaging materials such as molding compounds and some lid sealing materials. Decreasing design rules of VLSI devices make them more sensitive to this problem.[24,32] Packaging materials (particularly ceramics) will probably not be pushed below the 0.001 to 0.01-alpha particle/(cm^2 • h) level. Because alpha particles have low penetrating power in solids, low-alpha materials have been suggested as alpha-absorbing coatings on silicon chips.[31,32] An 0.001-alpha-particle/(cm^2 • h) emission rate, the lowest level anticipated, has been reported using silicone coatings.[32]

Molding compounds are typically filled with silica fillers to reduce the CTE, but a side effect is alpha-particle emission that can cause soft errors in memory devices assembled in plastic packages. Applying a silicone coating to the surface of the memory chip, prior to molding, reduces the alpha-particle absorption. Since the 1980s, the molding compounds used for encapsulating memory devices contain silica fillers, made by chemical vapor deposition (CVD), which do not emit alpha-particles. As a result surface coating of memory chips is no longer required for alpha-particle protection.

6.5.2.5 Assembly Environment.   The ever-decreasing feature sizes on VLSI chips down into the submicrometer range (Fig. 6.4) and the use of finer bond pad pitches impose stringent requirements on the control of cleanliness in the assembly environment. At a feature size of 1 μm, as shown in Fig. 6.75, the density for foreign particles of 1 μm or greater in a typical room environment is astronomical. Foreign particles not removed before molding of plastic packages, or lid sealing in ceramic packages, can degrade the reliability of the packaged device. Examples of such failures are electrical short-circuiting between adjacent interconnections by metallic foreign particles, and hard failures due to metal corrosion from foreign particles with high levels of contaminants. Critical assembly processes such as chip attachment, wire bonding and molding or lid sealing are currently being done, by many assembly houses, in clean rooms having a Class 10,000 or better environment.

6.5.3   Package to Second-Level Interconnection

Second-level interconnections for the four package technologies are compared in Table 6.12. Interconnections are identified as TH or SM types. Other parameters affecting level 2

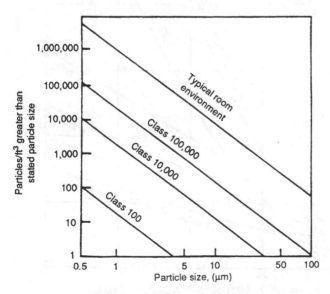

FIGURE 6.75   Particle-size distribution in typical room atmosphere and in three classes of clean environments. *(After Lewis and Berg,[61] © 1986, IEEE.)*

interconnections are the lead configuration (in line versus perimeter versus area array) and the shape or geometry of the lead (GW, J, TAB, side-brazed leads, solder balls, and FC). Details of the TH and SM processes are covered in Chap. 9.

6.6 SUMMARY AND FUTURE TRENDS

The packaging and interconnection of IC devices is summarized in Tables 6.9, 6.10, and 6.12. Four package technologies have emerged over the years, and each has enjoyed its share of success. Demand for higher pin counts will undoubtedly continue with ever-increasing silicon capability. Laminated Al_2O_3 ceramic packages have dominated the high-performance VLSI packaging technology up to several years ago, but factors such as its high dielectric constant, modest thermal conductivity and high cost has forced the industry to consider other packaging materials including laminated plastic packages.

Requirements for higher packaging density on the PWB level are driving package designs towards smaller perimeter lead pitches and array type packages. Figure 6.76 illustrates the area advantages of a ball grid array (BGA) package over a peripheral lead type such as a QFP. Above 100 I/Os the BGA, with solder balls on 0.050-in (1.27-mm) pitch and fully populated, uses less area on the PWB than the QFP with peripheral leads on 0.020-in (0.50-mm) pitch. At a peripheral lead pitch of 0.012 in (0.30 mm), the crossover

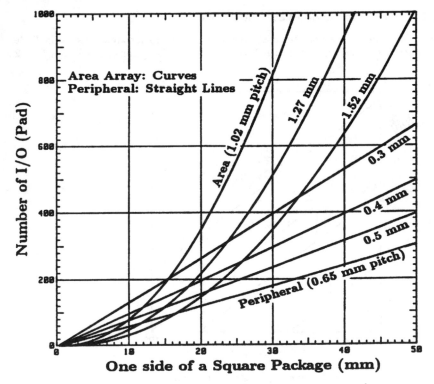

FIGURE 6.76 BGA and PQFP comparison: package pin counts (square). *(After Lau.[1])*

is at approximately 300 I/Os. The demand for BGAs has increased over the past few years because of their robust construction and ease of assembly to the PWB. BGAs are now challenging the fine pitch perimeter packages which are difficult to handle and have a greater board assembly yield loss than the array type packages.

Competing with traditional packaged devices will be the drive to direct chip attach (both perimeter and array types) and chip scale packaging where the ultimate in high density and performance can be achieved.

Newer packaging materials are being developed in order to improve material properties to achieve better performance, better reliability, and lower costs. Designers will make more use of package enhancement features, for example: heat spreaders in molded plastic packages,[62] and ground/power planes and heat spreading copper slugs in laminated plastic packages.[63,64]

The ever-decreasing feature sizes of IC chips will place demands on interconnect technology capabilities for connecting to very closely spaced bond pads. Wire bonding in particular will be challenged. If wire bonding is to survive, it will need improvements in placement accuracy to meet demands of tighter bond pad pitches. It will also be necessary to provide the capability of longer wire spans with controlled wire dressing to bridge the gap between the chip and the package bond pads while avoiding short circuits between adjacent wires in the dense wire fanout pattern. Wire bonding of high-I/O chips will need to be done in Class 10,000, or better, clean rooms to optimize automatic wire bonder performance and achieve acceptable bonding yields.

FC will be giving wire bonding serious competition as the future interconnection technology for IC packaging. FC has the advantage of very low power and ground inductance connections to the package and the best capability in terms of the density of interconnection (using an area array) to the device compared to wire or TAB.

The system design will depend more and more on a systematic optimization of the entire interconnection scheme to achieve the potential benefits of improved silicon capability. This optimization may lead to completely new requirements for assembly and packaging but, more than likely, it will lead to a sorting out of the various existing assembly technologies and packages.

REFERENCES

1. J. H Lau, *Ball Grid Array Technology,* McGraw-Hill, New York, 1995.

2. S. M. Sze, *VLSI Technology,* 2d ed., McGraw-Hill, New York, 1988, pp. 1–4.

3. C. Cohn, R. M. Richman, L. S. Saxena, and M. T. Shih, "High I/O Plastic Ball Grid Array Packages—AT&T Microelectronics Experience," in *Proc. 45th Electronic Components and Technology Conf.,* May 1995, pp. 10–20.

4. K. M. Striny, "Assembly Techniques and Packaging of VLSI Devices," in S. M. Sze (ed.), *VLSI Technology,* 2d ed., McGraw-Hill, New York, 1988, chap. 13, pp. 566-611.

5. R. Hannemann, "Physical Technology for VLSI Systems," in *Proc. IEEE Int. Conf. on Computer Design: VLSI in Computers,* 1986, pp. 48–53.

6. B. C. Johnson, "Overview," in *IEEE High Performance Integrated Circuit Packaging,* (IEEE Videoconf. Seminar via Satellite, Notes, Sept. 22, 1987), p. 23.

7. P. Hoffman, "TAB Implementation and Trends," *Solid State Technol.,* vol. 31, p. 86, June 1988.

8. M. F. Bregman and C. A. Kovac, "Plastic Packaging for VLSI-Based Computers," *Solid State Technol.,* vol. 31, p. 79, June 1988.

9. R. K. Traeger, "Hermeticity of Polymeric Lid Sealants," in *Proc. 25th Electronic Components Conf.,* 1976, p. 361.

10. F. N. Sinnadurai, *Handbook of Microelectronics Packaging and Interconnection Techniques,* Electrochemical Publ. Ltd., 1985, p. 4.

11. J. R. Howell, "Reliability Study of Plastic Encapsulated Copper Lead Frame Epoxy Die Attach Packaging System," in *Proc. Int. Reliability Physics Symp.,* 1981, p. 104.

12. F. J. Parker and J. W. Stafford, "Development of an 18-Pin Premolded Plastic DIP," *Semiconductor Int.,* pp. 49–61, Nov. 1979.

13. K. M. Striny and A. W. Schelling, "Reliability Evaluation of Aluminum-Metallized MOS Dynamic RAM's in Plastic Packages in High Humidity and Temperature Environments," *IEEE Trans. Components, Hybrids, Manuf. Technol.,* vol. CHMT-4, pp. 476–481, 1981.

14. Integrated Circuit Engineering Corp. Tech. Staff, "Surface Mount Packaging Report," *Semiconductor Int.,* pp. 72–77, June 1986.

15. M. Kakei, Y. Ikeda, and S. Koshibe, "Low Stress Molding Compounds for VLSI Devices," *Nikkei Electron.,* Spec. Issue on Microdevices 11, pp. 82–92, 1984.

16. M. Murtuza, J. C. Lee, R. Tan, W. H. Sehroen, and G. Bednarz, "Flux Penetration and Pressure Cooker Fail Mechanism in Plastic IC Packages," in *Proc. IEEE 36th Electronic Components Conf.,* May 1986, p. 620.

17. P. H. Singer, "Leak Detection in IC Packages," *Semiconductor Int.,* pp. 91–96, Nov. 1984.

18. "Kyocera Packaging," Publ. A-122E, Kyocera America Inc., San Diego, Calif.

19. B. Gunasekera, "Automatic Back-End Enhances Ceramic Quad Assembly," *Semiconductor Int.,* pp. 292–294, May 1987.

20. E. Philofsky, "Purple Plague Revisited," *Solid State Electron.,* vol. 13, p. 1391, 1970.

21. A. Sharan and L. Valero, "Application of Controlled Atmosphere IR Furnaces to Cerdip Sealing," *Semiconductor Int.,* pp. 91–95, Aug. 1987.

22. K. Eastman, "Ceramic Quad Package Meets High-Density SMT Needs," *Electron. Packag. Prod.,* pp. 70–71, Jan. 1987.

23. R. Corey R. Nordman, and A. Amalfitano, "Fine Forming," *Circuits Manuf.,* pp. 41–45, July 1988.

24. T. C. May and M. H. Woods, "A New Physical Mechanism for Soft Errors in Dynamic Memories," in *Proc. 16th Int. Reliability Physics Symp.,* Apr. 1978, pp. 33–40.

25. M. C. Loo, "Influence of Seal Glass on the Hermeticity of Cerdip Packages," *Solid State Technol.,* pp. 163–165, Aug. 1986.

26. Silicon Carbide Tech. Inf. RDP-TB0035, Kyocera America Inc., San Diego, Calif., 1989.

27. "Design Guidelines Multilayer Ceramic," Catalog CATD6, Kyocera America Inc., San Diego, Calif., 1995.

28. S. L. Spitz, "Ceramics Keep Circuits Cool," *Electron. Packag. Prod.,* pp. 36–41, July 1989.

29. "MetCeram Design Guidelines," MetCeram, Providence, R.I.

30. C. Cohn, N. V. Gayle, L. S. Saxena, "Design and Development of Plastic Pin Grid Array Packages," in *Proc. 1990 International Electronics Packaging Conf.,* September 1990, pp. 882–893.

31. K. Yamaguchi and M. Igaroshi, "Screen Printing Grade Polyimide Paste for Alpha-Particle Protection," in *Proc. IEEE 36th Electronic Components Conf,* May 1986, p. 100.

32. M. C. White, J. W. Serpiello, K. M. Striny, and W. Rosensweig, "The Use of Silicone RTV Rubber for Alpha Particle Protection on Integrated Circuits," in *Proc. Int. Reliability Physics Symp.,* 1981, p. 43.

33. G. L. Ginsberg, "Chip and Wire Technology: The Ultimate in Surface Mounting," *Electron. Packag. Prod,* pp. 78–83, Aug. 1985.

34. Citizen Watch Co., Ltd, Data sheets on flip-chip technology, 1996.

35. "JEDEC Registered and Standard Outlines for Semiconductor Devices," Publ. 95, Electronics Industries Assoc., Washington, D.C., 1985.

36. C. L. Cohen, "Japan's Packaging Goes World Class," *Electronics,* pp. 26–31, Nov. 11, 1985.

37. J. S. Braden, "Advanced Surface Mountable Packages for VLSI Devices," *Semiconductor Int.*, pp. 82--85, Apr. 1987.

38. L. W. Schaper, "The Impact of Inductance on Semiconductor Packaging," in *Proc. Int. Reliability Physics Symp.*, 1981, p. 38.

39. A. J. Rainal, "Transmission Properties of Various Styles of Printed Wiring Boards," *Bell Sys. Tech. J.*, vol. 56, p. 995, 1979.

40. J. A. DeFalco, "Reflections and Crosstalk in Logic Circuit Interconnections," *IEEE Spectrum*, p. 44, July 1970.

41. R. Kamikawai, M. Nishi, K. Nakanishi, and A. Masaki, "Electrical Parameter Analysis from Three-Dimensional Interconnection Geometry," *IEEE Trans. Components, Hybrids, Manuf. Technol., vol.* CHMT-2, p. 269, 1985.

42. F. F. Oettinger, "Thermal Evaluation of VLSI Packages Using Test Chips—A Critical Review," *Solid State Technol.*, p. 169, Feb. 1984.

43. G. K. Baxter, "A Recommendation of Thermal Measurement Techniques for IC Chips and Packages," in *Proc. Int. Reliability Physics Symp.*, 1977, p. 204.

44. R. J. Hannemann, "Microelectronic Device Thermal Resistance: A Format for Standardization," in *Heat Transfer Dig.*, vol. 20: *Heat Transfer in Electronic Equipment*, Am. Soc. Mechanical Eng., New York, p. 38, 1981.

45. J. C. Bolger and C. T. Mooney, "Die Attach in Hi-Rel P-DIPS: Polyimides or Low Chloride Epoxies?" *IEEE Trans. Components, Hybrids, Manuf. Technol.*, vol. CHMT-4, p. 394, 1984.

46. D. R. Olsen and H. M. Berg, "Properties of Die Bond Alloys Relating to Thermal Fatigue," in *Proc. 27th Electronic Components Conf,* 1977, p. 193.

47. P. J. Planting, "An Approach for Evaluating Epoxy Adhesives for Use in Hybrid Microelectronic Assembly," *IEEE Trans. Parts, Hybrids, Packag.*, vol. PHP-11, p. 305, 1975.

48. K. Otsuka and T. Usami, "Ultrasonic Wire Bonding Technology for Custom LSI's with Large Numbers of Pins," in *Proc. 23d Electronic Components Conf.*, 1981, p. 350.

49. S. Khadpe (ed.), "Semiconductor Packaging Update: Worldwide TAB Status," in *Proc. 1st Int. TAB Symp.*, Feb. 1989, pp. 1–38.

50. R. K. Shukla and N. P. Mencinger, "A Critical Review of VLSI Die-Attachment in High Reliability Applications," *Solid State Technol.*, pp. 67–74, July 1985.

51. M. L. White, K. M. Striny, and R. E. Sammons, "Attaining Low Moisture Levels in Hermetic Packages," in *Proc. 20th Int Reliability Physics Symp.*, 1982, p. 253.

52. K. M. Kearney, "Trends in Die Bonding Materials," *Semiconductor Int.*, pp. 84–88, June 1988.

53. F. K. Moghadam, "Development of Adhesive Die Attach Technology in Cerdip Packages: Material Issues," *Solid State Technol.*, pp. 149–157, Jan. 1984.

54. G. S. Selvaduray, "Die Bond Materials and Bonding Mechanisms in Microelectronic Packaging," *Thin Solid Films*, vol. 153, pp. 431–445, 1987.

55. *Bonding Handbook and General Catalog*, Kulicke and Soffa Industries, Inc., Willow Grove, Pa., 1990.

56. "Standardization and Implementation Requirements for Fine Pitch Technology with Emphasis on Tape Automated Bonding," IND-TAB-XXX, Inst. for Interconnecting and Packaging Electronic Circuits, 1988.

57. C. J. Bartlett, J. M. Segelken, and N. A. Teneketges, "Multichip Packaging Design for VLSI-Based Systems," *IEEE Trans. Components, Hybrids, Manuf Technol.*, vol. CHMT-10, pp. 647–653, Dec. 1987.

58. I. Memis, "Quasi-Hermetic Seal for IC Modules," in *Proc. 30th Electronic Components Conf,* 1980, p. 121.

59. R. W. Thomas, "Moisture, Myths, and Microcircuits," *IEEE Trans. Parts, Hybrids, Packag.*, vol. PHP-12, p. 167, 1976.

60. "Moisture Measurement Technology for Hermetic Semiconductor Devices," NBS Publ., p. 400, 1981.

61. G. L. Lewis and H. M. Berg, "Particulates in Assembly: Effect on Device Reliability," in *Proc. IEEE 36th Electronic Components Conf.,* May 1986, p. 100.

62. M. Aghazadeh and B. Natarajan, "Parametric Study of Heatspreader Thermal Performance in 48 Lead Plastic DIPs and 68 Lead Plastic Leaded Chip Carriers," in *Proc. IEEE 36th Electronic Components Conf,* May 1986, p. 143.

63. D. Mallik and B. K. Bhattaccharyya, "High-Performance PQFP," in *Proc. IEEE 39th Electronic Components Conf.,* May 1989, p. 494.

64. C. Patton, "Novel Arrangement of Contacts and Buses Shrinks 1-Mbit RAM," *Electron. Des.,* p. 42, June 27, 1985.

65. Tessera Inc., data sheet, "The Tessera® μBGA™ Package—A Low-Cost Alternative to High Pin Count Packaging," 1995.

CHAPTER 7

THE HYBRID MICROELECTRONICS TECHNOLOGY

Jerry E. Sergent
BBS PowerMod
Palmyra, New York

7.1 INTRODUCTION

The hybrid microelectronics technology is one branch of the electronics packaging technology, and it is differentiated from other branches by the manner in which the interconnection pattern is formed. A metallization pattern is formed on a ceramic substrate, and the circuit is completed by mounting and bonding active and passive devices as necessary. Another characteristic of the hybrid technology is the ability to fabricate passive components. The thick and thin film technologies, for example, can be used to manufacture resistors with parameters superior to the carbon resistors commonly used in conjunction with printed circuit boards.

The most commonly accepted definition of a hybrid circuit is one in which the substrate is metallized by one of the methods in Fig. 7.1, and that contains at least two components, one of which must be active. This definition is intended to exclude single-chip packages and circuits that contain only passive components, such as resistor networks. By this definition, a hybrid circuit may range from a simple diode-resistor logic gate to a circuit containing in excess of 100 integrated circuits (ICs).

This chapter describes the methods and properties of the various metallization approaches and assembly methods used to manufacture hybrid circuits. Also included are design guidelines, a discussion of reliability considerations, and applications of hybrid circuits.

7.2 THICK FILM TECHNOLOGY

Thick film circuits (see Fig. 7.2) are fabricated by screen printing a conductive, resistive, or insulating material in the form of a viscous paste onto a ceramic substrate and exposing

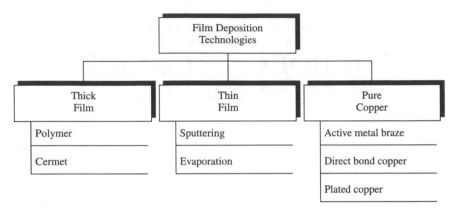

FIGURE 7.1   Film deposition methods.

FIGURE 7.2   Signal conditioning circuit fabricated on a double-sided thick film substrate with surface mount components. *(Courtesy of Mini-Systems, Inc.)*

the film to an elevated temperature to activate the adhesion mechanism that adheres the film to the substrate. In this manner, by depositing successive layers, as shown in Fig. 7.3, multilayer interconnection structures can be formed which may contain integrated resistors, capacitors, or inductors.

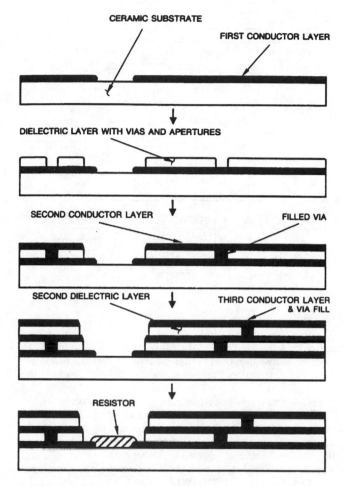

FIGURE 7.3 Thick film multilayer fabrication steps.

7.2.1 Thick Film Materials

All thick film pastes have two general characteristics in common.[1]

1. They are a viscous fluid with a non-newtonian rheology suitable for screen printing.

2. They are composed of two different multicomponent phases; a functional phase that imparts the electrical and mechanical properties to the finished film and a vehicle phase which imparts the proper rheology.

There are numerous ways of categorizing thick film pastes. One such way is shown in Fig. 7.4, which depicts three basic categories: polymer thick films, refractory thick films, and

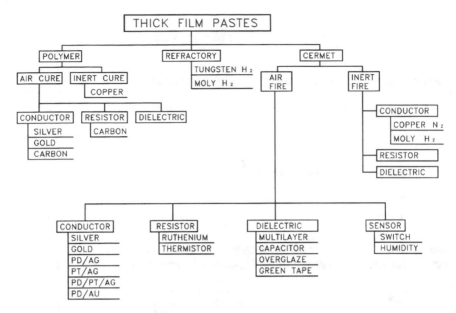

FIGURE 7.4 Thick film paste matrix.

cermet thick films. Refractory thick films are a special class of cermet thick films and are frequently categorized separately. These materials are designed to be fired at much higher temperatures (1500–1600°C) than conventional cermet materials, and are also fired in a reducing atmosphere.

Polymer thick films consist of a mixture of polymer materials with conductor, resistor, or insulating particles, and cure at temperatures ranging from 85 to 300°C. Polymer conductors are primarily silver, with carbon being the most common resistor material. Polymer thick film materials are more commonly associated with organic substrate materials as opposed to ceramic and therefore will not be considered in further detail.

Cermet thick film materials in the fired state are a combination of glass ceramic and metal and are designed to be fired in the range 850–1000°C.

A standard cermet thick film paste has four major ingredients: an active element, which establishes the function of the film; an adhesion element, which provides the adhesion to the substrate and a matrix that holds the active particles in suspension; an organic binder, which provides the proper fluid properties for screen printing; and a solvent or thinner, which establishes the viscosity of the vehicle phase.

The Active Element. The active element within the paste dictates the electrical properties of the fired film. If the active element is a metal, the fired film will be a conductor; if it is a conductive metal oxide, a resistor; and, if it is an insulator, a dielectric. The active element is most commonly found in powder form ranging from 1 to 10 μ (in size, with a mean diameter of about 5 μ). Particle morphology can vary greatly depending on the method used to produce the metallic particles. Spherical, flaked, or a circular shapes (both amorphous and crystalline) are available from powder manufacturing processes. Structural shape and particle morphology are critical to the development of the desired electrical performance, and extreme control on the particle shape, size, and distribution must be maintained to ensure uniformity of the properties of the fired film.

The Adhesion Element. There are two primary constituents used to bond the film to the substrate; glass and metal oxides, which may be used singly or in combination. Films which use a glass, or *frit*, are referred to as "fritted" materials and have a relatively low melting point (500–600°C). There are two adhesion mechanisms associated with the fritted materials, a chemical reaction and a physical reaction. In the chemical reaction, the molten glass chemically reacts with the glass in the substrate to a degree. In the physical reaction, the glass flows into and around the irregularities in the substrate surface. The total adhesion is the sum of the two factors. The physical bonds are more susceptible to degradation by thermal cycling or thermal storage than the chemical bonds and are generally the first to fracture under stress. The glass also creates a matrix for the active particles, holding them in contact with each other to promote sintering and to provide a series of three-dimensional continuous paths from one end of the film to the other. Principal thick film glasses are based on B_2O_3-SiO_2 network formers with modifiers such as PbO, Al_2O_3, Bi_2O_3, ZnO, BaO, and CdO added to change the physical characteristics of the film, such as melting point, viscosity, and coefficient of thermal expansion. Bi_2O_3 also has excellent wetting properties, both to the active element and to the substrate and is frequently used as a flux. The glass phase may be introduced as a prereacted particle or formed in situ by using glass precursors such as boric oxide, lead oxide, and silicon. Fritted conductor materials tend to have glass on the surface, making subsequent component assembly processes more difficult.

A second class of materials utilizes metal oxides to provide the adhesion to the substrate. In this case, a pure metal, such as copper or cadmium, is mixed with the paste and reacts with oxygen atoms on the surface of the substrate to form an oxide. The conductor adheres to the oxide and to itself by sintering, which takes place during firing. During firing, the oxides react with broken oxygen bonds on the surface of the substrate, forming a Cu or Cd spinel structure. Pastes of this type offer improved adhesion over fritted materials, and are referred to as "fritless," "oxide-bonded," or "molecular-bonded" materials. Fritless materials typically fire at 900–1000°C, which is undesirable from a manufacturing aspect. Ovens used for thick film firing degrade more rapidly and need more maintenance when operated at these temperatures for long periods of time.

A third class of materials utilizes both reactive oxides and glasses. The oxides in these materials react at lower temperatures, but are not as strong as copper. A lesser concentration of glass than found in fritted materials is added to supplement the adhesion. These materials, referred to as *mixed bonded* systems, incorporate the advantages of both technologies and fire at a lower temperature.

Organic Binder. The organic binder is generally a thixotropic fluid and serves two purposes: it holds the active and adhesion elements in suspension until the film is fired, and it gives the paste the proper fluid characteristics for screen printing. The organic binder is usually referred to as the "nonvolatile" organic since it does not evaporate but begins to burn off at about 350°C. The binder must oxidize cleanly during firing, with no residual carbon which could contaminate the film. Typical materials used in this application are ethyl cellulose and various acrylics.

For nitrogen-fireable films, where the firing atmosphere can contain only a few ppm of oxygen, the organic vehicle must decompose and thermally depolymerize, departing as a highly volatile organic vapor in the nitrogen blanket provided as the firing atmosphere, since oxidation into CO_2 or H_2O is not feasible due to the oxidation of the copper film.

Solvent or Thinner. The organic binder in the natural form is too thick to permit screen printing, which requires the use of a solvent or thinner. The thinner is somewhat more volatile than the binder, evaporating rapidly above about 100°C. Typical materials used for this application are terpineol, butyl carbitol, or certain of the complex alcohols into which the nonvolatile phase can dissolve. The low vapor pressure at room temperature is desirable to minimize drying of the pastes and to maintain a constant viscosity during printing. Additionally, plasticizers, surfactants, and agents that modify the thixotropic nature of the paste are added to the solvent to improve paste characteristics and printing performance.

To complete the formulation process, the ingredients of the thick film paste are mixed together in proper proportions and milled on a three-roller mill for a sufficient period of time to ensure that they are thoroughly mixed and that no agglomeration exists. There are three important parameters that may be used to characterize a thick film paste: fineness of grind, percent solids, and viscosity.

Fineness of grind is a measure of the particle size distribution and dispersion within the paste. A fineness-of-grind (FOG) gauge is a hard steel block with a tapered slot ground into one surface to a maximized depth, such as 50 μ, with a micrometer scale marked along the groove. The paste is placed in the deep end and drawn down the block toward the shallow end by a tapered doctor blade. At the point where the largest particles cannot pass under the gap between the groove and the doctor blade, the film will begin to form *streaks,* or areas with no paste. The location of the first streak with respect to the scale denotes the largest particle, and the point where approximately half of the width of the groove is composed of streaks is the mean value of the particle size. At some point, essentially all of the particles will be trapped, which represents the smallest particle.

The *percent solids* parameter measures the ratio of the weight of the active and adhesion elements to the total weight of the paste. This test is performed by weighing a small sample of paste, placing it in an oven at about 400°C until all the organic material is burned away, and reweighing the sample. The percent solids parameter must be tightly controlled in order to achieve the optimum balance between printability and the density of the fired film. If the percent solids content is too high, the material will not have the proper fluid characteristics to print properly. If it is too low, the material will print well, but the fired print may be somewhat porous or may lack in definition. A typical range for percent solids is about 85–92 percent by weight. By volume, of course, the ratio is somewhat lower due to the lower density of the vehicle as compared to the active and adhesion elements.

The *viscosity* of a fluid is a measure of the tendency of the fluid to flow and is the ratio of the shear rate of the fluid in sec^{-1} to the shear stress in force/unit area. The unit of viscosity is the *poise*, measured in dynes/cm^2-sec. Thick film pastes typically have the viscosity expressed in *centipoise* (cp), although the actual viscosity may be in the thousands of poise. An alternate unit of viscosity is the Pascal-sec. One Pascal-sec is equivalent to 0.001 centipoise.

In an ideal or "newtonian" fluid, the relationship between shear rate and shear stress is linear, and the graph passes through the origin. Newtonian fluids are not suitable for screen printing since the force of gravity assures that some degree of flow will always be present. As a basis for comparison, the flow properties of water approach those of newtonian fluids.

To be suitable for screen printing, a fluid must have certain characteristics.[1]

- The fluid must have a *yield point,* or minimum pressure required to produce flow, which must obviously be above the force of gravity. With a finite yield point, the paste will not flow through the screen at rest and will not flow on the substrate after printing.
- The fluid should be somewhat *thixotropic* in nature. A thixotropic fluid is one in which the shear rate /shear stress ratio is nonlinear. As the shear rate (which translates to the combination of squeegee pressure, velocity, and screen tension) is increased, the paste becomes substantially thinner, causing it to flow more readily. The corollary to this term is *pseudoplastic*. A pseudoplastic fluid is one in which the shear rate does not increase appreciably as the force in increased.

• The fluid should have some degree of hysteresis so that the viscosity at a given pressure depends on whether or not the pressure is increasing or decreasing. Preferably, the viscosity should be higher with decreasing pressure, as the paste will be on the substrate at the time and will have a lesser tendency to flow and lose definition.

The shear rate versus shear stress curve for a thixotropic paste that has these characteristics is shown in Fig. 7.5. A third variable (time) should also be considered in this figure. In practice, a finite and significant amount of time elapses between the application of the force and the time when the steady-state viscosity is attained. During the printing process, the squeegee velocity must be sufficiently slow to allow the viscosity of the paste to lower to the point where printing is optimum. After the print, sufficient time must be allowed for the paste to increase to nearly the rest viscosity (leveling). If the paste is placed in the drying cycle prior to leveling, the paste will become still thinner due to the increased temperature, and the printed film will lose line definition.

Under laboratory conditions, the viscosity of the paste may be measured with a cone-and-plate or a spindle viscometer. The cone-and-plate viscometer consists of a rotating cylinder with the end ground to a specific angle. A sample of paste is placed on a flat plate, and the cylinder is inserted into the paste parallel to the plate. The cylinder is rotated at a constant velocity and the torque required to maintain this velocity is converted to viscosity. The spindle method utilizes a cylinder of known volume which is filled with paste. A spindle of known size is rotated at a constant speed inside the cylinder, and the torque measurement is converted to viscosity as with the cone-and-plate. The spindle method provides a more consistent reading because the boundary conditions can be more tightly controlled, and it is the most common method of characterizing thick film pastes in a development or manufacturing facility.

It is important to understand the limitations of viscometers for laboratory use. These are not analytical tools in that a viscometer reading may be directly translated to the settings on the screen printer. Most viscometers of this type are designed to operate at two speeds, which would yield only two points on the curve in Fig. 7.5. This limits the utilization of the viscometer to simply comparing one paste with another. When used for this purpose, the measurement conditions must be identical; the same type spindle in the same volume of paste must be used, and the sample must be at the same temperature if the correlation is to be meaningful.

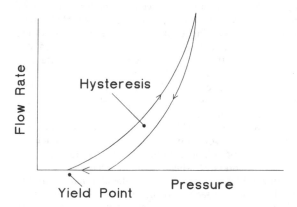

FIGURE 7.5 Viscosity curve for thick film paste.

Viscosity can be readily lowered by addition of the appropriate solvent. This is frequently required when the paste jar has been opened a number of times or if paste has been returned to the jar from the screen. If the original viscosity is recorded, the paste can be returned to the original viscosity with the aid of the viscometer. Increasing the viscosity is more difficult, requiring the addition of more nonvolatile vehicle, followed by remilling of the paste.

Cermet thick film pastes are divided into three broad categories: conductors, resistors and dielectrics. Each of these categories may have several subcategories that describe materials for a specific application.

7.2.2 Cermet Thick Film Conductor Materials

Thick film conductors must perform a variety of functions in a hybrid circuit.

- The most fundamental function is to provide electrically conductive traces between the nodes of the circuit.
- They must provide a means to mechanically mount components by reflow solder, epoxy, or by direct eutectic bonding.
- They must provide a means for the electrical interconnection of components to the film traces and to the next higher assembly.
- They must provide a means of terminating thick film resistors.
- They must provide electrical connections between conductor layers in a multilayer circuit.

Thick film conductor materials may be divided into three broad classes: air fireable, nitrogen fireable, and those that must be fired in a reducing atmosphere. Air fireable materials are made up of noble metals that do not readily form oxides. The basic metals are gold and silver, which may be used in the pure form or alloyed with palladium and/or platinum. Nitrogen fireable materials include copper, nickel, and aluminum, with copper being the most common. The refractory materials, molybdenum, manganese, and tungsten, are intended to be fired in a reducing atmosphere consisting of a mixture of nitrogen and hydrogen.

Gold is most often used in applications where a high degree of reliability is required, such as military and medical applications or where gold wire bonding is desirable for reasons of speed. The assembly processes (i.e., soldering, epoxy bonding, and wire bonding) used with gold thick films must be selected with care if reliability is to be maintained at a high level. For example, gold readily alloys with tin and will leach rapidly into certain of the tin-bearing solders, such as the lead-tin (Pb/Sn) alloys. Gold and tin also form intermetallic compounds which are brittle and which have a high electrical resistivity. Where Pb/Sn solders are to be used for component or lead attachment, gold must be alloyed with platinum or palladium to minimize leaching and intermetallic compound formation. Gold also forms intermetallic compounds with aluminum, which is commonly used as the contact material on semiconductor devices and for wire bonding. The diffusion coefficient of aluminum into gold is much higher than that of gold into aluminum, with the diffusion rate increasing rapidly with temperature. Consequently, when an Au-Al interface occurs, as when an aluminum wire is bonded to a gold thick film conductor, the aluminum will diffuse into the gold wire, leaving voids in the interface (Kirkendall voids) that weaken the bond strength and increase the electrical resistance. This phenomenon is accelerated at temperatures above about 170°C and represents a reliability risk. The addition of palladium alloyed with the gold lowers the rate of aluminum diffusion significantly and improves the reliability of aluminum wire bonds.

Silver is commonly used in commercial applications where cost is a factor. Like gold, silver leaches into Pb/Sn solders, although at a slower rate. Pure silver may be used in applications where the exposure to Pb/Sn solder in the liquidus state is minimized and may also be nickel plated to further inhibit leaching.

Silver also has a tendency to migrate when an electrical potential is applied between two conductors in the presence of water in the liquid form. As shown in Fig. 7.6, positive silver ions dissolve into the water from the positive conductor. The electric field between the two conductors transports the Ag ions to the negative conductor where they recombine with free electrons and precipitate out of the water onto the substrate as metallic silver. Over time, a continuous silver film grows between the two conductors, forming a conductive path. While other metals, including gold and lead, will migrate under the proper conditions, silver is the most notorious because of its high ionization potential.

Alloying silver with palladium and/or platinum slows down both the leaching rate and the migration rate, making it practical to use these alloys for soldering. Palladium/silver conductors are used in most commercial applications and are the most common materials found in hybrid circuits. However, the addition of palladium increases both the electrical resistance and the cost. A ratio of 4 parts silver to 1 part palladium is frequently used, which offers a good compromise between performance and cost.

Copper-based thick films were originally developed as a low-cost substitute for gold, but copper is now being selected when solderability, leach resistance and low resistivity are required. These properties are particularly attractive for power hybrid circuits. The low resistivity allows the copper conductor traces to handle higher currents with a lower voltage drop and the solderability allows power devices to be soldered directly to the metallization for better thermal transfer.

Copper thick film systems are known to exhibit the following problems:

- The requirement for a nitrogen atmosphere (<10 ppm oxygen) has created problems when scaling up from a prototype effort to high-volume production. Organic materials used in air-fireable systems combine with oxygen in the furnace atmosphere and "burn" off, while those used in copper paste systems "boil" off (or "unzip") and must be carried away by the nitrogen flow. It has proven difficult to maintain a consistent, uniform nitrogen blanket in the larger furnaces required for production, which has necessitated collaboration between furnace manufacturers, paste manufacturers, and users to come up with special, innovative furnace designs for copper firing. While

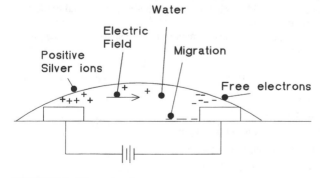

FIGURE 7.6   Silver migration.

some furnaces may be used for both nitrogen and air firing in prototype quantities, it is not practical to switch back and forth for production.

• Due to the large print areas normally required for dielectric materials, the problem of organic material removal is amplified when these materials are used to manufacture multilayer circuits. Consequently, multilayer dielectric materials designed to be used with copper are generally more porous than those designed for air-fireable materials, and, as a result, it has proven difficult to manufacture dielectric materials that are hermetic. This generally has required three layers of dielectric material between conductor layers to minimize short circuits and leakage, as opposed to the normal two required for air-fireable systems.

• Many resistor materials, particularly in the high ohmic range, have not proven to be as stable as air-fireable resistors when fired at temperatures below 980°C.

Refractory thick film materials, typically tungsten, molybdenum, and titanium, may also be alloyed with each other in various combinations. These materials are designed to be cofired with ceramic substrates at temperatures ranging up to 1600°C and are post-plated with nickel and gold to allow component mounting and wire bonding.

The properties of thick film conductors are summarized in Table 7.1.

TABLE 7.1 Thick Film Conductor Capabilities

	Au wire bonding	Al wire bonding	Eutectic bonding	Sn/Pb solder	Epoxy bonding
Au	Y	N	Y	N	Y
Pd/Au	N	Y	N	Y	Y
Pt/Qu	N	Y	N	Y	Y
Ag	Y	N	N	Y	Y
Pd/Ag	N	Y	N	Y	Y
Pt/Ag	N	Y	N	Y	Y
Pt/Pd/Ag	N	Y	N	Y	Y
Cu	N	Y	N	Y	N

7.2.3 Thick Film Resistor Materials

Thick film resistors are formed by adding metal oxide particles to glass particles and firing the mixture at a temperature/time combination sufficient to melt the glass and to sinter the oxide particles together. The resulting structure consists of a series of three-dimensional chains of metal oxide particles embedded in a glass matrix. The higher the metal oxide-to-glass ratio, the lower the resistivity and vice versa.

Referring to Fig. 7.7, the electrical resistance of a material in the shape of a rectangular solid is given by the classic formula

$$R = \frac{\rho_B L}{WT} \tag{7.1}$$

where R = electrical resistance in ohms

ρ_B = bulk resistivity of the material in ohms-length

L = length of the sample in the appropriate units

W = width of the sample in the appropriate units

T = thickness of the sample in the appropriate units

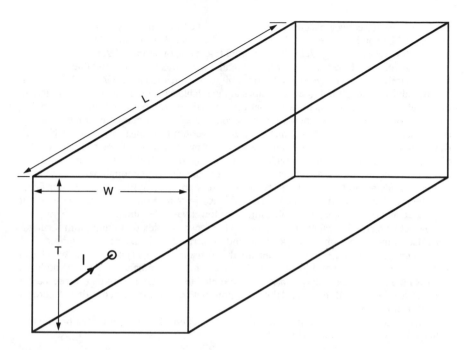

FIGURE 7.7   Resistance vs. temperature.

A "bulk" property of a material is one that is independent of the dimensions of the sample.

When the length and width of the sample are much greater than the thickness, a more convenient unit to use is the "sheet" resistivity, which is equal to the bulk resistivity divided by the thickness.

$$\rho_s = \frac{\rho_B}{T}$$ (7.2)

where ρ_s = sheet resistivity in ohms/square/unit thickness.

The sheet resistivity, unlike the bulk resistivity, is a function of the dimensions of the sample. Thus, a sample of a material with twice the thickness as another sample will have half the sheet resistivity although the bulk resistivity is the same. In terms of the sheet resistivity, the electrical resistance is given by

$$R = \frac{\rho_s L}{W}$$ (7.3)

For a sample of uniform thickness, if the length is equal to the width (i.e., the sample is a square), the electrical resistance is the same as the sheet resistivity independent of the actual dimensions of the sample. This is the basis of the units of sheet resistivity, "ohms/square/ unit thickness." For thick film resistors, the standard adopted for unit thickness is 0.001 in or 25 μ of *dried* thickness. The dried thickness is chosen as the standard as opposed to the fired thickness for convenience in process control. The dried thickness can be obtained in minutes, while obtaining the fired thickness can take as much as an hour. The specific units for thick film resistors are "$\Omega/\square/0.001$" of dried thickness, or simply "$\Omega/\square$."

A group of thick film materials with identical chemistries that are blendable is referred to as a "family" and will generally have a range of values from 10 $\Omega/\square$ to 1 M$\Omega/\square$ in decade values, although intermediate values are available as well. There are both high and low limits to the amount of material that may be added. As more and more material is added, a point is reached where there is not enough glass to maintain the structural integrity of the film. With conventional materials, the lower limit of resistivity is about 10 $\Omega/\square$. Resistors with a sheet resistivity below this value must have a different chemistry and often are not blendable with the regular family of materials. At the other extreme, as less and less material is added, a point is reached where there are not enough particles to form continuous chains, and the sheet resistance rises very abruptly. Within most resistor families, the practical upper limit is about 2 M$\Omega/\square$. Resistor materials are available to about 20 M$\Omega/\square$ but are not amenable to blending with lower value resistors.

The active phase for resistor formulation is the most complex of all thick film formulations due to the large number of electrical and performance characteristics required. The most common active material used in air-fireable resistor systems is ruthenium, which can appear as RuO_2 (ruthenium dioxide) or as $BiRu_2O_7$ (bismuth ruthenate). With the addition of *TCR* modifiers, these materials can be formulated to provide a temperature coefficient of resistance [as defined in Eq. (7.4)] of ±50 ppm with a stability of better than 1 percent after 1000 hours at 150°C.

Certain properties of thick film resistors as a function of ohmic value are predictable with qualitative conduction models.[2]

- High-ohmic-value resistors tend to have a more negative *TCR* than low-ohmic-value resistors. This is not always the case in commercially available systems due to the presence of *TCR* modifiers, but it always holds true in pure metal oxide-glass systems.
- High-ohmic-value resistors exhibit substantially more current noise than low-ohmic-value resistors as defined in MIL-STD-202. Current noise is generated when a carrier makes an abrupt change in energy levels, as it must when it makes the transition from one metal oxide particle to another across the thin film of glass. When the metal oxide particles are directly sintered together, the transition is less abrupt and little or no noise is generated.
- High-ohmic-value resistors are more susceptible to high-voltage pulses and static discharge than low-ohmic-value resistors. The high-voltage impulse breaks down the thin film of glass and forms a sintered contact, permanently lowering the value of the resistor. The effect of static discharge is highly dependent on the glass system used. Resistors from one manufacturer may drop by as much as half when exposed to static discharge, while others may be affected very little. This can be verified experimentally by heating a previously pulsed resistor to about 200°C. The value increases somewhat, indicating a healing of the glass oxide layer.

7.2.3.1 Electrical Properties of Thick Film Resistors
The electrical properties of resistors can be divided into two categories.

1. Time-zero (as-fired) properties
   a. Temperature coefficient of resistance (*TCR*)
   b. Voltage coefficient of resistance (VCR)
   c. Resistor noise
   d. High voltage discharge

2. Time-dependent (aged) properties
   a. High temperature drift
   b. Moisture stability
   c. Power handling capability

Time-Zero Properties.

Temperature coefficient of resistance (TCR).   Referring to Eq. (7.4), all real materials exhibit some change in resistance with temperature, and most are nonlinear to a greater or lesser degree. Figure 7.8 shows a graph of resistance versus temperature for a typical material. The *TCR* is a function of temperature, and is defined as the slope of the curve at the test temperature, *T*.

$$TCR(T) = \frac{dR(T)}{dT} \qquad (7.4)$$

Referring again to Fig. 7.8, the *TCR* is often linearized over a range of temperatures as depicted in Eq. (7.5).

$$TCR = \frac{\Delta R}{\Delta T} \qquad (7.5)$$

In general, this result is a small number expressed as a decimal with several preceding zeroes. For convenience, Eq. (7.5) is typically normalized to the initial value of resistance and is multiplied by one million to produce a whole number as shown in Eq. (7.6).

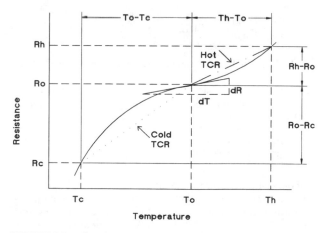

$$TCR\ (To) = \frac{dR}{dT}$$

$$TCR\ (Hot) = \frac{Rh - Ro}{Ro\ (Th - To)}$$

$$TCR\ (Cold) = \frac{Ro - Rc}{Ro\ (To - Tc)}$$

FIGURE 7.8 Resistance vs. temperature.

$$TCR = \frac{R(T_2) - R(T_1)}{R(T_1)(T_2 - T_1)} \times 10^6 ppm/°C \qquad (7.6)$$

where $R(T_2)$ = resistance at a temperature T_2

$R(T_1)$ = resistance at a temperature T_1

Most paste manufacturers present the *TCR* as two values:

- The average from 25°C to 125°C (the "hot" *TCR*)
- The average from 25°C to –55°C (the "cold" *TCR*)

An actual resistor paste carefully balances the metallic, nonmetallic, and semiconductive fractions to obtain a *TCR* as close to zero as possible. This is not a simple task, and the "hot" *TCR* may be quite different from the "cold" *TCR*. While linearization does not fully define the curve, it is adequate for most design applications.

It is important to note that the *TCR* of most materials is not linear and the linearization process is at best an approximation. For example, the actual *TCR* of most thick film resistor materials at temperatures below –40°C tends to drop very rapidly, and may be somewhat below the linearized value. The only completely accurate method of describing the temperature characteristics of a material is to examine the actual graph of temperature vs. resistance. The *TCR* for a material may be positive or negative. By convention, if the resistance increases with increasing temperature, the *TCR* is positive. Likewise, if the resistance decreases with increasing temperature, the *TCR* is negative.

In general, metals exhibit positive *TCR*s and nonmetals exhibit negative *TCR*s. In metals, the electron cloud is more disordered with increasing thermal energy and resistance increases. Nonmetals (or semiconductors), which have electrons firmly bonded to crystal locations, become more mobile with energy and are better conductors as temperature is increased. They have a negative *TCR*.

While the absolute *TCR* of a resistor is important, in some cases the ratio of resistance change between two resistors is more important. For example, assume that two resistors with a value of 1 kΩ have *TCR*s of +75 ppm and –40 ppm, respectively, at 0°C. After a change in temperature of +100°C, the respective values of resistance would be 1008 Ω and 996 Ω. The ratio of these two resistors at 100°C is

$$\frac{R_1}{R_2} = \frac{1.008}{996} = 1.012$$

which translates to a net *TCR* of 115 ppm/°C, or the algebraic difference of the *TCR*s of the individual resistors.

Now consider that the *TCR*s of the two resistors are 200 ppm/°C and 175 ppm/°C, which are substantially higher in magnitude than the previous example. For this case, R_1 = 1020 Ω and R_2 = 1018 Ω at 100°C, and the ratio is

$$\frac{R_1}{R_2} = \frac{1020}{1018} = 1.002$$

which is equivalent to a *TCR* of only 24.6 ppm/°C for the ratio.

It is therefore possible to have low *TCR* tracking with high absolute *TCR* if both resistors move in the same direction with temperature variations. In many circuit designs, this parameter is more important than a low absolute *TCR*. The *TCR* tracking of two resistors can be enhanced by following these guidelines.

- Resistors made from the same value paste will track more closely than resistors made from different decades.
- Resistors of the same length will track more closely than resistors of different lengths.
- Resistors printed with the same thickness will track more closely than resistors of different thicknesses.

Voltage coefficient of resistance (VCR). Certain resistor materials also exhibit a sensitivity to high voltages or, more specifically, to high electric fields as defined in Eq. (7.7). Note that the form of this equation is very similar to that of the TCR.

$$VCR = \frac{R(V_2) - R(V_1)}{R(V_2) - R(V_2 - V_1)} \times 10^6 \text{ ppm/°C} \tag{7.7}$$

where $R(V_1)$ = the resistance at V_1

$R(V_2)$ = the resistance at V_2

V_1 = the voltage at which $R(V_1)$ is measured

V_2 = the voltage at which $R(V_2)$ is measured

Due to the semiconductive component in resistor pastes, the VCR is invariably negative. That is, as V_2 is increased, the resistance decreases. Also, because higher resistor decade values contain more glass and oxide constituents, and are more semiconductive, higher paste values tend to have more negative VCRs than lower values.

Finally, VCR is dependent on resistor length. The voltage effect on a resistor is a gradient; it is the volts/mil rather than the absolute voltage that causes resistor shift. Therefore, long resistors show less voltage shift than short resistors, for similar compositions and voltage stress.

Resistor noise. On a fundamental level, noise occurs when an electron is moved to a higher or lower energy level. This change in energy of the electron is noise. The greater the potential difference between the energy levels, the greater the noise. Metals with many available electrons in the "electron cloud" have low noise, while semiconductive materials have fewer free electrons and exhibit higher noise.

There are two primary noise sources present in thick film resistors: thermal, or "white" noise, and current, or "pink" noise. The thermal noise is generated by the random transitions between energy levels as a result of heat and is present to a degree in all materials. Current noise occurs as a result of a transition between boundaries in a material, such as grain boundaries, where the energy levels may undergo abrupt changes from one region to another. In thick film resistors, the prime source of current noise is the thin layer of glass which may exist between the active particles.

The frequency spectrum of thermal noise is independent of frequency and is expressed in dB/Hz. The total noise is calculated by multiplying the noise figure by the bandwidth of the system. Current noise, on the other hand, has a frequency spectrum proportional to the inverse of the frequency, or $1/f$, and is expressed in units of dB/decade. The level of current noise in most applications is insignificant after the frequency exceeds 10–20 kHz.

In applications where the noise level is of significance, these guidelines may help to improve performance.

- High-value resistors exhibit a higher noise level than low value.
- Large area resistors exhibit a lower noise level.
- Thicker resistors exhibit a lower noise level.

Noise information is particularly important for low-signal applications as well as a quality check on processing. A shift in noise index, with constant resistor value, geometry and termination, indicates a process variation that must be investigated. For example, thin or underfired resistors generate higher than normal noise. The conductor/resistor interface can also be an important noise generator if it is glassy or otherwise imperfect. Finally, poor or incomplete resistor trimming also generates higher noise. A resistor noise test is an excellent method of measuring a resistor attribute not easily obtained by other methods.

Time-Dependent Properties.

High-temperature drift.   Thick film resistors in the untrimmed state exhibit a slight upward drift in value, primarily as a result of stress relaxation in the glasses that make up the body of the resistor. In properly processed resistors, the magnitude of the drift over the life of the resistor is measured in fractions of a percent and is not significant for most applications. At high temperatures, however, the drift is accelerated and may affect circuit performance in resistors that have not been properly fired or terminated or that are incompatible with the substrate.

To characterize the drift parameters, accelerated testing is frequently performed. A standard test condition is 125°C for 1000 hours at normal room humidity, corresponding to test condition S of MIL-STD-883C, method 1005.[4] This test is considered to be equivalent to end-of-life conditions. More aggressive testing conditions are 150°C for 1000 hours or 175°C for 40 hours.

Moisture stability.   Resistance drift in the presence of moisture is a discriminating and important test. The most common test condition is 85 percent relative humidity and 85°C. Past studies indicate that this condition accelerates failure in thick film circuits by a factor of almost 500, compared to normally stressed circuits in the field. Humidity testing of resistors and circuits is more costly than simple heat aging, but all the evidence indicates it is a good predictor of reliability.

Power-handling capability.   Drift due to high power is primarily due to internal resistor heating. It is different from thermal aging in that the heat is generated at the point-to-point metal contacts within the resistor film. When a resistor is exposed to elevated temperature, the entire bulk of the resistor is uniformly heated. Under power, local heating can be much greater than the surrounding area. Because lower value resistors have more metal and, therefore, many more contacts, low value resistors tend to drift less than higher value resistors under similar loads. For most resistor systems, the shape of the power aging curve is a "rising exponential" as shown in Fig. 7.9.

The most generally accepted power rating of thick film resistors to achieve a drift of less than 0.5 percent over the resistor life is 50 W/in^2 of active resistor area. If more drift can be tolerated, the resistor can be rated at up to 200 W/in^2, as catastrophic failure will not occur until this rating is exceeded by a factor of several times.

Typical properties of thick film resistors are shown in Table 7.2.

7.2.3.2 Process considerations for thick film resistors.

The process windows for printing and firing thick film resistors are extremely critical in terms of both temperature control and atmosphere control. Small variations in temperature or time at temperature can cause significant changes in the mean value and distribution of values. In general, the higher the ohmic value of the resistor, the more dramatic will be the change. As a rule, high ohmic values tend to decrease as temperature and/or time is increased, while very low values (<100 Ω/□) may tend to increase.

Thick film resistors are very sensitive to the firing atmosphere. For resistor systems used with air fireable conductors, it is critical to have a strong oxidizing atmosphere in the

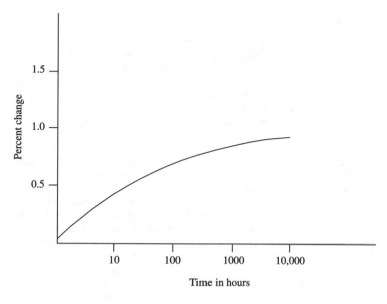

FIGURE 7.9   Resistance drift vs. time at 150°C.

firing zone of the furnace. In a neutral or reducing atmosphere, the metallic oxides which comprise the active material will reduce to pure metal at the temperatures used to fire the resistors, dropping resistor values sometimes by more than an order of magnitude. Again, high-ohmic-value resistors are more sensitive than low-ohmic-value ones. Atmospheric contaminants, such as vapors from hydrocarbons or halogenated hydrocarbons, will break down at firing temperatures, creating a strong reducing atmosphere. For example, one of the breakdown components of hydrocarbons is carbon monoxide, one of the strongest reducing agents known. Concentrations of fluorinated hydrocarbons in a firing furnace of only a few ppm can drop the value of a 100 kΩ resistor to below 10 kΩ. As a rule of thumb, no solvents should be permitted in the vicinity of a furnace used to fire thick film materials.

TABLE 7.2   Typical Thick Film Resistor Characteristics

Tolerances, as fired	± 10–$\pm 20\%$
Tolerances, laser trimmed	± 0.5–± 1
TCRs	
5–100 k$\Omega/\square$, $-55°$ to $+125°$C	± 100–± 150 ppm/°C
100 k$\Omega/\square$ to 10 M$\Omega/\square$, $-55°$ to $+125°$C	± 150–± 750 ppm/°C
Resistance drift after 1000 h at 150°C, no load	$+0.3$ to -0.3%
Resistance drift after 1000 h at 85°C, with 25 W/in^2	0.25–0.3%
Resistance drift, short-term overload (2.5 times rated voltage)	$<0.5\%$
Voltage coefficient	-20 ppm/V · in
Noise (Quan-Tech)	
100 $\Omega/\square$	-30 to -20 dB
100 k$\Omega/\square$	0 to $+20$ dB
Power rating	40–50 W/in^2

7.2.4 Thick Film Dielectric Materials

Thick film dielectric materials are used primarily as insulators between conductors, either as simple crossovers or in complex multilayer structures. Small openings, or *vias*, may be left in the dielectric layers so that adjacent conductor layers may interconnect. In complex structures, as many as several hundred vias per layer may be required. In this manner, complex interconnection structures may be created. Although the majority of thick film circuits can be fabricated with only three layers of metallization, others may require several more. If more than three layers are required, the yield begins dropping dramatically with a corresponding increase in cost.

Dielectric materials used in this application must be of the *devitrifying* or *recrystallizable* type. These materials in the paste form are a mixture of glasses that melt at a relatively low temperature. During firing, when they are in the liquid state, they blend together to form a uniform composition with a higher melting point than the firing temperature. Consequently, on subsequent firings they remain in the solid state, which maintains a stable foundation for firing sequential layers. By contrast, vitreous glasses always melt at the same temperature and would be unacceptable for layers to either "sink" and short to conductor layers underneath, or "swim" and form an open circuit. Additionally, secondary loading of ceramic particles is used to enhance devitrification and to modify the temperature coefficient of expansion (TCE).

Dielectric materials have two conflicting requirements in that they must form a continuous film to eliminate short circuits between layers while, at the same time, they must maintain openings as small as 0.010 in. In general, dielectric materials must be printed and fired twice per layer to eliminate pinholes and prevent short circuits between layers.

The TCE of thick film dielectric materials must be as close as possible to that of the substrate to avoid excessive *bowing*, or warpage, of the substrate after several layers. Excessive bowing can cause severe problems with subsequent processing, especially where the substrate must be held down with a vacuum or where it must be mounted on a heated stage. In addition, the stresses created by the bowing can cause the dielectric material to crack, especially when it is sealed within a package. Thick film material manufacturers have addressed this problem by developing dielectric materials that have an almost exact TCE match with alumina substrates. Where a serious mismatch exists, matching layers of dielectric must be printed on the bottom of the substrate to minimize bowing, which obviously increases the cost.

Dielectric materials with higher dielectric constants are also available for manufacturing thick film capacitors. These generally have a higher loss tangent than chip capacitors and utilize a great deal of space. While the initial tolerance is not good, thick film capacitors can be trimmed to a high degree of accuracy.

Another consideration for selecting a thick film dielectric is the compatibility with a resistor system. As circuits become more complex, the necessity for printing resistors on dielectric as opposed to directly on the substrate becomes greater. In addition, it is difficult to print resistors on the substrate in the proximity of dielectric several dielectric layers. The variations in the thickness affect the thickness of the substrate and, therefore, affect the spread of the resistor values.

7.2.4.1 Overglaze Materials.
Dielectric overglaze materials are vitreous glasses designed to fire at a relatively low temperature, usually around 550°C. They are designed to provide mechanical protection to the circuit, to prevent contaminants and water from spanning the area between conductors, to create solder dams, and to improve the stability of thick film resistors after trimming.

Noble metals, such as gold and silver, are somewhat soft by nature. Gold, in particular, is the most malleable of all metals. When subjected to abrasion or to scraping by a sharp

object, the net result is likely to be a bridging of the metal between conductors, resulting in a short circuit. A coating of overglaze can minimize the probability of damage and can also protect the substrates when they are stacked during the assembly process.

Contaminants that are ionic in nature, combined with water in the liquid form, can accelerate metal migration between two conductors. An overglaze material can help to limit the amount of contaminant which actually contacts the surface of the ceramic, and can help to prevent a film of water from forming between conductors. A ceramic substrate is a very "wettable" surface in that it is microscopically very rough, and the resulting capillary action causes the water to spread rapidly, creating a continuous film between conductors. A vitreous glass, being very smooth by nature, causes the water to "bead up," much like on a waxed surface, helping to prevent the water from forming a continuous film between conductors, thereby inhibiting the migration process.

When soldering a device with many leads, it is imperative that the volume of solder under each lead be the same. A well designed overglaze pattern can prevent the solder from wetting other circuit areas and flowing away from the pad, keeping the solder volume constant. In addition, overglaze can help to prevent solder bridging between conductors.

Overglaze material has long been used to stabilize thick film resistors after laser trim. In this application, a green or brown pigment is added to enhance the passage of a YAG (yttrium-aluminum-garnet) laser beam. Colors toward the shorter wavelength end of the spectrum, such as blue, tend to reflect a portion of the YAG laser beam and reduce the average power level at the resistor. There is some debate as to the effectiveness of overglaze in enhancing the resistor stability, particularly with high ohmic values. Several studies have shown that, while overglaze is undoubtedly helpful at lower values, it can actually increase the resistor drift of high-value resistors by a significant amount.

7.2.5 Conclusions

It is apparent that cermet thick film materials are complex structurally and electrically. More than 120 variables related to material properties and processes have been identified with thick film resistors. Clearly, it is impossible to control all these variables in a manufacturing environment. To achieve the best results, it is important that the user know how the various constituents interact and the role each of them plays in the final product. While the user need not know all the ingredients making up a particular paste, it is critical that the degree of compatibility with other pastes to be used in the circuit be assured. Very few hybrid circuits can be made with only one thick film paste, and if one or more of the pastes are incompatible with the others, unexpected reactions can occur that result in circuit failure. Usually, the only source of information, other than direct experimentation, is the paste manufacturer. A close working relationship with the manufacturer is a vital element toward creating a successful operation.

7.3 THIN FILM TECHNOLOGY

Thin film technology is a subtractive technology in that the entire substrate is coated with several layers of metallization and the unwanted material is etched away in a succession of photoetching processes. The use of photolithographic processes to form the patterns enables much narrower and more well defined lines than can be formed by the thick film process. This feature promotes the use the thin film technology for high-density and high-frequency applications.

Thin film circuits typically consist of three layers of material deposited on a substrate. The bottom layer serves two purposes: it is the resistor material and also provides the adhesion to the substrate. The middle layer acts as an interface between the resistor layer and the conductor layer, either by improving the adhesion of the conductor or by preventing diffusion of the resistor material into the conductor. The top layer acts as the conductor layer.

7.3.1 Thin Film Deposition Processes

The term *thin film* refers more to the manner in which the film is deposited onto the substrate as opposed to the actual thickness of the film. Thin films are typically deposited by one of the vacuum deposition techniques or by electroplating.

7.3.1.1 Sputtering   Sputtering is the prime method by which thin films are applied to substrates. In ordinary triode sputtering, as shown in Fig. 7.10, a current is established in a conducting plasma formed by striking an arc in a partial vacuum of approximately 10 μ pressure. The gas used to establish the plasma is typically an inert gas, such as argon, which does not react with the target material. The substrate and a target material are placed in the plasma with the substrate at ground potential and the target at a high potential, which may be ac or dc. The high potential attracts the gas ions in the plasma to the point where they collide with the target with sufficient kinetic energy to dislodge microscopically sized particles with enough residual kinetic energy to travel the distance to the substrate and adhere.

The adhesion mechanism of the film to the substrate is an oxide layer that forms at the interface. The bottom layer (first layer sputtered) must therefore be a material that oxidizes readily. Gold and silver, for example, being noble metals, do not adhere well to ceramic surfaces. The adhesion is enhanced by presputtering the substrate surface by random bombardment of argon ions prior to applying the potential to the target. This process removes several atomic layers of the substrate surface, creating a large number of broken oxygen bonds and promoting the formation of the oxide interface layer. The oxide formation is further enhanced by the residual heating of the substrate that results from the transfer of the kinetic energy of the sputtered particles to the substrate when they collide.

Ordinary triode sputtering is a very slow process, requiring hours to produce useable films. By utilizing magnets at strategic points, the plasma can be concentrated in the vicinity of the target, greatly speeding up the deposition process. The potential that is applied to the target is typically RF energy at a frequency of 13.56 MHz. The RF energy may be generated by a conventional electronic oscillator or by a magnetron. The magne-

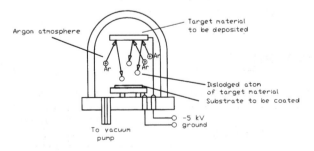

FIGURE 7.10   A dc sputtering chamber.

tron is capable of generating considerably more power with a correspondingly higher deposition rate.

By adding small amounts of other gases, such as oxygen and nitrogen to the argon, it is possible to form oxides and nitrides of certain target materials on the substrate. It is this technique, called *reactive sputtering,* which is used to form tantalum nitride, a common resistor material.

7.3.1.2 Evaporation.

7.3.1.2 Evaporation. The evaporation of a material into the surrounding area occurs when the vapor pressure of the material exceeds the ambient pressure and can take place from either the solid state of the liquid state. In the thin film process, the material to be evaporated is placed in the vicinity of the substrate and heated until the vapor pressure of the material is considerably above the ambient pressure. The evaporation rate is directly proportional to the difference between the vapor pressure of the material and the ambient pressure and is highly dependent on the temperature of the material.

Evaporation must take place in a relatively high vacuum ($<10^{-6}$ torr) for three reasons.

1. To lower the vapor pressure required to produce an acceptable evaporation rate, thereby lowering the required temperature required to evaporate the material.

2. To increase the mean free path of the evaporated particles by reducing the scattering due to gas molecules in the chamber. As a further result, the particles tend to travel in more of a straight line, improving the uniformity of the deposition.

3. To remove atmospheric contaminants and components, such as oxygen and nitrogen, which may react with the evaporated film.

At an ambient pressure of 10^{-7} torr, a vapor pressure of 10^{-2} torr is required to produce an acceptable evaporation rate. A table of common materials, their melting points, and the temperature at which the vapor pressure is 10^{-2} torr is shown in Table 7.3.

TABLE 7.3 Melting Points and $P_v = 10^{-2}$ Torr Temperatures of Some Common Metals Used in Thin Film Applications

Material	Melting point (°C)	Temperature $P_v = 10^{-2}$ torr (°C)
Aluminum	659	1220
Chromium	1900	1400
Copper	1084	1260
Germanium	940	1400
Gold	1063	1400
Iron	1536	1480
Molybdenum	2620	1530
Nickel	1450	2530
Platinum	1770	2100
Silver	961	1030
Tantalum	3000	3060
Tin	232	1250
Titanium	1700	1750
Tungsten	3380	3230

The "refractory" metals, or metals with a high melting point such as tungsten, titanium, or molybdenum, are frequently used as carriers, or *boats*, to hold other metals during the evaporation process. To prevent reactions with the metals being evaporated, the boats may be coated with alumina or other ceramic materials.

If it is assumed that the evaporation takes place from a point source, the density of the evaporated particles assumes a cosine distribution from the normal. The distance of the substrate from the source then becomes a compromise between deposition uniformity and deposition rate; if the substrate is closer (further away), the deposition is greater (lesser) and the deposition is less (more) uniform over the face of the substrate.

In general, the kinetic energy of the evaporated particles is substantially less than that of sputtered particles. This requires that the substrate be heated to about 300°C to promote the growth of the oxide adhesion interface. This may be accomplished by direct heating of the substrate mounting platform or by radiant infrared heating.

There are several techniques by which evaporation can be accomplished. The two most common of these are resistance heating and electron-beam (E-beam) heating.

Evaporation by resistance heating, as depicted in Fig. 7.11, usually takes place from a boat made with a refractory metal, a ceramic crucible wrapped with a wire heater, or a wire filament coated with the evaporant. A current is passed through the element and the heat generated heats the evaporant. It is somewhat difficult to monitor the temperature of the melt by optical means due to the propensity of the evaporant to coat the inside of the chamber, and control must be done by empirical means. There exist closed loop systems that can control the deposition rate and the thickness, but these are quite expensive. In general, adequate results can be obtained from the empirical process if proper controls are used.

The E-beam evaporation method takes advantage of the fact that a stream of electrons accelerated by an electric field tend to travel in a circle when entering a magnetic field. This phenomenon is utilized to direct a high-energy stream of electrons onto an evaporant source. The kinetic energy of the electrons is converted into heat when they strike the evaporant. E-beam evaporation is somewhat more controllable since the resistance of the boat is not a factor and the variables controlling the energy of the electrons are easier to measure and control. In addition, the heat is more localized and intense, making it possible to evaporate metals with higher 10^{-2} torr temperatures and lessening the reaction between the evaporant and the boat.

7.3.1.3 Comparison between sputtering and evaporation.

While evaporation provides a more rapid deposition rate, there are certain disadvantages when compared with sputtering.

1. It is difficult to evaporate alloys such as NiCr due to the difference between the 10^{-2} torr temperatures. The element with the lower temperature tends to evaporate somewhat faster, causing the composition of the evaporated film to be different from the composition of the alloy. To achieve a particular film composition, the composition of

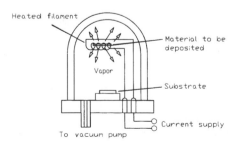

FIGURE 7.11 Thermal vacuum evaporation system.

the melt must contain a higher portion of the material with the higher 10^{-2} torr temperature and the temperature of the melt must be tightly controlled. By contrast, the composition of a sputtered film is identical to that of the target.

2. Evaporation is limited to the metals with lower melting points. Refractory metals, ceramics, and other insulators are virtually impossible to deposit by evaporation.

3. Reactive deposition of nitrides and oxides is very difficult to control.

The comparisons between sputtering and evaporation are summarized in Table 7.4.

TABLE 7.4   Comparison of Evaporation and Sputtering Processes for Nichrome

	Vacuum evaporation	Sputtering
Mechanism	Thermal energy	Momentum transfer
Deposition rate	Can be high (to 750,000 Å/min)	Low (20–100 Å/min) except for some metals (Cu = 10,000 Å/min)
Control of deposition	Sometimes difficult	Reproducible and easy to control
Coverage for complex shapes	Poor, line of sight	Good, but nonuniform thickness
Coverage into small blind holes	Poor, line of sight	Poor
Metal deposition	Yes	Yes
Alloy deposition	Yes (flash evaporation)	Yes
Refractory metal deposition	Yes (by e-beam)	Yes
Plastics	No	Some
Inorganic compounds (oxides, nitrides)	Generally no	Yes
Energy of deposited species	Low (0.1–0.5 eV)	High (1–> 100 eV)
Adhesion to substrate	Good	Excellent

7.3.1.4 Electroplating.   Electroplating is accomplished by applying a potential between the substrate and the anode, which are suspended in a conductive solution of the material to be plated. The plating rate is a function of the potential and the concentration of the solution. In this manner, most metals can be plated to a metal surface.

In the thin film technology, it is a common practice to sputter a film of gold that is only a few Å thick and to build up the thickness of the gold film by electroplating. This is considerably more economical and results in much less target usage. For added savings, some companies apply photoresist to the substrate and electroplate gold only where actually required by the pattern.

7.3.2   Photolithographic Processes

In the photolithographic process, the substrate is coated with a photosensitive material which is exposed with ultraviolet light through a pattern formed on a glass plate. The photoresist may be of the positive or negative type, with the positive type being prevalent due to its inherently higher resistance to the etchant materials. The unwanted material, which is not protected by the photoresist, may be removed by "wet," or chemical etching or by "dry" or sputter etching.

In general, two masks are required, one corresponding to the conductor pattern and one corresponding to a combination of both the conductor and resistor patterns—generally referred to as the "composite" pattern. As an alternative to the composite mask, a mask that contains only the resistor pattern plus a slight overlap onto the conductor to allow for misalignment may be used. The composite mask is preferred because it allows a second gold etch process to be performed to remove any bridges or extraneous gold to be removed which might have been left from the first etch.

Sputtering may also be used to etch thin films. In this technique, the substrate is coated with photoresist and the pattern exposed in exactly the same manner as with chemical etching. The substrate is then placed in a plasma and connected to a potential. In effect, the substrate acts as the target during the sputter etching process, with the unwanted material being removed by the impingement of the gas ions on the exposed film. The photoresistive film, being considerably thicker than the sputtered film, is not affected. Sputter etching has two major advantages over chemical etching.

1. There is virtually no undercutting of the film. The gas ions strike the substrate in approximately a cosine distribution with respect to the normal of the substrate. This means that virtually no ions strike the film tangentially, leaving the edges intact. This results in more uniform line dimensions, which further results in better resistor uniformity. By contrast, the rate of the chemical etching process in the tangential is the same as in the normal direction, which results in the undercutting of the film by a distance equal to the thickness.

2. The potent chemicals used to etch thin films are no longer necessary, with less hazard to personnel and no disposal problems.

7.3.3 Thin Film Materials

Virtually any inorganic material may be deposited by the sputtering process, but the outgassing of most organic materials is too extensive to allow sputtering to take place. A wide variety of substrate materials are also available, but these must contain an oxygen compound to permit adhesion of the film.

7.3.3.1 Thin film resistors.   Materials used for thin film resistors must perform a dual role in that they must also provide the adhesion to the substrate, which narrows the choice to those materials which form oxides. The resistor film begins forming as single points on the substrate in the vicinity of substrate faults or other irregularities which might have an excess of broken oxygen bonds. The points expand into islands that, in turn, join to form continuous films. The regions where the islands meet are called "grain boundaries," which are a source of collisions for the electrons. The more grain boundaries that are present, the more negative will be the *TCR*. Unlike thick film resistors, however, the boundaries do not contribute to the noise level. Furthermore, laser trimming does not create microcracks in the glass-free structure, and the inherent mechanisms for resistor drift are not present in thin films. As a result, thin film resistors have better stability, noise, and *TCR* characteristics than thick film resistors.

The most common types of resistor material are nichrome (NiCr) and tantalum nitride (TaN). Although NiCr has excellent stability and *TCR* characteristics, it is susceptible to corrosion by moisture if not passivated by sputtered quartz or by evaporated silicon monoxide (SiO). TaN, on the other hand, may be passivated by simply baking in air for a few minutes. This feature has resulted in the increased use of TaN at the expense of NiCr, especially in military programs. The stability of passivated TaN is comparable to that of passivated NiCr, but the *TCR* is not as good unless annealed for several hours in a vacuum to minimize the effect of the grain boundaries. Both NiCr and TaN have a relatively low maximum sheet resistivity on alumina—about 400 $\Omega/\square$ for NiCr and 200 $\Omega/\square$ for TaN. This requires complex patterns to achieve a high value of resistance, resulting in a large required area and the potential for low yield. Chrome disilicide has a maximum sheet resistance of $1000\Omega/\square$ and overcomes this limitation to a large extent.

The TaN process is more often used due to the inherently high stability. In this process N_2 is introduced into the argon gas during the sputtering process, forming TaN by reacting with pure Ta atoms on the surface of the substrate. By heating the film in air at about 425°C for 10 minutes, a film of TaO is formed over the TaN that is virtually impervious to further O_2 diffusion at moderately high temperatures, which helps to maintain the composition of the TaN film and stabilizes the value of the resistor. TaO is essentially a dielectric and, during the stabilization of the film, the resistor value is increased. The amount of increase for a given time and temperature is dependent on the sheet resistivity of the film. Films with a lower sheet resistivity increase proportionally less than those with a higher sheet resistivity. The resistance increases as the film is heated longer, making it possible to control the sheet resistivity to a reasonable accuracy on a substrate-by-substrate basis.

The properties of tantalum nitride and nichrome resistors are shown in Tables 7.5 and 7.6 respectively.

TABLE 7.5   Characteristics of Tantalum Nitride Resistors

Sheet resistance	20–150 $\Omega/\square$, 100 $\Omega/\square$ (typical)
Sheet resistance tolerance	±10% of nominal value
TCR	−75 ± 50 ppm/°C (typical), 0 ± 25 ppm/°C with vacuum anneal
TCR tracking*	<2 ppm
Resistance drift (1000 h at 150°C in air)	<1000 ppm
Ratio tracking	5 ppm
Resistor tolerance after anneal and laser trim	±0.10% standard, ±0.03% bridge trim
Noise (100 Hz to 1 MHz)	< −40 dB

* −55° to 125°C.

TABLE 7.6   Characteristics of Nichrome Resistors

Sheet resistance	25–300 $\Omega/\square$, 100–200 $\Omega/\square$ (typical)
Sheet resistance tolerance	±10% of nominal value
TCR	0 ± 50 ppm/°C, 0 ± 25 ppm/°C (with special anneal)
TCR tracking*	2 ppm
Resistance drift	<2000 ppm after 1000 h at 150°C <1000 ppm with special anneal <200 ppm, sputtered films with 350°C anneal
Ratio tracking	5 ppm
Resistor tolerance after anneal and laser trim	±0.1
Noise (100 Hz to 1 MHz)	−35 dB (maximum)

* −55 to +125°C.

7.3.4 Barrier Materials

When Au is used as the conductor material, a barrier material between the Au and the resistor is required. When gold is deposited directly on NiCr, the Cr has a tendency to diffuse through the Au to the surface, which interferes with both wire bonding and eutectic die bonding. To alleviate this problem, a thin layer of pure Ni is deposited over the NiCr. In addition, the Ni improves the solderability of the surface considerably. The adhesion of Au to TaN is very poor. To provide the necessary adhesion, a thin layer of 90Ti/10W may be used between the Au and the TaN.

7.3.5 Conductor Materials

Gold is the most common conductor material used in thin film hybrid circuits because of the ease of wire and die bonding and the high resistance of the gold to tarnish and corrosion. Aluminum and copper are also frequently used in certain applications. It should be noted that copper and aluminum will adhere directly to ceramic substrates, but gold requires one or more intermediate layers since it does not form the necessary oxides for adhesion.

7.4   COMPARISON OF THICK AND THIN FILM

Although the thin film process provides better line definition, smaller line geometry, and better resistor properties, it has several disadvantages compared to thick film.

1. Due to the added labor involved, the thin film process is almost always more expensive than the thick film process. Only in the case where a number of thin film circuits can be fabricated on a single substrate can thin film compete in price.
2. Multilayer structures are extremely difficult to fabricate. While they are possible with multiple deposition and etching processes, this is a very expensive and labor-intensive process and is limited to very few applications.
3. The designer is, in most cases, limited to a single sheet resistivity. This requires a large area to fabricate both large and small value resistors.

Table 7.7 summarizes the differences between thick and thin films. Design guidelines for the two technologies are shown in Figs. 7.12–7.28.

7.5   COPPER METALLIZATION TECHNOLOGIES

The thick film and thin film technologies are limited in their ability to deposit films with a thickness greater than 1 mil (25 µ). This factor directly affects the ohmic resistance of the circuit traces and affects their ability to handle large currents or high frequencies. The copper metallization technologies provide conductors with greatly increased conductor thickness which offer improved circuit performance in many applications. There are three basic technologies available to the hybrid designer: direct bond copper (DBC), active metal braze (AMB), and the various methods of plating copper directly to ceramic.

TABLE 7.7 Comparison of Design Rules for Pure Copper Metallization

Technology	Line width, in		Etch factors,* in	Registration,† in	Copper thickness,‡ in			Via diameters, in		Integral leads from edge of ceramic	Camber,§ inch per linear inch
	Min.	Typical			Min.	Typical	Max.	Min.	Typical		
Direct bond copper	0.015	0.020	0.004 to 0.008 with 0.020 pullback required from ceramic edge	±0.008	0.005	0.012	0.500	0.016	0.064	0.100 in sq. required on ceramic for integral leads	±0.004
Plated copper	0.002	0.005	0.0005	±0.005	0.001	0.002	0.005	0.005	0.025	Leads not possible	±0.003
Active metal bond copper	0.015	0.020	0.004 to 0.008 with 0.020 pullback required from ceramic edge	±0.010	0.008	0.010	0.012	Vias not possible		0.100 in sq. required on ceramic for integral leads	±0.004

*The width of the artwork pattern determines final conductor width. Etching will reduce the original artwork width by several mils, depending upon copper thickness. Therefore the artwork must have a wider dimension to compensate. This is called the etch factor, as shown in Fig. 5.3.
†The artwork is registered to a feature on the ceramic substrate, usually a corner of the ceramic surface.
‡Copper layers can be lapped to a lesser thickness after metallization.
§Assume that both ceramic sides have copper metallization. Then camber depends on differences in percentage of ceramic surface area coverage by copper.

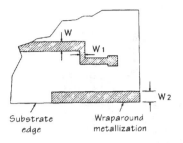

Dimension	Thick film, mils		Thin film, mils		Remarks
	Preferred	Minimum	Preferred	Minimum	
W	As required for current-carrying				*
	capability or electrical resistance				
W_1	10	5	5	2	
W_2	20	15	15	10	†

*I_{MAX} may be calculated from Eq. (11.74).
†Wraparound metallization for thick film printing requires special tooling. Do NOT use laser-scribed substrates for either thick film or thin film fabrication with this process unless the substrates have been properly annealed, as the metallization will not adhere well to the initial laser-scribed surface.

FIGURE 7.12   Conductor line widths.

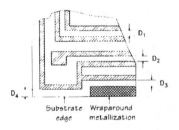

Dimension	Thick film, mils		Thin film, mils		Remarks
	Preferred	Minimum	Preferred	Minimum	
D_1	10	7.5	5	2	*
D_2	15	10	7.5	5	†
D_3	15	10	7.5	5	
D_4	10	10	7.5	7.5	‡

*Conductor lines ≤ mils in length.
†Conductor lines > 15 mils in length.
‡From minimum substrate size.

FIGURE 7.13   Conductor line spacing.

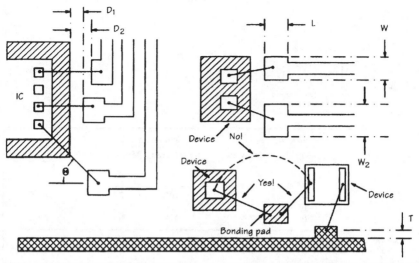

| Dimension | Thick film, mils | | Thin film, mils | | Remarks |
	Preferred	Minimum	Preferred	Minimum	
L	10	7.5	7.5	5	(1)
W	10	7.5	7.5	5	(1)
W_2	15	12.5	12.5	10	(2)
Θ	90	45	90	45	(3)
D_1	75	100 max.	75	100 max.	(4)
D_2	30	15	30	15	(5)
W_2	15	10	NA	NA	(6)
T	25	20	12.5	10	(7)

1. Use 10 × 10 minimum for automatic wire bonding.
2. Width for double wire.
3. Locating angle, die edge to center of wirebond site.
4. *Maximum* distance from edge of die to edge of pad.
5. Distance from mounting pad to wirebonding pad.
6. Distance from center of wirebond site to dielectric.
7. Wraparound metallization distance to wirebond site.

FIGURE 7.14 Bonding pad sizes.

7.5.1 Direct Bond Copper

Copper may be bonded to alumina ceramic by placing a film of copper in contact with the alumina and heating to about 1065°C, just below the melting point of copper, 1083°C. At this temperature, a combination of 0.39 percent O_2 and 99.61 percent Cu form a liquid that can melt, wet, and bond tightly to the surfaces in contact with it when cooled to room temperature. In this process, the copper remains in the solid state during the bonding process and a strong bond is formed between the copper and the alumina with no intermediate material required. The metallized substrate is slowly cooled to room temperature at a controlled rate to avoid quenching. To prevent excessive bowing of the substrate, copper must be bonded to both sides of the substrate to minimize stresses due to the difference in TCE between copper and alumina.

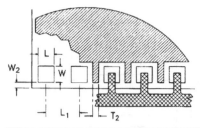

Dimension	Thick film, mils		Thin film, mils		Remarks
	Preferred	Minimum	Preferred	Minimum	
L	60	30	60	30	
W	60	50	60	50	
L_1	100	50	100	50	*
W_2	10	5	10	5	†
T_2	10	7.5	NA	NA	‡

*Pad spacing, center to center.
†Distance from edge of substrate.
‡Screened dielectric between pads.

FIGURE 7.15   Pads for lead frame attachment.

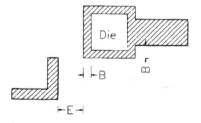

Dimension	Thick film, mils		Thin film, mils		Remarks
	Preferred	Minimum	Preferred	Minimum	
B	10	5	10	5	*,‡
B	10	10	10	10	†,‡
E	15	10	10	2	§

*Distance from edge of chip to edge of mounting pad; chip size≤100 mils.
†Chip size > 100 mils.
‡Distance=10 mils *maximum*.
§Pad distance to other metallization.

FIGURE 7.16   Mounting pads for semiconductor die.

In this manner, a film of copper from 5–25 mils thick can be bonded to a substrate and a metallization pattern formed by photolithographic etching. For subsequent processing, the copper is usually plated with several hundred microinches of nickel to prevent oxidation. The nickel-plated surface is readily solderable, and aluminum wire bonds to nickel is one of the most reliable combinations.[3] Aluminum wire bonded directly to copper is not as reliable and may result in failure on exposure to heat and/or moisture.[4]

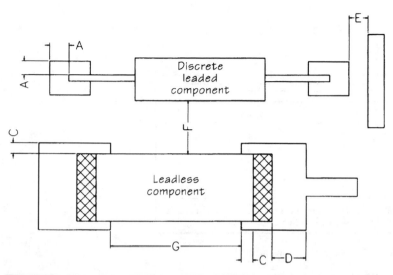

Dimension	Thick film, mils		Thin film, mils	
	Preferred	Minimum	Preferred	Minimum
A	10	10	10	10
C	10	5	10	5
D	20	10	20	10
E	15	15	15	15
F	40	40	40	40
G	AR	30	AR	30

1. Metallized pad overlap for leaded component.
2. Distance from edge of pad to chip metallization on leadless chip side and underneath.
3. Distance from edge of pad to chip metallization end.
4. Pad distance to other metallization.
5. Distance from leaded component.
6. Spacing between termination for leadless chip components.

FIGURE 7.17   Pads for discrete components.

Multilayer structures of up to four layers have been formed by etching patterns on both sides of two substrates and bonding them to a common alumina substrate. Interconnections between layers are made by inserting oxidized copper pellets into holes drilled or formed in the substrates prior to firing. Vias may also be created by using one of the copper plating processes.

Integral leads extending beyond the substrate edge may be created by bonding a copper sheet larger than the ceramic substrate and etching a lead frame at the same time as the pattern.

The line and space resolution of DBC are limited due to the difficulty of etching thick layers of metal without substantial undercutting. Special design guidelines must be followed to allow for this factor, as shown in Fig. 7.29. While the DBC technology does not

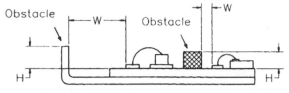

H, mils	W,* mils
20	40
40	45
60	50
80	55
100	65

*The W dimension may be smaller in thermosonic bonding if a tip with a smaller angle is used.

FIGURE 7.18   Pads for lead frame attachment.

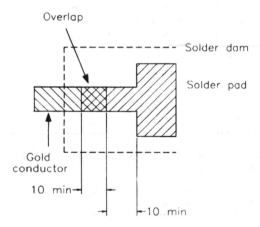

FIGURE 7.19   Wirebond distance from obstacle.

have a resistor system, the thick film technology can be used in conjunction with DBC to produce integrated resistors and areas of high-density interconnections.

Aluminum nitride can also be used with copper, although the consistency of such factors as grain size and shape are not as good as aluminum oxide at this time. Additional preparation of the AlN surface is required to produce the requisite layer of oxide necessary to produce the bond. This can be accomplished by heating the substrate to about 1250°C in the presence of oxygen.

Direct bond copper offers considerable advantages when packaging power circuits. The thick layer of copper can handle considerable current without excessive voltage drops and heat generation and allows the heat to spread rapidly outward from semiconductor devices, which lowers the thermal impedance of the system dramatically. The layer of copper on the bottom also contributes to heat spreading.

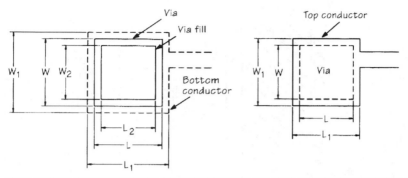

Dimension	Thick film, mils		Thin film, mils		Remarks
	Preferred	Minimum	Preferred	Minimum	
L	12.5	10	NA	NA	(1), (7), (8)
L_1	$L + 5$	L	NA	NA	(2)
L_2	L	L	NA	NA	(3), (9)
W	12.5	10	NA	NA	(4), (7), (8)
W_1	$L + 5$	L	NA	NA	(5)
W_2	W	W	NA	NA	(6), (9)

1. Via length.
2. Insert top and bottom.
3. Length of via fill.
4. Via width.
5. Conductor overlap, top and bottom.
6. Width of via fill.
7. It is preferred for inspection purposes that $L=W$.
8. It is preferred for inspection purposes that all vias on a substrate be the same size.
9. Via fills are required with more than two metallization layers. For more than two layers, the top via fill may be omitted.

FIGURE 7.20   Via and conductor pad sizes.

7.5.2   Plated Copper Technology

While a copper film can be mechanically bonded to a rough substrate surface, the adhesion is usually inadequate for most applications. The various methods of plating copper to a ceramic all begin with the formation of a conductive film on the surface. This film may be vacuum deposited by thin film methods, screen printed by thick film processes, or deposited with the aid of a catalyst. A layer of electroless copper may be plated over the conductive surface, followed by a layer of electrolytic copper to increase the thickness.

A pattern may be generated in the plated surface by one of two methods. Conventional photolithographic methods may be used to etch the pattern, but this may result in undercutting and loss of resolution when used with thicker films. To produce more precise lines, a dry film photoresist may be utilized to generate a pattern on the electroless copper film which is the negative of the one required for etching. The traces may then be electroplated to the desired thickness using the photoresist pattern as a mold. Once the photoresist pattern is removed, the entire substrate may be immersed in an appropriate etchant to remove the unwanted material between the traces. Plated copper films created in this manner may be fired at an elevated temperature in a nitrogen atmosphere to improve the adhesion.

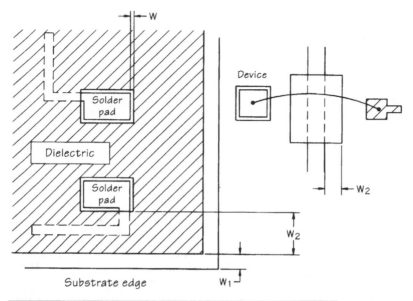

Substate edge

Dimension	Thick film, mils		Remarks
	Preferred	Minimum	
W	10	5	*, †
W_1	10	5	‡
W_2	10	5	§

*Distance to metallized pad for solder or epoxy attach.
†Low temperature dielectric preferred for solder attach, high temperature preferred for epoxy attach.
‡Distance to edge of substrate.
§Dielectric overlap when covering conductor.

FIGURE 7.21   Thick film dielectric spacing and size.

7.5.3   Active Metal Brazing Copper Technology

The active metal brazing (AMB) process utilizes one or more of the metals in the IV-B column of the periodic table, such as titanium, hafnium, or zirconium, to act as an activation agent with ceramic. These metals are typically alloyed with other metals to form a braze which can be used to bond copper to ceramic. One such example is an alloy of 70Ti/15Cu/15Ni which melts at 960—1000°C. Numerous other alloys can also be used.[2]

The braze may be applied in the form of a paste, a powder, or a film. The combination is heated to the melting point of the selected braze in a vacuum to minimize oxidation of the copper. The active metal forms a liquidus with the oxygen in the system which acts to bond the metal to the ceramic. After brazing, the copper film may be processed in much the same manner as DBC.

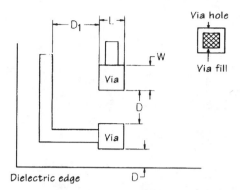

	Thick film, mils		
Dimension	Preferred	Minimum	Remarks
L	20	10	*
W	20	10	*
D	20	10	†
D_1	15	10	‡
D_2	10	10	§

*It is preferred for inspection purposes that all vias on a substrate be the same size.
†This also applies to vias on adjacent layers. Vias must be staggered unless used for heat dissipation purposes.
‡Via distance to metallization.
§Conductor distance to metallization.

FIGURE 7.22 Dielectric via sizes and locations.

7.5.4 Comparison of Copper Metallization Technologies

The DBC and AMB processes are advantageous in applications where a thicker copper layer is beneficial. In particular, these technologies are very suitable for use in power circuits where their ability to handle high currents and to aid in dissipating heat can be utilized. Plated copper is most applicable to applications where fine lines and precise geometries are required, such as RF circuits. Design guidelines for each of these technologies is given in Table 7.7.

7.6 SUMMARY OF SUBSTRATE METALLIZATION TECHNOLOGIES

The applications of the various substrate metallization technologies rarely overlap. Given a particular set of requirements, the choice is usually very apparent. Table 7.8 illustrates the properties of the different technologies.

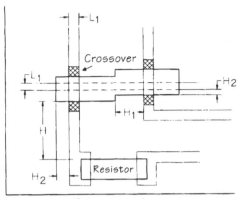

Substrate edge

| Dimension | Thick film, mils | | Remarks |
	Preferred	Minimum	
H	50	30	*
H_1	20	10	†
H_2	20	10	‡
L_1	15	10	§

*Distance to resistor.
†Conductor distance to dielectric step. Use staggered crossovers or H patterns when two or more adjacent lines must cross in the same direction. (See Fig. 11.40.)
‡Overlap distance on crossover.
§Conductor lines overlap.

FIGURE 7.23   Dielectric distance on crossovers.

7.7 RESISTOR TRIMMING

Thick film resistors in the as-fired state can be expected to hold a tolerance of ±20 percent and thin film resistors can be processed to a tolerance of ±10 percent. While these tolerances may be adequate for some applications, other applications require tighter tolerances. One of the major advantages of the hybrid technology is the ability to adjust the resistance to a predetermined value by laser or abrasive trimming. By removing a portion of the resistor, the geometry can be altered to produce a higher value. It is convenient to think of thick and thin film resistors as one-way potentiometers in which the value can only be increased. The resistors must therefore be designed to a target value such that all the resistors are below the desired final value.[2]

For laser trimming, a neodymium-doped yttrium aluminum garnet (Nd:YAG) laser operating at 1.06 μm wavelength is typically utilized. The intensity of the laser beam evaporates the resistor material at the point of contact, creating a narrow path, called a "kerf," in the resistor, as shown in Fig. 7.30. The laser acts as a closed-loop control system which continuously monitors the value of the resistor being trimmed and compares it to a preset value. When the resistor reaches the desired value, the laser stops cutting.

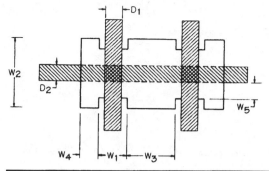

| Dimension | Thick film, mils | | Remarks |
	Preferred	Minimum	
D_1	10	5	(1)
D_2	10	5	(2)
W_1	$D_1 + 10$	$D_1 + 10$	(3)
W_2	$D_1 + 10$	$D_1 + 10$	(4)
W_3	10	5	(5)
W_4	15	10	(6)
W_5	10	5	(7)

1. Top conductor widths.
2. Bottom crossover width.
3. Width of crossover.
4. Overlap of bottom conductor.
5. Width of dielectric print between crossovers.
6. Width of crossover arm.
7. Extension of crossover past conductor.

FIGURE 7.24   H pattern crossovers.

Due to the large difference in temperature between the trim area and the surrounding resistor material, laser trimming may create tiny microcracks in the glass structure of thick film resistors which may propagate over the life of the resistor, resulting in a slight upward drift of the resistor value. If the trim parameters are tightly controlled and if the penetration of the laser cut into the resistor is restricted to 50 percent of the width, the drift can be minimized. A properly trimmed thick film resistor can be expected to maintain a tolerance of less than 0.5 percent over the life of the resistor without special processing. The tolerance can be reduced to less than 0.1 percent if the resistor is pretrimmed, thermally aged, and trimmed to the target value. Microcracking does not occur in thin film resistors since the microstructure is not glassy.

Abrasive trimming can also be used to trim resistors. Sand is forced through a small nozzle by compressed air and abrades away the resistor material. Abrasive trimming is not as precise as laser trimming, and is primarily restricted to thick film resistors. Abrasive trimming is slower and more difficult to automate than laser trimming, but is easier to set up and finds extensive use in fabricating prototypes. Abrasive trimming does not create microcracks, and resistors trimmed in this manner are inherently more stable than laser trimmed resistors under conditions of high power dissipation. The power hybrid industry

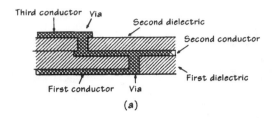

(a)

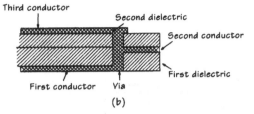

(b)

FIGURE 7.25 Via location on adjacent layers, (*a*) permitted and (*b*) not permitted. The only exception to this rule is when vias are used as heat sinks under components.

uses abrasive trimming on a large scale to trim power resistors to take advantage of this feature.

It is also possible to actively trim hybrid circuits with a laser. The circuit is connected to the appropriate power and signal sources and the desired outputs monitored during trim. This process is particularly advantageous for trimming active filters and similar circuits.

7.8 ASSEMBLY OF HYBRID CIRCUITS

There are two basic methods of assembling hybrid circuits; the "chip-and-wire" approach in which semiconductor devices in the unpackaged form are mechanically attached to the substrate metallization by epoxy, solder, or eutectic bonding, and electrically connected by wire bonding, and the "surface mount" approach in which packaged devices are soldered to the substrate, making both the electrical and mechanical connections simultaneously. These methods are not mutually exclusive; it is very common to combine these approaches on a single substrate. This chapter is primarily concerned with the chip-and-wire technology.

7.8.1 Chip-and-Wire Technology

Semiconductor devices in the unpackaged state are very delicate. The metallization patterns are extremely thin and easily damaged even with normal handling. The Input/Output (I/O) terminals are electrically open, and electrical charge has no path to ground, making the die extremely susceptible to damage by electrostatic discharge (ESD). ESD can be generated from a variety of sources, as shown in Table 7.9. The die must be handled with extreme care, using a vacuum pickup, as opposed to tweezers, which is designed to dissipate electrical charge. Operators must be properly grounded, and ionized air blowers may be necessary when especially susceptible devices are being mounted.

7.8.1.1 Direct eutectic bonding of semiconductor devices. Silicon forms a eutectic composition with gold in the ratio of 94 percent gold and 6 percent silicon by weight, which melts at 370°C. Gold is the only metal to which the eutectic temperature in combi-

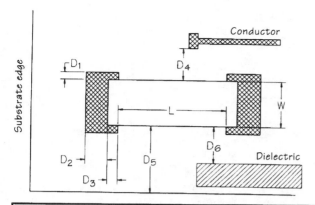

Dimension	Thick film, mils		Remarks
	Preferred	Minimum	
L	40	20	(1)
W	See Table 11.4		(2)
D_1	10	5	(3)
D_2	10	5	(4)
D_3	10	7.5	(5)
D_4	20	15	(6)
D_5	30	20	(7)
D_6	20	20	(8)

1. Aspect ratio $0.5 \leq L/W \leq 5$ preferred, $0.3 \leq L/W \leq 10$ maximum.
2. Depends on the more restrictive of tolerance and power.
3. Minimum excess conductor width.
4. Minimum excess conductor length.
5. Resistor overlap onto conductor.
6. Conductor distance from resistance.
7. Distance from edge of substrate.
8. Distance to multilayer or crossover dielectric.

FIGURE 7.26 Thick film resistor dimensions.

nation with silicon is sufficiently low to be practical. By contrast, the eutectic temperature of silicon and silver is nearly 800°C.

In the eutectic bonding process, the substrate is preheated to about 200°C to minimize thermal shock and then transferred to a heated stage at about 400°C. The silicon die is picked up by a heated collet, also to minimize thermal shock, which is sized to match the size of the die and connected to a small vacuum pump for the purpose of holding the die in place. The collet is also connected to a motor which is capable of providing mechanical scrubbing to assist in making the bond. The die is picked up by the collet and transported to the desired metallization pattern on the substrate. Simply placing the die in contact with the heated substrate will not automatically form the bond; the materials must be in the proper proportion. This is accomplished by mechanically scrubbing the die into the gold metallization. During the scrubbing process, the eutectic alloy will be formed at some random point along the interface, which will then become molten. This, in return, will rapidly

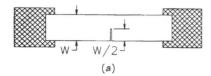

(a)

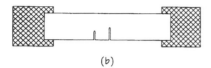

(b)

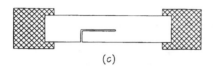

(c)

FIGURE 7.27 Laser trim cut modes. Do not penetrate more than halfway into the resistor. When trimming to less than 50% tolerance, use the double-plunge cut or L-cut mode. Trim to –5% or until *W*/2 is reached with a single-plunge cut, then switch to double-plunge or L cut.

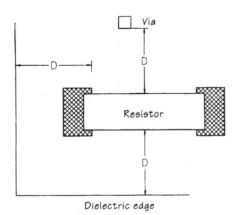

Dimension	Thick film, mils		Remarks
	Preferred	Minimum	
D	20	20	*, †, ‡

*Distance to edge of dielectric or via when resistors are printed on top of dielectric.
†The same dimensions for resistors in Fig. 11.43 also hold.
‡The use of resistors on dielectric requires an alternative set of resistor curves.

FIGURE 7.28 Dimensions of resistors on dielectric.

TABLE 7.8 Comparison of Hybrid Metallization Technologies

Technology	Ceramic selection	Metallization description	Adhesion mechanism	Geometry (typical)	Electrical and thermal	Hybrid assembly	Reliability	Cost
Thick film	Usually oxide, most types possible	Metal + glass	Chemical + mechanical	0.010 in (250 μm) lines and more than 0.0005 in (12 μm) thickness	Poor	Good, solder rework can be a problem	Good, well-understood	Variable; can use expensive precious metals
Thin film	All types	Pure metal, require adhesion layers	Mechanical + chemical	0.002 in (50 μm) lines and less than 0.0005 in (12 μm) thickness	Adequate	Adequate, not easily soldered	Good, well-understood	High for equipment, process, and materials
Direct bond copper	More selective, oxide based	Pure copper	Chemical + mechanical	0.020 in (400 μm) lines and 0.008 to 0.020 in (200 to 500 μm) thickness	Good	Good	Good, well-understood	Reasonable
Plated copper	All types	Pure copper, usually thin adhesion layer	Mechanical (chemical?)	0.004 in (100 μm) lines and 0.0005 to 0.005 in (12 to 125 μm) thickness	Good	Good	Less understood	Reasonable
Active metal braze copper	All types	Pure copper, with braze adhesion layer	Chemical + mechanical	0.020 in (400 μm) lines and 0.008 to 0.020 in+(200 to 500 μm+) thickness	Good	Good	Adequate	Reasonable

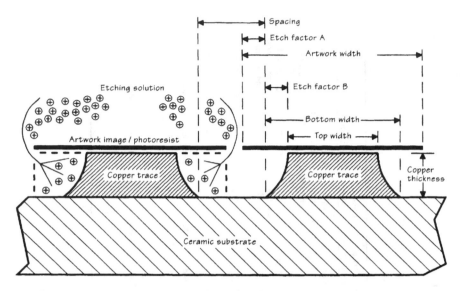

FIGURE 7.29 Design guidelines for photolithographic etch process.

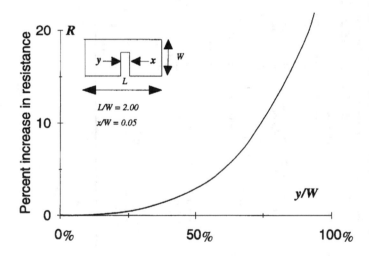

FIGURE 7.30 Increase in resistance with laser cut length.

dissolve more material until the entire interface is liquid and the bond is formed. Devices larger than 0.020×0.020 in tend to crack when mechanically scrubbed, and require a gold-silicon preform to make the bond.

The eutectic process makes a very good bond, mechanically, electrically and thermally, but the high bonding temperature is an extreme disadvantage for a variety of reasons, and this process is generally used only in single chip applications. Where a metallurgical bond is required, solder is the preferred process. Semiconductor devices

TABLE 7.9 Typical Electrostatic Voltages

Event	Relative humidity, %		
	10	40	55
Walking across carpet	35,000	15,000	7,500
Walking across vinyl floor	12,000	5,000	3,000
Motions of bench worker	6,000	800	400
Remove DIPs from plastic tubes	2,000	700	400
Remove DIPs from vinyl trays	11,500	4,000	2,000
Remove DIPs from Styrofoam	14,500	5,000	3,500
Remove bubble pack from PWBs	26,000	20,000	7,000
Pack PWBs in foam-lined box	21,000	11,000	5,500

SOURCE: From Ref. 43.

intended for solder bonding are typically metallized with a titanium/nickel/silver alloy on the bottom.

7.8.1.2 Organic bonding materials. The most common method of mechanically bonding active and passive components to a metallized substrate is with epoxy. Most epoxies for hybrid circuit applications have a filler added for electrical and/or thermal conductivity. The most common conductive filler is silver. Silver has a high electrical conductivity, and a smaller proportion of silver to epoxy is required to produce a given resistivity than other metals. Silver epoxy, therefore, has a higher mechanical strength since more epoxy is present. Other conductive filler materials include gold, palladium-silver for reduced silver migration, and tin-plated copper. Nonconductive filler materials include aluminum oxide, beryllium oxide, and magnesium oxide for improved thermal conductivity.

Epoxy may be dispensed by screen printing, pneumatic dispensing through a nozzle, and by the die transfer method. Screen printing requires a planar surface, while pneumatic dispensing and die transfer may be used with irregular surfaces. Where many die are to be mounted on a single substrate, screen printing is the preferred method because epoxy can be applied to all mounting pads in a single pass of the screen printer.

Where large surfaces are to be mounted, such as substrates, liquid epoxies are difficult to use. For this application, a large sheet of glass cloth can be impregnated with epoxy, which is then B-staged, or partially cured, to produce a solid structure of epoxy generally referred to as a "preform." The sheet may then be cut into smaller pieces and used to mount large components and substrates. Where high thermal conductivity is required, the mesh may be made from silver wire and silver epoxy used.

Epoxy as a die attach material suffers from two major disadvantages.

1. The thermal resistance is quite high compared to direct methods of bonding such as soldering or eutectic bonding.
2. The operating temperature is limited to about 150°C, as many epoxies begin breaking down and outgassing at temperatures above this figure.

These factors inhibit the use of epoxy in mounting power devices, in applications where a high processing temperature is utilized, or where performance at a high ambient temperature is required. One phenomenon that can occur when the epoxy is operated at elevated temperatures near or above the T_g, or glass transition temperature, is the settling of the filler away from the device interface, leaving a thin layer of resin at the bond line and increasing the electrical resistance of the bond to the point where circuit failure occurs. Another failure mechanism that can occur is silver migration, which can be accelerated by water absorption of a resin with a relatively high mobile ion content.

Other materials which can be utilized for die bonding are polyimide and certain of the thermoplastic materials. Polyimide is stable out to around 350°C, unless filled with silver, which acts as a catalyst to promote adhesion loss. Polyimides are typically dissolved in a solvent, such as xylene, which requires a two-step curing process. The first step, at a low temperature, evaporates the solvent, then the second step, at an elevated temperature, crosslinks the polyimide. Thermoplastic materials are particularly advantageous in high-volume applications, as the cure cycle is much shorter than that of either epoxy or polyimide.

7.8.1.3 Solder attachment.   In the soldering process, an alloy of two or more metals is melted at the interface of two metal surfaces. The molten solder dissolves a portion of the two surfaces and, when the solder cools, a junction, or "solder joint," is formed, joining the two metal surfaces.

For the solder joint to occur, both metal surfaces and the solder must be clean and free from surface oxides. Removal of the oxides is accomplished by the use of a "flux," an organically based acid. Fluxes are categorized by their strength and by the requirement for cleaning. Fluxes requiring solvent cleaning are rapidly being phased out due to environmental regulations. The so-called "no-clean" fluxes, which remain on the circuit after the soldering process, are seeing extensive use in commercial applications. Water soluble fluxes are still widely used in the printed circuit board industry. Surfaces which are heavily oxidized before soldering may require the use of a stronger flux such as an RMA (resin mildly activated) flux.

Solder materials are selected primarily for their compatibility with the surfaces to be soldered and their melting point. Solders are generally divided into two categories: "hard" solders, which have a melting point above about 500°C and are often referred to as "brazes," and "soft" solders, which have a lower melting point. Soft solders are used primarily in the assembly process, while hard solders are used in lead attachment and package sealing.

Soft solders may also be divided into two categories: eutectic solders and non-eutectic solders. Eutectic solders have the lowest melting point and are typically more rigid in the solid state than other solders with the same constituents. This is die in part to the fact that eutectic solders go directly from the liquid to the solid state without going through a "plastic" region. Their lower melting point makes them very attractive in many applications, and their use is quite common. Some of the more common solders and their characteristics are listed in Table 7.10.

The compatibility of the solder with the material in the metal surfaces must be a prime consideration when selecting a solder, in particular the tendency of the metal to "leach" into the solder and the tendency to form intermetallic compounds, which might prove detrimental to reliability. Leaching is the process by which a material is absorbed into the molten solder to a high degree. While a certain degree of leaching must occur to produce the solder joint, excessive leaching can cause the metallization pattern to vanish into the molten solder, creating an open circuit. Tin-bearing solders used in conduction with gold or silver conductors are especially prone to this phenomenon, since these materials have a strong affinity with tin. A thick or thin film gold or silver conductor will dissolve into a tin-lead solder in a matter of seconds. The addition of platinum and/or palladium to the thick film conductor materials will greatly enhance the leach resistance to tin-bearing solders, but it must still be a consideration. The leach resistance is proportional to the amount of Pt and Pd added, but this has the adverse effect of adding both to the cost and the electrical resistance. Leaching can be minimized by soldering at the lowest possible temperature for

TABLE 7.10 Processing Temperatures of Selected Organic and Metallic Attachment Materials

Organic Attachment Materials	
Material	Temperature, °C
Polyimide	250–350°C*
Epoxy	150°C

Metallic Attachment Materials		
Alloy†	Liquidus, °C§	Solidus, °C
In 52–Sn 48‡	118	118
Sn 62.5–Pb 36.1–Ag 1.4‡	179	179
Sn 63–Pb 37‡	183	183
In 60–Pb 40	185	174
Sn 60–Pb 40	188	183
Sn 96.5–Ag 3.5‡	221	221
Pb 60–Sn 40	238	183
Pb 70–Sn 27–Ag 3	253	179
Pb 92.5–Sn 5–Ag 2.5	280	179
Sn 90–Ag 10	295	221
Pb 90–Sn 10	302	275
Au 88–Ge 12‡	356	356
Au 96.4–Si 3.6‡	370	370
Ag 72–Cu 28‡	780	780

*Polyimide materials may also require a precure step at 70°C to remove solvents.
†Numerical values are percentages. In = indium, Sn = tin, Pb = lead, Ag = silver, Au = gold, Ge = germanium, Si = silicon, Cu = copper.
‡Eutectic composition.
§The processing temperature of most alloys is ≥20°C above the liquidus.

the shortest possible time consistent with forming the solder joint. The addition of a small amount of silver to the solder lowers the melting point slightly, and, with proper control of the soldering temperature, the silver already present in the solder partially saturates the solution, further inhibiting leaching. Soldering directly to gold requires the use of AuSn, PbIn or other solders that do not leach gold. The added Au in the AuSn combination inhibits leaching of the gold in the film.

Intermetallic compound formation is also a consideration. Certain compounds have a high electrical resistance and are susceptible to mechanical failure when exposed to temperature cycling or storage at temperature extremes. Tin forms intermetallic compounds with both gold and copper, and indium also forms intermetallic compounds with copper. The most commonly used solders are the tin-lead solders, which are used extensively on copper and the PdAg and PtAg alloys.

Lead-free solders, such as the tin/silver combinations, are becoming more widely used. Apart from environmental concerns, the tin/silver solders have proven to be more resilient to extensive temperature cycling conditions.

In the soldering process, it is desirable to minimize the exposure of the part to elevated temperatures. High temperatures accelerate the rate of chemical reactions which may be detrimental to the reliability of the circuit. Furthermore, excessive exposure of metal surfaces to liquid solder increases the rate of formation of intermetallic compounds and also increases leaching.

Solders for microelectronic applications are generally in the form of a paste, with the solder in powder form mixed with an appropriate flux and a dispensing vehicle. Solder in paste form may be screen printed or dispensed pneumatically. The part is placed in the wet solder paste prior to soldering by an automatic pick-and-place system or manually.

The most effective method of soldering is to place the part in conjunction with the solder paste into a tunnel furnace which has several heated zones. By controlling the speed of the belt and the temperature of the individual zones, a time-temperature relationship, or "profile," can be established which will optimize the soldering process. Heating of the part may be accomplished by resistance heating, by infrared (IR) heating, or by a combination of both.

The use of a nitrogen or forming gas blanket during the soldering process to prevent oxidation of the solder and/or the surfaces aids greatly in soldering, particularly to nickel, and improves the wetting of tin-bearing solders. The formation of a gas blanket can be easily accomplished by connecting a gas source to the furnace and by the use of baffles at the ends of the furnace to minimize the intrusion of air into the heated region. A typical profile has a duration of several minutes and has a plateau just below the melting point of the solder for a period of time followed by a rapid rise in temperature, or "spike," above the melting point for a short duration, and a gradual decline down to room temperature. In general, the duration of the spike should be held as short as possible consistent with good solder flow and wetting to minimize such effects as leaching of conductor materials and the effect of high temperatures on the components.

7.8.1.4 Wire bonding.

Ohmic contacts to semiconductor devices are typically made with aluminum, since that material diffuses well into the silicon structure at a moderate temperature. Wire bonding is used to make the electrical connections from the aluminum contacts to the substrate metallization or to a lead frame, from other components, such as chip resistors or chip capacitors, to substrate metallization, from package terminals to the substrate metallization, or from one point on the substrate metallization to another.

There are two basic methods of wire bonding, thermocompression wire bonding, which uses primarily gold wire, and ultrasonic wire bonding, which uses primarily aluminum wire. Thermocompression wire bonding, as the name implies, utilizes a combination of heat and pressure to form an intermetallic bond between the wire and a metal surface. In thermocompression bonding (Fig. 7.31) a gold wire is passed through a hollow capillary, generally made from a refractory metal such as tungsten, and a ball formed on the end by means of an electrical arc. The substrate is heated to about 300°C and the ball is forced into contact with the bonding pad on the device with sufficient force to cause the two metals to bond. The capillary is then moved to the bond site on the substrate, feeding the wire as it goes, and as the wire is bonded to the substrate by the same process, except that the bond is in the form of a "stitch," as opposed to the "ball" on the device.

The wire is then clamped and pulled to break just above the stitch and another ball formed as above. Thermocompression bonding is rarely used for a variety of reasons.

- The high substrate temperature precludes the use of epoxy for device mounting.
- The temperature required for the bond is above the threshold temperature for gold-aluminum intermetallic compound formation. The diffusion rate for aluminum into gold is much greater than for gold into aluminum. The aluminum contact on a silicon device is very thin, and, when it diffuses into the gold, voids called "Kirkendall voids" are created in the bond area, increasing the electrical resistance of the bond and decreasing the mechanical strength.
- The thermocompression bonding action does not effectively remove trace surface contaminants that interfere with the bonding process.

The ultrasonic bonding process (Fig. 7.32) uses ultrasonic energy to vibrate the wire against the surface to combine the atomic lattices together at the surface. Localized heat-

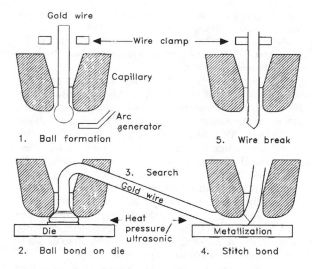

FIGURE 7.31 Typical ball bonding sequence.

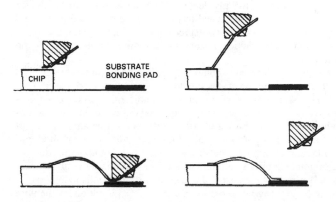

FIGURE 7.32 Aluminum ultrasonic wedge-wedge bonding.

ing at the bond interface caused by the scrubbing action, aided by the oxide on the aluminum wire, assists in forming the bond. The substrate itself is not heated. Intermetallic compound formation is not as critical as with the thermocompression bonding process, as both the wire and the device metallization are aluminum. Kirkendall voiding on an aluminum wire bonded to gold substrate metallization is not as critical, since there is substantially more aluminum available to diffuse than on device metallization. Ultrasonic bonding makes a stitch on both the first and second bonds. For this reason, ultrasonic bonding is somewhat slower than thermocompression, since the capillary must be aligned with the second bond site when the first bond is made. Ultrasonic bonding to package leads may be difficult if the leads are not tightly clamped, since the ultrasonic energy may be propagated down the leads instead of being coupled to the bond site.

The use of thermosonic bonding largely overcomes the difficulties noted with thermocompression bonding. In this process, used with gold wire, the substrate is heated to 150°C, and ultrasonic energy is coupled to the wire through the transducer action of the capillary, scrubbing the wire into the metal surface and forming a ball-stitch bond from the device to the substrate as in thermocompression bonding.

Thermosonic gold bonding is the most widely used bonding technique, primarily because it is faster than ultrasonic aluminum bonding. Once the ball bond is made on the device, the wire may be moved in any direction without stress on the wire, which greatly facilitates automatic wire bonding, as the movement need only be in the x and y directions.

By contrast, before the first ultrasonic stitch bond is made on the device, the circuit must be oriented so that the wire will move toward the second bond site only in the direction of the stitch. This requires rotational movement, which not only complicates the design of the bonder but increases the bonding time as well. Attempts at designing an aluminum ball bonder have not proven successful to date because of the difficulty in making a ball on the end of the wire. Aluminum bonding is commonly used in hybrids or chip carriers where the package-sealing temperature may exceed the threshold temperature for the formation of intermetallic compounds, in power hybrids where the junction temperature is high, and in applications where large wires are required. Gold wire is difficult to obtain in diameters above 0.002 in, while aluminum wire is available up to 0.022 in.

Automatic thermosonic gold bonding equipment now exists that can bond up to four wires per second under optimum conditions. The actual rate depends on the configuration of the bonding pattern and the number of devices to be bonded. The utilization of an automatic bonder with pattern recognition is strongly dependent on the accuracy of the device placement and on the quality of the substrate metallization. If the devices are not placed within a few mils of the bond site and within a few degrees of rotation, the pattern recognition system will not be able to locate the device, resulting in lost production time while the operator feeds in the coordinates.

The substrates must be kept very clean and should only be handled with finger cots during the assembly process. In storage, the substrates should be kept in a clean, dry area. Thin film metallization makes the best bonding surface, since it is uniform in thickness and is almost pure gold. There exists a considerable variation in bondability among the various thick film golds and bonding studies under a variety of conditions must be an integral part of the selection process of a thick film gold material. Thick film gold for use with aluminum wire must contain palladium to improve the strength of the bond under conditions of thermal aging. The same principle applies to aluminum bonding on palladium-silver materials. As the palladium-to-silver ratio is increased, the reliability of the bonds increases.

The integrity of the wire bonds may be tested by placing a small hook under the wire and pulling at a constant rate of speed, usually very slow, until the bond fails. The amount of force required is the "bond strength." The mode of failure is just as important as the actual bond strength. There are five points at which a bond can fail.

1. The ball (or stitch) at the device may fail.
2. The wire may break just above the stitch at the device end.
3. The wire may break in the center.
4. The wire may break just above the stitch at the substrate end.
5. The stitch at the substrate may lift.

Of the five options, number three is the most desirable, with two and four following in turn. If an excessive number of lifts occur, particularly at the device end, a serious bonding problem may be indicated. Some of the more common bonding problems and their possible causes are listed below:

- *Excessive bond lifts at the device.* This problem may be a result of one of three factors: improper bond setup, contamination on the device, or incomplete etching of the glass passivation on the device.
- *Excessive bond breaks just above the device.* This is probably caused by either a worn bonding tool or by excessive bonding pressure, both of which can crimp the wire and create a weak point.
- *Excessive bond breaks just above the substrate.* This is probably caused by excessive bonding pressure. This may result from improper bonding setup, which in turn may be due to contamination or to poor substrate metallization. A worn bonding tool may result in the wire being nicked, creating a possible failure point.
- *Excessive stitch lifts at the substrate.* This may be due to a worn bonding tool, poor bonder setup, or poor device metallization.

The wire size is dependent on the amount of current that the wire is to carry, the size of the bonding pads, and the throughput requirements. Table 7.11 illustrates the wire size for a specified current for both gold and aluminum wire. For applications where current and bonding size is not critical, 0.001 in wire is the most commonly used size. While 0.0007 in wire is less expensive, it is difficult to thread through a capillary and bond without frequent breaking and consequent line stoppages.

TABLE 7.11 Maximum Current for Wire Size

Material	Diameter, in	Maximum current, A	
		$L < 0.040$ in	$L > 0.040$ in
Gold	0.001	0.949	0.648
	0.002	2.683	1.834
Aluminum	0.001	0.696	0.481
	0.002	1.968	1.360
	0.005	7.778	5.374
	0.008	15.742	10.876
	0.012	28.920	19.981
	0.015	40.417	27.924
	0.022	71.789	49.600

7.9 PACKAGING

Packaging for individual devices or complete hybrid assemblies can be characterized as shown in Table 7.12. There are many driving forces and considerations in the choice of technology for packaging and the subsequent next level assembly technologies.

7.9.1 Hermetic Packages

A true hermetically sealed package would prevent intrusion of contaminants (liquid, solid, or gas) for an indefinite period of time. In practice, however, this is not realistic. Even in a perfectly sealed structure, diffusion phenomena will occur over time, allowing the smaller molecules, such as helium or water vapor, to penetrate the barrier medium and ultimately reach equilibrium within the package interior. A hermetic package is defined as one in

TABLE 7.12 Characterization of Package Types

Leads	Type
Through-hole	Hermetic
Plug-in	Metal
Dual In-line (DIP)	Ceramic
Pin grid array (PGA)	
TO styles	
Surface mount	Plastic
Small outline (SOIC, SOT)	
Chip carrier (leadless, leaded)	
Thin small outline (TSOP)	

which the leak rate of helium after pressurization is below a specified rate with reference to the package size as shown in Table 7.13.

TABLE 7.13 Definition of Hermetic Package

Package volume (cm^3)	Bomb Condition			Reject limit (atm-cm^3/s/He)
	PSIG	Exposure time (h)	Maximum dwell (h)	
$V < 0.40$	60 ± 2	$2 + 0.2, -0$	1	5×10^{-8}
$V > 0.40$	60 ± 2	$2 + 0.2, -0$	1	2×10^{-7}
$V > 0.40$	30 ± 2	$4 + 0.4, -0$	1	1×10^{-7}

A hermetic package must either be metal, ceramic, or glass. Organic packages, or packages with an organic seal, may initially pass the pressurization test described above but will allow water vapor to pass back and forth from the atmosphere to the package interior. Therefore, they are not truly hermetic. Interconnections through a metal package may be insulated by glass-to-metal seals utilizing glass that matches the coefficient of expansion to the metal.

A hermetic package allows the circuit mounted inside to be sealed in a benign environment, generally nitrogen, which is obtained from a liquid nitrogen source. Nitrogen of this type is extremely dry, with a moisture content of less than 10 ppm. As a further precaution, the open package with the enclosed circuit mounted inside is subjected to an elevated temperature, usually 150°C, in a vacuum, to remove absorbed and absorbed water vapor and other gases prior to sealing. For added reliability, the moisture content inside a package should not exceed 5000 ppm. This figure is below the dew point of 6000 ppm at 0°C, ensuring that any water that precipitates out will be in the form of ice, which is not as damaging as water in the liquid form.

A hermetic package adds considerably to the reliability of a circuit by guarding against contamination, particularly of the active devices. An active device is susceptible to a number of possible failure mechanisms, such as corrosion and inversion, and may be attacked by something as benign as distilled, deionized water, which can leach phosphorous out of the passivating oxide to form phosphoric acid which, in turn, can attack the aluminum bonding pads.

7.9.1.1 Metal packages.   The most common type of hermetic package is the metal package, which is fabricated primarily from ASTM F-15 alloy, $Fe_{52}Ni_{29}Co_{18}$ (known also as

Kovar®). The tub-type package is fabricated by forming a sheet of the F-15 alloy over a set of successive dies. Holes for the leads are then punched in the bottom for plug-in packages and in the side for flat packages. A layer of oxide is then grown over the package body. Beads of borosilicate glass, typically Corning 7052 glass, are placed over the leads and placed in the holes in the package body. A reactive glass-to-metal seal is formed by heating the structure above the melting point of the glass (approximately 500°C). The molten glass dissolves some of the oxides on the alloy (primarily iron oxide) which, upon cooling, provided the adhesion mechanism. The glass-to-metal seal formed in this manner has four distinctive layers.

1. Metal
2. Metal oxide
3. Metal oxide dissolved in glass
4. Glass

After the glass-to-metal seals have been formed, the oxide not covered by the glass must be removed and the metal surface plated to allow the package to be sealed and to allow the package leads to be soldered to the next higher assembly. The prime plating material is electrolytic nickel, although gold is frequently plated over the nickel to aid in sealing and to prevent corrosion. In either case, the package leads are plated with gold to allow wire bonding and to improve solderability. Although electroless nickel has better solderability, it tends to crack when the leads are flexed.

The glass-to-metal seal formed in this manner provides an excellent hermetic seal and the close match in TCE between the glass and the F-15 alloy (approximately $5.0 \times 10^{-6}/$°C) maintains the hermeticity through temperature cycling and temperature storage.

There are three types of lids commonly used on metal packages; the domed lid, the flat lid, and the stepped lid. These are also fabricated with ASTM F-15 alloy with the same plating requirements as the packages. The domed lid is designed for use with platform packages and may be projection welded or soldered. The flat lid is designed for use with the tub package and is primarily soldered to the package. The stepped lid is fabricated by photoetching a groove in a solid sheet of F-15 alloy such that the flange is about 0.004 in thick. This lid is designed to be seam welded to a tub package. When lids are designed for soldering, a preform of the desired solder material is generally attached to the outer perimeter of the bottom of the lid.

Methods of sealing metal packages.   A flat lid or a stepped lid may be soldered to the package by hand, by the use of a heated platen, or in a furnace. While the platen is somewhat faster, the metal package acts as a heat sink, simultaneously drawing heat away from the seal area and raising the temperature inside the package, unless the glass beads used for insulating the leads extend entirely around the periphery of the package, an obviously impractical requirement. In addition, leaks through the solder, or "blow holes," caused by a differential in pressure between the inside and outside of the package will occur unless the ambient pressure outside the package is increased at the same rate as the pressure inside created by heating the package. Because of the temperature rise inside the package, it is risky to use epoxy to mount components unless the glass beads extend around the periphery of the package as described above. Solder sealing may be accomplished in a conveyor-type furnace, which has a nitrogen atmosphere. The nitrogen prevents the oxidation of the solder and also provides a benign environment for the enclosed circuit. Furnace sealing requires a certain degree of fixturing to provide pressure on the lid.

Parallel seam welding is accomplished by the generation of a series of overlapping spot welds by passing a pair of electrodes along the edge of the lid. The alignment of the lid to the package is quite critical and is best accomplished outside the sealing chamber, with the lid being tacked to the package in two places by small spot welds. A stepped lid greatly facilitates the process and improves the yield, since it requires considerably less power than a flat lid of greater thickness. While the sealing process is relatively slow compared to other methods, a package sealed by parallel seam welding can be easily delidded by grinding the edge of the lid away. Since the lid is only about 0.004 in thick in the seal area, this may be readily accomplished in a single pass of a grinding wheel. With minimal polishing of the seal area of the package, another lid may be reliably attached.

Certain classes of packages with a flange on the package may be sealed by a process called projection, or one-shot, welding. In this process, an electrode is placed around the flange on the package and a large current pulse is passed through the lid and the package, creating a welded seam. Heavy-duty resistance welding equipment capable of supplying 500 pounds of pressure and 12,000 amperes per linear inch of weld is required for these packages. The major advantages of one-shot welding are process time and a less expensive package. The major disadvantage is the difficulty of removing the lid and repairing the circuit inside. Delidding a projection welded package is a destructive process, and the package must be replaced.

7.9.1.2 Ceramic packages. Ceramic packages in this context are considered to be structures that permit a thick or thin film substrate to be mounted inside in much the same manner as a metal package. Ceramic structures that have metallization patterns, which allow direct mounting of components are referred to as "multichip ceramic packages." Ceramic packages for hybrid circuits generally consist of three layers of alumina. The bottom layer may or may not be metallized depending on how the substrate is to be mounted. A ring of alumina is attached to the bottom layer with glass and a lead frame is sandwiched between this ring and a top ring with a second glass seal. The top ring may be metallized to allow a solder seal of the lid or may be left bare to permit a glass seal.

Methods of sealing ceramic packages. The most common method of sealing ceramic packages is solder sealing. During the manufacturing process, a coating of a refractory metal or combination of metals, such as tungsten or an alloy of molybdenum and manganese, is fired onto the ceramic surface around the periphery of the seal area. On completion, the surface area is successively nickel plated and gold plated. A lid made from ASTM F-15 alloy is plated in the same manner and soldered onto the package, usually with an alloy of $Au_{80}Sn_{20}$, in a furnace with a nitrogen atmosphere.

A less expensive, but also less reliable, method of sealing is to use a glass with a low melting point to seal a ceramic lid directly to a ceramic package. This avoids the use of gold altogether, lowering the material cost considerably. The glass requires a temperature of about 400°C for sealing, as opposed to about 300°C for the AuSn solder. The glass seal is somewhat susceptible to mechanical and thermal stress, particularly at the interface between the glass and the package.

These two techniques have a common problem; it is difficult to remove the lid for repair without rendering the package useless for further sealing. An alternative approach seeing increased use is to braze a ring of ASTM F-15 alloy, which has been nickel and gold plated as described above onto the sealing surface of the ceramic package. It is then possible to use parallel seam welding with its inherent advantages for repair. This approach is also frequently used for ceramic multilayer packages designed for multichip packaging.

7.9.2 Non-hermetic Packaging Approaches

The term "non-hermetic" package encompasses a number of configurations and materials, all of which ultimately allow the penetration of moisture and/or other contaminants to the circuit elements. Most techniques involve encapsulation with one or more polymer materials, with the most common being the molding and fluidized bed approaches.

Both injection and transfer molding techniques utilize thermoplastic polymers, such as acrylics or styrenes, to coat the circuit. In transfer molding, the material is heated and transferred under pressure into a closed mold in which the circuit has been placed while, in injection molding, the material is heated in a reservoir and forced into the mold by piston action.

The fluidized bed technique uses an epoxy powder, which is kept in a constant state of agitation by a stream of air. The circuit to be coated is heated to a temperature above that of the melting point of the epoxy and is placed in the epoxy powder. The epoxy melts and clings to the circuit, with the thickness controlled by the time and the preheat temperature.

Both methods are used to encapsulate hybrids and individual devices, and are amenable to mass production techniques. The overall process may be performed at a cost of only a few cents per circuit. The coatings are quite rugged mechanically, are resistant to many chemicals and have a smooth, hard surface suitable for marking.

7.9.3 Package Types

Plug-in packages. Plug-in packages have leads protruding from the bottom with a lead spacing of 0.100 in. A special case of plug-in packages is the DIP package (Fig. 7.33). Standard DIP packages are used to package individual die and have two rows of leads on 0.100 in centers separated with each row being 0.300 in apart. Other package types designed for hybrid use (platform or bathtub) often are found to have a lead configuration consistent with the DIP design for commonality with test mounting sockets.

Plug-in packages are designed for through-hole insertion into printed wiring boards. In addition to DIP packages, a single-in-line (SIP) packaging technique was developed for resistor and capacitor components or networks. For all plug-in packages, the leads provide a convenient method for ensuring clearance on assembly and providing a degree of compliance into the mechanical stress established by assembly or expansion coefficient mismatch between the package and the mounting substrate. DIP packages have been the mainstay in discrete device packaging while device complexity was low. However, with the advent of very large scale integration (VLSI) devices and high input/output (I/O) count, the 100 mil lead centers required development of very large packages. A standard 40-pin device required a length exceeding 2.00 in and higher pin counts became increasingly difficult to package in DIP form.

Small outline package. The small outline (SO) package is shown in Fig. 7.34. The leads on the SO package are on 0.050 in centers, as opposed to 0.100 in for the DIP pack-

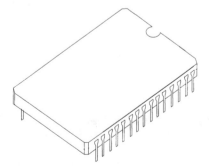

FIGURE 7.33 Dual in-line package (DIP).

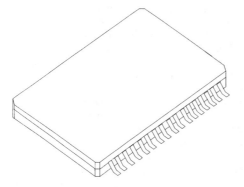

FIGURE 7.34 Small outline (SO) package.

age. The SO has a low profile and occupies less than 50 percent of the area of the DIP. It weighs about one-tenth that of a DIP.

The SO package family includes packaging of passive devices; packages that contain IC's, known as SOIC packages; and packages that contain transistors, known as SOT's. Both plastic and ceramic SO packages are available.

Ceramic chip carriers. A special case of the ceramic package is the hermetic chip carrier as shown in Fig. 7.35. The wire bonding pads are routed to the outside between layers of ceramic and are connected to semicircular contacts called "castellations." The most common material is alumina, which is metallized with a refractory metal during fabrication, and then successively plated with nickel and gold. Most multilayer chip carriers are designed to be sealed with solder, usually $Au_{80}Sn_{20}$.

Configurations of chip carriers for military applications have been standardized by JEDEC in terms of size, lead count, lead spacing and lead orientation, although nonstandard carriers can be used for specialized applications. The most common lead spacing is 0.050 in, with high lead count packages having a spacing of 0.040 in.

The removal of heat from chip carriers in the standard "cavity up" configuration has been a problem because the only path for heat flow is along the bottom of the carrier out to the edge of the carrier, where it flows down to the substrate through the solder joints. This problem can be alleviated to a certain extent by printing pads on the bottom of the carrier, which are soldered directly to the substrate. This lowers the thermal impedance by a factor of several times. If this does not prove adequate, carriers with the cavity pointing down can be utilized. In this configuration, the chip is mounted on the top of the carrier in the upside-down position, the lid is mounted in a recess on the bottom of the carrier, and a

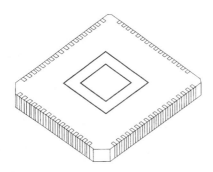

FIGURE 7.35 Leadless chip carrier (LLCC).

heat sink is mounted to the top to enable the heat to be removed by convection. Beryllia and aluminum nitride chip carriers are being used to package high-power devices,

Devices mounted in chip carriers can be thoroughly tested and burned in before mounting on a substrate or printed circuit board. This process can be highly automated and is frequently done in military applications. Sockets exist for the standard sizes, which make contacts to the castellations without the necessity for solder.

Chip carriers have proven to be a viable approach for packaging hybrid circuits in a minimum size. Although available with pin counts up to 128, chip carriers have proved to be a reliability risk when the pin count is greater than 84 because the net expansion of the carrier at temperature extremes, and, therefore, the stress on the solder joints, is proportional to its size. Furthermore, the temperature at which chip carriers may be used on PC boards is limited due to the difference in coefficient of expansion (TCE) between the carrier and board material.

As the solder joint is made higher, the difference in TCE becomes less significant. The highest solder column that can be made by ordinary means is about 0.007 in. Above this height, a molten solder column begins collapsing of its own weight. Power cycling, in which the device in the carrier is powered on and off at periodic intervals, has proven to be a serious reliability risk, even more than temperature cycling, when power devices are mounted. While the device in the carrier is being power cycled, the carrier and the board are in a non-equilibrium state with respect to temperature. This causes considerable stress on the solder joints, ultimately resulting in failure due to metal fatigue.

7.9.4　Packages for Power Hybrid Circuits

As the power requirements of package materials have become more demanding, ASTM F-15 alloy and alumina become less attractive due to their relatively low thermal conductivity. Copper, molybdenum, copper-clad materials, aluminum nitride, and beryllia have all been used to manufacture packages. This requires some innovation on the part of the package manufacturer to develop methods of sealing and through-hole connections, as copper is not amenable to seam welding or glass-to-metal sealing.

One power package uses a cavity machined from a solid block of copper, plated with nickel and gold, which has a stainless steel seal ring brazed around the perimeter. The leads are made from copper-cored Alloy 52 material with a ceramic insulator. The insulator is generally metallized on the outside with an alloy of MbMn which is successively plated with Ni and Au. The pins are individually soldered to the package body with 80Au/20Sn solder. This package is compatible with conventional seam welding, and offers the best thermal conductivity of the various configurations when oxygen free high conductivity (OFHC) copper is used. The copper must be processed such that the cold-worked mechanical properties are not destroyed to enable the package to withstand the constant acceleration test. This limits the temperature range which the package can see during processing and use.

An alternative method of attaching the leads to the package is to use a glass with a low melting point, such as potash soda barium glass, which has a different temperature coefficient of expansion (TCE) than the package material such that a compression seal is formed.

Another approach uses a copper base that has an ASTM F-15 lead frame brazed to it and is plated with Ni and Au. This package can use conventional glass-to-metal sealing in the manufacturing process, and is compatible with parallel seam welding. This package must be less than 1×1 in and must use a copper base less than 0.060 in thick due to the large TCE mismatch between copper and ASTM F-15 alloy.

7.10 DESIGN OF HYBRID CIRCUITS

The successful design of a hybrid microcircuit involves not only the design of a circuit that meets the technical requirements but that also meets the cost, reliability, and schedule requirements. By using accepted design guidelines along with a systematic design approach, the hybrid designer will have the best chances for success. A procedure that meets these criteria is shown in Table 7.14. It will result in a drawing package as outlined in Table 7.15.

7.10.1 Sizing

The approximate size of a hybrid circuit can be determined by using one of the empirical methods based on component count and the weighting factors for each component. The weighting factors are different for thin film and thick film and are based on the standard

TABLE 7.14 Hybrid Circuit Design Sequence

Operation	Output
Partitioning	Division of system into individual circuits
Initial concept review	Preliminary packaging concepts Circuit schematics Risk analysis
Circuit analysis	Verification of electrical design Design parameters centered Sensitivity analysis Voltage, current, and power levels in each component
Breadboard tests	Verification of circuit analysis
Component selection	Preliminary parts list
Preliminary thermal analysis	Indication of potential thermal problems
Sizing analysis	Approximate size of circuit
Technology selection	Determination of substrate technology
Process sequence	Selection of manufacturing process
Material selection	Selection of attachment materials
Circuit layout	Location of components and interconnection traces Layer drawings Assembly drawings
Detailed thermal analysis	Temperature profile of circuit
Preliminary design review	Review prior to prototype build Exceptions to design guidelines Risk analysis
Prototype build	Verification of performance Electrical performance Conformance to quality standards
Documentation release	Assembly drawings Process instructions Travelers
Detailed design review	Review of prototype build and documentation
Preproduction build	Verification of design and documentation
Production release review	Review of preproduction build prior to production
Production	Manufacturing of hybrid circuit

TABLE 7.15 The Design Package

Circuit schematic
Complete parts and materials lists
Complete process sequence
Process specifications
Layout with individual layer drawings
Assembly diagram and instructions
Electrical test procedure
Environmental test procedure
Troubleshooting procedure
Qualification procedure
Handling and packaging instructions
special instructions

line widths of 0.002 in for thin film and 0.010 in for thick film. For wider lines, the weighting factors must be adjusted proportionately. The weighting factors for different components are shown in Table 7.16.

TABLE 7.16 Weighting Factors for Components for Substrate Size

	Weighting factor	
Component	Thick film	Thin film
IC (chip-and-wire)	5	4
Transistor (small signal)	4	3
Diode (small signal)	4	3
Transistor (power)	5	5
Diode (power)	4	4
Resistor	3	3
Capacitor	2	2
Chip resistor	3	2
Inductor	3	2

Note: Do not load substrate more than 80 percent of available area.

The size of a hybrid is often determined by the pin count requirements. The maximum pin count is limited by the length around the periphery of the circuit and the pitch. If the size as determined by the empirical formula is not sufficient to accommodate the required pin count, a larger size must be used. For example, a 1.0×1.0 in circuit with 0.1 in pin spacing can accommodate 36 leads.

7.10.2 Selection of Technology

It is frequently possible for more than one substrate technology to be utilized to manufacture a given circuit. The selected technology must first meet the technical requirements. After these are satisfied, cost becomes the prime consideration. The technology selection process begins by assuming that the circuit will be fabricated with the least expensive process, or with the "standard processes" that exist within a given company. This assumption remains in effect until it is proven that the standard processes cannot meet the technical requirements. In this manner, the most economical design will result.

7.10.3 Layout

The layout is a graphical representation of each of the individual layers on the substrate, the location of the individual components on the substrate, and the location of the substrate with respect to the input/output pins. An important part of the layout process is to perform a worst-case analysis on the dimensional tolerances to determine whether a potential interference fit exists.

Virtually all layouts are created on a CAD system with software designed specifically for the purpose. In addition to the drafting function, most CAD systems have the following features:

- *Automatic checking.* With the schematic entered into the system, the CAD system can perform a point-to-point continuity check of the layout.
- *Component boundaries.* The computer prevents the designer from placing a component within a predetermined distance from another component to prevent interference.
- *Parts List.* A complete parts list can be created. When thick film circuits are being laid out, the area of each print can be calculated and the amount of paste required can be determined from an algorithm.
- *Visual aids.* The visual aids needed for fabricating the circuit can be created by the CAD system directly from the layout.
- *Resistor design.* The data from the resistor characterizations can be fed into the computer, which will then calculate empirical design equations allowing for power dissipation, tolerance, and termination effects.

It is important to realize that no CAD system to date can perform the complete layout of a chip-and-wire hybrid circuit without operator intervention. There are simply too many variations to consider.

A successful hybrid design requires extreme attention to detail. Each aspect of the design, including thermal, electrical, and mechanical aspects, must be thoroughly analyzed and reviewed by the design team before the design proceeds. There is an old saying among hybrid engineers: "The sooner the layout begins, the longer it will take, and the more mistakes will be made."

7.11 MULTICHIP MODULES

Multichip modules are an extension of the hybrid technology which permit a higher packaging density than can be attained with other approaches, allowing a silicon-to-substrate area ratio of greater than 30 percent. There are three branches of the MCM technology as depicted in Table 7.17;[5] MCM-L, based on a *laminated* printed circuit board structure, MCM-C, based on cofired *ceramic* structures, and MCM-D, which utilizes *deposited* conductors and dielectrics.

7.11.1 The MCM-L Technology

The MCM-L technology is based on the printed circuit board technology. Multilayer structures are formed by etching patterns in copper foil laminated to both sides of resin-

TABLE 7.17   Types of Multichip Modules

MCM-L	Substrates formed by *laminating* layers of printed circuit board material to form multilayer interconnection structures
MCM-C	Substrates formed by cofired *ceramic* or glass/ceramic structures, similar to the thick film process.
MCM-D	Interconnections formed by *depositing* alternate layers of conductors and dielectrics onto an underlying substrate, similar to the thin film process

based organic panels ("cores") which are laminated together with one or more layers of the basic resin in between to act as an insulator. Interconnections between layers may be formed by "through" vias, which extend all the way through the board, "blind" vias, which extend from the surface part way through the board, or "buried" vias, which connect only certain of the inner layers and do not extend to the surface in either direction. Through vias may be drilled and plated after laminating, while blind and buried vias must be drilled and plated prior to laminating.

There are numerous materials which may be used to fabricate MCM-L structures as shown in Table 7.18. The criteria for selection will vary with the application. Cyanate ester, for example, has a very low dielectric constant and excellent high-frequency characteristics.

Design guidelines for MCM-L substrates are presented in Table 7.19. Printed circuit boards for MCM applications must be suitable for wire bonding. This is accomplished by selectively plating nickel and gold on the copper traces as required. The difference in gold thickness is due to the different bonding mechanism between gold and aluminum wire. Gold wire bonding is accomplished by thermosonic bonding and is fundamentally a gold-to-gold bond, whereas aluminum wire bonding is accomplished by ultrasonic bonding and is an aluminum-nickel bond. Gold plating for aluminum bonding is thinner since it acts only to keep the nickel from oxidizing and interfering with the bonding process.

7.11.2   THE MCM-C TECHNOLOGY

MCM-C multilayer structures are fabricated from ceramic or ceramic/glass materials, with alumina being the primary base. There are two basic types of MCM-C substrates: high-temperature cofired ceramic (HTCC) and low-temperature cofired ceramic (LTCC). Both processes begin with thin sheets of unfired material approximately the consistency of putty which are referred to as *green tape*. Green tape is created by mixing the base powder with an organic vehicle and forming it into sheets by doctor blading or other means. Vias are punched in the green tape where interconnections between layers are to be made and filled with thick film paste designed specifically for via filling. The individual layers are printed with thick film paste to create the metallization patterns, aligned with the other layers, and laminated at elevated temperature and pressure. The laminated structures are then subjected to a lengthy bakeout cycle to remove the organic material and cofired at an elevated temperature to form a monolithic structure.

The HTCC and the LTCC processes differ in two primary ways: the firing temperatures and the thick film materials.[5] HTCC ceramics are designed to be fired at approximately 1600°C, which requires the use of refractory metals such as tungsten and molybdenum/manganese alloys as the conductors, and the firing process must take place in a reducing atmosphere to avoid oxidation of the metals. The top and bottom metallization layers are plated with nickel and gold to permit die and wire bonding. LTCC materials primarily

TABLE 7.18   Physical and Electrical Properties of Kapton 100 HN Film

Physical properties	Typical values		Test method
	23°C (73°F)	200°C (392°F)	
Ultimate tensile (MD)* strength, MPa (lb/in²)	231 (33,500)	139 (20,000)	ASTM D-882-83, method A†
Yield point (MD) at 3%, MPa (lb/in²)	69 (10,000)	41 (6,000)	ASTM D-882-81
Stress to produce (MD) 5% elongation, MPa (lb/in²)	90 (13,000)	61 (9,000)	ASTM D-882-81
Ultimate elongation (MD), %	72	83	ASTM D-882-81
Tensile modulus (MD), GPa (lb/in²)	2.5 (370,000)	2.0 (290,000)	ASTM D-882-81
Impact strength, kg • cm (ft • lb)	8 (0.58)		DuPont pneumatic impact test
Folding endurance‡, cycles	285,000		ASTM D-2176-69 (1982)
Tear strength (MD), propagating (Elmendorf), g	7		ASTM D-1922-67 (1978)
Density, g/cm³	1.42		ASTM D-1505-68 (1979)
Coefficient of friction, kinetic (film-to-film)	0.48		ASTM D-1894-78
Coefficient of friction, static (film-to-film)	0.63		ASTM D-1894-78
Refractive index (Becke line)	1.66		ASTM D-542-50 (1977)
Poisson's ratio	0.34		Avg 3 samples elongated at 5%, 7%, 10%
Low-temperature flex life	Pass		IPC-TM-650, method 2.6.18
Dielectric strength at 60 Hz, V/mil	7,700		ASTM D-149-81
Dielectric constant at 1 kHz	3.4		ASTM D-150-81
Dissipation factor	0.0018		ASTM D-150-81
Volume resistivity, Ω • cm	10^{17}		ASTM D-257-78

*MD—Machine direction

†Specimen size: 25 by 170 mm (1 by 6 in); jaw separation: 100 mm (4 in); jaw speed: 50 mm/min (2 in/min); ultimate refers to tensile strength and elongation measured at break.

‡Massachusetts Institute of Technology Test.

Source: From duPont.[12] Reprinted with permission.

consist of glass/ceramics and are designed to be fired at much lower temperatures—in the range 850–1050°C. This permits the use of standard thick film materials, such as silver and gold, which have much lower sheet resistivity than the refractory metals and do not require subsequent processing for assembly. A comparison of the HTCC and LTCC processes is shown in Table 7.20.

As a result of the organic burnout, the substrates shrink during firing. The shrinkage is very predictable, however, and may be accounted for during the design stage. This prop-

TABLE 7.19 Design Guidelines for MCM-L Substrates

Parameter	Value
Minimum metal spacing	0.003 in
Minimum trace width/maximum metal thickness	0.003/0/0018 in
Minimum Pth pad diameter	0.015 in
Surface layer nickel thickness (μ in) Aluminum/Gold wire bonding (μ in)	30+ 100–200
Surface layer gold thickness (μ in) Aluminum wire bonding (μ in) Gold wire bonding (μ in)	10+ 10–40 50–100
Minimum blind via diameter	0.006 in
Minimum finished hole diameter	0.010–0.000 in
Minimum buried via diameter	0.006 in
Minimum via pitch (no tracks)	0.015 in

TABLE 7.20 Electrical Properties of Nomex*

Nomex type	Thickness, mil	Dielectric strength, V/mil, ASTM D-149	Dielectric constant, at 60 Hz, ASTM D-150	Dissipation factor at 60 Hz, ASTM D-150	Volume resistivity, Ω • cm, ASTM D-257
410	3	540	1.6	0.005	10^{16}
411	5	230	1.2	0.003	—
414	3.4	530	1.7	0.005	10^{16}
418	3	730	2.9	0.006	10^{16}
419	7	325	2.0	—	—
992	125	380	1.7	0.020	10^{17}
993	120	540	2.6	0.015	10^{17}
994	250	250	3.5	0.010	10^{16}

*Unless otherwise noted, the Nomex properties are typical values measured by air under "standard" conditions (in equilibrium at 23°C, 50% relative humidity) and should not be used as specification limits. The dissipation factors of types 418, 419, 992, and 993 and all of the volume resistivities are measured under dry conditions. Nomex is a registered trademark of E.I. duPont de Nemours & Co. for its aramid products.

Source: From duPont Co.[11] Reprinted with permission.

erty is very critical when selecting a via fill material. The shrinkage of the via fill material must match that of the ceramic to prevent open circuits between layers.

Resistors and capacitors compatible with the LTCC green tape process may also be fabricated.[6] A distinction is made here between a *sandwiched* resistor, which is formed between two layers of green tape, and a *buried* resistor, which is printed on an alumina substrate and covered with layers of green tape. The resistor pastes developed for this process show a high degree of stability after several refirings at high temperatures and exhibit TCRs comparable to those of standard materials (<100 ppm/°C). Although the accuracy of untrimmed resistors is adequate for many digital circuit applications, buried resistors printed and fired on the substrate may be laser trimmed prior to lamination with green tape. These show excellent stability under conditions of high-temperature storage, temperature cycling, and harsh environments because they are covered with hermetic dielectric.

7.11.3 THE MCM-D TECHNOLOGY

The MCM-D technology utilizes processes similar to those used to fabricate thin film hybrid circuits. The conductors are primarily sputtered or plated metals, gold, aluminum, or copper, deposited on a variety of substrate materials, including ceramic and silicon. The dielectric materials are primarily used in the liquid state and are applied by spinning. Vias may be opened in the dielectric film by applying photoresist and etching, or by using photosensitive dielectric materials. A list of common dielectric materials is shown in Table 7.21.

TABLE 7.21 Properties of Fluoropolymers

	PTFE	FEP	PFA	ETFE	PVDF	PCTFE	ECTFE	PVF
% flourine	76.0	76.0	76.0	59.4	59.4	48.9	39.4	41.3
Melting point, °C	327	265	310	270	160	218	245	200
Upper use temperature, °C	260	200	260	180	150	204	170	110
Density, g/cm³	2.13	2.15	2.12	1.7	1.78	2.13	1.68	1.38
Oxygen index, %	>95	95	95	28–32	44	—	48–64	—
Arc resistance, seconds	>240	>300	>300	72	60	2.4	18	—
Dielectric constant	2.1	2.1	2.1	2.6	9–10	2.5	2.5	9
Dissipation factor	0.0002	0.0002	0.0002	0.0008	0.02–0.02	0.02	0.003	0.002
Tensile strength, lb/in²	5000	3100	4300	7000	6200	6000	—	—
Specific gravity	2.2	2.17	2.17	1.7	1.78	2.2	1.68	—
Water absorption, % 24 h	0	0	0.03	0.03	0.06	0	—	—
Electrical strength, V/mil	480	600	500	400	280	600	—	—

Source: From Harper.[7] Reprinted with permission.

MCM-D structures can be made much denser than the other types as a result of the photoetching process. Line widths of 10μ and via diameters of 15μ are common.

7.11.4 SUMMARY

Multichip modules are an important part of the repertoire of the packaging engineer for at least two reasons.

1. By utilizing the increased density available with this technology, more functionality can be incorporated into a smaller volume with all the advantages that this ability encompasses.
2. The variety of materials enables the substrate/interconnection structures to be more nearly tailored to a particular application.

A comparison of the various interconnection technologies is given in Table 7.22.

TABLE 7.22 General Characteristics of Thermoplastics

Material	Characteristic properties	Processing*	Electrical/electronic applications
Acrylics	Crystal clarity, good surface hardness, weatherability, chemical and environmental resistance, mechanical stability	1,2,3,4,5,6	Colored electronic display filters, conformal coatings
Fluoroplastics	Heat resistance, superior chemical resistance, low dielectric losses, zero water absorption, low friction coefficient	9,10,11,12,13 Some fluoroplastics can be molded by more conventional methods (2,7)	Wire and cable insulation, electrical components
Ketone plastics	Heat resistance, chemical resistance, high strength, resistance to burning, thermal and oxidative stability, excellent electrical properties, low smoke emission	1,2,8,13	Wire insulation, cable connectors
Liquid crystal polymers	High temperature resistance, chemical resistance, high mechanical strength, low thermal expansion	1	Chip carriers, sockets, connectors, relay cases
Nylon	Mechanical strength, tough, abrasion and wear resistance, low friction coefficient	1,2,3,4,6,8	Connectors, wire jackets, wire ties, coil bobbins
Polyamide-imide	High temperature resistance, superior mechanical properties at elevated temperature, dimensional stability, creep and chemical resistance, radiation resistance	1,2,7,9	Connectors, circuit boards, radomes, films, wire coating
Polyarylate	Ultraviolet stability, dimensional stability, heat resistance, stable electrical properties, flame-retardant, high arc resistance	1,2,3,4	Connectors, coil bobbins, switch and fuse covers, relay housings
Polycarbonate	Clarity, toughness, heat resistance, flame-retardant	1,2,3,4	Connectors, terminal boards, bobbins
Polyesters (PBT, PCT, PET)	Good electrical properties, chemical resistance, high temperature resistance, low moisture absorption	1,2	Connectors, sockets, chip carriers, switches, coil bobbins, relays

TABLE 7.22 General Characteristics of Thermoplastics (*Continued*)

Material	Characteristic properties	Processing*	Electrical/electronic applications
Polyetherimide	Good high temperature strength, dimensional stability, chemical resistance, long-term heat resistance, low smoke generation	1,2,3,7	Connectors, low-loss radomes, printed circuit boards, chip carriers, sockets, bobbins, switches
Polyolefins	Range of strength and toughness, chemical resistance, low friction coefficient, processability, excellent electrical properties	1,2,3,4,7,8	Wire and cable insulation
Polyimides	Superior high temperature properties, radiation resistance, flame resistance, good electrical properties	1,6,7	Insulation for electric motors, magnet wire, flat cable, integrated-circuit applications
Polyphenylene oxide	Low moisture absorption, good electrical properties, chemical resistance	1,2,3,4	Connectors, fuse blocks
Polyphenylene sulfide	Flame resistance, high temperature resistance, dimensional stability, chemical resistance, good electrical properties	1	Connectors
Polyphthalamide	Good combination of mechanical, chemical, and electrical properties	1	Connectors, switches
Styrenes	Range of mechanical, chemical, electrical properties depending on type of styrene polymer, low dielectric losses	1,2,3,4,8	Housings
Polysulfones	High temperature resistance, excellent electrical properties, radiation resistance	1,2	Circuit boards, connectors, TV components
Vinyls	Range of properties depending on type	1,2,3,4	Wire insulation, tubing, sleeving

*1 Injection molding   6 Casting   11 Dispersion coating
2 Extrusion   7 Compression molding   12 Fluidized bed coating
3 Thermoforming   8 Rotational molding   13 Electrostatic coating
4 Blow molding   9 Powder metallurgy
5 Machining   10 Sintering

Source: From Harper[3] and Greene.[4] Reprinted with permission.

REFERENCES

1. ISHM Modular Series on Microelectronics, *Handbook of Hybrid Microelectronics,* Jerry Sergent and Charles Harper, eds., McGraw-Hill, 1995.

2. George Harman, "Wire Bond Reliability and Yield," ISHM Monograph, 1989.

3. "Temperature Dependent Wear-Out Mechanism for Aluminum/Copper Wire Bonds," Craig Johnston, Robin A. Susko, John V. Siciliano, and Robert J. Murcko, *Proceedings of the ISHM Symposium,* 1991.

4. MIL-STD-883C, U.S. Government Printing Office, Washington, D.C.

5. J. Sergent, "Materials for Multichip Modules," *Semiconductor International,* June 1996.

6. H. Kanda, R.C. Mason, C. Okabe, J. D. Smith, and R. Velasquez, "Buried Resistors and Capacitors for Multilayer Hybrids," *Proceedings ISHM Symposium,* 1994.

CHAPTER 8

RIGID AND FLEXIBLE PRINTED WIRING BOARDS

Victor J. Brzozowski

Northrop Grumman Corp.
Electronic Systems and Sensors Division
Baltimore, Maryland

8.1 INTRODUCTION

Printed-wiring boards, commonly called PWBs, are sometimes referred to as the baseline in electronic packaging. Electronic packaging is fundamentally an interconnection technology and the PWB is the baseline building block of this technology. It serves a wide variety of functions. Foremost, it contains the wiring required to interconnect the component electrically and acts as the primary structure to support those components. In some instances it is also used to conduct away heat generated by the components. The PWB is the interconnection medium upon which electronic components are formed into electronic systems. Based on the critical importance of PWBs as the interconnecting link between individual component parts into the formulation of electronic systems, this chapter is devoted to presenting an understanding of the many important facets of PWBs that must be fully understood to achieve success in electronic systems.

8.1.1 Critical Factors in PWB Technology

The constant pressures for improvements in PWB technology arise in all aspects of this technology. Electrically, the increase of high-speed and high-frequency electronic systems creates demand for PWBs having lower electrical losses. This subject is further covered later in this chapter and in Chap. 11. In addition, higher operating voltages increasingly require PWBs with greater resistance to voltage breakdown, high-voltage tracking, and corona. These aspects are covered in Chap. 12.

Aside from the requirements for higher electrical performance of PWBs, higher electronic system functional densities and the resultant higher thermal densities create demand for lower thermal resistance of PWB materials. In an increasing number of applications, the old industry standard epoxy-glass PWB is thermally inadequate. This is discussed further in this chapter as well as in Chaps. 9 and 10.

New developments in component technology in the past 30 years, with the movement away from through-hole technology to the higher-density surface-mount technology

(SMT) using leadless chip carrier (LCC) components, have forced innovations in PWB materials and processes to control the end-item printed-wiring-assembly (PWA) coefficient of thermal expansion (CTE). SMT has caused the solder joint to act not only as the electrical interconnect between the component and the PWB but also as the main structural element attaching the component to the PWB. This PWA CTE control is required to reduce stresses imparted into the solder joint due to thermal operational excursions. Without it, solder joint failure occurs.

The constant trend toward higher-functionality integrated-circuit (IC) components with higher input-output (I/O) pin counts of the IC packages has resulted in increasing demand for finer-line PWBs. High I/O count package technology and fine-line PWB technology are often called fine-pitch technology (FPT) and are discussed further in Chaps. 5, 6, 9, and 10.

Basically, the problem is that advances in IC technology have caused significant increases in the complexity of component packaging technology, which many times are implemented without seeing what impacts may be imparted to PWB technology. PWB technology advances are often implemented after the fact instead of in conjunction with IC technology advances. With this trend, standard subtractive PWB technology often becomes inadequate, leading to new fine-line circuit interconnection forms such as additive PWB circuits, microwire interconnections, multichip modules, and chip-on-board. Again, the reader is referred to Chap. 10 for further discussion.

8.1.2 Glossary of Terms

For convenient reader reference a dictionary of terms for PWB technology and the complete subject of electronic packaging has been published.[1] Applicable industry and military standards are listed at the end of this chapter.

8.2 PRINTED-WIRING-BOARD SYSTEM TYPES

PWBs can be classified into several categories based on their dielectric material or their fabrication technique. This section describes PWBs fabricated using organic dielectric, ceramic dielectric, and discrete wire techniques. In addition, advances in the state of the art of printed wiring are leading to new interconnection techniques in very-high-density applications. These trends are discussed in Sec. 8.2.3.

8.2.1 Organic PWBs

These PWBs are fabricated using an organic dielectric material with copper usually forming the conductive paths. The organic-based boards can be subdivided into the following classifications: rigid, flexible, rigid-flex combining the attributes of both rigid and flexible boards in one unit, and molded. Each of these classifications, except for molded, can be further subdivided into single-sided, double-sided, or multilayer PWBs.

The circuit interconnection pattern, except for molded PWBs, is created by imaging the conductor pattern on copper sheets using a photoresist material and one of two image-transfer techniques—screen printing or photo imaging. The resist acts as a protective cover defining the conductor patterns while unwanted copper is etched away. Molded PWBs are usually nonplanar (three-dimensional) and consist of conductive materials selectively applied through printing conductive pastes or additively plating conductors to either extruded or injection molded thermoplastic resins.

These techniques can be used with a variety of dielectric materials to achieve various mechanical and electrical characteristics in the final product. Among the most common dielectric materials are epoxy/e-glass (electronic-grade glass) laminates used in the fabrication of rigid PWBs and polyimide film used in the fabrication of flexible printed wiring. The rigid-flex boards use a combination of these two materials. Molded PWBs use high-temperature thermoplastic resins. Section 8.3 addresses the wide range of materials and their characteristics that may be used for a given application. The following sections describe in detail the construction of each organic board type.

8.2.1.1 Rigid.   The rigid PWB is fabricated from copper-clad dielectric materials. The dielectric consists of an organic resin reinforced with fibers. The most commonly used fiber materials are paper and e-glass. Quartz, aramid and S2-glass are other fibers which have been used in specialized SMT and high speed applications. The fibers are either chopped (usually paper) or woven into a fabric. The organic media can be of a wide formulation and include flame-retardant phenolic, epoxy, polyfunctional epoxy, or polyimide resins. Built within the laminate structure may be low-CTE clad metals such as copper-invar-copper (CIC) or copper-molybdenum-copper (CMC) for decreasing the CTE of the overall PWB structure in SMT applications. Detailed information on base materials is included in Sec. 8.3.

As the name implies, rigid PWBs consist of layers of the organic laminates that are laminated through heat and pressure into a rigid interconnection structure. This structure is usually sufficiently rigid in nature to be able to support the components that are mounted to it. Specialized applications may require the PWB to be mounted to a support structure. The support structure may be used to remove heat generated by the components, decrease the movement of the PWB under extreme vibration, or decrease the CTE of the PWB in SMT applications (see Section 8.4.4).

The rigid PWB interconnection structure may be further subdivided by the number of wiring layers contained within the structure into the following three categories—single-sided, double-sided, or multilayered. Figure 8.1 shows a cross-sectional view of each type.

Single-Sided PWBs.   A single-sided PWB consists of a single layer of copper interconnection (usually on the component side of the PWB). The rigid dielectric material is fabricated from multiple layers of unclad laminate material pressed to the final end-use thickness. A single layer of copper cladding is applied to one of the outside layers during this process. In some instances double-sided copper cladding may be used, with the copper on one face being completely etched away during the processing.

The base laminate of single-sided boards can be of woven or paper (unwoven) materials with copper foil, usually of 1- or 2-oz weight, clad to one side. It should be noted here that copper cladding is most often referred to by its weight (1 oz/ft^2 equals 0.00137 in thick) rather than by its thickness. The raw clad laminate is first cut into working panels suited to the equipment, which will handle the subsequent operations. The panel is then drilled or punched to provide a registration system. Laminate flatness is important in achieving a good registration baseline. This is critical in an automated print and etch system because the panel tends to warp after the copper is removed during etching. This warping allows stresses built into the material during its fabrication to be relieved. Excessively warped panels may not register properly for subsequent operations.

The individual artworks which define the conductor patterns are then arranged or panelized so that one or more PWBs will be produced from a single panel. This is accomplished by stepping and repeating the patterns into a panel phototool. Once the panel layout is established, the panel can be drilled or punched to produce the final hole pattern. Holes required are either drilled in glass-reinforced products or punched in paper-reinforced products. Registration of the conductor pattern to holes is accomplished through either the right-angle edge of the panel or on pilot holes contained in opposite corners of

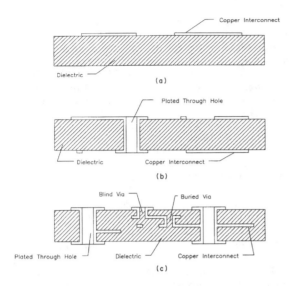

FIGURE 8.1 Cross-sectional views of organic PWBs: (*a*) single-sided, (*b*) double-sided, and (*c*) multilayer.

the panel. Drilling of holes is usually done after the panels are first cut; punching of holes is done as the last operation.

Following the drilling operation, the etch resist is applied and the circuit pattern formed. This pattern can be made by printing a liquid resist or photo imaging of a film or liquid. The next step is to etch away the unwanted copper from the laminate, leaving only the desired circuit pattern. Finally the resist is stripped and the single-sided board is complete in panel form. At this point additional processes such as platings or solder masks may be performed, or the individual boards may be sheared or routed from the panel.

While single-sided boards with their simplicity might seem doomed due to the emergence of modern complex electronics, they continue to have a substantial market, especially where cost is a strong driver. Their use with fairly complex IC devices can be seen in many places, such as watches, cameras, and automobiles. This will probably continue to be the case for the foreseeable future.

Double-Sided PWBs. From a historical perspective the double-sided board is probably the most often designed type of all PWBs. It retains much of the production simplicity of the single-sided board but allows circuit complexities far in excess of 2:1 over its simpler cousin. This is the case because it allows basic *x* and *y* routing of the circuit on its two outer faces, thus improving the routing efficiency and the circuit density. Interconnection of the two conductor patterns is accomplished through drilling and subsequent plating or filling the interconnection holes, called vias. The most widely used method is to plate the vias with copper. Silver-conductive ink is another process used in low-cost consumer products.

Double-sided boards are fabricated from laminates with copper (usually 1 oz) clad to both outside layers. The copper may be clad to a variety of materials, as discussed in Sec. 8.3. Here, as with the single-sided board, the material is purchased from a laminator who specializes in providing laminates to the electronic industry.

Once the raw laminate is cut into panels, the fabrication operation begins with the interconnection hole drilling. The via holes may also serve as mounting holes for the components to be soldered into. After the via hole pattern has been drilled, the holes may be filled with the conductive ink or the panel is copper plated by an electroless technique in preparation for subsequent plating by either of two methods—pattern plating or panel plating.

The conductor image is formed in a similar way as with single-sided boards, except that the photoresist application and imaging take place on both sides of the panel. Obviously, the registration of the photo images from one side of the panel to the other is critical. The circuit pattern on one side must be properly registered to the pattern on the other side, or the plated-through holes will not properly connect between the two sides.

The next step is to etch away the copper laminate, leaving only the desired circuit pattern. At this point additional processes such as resist stripping, platings, or solder masks may be performed and the individual PWBs then sheared or routed from the panel (see Fig. 8.1).

Multilayered PWBs. Multilayer boards are those PWBs having three or more conductive layers, including any pads-only layers. The typical modern multilayer board will have from 4 to 10 layers of circuitry, with some high-density applications requiring upward of 50 layers. Most multilayer boards are fabricated by laminating single- or double-clad, pre-etched, patterned sheets of thin (<0.005-in) laminate together using partially cured resin in a carrier fabric. The materials commonly used for this purpose are discussed at length in Sec. 8.3.

The single- or double-clad laminate material is processed similarly to the single- or double-sided PWB, except that the via or component holes are usually not drilled until after lamination. It should be noted here that the importance of registration is amplified as the layer count increases. Increased pad sizes may be required to minimize via hole breakout due to misregistration. The same requirement may limit the size of panels due to run out of the circuit features.

Following the fabrication of the individual layers or layer pair, a "book" of layers and their interposed bonding layers are stacked together in a particular sequence to achieve the required lay-up. This book is then laminated under heat and pressure to the appropriate thickness for the final board. The outer layers are not pre-etched so that the laminated book appears the same as a double-sided copper-clad laminate of comparable thickness. After lamination, the book is processed the same as a thick double-sided board. The book is drilled to add the via holes and then processed as if it were a double-sided board using plated-through holes. A cross-section of a typical multilayer board is shown in Fig. 8.1.

In some cases standardized layers, such as power or ground distribution, can be "mass"-laminated into the raw laminate. This is a very cost-effective means of achieving multilayer density at near double-sided board cost since the outer layer processing and via drilling are identical to that for double-sided PWB processing. As a result, four-layer boards are the most prevalent multilayer boards.

Where circuit density requirements override cost considerations, techniques such as blind or buried vias are used to increase the interconnection wiring density on a given layer. Where these techniques are used, the inner layer pairs are fabricated as double-sided boards, complete with plated vias, and then assembled into books for processing into multilayer boards. Thus the inner layers may be interconnected without a via hole through the entire board. Similarly, blind vias may connect to the first or subsequent buried layer on each side of the board without penetrating the entire board (see Fig. 8.1).

The multilayer board has now achieved a cost and reliability level that allows its use in any level of electronics. It is no longer the exclusive tool of the mainframe computer, telecommunications, or military electronics. It is often seen even in toys.

8.2.1.2 Flexible. As defined by the Institute for Interconnection and Packaging Electronic Circuits (IPC), flexible printed wiring is a random arrangement of printed wiring, utilizing flexible base material with or without cover layers. Interconnection systems consisting of flat cables, collated cable, ribbon cable and sometimes wiring harnesses are sometimes confused with flexible printed wiring. Flexible printed wiring is used in applications requiring continuous or periodic movement of the circuit as part of the end product function and those applications where the wiring cannot be planar and is moved only for servicing. Flexible wiring can be used as interconnect cabling harnesses between various systems circuit card assemblies and or connectors as well as to interconnect individual electrical components. Figure 8.2 shows a typical flexible printed wiring interconnect.

Visually, flexible printed wiring looks similar to rigid printed wiring. The main difference in the products is the base or dielectric material. Flexible printed wiring is manufactured using ductile copper foil bonded to thin, flexible dielectrics. The dielectric materials include polyimide (Kapton), polyester terephthalate (Mylar), random fiber aramid (Nomex), polyamide-imide TFE Teflon and FEP Teflon and polyvinyl chloride (PVC). Dielectric materials are discussed in further detail in Section 8.3.2.1.

As with their rigid printed wiring brethren, the flexible printed wiring may be manufactured in single-sided, double-sided or multilayer configurations (see Fig. 8.3). The conductor patterns are formed in a similar manner to rigid PWBs using either a screen printing or photo-imaging of a resist to form the conductor pattern and then etching of the unwanted copper. A variety of adhesive materials are used in their manufacture to bond the various layers together. These including acrylics, epoxies, phenolic butyrals, polyesters and polytetrafluoroethylene (PTFE). In addition, newer processes have been developed to laminate the conductor directly to the dielectric film without the use of an adhesive layer. Adhesive materials are discussed in Section 8.3.2.2.

On single- and double-sided flexible wiring, cover layers are often used to protect the etched copper circuitry. When a film cover layer is employed, the access holes to the circuitry are either punched or drilled in the adhesive-coated film. The cover layer is then mechanically aligned to the wiring and laminated in a platen press under heat and pressure. The adhesive systems must be of a "no" or "low" flow formulation to keep from contaminating the interconnection pads. Screen printable cover coats may also be used. These are usually ultraviolet curable and give a moderately pinhole-free insulating surface.

Unlike their rigid counterparts, the outline features of flexible wiring are not routed but cut using either soft or hard tooling. Soft tooling involves a rule die. In its simplest form, this can be a steel die and Exacto knife to cut the outline pattern. More complex dies may consists of steel strips imbedded in plywood. When pressed against the flexible wiring they act as cookie cutters to form the outline. Rule dies are dimensionally less accurate and less expensive than hard tools. Hard tools are punches, die plates and strippers manufactured from hardened steel. They can hold tolerances within the die of ±0.001 in.

FIGURE 8.2 Typical flexible printed wiring interconnect in connector assembly. *(Courtesy of Teledyne Electro-Mechanisms, Hudson, N.H.)*

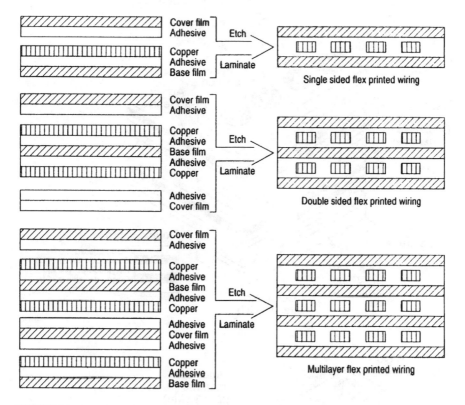

FIGURE 8.3 Anatomy of a flexible PWB.

Due to the extreme flimsiness of flexible wiring, when components are to be mounted, adequate reinforcement must be added to the flexible wiring to eliminate stress points at the component circuit interfaces. Reinforcements typically used are simple pieces of unclad rigid laminates or complex formed, cast or machined metals or plastics to which the flexible wiring is laminated.

8.2.1.3 Rigid-Flexible. Rigid-flex circuitry consists of single or multiple flexible wiring layers selectively bonded together using either a modified acrylic adhesive or an epoxy prepreg bond film. Cap layers of rigid core copper clad laminate may be bonded to the top and bottom surfaces of the circuit to add further stability to the bonded areas (See Fig. 8.4). This technique is in wide use in the industry today to provide rigid areas at connector interfaces (see Fig. 8.5). In other instances, PWBs have incorporated within the structure flexible layers to interconnect one PWB to another or to interconnect a PWB to a connector. The rigid-flexible wiring exhibits the lowest profile form factor of all the interconnect system types.

In multilayer applications, problems can arise in fabricating rigid flex assemblies with the kapton/acrylic film layers integrated into the hard board PWB section which would contain epoxy/e-glass materials. The high moisture absorption rate (4 percent) and CTE (400–600 ppm/°F) of the acrylic adhesive can lead to reliability problems in the finished product under thermal uses.

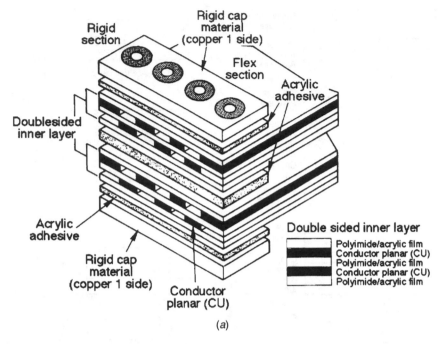

(a)

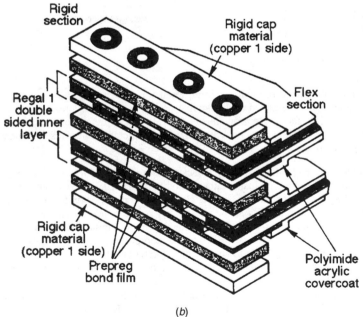

(b)

FIGURE 8.4 (a) Typical rigid-flexible multilayer, (b) REGAL I flexible multilayer. *(Courtesy of Teledyne Electro-Mechanisms, Hudson, N.H.)*

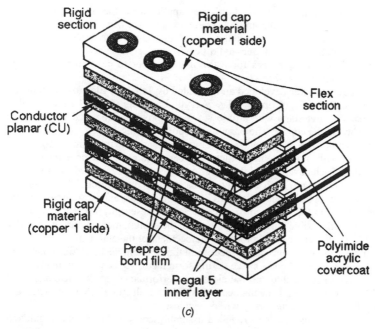

FIGURE 8.4 *(Continued)* (c) REGAL 5 flexible multilayer.

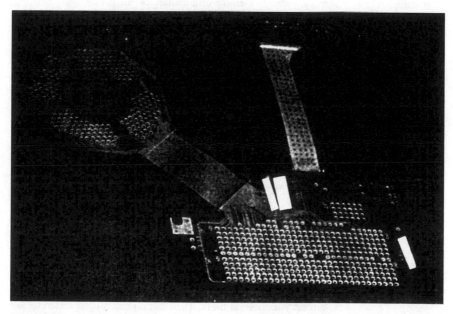

FIGURE 8.5 Typical rigid-flexible printed wiring. *(Courtesy of Teledyne Electro-Mechanisms, Hudson, N.H.)*

To allow for increased reliability in multilayer applications, the amount of acrylic adhesive in the rigidized area must be limited. There are a couple of methods in use in the industry today to accomplish this. One method involves using a polyimide/acrylic base with a prepreg covercoat in the rigid area and a polyimide acrylic selective cover coat in the flexible areas. This technique does force the manufacturer to develop tooling techniques to overlap the separate dielectrics.

A second technique that was developed and patented by Teledyne Electro-Mechanisms of Hudson, N.H is called a Rigid Epoxy Glass Acrylic Laminate (REGAL) Flex. In this technique, the manufacturer starts with a base stock of Epoxy prepreg clad with copper. The traces are etched in the copper. The base-stock epoxy prepreg is then encapsulated in the flex area with flexible dielectric to allow the circuit to bend. The traces in the rigid section are encapsulated with epoxy prepreg which has been windowed to remove the prepreg from the flex section. Each cover coated flex layer can be selectively bonded together with pre-windowed epoxy prepreg to form a rigidized area for subsequent through hole processing. A second construction technique called REGAL 5 removes the epoxy prepreg from the flex area. Figure 8.4 shows the various differences between these rigid-flex techniques.

The benefits of rigid-flexible wiring are apparent in the design, manufacturing, installation and assembly, quality control and product enhancement of the end item.

The designer has increased conceptual freedom in the end item design. Conformability, three dimensional interconnects and a space saving form factor are benefits. In many cases reduced interconnect length lead to optimal electrical performance. Mechanical and electrical interfaces are reduced and mechanical, thermal and electrical characteristics are more repeatable than with conventionally wired systems.

In manufacturing their use leads to reduced assembly costs within a totally unitized interconnect system. There are increased opportunities for automation. In addition, reduced system interconnect errors and improved system interconnect yields occur.

Installation and assembly benefits include elimination of miswiring and indexing errors encountered in discrete wired systems. A reduction in the installer skill and training, increased speed of installation and mounting simplification leading to reduced hardware requirements are other benefits.

Quality control benefits by the adaptability of the product to automated inspection, a simplification of error cause analysis and more effective error cause correction.

Products are enhanced through a reduced weight, volume and cost. Fewer interconnections are usually required leading to increased system reliability. Reduction in product inventory, maintenance and field service time and expense are also realized.

8.2.1.4 Molded.   One other PWB concept with many functional and design advantages in many application areas is the molded or three-dimensional PWB. As stated previously, these boards are usually nonplanar (three-dimensional) and consist of conductive materials selectively applied to either extruded or injection molded thermoplastic resins. Standard rigid PWB laminate structures produce formed circuitry only in two dimensions, by comparison. Representative molded three-dimensional circuits might be cases or covers which contain an electronic assembly, or a molded three-dimensional IC chip carrier, as shown in Fig. 8.6. This technique can be applied to any three-dimensional molded part onto which formed circuitry is beneficial. A single three-dimensional case with integral formed circuitry, for instance, could replace a two-part case and separate circuit board assembly.

High-temperature thermoplastics are commonly used in these applications due to the soldering of parts to the circuitry formed on them. The most common used materials are polyethersulfone, polyetherimide, or polysulfone. Other resins used include but are not limited to polyaryl esters, polyamides, polyphenylene sulfides, and various polyesters.

FIGURE 8.6 Chip carrier package made by two-step molding process for achieving three-dimensional circuit patterns. *(Courtesy of Pathtek Co., Rochester, N.Y.)*

Each resin has its own unique set of properties which must be matched to the functional and cost requirements of the end item.

The benefits of using this technology in a functional part are many and varied. The resin system's thermal and electrical properties are superior to standard epoxy/e-glass rigid laminate materials. Manufacturing tolerances of the finished part can be held to ±0.001 in. Holes can be rectangular, square, oval, or tapered. Features such as connectors, clips, bosses, and spacers can be molded into the finished part. This concept also offers cost advantages for sufficiently high-volume usage over conventional rigid laminates.

The formed three-dimensional circuitry can be applied in several ways, with two common application methods being (1) circuitry transfer process and (2) two-step molding process.

In the circuitry transfer process, the molded part is made as one step, and the circuit pattern is applied to a flat release sheet in another step. The pattern is applied by screen printing of a polymer thick-film material onto the release sheet. A polymer thick film as used here is basically a conductive powder such as copper or silver mixed into a polymer resin to a screenable form (see also Chap. 1). The release sheet is slit, punched, and so on, where required, so that it can be inverted onto the molded form in a contour that fits the contour of the molded part exactly. The circuit pattern is now pressed, or transferred from the release sheet, onto the three-dimensional contour of the molded part. This is done in a heated press in order to thoroughly bond the cured polymer thick-film circuitry to the molded part. The Konec[*] processes use polymer thick films to manufacture molded interconnections.[5]

---

*Konec is a registered trademark of Amoco Performance Products, Ridgefield, Conn.

The two-step molding process is performed by overmolding a separate initially molded part. This process usually involves the selective additive plating of copper to form the interconnect. In this process, the initially molded part (the first step of molding) is molded using a plastic material which is catalyzed so that it is a platable plastic. The platable plastic part is molded so that any ink or contact area that is to become circuitry protrudes from the rest of the molded part. This first molding of platable plastic is now placed in a second mold of the final desired part form, but with the protruding final circuitry areas contacting the mold walls. Thus in this second molding step, the final part is molded with an uncatalyzed, hence unplatable, plastic. When this piece is now removed from the mold, the areas of catalyzed plastic are exposed at the surfaces. Plating of the part will then plate only the catalyzed plastic surfaces. This forms the circuitry pattern. The rest of the part will not be plated. There are two variations on this type of process being used in the industry today.

In the photoselective plating process a molded part is chemically treated and then sensitized in a liquid non-noble metal photo initiator to promote copper adhesion. Contact printing is the preferred method of photo exposure to achieve the required through-hole and surface conductor patterns. The sensitized sites are reduced to metal under ultraviolet light, and then the part is selectively plated in an electroless copper bath. PCK Technology, Pathtek, Amp, Texas Instruments, and Additive Technologies are just a few of the companies producing products using this technology.

LeaRonal has developed a similar additive metallization process. In this process, a molded part is processed through an adhesion promotion process which includes a swell to prepare the part for etching, an etch to produce a microroughened surface, and a neutralization to solubilize the by-products of the etch. A permanent ultraviolet-curable epoxy resist is next applied to define the circuit pattern, prevent metallization of noncircuit areas, provide environmental protection, and provide thermal protection to the substrate during soldering. A unique process is then used to selectively activate and deposit a thin layer of electroless copper in the exposed activated substrate surface and holes. This is followed by a fully additive copper deposition to the final required copper thickness. This process has fabricated successfully a pin grid array adapter circuit.

8.2.2   Ceramic PWBs

These PWBs arc classified by their method of manufacture and type of metallization. There are four distinct types. *Thick film* uses alumina, beryllia, and similar materials as the substrate base material and fired thick-film dielectric paste (a glass-frit paste) as the dielectric. Conductors are formed from fired conductive noble metal pastes. *Thin film* uses ceramic, glass, quartz, silicon, or sapphire as the substrate base and deposits various metals by plating, sputtering, or vapor deposition. *Cofired* substrates can be broken into two distinct groupings. Cofired ceramic uses ceramic tape as the dielectric that is cofired with refractory metal pastes which form the conductors; cofired low-temperature tape uses a glass/ceramic tape dielectric that is cofired with noble metal pastes which form the conductors. *Direct-bond copper* directly bonds copper conductors to a ceramic substrate. All of these ceramic-based PWBs are most often referred to as substrates.

Ceramic boards do offer advantages, compared to organic boards. The ceramic dielectric is inherently much more rigid than organic material dielectrics. Flatness values of 0.002 to 0.003 in/in are normal and can be as low as 0.001 in/in. Component soldering (183 to 240°C) is usually performed above or near the glass transition temperature T_g of organic materials (100 to 240°C) and can lead to damaged PWBs when the process is controlled improperly. The 1600°C needed to fire ceramic is well above this soldering temperature.

Higher thermal conductivities available with ceramic materials offer improved thermal management over organic boards. When thermal vias are required, the smaller buried vias available with ceramic boards provide a low thermal resistance while sacrificing less routing area.

The CTE matching to hermetic component cases is available with a ceramic board and offers improved solder joint reliability in SMT applications.

Increased costs and design time are disadvantages to the use of ceramic boards. A weight penalty is usually paid when ceramic boards are used. The ceramic and noble metal materials used in ceramic boards are also more costly than their organic counterparts. The demand for these boards has usually been in low-volume military and avionics applications. This has led to a limited number of ceramic PWB facilities, which has also helped to maintain higher costs. The following sections describe in detail each of the four different ceramic substrate types.

8.2.2.1 Thick Film. This class of ceramic PWBs is manufactured by building up alternating layers of conductors and dielectric on a ceramic substrate. A thick-film substrate may be called a true printed circuit in that resistive elements may also be built into the substrate. Thick-film substrates have dielectric thicknesses of 0.0015 to 0.0025 in, metallization thicknesses of 0.0005 to 0.001 in, and resistor thicknesses of 0.001 to 0.0015 in. Each layer is pattern-printed onto the substrate using screen or stencil printing processes.

Several different ceramic materials can be used as the substrate base. These include alumina, beryllia, aluminum nitride, boron nitride, silicon carbide, and silicon nitride, depending on the end item requirements. Dielectric, conductor, and resistive inks (pastes) are printed and fired to build the interconnect structure. These materials are discussed in Chap. 7.

The manufacture of a thick-film ceramic PWB begins with the generation of artwork defining the following: conductor patterns, dielectric layers including via openings in multilayer applications, via fill patterns, and resistor networks when required. From this artwork a screen or stencil for each wiring, via, resistor, and dielectric layer is developed.

When a screen is to be used, its manufacturing begins with stainless-steel wire mesh stretched over a metal frame. Nylon and polyester mesh are sometimes used due to their lower costs. They can stretch and be affected by temperature and humidity and are not as durable as the stainless-steel screens. The mesh count (number of wires per linear inch of screen) and wire diameter determine the obtainable print resolution of the various layers. In general, mesh counts vary from 200 to 400 and wire diameters from 0.9 to 1.1 mil, with the lower mesh counts (200) and greater wire diameters (1.1 mil) being used for gross conductor patterns (>0.010-in line widths and spaces).

A photosensitive polyvinyl or polyimide emulsion is next applied to the screen, and the conductor, dielectric, via, or resistor pattern is photo-imaged on the emulsion under ultraviolet light using the artwork. Uncured resin is then washed away, leaving the final conductor pattern. The emulsion thickness determines the final wet print thickness of the various layers for a given mesh and wire diameter.

Stencil printing involves etching the patterns to be printed in a thin metal foil, usually nickel or brass. This once again uses photoresistive materials in a photoimaging operation to define the pattern and then etching away the unwanted metal similar to etching copper on a PWB laminate. The metal stencil is then mounted in a metal frame. The advantages of stencils over screen meshes are many. They offer more uniform print thicknesses, greater print resolution, reduced dimensioning capabilities, and easier process control.

The ceramic substrate is prepared by cutting to size using laser drilling, diamond scribing, or ultrasonic milling. The laser is by far the most prevalent method. Overlapping of the laser drill hole pattern can yield a smooth cut surface. Spacing of the holes yields a perforated surface, which can be used to define a number of substrates on a single ceramic panel. This "snapstrate" can be processed, and after the processing is completed, the individual substrates can be snapped along the perforation.

Cleaning of the ceramic to remove the melted ceramic or glass residue (slag) left after laser drill is called deslagging. A variety of methods may be used, including sandblasting with an alumina slurry followed by a cleaning in hot isopropyl alcohol. Subsequent heating in a furnace at 800 to 925°C is usually done to burn off any organic contaminants left during the previous processes.

Following substrate preparation, the metallization process begins. Conductive, dielectric, or resistive inks (pastes) contain the desired metals or conductors. These are combined with glass frits to allow bonding during firing and needed solvents to accomplish a definable print. Each layer is printed, dried to volatilize the solvents, and then fired in a furnace. This print, dry, fire sequence continues until the multilayer structure is complete.

Resistor layers are the last high-temperature firing (800 to 900°C) performed and are done together. Subsequent high firings can cause an oxidation of the resistive material and an unacceptable rise in resistance of the material. A low-temperature (425 to 525°C) glass encapsulant can be printed and fired over selective resistors and conductors as a protective overcoat or solder mask. Resistors are usually trimmed to a final value using a laser-trimming process. This requires the overcoat encapsulant to be composed of a glass that allows the laser to penetrate to the resistor.

The metallization buildup on the substrate can cause excessive warping to occur. Warping can be controlled through proper design and process control. These techniques include elimination of unnecessary layers through optimal routing, cofiring layers, and printing a layer of large CTE mismatch material on the opposite side of the substrate for every three to four topside layers.

Thick-film ceramic boards are used mainly as interconnect substrates in multichip modules. Their use as interconnection substrates for applications similar to rigid organic PWBs is usually limited to a maximum PWB size of 8 in.

8.2.2.2 Thin Film. Thin-film ceramic boards are normally limited to specialized designs or single-layer applications. They are more expensive and difficult to multilayer when compared to their thick-film counterparts. Their use requires the substrate surface to be very flat and smooth and causes higher-purity ceramics to be used. These include alumina, glass, quartz, silicon, or sapphire. Thin-film metallizations use noble metals (such as gold) and are used most often in microwave applications due to their improved electrical performance over thick-film substrates at higher frequencies.

Thin-film interconnections in multilayer applications are through buried vias, as is the case with all ceramic PWBs. The top and bottom metallization on a double-sided substrate can be connected using plated-through holes for electrical interconnection or improved thermal performance. Metallization patterning of thin-film ceramics is accomplished through the use of photo lithography, plating, etching, vapor deposition, and sputtering methods.

8.2.2.3 Cofired. This type of ceramic PWB requires the printing of pastes containing conductor metallization onto unfired tape (dielectric) materials. These layers are then stacked and cofired together in a furnace to form the interconnect structure. The unfired tape materials can be either ceramic or a low-temperature dielectric.

The ceramic tape system requires higher firing temperatures. This results in refractory metals such as tungsten, molybdenum, or tungsten copper to be used as the conductor within the paste. These metals have higher vaporization temperatures to withstand the firing, but lower thermal and electrical conductivities than the noble metals (gold, silver, or copper). Their lower conductivities typically limit the use of these substrates to digital applications.

The conductor paste is applied to the tape using a screen or stencil similar to the thick-film process. For multilayer applications, holes are punched in the dielectric prior to printing. The conductive paste fills the hole and later forms a buried via during the firing operation. After all layers have been printed, they are stacked in the proper sequence, laminated together under heat and pressure, and fired to solidify the ceramic.

Upon completion of the cofiring operation, the exposed refractory metals are electroplated with typically 0.000080 to 0.000350 in of nickel and 0.000050 to 0.000100 in of gold. The nickel acts as a barrier to intermetallic formations between the gold and tungsten and as a corrosion barrier. The gold serves as a wire-bondable or solderable surface for component attachment.

The dielectric tape systems are composed of lower-temperature reflow glasses similar to those found in thick-film pastes. The printing, stacking, and laminating operations fare the same as those used for the ceramic materials. The firing, however, occurs at a lower firing temperature, which allows the use of noble metal pastes. These tapes allow the substrates to be used in microwave applications. In addition, no additional platings are required upon postfiring.

Cofired PWBs offer distinct advantages over thin- or thick-film processed PWBs. Multilayering is limited only by the thickness limitation of the overall package. Each fired layer is 0.003 to 0.012 in thick, depending on the tape thickness used. Thermal vias may be more readily incorporated into the design using an array of vias punched in the dielectric and filled with conductive pastes. Cutting of the tape prior to stacking and firing can allow cavities to be formed in the final product to allow component mounting. Cofired substrates are the main manufacturing method used to produce leadless chip carrier and multichip module component bodies. The patterning of the conductor layers prior to firing allows for easier inspection and rework.

The main disadvantages are in the longer life-cycle time needed to develop the tooling required to produce the item.

8.2.2.4 Direct-Bond Copper.

As the name implies, a direct-bond copper (DBCu) board uses copper directly bonded to a ceramic dielectric. The most commonly used ceramic is alumina, although beryllia and aluminum nitride (which offer improved thermal performance) have been used successfully in the process. The DBCu structure offers improved thermal and structural performance compared with conventional thick- or thin-film technologies using alumina dielectrics.

The process involves oxidation of the surface of a copper foil, which is then placed against a ceramic substrate. The pieces are placed in a furnace which reflows the copper oxide and fuses it with the surface ceramic oxides. This directly bonds the two materials together.

Any currently available ceramic material thicknesses can be used (0.005 to 0.125 in). Copper foils of 0.001- to 0.080-in thickness have been used successfully. To prevent the substrate from warping or cracking, it is recommended that the copper foil thickness be less than or equal to the ceramic thickness.

The bonding process occurs at approximately 1000°C. During cooling, the copper contracts at a much higher rate than the ceramic due to its greater CTE. The cooling increases the tensile strength of the ceramic by an order of magnitude by placing it in compression. This allows thinner ceramic materials to be used and will decrease the overall assembly height and reduce the thermal resistance of the board. Coupling the reduction in thickness with the heat-spreading capabilities of copper allows a DBCu board to offer reductions in thermal resistance comparable to a minimal-thickness BeO thick-film board.

The copper interconnect features can be formed by punching the copper sheet prior to attachment to the ceramic or by photo-imaging techniques similar to those used in rigid

PWBs after bonding to the ceramic. The latter process allows finer line features. Typically, 0.015-in minimum line widths and spacings are used.

Alternately stacking layers of copper and ceramic can create multilayer interconnect structures. Three conductor layers are the limits achieved to date. Buried vias consisting of windows in the ceramic filled with copper spheres or particles are used to interconnect from layer to layer.

DBCu boards offer the advantages of improved structural strength, thermal management, and high thermal conductivity. In addition, the copper offers bondability when nickel- or gold-plated and is capable of excellent solderability. Limitations in dimensioning and multilayering are its disadvantages.

8.2.3 Developmental PWBs

Advances in the state of the art of integrated circuitry are leading to systems applications requiring higher packaging densities and improved thermal and electrical performance. In many instances this has led to the requirement for multichip modules to be designed to meet system electrical throughput and volumetric requirements. This has placed increasing emphasis on the interconnection substrate design within the module. Ceramic materials with their high dielectric constant impart unacceptable losses in throughput due to propagation delay for high-speed applications. In addition, limitations in wiring density on the substrate due to the printing techniques used cause unacceptably thick substrates because of the need for more wiring layers. New combinations of standard materials and processes have been combined to address these requirements. These are mainly in the form of a thin-film dielectric on silicon or ceramic.

The dielectric used by most companies is polyimide. A thin-film polyimide layer is applied through screen printing or spinning onto a silicon or ceramic substrate. In some instances the silicon or ceramic may have wiring routed within. The polyimide is then metallized with aluminum or copper and patterned using standard thin-film photo-imaging techniques. Via interconnects from layer to layer are done in an additive "pillar" process. The polyimide dielectric layers provide a thin, low-dielectric-constant material (<4.0), and the thin-film metallization allows very-high-density routing (0.001- to 0.002-in lines and spaces).

A second interconnection technique involving a thin-film polyimide process involves the "growing" of a circuit on top of the die in a multichip module. This technique, called high-density interconnect (HDI), was developed by General Electric. Here, die are mounted into wells inside a multichip package (MCP) such that the top of the die forms a planar surface with the inside surface of the module. Polyimide dielectric and thin-film metal layers are then fabricated on top of the die to perform the interconnection.

Another developing technology for multichip module applications is the silicon circuit board (Figs. 8.7 and 8.8) being developed by nCHIP in San Jose, Calif. This board uses silicon as a base substrate. The difference in this board over similar silicon interconnection techniques is the use of silicon dioxide as the dielectric material instead of polyimide. The manufacturer reports that material costs less as a base material and requires fewer processing steps than polyimide. It offers a second advantage in that a thin silicon dioxide layer between power and ground planes in the substrate gives an integral decoupling capacitor between the planes not obtainable with other technologies.

8.2.4 Discrete-Wired PWBs

Most discrete-wired boards use an organic rigid PWB as a base substrate, their primary difference being that the circuit is wired using discrete or individual wires. Several variet-

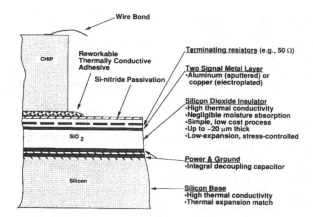

FIGURE 8.7   Cross-sectional view of silicon circuit board. *(Courtesy of nCHIP, San Jose, Calif.)*

FIGURE 8.8   High-speed RISC/SPARC processor using multichip design with silicon circuit board. *(Courtesy of PCK Technology, Melville, N.Y.)*

ies of these boards are in use today. The two resembling an organic rigid PWB most are the Multiwire[*] board and the Microwire board. Two other commonly used discrete wiring systems are the stitch-welded board and the Wire-Wrap[†] technique. These technologies are described in the following sections.

8.2.4.1 Multiwire.   Multiwire technology was developed as an alternative PWB method to conventional multilayer PWBs. The invention of Multiwire addressed a variety of concerns of multilayer PWB designers and fabricators. The time and cost required to produce the artwork films (especially for existing designs, which need to be quickly modified), layer-to-layer registration requirements of a large number of layers, and the inability to inspect or repair inner layers in a finished product are reduced or eliminated using this technology.

Multiwire most commonly uses a rigid epoxy/e-glass laminate as a base, although any base material (polyimide, metal core, or flexible material) used in the printed-circuit industry can be used. Power and ground planes are normally photo-imaged on each side of the base laminate from the copper cladding, as explained in the manufacture of double-sided rigid boards in Sec. 8.2.1. Overlaid on each side of the base is a proprietary adhesive.

The design of the wiring is done on a computer-assisted design (CAD) system. The CAD system generates the necessary instructions to drive a numerically controlled wiring machine. The machine is used to embed wires into the adhesive in the pattern of the designed circuit. The discrete wire runs use insulated no. 34 AWG (6.3-mil-diameter) or no. 38 AWG (4.0-mil diameter) magnet wire. Wiring can be done in an *x-y* orthogonal grid as well as at a 45° angle to intersect the component locations (Fig. 8.9).

After all wires have been routed, they are pressed more deeply into the adhesive and then encapsulated with an epoxy cover and copper foil. The board at this stage resembles a conventional multilayer PWB prior to the drilling step.

A via pattern is drilled at the component locations using conventional drilling processes, which cut into the wiring. The holes are cleaned using high-pressure water and prepared for copper plating in a proprietary alkaline permanganate solution. This solution

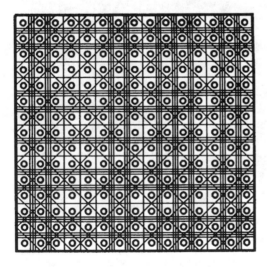

FIGURE 8.9   Multiwire typical wired circuit pattern using *x-y* and 45°C geometries. *(Courtesy of PCK Technology, Melville, N.Y.)*

---

*Multiwire and Microwire are registered trademarks of PCK Technology Division, Kollmorgan Corp., Melville, N.Y.

†Wire-Wrap is manufactured by the Gardner-Denver Co.

removes resin smear and microetches the hole wall to promote plating adhesion. In addition, it chemically removes some of the wire insulation at the point of wire entry into the drilled hole.

Hole plating to complete the interconnect is accomplished in a way similar to conventional PWB processing. The boards are catalyzed and plated using an appropriate electroless copper bath. Application of a dry-film photoresist defines the surface features. Exposed holes and surface features are then electroplated with copper and then tin-lead plated. The photoresist is stripped and the tin-lead plate acts as the resist for etching the background copper and completing the termination process. Figure 8.10 describes the final structure of a Multiwire board.

It should be noted that the Multiwire board use is aimed toward through-hole component (such as DIP and PGA) designs and low-density SMT designs. To address ultra-high-density SMT designs another product, called Microwire, was developed. This is discussed in the next section.

8.2.4.2 Microwire.   With the advent of high-speed VLSI circuitry, Microwire interconnection technology was developed to simultaneously address unique thermal, electrical, and density requirements typically associated with high-speed SMT devices.

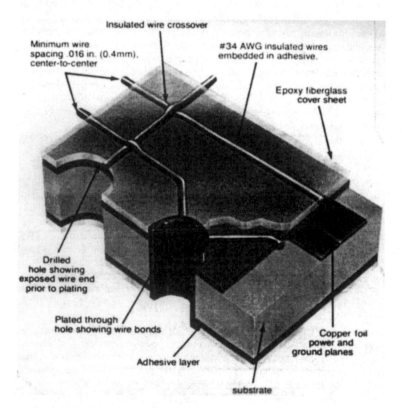

FIGURE 8.10   Cross-sectional view of typical multiwire board. *(Courtesy of PCK Technology, Melville, N.Y.)*

Microwire is similar to Multiwire in that discrete wires are once again used to form the PWB circuitry. This technology, however, uses a finer wire gauge for increased wiring density, a metal core base construction for improved thermal characteristics and CTE control when required, and laser-drilled vias to interconnect the bulk of the wiring.

The two basic constructions used in Microwire technology are solid core and twin-foil core. Since the two constructions differ only in the core makeup, the solid core is described, and the distinguishing characteristics of the twin-foil core are discussed subsequently.

In most instances the core material used is a CIC substrate in thicknesses ranging from 0.010 to 0.062 in. The core has a 20 percent copper, 60 percent invar, 20 percent copper ratio by thickness, yielding a substrate CTE of 5.2 ppm/°C and a weight of 0.305 lb/in^3, with a thermal conductivity that varies depending on the thickness of the core. The core thickness selection requires the designer to make trade-offs between many parameters. This core serves to control the CTE of the surface of the PWB when leadless components are used. If the CTE is not a concern in a design, other materials such as copper may be used. Since the epoxy compounds used in the subsequent processing steps have much higher CTEs, the interface between core and epoxy must be specially prepared to have good adhesion. This is accomplished through a mechanical abrading and chemical oxidation of the substrate surface. In very-high-density circuits requiring wiring on both sides of the core, the core would be drilled at the through-via sites and the holes backfilled with epoxy for subsequent drilling and plating operations.

To the core, 1-oz (0.0014-in) copper is bonded using 0.0035-in-thick epoxy prepreg. In most applications, the same copper-prepreg combination is bonded to both sides of the core to prevent substrate warpage. These subsequently serve as power and ground distribution planes. The planes are formed using standard photo-imaging and etching techniques. Small isolated copper pads are also etched into the planes directly under all sites that require the wires to be terminated. These "bounce" pads act as small mirrors to reflect the energy from a laser used in the subsequent laser-drilling operation. The planes are chemically oxidized to promote adhesion of subsequent dielectric and wiring layers.

Signal formation in the Microwire process is accomplished using no. 42 AWG (0.0025-in-diameter) C16200 copper cadmium wire with 0.001-in radial polyimide insulation. Wiring begins by first applying a liquid modified epoxy dielectric and wiring adhesive to the surface of the package. The liquid dielectric and 0.002-in-thick film adhesive are applied simultaneously by running the package through a set of precision steel rollers. The steel rollers allow a precise thickness of dielectric to be extruded. The epoxy resin is allowed to cure at room temperature to minimize any stresses.

As in the case of Multiwire, a numerically controlled wiring machine is driven by a CAD database containing all wiring instructions. Wiring can once again follow an x-y orthogonal and 45° wiring path. The wire is continuously bonded ultrasonically to the substrate, starting at a termination site and following a predetermined path, which places the wire directly over any termination sites that must be connected electrically to that net. The process continues until the end of the net is reached and the wire cut, and the next net routing begins. Figure 8.11 shows the wired substrate.

After wiring, the circuitry is encapsulated in a layer of modified epoxy and 0.001-in-thick plating adhesive to form a flat surface. This operation is similar in nature to the initial epoxy/adhesive layer application process. The entire interconnect layer from the top of the core copper plane to the surface of the board is 0.015 to 0.017 in thick. These wiring layers can be fabricated on both sides of the substrate when wiring density requires it.

Vias are formed through a laser-drilling technique. The substrate is mounted on a numerically controlled x-y table under the optics of a CO_2 laser. The substrate is positioned at a termination site and the laser energized for several milliseconds. The laser

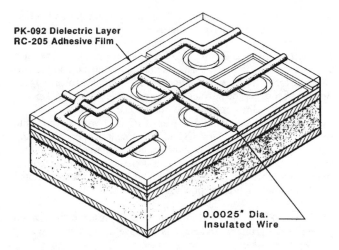

**PK-092 Dielectric Layer
RC-205 Adhesive Film**

**0.0025″ Dia.
Insulated Wire**

FIGURE 8.11   Microwire wired substrate. *(Courtesy of PCK Technology, Melville, N.Y.)*

evaporates the organic material in its path (including the 0.001-in polyimide wire insulation) and is reflected off the copper bounce pad, terminating the depth of the hole. The result is a 0.010-in-diameter blind via hole with a stripped wire passing through it. When wiring density requires wiring on both sides of the core, a conventional drilling process is used to interconnect the two wiring layers. Figure 8.12 shows the laser-drilled vias.

The termination to the hole is accomplished with electroplated copper and tin-lead. The holes are cleaned chemically to remove debris, and then the boards are seeded with a

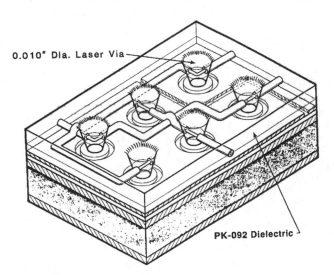

0.010″ Dia. Laser Via

PK-092 Dielectric

FIGURE 8.12   View of laser-drilled vias in Microwire board. *(Courtesy of PCK Technology, Melville, N.Y.)*

palladium catalyst. A thin layer of electroless copper is deposited in the holes and across the entire surface of the substrate. The surface pad patterns are defined using a dry-film resist and a standard photo-imaging technique. The panels then go into a modified electroplating bath where copper and then tin-lead are plated. The plating processes deposit a combined 0.0025-in copper plate. The photo mask is subsequently removed and the tin-lead plate acts as the final resist for etching away the background thin electroless copper plate, leaving only the component pads on the surface. Subsequent routing of the panel to the finished PWB size and electrical testing complete the operation.

The twin-core Microwire board replaces the solid metal CIC core with a composite laminate consisting of two layers of 0.006-in-thick CIC (12.5 percent copper, 75 percent invar, and 12.5 percent copper) having a CTE of 4.6 ppm/°C and a weight of 0.301 lb/in^3. These planes are separated by 0.005 in of epoxy/e-glass. This construction gives an equivalent CTE at a reduced weight compared to the solid-core CIC PWB. A cross-sectional view of a typical Microwire PWB construction is shown in Fig. 8.13.

8.2.4.3 Stitch-Weld. Stitch-weld PWBs are used throughout the industry mainly when prototyping electronic systems. Their primary advantage is the ease of conducting wiring changes.

An organic rigid double-sided PWB is once again used as a base. Usually the double-sided epoxy/e-glass PWB is designed with power and ground layers conventionally photo-imaged and etched onto the outer layers. A standard hole pattern is drilled to accommodate a wide variety of component types (such as DIP, SIP, and PGAs).

A tightly toleranced hole is drilled into which a specialized pin is force-fitted. The pin contains a well with a spring contact into which a component lead may be inserted. The spring contact serves to hold the component in place as well as to make electrical contact between the component lead and pin.

The pin has a closed, flattened bottom, which allows a wire to be welded to this surface. The wires are stitched and welded from pin to pin to form the appropriate interconnect circuitry. Connection to the power and ground planes is either by pressing the pins

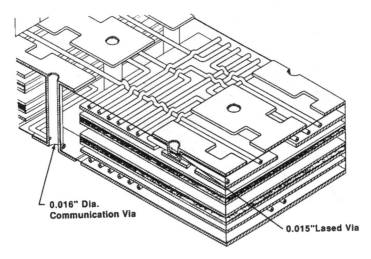

0.016" Dia. Communication Via

0.015" Lased Via

FIGURE 8.13 Cross-sectional view of typical Microwire board using twin CIC foil construction. *(Courtesy of PCK Technology, Melville, N.Y.)*

into plated-through holes connected to the plane or by using a washer or a specialized springed wire shaped similar to a cotter pin. The springed-wire circular wire head is sprung around the outside diameter of the pin and held in place by a spring force. The legs extend from the pin onto the ground or power plane. Soldering is used to complete the connection. The washer fits snugly around the pin head and is again soldered in place. The wired assembly is coated using a conformal coating material to contain the loose "rats' nest" of circuit wires. The primary advantage of the stitch-wired board is its ease of change or redesign. Using a single, universal part layout, one can wire many different circuits based on the same artwork. The disadvantage of this type of board is its inherent need for more volume due to the discrete wires. Figure 8.14 shows a typical stitch-welded board.

8.2.4.4 Wire-Wrap.   Once more an organic rigid PWB is most often used as a base panel. The PWB usually has at least two photo-imaged planes within or on the surface, which carry the power and ground distribution. Vias are drilled at appropriate sites to form a pattern to which components may be attached or, more often, to form a connector pattern into which other PWBs may be plugged. Each via hole in a wire-wrapped board receives a pin that is either pressed or soldered into place. The tail or wiring end of the pin has a square cross section around which an insulated wire is tightly wrapped. The tightly wrapped wire develops a hoop stress which maintains a gas-tight electrical connection on the corners of the square pin. The wrapped wires are used to carry signals from pin to pin. Figure 8.15 shows a typical wire-wrapped board.

The wire-wrapped board has similar advantages and disadvantages as the stitch-welded board. When used as a mother board into which daughter boards are plugged, the wire-wrapped board has a distinct advantage over other PWBs in that its circuit is easily reconfigured by clipping the terminus of a connection and wrapping a new circuit wire in its place.

FIGURE 8.14   Typical stitch-welded PWB. *(Courtesy of Stitch Wire, Inc., Newbury Park, Calif.)*

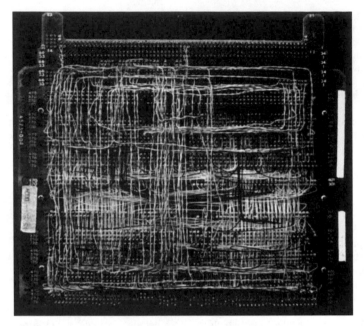

FIGURE 8.15   Typical wire-wrapped PWB.

8.3   MATERIALS FOR PRINTED-WIRING BOARDS

As Sec. 8.2 delineated, PWBs are designed in various sizes and shapes, use a variety of processes and materials, and perform a variety of electrical, structural, and sometimes thermal functions. Paramount to achieving a PWB which performs its intended function reliably, is producible, and is fabricated for the lowest cost possible, the designer must have a fundamental knowledge of the materials used in the end item. However, the knowledge must not be limited to knowing the end-item material's electrical, mechanical, thermal, and chemical properties. Materials knowledge must include environmental effects (thermal, mechanical, and humidity) on properties and their impacts on the material's performance in a given design. In addition, manufacturing-process-related stresses must also be considered. The following sections address these aspects in materials used in PWBs.

8.3.1   Rigid PWB Materials

As delineated in Sec. 8.2.1.1, organic rigid PWBs consist of a dielectric material onto which is patterned some form of metallization, which creates the actual circuit.

Fiber-reinforced resin dielectric materials, referred to as laminates, clad with copper sheets, are most commonly used for rigid PWB applications. The following sections describe the typical laminate anatomy, the materials that comprise typical laminates, the variety of laminates in use today, the specifications that govern their fabrication and end-item properties, and the problems one may encounter with laminate materials.

8.3.1.1 Anatomy of a PWB Laminate.   The laminate properties are related to the constituents in the composite laminated structure, that is, to the anatomy of the laminate. In this section, the nature of the PWB laminate construction is explained and the rules of the various constituents are discussed. An organic rigid PWB laminate consists of three major elements and some auxiliary elements. The major elements are the fabric, the resin (which combined comprise the dielectric), and the metal foil (usually copper). The auxiliary elements are the adhesion promoters or treatments that are applied to the fabric and to the foil to assure maximum adhesion of the resin to the fabric and to the foil.

The manufacture of a copper-clad PWB laminate begins in a machine called a treater. Fundamentally, the operational sequence is that the fabric is fed off the fabric roll and through a dip pan containing resin. The resin-curing agent mixture in the dip pan is called "A-stage," a term used to describe totally unreacted resin.

The resin impregnates the fabric, is passed through a set of metering rolls to control the thickness, and then passes through a treating oven for partial cure (polymerization) of the resin onto the fabric. After the resin-soaked fabric is partially cured in the treater oven, the fabric-resin combination is called "B-stage" or prepreg. These two terms are used to describe the partially cured resin. Finally, the B-stage coated fabric is cut into predetermined sizes for laminating (as large as 4 ft by 8 ft).

The B-stage is especially critical since it can be undercured (understaged) or overcured (overstaged). Ideally, the B-stage will be dry to the touch, but capable of reflow and optimized bonding in the laminating press. If the B-stage sheets are undercured, they may not be dry to the touch, and the resin will flow out of the laminate during processing under heat and pressure in the laminating press. Dry spots will then exist in the laminate. On the other hand, if the B-stage sheets are overcured, poor bonding will occur in the lamination processing. This will result in poor bond strength, both between layers of fabric and to the copper foil. Delamination and poor copper peel strength will occur as problems in the final PWB laminate. Hence the B-stage must be closely controlled for optimum PWB laminates to be produced.

Two other important factors in handling B-stage material are (1) resin aging and (2) moisture layers on the B-stage sheets. The resins used to impregnate the fabric are an organic polymer. The nature of polymer reactions is such that resin curing, or polymerization, will slowly continue at all times, the reaction rate being a function of temperature. Therefore, since an optimum B-stage is only partly cured, the curing will continue toward overcure, especially in warm or hot conditions, such as summer shipping and storage.

Under any given set of storage temperature conditions, a specific useful life, or shelf life, will exist for any given B-stage sheet. Thus cool shipment and storage conditions are usually recommended for B-stage stacks. The B-stage sheets must also be shipped and stored in dry conditions, since moisture films can be condensed onto cool sheets. When B-stage sheets with invisible moisture films are laminated together into a PWB laminate, moisture entrapment will result between layers in the cured laminate. During subsequent soldering operations on the PWB, this entrapped moisture will explode into small entrapped delamination spots. These white spots, known as blisters, can be sufficiently large or dense to affect PWB performance and reliability. Often it is necessary to dry boards in an oven before assembly and soldering to drive out any moisture and, thereby, prevent blisters.

The overall important points, then, are that B-stage sheets should be kept cool and dry (perhaps in sealed plastic bags) during shipping and storage. The laminate or B-stage sheet supplier can advise on optimum storage and shelf life conditions. The end-item prepreg critical factors are the maintenance of the ratio of fabric to resin, final material thickness, and degree of cure (polymerization) of the resin. The critical machine parameters to be monitored are the metering roll thickness, speed of the fabric through the treater, and the air velocity and oven temperature within the treater.

The final cured laminate is referred to as "C-stage" laminate and is achieved by pressure and heat in a laminating press. A final copper-clad C-stage laminate of a given thickness is made up of a number of thin B-stage laminates. This complete stack, including the copper foil, is pressed together and heated between flat plates, or platens, in a heated laminating press for the time required to completely polymerize the epoxy resin at the selected press temperature. Laminates used in the manufacture of single- and double-sided PWBs are usually thick laminates (>0.030 in) and are made up of a number of thin unclad laminate B-stage plies. Multilayer PWBs use thin clad C-stage laminate plies bonded together with thin B-stage plies. Each ply thickness is typically in the 0.004- to 0.008-in-thick range. As an example, 0.062-in-thick double-sided NEMA grade FR-4 epoxy-glass laminate would contain enough layers of B-stage epoxy-resin-soaked glass sheets to achieve the desired final pressed thickness, with copper foil on both the top and the bottom of the stack. The final product would be processed to conform to NEMA or military specification standards.

Prior to discussing the various laminates and their properties and uses, the major elements, that is, fabric, resin, and metal, will be highlighted.

8.3.1.2 Fabric Materials and Construction.

There are five materials that usually constitute the base fabric of the PWB laminate, namely, paper, E-glass, S2-glass, quartz, and aramid (aromatic polyamide polymer) fiber. The aramid fiber is manufactured by DuPont under the trade name Kevlar. Table 8.1 shows the properties of the various materials. In some instances a hybrid mixture of these materials is used to achieve certain end-item properties.

Paper-based materials are used with flame-retardant resins in low-cost PWBs, where laminate dimensional stability is not critical and where holes are usually punched in the material. Their use is mainly limited to single- and double-sided laminates for consumer electronics such as toys, calculators, and radios.

The most widely used fabric material in PWB applications is E-glass, a borosilicate type. Both randomly oriented glass-fiber mattes and woven glass-fiber fabrics are used. Its material properties satisfy the electrical and mechanical needs of most applications. It is used in single-sided, double-sided, and multilayer applications for aerospace, computer, communication, and industrial control applications.

S2-glass contains a higher percentage of silicon dioxide and aluminum oxide than E-glass giving it greater strength. It requires special tooling to manufacture and has a higher melting point than E-glass making it more costly to manufacture. Large minimum purchases relative to PWB program usage requirements are the norm. Due to the higher per-

TABLE 8.1 PWB Fiber Material Properties

	e-glass	s-glass	Quartz	Aramid
Specific gravity, g/cm³	2.54	2.49	2.20	1.40
Tensile strength, kg/mm	350	475	200	400
Young's modulus, kg/mm	7400	8600	7450	13,000
Maximum elongation, %	4.8	5.5	5.0	4.5
Specific heat, cal/(g · °C)	0.197	0.175	0.230	0.260
Thermal conductivity, W/(m · °C)	0.89	0.9	1.1	0.5
CTE, ppm/°C	5.0	2.8	0.54	−5.0*
Softening point, °C	840	975	1420	300
Dielectric constant at 1 MHz	5.8	4.52	3.5	4.0
Dissipation factor at 1 MHz	0.0011	0.0026	0.0002	0.001

*CTE given is along fiber axis; radial CTE = 60 ppm/°C.
Source: From Senese.[9]

centage of Silicone Dioxide relative to E-glass, the S2-glass exhibits a lower CTE, dielectric constant and dissipation factor than E-glass. Its high cost relative to the incremental improvement in electrical and CTE performance limit its use in today's PWBs to very special applications.

Quartz, a nearly pure silica, when used as the fiber in a PWB behaves typically like E-glass. Its advantages are that it has a lower dielectric constant, dissipation factor and CTE compared to E-glass making it attractive for microwave applications and high reliability SMT applications. These however must be weighted against its disadvantages. The fabric is very expensive to produce due to its higher melting temperature. In addition, during the drilling operation of PWB manufacture, the hardness of the quartz limits carbide drill bit hits to 25 or less.

Aramid-based fabrics have been used since the mid 1980s in PWB SMT applications. Their negative in-plane CTE combined with the high modulus (see Table 8.1) compared to that of E-glass along the fiber allows for a lower overall PWB CTE when processed with a suitable resin and a proper fiber-to-resin ratio. This is required in SMT applications to improve solder joint reliability. The material is also lower in density and has a lower dielectric constant than E-glass. The raw material costs for these materials are greater than for E-glass. Extra costs are incurred due to the difficulty in processing laminates made from these materials throughout the PWB fabrication cycle. The laminate typically has a lower dimensional stability compared to typical E-glass laminates. It is difficult to machine relative to E-glass and must be plasma etched prior to plating. The laminate also soaks up moisture and processing solutions very readily. The fiber's high radial coefficient of thermal expansion compared to the other fiber materials can decrease plated through hole reliability in high aspect ratio boards. Style 120 fabric has been the mainstay in the industry for Aramid boards. Recently DuPont developed a nonwoven aramid fabric with the trade name Thermount. This product has shown improved dimensional stability and reduced cost with a CTE comparable to 120 cloth style fabrics. Their applications to date have been limited to high-reliability military and aerospace applications.

Numerous types of base fabrics both nonwoven and woven are manufactured using these materials. Short randomly oriented glass, paper or aramid fibers known as glass matte, electrical-grade papers, and others make up a small percentage of the industry. Woven continuous fiber fabrics are the most common materials (Table 8.2). As Table 8.2 shows, cloths are designated by three- or four-digit style numbers indicating weight, thickness, and thread count, with an increase in the number showing an increase in the fabric thickness. The choice of fabric used in a particular laminate is dependent on electrical requirements, PWB mechanical properties (CTE in particular), dielectric thickness and tolerance, and circuit filling needs during processing.

Fabrics are manufactured by weavers on looms. Thousands of individual strands of fiber are placed on a master roll and fed into the loom. This is known as the "warp" or "machine" direction. "Fill" yarn is fed crosswise in the loom (the fill direction) which creates the fabric as the warp yarn is fed into the machine. Due to this weaving operation, the number of plies or strands of yarn per inch in each direction do vary along with the tension on the fabric strands. This does cause slight differences in material properties and dimensional stability in the two in-plane material directions. In finished prepreg or laminate product, by convention, the warp is typically parallel to the long side of a cut panel.

Prior to weaving, the warp fibers are run through various treatments called sizing. Sizing is often waxlike and provides lubricity to the fibers during the mechanical weaving operation. This treatment must be removed from the woven glass fabric, otherwise poor adhesion of the resin to the fabric will occur. This removal process is known as scouring. Removal of this sizing treatment is especially difficult at the fiber intersections in the woven fabric. Failure to remove the sizing agent from these intersections can result in poor

TABLE 8.2 PWB Fabric Cloth Makeup

Fabric range	Style: plain weave	Thickness,*† in	Thread count per inch ±3* ($W \times F$)	Weight ±5%,* oz/yd²
G1 (glass)	104	0.0010	60 × 52	0.56
	106	0.0012	56 × 56	0.73
	1070	0.0014	60 × 35	1.05
	107	0.0015	60 × 35	1.05
	1080	0.0020	60 × 47	1.40
	108	0.0022	60 × 47	1.40
G2 (glass)	2112	0.0030	40 × 39	2.05
	112	0.0036	40 × 39	2.10
	2113	0.0029	60 × 56	2.30
	2313	0.0030	60 × 64	2.40
	113	0.0032	60 × 64	2.43
	2125	0.0035	40 × 39	2.60
	1125	0.0039	40 × 39	2.60
	2116	0.0036	60 × 58	3.10
	116	0.0038	60 × 58	3.10
	1675	0.0040	40 × 32	2.90
	2119	0.0034	60 × 46	2.80
	119	0.0038	54 × 50	2.80
	2165	0.0040	60 × 52	3.55
	1165	0.0042	60 × 52	3.55
S2 glass	6106	0.0013	56 × 56	0.72
	6080	0.0020	60 × 47	1.45
	6180‡	0.0022	55 × 55	1.20
	6212	0.0032	40 × 39	2.04
	6313	0.0033	60 × 64	2.40
	6120‡	0.0035	60 × 60	3.10
	6216	0.0035	60 × 58	3.09
	6116	0.0036	60 × 58	3.10
	6628	0.0067	44 × 32	6.03
G3 (glass)	7628	0.0067	44 × 32	6.00
	7629	0.0071	44 × 34	6.18
	1528	0.0065	42 × 32	5.95
	7637	0.0089	44 × 22	7.00
	7642	0.0099	42 × 20	6.70
	6628	0.0067	44 × 32	6.03
	6628	0.0067	44 × 32	6.03
A (aramid)	120	0.0040	34 × 34	1.70
	108	0.0020	60 × 60	0.80
	177	0.0030	70 × 70	0.93
	3080	0.0020	60 × 60	0.90
	3081	0.0020	60 × 60	0.90
	3500	0.0040	34 × 34	1.80
	3511	0.0040	33 × 33	1.71
Q (quartz)	503	0.0050	50 × 40	3.30
	525	0.0030	50 × 50	2.00

*Based on finished-goods state in which heat cleaning and finishing have been applied.
†These values should not be used for computation of dielectric thickness in board design or layout. Tolerance is 20% on fabric ranges G1 and G2 and 10% on all others.
‡4- or 8-harness satin weave.
Source: From MIL-S-13949 and Senese.[5]

adhesion of resin at the intersections, usually observable as small white crosses, known as measling or crazing, in the final cured laminate. Scouring is followed by a high-temperature oven bake to ensure all sizing and lubricants are removed.

Many fabrics, especially glass and aramid, do not bond well to the resins which are used to impregnate them. Thus to assure a strong laminate, it is necessary to treat the fabric with an organosilane adhesion-promoting treatment. The optimum treatment will vary, the treatment being selective to the fabric-resin combination being used. Inadequate bonding of the fabric-resin interface can result in the migration of water, plating solutions, copper etchants, and other liquids into the laminate during PWB fabrication processes. For further detail see laminates and laminate problems, as discussed in Sec. 8.3.1.9.

In order to achieve smoother surface finishes in laminates for fine line PWB applications, laminators tend to supply to PWB manufacturers laminates with finer weave fabrics. A recent development in fabric construction has been a product called Surface Enhanced (SE) glass fabric. In this product, used on heavier weave styles such as 7628 or 7629, the fabric is manufactured using a process that tends to flatten the glass bundles and hence lead to a laminate product with a smoother surface. These heavier weaves tend to be less expensive to produce than the finer weave materials and hence will save cost to the PWB manufacturer. Surface Enhanced glass fiber product will usually be designated with an SE suffix (i.e., 7628SE).

8.3.1.3 Resin Systems.   A wide range of resin materials are in use in today's organic PWBs, with new formulations being brought to market continually. As previously mentioned, most resin systems used in organic PWB laminates are a thermosetting plastic, with thermoplastic materials used primarily in microwave applications and in molded PWB applications. Thermoplastics are those plastics that will remelt upon heating to some melting temperature. Thermosetting plastics are those plastics that will not remelt upon heating. Thermosetting plastics will soften upon application of heat, however, as will thermoplastics. Both will soften dramatically at the glass transition temperature T_g of the plastic.

T_g is the temperature at which the plastic changes from a rigid or harder material to a softer or glassy-type material. It is a definite characteristic of all plastic materials, but not a property of the resin system where molecular bonds are broken. T_g is the point where the physical properties of the resin change due to a weakening of the resin system's molecular bonds. The physical properties (such as CTE, flexural strength, and tensile strength) as a function of temperature undergo significant changes in the slope of the property versus temperature curve above and below the T_g point (Fig. 8.16). There is a transition region approximately 25°C below and above T_g, where gradual changes in the material properties occur. T_g is determined by the intersection of the property curves above and below this transition region. The glass transition temperature is normally determined by measuring the change in the slope of the volume expansion of the resin using differential scanning calorimetry (DSC) or thermomechanical analysis (TMA).

There are a wide variety of resin systems used in today's laminates based on the following: epoxies, polyimides, polytetrafluoroethylene (PTFE) and cyanate ester. Most manufacturers use one of these base systems and add modifiers to improve properties such as T_g.

The most common resin systems in use today are the standard epoxies. These are used in NEMA grade G-10 and FR-4 laminates and have a relatively low T_g (105 to 125°C). The G-10 epoxy is a general-purpose bisphenol A difunctional epoxy, while the FR-4 epoxy is a brominated bisphenol A difunctional epoxy. The bromines make the FR-4 epoxy flame-retardant. These epoxies are easy to process and B-stage, and they have excellent adhesion to copper at room temperature. The low T_g causes a large percent expansion, which weakens the copper adhesion during soldering temperatures.

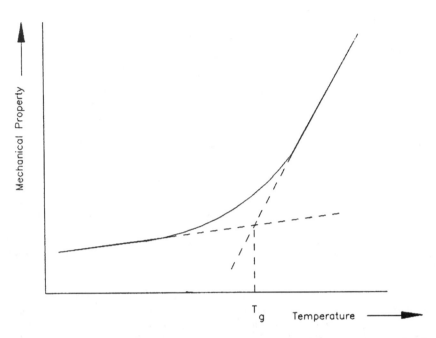

FIGURE 8.16 Determination of resin glass transition temperature T_g where physical properties of resin exhibit significant change.

The high-performance epoxies are modifications to these difunctional epoxies formulated with many materials including bismaleimide triazine (BT), polyimide, and cyanate ester to name a few. They are usually made to increase the base epoxy resin T_g and improve the chemical and thermal stress resistance of the bisphenol A epoxies. The type of modification and the percentage can cause T_g to vary between 125 and 200°C. These higher performance resins fall into roughly three categories based on the T_g: Difunctional of 110–130 °C, Tetrafunctional of 135–160 °C and Multifunctional of 170–200 °C. Raising the epoxy resin T_g usually leads to a resin system that is more brittle than the base epoxy. This can make the new resin more difficult to process and can lead to laminate reliability problems in harsh service environments.

Polyimides are the second major type of resin system in use today for organic PWB laminates. Until the 1980s, the polyimide used in PWB applications was Kerimid 601 from a European source. Concerns about carcinogenic effects of this compound led to the development of variations of this resin. The polyimide resins exhibit T_g values of over 200°C. In addition to the higher T_g, they exhibit superior adhesion to copper at soldering temperatures and have a lower CTE than epoxies. Their disadvantages are that they are quite brittle, and hence more care needs to be followed during their processing, they cost more, and they have a higher moisture absorption and a lower flammability rating. Modifications to polyimides are usually done with epoxies to improve their processibility, reduce laminate moisture absorption, and improve their adhesion characteristics.

Many new polyimide formulations are coming to market today due to their use in multichip packaging as a thin-film dielectric. The formulations strive to decrease the material dielectric constant, improve the resin ductility, lower the resin moisture absorption, and improve the processibility.

To improve electrical performance (lower signal propagation delay) in high-speed applications, the laminate dielectric constant must be reduced. Though S2-glass, quartz and aramid fibers do have lower dielectric properties than E-glass (see Table 8.1), the lowering of the resin dielectric constant is the main driver in achieving lower dielectric constant laminate materials. In that regard, most low dielectric constant laminates on the market today use PTFE as the resin system.

PTFE is an ultra high molecular weight molecule consisting of symmetrically oriented fluorine atoms around a carbon center giving it a non-polar nature. It has strong primary chemical bonds making it thermally stable. The non-polar nature and unique molecular nature of PTFE are responsible for the following:

PTFE resins are non-hygroscopic

Unlike other plastics, there are no reactive groups in the molecule which would cause further cross-linking to occur during the lamination process.

Glass transition temperature does not strictly apply to PTFE. Its properties are maintained throughout its useful temperature range of well below 0 to 327°C.

The above 3 unique PTFE characteristics lead to PTFE resins exhibiting stable dielectric constant and dissipation factor properties under varying frequency, temperature and humidity conditions.

A number of manufacturers have laminated materials on the market today using PTFE resins. The main drawback to PTFE based laminates is their low modulus due to the low PTFE modulus and the high lamination processing temperatures required compared to conventional epoxies or polyimides. Applications of PTFE resin system laminates tend to be limited to relatively specialized small microwave substrates.

Cyanate ester resin systems are the fourth major base resin system in use today. They have a dielectric constant, 2.8, approaching that of PTFE, 2.2. Cyanate ester systems have been available for the past 15 years and have been known as triazine. In the past, they have been added as modifiers to epoxy resins. Recent advances by Dow Chemical and others have improved the cycloaliphatic backbone structure, dicyclopentadiene. This has improved the dielectric properties and moisture resistance while maintaining standard processes in both the laminate and PWB operations.

Table 8.3 lists some of the major physical properties for various resin systems. They have been grouped into epoxy, polyimide, PTFE and cyanate ester. The actual properties vary somewhat about these norms due to the differences in manufacturers formulation. The user should consult the laminate manufacturer for the exact physical properties of the resin used in a given laminate.

Regarding thermoplastic substrates, where they are used, they must be made of high-performance thermoplastics since other thermoplastics will soften excessively or melt at

TABLE 8.3 Typical Resin System Physical Properties

Type	T_g, °C	Elastic modulus, 10^6 lb/in^2	Poisson's ratio	CTE, ppm/°C	Thermal conductivity, Btu/) (h · ft · °F)	Dielectric constant at 1 MHz
Epoxy	130	0.5	0.35	58	0.14	4.5
Polyimide	260	0.6	0.33	49	0.14	4.3
Cyanate ester	260	0.5	0.35	55	0.14	3.8
PTFE	—	0.05	0.46	99	0.14	2.6

temperatures required for PWB soldering operations. PTFE and polysulfones are the two most commonly used thermoplastic resin systems. Readers are referred to Chap. 1 for information on these high-performance thermoplastics.

8.3.1.4 Metals in Laminates. The third major component of a PWB laminate is the metal used to create the interconnection circuitry. Copper is used almost exclusively. In thin-film multichip module applications, aluminum is sometimes used as the interconnect metal. This chapter does not discuss aluminum applications. Certain SMT applications use a clad metal made of CIC or CMC as a constraining core. This is covered briefly at the end of this subsection. In addition planar resistor technology may be manufactured into the PWB foil. This is also discussed briefly at the end of this section.

Copper foils used in PWB applications are covered in IPC MF-150. Two types of copper foil are used, either electrodeposited (E), the most prevalent, or wrought (W). The wrought foils are usually limited to special applications such as flexible printed circuitry where high ductility is essential. The electrodeposited and wrought types are further subdivided into classes to reflect functional performance and testing properties. Table 8.4 covers the eight IPC and 14 Military class descriptions, Table 8.5 the minimum mechanical properties for each class, and Table 8.6 is an application guide for copper foil usage.

Class 1, 2, 3, and 4 electrodeposited foils are used predominantly in laminates. The class 1 and 2 foils are more brittle and are not generally used in high-performance laminate applications where substantial thermal stress ranges are to be incurred. In class 1 foils, fracture without deformation will occur under relatively low stress levels. This fracturing occurs under a combination of thermal stress imposed on a copper with degraded grain boundary strength. This degradation is due to such factors as codeposited organics, which can cause easy separation of grain boundaries at elevated temperatures. Class 3 and 4 foils are much more ductile at elevated temperatures and as such are used in the higher-

TABLE 8.4   Metal Foil Designations

Class IPC-MF-150)	Designation (MIL-S-13949)	Description
1	C	Copper, standard electrodeposited (STD, type E)
2	G	Copper, high ductility, electrodeposited (HD, type E)
3	H	Copper, high temperature, elongation electrodeposited (HTE type E)
4	J	Copper, annealed, electrodeposited (ANN, type E)
5	A	Copper, as rolled, wrought (AR, type W)
6	K	Copper, light cold-rolled, wrought (LCR, type W)
7	L	Copper, annealed, wrought (ANN, type W)
8	M	Copper, as rolled, wrought, low-temperature annealable (ARLT, type W)
—	B	Copper, rolled (treated)
—	D	Copper, drum side out (double treated), electrodeposited
—	N	Nickel
—	U	Aluminum
—	Y	Copper-Invar-copper
	O	Unclad

Source: From IPC-MF-150 and MIL-S-13949.

TABLE 8.5 Copper Foil Minimum Mechanical Properties*

Class	Weight,‡ oz	Tensile strength at room temperature (23°C)		Tensile strength at elevated temperature (180°C)	
		10^3 lb/in^2	MPa	10^3 lb/in^2	MPa
		Type E			
1	$^1/_2$	30	207	N/A	N/A
	1	40	276	N/A	N/A
	2+	40	276	N/A	N/A
2	$^1/_2$	15	103	N/A	N/A
	1	30	207	N/A	N/A
	2+	30	207	N/A	N/A
3	$^1/_2$	30	207	15	103
	1	40	276	20	138
	2+	40	276	20	138
4	1	20	138	15	103
	2+	20	138	15	103
		Type W			
5	$^1/_2$	50	345	—	—
	1	50	345	20	138
	2+	50	345	40	376
6	1	25–50	177–345	N/A	N/A
	2+	25–50	177–345	N/A	N/A
7	$^1/_2$	15	103	—	—
	1	20	138	14	97
	2+	25	172	22	152
8†	$^1/_2$	15	103	N/A	N/A
	1	20	138	N/A	N/A
	2	20	138		

*The copper foil shall conform to the tensile requirements when tested in both longitudinal and transverse directions.

†Properties given are following a time-temperature exposure of 15 min at 177°C (350°F).

‡Minimum properties for testing copper weights less than $^1/_2$ oz shall be agreed to between user and vendor.

Source: From IPC-MF-150.

thermal-stress environments. It is apparent from Table 8.5 that at room temperature, the mechanical properties of the foils overlap, and it would appear that they are equivalent. At elevated temperatures, however, the ductilities of the classes are significantly different. Testing the mechanical properties of foils at elevated temperatures to distinguish between classes can be done using a hot rupture test methodology developed by Zakraysek of General Electric.[10]

As mentioned, copper foil thicknesses are described in terms of area weight (ounces per square foot). The various common foil thicknesses are shown in Table 8.7.

The side of the copper foil, which is to be bonded to the base laminate, must undergo a treatment to promote better adhesion. The type of treatment is designated by a letter, as shown in Table 8.8. The copper foil treatment is an adhesion promoter selective for the resin being used. Existing treatments used in the industry today are black, brown, and

TABLE 8.6 Application Guide for Copper Foil

Type	Grade	Class	Handling	At room temperature			At elevated temperature (180°C)[c]		
				Flex to install		Continuous flexing, high cycle fatigue[b]	Flex to install		Continuous flexing, high cycle fatigue[b]
				Single bend	Low cycle fatigue[b]		Single bend	Low cycle fatigue[b]	
E	1	All	Good	30/3.3	7.1/38	0.19/1470	Not recommended for applications requiring foil flexing or bending		
E	2	All	Good	20/5.6	5.3/52	0.18/1560	Not applicable		
E	3	All	Good	50/1.4	10.3/26	0.21/1340	15/7.9	4.2/63	0.17/1610
E	4	All	Caution[h]	30/3.3	7.5/36	0.32/870	30/3.3	6.9/39	0.15/1230
W	5[d]	All	Good	65/0.8	13.1/20	0.32/870	15/7.9	4.2/63	0.17/1610
W	5[e]	All	Good				TBD	TBD	TBD
W	6[f]	All							
W	7	All	Caution[h]	65/0.8	12.5/21	0.32/870	45/1.7	9.5/28	0.20/1370
W	8[g]	All	Good	25/4.0	6.2/44	0.15/1890	TBD	TBD	TBD

[a]Larger maximum strain range and smaller minimum bend diameter values indicate superior performance for a given strain mode.

[b]Low cycle fatigue < 500 cycles to failure; high cycle fatigue > 10^4 cycles to failure. The values given here have been calculated for 20 and 10^5 fatigue cycles for low and high cycle fatigue, respectively. For the calculations the minimum mechanical properties for 34-μm [1-oz] copper foil given in the respective slash sheets have been used. Typical property values can be considerably higher. (See IPC-TR-484, "Results of Copper Foil Ductility Round Robin Study" April 1986.)

[c]The values for elevated-temperature applications should primarily be used for qualitative purposes, since unproven assumptions were necessary for their calculation.

[d]W5 foil is highly anisotropic due to the rolling process; in the rolling direction the performance values can surpass those given for W7 foil.

[e]The values given apply in the foil rolling direction only. For cross-machine direction values see previous line.

[f]Choice of temper allows tradeoff in handling/high cycle fatigue versus high strain/low cycle fatigue properties. Grades 5 and 7 are the limiting tempers.

[g]Handling characterization is for as-received foil; maximum strain range/minimum bend diameter values apply to W8 foil after low-temperature annealing at 177°C (350°F) for 15 min.

[h]Handling difficulties normally occur only with thinner foils.

Source: From IPC-MF-150.

TABLE 8.7 Copper Foil Thickness Requirements

Designator	Thickness by weight* oz/ft^2	g/m^2	Thickness by gauge (for reference only) in nominal	mm nominal	Tolerance Type E	Type W
E	0.146 (1/8)	44.57	0.0002	0.005	N/A	N/A
Q	0.263 (1/4)	80.18	0.0004	0.009	N/A	N/A
T	0.350 (3/8)	106.9	0.0005	0.012	N/A	N/A
H	0.500 (1/2)	153	0.0007	0.018	↑	↑
M	0.750 (3/4)	229	0.0010	0.025	+10%	+5%
1	1	305	0.0014	0.035	of	of
2	2	610	0.0028	0.071	thickness	thickness
3	3	915	0.0042	0.106	or weight	or
4	4	1221	0.0056	0.142		weight
5	5	1526	0.0070	0.178	↓	↓
6	6	1831	0.0084	0.213	↓	↓
7	7	2136	0.0096	0.249	↓	↓
10	10	3052	0.0140	0.355	↓	↓
14	14	4272	0.0196	0.496	↓	↓

*Fractions in parentheses represent common industry terminology.
Source: From IPC-MF-150.

TABLE 8.8 Copper Foil Treatment Designations

Designation	Description
D	No treatment, no stain proofing
P	No treatment, with stain proofing
S	Single-sided treatment, with stain proofing
D	Double-sided treatment, with stain proofing

Source: From IPC-MF-150.

white oxide treatment. Black oxide is a controlled coating of a cupric copper oxide that develops long crystals to promote adhesion. Brown oxide has shorter crystals. It is used in flexible and rigid flexible boards since it is more resistant to flexing and higher lamination temperatures. The white oxide treatment was developed by DuPont and named Dura-BOND. This process applies a 0.1 μm coating of tin-tin oxide on the copper surface followed by a silane coupling agent application which forms covalent chemical bonds with both the tin oxide and laminate resin.

While the previously mentioned processes are applied by the PWB manufacturer, a second treatment called *double treat* copper can be applied by the foil manufacturer. This involves an oxide bonding treatment followed by a thermal barrier of brass, zinc or nickel and a final chrome based stabilizer.

Oxidized materials are generally known to provide optimized bond strength for most bonding systems. Failure to achieve an optimum bond of the copper foil results in poor bond strength of the etched copper circuits, a very important factor in PWB performance and reliability. In the resin treatment and lamination operation, great care must be taken to optimize resin bond strength to the copper, to the fabric, and between the thin sheets of resin-soaked fabric that make up the laminate. Failure in any of these can result in very difficult end-product problems. Even with the very best adhesion between sheets (known as interlayer bond strength), there is always a weakness between layers. This is clearly indicated by the fact that voltage breakdown occurs 20 to 25 times faster parallel to these layers than perpendicular to them. See Chap. 15 for further discussion on this subject.

The requirements for finer line circuitry, thin weight copper (less than 1/2 oz) and thin laminates have led to the development of improved electrodeposited copper foil products. Three such products are JTCAM from Gould, SQ-VLP from Mitsui-Kinzoku and ML-MLS from OAK-Mitsui. In standard copper foils, a microstructure with long columnar grains and hence a jagged sharp peaked surface (tooth) is manufactured. Each of these products is developed around a copper foil with a smaller grain microstructure and a low profile and more rounded surface (tooth). Due to their unique structure, these foils exhibit a combination of higher tensile strength, very high hardness and higher elongation at elevated temperature than conventional foils. The higher tensile and hardness properties lead to less yield loss during the handling of thin foils and a reduction in pits, dents and wrinkles during lamination. The finer grain structure and lower tooth profile are instrumental in improving the yield of fine etched circuitry with more vertical side walls being produced after etch and in reducing the etch time required.

SMT applications requiring a controlled PWB CTE have led to the use of clad metals in PWB applications. These metals used are either invar or molybdenum clad on their outer surface with copper. The thickness ratio of the invar or molybdenum to copper controls the CTE of the clad metal. The ratio of the thickness of the clad metal to the thickness of the laminate dielectric as well as their respective CTEs and elastic moduli determine the CTE of the overall PWB.

Laminate manufacturers have developed CIC- and CMC-clad dielectrics as cores for PWBs. These cores usually consist of an epoxy/E-glass or polyimide/E-glass thin dielectric (0.004 to 0.006 in thick) sandwiched between thin CIC or CMC layers (usually 0.006 in thick). One or two cores may be used, depending on the application. In addition, the CIC or CMC layers can serve as power or ground planes.

The CIC foil requirements are currently defined in IPC MF-152. Table 8.34 delineates some of the important mechanical properties for both CIC and CMC.

Planar resistor technology (PRT)[14] has been used in electronics applications for over 21 years and is referred to as Ohmega-ply material. PRT is a thin film resistive material made from 0.1 to 0.4 μm of nickel alloy which has been deposited to the rough side of the copper foil to be used in the PWB laminate. The deposited material onto the copper foil is identified as resistive conductive material (RCM). The RCM material is available on both single and double treat copper, Class 1 and Class 3 copper foils and in copper weights from 1/4 ounce to 2 ounces. It can be supplied on almost any dielectric material by a laminator willing to laminate the RCM material.

Sheet resistances of the material come in the following standards, 25, 100 and 250 Ω per square with others being developed. To design a specific resistor, thin film resistor design criteria are used in that the square size of the resistor is a dimensionless number. The ratio of the length to width of the resistor multiplied by its sheet resistivity gives the resistor value.

The PRT laminate is processed by PWB manufacturers using a print and etch process to form the circuitry. An extra print and etch step is required to form the resistor. Resistor tolerances of 10 percent are typical after the etch process. Tighter tolerances can be achieved using thin film resistor trim technology of either laser, mechanical, or abrasive trimming.

The PRT technology has been used in supercomputer, parallel processor mainframe and VLSI testing applications. The resistors can either be buried within the PWB or on the surface.

8.3.1.5 Laminate Types and Specifications. As was mentioned above, organic PWB laminates are by far the largest group of materials used for PWB applications in the electronics industry. Combining materials from the three major groups discussed above; fabric, resin and metal; leads to a wide variety of available laminates whose properties are tailored to meet specific PWB application requirements. There are strong industry standards for the laminates and the printed wiring boards made using these laminates, as explained below.

The major groups which issue these standards are National Electrical Manufacturers Association (NEMA), Department of Defense (DOD) for Military (MIL) Specifications and Institute for Interconnecting and Packaging Electronic Circuits (IPC). Laminates defined by NEMA are covered in document LI 1-89. This document is currently undergoing extensive revision and should be published in the spring of 1996. Laminates meeting DOD and MIL requirements are specified in MIL-S-13949, currently at revision H. It contains specifications for both clad-metal C-stage as well as unclad B-stage laminate materials. The laminate materials meeting IPC specifications are divided across five documents. ANSI/IPC-L-108 covers the requirements for metal clad thin (>0.001 in.) base material laminates for multilayer PWBs. ANSI/IPC-L-109 covers the requirements for B-stage laminate materials used in bonding the layers of high-temperature multilayer PWBs. ANSI/IPC-L-112 covers the requirements for metal clad laminated polymeric sheets containing 2 or more types of reinforcing materials for PWBs. In this document, the basic laminate thicknesses are from 0.031 in and greater. ANSI/IPC-L-115 defines the laminate materials used for manufacturing single- and double-sided PWBs of 0.020 in and greater in thickness. ANSI/IPC-L-125 covers the requirements for high-speed/high-frequency performance laminates both clad and unclad (<0.002 in thick) for use in microwave and other high-speed electrical applications.

Basically, PWB laminators manufacture PWB laminates to NEMA and or IPC standards, as explained below. Many of these laminates are qualified to the more rigid DOD requirements for military electronic systems. IPC is an important major industry association whose documents include test standards, workmanship standards, PWB operations standards and much more. There are, of course, numerous other important industry groups for specific objectives, specific geographical technology transfer, and other purposes. It should be understood that the members of the industry, both producers and suppliers, make up the bodies of both NEMA and IPC and are responsible for inputting the technical data within these material specifications that are issued through these industry groups. The DOD uses the inputs from the industry in updating and modifying MIL-S-13949 but it is the DOD that controls and issues this document.

The major NEMA and Military standard grades for PWB laminates are shown in Tables 8.9, 8.10, and 8.11 respectively. These standards are based on the type of resin and type of fabric used in the laminate, and they address copper clad laminates as well as prepreg.

The NEMA standard grades are used in a variety of commercial applications. FR-2, FR-3, CEM-1, CEM-3, FR-4, and FR-5 are most widely used. The paper base FR-2 laminates are used in low-cost consumer products such as toys, TV games and calculators. The FR-3 laminate with its higher electrical and physical properties is used in televisions, computers and communications equipment. CEM-1 laminate has punching properties similar to FR-2 and 3 but with electrical properties approaching FR-4. It is used in industrial electronics, automobiles and smoke detectors. CEM-3 is higher in cost than CEM-1 and is

TABLE 8.9   NEMA Copper-Clad Laminate Requirements*

LI-1-1989 grades	XXXP	XXXPC	FR-1	FR-2	FR-3	FR-4	FR-5	FR-6	CEM-1	CEM-3	G-10	G-11
Fabric description	Paper base	Paper base	Paper base	Paper base	Paper base	E-glass fabric base	E-glass fabric base	E-glass matte	E-glass cloth surf cellulose paper core	E-glass cloth surf nonwoven glass core	E-glass fabric base	E-glass fabric base
Resin description	Phenolic	Phenolic	Phenolic	Phenolic	Epoxy, flame-resistant	Epoxy, flame-resistant	Epoxy, flame-resistant	Polyester, flame-resistant	Epoxy, flame-resistant	Epoxy, flame-resistant	Epoxy, general-purpose	Epoxy, temperature-resistant
Copper peel, min. lb/in 1 oz after solder float	6.00	6.00	6.00	6.00	8.00	8.00	8.00	7.00	7.00	8.00	8.00	8.00
Volume resistivity, min. $M\Omega \cdot cm$	10^4	10^4	10^3	10^4	10^4	10^6	10^6	10^6	10^6	10^6	10^6	10^6
Surface resistivity, min. $M\Omega$	10^3	10^3	10^2	10^3	10^3	10^4	10^4	10^4	10^4	10^4	10^4	10^4
Water absorption, max. %												
0.031 in thick	1.30	1.30	5.60	1.30	1.00	0.50	0.50	—	0.50	0.50	0.50	0.50
0.062 in thick	1.00	0.75	3.60	0.75	0.65	0.25	0.25	0.40	0.30	0.25	0.25	0.25
Dielectric breakdown, min. kV parallel to lam.	15.00	15.00	5.00	15.00	30.00	40.00	40.00	30.00	40.00	40.00	40.00	40.00
Permittivity (dielectric constant) at 1 MHz, max. average	4.80	4.80	6.00	4.80	4.80	5.40	5.40	4.30	5.20	5.40	5.40	5.40
Loss tangent at 1 MHz max. average	0.040	0.040	0.060	0.040	0.040	0.035	0.035	0.030	0.035	0.035	0.035	0.035
Flexural strength, 10^3 lb/in²												
Lengthwise	12.00	12.00	12.00	12.00	20.00	60.00	60.00	15.00	50.00	50.00	60.00	60.00
Crosswise	10.50	10.50	10.00	10.50	16.00	50.00	50.00	—	40.00	40.00	50.00	50.00
Flammability, min.	N/A	N/A	UL94 V-1	UL94 V-1	UL94 V-1	UL94 V-1	UL94 V-1	—	UL94 V-0	UL94 V-0	UL94 V-0	UL94 V-0

*All values are for 0.031-in-thick material except for: (1) FR-6 properties are for 0.062 in thick; (2) volume and surface resistivities are for 0.062-in thick; (3) water absorption gives both 0.031- and 0.062-in thicknesses.

Source: From NEMA LI-1-1989.

TABLE 8.10 Copper-Clad Laminate Requirements

MIL-S-13949, sheet #	2	4	5	6	7	8	9
Type designation	GB	GF	GH	GP	GR	GT	GX
Fabric description	Woven E-glass	Woven E-glass	Woven E-glass	Nonwoven E-glass	Nonwoven E-glass	Woven E-glass	Woven E-glass
Resin description	Polyfunct. epoxy	Difunct. epoxy	Polyfunct. epoxy	PTFE	PTFE	PTFE	PTFE
Copper peel, min. lb/in							
1 oz as received	8.0	6.0	6.0	5.0–7.0	5.0–7.0	5.0	5.0
1 oz at elevated temperature	3.0	5.0	5.5	4.0–5.0	4.0–5.0	2.0	2.0
Volume resistivity, min. M$\Omega \cdot$cm	1.00E + 06	1.00E + 06	1.00E + 07	1.00E + 06	1.00E + 06	1.00E + 06	1.00E + 06
	1.00E + 05	1.00E + 03	1.00E + 05	1.00E + 05	1.00E + 05	1.00E + 05	1.00E + 05
Surface resistivity, min. MΩ	1.00E + 04	1.00E + 04	1.00E + 08	1.00E + 04	1.00E + 04	1.00E + 04	1.00E + 04
	1.00E + 03	1.00E + 03	1.00E + 05	1.00E + 03	1.00E + 03	1.00E + 03	1.00E + 03
Water absorption, max. %							
0.031 in thick	0.80	0.80	0.80	0.30	0.10	0.20	0.20
0.062 in thick	0.35	0.35	0.35	0.20	0.10	0.10	0.10
Dielectric breakdown, min. kV parallel to laminate	40	40	40	30	30	20	20
Electric strength, min. v/mil	—	750	—	—	—	—	—
Permittivity (dielectric constant)							
at 1 MHz	5.4	5.4	4.5	2.4	2.20–2.40	2.8	2.40–2.60
at 50 MHz	5.4	—	4.5	—	at X band	—	at X band
Loss tangent							
at 1 MHz	0.03	0.03	0.2	0.001	0.0015 max	0.005	0.0022
at 50 MHz	0.03	—	0.2	—	at X band	—	at X band
Flexural strength, 10^3 lb/in^2							
Lengthwise	60	60	60	8	3.5	12	12
Crosswise	50	50	50	6	3.5	10	10
Arc resistance, min. seconds	60	60	90	120	180	180	180
Flammability, max. burn time, seconds	N/A	15	15	N/A	N/A	N/A	N/A

*Sheets 1 and 3 were canceled June 1988.

Source: From MIL-S-13949.

TABLE 8.10 Copper-Clad Laminate Requirements *(Continued)*

10	14	15	19	22	24	25	27	29	31
GI	GY	AF	QI	BF	GM	CF	SC	GC	BI
Woven E-glass	Woven E-glass	Woven aramid	Woven quartz	Nonwoven aramid	Woven E-glass	Nonwoven poly-glass	Woven S2-glass	Woven E-glass	Nonwoven aramid
Polyimide	PTFE	Polyfunct. epoxy	Polyimide	Epoxy	Triazene or BT epoxy	Polyfunct. epoxy	Cyanate ester	Cyanate ester	Polyimide
5.0	5.0	5.0	5.0	3.0	6.0	10.0	.5	4.5	—
5.0	2.0	4.0	5.0	2.5	3.0	1.0	4.5	4.5	2.0
6.00E+04	1.00E+06	1.00E+06	6.00E+04	1.00E+06	1.00E+06	1.00E+06	1.00E+06	1.00E+06	1E6
6.00E+04	1.00E+05	1.00E+03	6.00E+04	1.00E+03	1.00E+05	1.00E+03	1.00E+04	1.00E+04	1E3
1.00E+04	1.00E+04	1.00E+04	1.00E+04	1.00E+04	1.00E+06	1.00E+05	1.00E+06	1.00E+06	1E4
6.00E+04	1.00E+03	1.00E+03	6.00E+04	1.00E+03	1.00E+05	1.00E+03	1.00E+04	1.00E+04	1E3
1.00	0.20	2.00	1.00	2.00	0.80	—	1.00	1.00	3.5
0.50	0.10	2.00	0.50	2.00	0.35	—	1.00	1.00	3.5
40	20	40	40	40	40	40	40	40	40
750	—	750	750	1500	750	750	750	750	750
5.4 —	2.15–2.40 at X band	4.5	3.4	4.0 max.	4.8	5.4	4.3	4.5	4.5 max.
0.025 —	.0015 at X band	0.035	0.01	—	0.02	0.03	0.015	0.015	0.035 max.
60	6	50	60	50	60	6	50	50	30
45	5	40	45	40	50	5.5	40	40	30
120	120	60	120	60	60	60	120	120	60
—	N/A	15	—	15	15	15	25–30	25–30	N/A

TABLE 8.11 Prepreg Requirements

MIL-S-13949, sheet #	11	12	13	16	18	21	23	26	28	30	32
Type designation	GE	GF	GI	AF	GH	AI	BF	GM	SC	GC	BI
Fabric description	Woven E-glass	Woven E-glass	Woven E-glass	Woven aramid	Woven E-glass	Woven aramid	Nonwoven aramid	Woven E-glass	Woven S2-glass	Woven E-glass	Nonwoven aramid
Resin description	Difunct. epoxy	Difunct. epoxy	Polyimide	Polyfunct. epoxy	Polyfunct. epoxy	Polyimide	Epoxy	Triazene or BT epoxy	Cyanate ester	Cyanate ester	Polyimide
Volatiles content, %	0.75	0.75	2	1.5	0.75	1.5	1.5	2	2	2	2.5
Electric strength, min. V/mil	750	750	750	750	750	750	1500	750	750	750	750
Permittivity at 1 MHz	5.4	5.4	5.4	4.5	5.4	4.2	4	4.3	4.3	4.5	4.5
Loss tangent at 1 MHz	0.035	0.035	0.025	0.035	0.035	0.035	0.025	0.02	0.015	0.015	0.035
Flammability, seconds	—	15	—	15	15	15	15	15	15	15	N/A

Source: From MIL-S-13949.

more suited to plated through hole applications. It is used in automobiles and home computers. FR-4 is the most widely used laminate material due to its excellent physical, electrical and processing properties. It is used in aerospace, computers, automotive and industrial control applications. FR-5 is used in applications requiring higher heat resistance than attainable with FR-4.

Chapters 5, 6, 9, and 10 discuss in detail all of the important trends in advanced electronic packaging and interconnecting. Many of these trends require higher-performance PWB laminate properties than are available with the industry standard NEMA laminates discussed. Many of the newer high-performance laminates are covered in MIL-S-13949, specification sheets 6-32. Other laminates are continually being developed and will no doubt be added to these documents in the near future.

Three major advanced PWB laminate requirements predominate. They are:

1. Higher thermal stability laminates (higher glass transition temperature, T_g, especially) to withstand increasing system thermal densities and to stabilize critical physical properties, particularly thermal expansion (CTE).

2. Lower thermal expansion laminates, especially x-y axis CTE, to eliminate or minimize solder joint cracking in surface mount packages.

3. Lower dielectric constant laminates are used to minimize signal propagation delay and losses in high-speed digital or microwave applications.

For varying electronic system requirements, any one or any combination of the above requirements might exist. Most of the advanced PWB laminates have been developed to achieve the higher glass transition requirement, many have been developed to meet the lower x-y axis CTE requirement, and a smaller number have been developed to meet the lower dielectric constant requirement. Basically, in the anatomy of a PWB laminate, the glass transition requirement is largely controlled by the resin component of the laminate, the x-y axis (in-plane) CTE is largely controlled by the fabric constituent of the laminate, the z axis (out of plane) CTE is controlled by both the fabric and resin, and the dielectric constant is largely controlled by the resin component—although selection of the fabric can have a substantial influence on dielectric constant.

To achieve higher glass transition temperature PWB laminates, several major higher temperature resin systems have been developed. The resin systems were discussed in Section 8.3.1.3. Laminates made using these resin systems include NEMA grades G-11 and FR-5 and military, MIL-S-13949, grades GI, GH, GM, AI, BI, QI and CF.

To achieve lower *in-plane (x-y)* axis PWB CTE, three fibers have predominated, mainly, aramid fiber (such as DuPont's Kevlar), S2-glass and quartz fiber. These fibers were discussed in Sec. 8.3.1.2. Laminates using these fibers have been made and qualified to military grades AE, AI, BI, SC and QI (see Table 8.10).

Finished PWBs using aramid-based laminates have two problems not associated with glass-based laminates. First, a phenomenon known as resin microcracking will occur during subsequent environmental thermal cycling. The microcracking is small cracks in the resin, usually at the PWB surface and sometimes within the finished laminate cross section. These cracks are caused by high stresses built up in the resin due to the high radial CTE of the aramid fabric. They occur at the crossover points of the fabric. Microcrack severity is a function of the resin system used, being most severe in brittle polyimide systems. Modified polyimide and epoxy systems can limit microcrack occurrence somewhat. The microcracking can cause failure in copper traces, especially when low-ductility (class 1 and 2) copper in combination with fine-line (<0.008-in-wide) copper traces is used in a design. The Thermount material developed by DuPont discussed in 8.3.1.2 is less susceptible to this phenomenon. Figure 8.17 shows typical surface resin microcracks.

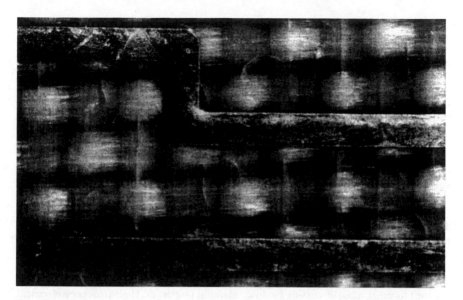

FIGURE 8.17 Typical surface resin microcracking in PWB manufactured from Kevlar-based laminate.

The second problem which develops in PWBs manufactured with laminates using aramid fabrics is premature failure in the barrel of copper PTHs during subsequent environmental thermal cycling. This is once again due to the high radial CTE of the fabric, which imposes a high-z axis (out-of-plane) CTE in the PWB. It should be pointed out that this phenomenon is not limited to aramid applications, but will occur whenever a material is used to constrain the x-y axis (in-plane) PWB CTE. This usually will cause an increase in the z axis CTE. The onset of cracking is determined by the thermal cycling temperature extremes, copper ductility in the barrel of the hole, aspect ratio of the PTH (aspect ratio is the ratio of board thickness to finished-hole diameter), and copper plating thickness. Barrel cracking can be controlled through the use of ductile copper plating (class 3), limiting the PTH aspect ratio (<3.5:1), increasing the copper thickness in the hole barrel from 1 mil minimum to >1.5 mil minimum, or applying an electroless or electrolytic nickel plate of 0.3 to 0.5 mil over the copper in the barrel of the hole.

Microwave applications in aerospace and telecommunications equipment would require the use of NEMA GT and GX or MIL grades GP, GR, GT, GX, or GY materials. Most of these laminates consist of PTFE/E-glass laminate composites.

The Type GP and GR laminates are produced from a mixture of randomly oriented E-glass fibers and PTFE resin yielding a low-loss printed circuit material with many unique properties. The fibers used are either a nominal 0.5 μm diameter e-glass microfiber in short lengths or a long length e-glass fiber. These materials resist water and chemical absorption resulting in excellent maintenance of their electrical properties. Holes may be easily machined or punched due to the very fine, randomly dispersed e-glass fibers. The product is very flexible and can be bent and used in conformable PWB or antenna configurations. It maintains excellent dimensional stability and physical strength. Applications for this product are in high-frequency analog and digital communications systems, missile and radar components, electronic warfare and countermeasures and satellite communication equipment.

Type GT, GX and GY laminates are produced from a mixture of woven fabric E-glass and PTFE resin. They are manufactured to closely controlled thickness tolerances and precise dielectric constant values. MIL-S-13949 designates that Type GT materials have dielectric constants less than 2.8 at 1 MHz, Type GX materials have dielectric constants from 2.4 to 2.6 at X-band and Type GY materials have dielectric constants from 2.15 to 2.40 at X-band. Applications for these products exist in many areas of microwave technology. These include ILS and MLS radar systems, phased array antennas, satellite communication equipment, TVRO and DBS low-loss circuits for LNA, LNB and BDC circuits, cellular communication systems, microwave transmission devices and a variety of R.F. componentry including power dividers, mixers, amplifiers, couplers and antennas.

Specialized woven e-glass PTFE laminate materials filled with an alumina particle to produce a high dielectric constant (10) laminate as a substitute for alumina ceramic substrates are also manufactured. They are used in applications which may see high vibration or stress and is more rugged than using conventional alumina ceramic substrates.

8.3.1.6 Electrical Properties of PWB Laminates. There are six laminate electrical properties the designer must be concerned with. These are the permittivity (dielectric constant), dissipation factor, current-carrying capacity, dielectric strength, dielectric breakdown, and insulation resistance.

Permittivity and Dissipation Factor. These two material properties are commonly grouped together when discussing laminate material properties. High-frequency electrical applications such as microwave, computer, and radar signal-processing applications often require PWB laminates or substrates with a lower permittivity or dissipation factor than found in standard FR-4 laminates. These are often referred to as low-loss laminates.

The permittivity is 'more commonly known as the dielectric constant. The dielectric constant of a material is a dimensionless value, which is defined as the ratio of the capacitance of a capacitor with a given dielectric to the capacitance of the same capacitor with air as the dielectric. It is the electrostatic energy storage capability of an insulating material. The dielectric constant plays a major role in high-frequency digital and microwave applications in determining the electrical propagation delay as well as the impedance and capacitance associated with the PWB. For high-frequency applications, the designer prefers a low-dielectric-constant laminate to decrease the signal propagation delay. The dielectric constant of a vacuum is 1.

The dissipation factor (sometimes referred to as loss tangent) is the second dimensionless property. It is defined as the ratio of the total power loss in a dielectric material to the product of the voltage and current in a capacitor in which the material is a dielectric. It gives a measure of the total signal transmitted power that will be lost as electrons dissipate into the laminate material. It varies as a function of frequency and is usually measured at 1 MHz per ASTM D-150 or IPC TM-650, method 2.5.5.3.

The major NEMA and military grades for these materials used in microwave applications are GP, GR, GY, GT, and GX. Other often used ungraded low-loss laminates include polystyrene and cross-linked polystyrene, polyethylene and polypropylene, polyphenylene oxide, polysulfones, and other low-loss high-thermal-performance thermoplastics.

The dielectric constant and dissipation factor for a given laminate are determined by the dielectric constants and dissipation factors of the constitutive fabric and resin materials and the fabric-to-resin ratio of the laminate. They will also vary with signal frequency as well as ambient temperature, and humidity. They are usually measured at 1 MHz using either ASTM D-150 or IPC TM-650, method 2.5.5.3. Table 8.12 gives typical values for the permittivities and dissipation factors of some important laminates and substrates. Figures 8.18 and 8.19 provide data on variations in the dielectric constant due to variations in fabric to resin ratio and constitutive materials.

TABLE 8.12 Typical Laminate Permittivity and Loss Tangent

MATERIAL		PERMITTIVITY	LOSS TANGENT
FABRIC	RESIN	@ 1 MHz	@ 1 MHz
E-GLASS	POLYIMIDE	4.3-4.5	.013-.020
E-GLASS	EPOXY	4.1-4.8	.015-.025
E-GLASS	CYANATE ESTER	3.8	0.01
S2-GLASS	CYANATE ESTER	3.5	0.01
ARAMID	EPOXY	3.9	0.027
THERMOUNT	POLYIMIDE	3.9-4.2	0.015-.022
THERMOUNT	EPOXY	3.8-4.1	.015-.022
QUARTZ	POLYIMIDE	3.5-3.7	.010
CERAMIC	PTFE	10.2-10.8	0.002
CERAMIC/E-GLASS SPHERES	PTFE	2.80-2.95	0.001
WOVEN E-GLASS/CERAMIC	PTFE	10	0.003
NON-WOVEN E-GLASS/CERAMIC	PTFE	3.50-4.50	0.0026
NON-WOVEN E-GLASS	PTFE	2.17-2.33	.0009-.0013
WOVEN E-GLASS	PTFE	2.17-3.20	.0013-.009

Current-Carrying Capacity. The current-carrying capacity (in amperes) of a copper conductor in a PWB is usually limited by the allowable temperature rise in the conductor or by the amount of voltage drop that can be allowed in the circuit path. This is a function of the current, trace thickness and width, environmental conditions, conductor density, laminate T_g, and whether the trace is within the PWB or on the PWB external surface. Figure 8.20 can be used to estimate temperature increases above ambient versus current. The curves include a 10 percent derating factor on a current basis. It is also best to further derate the values another 15 percent when the PWB is less than 0.032 in thick or the conductors are of greater than 2-oz thickness.

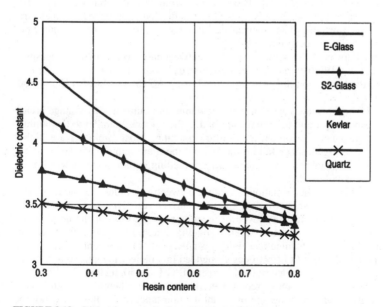

FIGURE 8.18 Dielectric constant of cyanate ester resin system.

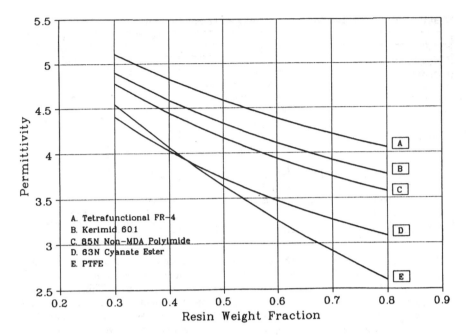

FIGURE 8.19 Permittivity vs. resin content, e-glass systems. *(Reprinted with permission from Guiles.[16])*

Dielectric Strength and Breakdown. The dielectric strength is the ability of an insulating material to resist the passage of a disruptive electric discharge produced by an electric stress. It is given in volts per mil and is measured through the thickness of a laminate per ASTM D-149 or IPC TM-650, method 2.5.1.

Dielectric breakdown is the value of the disruptive charge (in kilovolts) measured between two Pratt and Whitney no. 3 taper pins inserted in the laminate on 1-in centers perpendicular to the lamination. It is also measured using ASTM D-149 or IPC TM-650, method 2.5.6.

Fundamentally, high voltages can cause severe damage to laminates by voltage breakdown (which is especially rapid for voltage applied in the *x-y* axis, parallel to the laminations), high-voltage puncture, corona, and carbon tracking. Most laminates are highly subject to all of these high-voltage failure modes, but there are numerous ways to minimize these destructive problems. This subject area, including high-voltage material properties, is covered in detail in Chap. 15.

Insulation Resistance. The insulation resistance between two holes or conductors in a laminate or PWB is composed of two components—surface resistivity and volume resistivity. The surface resistivity (usually given in megohms) is the ratio of the dc potential applied between two conductive points on an insulating material surface to the total current between the points. The volume resistivity (usually given in ohm-centimeters) is the ratio of the dc potential applied to electrodes embedded in a dielectric material to the current between them. They may be determined by using IPC TM-650, method 2.5.17.1.

These properties are usually measured at specific temperature-humidity conditions to evaluate their degradation. The tests are of greatest importance when the test condition matches the end-use environmental conditions.

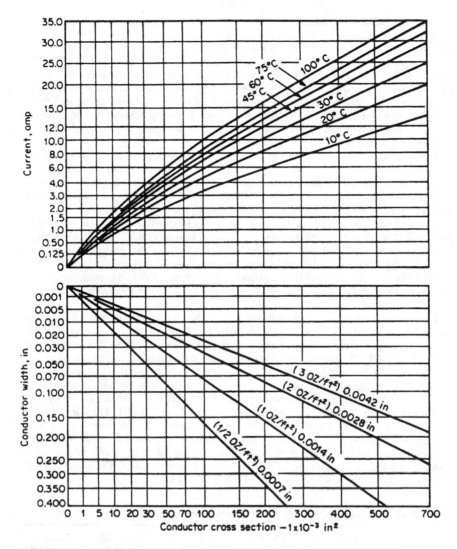

FIGURE 8.20 Conductor cross section and current-carrying capacity. *(From MIL-STD-275.)*

The volume resistivity is very high, above 10^{15} $\Omega \cdot$ cm, at room temperature and low humidity for most laminate materials. However, the resistivity values drop very rapidly as temperature or humidity, or both are increased. It is not uncommon for volume resistivity values to decrease to 10^{6} $\Omega \cdot$ cm or lower in difficult temperature-humidity environments. For high performance, it is desirable that volume resistivity values remain above 10^{9} or 10^{10} $\Omega \cdot$ cm. Reduced values will recover when the temperature or humidity condition is removed. Improvement in retaining high resistivity in humid environments is achieved by conformal coatings or low surface energy treatments being applied to the complete PWB assembly. This is discussed in Sec. 8.3.5.

8.3.1.7 Physical and Mechanical Properties of PWB Laminates. Certain physical and mechanical properties of PWB laminates are very important in the design of PWB assemblies. Some important data on these factors are presented at this point.

Laminate Thickness. Laminate material thicknesses and tolerances are set per IPC L-108 and MIL-S-13949 for thin-clad laminates used in multilayer applications and IPC L-115 for single- and double-sided applications. Thickness is measured after removal of the copper cladding, using a mechanical micrometer, or by microsection, using the closest point between metal claddings for the value. Data from the publications are given in Tables 8.13 and 8.14. The military specification does not allow double-sided laminates with thicknesses below 0.0035 in between the clad metal.

With the explosion in electronics in the consumer and telecommunications fields, ultra thin laminates have been developed by manufacturers to meet the growing needs of packaging multichip module laminate (MCM-L), PCMCIA and thin light cards (TLCs). Minimum dielectric thicknesses of 0.002 in, multilayer builds of 0.014 in and copper cladding down to 9 μm are some of the challenges being met by the laminate industry.

Bow and Twist. Bow and twist of laminate panels is a measure of the deviation of the planarity of the laminate or PWB panel, expressed in inches per inch. Acceptable limits of bow and twist per MIL-S-13949 are given in Table 8.15. Laminate materials specified in IPC L-115 typically have values of bow and twist between 0.010 and 0.025 in/in. These values differ by material, panel size, and thickness. The reader should consult the applicable IPC material specification sheet. Panel bow and twist can be measured per IPC TM-650, method 2.4.22.

Flexural Strength. The laminate flexural strength is a measure (in pounds per square inch) of the load that a beam will withstand without fracturing when loaded in the center and supported on the ends. The test methods are covered in ASTM D-790 and IPC TM-650, method 2.4.4. The flexural strengths of typical PWB materials are given in Table 8.16.

Coefficient of Thermal Expansion (CTE). The CTE of a material is the change in linear length per unit length per degree change in temperature. Both x-y axis (in-plane) and z axis (out-of-plane) CTEs are important parameters. CTE test methods are covered by ASTM D-696 as well as by IPC TM-650, method 2.4.41 or 2.4.41.1. The laminate CTE varies due to the CTE of the constitutive fiber and resin materials, the fiber weave and orientation, and the ratio of fiber to resin in the laminate. In a finished PWB, the percent copper loading in a multilayer PWB and the mounting method of the PWB to any carrier plates or heat sinks will affect the CTE of the overall laminate. CTEs of some typical laminate materials are given in Table 8.17 and Figs. 8.21 and 8.22.

TABLE 8.13 Laminate Thicknesses and Tolerances, in mil

Base thickness	Class 1*	Class 2*	Grade A†	Grade B†	Grade C†
1– 4.5	±1	±0.75	—	—	—
4.6– 6.0	±1.5	±1	—	—	—
6.1– 12.0	±2	±1.5	—	—	—
12.1– 19.9	±2.5	±2	—	—	—
20.0– 30.9	—	—	± 3	± 2.5	±1.5
31.0– 40.9	—	—	± 6.5	± 4	±2
41.0– 65.9	—	—	± 7.5	± 5	±2
66.0–100.9	—	—	± 9.0	± 7	±3
101.0–140.9	—	—	±12.0	± 9	±3.5
141.0–250.0	—	—	±22.0	±12	±4

*From IPC L-108.
†From IPC L-115.

TABLE 8.14 Laminate Thicknesses and Tolerances,* in mil

Nominal thickness of base laminate without cladding	Class 1 PX, paper base only	Class 1 reinforced	Class 2 reinforced	Class 3† reinforced	Class 4 for microwave application, GR, GX, and GY	Class 5† reinforced	
						−	+
1.0– 4.5	—	± 1.0	± 0.7	±0.5	—	0.5	1.0
4.6– 6.5	—	± 1.5	± 1.0	±0.7	—	0.7	1.2
6.6– 12.0	—	± 2.0	± 1.5	±1.0	±0.75‡	1.0	1.5
12.1– 19.9	—	± 2.5	± 2.0	±1.5	±1.0	1.5	2.0
20.0– 30.9	—	± 3.0	± 2.5	±2.0	±1.5	2.0	2.5
31.0– 40.9	± 4.5	± 6.5	± 4.0	±3.0	±2.0	3.0	3.5
41.0– 65.9	± 6.0	± 7.5	± 5.0	±3.0	±2.0	3.0	3.5
66.0–100.9	± 7.5	± 9.0	± 7.0	±4.0	±3.0	4.0	4.5
101.0–140.9	± 9.0	±12.0	± 9.0	±5.0	±3.5	5.0	5.5
141.0–250.0	±12.0	±22.0	±12.0	±6.0	±4.0	6.0	6.5

*Tolerance value is determined by nominal base thickness (less cladding). Tolerance is applied over base plus cladding, with no additional tolerance for cladding thickness allowed. Tolerance of class 5 materials is applied to base thickness.

†These tighter tolerances are available only through product selection on most material types.

‡For some base materials, materials below certain base thickness are not covered by this specification. For example, types GT, GX, and GY are not covered for less than 10-mil core thickness and core thicknesses below 3.5 mil are not currently covered for any double-sided laminate.

Source: From MIL-P-13949.

TABLE 8.15 Laminate Bow and Twist, Maximum Total Variation (on basis of 36-in dimension),* in percent

Thickness,† in	Class A			Class B		
	All types, all weights, metal (one side)	All types, all weights, metal (two sides)		All types, all weights, metal (one side)	All types, all weights, metal (two sides)	
		Glass	Paper		Glass	Paper
0.20 and over	—	5	—	—	2	—
0.030 or 0.031	12	5	6	10	2	5
0.060 or 0.062	10	5	6	5	1	2.5
0.090 or 0.093	8	3	3	5	1	2.5
0.120 or 0.125	8	3	3	5	1	2.5
0.240 or 0.250	5	1.5	1.5	5	1	1.5

*Values apply only to sheet sizes as manufactured and to cut pieces having either dimension not less than 18 in.

†For nominal thicknesses not shown in this table, the bow or twist for the next lower thickness shown shall apply.

Source: From MIL-P-13949.

TABLE 8.16 Typical Laminate Flexural Strengths, Minimum Average, in lb/in^2

Material	Lengthwise	Crosswise
XXXPC	12,000	10,500
FR-2	12,000	10,500
FR-3	20,000	16,000
FR-4	60,000	50,000
FR-5	60,000	50,000
FR-6	15,000	15,000
G-10	60,000	50,000
G-11	60,000	50,000
CEM-1	35,000	28,000
CEM-3	40,000	32,000
GT	15,000	10,000
GX	15,000	10,000
GI	50,000	40,000

Source: From Coombs.[11] Reprinted with permission.

Regarding the CTEs shown in Table 8.17, most PWB laminates have CTE values of 7 to 25×10^{-6} cm/(cm · °C) in the x and y axes (in-plane). The z axis CTE values (perpendicular to the plane of the laminate) are much higher, often 40 to 110×10^{-6} cm/(cm · °C). The reason for this great difference is that the x-y axis expansion is limited by the low CTE values of the fabric in the laminate (see Table 8.1), while the z axis expansion is largely controlled by the resin component in the laminate (except when aramid fabric is used); see Table 8.3.

Thermal Conductivity. The thermal conductivity values of PWB laminates are all very low [about 0.14 Btu/(h · ft^2 · °F)], as is the case for nearly all electrically insulating materials. Thus heat cannot be removed from PWB assemblies through the laminate materials.

TABLE 8.17   Typical Laminate CTE, ppm/C

MATERIAL		X, Y AXIS	Z AXIS
FABRIC	RESIN		
E-GLASS	POLYIMIDE	12-15	55-75
E-GLASS	EPOXY	12-17	60-90
E-GLASS	CYANATE ESTER	12-15	55-75
S2-GLASS	CYANATE ESTER	7-10	45-55
ARAMID	EPOXY	6.5-7.5	90-110
THERMOUNT	POLYIMIDE	6-10	85-95
THERMOUNT	EPOXY	7-9	100-110
QUARTZ	EPOXY	8-11	60-80
QUARTZ	POLYIMIDE	8-11	55-75
CERAMIC	PTFE	45-50	45-50
CERAMIC/E-GLASS SPHERES	PTFE	30	30
WOVEN E-GLASS/CERAMIC	PTFE	14-16	36
NON-WOVEN E-GLASS/CERAMIC	PTFE	30-50	110-300
WOVEN E-GLASS	PTFE	13-34	150-250

Other thermal management methods must be used. The reader is referred to Chaps. 1 and 2 for further information on this subject.

Flammability. With respect to flammability, the ratings of Underwriters Laboratories are most commonly used. These ratings for various laminates are shown in Table 8.18. The requirements for the various UL ratings are as follows:

94V-0: Specimens must extinguish within 10 s after each flame application and a total combustion of less than 50 s after 10 flame applications. No samples are to drip flaming particles or have glowing combustion lasting beyond 30 s after the second flame test.

94V-1: Specimens must extinguish within 30 s after each flame application and a total combustion of less than 250 s after 10 flame applications. No samples are to drip flaming particles or have glowing combustion lasting beyond 60 s after the second flame test.

94V-2: Specimens must extinguish within 30 s after each flame application and a total combustion of less than 250 s after 10 flame applications. Samples may drip flame particles, burning briefly, and no specimen will have glowing combustion beyond 60 s after the second flame test.

94HB: Specimens are to be horizontal and have a burning rate less than 1.5 in/min over a 3.0-in span. Samples must cease to burn before the flame reaches the 4-in mark.

TABLE 8.18   UL Flammability Ratings for Typical Laminates

Grade	UL classification	Grade	UL classification
XXXPC	94HB	G-10	94HB
FR-2	94V-1	FR-4	94V-0
FR-3	94V-0	G-11	94HB
CEM-1	94V-0	FR-5	94V-0
CEM-3	94V-0	FR-6	94V-0

Source: From Coombs.[11] Reprinted with permission.

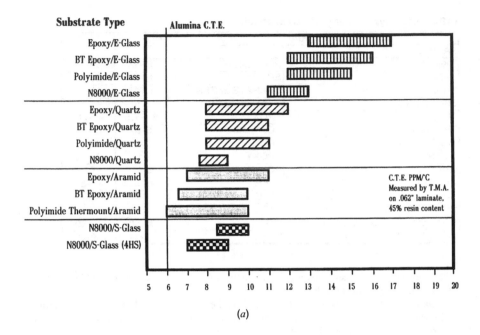

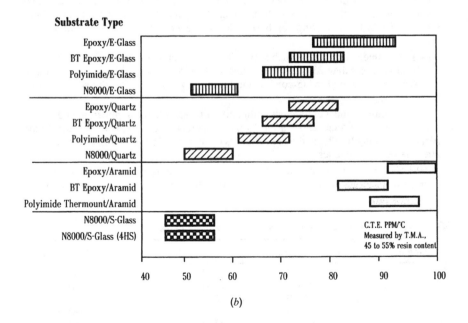

FIGURE 8.21 *(a)* x- and y-axis CTE and *(b)* z-axis CTE, (−55 to 95°C). *(Reprinted with permission of NELCO International Corp.)*

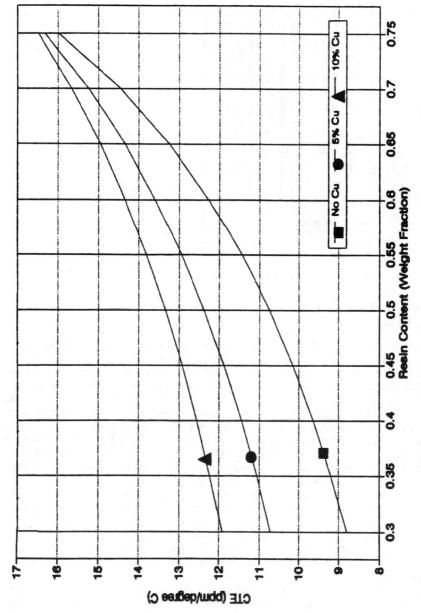

FIGURE 8.22 CTE as f ((resin and Cu volume), cyanate ester, S2 glass system. *(Reprinted with permission from Guiles.[16])*

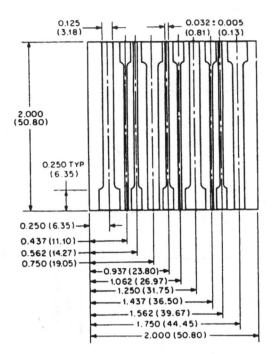

FIGURE 8.23   Typical peel-strength test pattern. *(From MIL-13949.)*

Copper Peel Strength. The peel strength of the copper circuitry is very important in PWB assemblies. The basic peel strength pattern is shown in Fig. 8.23. The wide end of each copper strip is peeled back from the edge of the specimen and gripped in a machine in which the force required to pull the copper away from the laminate may be measured. The foil is pulled perpendicular to the specimen surface at a rate of 2 in/s minimum and the force recorded. The force is converted to pounds per inch width of peel. Initially, peel strengths of most laminates are very good, with peel values of 5 to 12 lb per inch of conductor width, as measured using test patterns of NEMA, IPC, or MIL-S-13949 standards. Additional peel tests are usually done on laminate copper traces to simulate peel during soldering, after soldering, and after plating. It is not uncommon for the initial peel values to be reduced by 50 percent after or during these operations.

Copper Surface. Copper surface standards in laminates are controlled by IPC-L-108 and MIL-S-13949 for thin clad laminates used in multilayer applications and IPC-L-115 for single- and double-sided applications. Foil indentations, wrinkles and scratches are the three criteria delineated in these documents. Foil indentations are specified on the basis of a point count per square foot of surface area. Wrinkles visible to the naked eye are not allowed. Scratches longer than 4 in and deeper than 5 percent of the nominal foil thickness are not allowed.

Thermal Stress.   Laminates are subjected to an elevated thermal stress test to ensure good adhesion of copper and fabric to the resin as well as to ensure that an excessive level of volatiles is not trapped in the laminate. The test involves floating a specimen (usually 2 by 2 in) in a solder pot for 10 s. IPC TM-650, method 2.6.8.1, is used. Any signs of surface blistering, interlayer delamination, blistering, or crazing is cause for rejection.

Water Absorption. Water absorption is defined as the ratio of the weight of water absorbed by the laminate material to the weight of the dry material expressed as a percent. IPC TM-650, method 2.6.2.1, can be used to measure this property. Cross-linked polystyrenes and GX and GT materials will have values of 0.01 to 0.2 percent, paper-based XXXP and XXPC products range from 0.65 to 1.3 percent, FR-3, FR-4, and G-10 epoxy laminates range from 0.2 to 0.5 percent, and GI materials range from 1 to 1.5 percent.

Chemical Resistance. The processing to which a laminate material is subjected during PWB fabrication incorporates a wide variety of solvents and aqueous solutions. These solutions are absorbed to some degree by the laminate. This absorption can damage the laminate in a number of ways. The copper bond strength to the resin may be degraded, and wicking of the solution along the fiber bundle can lead to material discoloration and a degradation of dielectric strength and volumetric resistance. IPC TM-650, method 2.3.4.2, is used to test the resistance of a laminate to various chemicals used in PWB processing.

Fungus Resistance. Epoxy, polyimide and cyanate ester based laminates are generally considered fungus-resistant, and phenolic laminates are not.

Dimensional Stability. Dimensional stability is the resistance of a laminate to planar dimensional changes through the processing. The changes may be either expansion or shrinkage. It is one of the critical material parameters in multilayer PWB processing. Its value is given in inches per inch of laminate length, and it is measured using IPC TM-650, method 2.4.39.

Laminate properties which determine the dimensional stability are the type of resin, resin content, fabric construction including fabric type, weave, and thickness, and the lamination cycle used in fabricating the laminate. Improvements in dimensional stability occur by lowering resin contents, lowering lamination pressures, and using heavier fabric materials.

8.3.1.8 Laminate Problems. The complexity of the overall PWB life cycle is such that complete freedom from laminate problems is usually unavoidable. Problems can occur in laminate manufacture or in PWB manufacturing processes. Certain types of problems are common ones, and some of these problems and the corrective actions are listed in Tables 8.19 to 8.23.

8.3.1.9 Troubleshooting Material Problems in PWB Laminates. In the manufacture of the base laminate, and in the processing of the laminate into a finished PWB assembly, there are certain common problems that occur often. An important group of these problems, their probable sources, and possible corrective actions were given in Coombs and are listed here.[11]

1. Resin starvation (bond line appears white, glass weave is visible) occurs because of too few plies of prepreg in the laminate. *Corrective action:* If not obvious, section the laminate or perform burn-out in a muffle furnace to determine the amount of prepreg actually used.

2. Resin starvation may occur because prepreg has too low a resin content, too long a gel time, or too high a resin flow. *Corrective action:* Any combination of the conditions may be accommodated by adjusting the heat-up rate by varying the amount of kraft paper or other insulation material between platen and laminating fixture.

3. Voids may be caused by excessive volatile content. *Corrective action:* Place prepreg in a vacuum chamber for 2 h at 27 in Hg, or perform lamination in a vacuum lamination press.

TABLE 8.19 Measling and Blistering

Cause	Corrective action
Entrapped moisture	Check drill hole quality; check for delamination during drilling
Excessive exposure to heat during fusing and hot-air leveling	Monitor equipment for proper voltage regulation and temperature
Laminate weave exposure	Ensure that there is sufficient butter coat; that all wet processes are checked when materials are changed
Measles related to stress on boards with heavy ground planes	Postbake panels prior to drilling; ensure that laminate has balanced construction; use optimum preheat conditioning before IR fusing or hot-air leveling
Handling when laminate temperature exceeds its glass transition temperature	Allow boards to cool to ambient temperature before handling
Measling when large components or terminals are tight enough to cause excessive stress when heated	Check tooling for undersized holes; loosen tight terminals

Source: From Coombs.[11] Reprinted with permission.

TABLE 8.20 Dimensional Stability

Cause	Corrective action
Undercured laminate	Check laminate vendor for glass transition temperature
Distorted glass fabric (FR-4)	Examine etched panels for yarns parallel to warp and fill direction
Dense hole patterns with fine lines and spaces	Prebake laminate in panel form before drilling
Dimensional change parallel to grain differs from that of cross grain	Have laminator identify grain direction.

Source: From Coombs.[11] Reprinted with permission.

4. Surface dents and resin on the surface of laminate may be caused by particles of resin left on lamination fixtures, by particles of dust or prepreg collecting on release sheets, or by creases in release sheets. *Corrective action:* Clean lamination fixtures with a plastic scraper and solvent. Perform lay-up in a filtered-air positive-pressure room to reduce the amount of airborne particulates in the layup area.

5. Voids and a thick laminate may occur if the press temperature is too low. *Corrective action*: Verify the press temperature during lamination by inserting a thermocouple in the edge of the lay-up. Check the platen temperature profile with thermocouples and data logger.

6. Placement of two or more lamination fixtures side by side that are more than 0.008 in different in thickness can cause "wedging" or circuit shift. *Corrective action:* When placing lamination fixtures side by side, be sure that the "book" in each fixture is made up of the same number of MLB lay-ups of the same part number.

TABLE 8.21 Warp and Twist

Cause	Corrective action
Improper packing	Ensure that packing skids are flat and have sufficient support
Storage	Stack material horizontally rather than vertically
Distorted glass fabric (FR-4)	Examine etched panels for yarns parallel to warp and fill direction
Excessive exposure to heat	Maintain proper temperature and exposure times in all heat-related fabricating processes (IR fusing, hot-air leveling, and solder mask applications)
Improper handling after exposure to heat applications	Cure solder mask horizontally; use proper cooling techniques after IR fusing and hot-air leveling
Unbalanced laminate construction	Work with laminator to get balanced construction

Source: From Coombs.[11] Reprinted with permission.

TABLE 8.22 Copper Foil

Cause	Corrective action
Fingerprints due to improper handling	Handle copper cladding with gloves
Oils from punching, blanking, or drilling	Degrease with proper solvent
Poor solderability on print-and-etch boards	Degrease to remove contaminates; use highly activated flux; check laminator for procedure to remove excessive antistaining compound

Source: From Coombs.[11] Reprinted with permission.

7. Insufficient prepreg plies can cause voids or resin starvation. *Corrective action:* Make certain that at least two plies of 108 glass prepreg are used, adding third or fourth plies if the inner layers are 2-oz copper.

8. Board warp and twist may be caused by nonsymmetrical lay-ups or designs. It can also be caused by cross-laying the laminate weaves. *Corrective action:* Make sure that design rules mandate that all multilayers containing ground and supply planes be symmetrically placed about the z axis of the MLB and that partial planes be avoided. When laying up, make sure that the strong direction (machine direction or warp thread) of the laminate material is laid in the same direction. Require all material to be marked in the strong direction with an arrow by the vendor. Some fabricators insist that the prepreg also be laid with its warp thread in the same direction as the inner-layer laminate.

8.3.1.10 Contamination and Contamination Control of PWB Assemblies. While the reader is referred to Tautscher[12] for detailed review of this subject, some important funda-

TABLE 8.23 Copper Bond Strength

Cause	Corrective action
Pad or trace lifting during wave soldering	Check for undercut due to overexposure during etch; check with laminator that solvents used in circuit manufacturing process are compatible with laminate
Pad or trace lifting during hand soldering	Hand-soldering device is too hot or wattage is too high for application; device operator is applying heat too long to soldered pad
Pad or trace lifting during processing	Check laminate supplier for bond strength properties of lot of material in process; check for undercut during etch; if pads lift after solder leveling, check fuser voltage or hot-air leveling equipment temperature

Source: From Coombs.[11] Reprinted with permission.

mentals will be highlighted at this point. There are three types of contaminants that can occur, and which must be controlled. They are either water soluble, water insoluble, or totally insoluble.

Water-soluble contaminants usually occur from residues of plating, etching, and similar operations. Since the acid and alkaline solutions used in these operations are water-based solutions, the resulting residues are slow water-soluble. This class of inorganic contamination is also called ionic (since water solutions form electrically charged particles called ions) and polar (since the charged particles are electrically polar, either positive or negative). These contaminants must be removed with a liquid that will dissolve and completely remove them, such as deionized water or alcohol.

Water-insoluble contaminants are those that will not dissolve in water, such as oils, greases, and waxes. These contaminants must be removed with a solvent that will dissolve and completely remove them, such as organic solvents. Typical organic solvents are chlorinated or fluorinated solvents. These may be used in liquid form, but are more effective in their vapor form, as in vapor degreasers. This is because the solubility of contaminants increases with temperature, and easier penetration, for example into crevices, occurs with vapors. These water insoluble organic contaminants are also called nonionic or nonpolar, since they do not ionize.

Totally insoluble contaminants may be the worst problem, since these contaminants are not soluble in any cleaning agent and must be physically removed, if they can be removed at all. Examples of these are dust, dirt, embedded pumice, and insoluble salts.

The water-soluble and water-insoluble contaminants may be cleaned separately, or they may be cleaned using cleaner combinations which simultaneously remove both types of contaminants. Leading suppliers of cleaning agents and solvents market such combined-action solvents, often for use in vapor degreasers.

One major problem with cleaning operations is to know when adequate cleaning has been achieved. There are some widely accepted standard tests, which are based on measuring the resistivity or conductivity of the cleaning solutions that have been used for the PWB cleaning. Testing equipment and test procedures for these solvent extract resistivity tests are available through the IPC. The test methods are contained in IPC TM-650 and include the following: method 2.3.25 and 26 for ionic contamination, method 2.3.38 for organic contaminants, and method 2.3.39 for organic nonionic contaminates. Again, readers are referred to Tautscher[12] for a detailed review of this subject.

One other key point which should be mentioned about cleaning solvents is that chlorofluorocarbon (CFC) solvents have been found responsible for atmospheric ozone depletion. A major international agreement, known as the Montreal Protocol, has been reached to eliminate CFC materials from use in all applications. Consequently, numerous alternatives are emerging, and their effectiveness and trade-offs should be considered at any given time. Three prime alternatives are (1) new chlorinated/fluorinated solvents which do not deplete ozone, (2) other types of solvents such as terpenes, and (3) noncorrosive or "clean" fluxes.

8.3.2 Flexible PWB Materials

The materials and the anatomy of flexible printed wiring (FPW) are quite different from those of the various rigid PWB constructions discussed in this chapter. As was mentioned elsewhere in this chapter, these flexible circuitry forms are used in the many product designs and applications where the physical flexibility and formability of these circuitry forms offer advantages. These applications and their advantages are many, and are constantly increasing. FPW is etched or formed to the desired circuit pattern, and thus is similar to rigid PWBs except that it is flexible in form. The anatomy of the flexible conductor system in both flex and rigid flex is similar in cross-section, however, and these cross-sections are shown in Figs. 8.3 and 8.4.

As seen in Figs. 8.3 and 8.4, the flexible circuit in both flexible and rigid-flex consists of three important materials, namely, the base film, the conductor and the adhesive. The metal foil is bonded and subsequently etched to the desired pattern on the adhesive coated base film. A second adhesive coated film is bonded on top of the formed circuit pattern. This top adhesive coated film is usually the same material construction as the bottom, or base, adhesive coated film. The top adhesive coated film is known as a cover coat film. It protects the circuitry much as conformal coatings protect circuitry on rigid PWB circuits. This cover coat is usually laminated to the base film, under heat and pressure, after the circuit pattern has been formed and any cleaning operations completed. Although many types of films and adhesives can be used, thermal stability must be considered in their selection, especially where parts will be soldered to the circuitry or when vapor degreasing operations are used for cleaning.

Completed FPW products are covered by a number of government and industry documents. The military delineate the requirements for the design of FPW and rigid FPW in MIL-STD-2118 and the fabrication of FPW and rigid FPW in MIL-P-50884. The IPC covers FPW in three documents; single-side and double-sided FPW design standards in IPC-D-249, performance specifications for single- and double-sided FPW in IPC-FC-250A and performance specifications for rigid FPW in IPC-RF-245.

The following sections discuss in detail the constitutive materials used in the fabrication of flexible printed wiring.

8.3.2.1 Dielectric Films. There are four IPC documents covering the dielectric films used in the fabrication of flexible printed wiring. IPC-FC-231B covers the basis dielectric film material. IPC-FC-232B covers adhesive coated dielectric films used as cover sheets in FPW. IPC-FC-241B is the specifications covering metal-clad FPW dielectrics and IPC-FC-FLX is a listing of specification sheets for the various basis dielectric film and adhesive materials used in FPW.

For the reasons mentioned above, and for the general good stability and other good material characteristics which they possess, the films used for flexible printed wiring are usually either polyimide or polyester. While there are other sources of these materials, perhaps the most widely used are the DuPont polyimides (Kapton H) and the DuPont polyesters (Mylar).

Polyimides are dark yellow films with thermal stabilities of over 200–250°C. Polyimide films can withstand soldering temperature. They are not attacked by any known organic solvent and cannot be fused. Standard film thickness are 0.0005, 0.001, 0.002, 0.003 and 0.005 in. There are variations of the basic Kapton H film to develop special properties. Kapton XT is a polyimide film that has been developed with alumina incorporated into the polymer to increase the thermal conductivity. Kapton XC has been developed with increased electrical conductivity for special applications. Polyimides are considerably higher in cost than polyesters (about 20×).

Polyesters are transparent films containing no plasticizers. They do not become brittle under normal conditions. Their low service temperature range of –70°C to 150°C limits their use in applications requiring soldering. Thus, polyesters, sometimes identified by their chemical class (polyethylene terephthalate), are best used when mechanical rather than soldered interconnections are employed for component attachment.

Other types of films which are sometimes used are aramids (DuPont Nomex), polyester-epoxies (Rogers Bend/Flex), Upilex^R S polyimide film (Gore-Clad^R), fluorocarbons (DuPont Teflons) and polyetherimides (General Electric Ultem).

Nomex is a high-temperature paper material made from random fiber aramid materials. Its very hygroscopic and tends to readily absorb processing chemicals. It has fairly low tear strength and a dielectric constant about half that of Kapton H.

Bend-Flex is manufactured using a nonwoven mat of Dacron polyester and glass fibers saturated in a B-stage epoxy. Copper foils are directly clad to the B-staged material. The dielectric material is available in 0.005 to 0.030 in thicknesses.

Gore-Clad^R is a new material developed by W.L Gore and Associates for use in rigid-flex applications. This system uses UpilexR S polyimide film. This film exhibits superior dimensional stability, chemical resistance and moisture resistance when compared to traditional Kapton H material systems. The adhesive system is a thin cyanate ester modified with PTFE. The material is not recommended for dynamic flex applications.

Material properties of the various dielectric films are given in Table 8.24.

8.3.2.2 Adhesives Two IPC documents delineate the requirements for adhesive films used in the fabrication of flexible printed wiring products. IPC-FC-233A covers the adhesive material (see Table 8.25) and IPC-FC-FLX gives is a specification sheet listing of the materials.

The adhesive types used to bond the metal foil to the dielectric film are most commonly epoxies, polyesters or acrylics. New product developments have resulted in adhesiveless material systems for bonding between the copper and dielectric. In addition, an electrically conductive adhesive system has been introduce to eliminate plated through hole interconnections.

Epoxy systems include modified epoxies known as phenolic butyrals and nitrile phenolics. These systems are widely used and are generally lower in cost than acrylics but higher in cost than polyesters. Epoxies have good high-temperature resistance, a lower moisture absorption rate than acrylics, a lower coefficient of thermal expansion than acrylics and remains in good condition in all approved soldering systems. They also have very long-term stability at elevated temperatures in environmental conditions up to 250°F.

Acrylic systems are most often used in high-temperature soldering applications. They exhibit excellent adhesion to both the copper interconnect as well as polyimide films. Typical bond strengths to copper are 8–15 lb/in at room temperature. These adhesives have very controlled flow when bonding cover sheets to the base laminate. They are easy to process and exhibit excellent batch-to-batch process consistency. There are a number

TABLE 8.24 Material Properties of Flexible Printed-Wiring Dielectric Film

IPC-FC-231, specification sheet	Polyimide 1	FEP 2	Polyester 3	Epoxy polyester 4	Aramid paper 9
Tensile strength, min. lb/in^2	24000	2500	20000	5000	4000
Elongation, min. %	40	200	90	15	4
Initial tear strength, min. grams	500	200	800	1700	—
Density, g/cm^3	1.40	2.15	1.38	1.53	0.65
Dimensional stability, max. %	0.15	0.3	0.25	0.2	0.30
Flammability, min. % oxygen	30	30	22	28	24
Dielectric constant, max., at 1 MHz	4.00	2.30	3.40	—	3.00
Dissipation factor at 1 MHz	0.0120	0.0007	0.0070	—	0.01
Volume resistivity (damp heat), $\Omega \cdot cm$	10^6	10^7	—	10^5	10^6
Surface resistivity (damp heat), $M\Omega$	10^4	10^7	—	10^3	10^5
Dielectric strength, min. V/mil	2000	2000	2000	150	390
Moisture absorption, %	4.00	0.10	0.80	1.00	13.00
Fungus resistance	Non-nutrient	Non-nutrient	Non-nutrient	Non-nutrient	Non-nutrient

Source: From IPC-FC-231B, class 3 properties.

TABLE 8.25 Material Properties of Flexible Adhesive Bonding Films

IPC-FC-233, specification sheet #	1 Acrylic	2 Epoxy	3 Polyester
Peel strength, lb/in width			
As received	8.00	8.00	5.00
After solder float	7.00	7.00	N/A
After temperature cycling	8.00	8.00	5.00
Flow max, squeeze out mil/mil	5.00	5.00	10.00
Volatiles content, %	2.00	2.00	1.50
Flammability, min. % oxygen	15.00	20.00	20.00
Solder float, (IPC-TM-650, method 2.4.13 B)	Pass	Pass	N/A
Chemical resistance, % (IPC-TM-650, method 2.3.2 B)	80	80	90
Dielectric constant, max., at 1 MHz	4.00	4.00	4.60
Dissipation factor, max., at 1 MHz	0.05	0.06	0.13
Volume resistivity, min. $M\Omega \cdot cm$	10^6	10^8	10^6
Surface resistance, min. $M\Omega$	10^5	10^4	10^5
Dielectric strength, V/mil	1000	500	1000
Insulation resistance, min. $M\Omega$	10^4	10^4	10^5
Moisture absorption, max. %	6	4	2
Fungus resistance (IPC-TM-650, method 2.6.1)	Nonnutrient	Nonnutrient	Nonnutrient

of properties that will adversely effect multilayer products. Their high moisture absorption rate of 3 percent make products not properly dried prior to soldering delaminate. They have a low T_g, 90°F, and a high coefficient of thermal expansion, 40–600 ppm/°F. In multilayer applications, these properties can cause delamination and barrel cracking of plated through holes when the product is subjected to extreme thermal environments.

Polyesters are the lowest-cost adhesives used and the only adhesives which can be used properly with polyester films for base laminate and polyester cover film. The major drawback of this system is low heat resistance.

New material systems have been developed recently to provide designer with thinner films and laminates through the use of adhesiveless flex circuits. Flex Products, Inc. of Santa Rosa, Calif., has developed a product called UltraFlex™. This product uses advanced vacuum deposition technology to bond copper to polyimide. Parlex, in Methuen, Mass., has developed PALFlex as an adhesiveless flex laminate. This process uses specially treated polyimide film to which an ultra thin coating of a barrier metal is applied to promote copper adhesion. The copper is then electrodeposited onto the film. Rogers Inc., Chandler, Az., uses a cast on process to apply copper to the base dielectric film in its Flex-I-Mid 3000 material. Sheldahl Inc., of Northfield, Minn., produces a product called Nova-clad™ which uses a semi-additive process to deposit copper on to the base film. These material systems provide for applications which may use thinner copper for finer line widths, they have improved dimensional stability, improved flame retardency and have extended the dynamic flexible life of the completed FPW product.

A new adhesive system recently introduced by Sheldahl Inc. is Z-link. Z-link is an anisotropic adhesive system that is impregnated with conductive particles dispersed uniformly throughout the adhesive layer. The material system has two important attributes that determine its performance. First, a solder alloy is used as the conductive spheres. When laminated, the spheres fuse to the mating surfaces creating a low-resistance, high-current-carrying interface. Secondly, the adhesive thermosetting formulation give it the mechanical strength to withstand severe thermal conditions. This material is being used in multilayer applications to replace plated through hole interconnections.

8.3.2.3 Foils.   Regarding metal foils used for conductors in flexible circuitry, these are most commonly electrodeposited copper or rolled annealed copper as defined in IPC-MF-150. The electrodeposited foil is economical, has a rougher base bonding surface which enhances bonding to the base dielectric and a vertical grain structure. Rolled annealed copper foil is normally supplied with a bonding enhancing treatment due to its smoother surface and has a horizontal grain structure. This grain structure makes it better suited for flexing applications which makes it the material of choice in dynamic flex applications. Other conductors may be used for special cases. For copper and other foils, important features to specify are hardness, flexibility and bondability to the adhesive being used.

The materials and the anatomy of flexible printed wiring (FPW) are quite different from those of the various rigid PWB constructions discussed in this chapter. As was mentioned, these flexible circuitry forms are used in the many product designs and applications where their physical flexibility and formability offer advantages. These applications and their advantages are many and are increasing constantly. FPW is etched or formed to the desired circuit pattern, and thus is similar to rigid PWBs except that it is flexible in form. The anatomy of the flexible conductor system is similar in cross section, however, as shown in Fig. 8.18.

8.3.3 Ceramic PWB Materials

An in-depth discussion of ceramic materials is contained in Chap. 7.

8.3.4 Solder Masks

In PWB assemblies, there are many instances when it is desired to protect patterned conductor surfaces on the laminate, such as etched copper circuitry from being coated by solder during either the soldering or tinning processes. In these instances, a solder resist (more commonly known as a solder mask) is applied to the surface of the PWB laminate, usually after the conductor pattern is formed and the PWB panel is cleaned. After the application of the solder mask, solder can then be selectively applied to the areas of PWB holes, pads and conductor line areas to which component parts are to be attached in soldering operations. The absence of solder under the PWB coating not only saves solder, but eliminates the many problems which can occur when solder is entrapped under coatings.

8.3.4.1 Functions and Design Considerations While the primary function of solder masks is to prevent solder movement from the solder land areas to attached conductors and PTHs, solder masks actually serve many other functions. These are summarized below:

Reduce solder bridging and electrical short circuits

Reduce the volume of solder pickup to obtain cost and weight savings

Reduce solder pot contamination (copper and gold)

Protect PWB circuitry from handling damage, such as dirt, fingerprints

Provide an environmental barrier

Fill the space between conductor lines and pads with material of known dielectric characteristics

Provide an electromigration barrier for dendritic growth

Provide an insulation or dielectric barrier between electrical components and conductor lines or via interconnections when components are mounted directly on top of these features.

The following design factors influence decisions to use solder masks:

Criticality of system performance and reliability

Physical size of the PWB

PWB metallizations (tin-lead, copper, gold)

Density of line widths and spacings

Amount and uniformity of the metallization

Size and number of drilled PTHs

Annular ring tolerance for PTH

Component placement on one or both sides of the PWB

Need to have components mounted directly on top of a surface conductor

Need to tent via holes in order to keep molten solder out of selected holes

Need to prevent the flow of solder up via holes that have components situated on top of them

Likelihood of field repair or replacement of components

Need to contain solder on the lands in order to maintain the proper volume of solder required for the component lead solder joint

Choice of specifications and performance class that will give the solder mask the properties necessary to achieve the design goals

It should be pointed out that solder masks, not covering the entire board and component assembly, do not serve the same functions as conformal coatings discussed in Section 8.3.5.

8.3.4.2 Specifications and Selections.   The use of solder masks is covered in specification IPC-SM-840, entitled "Qualification and Performance of Permanent Polymer Coating (Solder Mask) for Printed Wiring Boards."

The IPC-SM-840 specification calls out three classes of performance which the designer can specify:

Class 1: Consumer—Noncritical industrial and consumer control devices and entertainment electronics

Class 2: General industrial—Computers, telecommunication equipment, business machines, instruments, and certain noncritical military application

Class 3: High reliability—Equipment where continued performance is critical; aerospace and military electronic equipment

In addition to calling out the performance classes, the specification assigns responsibility for the quality of the solder mask to the materials supplier, the PWB fabricator, and finally, the PWB user.

Solder masks may be temporary or permanent. The following sections will concentrate on permanent solder mask technology. The various types are illustrated in the technology chart shown in Fig. 8.24. It can be seen that various chemistries, uncured forms, application technologies and curing methodologies exist. A selection guide for choosing among the permanent solder masks is shown in Table 8.26. Permanent solder masks, become a functional part of the PWB assembly, and hence their performance in the system life cycle must be considered.

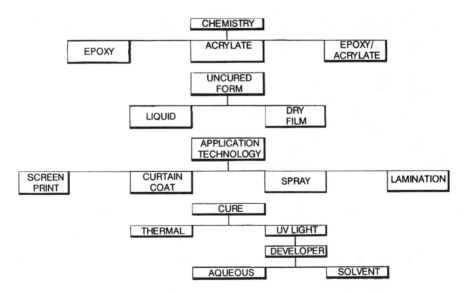

FIGURE 8.24   Permanent solder mask technology chart.

TABLE 8.26 Selection Guide for Permanent Solder Masks*

Feature	Screen print		Dry film		Liquid photoresist solvent
	Thermal	UV	Aqueous	Solvent	
Soldering performance	1	1	1	1	1
Ease of application	1	1	2	2	2
Operator skill level	2	2	2	2	2
Turnaround time	2–3	2–3	2	2	2
Inspectability	2–3	2–3	3	3	3
Feature resolution	3	3	1	1	1
Adhesion to Sn-Pb	1	3	1–2	1–2	1
Adhesion to laminate	1	1–2	1	1	1
Thickness over conduct or lines	3–4	3–4	1–2	1–2	2
Bleed or residues on pads	3–4	3	1	1	1
Tenting or plugging of selected holes	4	4	1	1	4
Handling of large panel size with good accuracy	3	3	1	1	1
Meeting IPC SM-840 class 3 specification	3–4	3–4	1–2	1	1
Two-sided application	4	4	1	1	3–4
Capital equipment cost	4	4	1–2	1–2	1

*1 = good or high; 2 = moderate; 3 = fair; 4 = poor or low.

Source: From Coombs.[5] Reprinted with permission.

The demand for permanent coatings on PWBs has greatly increased as the trend toward surface mounting and higher circuit density has increased. When the conductor line density was low, there was little concern about solder bridging, but as the density increased, the number and complexity of the components increased. At the same time, the soldering defects, such as line and component shorts, greatly increased. Inspection, testing, and rework costs accelerated as effort went into locating and repairing the offending solder defects. The additional cost of the mask on one or both sides of the PWB was viewed as a cost-effective means to offset the higher inspection, testing, and rework costs. The addition of a mask also had the added value for the designer of providing an environmental barrier on the PWB. This feature was important and dictated that the materials considered for a permanent solder mask should have similar physical thermal, electrical, and environmental performance properties that are in the laminate material.

There are four major types of permanent solder masks available on the market today, screen printable thermally cured liquids, photoimagable liquids, photoimagable dry films and photoimagable liquid/dry films.

The screen printable thermally cured solder masks have been in existence the longest. These masks are usually epoxy based. Application of these masks is usually through the use of a screen printer using either a stainless steel or polyester mesh. The screen has openings which are aligned to the PWB features and define the mask pattern to be applied to the PWB. Curing of the mask occurs at elevated temperature (usually about 125°C).

These masks have strong hard surfaces and excellent adhesion to the PWB surface. They usually are used on PWBs having relative coarse feature patterns (>0.025 in). Fine mask features (<0.025 in) are difficult to maintain in older epoxy formulations due to limitations in screen printer technology, smearing of the liquid and bleeding of solvents from the mask during cure. A further limitation of their use was in the tendency of the mask to retain residues during further processing steps.

The advent of new screen printing technologies coupled with a new generation of liquid polymers is causing a reevaluation of this solder mask technology for finer pitch SMT Technology. Many of the new thermal cure liquids on the market today are based on poly-functional epoxies which cure to a high cross-link density providing a low porosity coating. In addition, advances have been made to reduce their tendency to smear during printing and bleed during cure.

With regard to the photoimagable solder masks, three types are used; liquid, dry film and liquid dry film. The liquid photoimagable solder masks are formulated from either epoxy, acrylate or epoxy-acrylate chemistries. The dry films are formulated from either acrylate or epoxy-acrylate chemistries. The liquid/dry film masks use a combination of both mask technologies. The mask chemistry can be either solvent or aqueous developable. Two different polymerization formulations are available; bulk and surface.

Bulk polymerization, the most widely used, provides fairly straight side walls after development. Liquid bulk polymerized masks yield the narrowest finished feature sizes, 0.001 to 0.003 in. The smallest dry film bulk polymerized mask feature size is limited to 0.006 in. Disadvantages to bulk polymerization are shoot-through on double-sided PWBs and light scattering. On shoot-through, the UV light travels through the PWB and partially cures the mask on the far side of the PWB with the image being developed on the near side. The scattering of the UV light source may cause exposure of the mask inside the barrel of the PTH or partial exposure on the solder lands.

Surface polymerization masks exhibit undercutting of the mask during development of about 0.0005–0.003 in. This prevents feature sizes of less than 0.004–0.007 in in size. Pin holes may develop in these masks when foreign material is present between the solder mask surface and the UV light source.

Dry film solder masks are supplied by their manufacturers in rolls. The thickness is usually 0.002 to 0.004 in. The major advantage in using a dry film mask over a liquid mask is the ability of the dry film mask to tent holes of diameters up to 0.030 to 0.040 in. Liquid solder masks while lacking the hole tenting characteristics of the dry films tend to give a finished mask thickness less than the dry film; usually 0.001 to 0.002 in thick over the laminate and 0.0007 to 0.001 over metalized traces.

Pencil hardness is a value assigned solder masks which defines its ability to resist abrasion or its scratch resistance. Current solder mask formulations have pencil hardness values of 3H to 8H, with 8H being the most scratch resistant. Dry film masks tend to have values in the 3H to 4H range while liquid photoimagable masks have values from 5H to 6H. Table 8.27 presents typical material property data on photoimagable dry film solder masks and Table 8.28 presents material property data on photoimagable liquid solder masks.

8.3.4.3 Processing Considerations. The major processing steps for solder mask applications are cleaning and preparation of the PWB surface, solder mask application, and solder mask development and cure.

Cleaning and Preparation of the PWB Surface. Optimal solder mask appearance and adhesion will not occur unless the PWB surface has been carefully prepared. The main objectives in this process step are to remove all surface contaminants and roughen up the surface and thoroughly dry the prepared surface to promote mask adhesion.

If only bare copper circuitry or other hard metallizations such as nickel are present on the surface conductors prior to mask application, the following steps are usually employed. An acid pre-clean is usually first used to remove oils and oxidation on the surface. This is followed by a mechanical scrubbing to remove further residues and roughen

TABLE 8.27 Material Property Data for Typical Dry-Film Photo-Imageable Solder Masks*

Type	Dynamask KM Epoxy	Conformask 1000	Conformask 2000 Epoxy	Laminar DM	Vacrel 8000	Vacrel 8100	Valu 8000†	Valu 8200†
Solids content, %	100	100	100	100	100	100	100	100
Cured film thickness, in	0.003 or 0.004	0.0023	0.0023	0.003 or 0.004	0.003 or 0.004	0.003 or 0.004		
UL rating	94V–0	94V–0	94V–0	94V–0	94V–0	94V–0	94V–0	94V–0
IPC-SM-840A rating	Class III	Class III	Class III	Class III	Class III	Class III	Class III	Class III
Pencil hardness	3H	2H	3H	2H–3H 2H–3H	4H–5H	4H	4H–6H	4H–6H
Insulation resistance, Ω	4.0×10^{12}	3.3×10^{12}	1.1×10^{11}	2.0×10^{14} 1.0×10^{14}	5.0×10^{14}	5.0×10^{14}	3.0×10^{14}	3.0×10^{14}
Dielectric strength, V/mil	3434 3386	1952	2969	1366 1600	3506	2829	3517	4310
Dielectric constant at 1 MHz	4.23 3.41	—	—	—	3.6	3.8	3.5	3.2
Dissipation factor at 1 MHz	0.031 3.41	—	—	—	0.033	0.042	0.027	0.027
Volume resistivity, $\Omega \cdot cm$	6.0×10^{15} 6.7×10^{15}	—	—	—	3.0×10^{14}	9.0×10^{14}	3.0×10^{16}	1.5×10^{16}
Surface resistivity, Ω	1.0×10^{17} 3.0×10^{17}	—	—	—	10^{15}	3.6×10^{13}	8.0×10^{13}	7.0×10^{13}
Solvent resistance	Pass	Pass	Pass	Pass	Pass	Pass	Pass	Pass

*Where two values are given, the upper value is for 0.003-in-thick film, the lower value for 0.004-in-thick film.
†Value system by duPont is a combination liquid and dry-film system.

TABLE 8.28 Material Property Data for Typical Liquid Photo-Imageable Solder Masks

	Lea Ronal OP SR 5500	Coates Imagecure AQ	Chemline Photobond 3000	WR Grace AM-300	Lackwerke Peters SD-2461-AE	M and T Photomet 1001	Ciba-Geigy Probimer 52	Dynachem EPIC SP-100
Type	Epoxy/acrylate	Epoxy/acrylate	Acrylate	Epoxy/acrylate	Epoxy	Acrylate	Epoxy	Epoxy
Solids contents, %	70	68	65	100	80	100	38	76
Cured film thickness, in	0.0005–0.0015	0.0007–0.002	0.0008–0.0015	0.001–0.0055	0.0007–0.002	0.002–0.006	0.0005–0.0017	0.0005–0.0017
UL rating	94V-0	94V-0	94V-0	94V-0	94V-0	94V-0	94V-0	94V-0
IPC-SM-840A rating	Class III	Class III	Class III	Class III	Class III	Class III	Class III	Class III
Pencil hardness	5H	7H	4H	4H	6H	5H	4H	6H
Insulation resistance, Ω	9.3×10^{12}	1.0×10^{12}	4.0×10^{13}	1.4×10^{10}	—	2.0×10^{14}	2.0×10^{11}	1.2×10^{12}
Dielectric strength, V/mil	2500	2080	1703	3500	1778	3000	3048	3018
Dielectric constant at 1 MHz	3.5	2.8	3.6	3.5	3.6 at 0.8 MHz	4.3	3.7	4.07
Dissipation factor at 1 MHz	0.042	—	0.015 at 0.5 MHz	0.011	0.040 at 0.5 MHz	0.007	0.002	0.035
Volume resistivity, $\Omega \cdot$ cm	10^{13}	10^{15}	—	10^{15}	10^{13}	9.0×10^{14}	10^{13}	2.7×10^{15}
Surface resistivity, Ω	2.0×10^{13}	10^{13}	4.0×10^{12}	10^{10}	10^{13}	4.0×10^{13}	10^{10}	3.7×10^{15}
Solvent resistance	Pass	Pass	Pass	Pass	Pass	Pass	Pass	Pass

the surface. Bristle brushes, composite brushes or the combination of a pumice slurry and bristle brushes are used in this step.

The critical parameter to be maintained in this process is the brush pressure. Too little pressure and contaminants will remained trapped on the PWB or the surface will not be roughened sufficiently to promote adhesion of the mask. Too high a brush pressure and fine circuit features may be broken or grooves may be made in the surface under the solder mask which may promote entrapment of developing solutions.

Where tin-lead is present on the PWB prior to solder mask application, mechanical scrubbing and pumice scrubbing are not normally used. This is due to smearing of the malleable tin-lead or imbedding of the pumice material in the tin-lead during this operation. Usually the pre-cleaning of PWBs with tin-lead features involves a solvent degreasing in Freon or 1,1,1-trichloroethane followed by a light chemical cleaning.

The final steps in this process are a rinse in deionized water to remove any remnants from the scrubbing or cleaning processes followed by a thorough drying of the PWB surface. In some instances particularly when dry film solder masks are used, an oven bake of the PWB prior to solder mask application is recommended to remove any entrapped moisture in the laminate.

Solder Mask Application. Application of the solder mask is usually accomplished through one of 4 procedures; screen printing, curtain coating, spray coating or dry film application.

Screen printing of oven cured and some liquid photoimagable masks uses stainless steel or polyester mesh screens which define the finished solder mask features. A manual or semiautomated screen-printing machine mounts the mesh screen and is used to align the screen to the PWB. The liquid mask material is squeegeed through the openings in the screen on to the PWB.

Parameters needed to be controlled to optimize this printing process include choosing the optimal screen mesh size (110–200 mesh polyester monofilament is typical), optimal screen mesh tension and emulsion, maintaining the proper mask liquid viscosity and using a proper durometer rubber squeegee (60–70 durometer, well sharpened and 10°–20° from vertical is typical).

Photoimagable liquid solder masks can be applied by the use of screen printing using a screen to coat the PWB completely with the mask material but are also applied through the use of either curtain coating or spray coating equipment.

Curtain coating equipment applies the mask material on one-side of the PWB at a time by passing it horizontally through a flowing curtain of the liquid mask material. The material is then oven dried to allow application of the material to the second side of the PWB.

Critical parameters to be controlled in this process to ensure a consistent mask deposit and coverage are maintenance of a consistent solder mask viscosity and temperature and using a proper conveyor speed through the equipment to ensure a thin uniform coat is applied. Some formulations of liquid photoimagable materials require the PWB to be preheated prior to application. The oven drying cycle must remove a sufficient amount of the solvents from the mask to allow direct contact with the equipment conveyor system. Typically, 85 to 90 percent of the solvents must be removed during this step.

Spray coating equipment has the advantage of applying the mask material to both sides of the PWB prior to oven dry. The spray coating process uses an electrostatic spray system for application. In this system, the PWB is grounded and processed vertically between a grounded back panel and spray head. The spray head atomizes negatively charged masked material which is attracted by the grounded PWB. The equipment automatically sprays one side of the PWB and then flips the PWB to coat the second side.

There are a few drawbacks to this process. Some waste of mask material occurs. Solder mask "wrap around," a buildup of material on the panel edge, can occur due to charge buildups. Solder mask "skipping," a lack of mask coverage, can occur due to like charges building up on adjacent traces which repel the mask material.

Dry film mask materials are supplied as 0.002 to 0.004 in thick films sandwiched between protective film layers. The mask is cut to size according to the size of the PWB panel to be masked. One side of the protective film is removed and the mask is applied to the panel through the use of specialized lamination equipment. The second protective film coat is removed just prior to film development. The lamination equipment process uses heat and a vacuum to cause the solder mask to flow and conform to the panel surface with no entrapped air pockets.

Temperature is the important process parameters to be maintained. Too low a temperature results in improper flow and air entrapment while too high a temperature causes thin coverages on circuitry and flow of material into PTH leading to difficulties in later development. Fine line closely spaced circuitry can be difficult for dry film masks to conform to.

The liquid/dry film systems use a specialized machine which applies a thin coating of the liquid solder mask to the PWB and floods closed the PWB PTHs. Then a thin dry film mask is immediately applied over the liquid to trap the liquid in place.

Solder Mask Development and Cure. Development of both liquid and dry film photoimagable masks occurs under UV light in specially designed equipment. In the UV curing process, the mask contains components that under go polymerization and cross-linking by photogenerated radical and cationic species which occur during the exposure to a UV light source. The UV light exposure causes almost instantaneous hardening of the material into a cured film. A phototool is used with the final solder mask pattern in a negative UV lightblocking image (the areas to keep solder mask are exposed). This tool is aligned to the PWB in a frame which is then drawn down on the panel in a vacuum to eliminate space between the panel and mask surface. It is then exposed to a UV light source which produces an initial cure of the exposed mask areas.

Removal of the photo tool occurs next and the final image is developed in either an aqueous or solvent chemistry. It should be noted that aqueous chemistries are being favored by more users due to imposition of stricter EPA guidelines on the storage, use and disposal of toxic solvents. This development removes the unwanted, non-cured mask film.

Developing critical parameters include the chemistry temperature and concentration, spray pressures, conveyor speed, rinsing and drying. Fine line mask features may not adhere well to the surface if exposed to too high spray pressures.

Acrylate and epoxy-acrylate masks usually require a secondary UV "bump" to complete the cure. This bump completes the cross-linking and makes the mask more robust. Epoxy based masks usually require a thermal cure to complete the cross-linking.

8.3.4.4 Solder Mask Over Bare Copper (SMOBC).

One major solder mask application technology is called SMOBC. This name stands for the application of "solder mask over bare copper." A problem for conventional copper-tin-lead electroplated PWBs is the flow of tin-lead solder under the solder mask during the wave or vapor phase or infrared soldering process. This flow of molten metal underneath the mask can prevent the mask from adhering to metal or laminate. If the mask fractures because of this hydraulic force, the surface integrity is lost and the effectiveness of the mask as an environmental or dielectric barrier can be severely impaired. In fact, such breaks can actually trap moisture, dirt, and soldering flux and serve as a conduit to direct liquids down to the mask-laminate interface. This situation could lead to serious reliability and/or performance concerns.

The SMOBC process addresses the tin-lead flow problem by eliminating the use of tin-lead electroplating on the conductor lines under the solder mask. An all-copper PTH printed wiring board is often produced by a "tent-and-etch" process. This process is one in which the PWB is drilled and plated with electroless copper. This is immediately followed by copper panel plating with the full-thickness copper required for the PTHs. A dry-film resist process is used with a negative phototool (clear conductor lines and pads) to polymerize the dry-film resist in only those clear areas of the phototool. This polymerized resist will protect the lines and PTHs during a copper-etching process which will remove all the unwanted background copper. The photoresist is then stripped off, and the solder mask material is applied and processed through curing. Tin-lead is next added to the open component pads and PTHs by the hot-air leveling process.

An alternative process uses conventional procedures to create a pattern-plated board. After etching, however, the metal etch resist is removed chemically, leaving the underlying copper bare. The primary function of tin-lead in the PTHs and on the component pads is to improve solderability and appearance. It is important to demonstrate the solderability of the holes and pads on the SMOBC panel. One method of accomplishing this is by a hot-air leveling process which places a thin coating of molten tin-lead on only those copper areas of the PTH that have not been covered by the solder resist. This hot-air leveling process proves the ability of the copper surface to be soldered and also improves the appearance and solderability after longer-term storage of the PTHs and pad surfaces.

Since there is no flowable metal under the solder mask during the hot-air leveling step or later during component soldering, the mask maintains its adhesion and integrity.

The lower metallization height of the conductor lines allows the use of a thinner dry-film resist and also makes the liquid and screen-printing application somewhat easier.

One variation on the basic SMOBC process is to produce the PWB by the conventional pattern-plate copper and tin-lead process followed by etching of the background copper. Then another photoresist step is used to tent the holes and pads so that the tin-lead can be selectively stripped from the conductor lines. This would be followed by infrared or oil reflow, cleaning, and application of solder mask.

A second process variation strips off the tin-lead plating completely and is followed by cleaning, solder mask application, and hot-air leveling. There are still other PWB fabricators who do not like either of the above processes and are opting to use a nonflowable, copper-etchant-resistant metal like tin-nickel under the solder mask. The major shortcoming to tin-nickel is that it is considerably more difficult to solder with low-activity soldering fluxes.

8.3.5 Conformal Coatings for PWB Assemblies

Systems of synthetic resins usually dissolved in solvent vehicles are applied to completed printed wiring assemblies (PWA) in many applications. When cured they form a secure plastic-film encapsulation around PWAs called a conformal coating. Their purpose is to seal out environmental contaminants that the assembly may be exposed to during its lifetime. However, they may also seal in contamination products from the manufacturing process not removed by a proper pre-cleaning.

Conformal coatings were originally developed to protect sensitive electronic assemblies from the hostile environments experienced in the military, aerospace and marine environments. As the level of integration has increased in the electronics industry with the use of surface mount technology and the inherent finer lead pitches associated with VLSI circuitry, the use of conformal coatings has spread into the commercial arena.

Conformal coatings usage occurs for a variety of reasons. In humid and/or marine salt laden atmospheres, the insulation resistance of PWB laminates often drops very significantly due to ionic contaminants that form conductive solution paths between circuits. Conformal coatings provide a major slowing of this degradation and in addition can prevent fungal attack. Abrasion related failures due to mishandling and environmental dirt can be averted with conformal coatings. Corrosion of metallizations due to chemicals such as jet fuel and engine fluids contaminants in the environment can also be reduced. Coatings can provide added insulation protection between conductors against increased voltage levels. If the coating is formulated to contain an alumina trihydrate powdered filler, then improved resistance to carbon tracking results at high-voltage arcing conditions.

Since most conformal coatings were developed in response to the harsh environments experienced in the aerospace and military applications, many coatings on the market today are identified in Military Specification MIL-I-46058. Furthermore, coatings selected from the Qualified Products List (QPL) for this specification will have already been qualified to high-performance standards.

Two other organizations have documentation on conformal coatings. The IPC covers the qualification and performance of PWA conformal coatings in IPC-CC-830. Underwriters Laboratory has qualified three groupings of conformal coatings; acrylics, solvent based and water based polyurethanes and high-temperature silicones. These have passed rigorous UL746C and 94V test qualification and are being used by the consumer electronics manufacturers to improve the reliability of their products.

The five classes of coatings listed in MIL-I-46058 are acrylic (Type AR), polyurethane (Type UR), epoxy (Type ER), silicone (Type SR) and polyparaxylylene (Type XY). Some comments on each of these coating types, along with polyimides and diallyl phthalates, are presented below. In addition, with the advent of chip on board packaging technology silicon nitride as a conformal coating is being proposed by a number of companies. This is also discussed at the end of this section. Chapter 1 covers in greater depth the chemistry of each of the coating types.

8.3.5.1 Acrylic Coatings.

Acrylics are excellent coating systems from a production standpoint because they are relatively easy to apply. Furthermore, application mistakes can be corrected readily, because the cured film can be removed by soaking the printed circuit assembly in a chlorinated solvent such as trichloroethane or methylene chloride. Spot removal of the coating from isolated areas to replace a component can also be accomplished easily by saturating a cloth with a chlorinated solvent and gently soaking the area until the cured film is dissolved.

Since most acrylic films are formed by solvent evaporation, their application is simple and is easily adaptable to manufacturing processes. Also, they reach optimum physical characteristics during cure in minutes instead of hours.

Acrylic films have desirable electrical and physical properties, and they are fungus-resistant. Further advantages include long pot life, which permits a wide choice of application procedures; low or no exotherm during cure, which avoids damage to heat-sensitive components; and no shrinkage during cure. The most obvious disadvantage of the acrylics is poor solvent resistance, especially to chlorinated solvents.

8.3.5.2 Polyurethane Coatings.

Polyurethane coatings are available as either single- or two-component systems. They offer excellent humidity and chemical resistance and good dielectric properties for extended periods of time.

In some instances the chemical resistance property is a major drawback because rework becomes more costly and difficult. To repair or replace a component, a stripper compound must be used to remove effectively all traces of the film. Extreme caution must be exercised when the strippers are used, because any residue from the stripper may corrode metallic surfaces.

In addition to the rework problem, possible instability or reversion of the cured film to a liquid state under high humidity and temperature is another phenomenon which might be a consideration. However, polyurethane compounds are available to eliminate that problem.

Although polyurethane coatings systems can be soldered through, the end result usually involves a slightly brownish residue which could affect the aesthetics of the board. Care in surface preparation is most important, because a minute quantity of moisture on the substrate could produce severe blistering under humid conditions. Blisters, in turn, lead to electrical failures and make costly rework mandatory.

Single-component polyurethanes, although fairly easy to apply, require anywhere from 3 to 10 days at room temperature to reach optimum properties. Two-component polyurethanes, on the other hand, provide optimum cure at elevated temperatures within 1 to 3 h and usually have working pot lives of 30 min to 3 h.

8.3.5.3 Epoxy Coatings. Epoxy systems are available, as two-component compounds only, for coating electronic systems. Epoxy coatings provide good humidity resistance and high abrasive and chemical resistance. They are virtually impossible to remove chemically for rework, because any stripper that will attack the coating will also attack epoxy-coating or potted components and the epoxy-glass printed board as well. That means that the only effective way to repair a board or replace a component is to burn through the epoxy coating with a knife or soldering iron.

When epoxy is applied, a buffer material must be used around fragile components to prevent fracturing from shrinkage during the polymerization process. Curing of epoxy systems can be accomplished either in 1 to 3 h at elevated temperature or in 4 to 7 days at room temperature. Since epoxies are two-component materials, a short pot life creates an additional limitation in their use.

8.3.5.4 Silicone Coatings. Silicone coatings are especially useful for high-temperature service (approximately 200°C). The coatings provide high humidity and corrosion resistance along with good thermal endurance, which makes them highly desirable for PWAs that contain high heat-dissipating components such as power resistors.

Repairability, which is a prime prerequisite in conformal coating, is difficult with silicones. Because silicone resins are not soluble and do not vaporize with the heat of a soldering iron, mechanical removal is the only effective way to approach spot repair. That means the cured film must be cut away to remove or rework a component or assembly. In spite of some limitations, silicone coatings fill a real need because they are among the few coating systems capable of withstanding temperatures of 200°C.

In general silicones have other disadvantages. Their pot life is usually limited. Bonding to certain surfaces may require a priming agent. They have a very high coefficient of thermal expansion (CTE). This high CTE can lead to failure of fragile solder joints under environment thermal excursions if the application procedure is not controlled to prevent a thick non uniform coating.

8.3.5.5 Polyparaxylene Coatings. Polyparaxylylene coatings are commonly known as Parylene[*] Coatings. This is the trade name given to these coatings by its developer, Union Carbide Company.

Parylenes are unique in several important ways. First, they are applied in a vapor deposition process under a 0.1 torr vacuum. The deposition occurs at room ambient in a solvent free atmosphere which places no thermal, mechanical or chemical stress on the PWA. This vacuum chamber process does require special equipment.

---

*Parylene is a registered trademark of Union Carbide Company

A second important unique characteristic is that the highly active gaseous monomer molecules within the vacuum chamber are not hindered by line of sight application procedures such as spray deposition. A thin uniform coating occurs on all exposed surfaces including sharp corners and edges of components and solder joints, tiny crevices and under components. Typical coating thicknesses are 0.5 to 2 mil. This outstanding feature produces a coated PWA which offers better protection against humid and contaminated environments than most other coatings. Tests have shown Parylene to provide the most effective protective barrier in nuclear-biological-chemical (NBC) warfare environments. They have also been shown to withstand and protect PWAs from the damaging effects of decontamination with highly caustic agents such as DS_2.

The requirement for special vacuum deposition equipment is often a major disadvantage, but there are numerous companies which will provide coating services. Since Parylene coats all exposed surfaces, the masking procedures of parts such as connector pins requires greater care and process development. The tenacious nature of the film causes a more difficult Repairability procedure to be implemented. Removal is usually accomplished through mechanical means or by burning off the film using a solder iron. Reapplication of Parylene to the repaired area must occur in the vacuum chamber. In many instances a repair of a Parylene coated assembly may be through the use of another type of coating material.

8.3.5.6 Solder Resists.   Solder resists represent a special set of coatings, having special functions. They do not function as conformal coatings. This is sometimes a point of confusion. Solder resist coatings are covered in detail in Sec. 8.3.4.

8.3.5.7 Polyimide Coatings.   Polyimide coating compounds provide high-temperature resistance and also excellent humidity and chemical resistance over extended periods of time. Their superior humidity resistance and thermal range qualities are offset by the need for high-temperature cure (from 1 to 3 h at 200 to 250°C). High cure temperatures limit the use of these coating systems on most printed circuit assemblies. Because the polyimides were designed for high-temperature and chemical resistance, chemical removal and burn-through soldering cannot be successful.

8.3.5.8 Diallyl Phthalate Coatings.   Diallyl phthalate (DAP) varnishes also require high-temperature cure (approximately 150°C), which limits their use on printed circuit assemblies. Furthermore, their removal with solvents or with a soldering iron is difficult, owing to their excellent resistance to chemicals and high temperatures (350°F).

8.3.5.9 Silicon Nitride[17].   Silicon Nitride has been in use as a passivation coating on integrated circuits since the early 1970s. With the advent of chip on board (COB) packaging for use in both military and harsh consumer environments, this coating has begun to be looked at as a conformal coating for board level COB applications. A room temperature plasma deposition system has been developed by Ionic Systems that allows a low stress silicon nitride coating to be applied over components on a circuit assembly. Environmental testing of assemblies fabricated with this coating have shown that Silicon Nitride coated assemblies can pass the applicable parts of MIL-STD-883 for reliability. The silicon nitride coating thickness is approximately 0.5 μm.

8.3.5.10 Coating Selection.   A designer has many considerations to be aware of in selecting the optimal conformal coating for a particular application. There is not one universal conformal coating which is right for all applications.

Manufacturers have developed a vast variety of coatings under each of the coating types mentioned above. The QPL for MIL-I-46058 dated January 21, 1992, lists 21 Type

AR, 10 Type ER, 19 Type SR, 34 Type UR, and 5 Type XR coating materials distributed by 21 different manufacturers. These various formulations are developed through the addition of accessory chemicals and solvent vehicles to the basic resin systems to optimize the cured films desired properties. As an example, dyes are added as an aid in identification and fluoresces added to aid inspection for coating uniformity under U V light. Increase film flexibility occurs through the addition of plasticizers and adhesion enhancements occur through the use of wetting agents.

There are a number of important attributes that are desirable when selecting a conformal coating. When uncured, it should have a long pot life to allow adequate manufacturing application procedures, it should be easy to apply, adhere to the PWA securely and be environmentally safe. When cured, it should not excessively shrink and induce damaging mechanical stress on component solder joints. The coating should be a long term effective barrier to moisture and chemical contaminants, provide adequate electrical insulation resistance, not crack under its intended thermal environment and be repairable. Table 8.29 gives a listing of coating selection attributes. Table 8.30 lists important physical characteristics of the various coating materials.

8.3.5.11 Coating Process. Proper preparation, application and curing procedures must be followed to ensure that conformal coatings will perform their intended function.

First, through cleaning of completed PWAs is essential. Conformal coatings once applied and cured will do as good a job of sealing in contaminants and causing later reliability problems as preventing contamination from outside sources if the PWA is improperly cleaned.

The following are typical procedures used after the PWA has undergone the visual, physical and electrical testing to prepare for the coating process:

1. Vapor decrease or ultrasonic clean in a solvent to remove flux residues and other contaminants from the previous processes.

2. Remove ionic salts and other contaminants not soluble in solvents by rinsing in deionized water or alcohol.

3. Dry for two hours at 66°C to remove any last traces of cleaners.

4. Handle with gloves and store in environmentally controlled areas prior to coating.

TABLE 8.29 Conformal Coating Selection Chart

	Acrylic	Urethane	Epoxy	Silicone	Parylene	Polyimide	DAP
Application	A	B	C	C	C	C	C
Removal (chemically)	A	B	—	C	—	—	—
Removal (burn through)	A	B	C	—	C	—	—
Abrasion resistance	C	B	A	B	A	A	B
Mechanical strength	C	B	A	B	A	B	B
Temperature resistance	D	D	D	B	C	A	C
Humidity resistance	B	A	C	B	A	A	A
Potlife	A	B	D	D	—	C	C
Optimum cure	A	B	B	C	A	C	C
Room temperature	A	B	B	C	A	—	—
Elevated temperature	A	B	B	C	—	C	C

Property ratings (A–D) are in descending order, A being optimum.

TABLE 8.30 Conformal Coating Physical Characteristics

Properties	Acrylic	Urethane	Epoxy	Silicone	Parylene	Polyimide	DAP	Silicon nitride
Volume resistivity, $\Omega \cdot$ cm (50% RH, 23°C)	10^{15}	2×10^{11} 11×10^{14}	10^{12} 10^{17}	2×10^{15}	6×10^{16}	10^{16}	1.8×10^{16}	10^{15}–10^{17}
Dielectric strength, V/mil					5600			12,700
Dielectric constant								
60 Hz	3–4	5.4–7.6	3.5–5.0	2.7–3.1	3.15	3.4	3.6	—
1 kHz	2.5–3.5	5.5–7.6	3.5–4.5	—	3.10	3.4	3.6	—
1 MHz	2.2–3.2	4.2–5.1	3.3–4.0	2.6–2.7	2.95	3.4	3.4	5.5–10*
Dissipation factor								
60 Hz	0.02–0.04	0.015–0.048	0.002–0.010	0.007–0.010	0.020	—	0.010	—
1 kHz	0.02–0.04	0.043–0.060	0.002–0.020	—	0.019	0.002	0.009	—
1 MHz	0.10–0.040	0.050–0.070	0.030–0.050	0.001–0.002	0.013	0.005	0.011	—
Thermal conductivity, 10^{-4} cal/(s · cm · °C)	3–6	1.7–7.4	4–5	3.5–7.5	—	—	4–5	—
Thermal expansion, 10^{-5}/°C	5–9	10–20	4.5–6.5	6–9	3.5	4–5		0.3–0.5
Tensile strength, lb/in²					10,000			†
Yield strength, lb/in²					8,000			
Elongation to break, %					200			
Yield elongation, %					2.9			
Resistance to heat, °F continuous	250	250	250	400	—	500	350	>500
Effect of weak acids	None	Slight to dissolve	None	Little or none	None	Resistant	None	None
Effect of weak alkalides	None	Slight to dissolve	None	Little or none	None	Slow attack	None	None

* Process dependent, can be held to ±2 percent with range cited.
† Elastic modulus 0.5 to 2.5 × 10^6 psi @ 1000Å thickness.

The coating process itself is normally accomplished in one of five different ways: spraying, dipping, flow coating, brushing, and vapor deposition.

Spray coating. Spray coat processes are very amenable to automated, high-volume conformal coating applications. The process variables which must be optimized to ensure repeatable results include the material viscosity and solids content, spray pressure, nozzle design (hole sizes as well as patterns within the nozzle), distance of the nozzle from the PWA and spray pattern across the PWA (velocity of the spray nozzle across the PWA, number of passes used and spray angles used during the process). Advance computer controlled machinery are available from manufacturers to accomplish this. In low volume, manual spraying is used.

The main drawback to the spray process is that shadowing due to varying component heights and components leads makes applying a uniform coat difficult. Applying coating under components is very difficult with this process.

Dipping. Dipping is a second method which is used. This process can ensure even uniform coatings across a PWA. To ensure all voids are properly coated, the main process variables that must be controlled are the coating material viscosity, and the immersion and withdrawal speeds of the equipment used. Evaporation of the solvents in the coating bath during this process is an important contributor to changing material viscosities and hence must be monitored. Typical immersion speed are 2 to 12 in/min. Automated equipment for this process is available in the industry.

Masking of areas in which coating material is not wanted makes this part of the process more critical to control than when using the spraying process.

Flow coating. Flow coating, often referred to as curtain coating, applies the material to the PWA by passing it through a "curtain" of flowing material. Material viscosity, curtain flow rate and the velocity of the PWA as it passes through the material are important process variables. Programmable deposition equipment is available for this process. Some equipment can be programmed to vary or even stop the flow when areas which are to be left uncoated pass within the coating curtain window.

Brushing. This operation, which is manual in nature, is normally used only when repair of conformal coatings is required. This can be due to either reworking of components on the PWA or due to improper manufacturing processes that have left small void areas in the coating.

Vapor Deposition. This process is normally done only with the Parylene coatings and silicone nitride.

Curing of most material coating systems is through either letting the coated assemblies remain at room temperature for significant period of time (usually 24 h but sometimes days) or through the cure of the material at an elevated temperature (usually from 15 minutes to 2 h at temperatures of about 170°F).

A recent development in conformal coating technology has been the implementation of UV curable coatings. There are currently 10 different UV curable coatings recognized in MIL-I-46058, 3 Type AR, 3 Type ER, 3 Type UR and 1 Type SR from six different manufacturers. All but one material consist of 100 percent solids.

The UV curing materials offer many advantages. Conventional thermoplastic and thermoset chemistry material properties can be formulated though the use of the UV curable coatings. During the manufacturing process, shorter cure cycles are needed; for the most part the cure occurs in less than 1 min. Little or no solvent emissions occur due to the 100 percent solids content of most of the systems. Reduced space requirements and reduced energy consumption are other advantages achievable.

In the UV curing process, the liquid UV monomer, oligomer and polymer components under go polymerization and cross-linking by photogenerated radical and cationic species which occur during the exposure to a UV light source. The UV light exposure causes almost instantaneous hardening of the material into a cured film. The most critical item which must be optimized in this process is the development of a correct coating formulation matched to a particular UV light source.

There are a few less than optimal features of the UV systems. Shadowing of the UV light source due to component variations and component leads causes a secondary cure to be imposed on most of the UV systems on the market today. This cure ranges from 24 hours at room temperature to as little as 10 to 15 minutes but at temperatures elevated to 250–300°F. The materials are also difficult to repair and rework.

8.4 DESIGN CONSIDERATIONS

Now that the different types of printed-wiring techniques and their materials have been discussed, the printed-wiring designer has a large number of other factors to consider in order to develop a board design that is optimized for economic fabrication and reliable end usage. This section will not delve into the trade-offs to be made in choosing the type of board to be fabricated. It will concentrate on PWB mechanical layout design considerations with some general electrical considerations. Electrical considerations for high-speed applications are covered in depth in Chap. 11. The scope of this section will cover both organic rigid applications as well as flexible and rigid-flex PWB design considerations.

8.4.1 Connectivity

One of the first tasks a designer must complete is setting up the PWB interconnection system. This system consists of four parts (Fig. 8.25): PTH (via) grid, channel space (space remaining between pads associated with the PTHs), wiring channels (number of wire runs within each PTH grid channel space), and number of wiring layers. As more and more functions are combined into single devices, device I/O counts are increasing. In the 1970s,

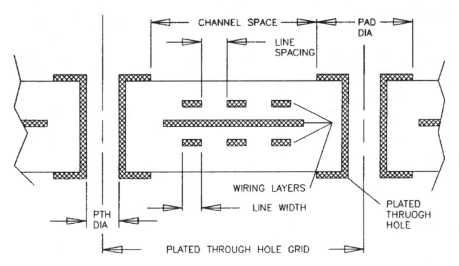

FIGURE 8.25 Elements of PWB grid system.

the most common devices were packaged in 14-, 16-, and 20-I/O dual-in-line packages (DIPs) with lead spacings on 0.100-in pitch off two sides of the package. With the advent of programmable logic and gate arrays in the 1980s, SMT packaging evolved with package I/O counts growing to 100 to 300 I/Os with peripheral lead pitches off four sides of the component as low as 0.020 in. Packaging trends in the 1990s are towards using single chip ASIC and/or multichip packaging. These are requiring packages with 700 I/Os with a switch from peripheral leads to pin grid array (PGA) or ball grid array (BGA) packaging on I/O down to 0.050-in pitch.

While the component I/O requirements have grown, this has caused a corresponding increase in the interconnection PTH and circuit wiring density requirements in the PWB. PTH density requirements in the 1970s for DIP technology were 60 to 70 PTH/in^2 Early SMT PTH density with 14 to 20 I/O devices increased to 80 to 100 PTH/in^2. The 1980s saw large I/O SMT gate-array components, further increasing the density to 120 to 150 PTH/in^2. The 1990s are beginning to have board designers consider PTH density requirements of up to 400 PTH/in^2.

The increase in component I/Os caused a corresponding increase in the circuit wiring density required to interconnect a given PWB. Multilayer boards became more prevalent; their layer count increased from four layers common in the early 1970s to 10 to 50 layers in today's high-speed computer and processor systems. Not only did the layer count increase, but so did the wiring density per layer through the use of finer conductor widths and spaces.

The circuit wiring density for a given PWB is often referred to in terms of inches of actual routed wiring per square inch of board surface area. A simple equation based on Rent's rule for connectivity was developed to estimate the total wiring per square inch required to interconnect a given number of I/O pins per square inch in a PWB. This is shown by the equation

$$L = 2.25N/2 \qquad (8.1)$$

where L is the total length of wiring per square inch of PWB and N is the total number of I/Os per square inch.

For a given PWB, the circuit wiring capacity per square inch of PWB area is determined by the PTH grid, the number of wiring channels between each set of PTHs, and the number of layers in the PWB. Due to the fact that in an actual design 100 percent utilization of the wiring capacity in a PWB is rarely achieved, a utilization factor must be applied to Eq. (8.1) (by division) to estimate the actual circuit capacity required for a given design. Utilization factors are difficult to determine and will vary for a number of reasons. Component placement, system architecture, component I/O pin arrangements, and electrical considerations such as crosstalk all affect the utilization factor. Different areas of a given PWB will have vastly different utilization factors. In designs such as computer applications, where large numbers of data buses are routed between components whose pins are easily aligned to allow straight wiring runs between the components, utilization factors of 70 percent or higher may be achieved. In random logic interconnection architectures, utilization factors below 50 percent are not uncommon.

Free vias are an important factor in increasing the utilization factor in a PWB. A free via is a PTH in a PWB not assigned to a component signal, power, or ground pin. There are three free via types—through-hole vias, buried vias, and blind vias. A buried via is a PTH connecting two or more internal layers in a PWB which does not go to the board surface (see Fig. 8.1c). A blind via is used to connect the board surface to an internal layer without going all the way through the PWB (see Fig. 8.1c). In all instances, free vias are used to route a given wiring run on the PWB from layer to layer. This jumping between layers of a given signal conductor through vias is used by a designer to route around blockages to a given routing path caused by other wiring runs. The use of buried and blind vias has been shown to decrease the required layer count in a multilayer PWB by 20 to 30 percent. Their use however does drive up the cost of the board over conventional multi-layer boards. The primary factors that determine the PTH grid and the number of wiring channels between PTH are discussed in the next sections.

8.4.2 PTH Grid

In through-hole component applications (where the lead of a component is placed into a hole in the board and then soldered to the hole) the component lead spacing for the most part determines the PTH grid. Due to the fact that IC devices for through-hole applications are packaged in DIPs, in most instances this will be a 0.100-in grid.

In SMT applications, the component lead is not soldered to a PTH, but to a land on the board surface. This land is interconnected to PWB vias and the designer has a wider lati-tude in choosing the grid. The choice of SMT PTH grid is influenced primarily by compo-nent density and wiring channel requirements. Designs having many components tightly packaged together may require that all PTHs associated with a given component be tightly bunched around the component periphery or under the component. The PTH grid must also be integrated with the lead solder land design to ensure that all solder lands can be connected to an interconnection PTH. Requirements such as testability, both bare board and functional testing, need to be considered. The SMT PTH grid systems used primarily fall between 0.050 and 0.100 in, with 0.050, 0.070, 0.075, and 0.100 in being the most common.

A common mistake made by novice designers is the failure to use a common grid sys-tem throughout a given PWB, especially when CAD automated routers are used. Place-ment of component interconnection patterns off grid or mixing different grid systems on the same PWB makes routing of the PWB more difficult and may even cause a need for extra layers to complete the route.

In conjunction with choosing a grid system, the PTH diameters and PTH land sizes to be used are determined. These are influenced by component lead type (through hole only) and the wiring channel requirements. Each of these has requirements that must be met to ensure a design that is producible and will operate as intended in its service environment. The following sections discuss these trade-offs and design requirements.

8.4.2.1 PTH Diameter. In a through-hole design, the finished PTH diameter chosen is a function of the component lead diameter or lead diagonal width for square or rectangular cross-sectioned leads. For an unsupported hole, the hole diameter usually is not less than 0.006 in greater than the lead diameter (diagonal width) and not more than 0.020 in greater than the nominal lead diameter (diagonal width). For a PTH the hole size is usually chosen to be no less than 0.006 in in diameter larger than the maximum lead diameter and no more than 0.032 in in diameter larger than the minimum lead diameter.

In an SMT design the designer has a wider latitude in the choice of PTH diameter. This diameter can be more readily traded off with a variety of PTH grids and wiring channel requirements. A more important trade-off usually made is the aspect ratio of the PTH (aspect ratio is the ratio of PWB thickness to PTH hole diameter). The aspect ratio plays an important part in the ability of the manufacturer to plate the barrel of the PTH. It also is one of the leading PTH reliability drivers when controlled-expansion laminate materials (aramid fibers and CIC planes) are used. The greater the aspect ratio, the more difficult it is to plate the PTH and the less reliable the hole will be in extreme thermal environmental conditions. The smaller the hole diameter, the more difficult hole drilling and cleaning also become. In general, SMT designs usually use drilled hole diameters ranging from 0.0135 to 0.025 in in diameter. Aspect ratios smaller than 5:1 are commonly used throughout the industry. Many multilayer fabricators for backplane applications are producing boards with PTH aspect ratios of 10:1.

8.4.2.2 PTH Land Size.   Land sizes associated with PTHs are a function of the manufacturer's capability of meeting minimum annular ring requirements and the number of wiring channels between the PTHs. The annular ring is the amount of copper between the edge of the finished pad and the edge of the hole (Fig. 8.26). Annular ring requirements for military applications are listed in Table 8.31. The size of the PWB, the number of wiring layers, the dimensional stability of the laminate material during processing, the accuracy of the phototool setup, and the manufacturer's processes for aligning the phototools during etching of the artwork features and laminating of the layers for multilayer applications all determine whether annular ring requirements will be met. In general, the most producible designs will maintain a land diameter at least 0.020 to 0.025 in in diameter greater than the drilled hole diameter. For buried or blind via applications, where adjacent layer pairs are being interconnected and tolerances due to movement during lamination need not be considered, land diameters 0.012 to 0.015 in in diameter greater than the drilled hole size may be used. Designers using a minimal land diameter will often bump the land (see Fig. 8.26)

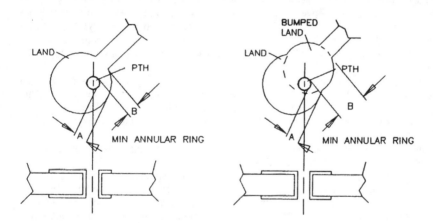

FIGURE 8.26   PTH land annular ring requirements. *(From MIL-STD-275E.)*

TABLE 8.31   PTH Land Annular Ring Requirements (in)

	*A**	*B**
Unsupported holes	0.015	0.015
External PTH	0.002	0.005
Internal PTH	0.002	0.005

*A—measured at any point around land circumference other than where conductor enters land; *B*—measured at point where conductor enters land (see Fig. 8.21).
Source: From MIL-STD-275E.

to help meet the annular ring B dimension. This bumped land is a second land offset from the PTH land in the direction that a conductor enters the PTH land.

The land diameter is invariably influenced to the greatest extent by the number of wiring channels required between PTHs. The wiring channel considerations are discussed in the next section.

8.4.3 Wiring Channel

The number of wiring channels used (see Fig. 8.25) between PTHs is primarily a function of the allowable wiring density within each wiring layer and the number of wiring layers allowed. Wiring channel selection requires trade-offs in PTH size and associated pad diameter. It can further be influenced by electrical limitations due to characteristic impedance and crosstalk requirements in high-speed designs. The PWB manufacturer's capability of producing a board of a given line width and spacing, the PTH drilled diameter to pad diameter ratio, and the layer count also have to be taken into consideration.

The choice of line width can be influenced by the current-carrying capability or electrical signal requirements, such as a controlled characteristic impedance. Minimal conductor spacing requirements are based on two factors—the voltage used and the level of electrical signal distortion that can be tolerated due to factors such as crosstalk. Table 8.32 gives the minimal spacing allowed between adjacent conductors due to differences in voltage. Line width and spacing limitations for military products are limited by MIL-STD-275E to a minimum of 0.004 in line widths and 0.004 in conductor-to-conductor spacings as long as the voltage requirements in Table 8.32 and the current-carrying capabilities of Fig. 8.20 are met.

The number of wiring channels required to achieve a given interconnect density also influences the PTH land sizes used. Greater wiring capacity requirements force the designer to add more wiring channels. The addition of wiring channels often will cause smaller land sizes to be used in order to maintain proper electrical clearances conductor to conductor and conductor to land.

In most designs, once the PTH grid system is chosen, the number of wiring channels between PTHs will usually be 1, 2, or 3. This is due to the fact that most PTH grid systems fall within a spacing of 0.050 to 0.100 in. Line width and spacing requirements coupled with PTH pad diameter requirements are usually the factors that limit this to three channels. When more than three channels are used, the lack of free vias in a design begins to bring the utilization factor down, drives the layer count up, and makes this a less attractive design choice.

TABLE 8.32 Minimum Conductor Spacing Requirements

Voltage between conductors, V dc or ac peak	Surface layers*	Internal layers
0–100	0.005 in	0.004 in
101–300	0.015 in	0.008 in
301–500	0.030 in	0.010 in
>500	0.00012 in/V	0.0001 in/V

*Assembly is assumed conformally coated.
Source: From MIL-STD-275E.

8.4.4 Structural Core Materials in PWAs

In many printed wiring assembly designs, in particular SMT applications, the finished PWB is mounted to another material usually for one of three reasons:

1. The second material functions as a structural stiffener to provide sufficient rigidity to the PWB to survive environmental shock and vibration inputs.
2. The material is used to reduce the CTE of the PWB to match the CTE of components such as ceramic leadless chip carriers. This in turn reduces the stresses imposed on solder joints during environmental thermal cycling the PWA may experience during its functioning lifetime.
3. The material is used to conduct heat being dissipated by the electrical components away from the PWA and into a heat exchanger to provide component reliability.

In many instances particularly in aerospace and military applications, this material performs all three functions.

The following paragraphs discuss the variety of material options available to the PWA designer today, compares their material properties, provides manufacturing incites and presents lists of applications which may require a particular material use.

Table 8.33 presents basic material properties of the constitutive materials which either by themselves or in combination constitute the PWA cores in use in today's electronics. Table 8.34 lists the properties of the core materials which are a combination of the basis materials given in Table 8.33.

Aluminum.   Aluminum materials have been used for many years in both military and commercial application for both SMT and Through Hole applications. Aluminum material properties are the baseline properties usually used to compare the advantages and disadvantages of new material products.

Two series of aluminums are predominantly used, 1100 usually H14 and 6061 usually T6 although 6063 T6 grades with their higher thermal conductivities are being considered. The 1100 series has a slightly higher thermal conductivity than the 6061 series. The 6061 series has a higher structural modulus. The 1100 series is most easily manufactured by punching. The 6061 series is easily machinable and both can be cast. Aluminum materials are the least expensive from a raw material cost when compared to all other materials presented in this chapter. The 1100 series predominates in through hole applications while the 6061 series predominates in SMT applications.

Aluminums have a relatively low density and are used in many applications where weight is critical. Their structural modulus provides a relatively stiff assembly though the thickness of the material must be increased to provide very high structural natural frequencies. The increased thickness tends to drive the assembly weight in very critical weight application to unacceptably high levels. This drives the designer to use materials with greater stiffness to weight ratios.

Aluminum has a very high CTE, 24×10^{-6} ppm/°C, when compared to the CTE of Alumina components, 6×10^{-6} ppm/°C. PWBs bonded to aluminum structures must use one of two methods to prevent solder joint failures during environmental thermal cycling. The first method is to use components with leads of sufficient structural compliance to reduce solder joint strain. In the second method, the PWB would be soft bonded to the aluminum using a silicone or gum rubber adhesive of sufficient thickness to uncouple the PWB structurally from the aluminum.

Copper. Copper may be used in applications requiring removal of high wattage from a PWA. It has a thermal conductivity 1.8 times that of aluminum. Its biggest drawback is its weight density, being 3.3 times that of aluminum.

TABLE 8.33 Typical Mechanical Properties of PWA Core Constitutive Material

Material	Density, g/cm³	CTE, 10⁻⁶ ppm/°C	Thermal conductivity, W/(m·°C)	Specific conductivity [W/(m·°C)]/(g/cm³)	Young's modulus, 10⁶ lb/in²	Yield strength, 10³ lb/in²	Ultimate strength, 10³ lb/in²	Elongation, %
Aluminum 1100H14			218		10.0			
Aluminum 6063T6	2.68	23.4	201	74.86	10.0	31.0	35.00	
Aluminum 6061T6	2.72	23.6	156	57.51	10.0	40.0	45.00	17
Copper	8.95	16.7	390	43.58		30.0		
Invar	8.03	1.7	12	1.49	20.0			
Molybdenum	10.28	4.5	142	13.81	47.0			
Beryllium S-200F	1.85	11.4	216	116.76	44.0	35.0	47.00	2
Beryllium oxide	3.06	6.4	250	81.70	56.0			
Graphite IG-610	1.85	4.5	128	69.02	1.7	4.5		
Graphite P100	2.20	-1.6	520	236.36				
Graphite P120	2.10	-1.6	640	304.76	120.0			
Graphite K1100	2.15	-1.6	1050	488.37	130.0	400.0		

TABLE 8.34 Typical Mechanical Properties of PWA Core Final Material

Material	Density, g/cm³	CTE, 10⁻⁶ ppm/°C		Thermal conductivity, W/(m·°C)		Specific conductivity [W/(m·°C)]/(g/cm³)	Young's modulus, 10⁶ lb/in²	Yield strength, 10³ lb/in²	Ultimate strength, 10³ lb/in²	Elongation, %
Aluminum 6061T6	2.72		23.6		156	57.35	10.0	40.0	45	17
Cu/Invar/Cu, 12.5/75/12.5	8.34	x–y	45.–4.7	x–y / z	100 / 19	11.99	20.3	55–70		
Cu/Invar/Cu, 20/60/20	8.45	x–y	5.1–5.3	x–y / z	167 / 24	19.76	19.6	45–60		
Cu/moly/Cu, 5/90/5	10.08	x–y	5.1	x–y / z	96	9.52	44.0	100.0		
Cu/moly/Cu, 13/74/13	9.89	x–y	5.7	x–y / z	120	12.13	39.0	87.0		
Cu/moly/Cu, 20/60/20	9.70	x–y	6.5	x–y / z	140	14.44	35.0	75.0		
Cu/moly/Cu, 25/50/25	9.56	x–y	7.9	x–y / z	155	16.22	32.0	62.0		
Moly/graphite/moly, 10/80/10	3.54	x–y	4.8	x–y / z	130	36.77	—	>2.0		
Beryllium S-200F	1.85		11.4		216	116.76	44.0	35.0	47	2
Be-BeO, E20	2.07		8.7		205	110.14	44.0	44.0	40	0.2
Be-BeO, E40	2.29		7.5		215	101.31	46.0	46.0	35	0.1
Be-BeO, E60	2.55		6.1		220	94.12	48.0	48.0	30	0.05
Beryllium-aluminum, AlBeMet 140	2.26		16.3		200	88.50	20.0	30.0	40	15
Beryllium-aluminum, AlBeMet 162	2.10		13.9		210	100.00	26.0	45.0	60	5
LECMCX-693 SiCp/Al	3.00		6.6		180	—	34.0	—	35	—

Copper is easy to machine. To prevent corrosion of the copper, protective coatings of nickel plate are usually used. Uses for this have been in high watt density through hole applications and surface mount applications where weight is not a consideration.

Clad Metal and Graphite Materials. Two clad metal materials predominate, copper-invar-copper and copper-molybdenum-copper. Both metal systems use outer copper surfaces clad to either invar or molybdenum on the interior. The thickness ratio of the invar or molybdenum to copper controls the CTE of the clad metal. Both of these materials are used when SMT applications require constraining materials to lower the CTE of the PWB to match that of the alumina chip carrier components.

Detailed explanation of CIC and CMC foils used within the PWB is given in Section 8.3.1.4. While some designs use internal PWB CIC or CMC foils, other designers have hard bonded completed PWB structures to either CIC or CMC structural members. The choice of which method to use is normally decided on two factors. To use thin clad metal cores integral to the PWB, the manufacturer must have a PWB facility with the technological expertise in PWB processing with these materials. PWBs using clad metal internal foils require expertise in etching, plating, laminating and drilling which vary from standard FR-4/copper clad processes. The second factor is weight. Both materials have very high weight densities which limit their use in many applications. A lighter weight assembly could be made with thin clad metal inner cores bonded to a light weight structure rather than a thin conventional PWB bonded to a thick CIC or CMC structural core.

In comparing these two clad metal systems, CIC has seen the most industry usage particularly in military applications. CIC is about 21 percent lighter, slightly less expensive and is easier to machine than CMC. CMC has about double the strength of CIC and is better able to resist warpage and remains relatively flat and stable compared to CIC. CMC is slightly more expensive than CIC.

Texas Instruments and Metalwerk Plansee of Austria have recently developed a new clad heat-sink core material consisting of 0.010 in molybdenum sheets bonded by a high-temperature vacuum brazing process to a lightweight central graphite plate. This molybdenum-graphite-molybdenum (MGM) composite when compared to conventional CIC or CMC clad metals offers a material that is less than 50 percent the density (although it is still 16 percent greater than aluminum), has a lower CTE and has better structural damping characteristics.

Existing plate sizes are limited to 12 by 18 in due to brazing furnace limitations. It is being produced in thicknesses ranging from 0.040 to 0.125 in with a tolerance of ±0.003 in. Flatness can be held to 0.002 in/in with a 5 μm surface finish.

Finished products can be machined using carbide drilling tools for hole drilling and profiling using carbide end mills or with a Nd:Yag laser.

Beryllium (Be). For applications where weight is extremely critical and high structural rigidity is required, beryllium is the material of choice based on technical merit. Pure beryllium is 68 percent the density of aluminum, has a structural modulus 4.4 times greater than aluminum, has a CTE 47 percent lower than aluminum and a thermal conductivity 23 percent greater than aluminum.

Beryllium has 3 major drawbacks; lack of multiple major suppliers, extremely high cost and potential health hazards. Brush Wellman, Inc., is the major supplier of beryllium with proven ore reserves of 73 years using 1992 consumption rates. Its high raw material cost and difficulty in machining, joining and forming make beryllium structures very expensive. It should be noted that 2 to 6 percent of the worlds population can develop allergic reactions to inhaled beryllium particles. This is known as berylliosis. Three factors are required for the person to contract the disease; they must be allergic, they must inhale particle sizes finer than 10 μm in diameter and they must be exposed to concentrations of >2 $\mu m/m^3$ over an eight-hour period. For this reason, manufacturing operations using beryllium based materials are carried out in well ventilated facilities with strict OSH policies regarding the disposal of this material. Solid pieces of beryllium are not a health hazard and can be handled without extreme safety precautions.

Be-BeO Composites.   One of the latest developments in electronic composite materials are metal-matrix-composites (MMCs) consisting of single crystal beryllium oxide (BeO) particles dispersed in a metallic beryllium matrix. The Be-BeO composites exploit the high thermal conductivity and chemical stability of BeO and the light weight and high modulus of beryllium. The resultant composites exhibit a high modulus, good thermal conductivity, low density and a low CTE relative to aluminum.

The composites are manufactured using powder metallurgy techniques. Machining of the composite material into a final product is difficult though advances in ultrasonics, CBN and diamond tooling are being used today to produce end item products from powder metallurgy blanks. Dry/cold pressing procedures to produce near-net shape and net shape products are being developed. Under development are also injection molding processes. Plating, alodining and anodizing can be used to prepare the surface for further processing or as a protective measure. Three forms are in use today depending on the specific material properties needed; E20, E40 and E60. The E designation refers to "Electronic Grade" products and the number refers to the percent loading of BeO in the product; 20 being 20 percent BeO. Under development are heat sink products for SMT applications using these alloys. Most application for this material revolve around housings for components such as transmit-receive modules for advance radar applications.

Aluminum-Beryllium Alloys.   An aluminum-beryllium alloy called Lockalloy was developed by Lockheed Aircraft Corp. some 30 years ago as a lightweight, structural material used in the folding Ventral fin of the SR-71 Blackbird aircraft. Recently, Brush Wellman has developed a new group of aluminum-beryllium alloys based on the Lockalloy process and has trademarked these products as AlBeMet.

The AlBeMet products are made using refinements to the basic powder metallurgical processing techniques used to produce Lockalloy. Current products are made from atomized powder consolidated by extrusion or hot isostatic press (HIP). AlBeMet consists of finely dispersed beryllium particles in an aluminum matrix. The beryllium particles act as reinforcing agents. The alloys developed have the stiffness of steel with a density 25 percent less than aluminum.

AlBeMet product designations are AlBeMet-XXX, with the first "X" referring to the aluminum matrix "1" being pure 1100 aluminum, a "2" would refer to 2024 aluminum; etc. The second "X" designates the minimum percentage weight of beryllium in the alloy; 6 being 60 percent. The third "X" designates alloys within a class.

A wide variety of manufacturing techniques can be used to produce end item products with AlBeMet materials. Health and safety procedures associated with beryllium-containing materials must be observed. The material machines much like magnesium with fully annealed material easier to machine than as-extruded stock. It does not require an etching treatment after machining like beryllium requires to remove machine marks and damage. Electron beam and TIG welding techniques can be used and the material can be welded to some aluminum grades. AlBeMet can be brazed and epoxy bonding can be used. Platings such as alodine, nickel and paints adhere well. Cost of the raw material is still very expensive compared to aluminum. Heat sinks for SMT applications are under development using this material as well as structural housings and covers for small electronic modules.

Aluminum Silicon-Carbide Composites.   Another composite material developed by LANXIDE Corporation for heat sink application is an Aluminum Silicon Carbide metal matrix composite. These were developed to take the place of CIC and CMC metal composites. By tailoring the ratio of silicon carbide (SiCp) to aluminum an isotropic composite with many of the characteristics of aluminum, light weight, good thermal and electrical conductivity is formed. In addition, a lower CTE from 6 ppm/C to 18 ppm/C is achieved as well as a higher Specific stiffness. The lower CTE is desirable to match the CTE of alumina chip carriers soldered to PWBs while the higher stiffness results in desirable higher resonant module frequencies under vibration.

Manufacturing techniques used to make this product include aluminum infiltration of tape cast and green machined SiCp pressed blocks, die casting investment casting, and cold injection molding. The material is difficult to machine due to the SiCp.

Graphite Matrix Composites. These composites using high-strength graphite fibers impregnated into a metal or plastic matrix have been under development for a number of years. A number of the major manufacturers; DWA, Sparta Inc., Americom, Courtalds Inc., AMOCO and Advanced Composite Products to name a few; have been involved in a number of research projects to develop these composites for electronics applications. The main thrust areas for this development revolve around reducing the weight and CTE of the resultant composites while increasing the thermal conductivity. For PWA applications, the most work has been done in developing thermal planes for SMT applications.

A number of high-strength pitch graphite fibers have been used (P75, P100, P120 and K1100X). The graphite fibers are produced in sheets with the fiber oriented in one direction (X). Along the X axis, the sheet exhibits a high value of thermal conductivity and excellent XZ rigidity. Thermal conductivity in the Y and Z directions is relatively poor as is the YZ rigidity.

The two most prevalent metals in use today as the base material are aluminum and copper. Aluminum serves as the base material for structures requiring light weight. Copper serves as the base material for structures requiring superior thermal conductivity.

Epoxies have served for the most part as the plastic base matrix. The graphite-plastic plates have been shown to provide a structure 30 percent lighter in weight than aluminum, provide a structure with much superior rigidity compared to aluminum, provide an in-plane CTE nearly matching ceramic and provide superior in-plane thermal conductivity.

These composite products are made by laying up alternating sheets of the base material (aluminum or copper or plastic) and graphite. Varying the percentage loading of graphite to base material and the ply orientation of the graphite sheets allow a wide variety of composite properties to be tailored to a particular application. For example, to maintain near isotropic in-plane properties, sheets of graphite fiber need to be laid-up or plied in alternating *x-y* orientations. Most products under development today use fiber ply orientations varying between 15° and 90° and percentage loading of graphite to base material by volume of from 20 to 60 percent. Isostatic pressure and heating of the laid up stack of materials fuses the graphite and base material together into the composite structure.

For the most part plates of the composite material are easily machinable. It should be noted however that machining exposes the graphite fibers which must then be protected to prevent corrosion of the composite. The most common means of protection for graphite-aluminum composites is by ion vapor deposition of 0.002 to 0.003 in of aluminum and for graphite-copper or graphite-plastic composites is 0.001 in of nickel plating.

It should be noted that graphite-plastic structures are susceptible to cracking of the plastic base material due to the extreme mismatch of the graphite fiber CTE and the plastic material CTE. Care needs to be taken in choosing plastic materials that are somewhat compliant. Brittle polyimides would be more susceptible to the cracking than many epoxies. Care needs to be taken in developing the composite manufacturing process and in particular the heat-up/cool-down cycle. Cracking may be induced during the initial processing by a non-optimal heat-up/cool-down cycle through the development of high stresses in the plastic. Subsequent thermal cycling of the product has shown that the initial cracks will grow in length and in extreme cases, cause fracture of the graphite-fibers within the structure. This could lead to a degradation in the initial CTE, thermal conductivity or structural moduli properties.

8.4.5 Flexible and Rigid-Flexible Design Considerations

Flexible and rigid-flexible circuits have certain design rules that are different from those of their rigid organic cousins. A number of these are given in this section. For more detailed information, the reader is referred to Ringling[2] and Gurley.[3]

1. The flexible film materials are dimensionally less stable than the rigid organic materials. Consequently, artwork used in the fabrication of the flexible interconnect is compensated to a greater extent to allow for shrinkage of the base material after etching of the copper. In addition, the designer usually tries to leave as much copper as possible and to distribute the copper evenly over the surface of the flexible base film.

2. The designer is working in a three-dimensional form and as such usually uses mockups of the interconnection system or detailed three-dimensional CAD layouts to help determine the critical length and bend dimension.

3. Applications for flexible circuits include both static and dynamic cases. Dynamic applications require that high-ductility rolled copper be used, the copper be at the neutral axis of the flex bend, and generous bend radii be used. Single-sided circuits should have a minimum bend radius of three times the total thickness, double-sided circuits a minimum of 10 times the total thickness.

4. All traces that cross a bend radius should cross at 90° to the bend for longest service life.

5. Internal corners should include generous bend radii of at least 0.030 in to limit tearing of the dielectric film. Tear guards consisting of a copper conductor 0.015 to 0.030 in wide should be used at least 0.030 in from the circuit edge.

6. Large lands should be used at interconnection holes and the cover coat overlap the land edges a minimum of 0.005 in to provide for added resistance for pad lifting during subsequent soldering operations. When this is not possible, tie-down tabs should be added to the pad opposite the side where the conductor enters to provide this capability.

7. When multiple layers of flex are bent, the layers to the outside of the bend radius should be designed with extra length to prevent buckling of the internal layers. This extra designed-in length is called a *bookbinder effect.*

REFERENCES

1. M. B. Miller, *Dictionary of Electronic Packaging, Microelectronic, and Interconnection Terms,* Technology Seminars, Lutherville, Md., 1990.

2. W. Ringling, *Rigid-Flex Printed Wiring Design for Production Readiness,* Marcel Dekker, New York, 1988.

3. S. Gurley, *Flexible Circuits, Design and Application,* Marcel Dekker, New York, 1984.

4. M. Nadel, "A Novel Approach to Manufacturing Printed Circuit Boards," in *Proc. IEPS Tech. Conf.* (1989).

5. F. Brewster and D. Fried, "Resistless Imaging of Photoselective Plating of Circuitry on Molded Plastic Parts," in *Proc. IEPS Tech. Conf.* (1989).

6. M. Precht et al., *Electronic Packaging Design Handbook,* Marcel Dekker, New York, 1990.

7. M. Fitzgibbon et al., "Multiwire Interconnection Boards," Kollmorgen Corp., unpublished.

8. T. Buck and M. Motazedi, "Microwire Interconnection Technology," PCK Technol. Div., Kollmorgen Corp., July 1989.

9. T. Senese, "N-8000/s-Glass for CTE Matching to Leadless Ceramic Chip Carriers, NELCO Laminates," presented at the Smart VI Conf., Jan. 16, 1990.

10. L. Zakraysek, "Rupture Testing for the Quality Control of Electrodeposited Copper Interconnections in High-Speed, High-Density Circuits," publication unknown.

11. C. F. Coombs, *Printed Circuits Handbook,* 3d ed., McGraw-Hill, New York, 1988.

12. C. J. Tautscher, *Contamination Effects on Electronic Products,* Marcel Dekker, New York, 1991.

13. D. P. Seraphim, R. C. Lasky, and C. Li, *Principles of Electronic Packaging,* McGraw-hill, New York, 1989.

14. Planar Resistor Technology, TYCO Technical Times, January, 1996, John Vesce editor, North American Printed Circuits, Stafford, Ct.

15. C. A. Harper, Electronic Materials and Processes Handbook, McGraw Hill, 1993.

16. C. L. Guiles, Everything You Always Wanted To Know About Laminates, Arlon Materials For Electronics, Fifth Edition, January, 1993.

17. R. L. Kubacki, Low Temperature Plasma Deposition of Silicon Nitride to Produce Ultra-Reliable, High Performance, Low Cost Sealed Chip-On-Board (SCOB)Assemblies, IEEE International Electronics Manufacturing Technology Symposium, Sept. 12–14, 1994, La Jolla, California.

INDUSTRY SPECIFICATIONS

ASTM-D-149 Test Method For Dielectric Breakdown Voltage and Dielectric Strength of Solid Electrical Insulating Materials at Commercial Power Frequencies

ASTM-D-150 Test Methods for AC Loss Characteristic and Permittivity Dielectric Constant of Solid Electrical Insulating Materials

ASTM-D-696 Test Method for Coefficient of Linear Thermal Expansion of Plastics

ASTM-D-790 Test Method For Flexural Properties of Unreinforced and Reinforced Plastics and Electrical Insulating Materials

IPC-T-50 Terms and Definitions for Interconnecting and Packaging Electronic Circuits

IPC-L-108 Specification for Thin Metal Clad Base Materials for Multilayer Printed Boards

IPC-L-109 Specification for Resin Preimpregnated Fabric (Prepreg) for Multilayer Printed Boards

IPC-L-112 Standard for Foil Clad, Composite Laminate

IPC-L-115 Specification for Rigid Metal-Clad Base Materials for Printed Boards

IPC-L-125 Specification for Plastic Substrates, Clad or Unclad, for High Speed/High-Frequency Interconnections

IPC-EG-140 Specification for Finished Fabric Woven From "E" Glass For Printed Boards

IPC-SG-141 Specification for Finished Fabric Woven From "S" Glass For Printed Boards

IPC-A-142 Specification for Finished Fabric Woven From Aramid For Printed Boards

IPC-QF-143 Specification for Finished Fabric Woven From Quartz (Pure Fused Silica) For Printed Boards

IPC-CF-148 Resin Coated Metal For Printed Boards

IPC-MF-150 Metal Foil For Printed Wiring Applications

IPC-CF-152 Composite Metallic Material Specification for Printed Wiring Boards

IPC-FC-231 Flexible Bare Dielectrics For Use in Flexible Printed Wiring.

IPC-FC-232 Specifications For Adhesive Coated Dielectric Films For Use As Adhesive Coated Cover Sheets For Flexible Printed Wiring.

IPC-FC-241 Metal-Clad Flexible Dielectrics For Use In Flexible Printed Wiring.

IPC-RF-245 Performance Specification For Rigid-Flex Printed Boards

IPC-D-249 Design Standard For Flexible Single- And Double-Sided Flexible Printed Wiring

IPC-FC-250 Specification for Single- And Double-Sided Flexible Printed Wiring.

IPC-D-275 Design Standard For Rigid Printed Boards And Rigid Printed Board Assemblies

IPC-RB-276 Qualification And Performance Specification For Rigid Printed Boards

IPC-D-300G Printed Board Dimensions and Tolerances

IPC-HF-318 Microwave End Product Board Inspection and Test

IPC-D-322 Guidelines For Selecting Printed Wiring Board Sizes Using Standard Panel Sizes

IPC-MC-324 Performance Specification For Metal Core Boards

IPC-D-325 Documentation Requirements For Printed Boards

IPC-TM-650 Test Methods Manual

IPC-CC-830 Qualification and Performance of Electrical Insulating Compound For Printed Board Assemblies

IPC-SM-840 Qualification and Performance of Permanent Polymer Coating (Solder Mask) For Printed Boards

UL796 Safety Standard For Printed Wiring Boards

MILITARY SPECIFICATIONS

MIL-STD-275 Design Standard For Rigid Printed Boards and Rigid Printed Board Assemblies (This document has been cancelled.)

MIL-S-13949 General Specification For Plastic Sheet, Laminated, Metal-Clad (For Printed Wiring)

MIL-STD-2000 Soldering Requirements for Soldered Electrical and Electronic Assemblies

MIL-I-46058 Insulating Compound, Electrical (For Coating Printed circuit Assemblies)

MIL-P-50884 Printed Wiring, Flexible

MIL-P-55110 General Specification For Printed Wiring Boards

CHAPTER 9

SURFACE MOUNT TECHNOLOGY

J. K. "Kirk" Bonner

Jet Propulsion Laboratory
California Institute of Technology
Pasadena, California

9.1 INTRODUCTION

Surface mount technology (SMT) is that packaging technology where components are mounted directly on the substrate surface rather than inserted into plated through holes (PTHs). This latter technology is an older one and is referred to as through hole technology (THT), and components whose leads are inserted into PTHs are referred to as through hole components (THCs). The trend is for new designs to incorporate either pure SMT or a combination of SMT/THT (so-called mixed technology). (See Fig. 9.1.)

SMT has been an important packaging technique for producing a wide variety of printed wiring assemblies (PWAs). It is used because of the numerous packaging benefits that it provides. Among these are:

- Increased functionality per substrate area
- Feasibility of using high input/output (I/O) count integrated circuit (IC) packages
- Very suitable for automated techniques
- Ideal for overall system size reduction
- Increased signal speed and enhanced electrical performance of circuitry
- Potentially higher quality due to consistent, repeatable processes

9.1.1 The Origins of Surface Mount Technology

Surface mount technology springs chiefly from two sources: (1) microelectronics thick film and thin film technology and (2) the flat pack technology of the mid-1960s. These two technologies fostered the placement of components by directly mounting the leads on the substrate surface. In the traditional through hole technology, which was prevalent for

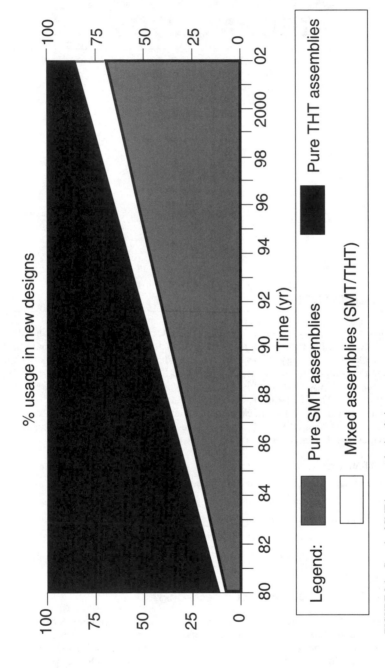

FIGURE 9.1 Growth of SMT in new packaging designs.

organic substrates, the component leads are inserted through plated through holes (PTHs). The metallic interconnects are then formed using a mass soldering operation such as wave soldering or dip soldering. In the case of SMT, mass soldering is generally accomplished using either forced convection heating or radiant energy or a combination of the two.

9.2   CONCURRENT ENGINEERING

This chapter should be used from a concurrent engineering point of view. The term *concurrent engineering* is used to express the integration of the different activities making up the product development and build cycles. To implement concurrent engineering, a team approach is required.

There are a number of critical issues to be addressed at the design/packaging stage that cannot be taken up after the fact without either seriously compromising the product goals or costing more than was anticipated. Simply put, concurrent engineering means "do not work in a vacuum." All relevant engineering disciplines should be brought into play by forming an SMT work team at the conceptual design stage before any of the critical design decisions have been made.

The relevant disciplines to be included in a concurrent engineering work team approach are:

- Logic/circuit
- Packaging
- Manufacturing (sometimes known as production)
- Test
- Quality assurance
- Reliability
- Safety/environmental
- Maintenance/facilities
- Systems
- Procurement

The concurrent engineering concept is illustrated in Fig. 9.2.

9.2.1   A New Approach to Manufacturing

The concurrent engineering team cuts across functional disciplines. The team concept is distinct from a committee. Three items are crucial to ensure success:

1. Designate a team leader.
2. Create a clear-cut written plan for launching SMT and providing ongoing support.
3. Obtain top management support.

The team leader should be a person highly skilled in diverse technical areas although not necessarily a technical specialist in any particular area. The team leader should have sufficient depth to understand the technical specialist's jargon and point of view. Just as important as this technical expertise and know-how, the team leader must also possess highly developed people skills and must be able to communicate well.

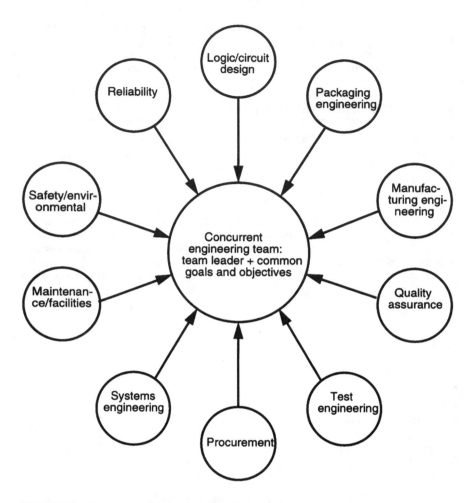

FIGURE 9.2   Concurrent engineering team concept.

The team leader should thoroughly understand the organization's goals and be well versed in total quality management (TQM) concepts and know how to put them into practice effectively. The team leader must be very clear about the organizations's goals and product requirements and should participate in planning.

To introduce SMT into an organization, a development team should be formed with the specific charter of setting up and integrating SMT onto the production floor. All the functional disciplines must be involved since introducing and implementing a new technology will have repercussions throughout the entire enterprise.

The workers should not be neglected. Introducing a new technology such as SMT will most likely entail training people in new skills and techniques. The workers may have to learn how to perform routine maintenance, recognize problems and react properly to them, and recognize product quality problems and take proactive steps.

9.3 THE ELECTRONICS PACKAGING HIERARCHY

9.3.1 The Packaging Hierarchy

There are six generally recognized levels of electronic packaging. Figure 9.3 shows the packaging hierarchy described. The six levels are:

1. *Level 0: Bare semiconductor (unpackaged).*
2. *Level 1: Packaged semiconductor or packaged electronic functional device.* The electronic device can be active, passive, or other (e.g., electromechanical). There are two cases to be distinguished regarding packaged IC devices. The first case entails a single semiconductor microcircuit within a suitable package. The package is either a hermetic sealed microcircuit (HSM) or a plastic encapsulated microcircuit (PEM). A single semiconductor microcircuit packaged in this manner can be referred to as a single chip module (SCM). The second case entails several semiconductor microcircuits plus discrete chips on a suitable substrate. This entire package is generally referred to as a multichip module (MCM) (see Sec. 9.5.6).

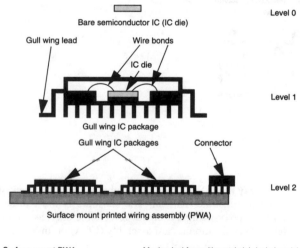

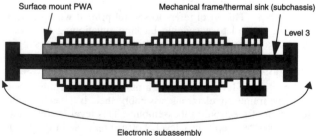

FIGURE 9.3 Packaging hierarchy.

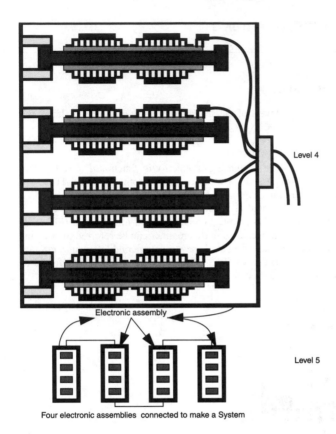

Level 4

Electronic assembly

Level 5

Four electronic assemblies connected to make a System

FIGURE 9.3 Packaging hierarchy. *(Continued.)*

3. *Level 2: Printed wiring assembly (PWA).* This level involves joining the packaged electronic devices to a suitable substrate material. The substrate is most often an organic material such as FR-4 epoxy-fiberglass board, or ceramic such as alumina. Level 2 is sometimes referred to as the circuit card assembly (CCA) or, more simply, the card assembly.

4. *Level 3: Electronic subassembly.* This level refers to several printed wiring assemblies (PWAs), normally two, bonded to a suitable backing functioning both as a mechanical support frame and a thermal heat sink. Sometimes this backing, or support frame, is called a subchassis. In some packaging hierarchies, e.g., computer packaging, Level 3 is the electronic assembly, also called the electronic box. See Fig. 9.4.

5. *Level 4: Electronic assembly.* This level consists of a number of electronic subassemblies mounted in a suitable frame. An electronic assembly, then, is a mechanically and thermally complete system of electronic subassemblies. This level is sometimes referred to as the electronic box or simply box level. For example, see Fig. 9.5. In some packaging hierarchies, e.g., computer packaging, Level 4 is the system.

6. *Level 5: System.* This refers to the completed product.

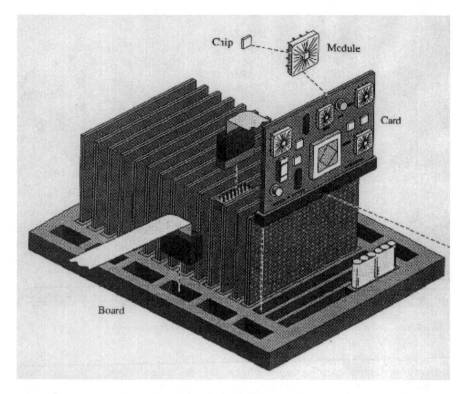

FIGURE 9.4 Level 3 packaging. *(Courtesy of Seraphim et al., 1989.)*

The trend in electronic packaging is to simplify and/or reduce the number of packaging levels. For example, the chip-on-board technology (COB), where a bare integrated circuit die (sometimes also called a chip) is placed directed on a printed wiring board and bonded to the board, eliminating the first level of packaging by going directly from the zeroth level to the second level. COB technology is a particular example of direct chip attach (DCA).

The packaging hierarchy given above is not universal. For computer packaging, for example, Level 3 entails a number of PWAs plugged into a backplane board and supported in a suitable chassis. See Fig. 9.5.

9.3.2 System Packaging Cycle

There are a number of distinct steps in the overall system packaging cycle. The SMT work team should be fully cognizant of all these steps, and the team leader should take steps to ensure that each step in this cycle is properly addressed and resolved. These steps are:

- Review the system specifications. The system specifications may be transformable into a suitable qualification test.
- Review all contractual requirements and commitments.
- Establish a performance schedule and milestone points.

FIGURE 9.5   SMAs in a chassis—Level 4 packaging. *(Courtesy of Litton Amecom.)*

- Determine the circuit design approach, special requirements, logic family, and constraints.
- Establish component requirements and a components listing.
- Perform system partitioning, and define system, subsystem, and electronic subassembly breakdowns. From this breakdown, determine the number and types of electronic surface mount printed wiring assemblies (PWAs) required.
- Establish and evaluate various packaging approach alternatives.
- Conduct a preliminary design review.
- Prepare detailed layouts of all major printed wiring assemblies and electronic subassemblies.
- Conduct a manufacturing analysis and prepare detailed flow process charts for each type of surface mount PWA.
- Prepare experimental models of critical printed wiring assemblies. Conduct tests to define performance characteristics and examine production, inspection, and test problems.
- Conduct a critical design review.
- Finalize the design and optimize it as much as possible at this point.
- Initiate preparation of detailed manufacturing drawings.
- Prepare breadboard, prototype, or engineering models as required.
- Conduct performance evaluation tests.
- Revise design and drawing set as required.
- Release to manufacturing.

- Support manufacturing during fabrication, assembly, and test of prototype PWAs.
- Revise design and drawings as a result of manufacturing and test.
- Release drawings for final product.
- Support manufacturing during fabrication, assembly, and test of PWAs.
- Perform final tests on electronic assemblies and system.

9.4 GENERAL PRINCIPLES AND GUIDELINES OF SMT

The purpose of this section is to give an overview of SMT before proceeding to a more detailed examination.

9.4.1 SMT vs. THT

Changing the placement method of components from THT to SMT entails a number of distinct changes:

- The packaging design must change to accompany new land pattern configurations and new component styles having different lead types and shapes.
- The manufacturing processes change to accompany the new component styles. Surface mount components (SMCs) are generally more amenable to placement by automated equipment, thus resulting in faster throughput. SMCs are normally soldered using a re-flow soldering process.
- Inspection processes and infrastructure changes to accommodate the above changes.
- Testing changes also to accommodate the above changes.

Table 9.1 shows some special points regarding changes due to SMT.

TABLE 9.1 Special Notes and Cautions Regarding Changes Due to SMT

NOTES AND CAUTIONS

Surface mount components (SMCs) are typically 2 to 10 times smaller than their corresponding through hole counterparts. This reduction in size ensures that the assembly size will get smaller or that the functionality per unit area will increase.

Surface mount technology very often entails an increased number of assembly steps. This is especially true for mixed technology assemblies: Type 1C and Type 2C. It is very important to have a design for manufacturability (DfM) scheme.

Because SMCs are smaller, they may be more difficult to inspect. Also, the use of statistical process control (SPC) techniques becomes much more important to ensure good yields when producing surface mount printed wiring assemblies. A design for inspectability (DfI) system is important.

In many cases testing surface mount PWAs becomes more difficult. A design for testability (DfT) system is important.

All of the above caveats imply the increased need for concurrent engineering practices and an adequate infrastructure to handle this newer packaging technology.

9.4.2 Surface Mount Classification

There are two widely used classification schemes for surface mount PWAs. These assemblies are often simply called surface mount assemblies, or SMAs. The first classification follows IPC-CM-770, Component Mounting Guidelines for Printed Boards. This classification utilizes two categories: assembly types and assembly classes. According to this classification, there are two assembly types:

1. Type 1 assembly, which has components only on its top side.
2. Type 2 assembly, which has components on both its top side and its bottom side.

According to the IPC classification, there are three assembly classes:

1. Class A assembly is entirely through hole technology (THT). A Type 1 Class A assembly is depicted in Fig. 9.6.
2. Class B assembly is entirely surface mount technology (SMT).
3. Class C assembly is a combined THT and SMT assembly (mixed technology).

A Class C assembly is often referred to as a mixed technology assembly. *Note:* Throughout this chapter, the assembly class will be denoted by a letter alone. For example, a Type 1 Class B assembly will simply be designated as: Type 1B.

Based on the IPC classification, the types of SMT assemblies are:

- Type 1B (single-sided pure SMT assembly). See Fig. 9.7.
- Type 2B (double-sided pure SMT assembly—no adhesive). See Fig. 9.8.
- Type 2B (double-sided pure SMT assembly—adhesive). See Fig. 9.8.
- Type 1C (single-sided with a mixture of THCs and SMCs). See Fig. 9.9.
- Type 2C (S) (simple) (double-sided with THCs only on top; small, discrete SMCs on the bottom). See Fig. 9.10.

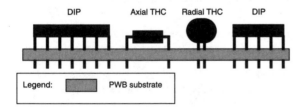

FIGURE 9.6 Type 1A assembly (all through hole).

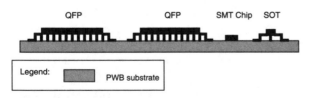

FIGURE 9.7 Type 1B assembly (single-sided pure SMT).

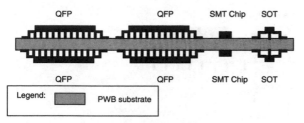

FIGURE 9.8 Type 2B assembly (double-sided pure SMT).

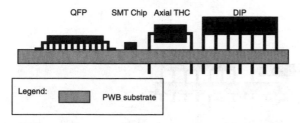

FIGURE 9.9 Type 1C assembly (single-sided mixed technology SMT)

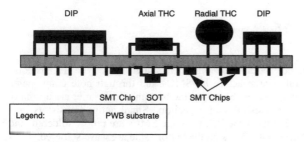

FIGURE 9.10 Type 2C (S) assembly.

- Type 2C (C) (complex) (double-sided with large SMCs and THCs on the top; small, discrete SMCs on the bottom). See Fig. 9.11.
- Type 2C (VC) (very complex) (double-sided with large SMCs and THCs on the top; large and small SMCs on the bottom). See Fig. 9.12.

Figures 9.6 through 9.12 are schematic drawings depicting the different types of surface mount PWAs. Figure 9.13 is a photograph of a Type 1B SMA, and Fig. 9.14 is a photograph of a Type 2B SMA.

The second classification scheme for surface mount PWAs is based on the soldering technology employed. This older classification involves only one category: assembly types.

A Type I assembly experiences only reflow soldering; it is a pure surface mount assembly. A Type IA assembly has SMCs only on its top side. A Type IB assembly has SMCs on both its top (primary) and bottom (secondary) sides.

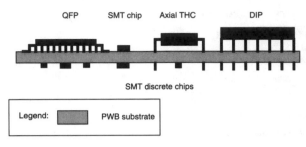

FIGURE 9.11   Type 2C (C) assembly.

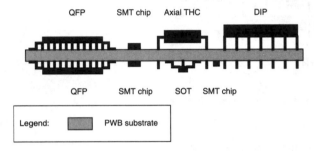

FIGURE 9.12   Type 2C (VC) assembly.

A Type II assembly experiences both reflow and wave soldering. A Type IIB has components on both sides—generally large active IC SMCs and through hole components (THCs) on the top side and small, discrete SMCs on the bottom. These latter are normally small passive SMCs such as chip resistors and capacitors. Small active transistors may also appear on the bottom side. For the Type IIB, the large SMCs are reflow soldered, and the THCs on the top side and the small SMCs on the bottom side are wave soldered.

Type III assembly is subjected only to wave soldering. A Type III assembly almost inevitably has through hole components (THCs) on its top side and small, discrete surface mount components (SMCs) on its bottom side.

The classification scheme followed in this chapter is the IPC-CM-770 classification. The scheme based on the soldering technologies is sometimes placed in brackets directly following the IPC classification if it contributes to clarity. Table 9.2 contrasts the IPC classification scheme with the older classification scheme based on the soldering technology used.

Note: The older classification scheme is mentioned here because it is still encountered in the SMT literature.

9.4.3   Producibility of Different Surface Mount PWAs

The order of producibility of the different types of surface mount PWAs begins from the simplest and progresses to the most difficult.

FIGURE 9.13   Type 1B surface mount PWA. *(Courtesy of Litton Amecom.)*

FIGURE 9.14   Type 2B surface mount PWA. *(Courtesy of Honeywell Space Systems.)*

1	Type 1B	Increasing complexity
2	Type 2B (no adhesive)	
3	Type 2C (S) (adhesive)	
4	Type 2B (adhesive)	
5	Type 1C	
6	Type 2C (C)	
7	Type 2C (VC)	▼

Not all components are available in a surface mount format. Some through hole components may be required because the SMT version does not yet exist.

TABLE 9.2 Surface Mount Assemblies—IPC and Former Classification Scheme

No.	IPC classification	Alternate (older) classification
1	Type 1B	Type IA
2	Type 2B (no adhesive)	Type IB
3	Type 2B (adhesive)	Type IB
4	Type 1C	Type IIA
5	Type 2C (S)	Type IIIB
6	Type 2C (C)	Type IIB
7	Type 2C (VC)	Type IIB

Note: Yield, rework, and therefore cost, will be affected by the assembly type chosen. Type 2C (VC) assemblies are not recommended because of their inherent complexity.

9.4.4 SMT Process Flow—Overview

Surface mount components are placed directly on the substrate surface. To create a metallurgical connection between the SMC and the substrate, solder paste is first deposited on the component lands. The SMCs are then mounted on the lands using a suitable placement method. The printed wiring boards (PWBs) plus SMCs are then reflow soldered, forming a surface mount PWA, or SMA. The PWAs are then cleaned and tested. For producing surface mount PWAs with components on both sides, the application of a suitable adhesive may be involved. For producing mixed technology (some THCs and some SMCs) PWAs, wave soldering is involved. In a nutshell, these are the major steps for producing surface mount PWAs:

- Solder paste application on the lands of a suitable substrate (e.g., a PWB)
- Adhesive deposition (not always required)
- Component preparation (if required)
- Component placement
- Soldering
- Cleaning
- Inspection
- Rework (when necessary)/hand assembly
- Clean prior to conformal coat (if required)

■ Conformal coat (if required)

■ Test

9.5 SMT COMPONENTS

There are several ways of approaching SMT components. The package style is one way; the type of component (active, passive, etc.) another. Table 9.3 shows the most common surface mount components and Table 9.4 the newer ones.

9.5.1 SMT Package Styles

There are several surface mount package styles. First, consider active devices. These have either an IC or a transistor inside the package if it is a single chip module. A multichip module (MCM) contains several ICs and other devices. For more details, see Sec. 9.5.3:

Active surface mount devices can be found in three distinct styles:

1. Leads around the periphery—all four sides. These are generally known as quad flat packs (QFPs) or chip carriers (CCs). In the case of the so-called leadless CCs, the term "termination" is more correct than "lead." Sometimes both QFPs and CCs are called peripheral array packages (PAPs).

2. Leads around the periphery—two sides only. These are generally known as small outline (SO) packages (0.050 in pitch) or shrink small outline (SSO) packages (0.025 in pitch).

3. Leads covering the entire, or almost entire, bottom area of the component. These are known as area array packages (AAPs). Ball grid arrays (BGAs) are a subset of AAPs.

Note: If the leads cover the entire bottom area of the component, they are full area array packages (FAAPs). If the leads cover only part of the area, they are depopulated area array packages (DAAPs). Note that a PAP ≠ DAAP.

Figure 9.15 sums up the principal different surface mount component package styles in a tree diagram format.

Note: It is beyond the scope of this chapter to describe in detail the internal fabrication of surface mount components, such as die fabrication, die attach, die bonding, encapsulation, and so on. Consult the literature, for example, see M. G. Pecht et al. (1995).

Another major distinction between active components is whether the component is hermetic or nonhermetic. Hermetic components have bodies of ceramic, whereas nonhermetic components typically have molded plastic bodies. The most common plastics used for nonhermetic components are epoxy and silicone. Most high reliability applications (Class 3) have required hermetic packages. It is an open question as to what extent nonhermetic components can be used for such applications. This is still an active area of investigation.

Closely connected with the package style is the lead (termination) type, lead configuration, and lead pitch. The pitch is defined as the distance from one lead center to an adjacent lead center. In the case of LCCs, the pitch refers to the distance from one termination center to an adjacent termination center.

Note: Standard pitch surface mount components' pitch is 0.050 in ≥ pitch > 0.025 in. Fine pitch surface mount components' pitch is 0.025 in ≥ pitch ≥ 0.020 in. Ultrafine pitch surface mount components' pitch is pitch < 0.020 in. Be aware that a 0.5 mm QFP has a pitch less than 0.020 in; hence, it is an ultrafine pitch SMC.

TABLE 9.3 Most Common Surface Mount Components

Component	Abbreviation	Type of component	Type of lead	Packaging + tape width (where applicable)
Rectangular chip component	R (resistor) or C (capacitor)	Passive	3-sided (wrap-around) or 5-sided termination	8mm or 12mm
Molded plastic capacitor	C (capacitor)	Passive		8mm or 12mm
Metal electrode leadless face	MELF	Passive	Cylindrical plated termination	8mm or 12mm
Transistor (or diode)	SOT23, SOT89, SOT143	Active	Gull wing	8mm
Leadless chip carrier (ceramic)	LCC	Active	Leadless with castellation	16mm, 24mm, 32mm, 44mm, matrix tray
Ceramic leaded chip carrier (J-Cerquad)	CLCC	Active	J-lead	Same as LCC
Ceramic quad flat pack (gull wing Cerquad)	CQFP	Active	Gull wing	Matrix tray
Small outline integrated circuit	SOIC	Active	Gull wing	12mm, 16mm, 24mm
Small outline J-lead integrated circuit	SOJIC	Active	J-lead	24mm
Small outline large integrated circuit	SOLIC	Active	Gull wing	16mm, 24mm
Plastic leaded chip carrier	PLCC	Active	J-lead	16mm, 24mm, 32mm, 44mm, 52mm, 68mm
Plastic quad flat pack (bumpered or unbumpered)	PQFP	Active	Gull wing	Matrix tray

TABLE 9.4 Newer Surface Mount Components

Component	Abbreviation	Type of Component	Type of lead	Packaging + tape width (where applicable)
Shrink small outline integrated circuit (PAP)	SSOIC	Active	Gull wing—2 sides (longer sides—similar to a Type II TSOP)	Tape and reel
Thin small outline package-Type I (PAP)	TSOP-I	Active	Gull wing—2 sides (shorter sides)	Tape and reel
Thin small outline package-Type II (PAP)	TSOP-II	Active	Gull wing—2 sides (longer sides)	Tape and reel
Thin quad flat pack (PAP)*	TQFP	Active	Gull wing—4 sides (periphery)	Matrix tray
Ball grid array—ceramic (CBGA) (AAP)	CBGA	Active	Solder balls (area array)	Matrix tray
Ball grid array—plastic (PBGA) (AAP)	PBGA	Active	Solder balls (area array)	Matrix tray
Land (pad) grid array (AAP)†	LGA or PGA‡	Active	Solder lands (area array)	Matrix tray
Column grid array—ceramic (CCGA) (AAP)	CGA	Active	Solder columns (area array)	Matrix tray

* PAP = peripheral array package, leads on all four sides
† AAP = area array package, leads (terminations) covering most or all of the bottom area
‡ Because the abbreviation for pin grid array is also PGA, the expression "land grid array" (LGA) is preferable to "pad grid array."

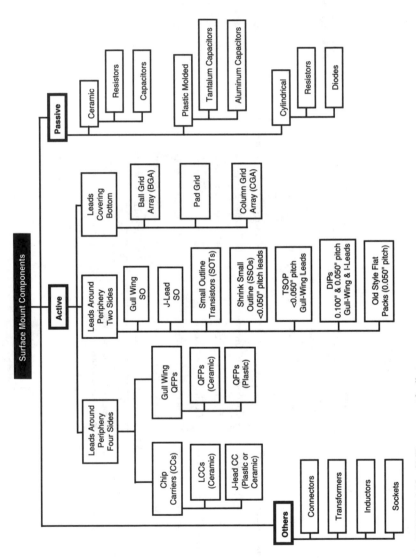

FIGURE 9.15 SMT component family tree.

The chief lead configurations are shown in Fig. 9.16.

There are two principal types of terminations for surface mount active devices: leadless and leaded. Leadless components are typically made of ceramic; leaded components may be of either ceramic or plastic. Leadless components having internal integrated circuit die are referred to as leadless ceramic chip carriers (LCCCs) or, more simply, leadless chip carriers (LCCs). LCCs have metallized castellations on their perimeters for solder fillet formation. Figure 9.17 shows LCC side castellations, and Fig. 9.16 shows the shape of a solder fillet formed with an LCC component.

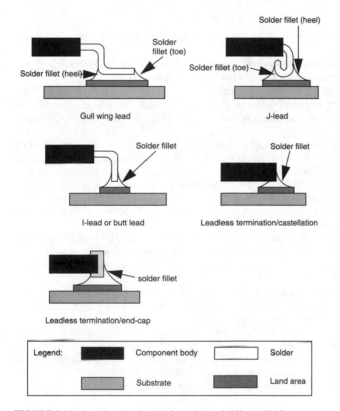

FIGURE 9.16   Lead/termination configurations of different SMCs.

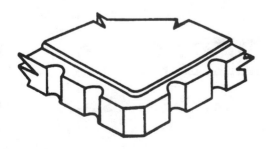

FIGURE 9.17   Leadless component showing side castellations.

Leadless passive components are generally referred to as ceramic chip components or simply chip components. The have end-cap terminations. See Fig. 9.16.

Note: A single IC with its planar circuitry on a substrate is called a die. Generally, the substrate is silicon (Si), but it may also be gallium arsenide (GaAs) or silicon-on-sapphire (SoS). (The word "die" has several acceptable plurals: *die, dies,* and *dice.*)

The three chief component lead configurations are:

1. Gull wing
2. J-lead
3. I-lead or butt lead

Figure 9.16 shows the shapes of these three lead configurations and the solder fillet formed from each.

Each of the lead configurations has certain characteristics shown in Table 9.5. For purposes of comparison, castellated LCCs are also included in this table.

I-lead or butt lead SMCs are generally formed from through hole dual in-line package (DIP) components having a pitch of 0.100 in. Although the leads are surface mountable, I-lead SMCs having a 0.100 in pitch are a special case of SMC. Some I-lead SMCs have a 0.050 in pitch.

Another important issue is coplanarity. Coplanarity is measured by reference to two (imaginary) planes passing through the component. If a component were completely coplanar, the distance between the two planes would be zero. The more a component lead is skewed from one of these reference planes, the greater the distance becomes. See Fig. 9.18 depicting the coplanarity of a gull wing SMC. For standard pitch SMCs, the recommended maximum allowable distance from true coplanarity is 0.004 in. For fine pitch and ultrafine pitch SMCs, it is 0.002 in.

In addition to coplanarity, lead fragility is also an issue. This issue is addressed in Table 9.5. From this table, it is evident that the gull wing lead configuration is the most delicate and requires the most careful handling. Leads can be bent and skewed and still exhibit good coplanarity.

Caution: Gull wings are especially vulnerable to damage. They must be handled extremely carefully prior to and during placement all the way through reflow. They must be placed in a suitable container while being transported.

Active devices also can be divided into those with terminations or leads on the periphery of the component or those with terminations (either pads or solder bumps) over much of the bottom component area. As mentioned, the former are called peripheral array packages, and the latter are known as area array packages.

There are three principal styles among surface mount passive devices:

1. Ceramic chip
2. Molded plastic
3. Cylindrical

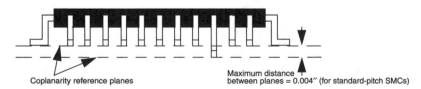

Coplanarity reference planes

Maximum distance between planes = 0.004″ (for standard-pitch SMCs)

FIGURE 9.18 Coplanarity in a gull wing SMC.

TABLE 9.5 SMC Lead/Termination Configuration Characteristics

Characteristic	Gull wing	J-lead	I-lead or butt lead	Castellated leadless	Ball of AAP
Lead rigidity and/or ruggedness (DfM)	P	G	E	E+*	E
Ability to self-align during reflow (DfM)	F	P	P	G	E
Suitability with various soldering methods (DfM)†	E	G	G	G	G
Ease of inspectability after soldering (DfI)	E	P	G	G	P
Ease of cleaning after soldering (DfM)‡	E	G	G	P–G	F
Printed wiring board real estate efficiency	P	G	G	G	E
Compatibility with future multipin packaging trends	E	P	P	P	E
General reliability of formed solder joint (DfR)**	E	G††	P–F	P–F**	P–G**
Suitable for fine pitch SMCs (DfM)	E	P	P	P	G

E = excellent, G = good, F = fair, P = poor

* Leadless carriers do not have leads, so there is very little possibility of termination damage.
† None of these components should be wave soldered, but some small SOICs and SOJICs are processed this way.
‡ Cleanability is largely directly proportional to the amount of standoff (distance between the bottom of the component and the top of the component).
**Solder joint reliability relates greatly to the CTE mismatch between the component and substrate. Careful CTE matching will increase SJ reliability.
††This assumes the lead is compliant.

These different styles are more carefully delineated in Sec. 9.5.2.

9.5.2 Surface Mount Components—Passive

Passive components are defined as components that change electronic signals but do not amplify, rectify, or switch the signal. This is in contradistinction to active components, which do function to amplify, rectify, or switch the electronic signal.

9.5.2.1 Resistors There are several types of surface mount resistors. The two most common types are discrete ceramic chip resistors and metal electrode leadless face (MELF) devices. Discrete ceramic resistors come in several sizes. The 1206 is the most common size, but smaller chip components such as 0603s and 0402s are rapidly becoming popular. MELFs are cylindrical in shape and also come in several sizes. Discrete ceramic resistors and MELFs have two terminations, one at each end. Surface mount resistor networks, known as R-packs, are typically packaged in an SOIC package format (see Sec. 9.5.3.2).

Figure 9.19 shows a ceramic chip resistor (RC 1206) and its associated land pattern, Fig. 9.20 shows a MELF, and Fig. 9.21 shows a MELF and its associated land pattern.

Chip Resistors. There are several common sizes for chip resistors. These are 0805, 1206, 1210, 2010, and 2512. Smaller resistors, such as 0603s and 0402s, are also fairly common. Surface mount chip resistors normally do not have value markings. The chip's number indicates its length and width dimensions in 0.001 in units; that is, an 0805 chip has a length of (roughly) 0.080 in and a width of (roughly) 0.050 in.

(a)   (b)

FIGURE 9.19 Ceramic chip resistor (RC 1206); (a) chip resistor and (b) footprint.

FIGURE 9.20 Metal electrode lead-less face (MELF) component.

(a)   (b)

FIGURE 9.21 Additional MELF example; (a) MELF device and (b) footprint.

An alumina substrate is used in making chip resistors. The resistive element is either sputtered (thin film) or screened (thick film) onto this substrate, and it is then trimmed to size. The end terminations of these components are either on three sides: top, bottom, and end (three-sided termination or wrap-around termination), or they cover five sides: top, bottom, end, and sides (five-sided termination).

Figure 9.22 shows the internal structure of a typical surface mount chip resistor. Figure 9.23 (reference drawing) shows the top and side view of a chip resistor. Chip resistor component dimensions are shown in Table 9.6. Also see Fig. 9.19, which shows a chip resistor (RC 1206) and its associated land pattern.

Resistor parameters that must be adequately specified are: resistance ± tolerance, power rating, thermal coefficient of resistance (TCR) in ppm, and the 10,000 hour stability.

Cylindrical Resistors. These are often referred to as metal electrode leadless face or MELF components. The MELF component style can be used for resistors, capacitors, and diodes, but this style is most commonly used for resistors. As resistors, MELFs are essentially like axial through hole resistors without the axial leads.

There are two common sizes, MLL 34 and MLL 41. MELFs have a tendency to roll off the surface of the PWB after placement. The MELF land pattern should have an indentation in which the MELF can fit snugly. Figure 9.24 shows a typical MELF being placed on

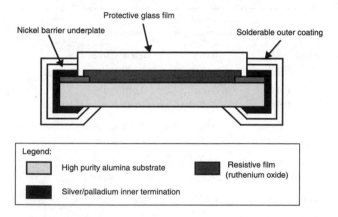

FIGURE 9.22 Internal structure of a typical SMT chip resistor.

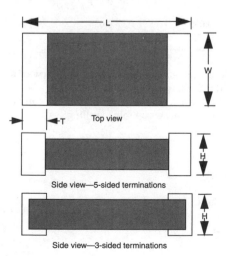

FIGURE 9.23 Top and side view of a chip component (RC or MLC)

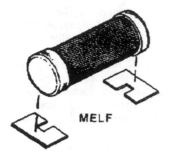

MELF

FIGURE 9.24 MELF being placed on its lands (courtesy of Traister, 1990).

TABLE 9.6 Typical Chip Resistor Component Dimensions, inches (mm)

Size	Length, L	Width, W	Height (maximum), H	Terminal, T
0402	0.039–0.043	0.019–0.024	0.016	0.008–0.012
	(1.00–1.10)	(0.48–0.60)	(0.40)	(0.20–0.30)
0603	0.059–0.065	0.028–0.035	0.024	0.010–0.016
	(1.50–1.65)	(0.70–0.90)	(0.60)	(0.25–0.40)
0805	0.075–0.083	0.045–0.053	0.026	0.018–0.026
	(1.90–2.10)	(1.14–1.35)	(0.65)	(0.45–0.65)
1206	0.118–0.126	0.057–0.067	0.030	0.022–0.030
	(3.00–3.20)	(1.45–1.70)	(0.75)	(0.55–0.75)
1210	0.118–0.134	0.094–0.110	0.030	0.022–0.030
	(3.00–3.40)	(2.39–2.79)	(0.75)	(0.55–0.75)
2010	0.191–0.203	0.094–0.110	0.030	0.025–0.034
	(4.85–5.15)	(2.39–2.79)	(0.75)	(0.65–0.85)
2512	0.242–0.254	0.120–0.132	0.030	0.025–0.034
	(6.15–6.45)	(3.05–3.35)	(0.75)	(0.65–0.85)

its lands on the PWB. Figure 9.25 (reference drawing) shows the front and side view of a MELF. Table 9.7 gives the MELF component dimensions.

MELFs are difficult to assemble because they have a tendency to roll. MELFs can be purchased having square-cap terminations. These are not susceptible to rolling.

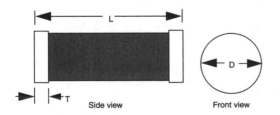

Side view   Front view

FIGURE 9.25 Front and side views of a MELF.

TABLE 9.7 Typical MELF Component Dimensions, inches (mm)

Size	Length, L	Diameter, D	Terminal, T
MLL 34	0.130–0.146	0.059–0.067	0.012–0.022
	(3.30–3.70)	(1.50–1.70)	(0.30–0.55)
MLL 41	0.189–0.205	0.096–0.100	0.014–0.020
	(4.80–5.20)	(2.44–2.54)	(0.35–0.50)

9.5.2.2 Capacitors   There are also several distinct types of surface mount capacitors. The most common type is the multilayer ceramic capacitor (MLC) which come in discrete packages very similar to the ceramic resistors. Ceramic capacitors also have two terminations (end-caps), one at each end. Ceramic resistors and ceramic capacitors are often called discrete monolithic chip components or more simply chip components. Molded plastic tantalum capacitors which are surface mountable are also available. These also have two leads for surface mounting to the PWB substrate.

Figure 9.26 shows a ceramic capacitor (MLC 1206) and its associated land pattern, and Fig. 9.27 shows a molded plastic tantalum capacitor and its associated land pattern.

(a)   *(b)*

FIGURE 9.26   Multilayer ceramic chip capacitor (MLC 1206); (*a*) chip capacitor and (*b*) footprint.

(a)   *(b)*

FIGURE 9.27   Molded plastic tantalum capacitor; (*a*) tantalum capacitor and (*b*) footprint.

Chip Capacitors.   These are also known as multilayer chip capacitors (MLCs) or monolithic chip capacitors. They look very similar to chip resistors, but they always have a five-
sided termination: top, bottom, end, and both sides. Refer back to Fig. 9.23 for the top and
side view of a chip capacitor. They have a complex internal structure (see Fig. 9.28). Chip
capacitor component dimensions are shown in Table 9.8.

TABLE 9.8   Typical Molded Capacitor Component Dimensions, inches (mm)

Case size	Length, L	Width, W	Height, H	Terminal length, L_T	Terminal width, W_T	Terminal height (minimum), H_T
A	0.118–0.134	0.050–0.070	0.055–0.070	0.020–0.043	0.043–0.051	0.020
	(2.99–3.40)	(1.27–1.78)	(1.39–1.78)	(0.50–1.10)	(1.10–1.30)	(0.70)
B	0.130–0.146	0.102–0.118	0.067–0.083	0.020–0.043	0.083–0.090	0.028
	(2.54–3.71)	(2.59–2.99)	(1.70–2.11)	(0.50–1.10)	(2.10–2.30)	(0.70)
C	0.224–0.248	0.114–0.138	0.087–0.110	0.020–0.043	0.083–0.090	0.040
	(5.69–6.30)	(2.89–3.50)	(2.21–2.79)	(0.50–1.10)	(2.10–2.30)	(1.00)
D	0.268–0.299	0.157–0.181	0.098–0.122	0.020–0.043	0.090–0.098	0.040
	(6.81–7.59)	(3.99–4.60)	(2.49–3.10)	(0.50–1.10)	(2.30–2.50)	(1.00)

There are three dielectric types used with MLCs. These are:

1. Z5U

2. X7R

3. NPO or COG

X7R and Z5U are low in cost. NPO MLCs give excellent stability over a wide range of
temperatures, frequencies, and voltages although they are somewhat more expensive.

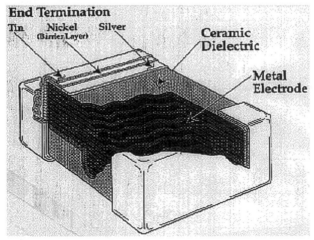

FIGURE 9.28   Internal structure of an MLC capacitor. *(Courtesy of
Kemet Electronics.)*

Capacitor parameters that must be adequately specified are capacitance ± tolerance, power rating and the 10,000 hour stability.

Molded Capacitors. These capacitors are often called plastic molded tantalum capacitors. Because of their shape, they are sometimes referred to as "brick" capacitors. Unlike chip capacitors, molded capacitors are polarized. They are also extremely stable; they are the capacitor of choice where capacitance stability is of prime importance. It is very important that these components be oriented the correct way on the PWB; otherwise, they will malfunction and may actually be hazardous. Some have been known to explode if oriented incorrectly and then energized. Their capacitance value varies from 0.1 mF to 100 mF and from 4 to 50 Vdc. There are four case sizes: Size A, Size B, Size C, and Size D (see Table 9.9). Sometimes other materials such as aluminum are used rather than tantalum. Table 9.9 still applies.

TABLE 9.9 Typical Molded Capacitor Component Dimensions, inches (mm)

Case size	Length, L	Width, W	Height, H	Terminal length, L_T	Terminal width, W_T	Terminal height (minimum), H_T
A	0.118–0.134 (2.99–3.40)	0.050–0.070 (1.27–1.78)	0.055–0.070 (1.39–1.78)	0.020–0.043 (0.50–1.10)	0.043–0.051 (1.10–1.30)	0.020 (0.70)
B	0.130–0.146 (2.54–3.71)	0.102–0.118 (2.59–2.99)	0.067–0.083 (1.70–2.11)	0.020–0.043 (0.50–1.10)	0.083–0.090 (2.10–2.30)	0.028 (0.70)
C	0.224–0.248 (5.69–6.30)	0.114–0.138 (2.89–3.50)	0.087–0.110 (2.21–2.79)	0.020–0.043 (0.50–1.10)	0.083–0.090 (2.10–2.30)	0.040 (1.00)
D	0.268–0.299 (6.81–7.59)	0.157–0.181 (3.99–4.60)	0.098–0.122 (2.49–3.10)	0.020–0.043 (0.50–1.10)	0.090–0.098 (2.30–20.50)	0.040 (1.00)

Note: For high capacitance or voltage requirements, there are four additional molded capacitor sizes known as extended range molded capacitors. Consult the literature. For example, see Hollomon (1989).

There are different techniques for attaching the terminations to the molded part. One technique is the welded stub, but it is being phased out due to placement problems on the PWB. Figure 9.29 shows the internal structure of a typical molded plastic capacitor. Figure 9.30 (reference drawing) shows the front and side view of a molded plastic "brick" capacitor. Table 9.9 shows the dimensions of the molded capacitors.

9.5.3 Surface Mount Components—Active

There are a number of active surface mount components. An active component displays electrical gain or control by amplifying, rectifying, or switching electronic signals. Except for discrete active components such as transistors and diodes, almost all active components contain an IC die inside the component package.

9.5.3.1 Small Outline Transistors (SOTs). These surface mount components (SMCs) typically have either three leads (SOT 23 and SOT 89) or four leads (SOT 143).

Figure 9.31 shows an SOT 23 and its associated land pattern, and Fig. 9.32 shows an SOT 89 and its associated land pattern.

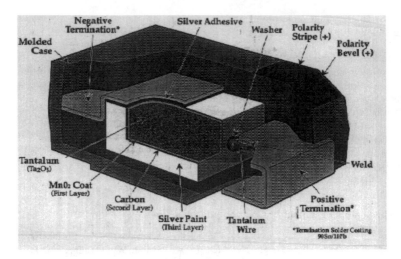

FIGURE 9.29 Internal structure of a molded plastic capacitor. *(Courtesy of Kemet Electronics.)*

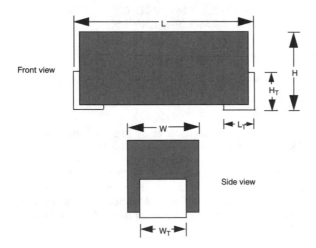

FIGURE 9.30 Front and side view of a molded plastic capacitor.

There are several important surface mount types of SOTs:

1. SOT 23
2. SOT 89
3. SOT 143
4. DPAK

The SOT 23, the SOT 89, and the DPAK have three leads whereas the SOT 143 has four leads. The SOT 23 is the most commonly used package; its leads are gull wing. Fig-

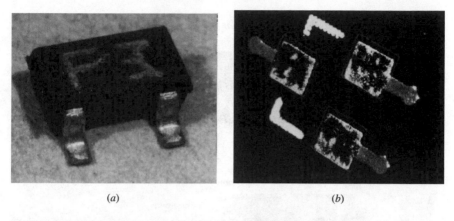

(a)   *(b)*

FIGURE 9.31   Small outline transistor 23 (SOT 23); *(a)* SOT and *(b)* footprint.

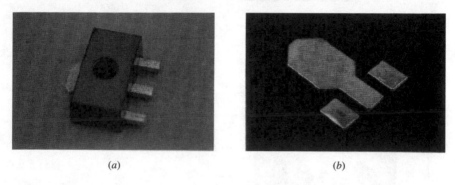

(a)   *(b)*

FIGURE 9.32   Small outline transistor 89 (SOT 89); *(a)* SOT and *(b)* footprint.

ure 9.33 (reference drawing) shows a top and side view of an SOT 23. The SOT 89 has planar leads. Figure 9.34 (reference drawing) shows a bottom and side view of an SOT 89. Except for having four leads rather than three, the SOT 143 looks very similar to the SOT 23 in all other respects. It has gull wing leads similar to the SOT 23. Therefore, only a top view of an SOT 143 is shown in Fig. 9.35 (reference drawing). The DPAK has planar leads. Figure 9.36 (reference drawing) shows a top and side view of a DPAK. The component dimensions of these four components (SOT 23, SOT 89, SOT 143, DPAK) are given in Table 9.10.

Note: It is also possible to package diodes in SOT packages. The DPAK is a higher wattage version of an SOT 23. For gull wing leaded components, the tip-to-tip width, $W_{T\text{-}T}$, is the same as the toe-to-toe width.

There are three profiles for SOT packages: low, medium, and high. See Fig. 9.37. The low profile is best when the component is to be glued to the PWB surface, as during wave soldering. However, if adequate cleaning is the objective, the high profile should be used.

9.5.3.2 Small Outline Integrated Circuits (SOICs).   These components typically have up to 28 leads, and the lead configuration is normally gull wing. Generally SOICs contain

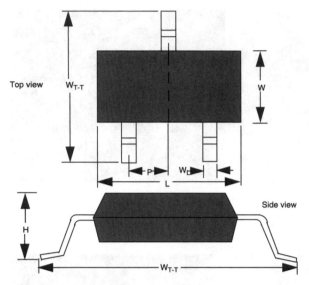

FIGURE 9.33 Top and side view of an SOT 23.

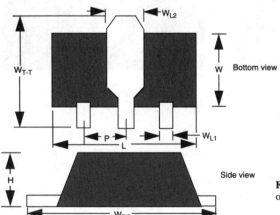

FIGURE 9.34 Bottom and side view of an SOT 89.

an IC chip and are called small outline integrated circuits (SOICs). They are rectangular in shape, with the leads running along the two long sides. The pitch is 0.050 in. Some SOICs, especially those used for memory chips, utilize a J-lead configuration. SOICs also come in different component widths. The shrink small outline integrated circuit (SSOIC) is similar to the SOIC except that it has a 0.025 in pitch.

They are similar in appearance to the through hole dual in-line package (DIP) except that SOICs are half the body wide of a DIP and half the pitch. The standard SOIC has a body width of 0.150 in as opposed to 0.300 in for a DIP; the standard SOIC has a pitch of 0.050 in as opposed to 0.100 in for a DIP. Like DIPs, SOICs have leads on two sides only. Another major difference, of course, is that DIPs are THCs and SOICs are SMCs.

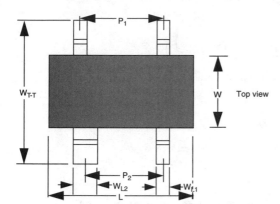

FIGURE 9.35 Top view of an SOT 143.

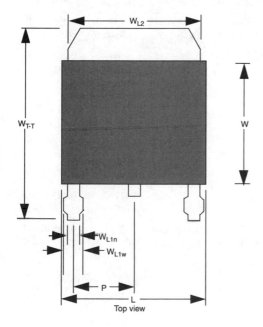

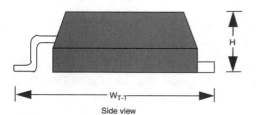

FIGURE 9.36 Top and side view of a DPAK.

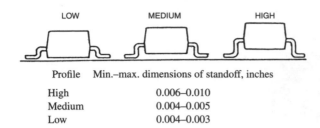

Profile	Min.–max. dimensions of standoff, inches
High	0.006–0.010
Medium	0.004–0.005
Low	0.004–0.003

FIGURE 9.37   Comparison of different profiles for the SOT 23 package.

TABLE 9.10   Typical SOT and DPAK Component Dimensions, inches (mm)

Component	Length, L	Width, W	Height, H	Tip-to-tip width, $W_{T\text{-}T}$	Lead width, W_L	Pitch, P
SOT 23	0.110–0.120 (2.80–3.05)	0.048–0.055 (1.20–1.40)	0.033–0.048 (0.85–1.20)	0.083–0.098 (2.10–2.50)	0.015–0.019 (0.36–0.46)	0.035–0.040 (0.90–1.00)
SOT 89	0.174–0.182 (4.40–4.60)	0.090–0.102 (2.29–2.60)	0.056–0.062 (1.40–1.60)	0.156–0.167 (3.94–4.25)	W_{L1} 0.015–0.019 (0.38–0.48) W_{L2} 0.064–0.071 (1.63–1.80)	0.057–0.060 (1.45–1.52)
SOT 143	0.110–0.122 (2.80–3.10)	0.048–0.067 (1.20–1.70)	0.033–0.048 (0.85–1.20)	0.083–0.102 (2.10–2.60)	W_{L1} 0.015–0.019 (0.38–0.48) W_{L2} 0.031–0.035 (0.80–0.90)	P_1 0.071–0.079 (1.80–2.00) P_2 0.063–0.071 (1.60–1.80)
DPAK	0.250–0.265 (6.35–6.73)	0.235–0.245 (5.97–6.22)	0.087–0.098 (2.20–2.50)	0.372–0.409 (9.45–10.40)	W_{L1n} 0.027–0.035 (0.70–0.90) W_{L1w} 0.030–0.045 (0.75–1.15) W_{L2} 0.205–0.215 (5.20–5.45)	0.085–0.095 (2.15–2.40)

Note: There are SOICs with a wider body, the so-called SOLIC having a body width (W) 0.300 in, the same as a DIP. The pitch of the SOLIC is still 0.050 in.

Figure 9.38 (reference drawing) shows the top and side view of a SOIC with 16 leads (gull wing), and the component dimensions are given in Table 9.11. Also see Fig. 9.39 showing a 14-lead SOIC with its associated land pattern, Fig. 9.40 showing a 16-lead SOIC with its associated land pattern, and Fig. 9.41 showing a different 16-lead SOIC with its associated land pattern.

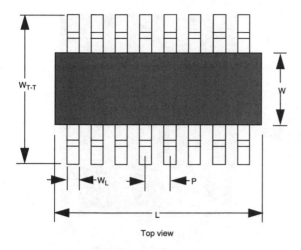

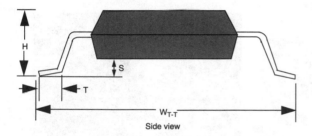

FIGURE 9.38 Top and side view of a gull wing SOIC.

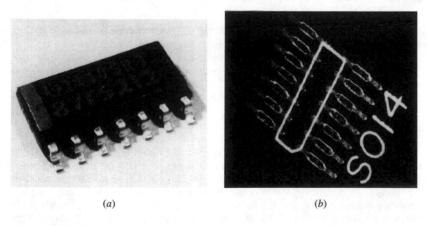

(a)   (b)

FIGURE 9.39 Small outline IC with 14 leads (SOIC 14); (a) SOIC and (b) footprint.

(a)   *(b)*

FIGURE 9.40   Small outline IC with 16 leads (SOIC 16); *(a)* SOIC and *(b)* footprint.

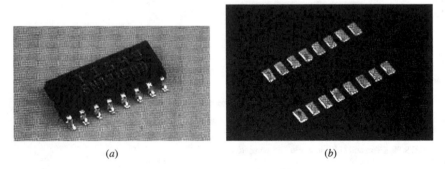

(a)   *(b)*

FIGURE 9.41   Additional SOIC with 16 leads (SOIC 16); *(a)* SOIC and *(b)* footprint.

A small outline package having J-leads package is generally designated as SOJIC or sometimes simply as SOJ. It is typically used to package memory integrated circuits. Figure 9.42 shows an SOJIC, and Fig. 9.43 (reference drawing) shows the top and side view of a SOJ with 16 leads (J-lead). There are two body widths (W) of an SOJ, namely, 0.300 in (nominal) and 0.350 in (nominal). Table 9.12 gives the component dimensions for the SOJ/0.300 in, and Table 9.13 gives the component dimensions for the SOJ/0.350 in. Two dimensions of an SOJ differing from the SOIC are the length (L) and the tip-to-tip width (W$_{T-T}$). Some SOJs may purposely have several center pins missing. This is to allow easier bussing of wider power traces.

Thin small outline packages (TSOPs) are a type of small outline package (SOP). These packages are rectangular shaped and have a very low profile. For example, see Fig. 9.44, comparing the lead height of a quad flat pack (QFP) to a thin small outline package (TSOP). Their pitch is normally 0.025 in or 0.020 in, with leads in the gull wing configuration running along either (1) the two shorter sides of the rectangle (Type I TSOP) or (2) along the two longer sides (Type II TSOP). The package material is typically plastic. Figure 9.45 shows a 32-lead Type I TSOP and its associated land pattern.

TABLE 9.11 Typical SOIC Component Dimensions, inches (mm)

No. of leads	Length, L	Width, W	Height, H	Tip-to-tip width W_{T-T}	Lead width W_L	Component pad length, T	Pitch, P (nominal)
8	0.189–0.197	0.150–0.157	0.053–0.069	0.228–0.244	0.013–0.020	0.035–0.040	0.050
	(4.80–5.00)	(3.80–4.00)	(1.35–1.75)	(5.80–6.20)	(0.33–0.51)	(0.89–1.02)	(10.27)
14	0.337–0.344	0.150–0.157	0.053–0.069	0.228–0.244	0.013–0.020	0.035–0.040	0.050
	(8.55–8.75)	(3.80–4.00)	(1.35–1.75)	(5.80–6.20)	(0.33–0.51)	(0.89–1.02)	(1.27)
16	0.386–0.394	0.150–0.157	0.053–0.069	0.228–0.244	0.013–0.020	0.035–0.040	0.050
	(9.80–10.00)	(3.80–4.00)	(1.35–1.75)	(5.80–6.20)	(0.33–0.51)	(0.89–1.02)	(1.27)
16L[*]	0.398–0.413	0.291–0.299	0.093–0.104	0.394–0.419	0.013–0.020	0.035–0.040	0.050
	(10.10–10.50)	(7.40–7.60)	(2.35–2.65)	(10.00–10.65)	(0.33–0.51)	(0.89–1.02)	(1.27)
20L	0.496–0.512	0.291–0.299	0.093–0.104	0.394–0.419	0.013–0.020	0.035–0.040	0.050
	(12.60–13.00)	(7.40–7.60)	(2.35–2.65)	(10.00–10.65)	(0.33–0.51)	(0.89–1.02)	(1.27)
24L	0.612–0.624	0.291–0.299	0.093–0.104	0.405–0.419	0.014–0.020	0.035–0.040	0.050
	(15.54–15.85)	(7.40–7.60)	(2.35–2.65)	(10.29–10.65)	(0.36–0.51)	(0.89–1.02)	(1.27)
28L	0.712–0.724	0.291–0.299	0.093–0.104	0.405–0.419	0.014–0.020	0.035–0.040	0.050
	(18.08–18.39)	(7.40–7.60)	(2.35–2.65)	(10.29–10.65)	(0.36–0.51)	(0.89–1.02)	(1.27)
32L	0.812–0.824	0.291–0.299	0.093–0.104	0.405–0.419	0.014–0.020	0.035–0.040	0.050
	(20.62–20.93)	(7.40–7.60)	(2.35–2.65)	(10.29–10.65)	(0.36–0.51)	(0.89–1.02)	(1.27)
36L	0.912–0.924	0.291–0.299	0.093–0.104	0.405–0.419	0.014–0.020	0.035–0.040	0.050
	(23.16–23.47)	(7.40–7.60)	(2.35–2.65)	(10.29–10.65)	(0.36–0.51)	(0.89–1.02)	(1.27)

* L signifies SOL; that is, the body width is a nominal 0.300 in.

9.5.3.3 Leadless Chip Carriers (LCCs). These are sometimes known as leadless ceramic chip carriers (LCCCs). Based on the classification scheme presented above in Fig. 9.15, these are classified as peripheral array packages with leads (or terminations) around all four sides. These components typically have fluted or grooved terminations on the four sides and foot pads on the bottom. The grooved side terminations are known as castellations. These castellations are metallized. They are almost always tungsten/nickel/solder plated over with gold. Component pitch is typically 0.050 in, although 0.040 in is also found.

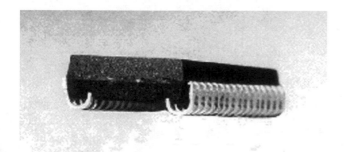

FIGURE 9.42 Small outline IC with J-leads (SOIC).

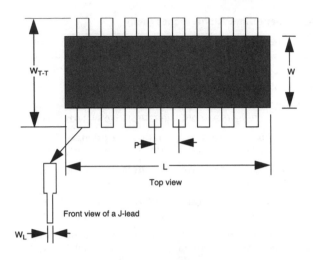

Top view

Front view of a J-lead

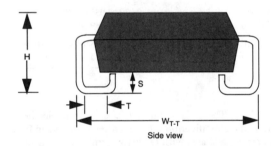

Side view

FIGURE 9.43   Top and side view of a J-lead SOJ/SOJIC.

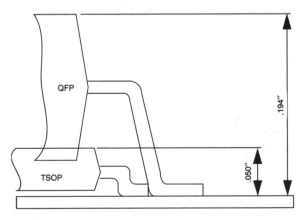

FIGURE 9.44   Component height comparison of a QFP and a TSOP.

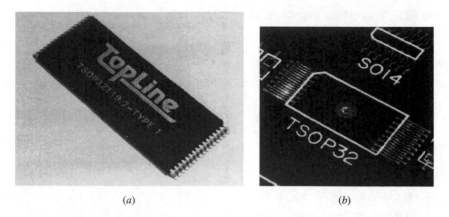

(a)   *(b)*

FIGURE 9.45   Type I thin small outline package with 32 leads; (*a*) TSOP and (*b*) footprint.

TABLE 9.12   Typical SOJ/0.300 Component Dimensions, Inches (mm)

No. of leads	Length, L	Width, W	Height, H (max.)	Tip-to-tip width, W_{T-T}	Lead width, W_L	Component pad length, T	Pitch, P (nominal)
14	0.380–0.392 (9.65–9.96)	0.295–0.305 (7.49–7.75)	0.148 (3.75)	0.330–0.345 (8.38–8.76)	0.015–0.020 (0.38–0.51)	0.035–0.040 (0.89–1.02)	0.050 (1.27)
16	0.430–0.442 (10.92–11.23)	0.295–0.305 (7.49–7.75)	0.148 (3.75)	0.330–0.345 (8.38–8.76)	0.015–0.020 (0.38–0.51)	0.035–0.040 (0.89–1.02)	0.050 (1.27)
18	0.480–0.492 (12.19–12.50)	0.295–0.305 (7.49–7.75)	0.148 (3.75)	0.330–0.345 (8.38–8.76)	0.015–0.020 (0.38–0.51)	0.035–0.040 (0.89–1.02)	0.050 (1.27)
20	0.530–0.542 (13.46–13.77)	0.295–0.305 (7.49–7.75)	0.148 (3.75)	0.330–0.345 (8.38–8.76)	0.015–0.020 (0.38–0.51)	0.035–0.040 (0.89–1.02)	0.050 (1.27)
22	0.580–0.592 (14.73–15.04)	0.295–0.305 (7.49–7.75)	0.148 (3.75)	0.330–0.345 (8.38–8.76)	0.015–0.020 (0.38–0.51)	0.035–0.040 (0.89–1.02)	0.050 (1.27)
24	0.630–0.642 (16.00–16.31)	0.295–0.305 (7.49–7.75)	0.148 (3.75)	0.330–0.345 (8.38–8.76)	0.015–0.020 (0.38–0.51)	0.035–0.040 (0.89–1.02)	0.050 (1.27)
26	0.680–0.692 (17.27–17.58)	0.295–0.305 (7.49–7.75)	0.148 (3.75)	0.330–0.345 (8.38–8.76)	0.015–0.020 (0.38–0.51)	0.035–0.040 (0.89–1.02)	0.050 (1.27)
28	0.730–0.742 (18.54–18.85)	0.295–0.305 (7.49–7.75)	0.148 (3.75)	0.330–0.345 (8.38–8.76)	0.015–0.020 (0.38–0.51)	0.035–0.040 (0.89–1.02)	0.050 (1.27)

Figure 9.17 shows an illustration of a leadless component with its castellations. Figure 9.46 shows a 20-termination LCC (LCC 20) with castellations and its associated land pattern. The view is from the top; its lid is clearly visible. Figure 9.47 shows a 68-termination LCC (LCC 68) with castellations and its associated land pattern. The view is from the bottom; pin one having an elongated pad is clearly visible.

TABLE 9.13   Typical SOJ/0.350 Component Dimensions, inches (mm)

No. of leads	Length, L	Width, W	Height H, (max.)	Tip-to-tip width, W_{T-T}	Lead width, W_L	Component pad length, T	Pitch, P (nominal)
14	0.380–0.392 (9.65–9.96)	0.345–0.355 (8.76–9.02)	0.148 (3.75)	0.380–0.395 (9.65–10.03)	0.015–0.020 (0.38–0.51)	0.035–0.040 (0.89–1.02)	0.050 (1.27)
16	0.430–0.442 (10.92–11.23)	0.345–0.355 (8.76–9.02)	0.148 (3.75)	0.380–0.395 (9.65–10.03)	0.015–0.020 (0.38–0.51)	0.035–0.040 (0.89–1.02)	0.050 (1.27)
18	0.480–0.492 (12.19–12.50)	0.345–0.355 (8.76–9.02)	0.148 (3.75)	0.380–0.395 (9.65–10.03)	0.015–0.020 (0.38–0.51)	0.035–0.040 (0.89–1.02)	0.050 (1.27)
20	0.530–0.542 (13.46–13.77)	0.345–0.355 (8.76–9.02)	0.148 (3.75)	0.380–0.395 (9.65–10.03)	0.015–0.020 (0.38–0.51)	0.035–0.040 (0.89–1.02)	0.050 (1.27)
22	0.580–0.592 (14.73–15.04)	0.345–0.355 (8.76–9.02)	0.148 (3.75)	0.380–0.395 (9.65–10.03)	0.015–0.020 (0.38–0.51)	0.035–0.040 (0.89–1.02)	0.050 (1.27)
24	0.630–0.642 (16.00–16.31)	0.345–0.355 (8.76–9.02)	0.148 (3.75)	0.380–0.395 (9.65–10.03)	0.015–0.020 (0.38–0.51)	0.035–0.040 (0.89–1.02)	0.050 (1.27)
26	0.680–0.692 (17.27–17.58)	0.345–0.355 (8.76–9.02)	0.148 (3.75)	0.380–0.395 (9.65–10.03)	0.015–0.020 (0.38–0.51)	0.035–0.040 (0.89–1.02)	0.050 (1.27)
28	0.730–0.742 (18.54–18.85)	0.345–0.355 (8.76–9.02)	0.148 (3.75)	0.380–0.395 (9.65–10.03)	0.015–0.020 (0.38–0.51)	0.035–0.040 (0.89–1.02)	0.050 (1.27)

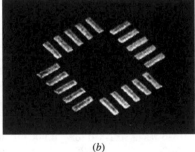

(*a*)   (*b*)

FIGURE 9.46   Leadless chip carrier with 20 terminations (LCC 20); (*a*) TSOP and (*b*) footprint.

LCCs are square in shape and have terminations on all four sides. Hence, they can be considered as peripheral array packages (PAPs). The pin number one mark is generally in the center of one of the sides. However, the manufacturer should always be consulted regarding the specific location of pin number one. Figure 9.48 (reference drawing) shows a top, bottom, and side view of a leadless chip carrier. Table 9.14 gives the component dimensions.

9.5.3.4 Leaded Chip Carriers. These can have either plastic or ceramic bodies. The plastic leaded chip carrier (PLCC) generally has leads in the J-lead configuration at 0.050

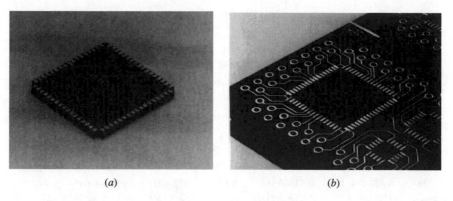

(a) *(b)*

FIGURE 9.47 Leadless chip carrier with 68 terminations (LC 68); (*a*) LCC, bottom view, and (*b*) footprint.

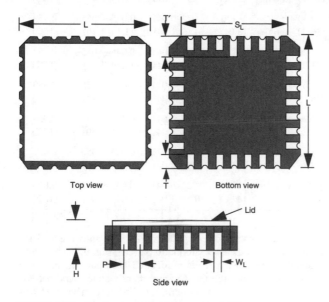

FIGURE 9.48 Top, bottom, and side view of an LCC.

in or 0.040 in pitch. The number of leads normally does not exceed 84. The ceramic leaded chip carrier (CLCC) also has J-leads. It is often referred to as a J-lead Cerquad. In both cases (PLCC and CLCC) leads are found on all four sides. Based on the classification scheme in Fig. 9.15, these are also classified as peripheral array packages with leads around all four sides.

The pin numbering system is different from that of the SOIC packages; the pin number one mark is typically in the center of one of the sides. However, the manufacturer should always be consulted regarding the specific location of pin number one. Figure 9.49 (reference drawing) shows a top and side view of a J-lead ceramic chip carrier. Table 9.15 gives the component dimensions. Also see Fig. 9.50 showing a 20-pin (lead) plastic leaded chip

TABLE 9.14   Typical LCC Component Dimensions, inches (mm)

No. of leads	Length/ width, L	Side length/side width, S_L	Height H, (max.)	Termination width, W_L	Component pad length, T	Pitch, P (nominal)
16	0.292–0.308 (7.42–7.82)	0.176–0.203 (4.46–5.16)	0.100 (2.54)	0.022–0.041 (0.56–1.04)	0.045–0.055 (1.14–1.40)	0.050 (1.27)
20	0.342–0.358 (8.69–9.09)	0.233–0.253 (5.91–6.43)	0.100 (2.54)	0.022–0.041 (0.56–1.04)	0.045–0.055 (1.14–1.40)	0.050 (1.27)
24	0.395–0.410 (10.04–10.41)	0.286–0.306 (7.26–7.76)	0.100 (2.54)	0.022–0.041 (0.56–1.04)	0.045–0.055 (1.14–1.40)	0.050 (1.27)
28	0.442–0.458 (11.23–11.63)	0.332–0.353 (8.45–8.97)	0.100 (2.54)	0.022–0.041 (0.56–1.04)	0.045–0.055 (1.14–1.40)	0.050 (1.27)
44	0.640–0.660 (16.26–16.76)	0.531–0.554 (13.48–14.08)	0.120 (3.04)	0.022–0.041 (0.56–1.04)	0.045–0.055 (1.14–1.40)	0.050 (1.27)
52	0.739–0.761 (18.78–19.32)	0.630–0.655 (16.00–16.64)	0.120 (3.04)	0.022–0.041 (0.56–1.04)	0.045–0.055 (1.14–1.40)	0.050 (1.27)
68	0.938–0.962 (23.83–24.43)	0.829–0.856 (21.05–21.74)	0.120 (3.04)	0.022–0.041 (0.56–1.04)	0.045–0.055 (1.14–1.40)	0.050 (1.27)
84	1.14–1.16 (28.83–29.59)	1.03–1.06 (26.05–26.88)	0.120 (3.04)	0.022–0.041 (0.56–1.04)	0.045–0.055 (1.14–1.40)	0.050 (1.27)
100	1.34–1.36 (34.02–34.56)	1.23–1.26 (31.24–31.88)	0.160 (4.06)	0.022–0.041 (0.56–1.04)	0.045–0.055 (1.14–1.40)	0.050 (1.27)
124	1.64–1.66 (41.64–42.18)	1.53–1.56 (38.86–39.50)	0.160 (4.06)	0.022–0.041 (0.56–1.04)	0.045–0.055 (1.14–1.40)	0.050 (1.27)
156	2.04–2.06 (51.80–52.34)	1.93–1.96 (49.02–49.66)	0.160 (4.06)	0.022–0.041 (0.56–1.04)	0.045–0.055 (1.14–1.40)	0.050 (1.27)

carrier (PLCC 20) with J-leads and its associated land pattern; Fig. 9.51 showing a 44-pin plastic leaded chip carrier (PLCC 44) with its associated land pattern; Fig. 9.52 showing a 68-pin plastic leaded chip carrier (PLCC 68) with its associated land pattern; Fig. 9.53 showing a 68-pin ceramic leaded chip carrier (CLCC 68) and its associated land pattern. In this case, the CLCC is a cavity down type component, and the bottom view clearly shows the die cavity. The die cavity is empty in this example. Normally, it would contain an integrated circuit wire bonded to the pads surrounding the inside periphery of the cavity. The component would also have a lid made of a suitable material sealing the cavity.

9.5.3.5 *Quad Flat Packs (QFPs).*   Again, based on the classification scheme in Fig. 9.15, these also are classified as peripheral array packages with leads around all four sides. Quad flat packs (often referred to as "quad packs" or "Q-packs") have leads in the gull wing configuration. They come in either a ceramic (CQFP) or plastic body (PQFP). The CQFP is also commonly called a gull wing Cerquad.

As pointed out, the gull wing leads of QFPs are very delicate and easily damaged. In general, the finer the pitch, the thinner the lead and the more susceptible it is to damage. Damage is particularly likely to occur during (1) shipping/handling and (2) component placement.

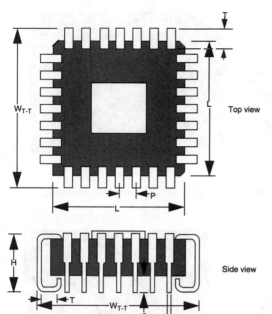

FIGURE 9.49 Top and side view of a J-lead CLCC.

TABLE 9.15 Typical Leaded CC Component Dimensions, inches (mm)

No. of leads	Length/ width, L	Height, H (maximum)	Tip-to-tip width, W_{T-T}	Lead width, W_L	Component pad length, T	Pitch, P (nominal)
20	0.350–0.356 (8.89–9.04)	0.180 (4.57)	0.385–0.395 (9.78–10.03)	0.013–0.021 (0.33–0.53)	0.035–0.040 (0.89–1.02)	0.050 (1.27)
28	0.450–0.456 (11.43–11.58)	0.180 (4.57)	0.485–0.495 (12.32–12.57)	0.013–0.021 (0.33–0.53)	0.035–0.040 (0.89–1.02)	0.050 (1.27)
44	0.650–0.656 (16.51–16.66)	0.180 (4.57)	0.685–0.695 (17.40–17.65)	0.013–0.021 (0.33–0.53)	0.035–0.040 (0.89–1.02)	0.050 (1.27)
52	0.750–0.756 (19.05–19.20)	0.200 (5.08)	0.785–0.795 (19.94–20.19)	0.013–0.021 (0.33–0.53)	0.035–0.040 (0.89–1.02)	0.050 (1.27)
68	0.950–0.958 (24.13–24.33)	0.200 (5.08)	0.985–0.995 (25.03–25.27)	0.013–0.021 (0.33–0.53)	0.035–0.040 (0.89–1.02)	0.050 (1.27)
84	1.150–1.158 (29.21–29.41)	0.200 (5.08)	1.185–1.195 (30.10–30.35)	0.013–0.021 (0.33–0.53)	0.035–0.040 (0.89–1.02)	0.050 (1.27)
100	1.350–1.358 (34.29–34.49)	0.200 (5.08)	1.385–1.395 (35.18–35.43)	0.013–0.021 (0.33–0.53)	0.035–0.040 (0.89–1.02)	0.050 (1.27)
124	1.650–1.658 (41.91–42.11)	0.200 (5.08)	1.685–1.695 (42.80–43.05)	0.013–0.021 (0.33–0.53)	0.035–0.040 (0.89–1.02)	0.050 (1.27)

(a)   (b)

FIGURE 9.50   Plastic leaded chip carrier with 20 leads (PLCC 20); (a) PLCC and (b) footprint.

(a)   (b)

FIGURE 9.51   Plastic leaded chip carrier with 44 leads (PLC 44); (a) PLCC and (b) footprint.

Regarding pin number one, the manufacturer should always be consulted regarding the specific location. Figure 9.54 (reference drawing) shows a top and side view of a gull wing ceramic quad flat pack. Table 9.16 gives the component dimensions.

Figure 9.55 shows (1) a flat Cerquad (CQFP) with its carrier ring (also called a guard ring or tie bars), (2) its carrier ring excised with the leads extending straight out from the package body, (3) the Cerquad with its leads formed in the shape of a gull wing, and (4) its corresponding footprint (land pattern configuration) on the printed wiring board surface. The pitch is 0.050 in. Figure 9.56 shows a PQFP having protruding bumpered ends to protect the leads during handling. It has 132 leads (BQFP 132). Its pitch is 0.025 in. Figure 9.57 shows a different PQFP (Japanese make) without the protective bumpers. It has 208 leads (PQFP 208). Its pitch is 0.0197 in (0.500 mm). Figure 9.58 shows a CQFP having 256 leads (CQFP 256). Its pitch is 0.020 in.

If a carrier ring is used (see Fig. 9.55), damage can be minimized during shipping/handling. But once the leads are formed, they are subject to damage. In many cases, ceramic QFPs (CQFPs) are supplied in a flat configuration with a carrier ring to protect the leads. To use, the ring must be excised and the leads formed.

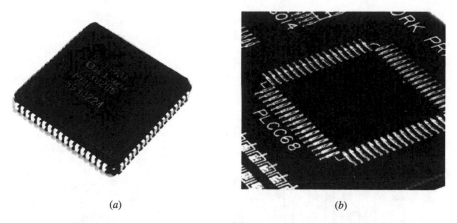

FIGURE 9.52 Plastic leaded chip carrier with 68 leads (PLC 68); (*a*) PLCC and (*b*) footprint.

FIGURE 9.53 Ceramic leaded chip carrier with 68 leads; (*a*) CLCC top, (*b*) CLCC bottom, cavity down, and (*c*) footprint.

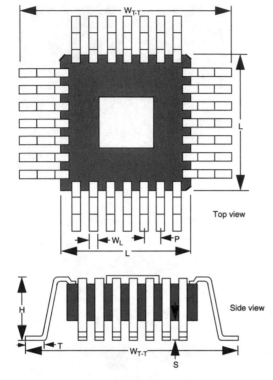

Top view

Side view

FIGURE 9.54 Top and side view of a gull wing QFP.

PQFPs are normally supplied in matrix trays with their leads already formed. In the U.S., the approved JEDEC version employs a protruding bumper to protect the leads from damage. See Fig. 9.56. The Japanese version of the PQFP does not have the protective bumpers. See Fig. 9.57.

A word should be said about metric units. Figure 9.57 shows a PQFP with a pitch of 0.500 mm. That is, its primary reference dimensions are in metric, or SI, units, not English, or inch-pound (IP), units. Generally, metric units can not be directly converted into round English units. A pitch of 0.500 mm is only approximately equal to 0.020 in. It is, more exactly, equal to 0.0197 in.

Note: If packages are referenced in metric units, this may have a serious impact if the land pattern dimensions are referenced in English units.

One of the great advantages of QFPs is their ability to accommodate a large number of leads. This means they can have a high I/O or high pin count. In particular, refer to Figs. 9.56–9.58. High pin count QFPs are classified as fine or ultrafine pitch SMCs. That is, their pitch is equal to or less than 0.025 in; often it is less than 0.020 in. See Sec. 9.5.7, which deals with high lead count packages.

9.5.3.6 Area Array Packages (AAPs). Examples of these are ball grid arrays (BGAs), land grid arrays (LGAs), and column grid arrays (CGAs). Unlike LCCs, leaded chip carriers, and quad packs, in which the terminations or leads are arranged around the periphery of the component, area array packages have terminations arranged not only along the component periphery but also on the bottom of the surface area. Hence the term "area array."

TABLE 9.16 Typical QFP Component Dimensions, inches (mm)

No. of leads	Length/ width, L	Height, H (max.)	Tip-to-tip width, W_{T-T}	Lead width, W_L	Component pad length, T	Pitch, P (nominal)
28	0.360–0.400 (9.15–10.16)	0.090 (2.30)	0.567–0.583 (14.40–14.80)	0.008–0.013 (0.20–.033)	0.035–0.040 (0.89–1.02)	0.050 (1.27)
36	0.460–0.500 (11.69–12.70)	0.195 (4.95)	0.675–0.685 (17.15–17.39)	0.008–0.013 (0.20–.033)	0.035–0.040 (0.89–1.02)	0.050 (1.27)
44	0.560–0.600 (14.23–15.24)	0.195 (4.95)	0.775–0.785 (19.69–19.93)	0.008–0.013 (0.20–.033)	0.0350–.040 (0.89–1.02)	0.050 (1.27)
52	0.660–0.700 (16.77–17.78)	0.195 (4.95)	0.875–.885 (22.23–22.47)	0.008–.013 (0.20–.033)	0.035–.040 (0.89–1.02)	0.050 (1.27)
68	0.860–0.900 (21.85–22.86)	0.195 (4.95)	1.075–1.085 (27.31–27.55)	0.008–0.013 (0.20–.033)	0.035–0.040 (0.89–1.02)	0.050 (1.27)
84	1.060–1.100 (26.93–27.94)	0.195 (4.95)	1.275–1.285 (32.39–32.63)	0.008–0.013 (0.20–.033)	0.035–0.040 (0.89–1.02)	0.050 (1.27)
100	1.260–1.300 (32.01–33.02)	0.195 (4.95)	1.475–1.485 (37.47–37.71)	0.008–0.013 (0.20–.033)	0.035–0.040 (0.89–1.02)	0.050 (1.27)
132	0.935–0.960 (23.75–24.38)	0.140 (3.55)	1.075–1.085 (27.31–27.56)	0.006–0.015 (0.15–0.38)	0.040–0.045 (1.02–1.14)	0.025 (0.635)
148	1.110–1.130 (28.19–28.71)	0.120 (3.05)	1.325–1.335 (33.65–33.91)	0.005–0.010 (0.12–0.25)	0.040–0.045 (1.02–1.14)	0.025 (0.635)
164	1.135–1.155 (28.83–29.34)	0.130 (3.30)	1.325–1.335 (33.65–33.91)	0.005–0.010 (0.12–0.25)	0.040–0.045 (1.02–1.14)	0.025 (0.635)
196	1.330–1.350 (33.78–34.30)	0.135 (3.43)	1.410–1.420 (35.81–36.07)	0.005–0.010 (0.12–0.25)	0.040–0.045 (1.02–1.14)	0.025 (0.635)
256	1.130–1.150 (28.70–29.21)	0.100 (2.54)	1.350–1.360 (34.29–34.54)	0.004–.008 (0.10–0.20)	0.045–.050 (1.14–1.27)	0.020 (0.508)

BGAs have small balls of solder on the bottom of the component; LGAs have pads of solder on their bottom side; CGAs have small columns of solder (about 0.090 in long) on their bottom side. They are the package analog of the flip-chip technology developed by IBM for denser interconnections. The balls, pads, or columns are currently at a pitch of 0.060 in (1.5 mm), 0.050 in (1.27 mm), or 0.040 in (1.0 mm). A plastic ball grid array is shown in Fig. 9.59.

As the complexity of the integrated circuit increases, so the number of external terminations (I/Os) also increases. This trend is pushing peripheral leaded components to pitches under 0.020 in. Since processing such ultrafine pitch components is very difficult, an alternative is to use area array packages to achieve increased pin count without the attendant processing difficulties of ultrafine pitch QFPs. The chief limitations of area array packages are (1) lack of direct visual inspectability and (2) possible difficulties in rework/repair. But it is probably easier to rework area array packages than ultrafine pitch QFPs.

FIGURE 9.55 Ceramic quad flat pack with 68 leads (CQFP 68); (*a*) CQFP w/carrier ring, (*b*) CQFP w/o carrier ring, (*c*) CQFP w/leads formed, and (*d*) footprint.

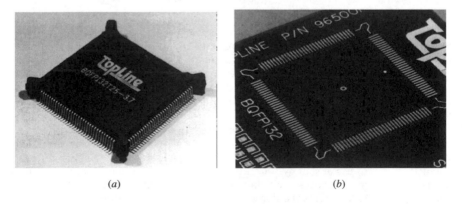

FIGURE 9.56 Bumpered plastic flat pack with 132 leads (BQFP 132); (*a*) BQFP and (*b*) footprint.

(*a*)   (*b*)

FIGURE 9.57   Plastic quad flat pack with 208 leads (PQFP 208); (*a*) PQFP and (*b*) footprint.

(*a*)   (*b*)

FIGURE 9.58   Ceramic quad flat pack with 256 leads (CQFP); (*a*) CQFP and (*b*) footprint.

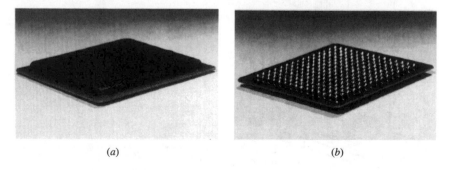

(*a*)   (*b*)

FIGURE 9.59   Ball grid array (BGA); (*a*) top and (*b*) bottom.

9.5.4 Other Surface Mount Components

9.5.4.1 Connectors. Surface mount connectors are becoming more prevalent, but it must be kept in mind that connectors can experience forces that other surface mount components are not subjected to, such as entry and withdrawal forces. It is best to consider zero-insertion force connectors when considering surface mount connectors. Otherwise, if a surface mount connector is used, it may be necessary to mechanically fasten the connector to the printed wiring board to avoid rupturing the solder joints during insertion and withdrawal of its mating connector.

9.5.4.2 Inductors. Inductors have fine wire wound around a ceramic or ferrite core. There are two styles for surface mount inductors. These are

1. Vertical windings—(*a*) in Fig. 9.60
2. Horizontal windings—(*b*) in Fig. 9.60

9.5.5 Plastic Encapsulated Microcircuits

Several of the surface mount components discussed above are found in plastic bodies; these are often referred to as plastic encapsulated microcircuits (PEMs). Among these are: SOICs and SOJICs. PLCCs and PQFPs are also plastic packages as are many of the AAP packages (ball grid arrays).

There are two principal methods of packaging semiconductor microcircuits to protect them from the ambient atmosphere to which they are exposed:

- Hermetically sealed microcircuits (HSMs)
- Plastic encapsulated microcircuits (PEMs)

Figure 9.61 shows the internal construction of a typical HSM, and Fig. 9.62 shows the construction of a typical PEM.

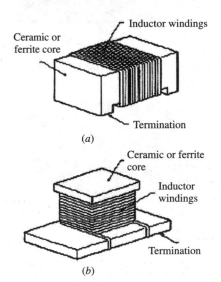

FIGURE 9.60 Surface mount inductors, two styles. *(Courtesy of Coombs, 1988.)*

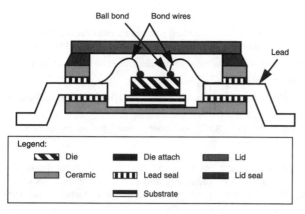

FIGURE 9.61 Hermetic sealed microcircuit (HSM).

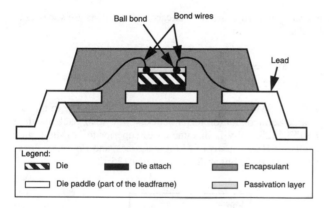

FIGURE 9.62 Plastic encapsulated microcircuit (PEM).

Historically HSMs have found wide use in military, avionics, space, under-the-hood automotive, and other situations where reliability and outstanding performance are the chief concerns. PEMs, on the other hand, have found wide use in telecommunications, computer, industrial, under-the-dashboard automotive, and commercial electronics applications where the service environment is more benign and/or the application is not as critical as those employing HSMs.

The ever shrinking military market, increasing expectations of higher product performance, and greater emphasis on cost savings have led to a renewed interest in PEMs for many applications where their use previously was categorically disallowed. The long-held perception is that PEM components are inherently unreliable due to the potential of moisture ingression followed by subsequent deterioration of performance.

However, during the last decade, numerous advancements have been made in the design and materials of PEMs resulting in much more reliable packages. These advancements have been incorporated into best commercial practice parts (BCPs). These advancements include:

- Improved leadframe design
- Increased plastic purity including much lower residual ionics and better matching of the plastic's coefficient of thermal expansion (CTE) to those of the die and lead frame
- Improved adhesion of plastic encapsulant material to the die and lead frame materials
- Plastic having much lower moisture absorption properties
- High quality microcircuit passivation (e.g., spun-on glass [SOG]) with better adhesion to the die
- Application of statistical process control (SPC) and continuous process improvement (CPI) to PEM manufacture and assembly

Note: For high reliability applications, best commercial practice (BCP) parts should be specified. Not all PEMs are equal in terms of reliability and performance. Always carefully investigate the particular PEM vendor under consideration. BCP parts are not the same as high-volume commercial parts.

PEMs offer several distinct advantages over HSMs. These are:

- They offer lower cost.
- Because the die and its internal bond wires are fully encapsulated, PEMs are superior to HSMs for mechanical shock and vibration.
- They are lighter in weight.
- Many integrated circuits are available as PEMs only.
- Component CTE more closely matches CTE of organic substrate.

The gap in reliability between HSMs and PEMs has decreased dramatically in the last decade. Data from several large semiconductor manufacturers who also package their semiconductors indicate that PEM reliability is in some cases superior to HSMs. Figure 9.63 shows operating life results for HSM versus PEM logic devices, and Fig. 9.64 shows operating life results for HSM versus PEM linear devices, both at 125°C.

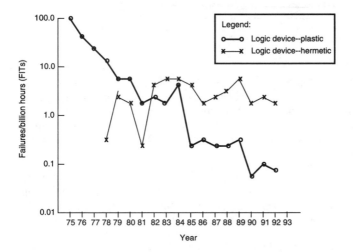

FIGURE 9.63 Plastic vs. hermetic reliability, logic devices. *(Courtesy of Texas Instruments.)*

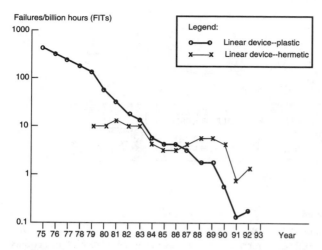

FIGURE 9.64   Plastic vs. hermetic reliability, linear devices. *(Courtesy of Texas Instruments.)*

There are two principal failure mechanisms in the case of plastic encapsulated microcircuits (PEMs).

- Thermomechanical induced failure
- Moisture induced failure

Thermomechanically induced failure is due chiefly to the differences in coefficient of thermal expansion (CTE) between the plastic encapsulant material, the die (silicon), and the leadframe material (for SMCs, the leadframe is normally a copper alloy, but may be Kovar or Alloy 42). Greatly improved plastic material along with adjustment of the filler material has largely corrected this failure mechanism.

Moisture absorption along the leadframe or directly through the bulk plastic gives rise to corrosion of the aluminum metallization, corrosion of the wire bonds (if aluminum is used), and increased ionic mobility. In addition, moisture absorbed in the plastic can lead to a phenomenon called "popcorning" during reflow soldering. Popcorning refers to the noise made as the absorbed moisture turns to steam during reflow, sounding much like popcorn popping in a pot. This often results in component cracking. Figure 9.65 illustrates this effect.

Improved plastic molding compounds having very low moisture absorption and ion scavengers results in low moisture absorption.

If PEMs are to be used, they ought to be stored in an inert atmosphere such as dry nitrogen (N_2) to prevent moisture absorption. Within one hour prior to assembly, the PEMs should be baked to remove moisture. If a vacuum oven is available, the following bakeout schedule must be followed:

Temperature 75°C ± 5°C, duration 24 h ± 0.5 h

If a vacuum oven is not available, the following bakeout schedule should be followed:

Temperature 100°C ± 5°C, duration 24 h ± 0.5 h, oven humidity < 50 percent RH

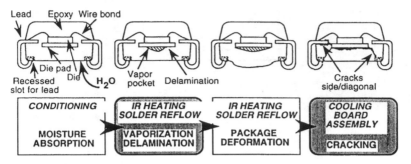

FIGURE 9.65   "Popcorn" effect with PEMs.

9.5.6 Multichip Modules

The driver behind the push to MCM technology is the continuing increase in semiconductor technology. See Fig. 9.66. MCMs represent a technique whereby very high pin counts can be achieved on the MCM substrate; the final MCM package will have fewer pins (or leads) than are found internally.

There are a number of techniques for attaching the die to the substrate in MCM technology. These are:

- Wire bonds
- Flip chip solder bumps
- Tape automated bonding (TAB) leads

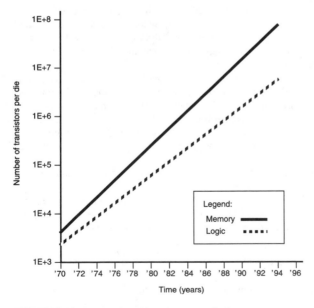

FIGURE 9.66   Increase in semiconductor complexity.

Figure 9.67 illustrates the different bonding techniques for MCMs along with the concept of a multilayer substrate.

The MCM packaging concept evolved out of that of hybrid microcircuit electronics; it entails placing a number of integrated circuit dice and supporting discrete chips on a single substrate. For example, see Fig. 9.68. This substrate is often a multilayer. This entire ensemble is then packaged using the packaging techniques common to single chip modules (SCMs), namely, it is packaged as either a hermetically sealed microcircuit (HSM) or a plastic encapsulated microcircuit (PEM). For PEMs see Sec. 9.5.5.

MCMs can be classified by substrate type. There are three classifications. They are:

1. *MCM-L.* Substrate based on laminated, multilayer printed wiring board technology. Polyimide and bismaleimide-triazine (BT) are often the laminate substrate materials used.

2. *MCM-C.* Substrates based on cofired ceramic or glass-ceramic technology.

3. *MCM-D.* Interconnection pattern created by depositing dielectrics and conductors on a base substrate, typically by thin film processes.

MCM-L is an extension of chip-on-board (COB) technology where the bare IC dies are mounted directly on a suitable organic substrate material. A typical material is BT-epoxy/fiberglass laminate using a multilayer construction. Figure 9.68 shows an example of this kind of MCM. In this case the packaging technology is PEM.

MCM-C is an extension of thick film hybrid technology. The package style is generally a ceramic body, hermetically sealed, that is, an HSM.

MCM-D involves vapor deposition processes to deposit layers of dielectric material. For example, a plasma enhanced chemical vapor deposition process (PECVD) has been used successfully to deposit layers of silicon dioxide. Organic dielectrics such as polyimide and benzocyclobutene can be deposited using chemical vapor deposition (CVD) techniques. The fine line lithographic technique (10–25 mm; 1 mil = 25 mm) used in MCM-D technology can produce very high density interconnections.

9.5.7 High Lead Count Packages

The trend in electronics is to continually decrease the feature size of the microelectronic devices (ICs), which are subsequently packaged to provide protection from a hostile envi-

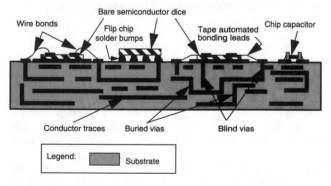

FIGURE 9.67 Different die attach techniques on an MCM substrate.

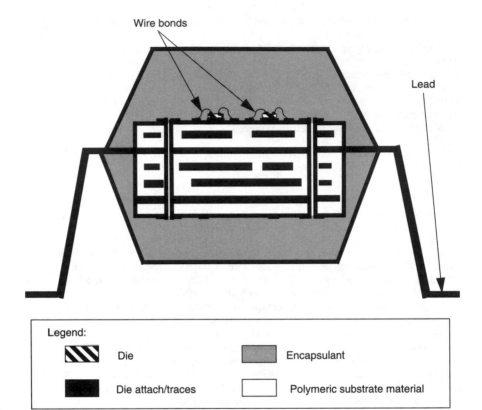

FIGURE 9.68 Seven-layer MCM-L packaged as a gull wing PEM.

ronment. As the feature size decreases, many more transistors can be placed on the micro-electronic substrate. For example, there are today IC devices where each individual IC contains several million transistors. This increases the pin count (I/O count, lead count).

There are essentially two ways to increase the number of pins without unduly increasing the package size:

1. Decrease the distance between the center of a lead and that of an adjacent lead, that is, decrease the lead pitch.

2. Cover the entire bottom area, or most of the area, of the component with leads.

In fact, both of these techniques are presently being used as packaging strategies to increase the number of available I/Os. The first strategy leads to ultrafine pitch quad flat packs (UFP QFPs) defined as having a lead pitch of 0.500 mm (0.01969 in) and under. The second strategy leads to area array packages, such as ball grid arrays, which were mentioned above.

Both strategies have advantages and drawbacks. The principal advantage of both is the increased pin size. The chief drawback of the UFP QFPs is that the leads, because they are very close together, are very delicate and hence easily damaged. Therefore, they are very difficult to process. AAPs, including BGAs, can be processed using standard SMT techniques. Their chief drawbacks are:

- One cannot visually inspect the leads under the component.
- Reworking a single lead after attachment is not possible; the entire component must be removed and then resoldered. This may compromise its reliability.

The main reason that BGAs can be processed using conventional SMT manufacturing techniques is that the lead pitch is generally 0.060 in, 0.050 in, and in some cases 0.040 in. Newer BGAs are being developed having a greatly reduced pitch, e.g., 0.500 mm or less. However, the balls used as the leads in BGAs are in no way as delicate as the gull wing leads used in UFP QFPs. Hence, they will be much less susceptible to damage and easier to process.

Leads between 300 to 600 are feasible with 0.050 in pitch BGAs; in the case of BGAs with a 0.500 mm pitch, more than 1000 leads is possible. It presently appears that the optimum number of leads used in conjunction with UFP QFPs is 200–350. Hence, over this number some type of AAP would be necessary.

It is appropriate to mention that one method for increasing the lead count is to dispense entirely with the package of the IC die and mount the die directly on the substrate surface. This is known as direct chip attach (DCA). There are three common ways of attaching the die directly to a substrate:

1. Wire bonding from die pads to bonding pads on the substrate (chip on board or COB)
2. The flip-chip technique where the die itself has small bumps of solder and these small solder bumps are then mounted on top of pads on the substrate (flip-chip DCA)
3. TAB (tape automated bonding)

Figure 9.67 shows different die attach techniques.

To protect the die, it is normal operating procedure to encapsulate the die using a suitable polymeric material such as an epoxy. This is often called glob-topping.

9.6 SYSTEM REQUIREMENTS

9.6.1 Functional Requirements

In a very broad sense, the goals and objectives of the overall system, whatever it may be, should drive the detailed design, packaging, and development requirements. The electronics hardware must meet the required use conditions and use environment, and it must function as intended for the entire defined life of the product, whatever it may be—from radio through complex spacecraft.

The system goals and objectives address the overall system design and development. The electronics hardware constitutes an integral part of many of the subsystems composing the system.

For a given system, the following packaging considerations must be determined:

- Definition and years of expected service
- Service environment and service environment characteristics
- Expected thermal environments, including power cycling
- Expected system reliability

9.6.2 System Reliability

System reliability means that the system performs as it is intended to perform over its intended life in its expected service environment. Strictly speaking, reliability means the probability that the system will perform as intended over its intended life in its expected service environment. If $R(x)$ is the reliability function, then $R(x)$ has two commonly acceptable meanings:

1. $R(x)$ is the probability that a random unit drawn from a population will still be operating after x hours.
2. $R(x)$ is the fraction of all units in the population that will survive at least x hours.

There are several important functions performed by reliability engineering. These are shown in Fig. 9.69. Design for X (DfX), the general concept, is addressed in Sec. 9.7.8. Another important concept related to reliability is the way failure can occur. There are two principal ways it can occur:

1. Catastrophic or overstress failure
2. Wearout failure

Catastrophic failure sometimes occurs when the product is subjected to accelerated stress testing. This is done to flush out those products that would fail due to overstress and to determine the wearout point. Wearout typically occurs much later in a product's life cycle.

There is a curve associated with overstress known as the infant mortality curve. A generalized infant mortality curve for a particular failure mode is shown in Fig. 9.70. There is also a curve associated with wearout. A generalized wearout curve for a particular failure mode is shown in Fig. 9.71. These two curves can be combined to produce the overall total failure curve for the product, shown in Fig. 9.72. Sometimes this curve is referred to as the "bathtub" curve because of its overall shape.

Note: The wearout point in Fig. 9.72 is the point where the failure curve no longer exhibits a constant failure rate.

Most reliability engineers do not believe the failure rate to be truly constant; it is assumed to be constant to a first approximation. Knowing the wearout point allows reliability engineers to make a useful prediction of a product's life.

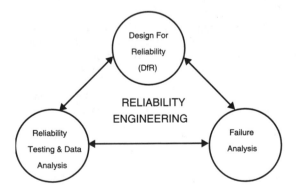

FIGURE 9.69 Different functions performed by reliability engineering.

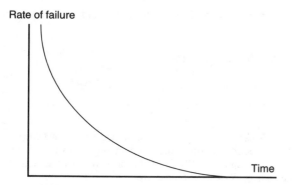

FIGURE 9.70   Generalized infant mortality failure curve.

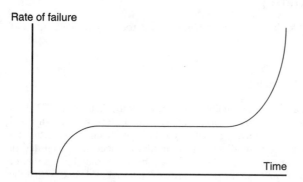

FIGURE 9.71   Generalized wearout failure curve.

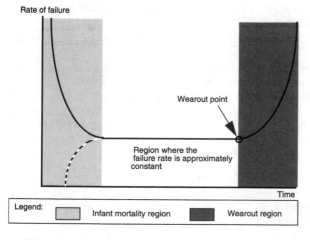

FIGURE 9.72   Generalized failure curve.

Infant mortality occurs early in a product's life. The causes of early failure include:

- Inadequate quality control
- Poor manufacturing, storage or transportation processes
- Bad design
- Faulty materials
- Inadequate burn-in

The failure rate during a product's useful life is generally assumed to be constant (see Fig. 9.72). However, after a certain period of time, the product begins to experience wearout failures. For surface mount technology (SMT) at the printed wiring assembly level, the two chief items that are subject to wearout failure are:

1. Solder joints
2. Vias/plated through holes (PTHs)

Note: Fatigue/creep of the solder joints is the predominant wearout failure mechanism of surface mount PWAs.

9.6.3 Physics of Failure Approach

Physics of failure (PoF) is an approach to reliability and design for reliability (DfR). This approach seeks to understand failures by a careful analysis of failure mechanisms and where they are likely to occur. Ultimately, the goal of the PoF approach is to construct a failure model whereby failures can be predicted. If the model is successful, then design rules can be formulated such that either the failure mechanisms are avoided altogether or their specific effects can be calculated and the product life cycle usage bounded so that the product is assured to last as long as it is intended within its intended service environment.

Four concepts are of especial importance for understanding the PoF approach. These are:

1. *Load.* The electrical, mechanical, or chemical stresses that the product is expected to experience during its entire life cycle.
2. *Failure site.* The particular site where the failure occurs. The failure site is geometry dependent.
3. *Failure mode.* The macroscopic manifestation of the failure mechanism. This normally is a visible phenomenon such as delamination or a crack in a solder joint. It also can be a measurement such as a significant change in resistance.
4. *Failure mechanism.* The hypothesized sequence of causes and effects leading to the failure mode. The initial cause of the entire chain of events leading to failure is the stress (load) on the system. The failure mechanism typically proceeds at the submicroscopic level.

Four examples are given here to illustrate these important concepts.

Example 1. A change in relative humidity (ΔRH) can lead to dendritic growth on the printed wiring board surface between adjacent conductor traces. The problem is exacerbated by narrower spaces between conductor traces.
- Load: change in ΔRH.
- Failure site: surface on PWB between two conductor traces.

- Failure modes: intermittent signals, leakage currents.
- Failure mechanism: dendritic growth between traces.

Example 2. A too rapid change in temperature during the soldering process leads to delamination of a multilayer PWB.
- Load: rapid change in temperature ($dT/dt \gg 0$).
- Failure site: internal layers of MLB.
- Failure mode: physical separation of the PWB's internal layers.
- Failure mechanism: absorbed water turning into steam and expanding, causing interfacial delamination.

Example 3. Ionic residue (contamination) remaining on the assembly surface near the recently soldered lands of some of the SMCs results in corrosion products and ultimately the disappearance of a small portion of the conductor trace due to the formation of corrosion products.
- Load: chemical contamination.
- Failure site: a conductor trace adjacent to lands having ionic contaminants remaining after flux/soldering.
- Failure mode: electrical open; evidence of corrosion products (e.g., white powder).
- Failure mechanism: mobile ionic species; moisture induced ionic corrosion.

Example 4. Excessive growth of intermetallic compounds (IMCs) in solder joints (SJs) due to too high a temperature excursion in the reflow tunnel over too long a time leads to brittle fracture of the SJs.
- Load: large ΔT, large t.
- Failure site: base of SJs.
- Failure mode: SJs visibly cracked; opens detected by electrical continuity checker.
- Failure mechanism: IMCs form and continue to grow due to excessively high t and long t.

At the PWB/PWA level, Table 9.17 gives a matrix of different failure modes associated with different loads and Table 9.18 gives a corresponding matrix of different failure sites associated with different failure modes.

TABLE 9.17 Typical Failure Modes/Loads

Failure mode	Contamination	ΔRH (RH_{min} and RH_{max})	ΔT (T_{min} and T_{max})	ΔCTE	Thermal shock	Vibration/ shock	High voltage	Time
Bow and twist		X	X	X	X	X		
Interfacial delamination	X	X	X	X	X			
Loss of mechanical rigidity			X		X			
Creep								X
Fatigue crack			X	X				
Brittle fracture						X		
Ductile fracture			X	X				
Dendritic growth	X	X					X	
Corrosion	X	X	X					
Vesication*	X	X						
Cross talk	X	X					X	
Dielectric breakdown	X	X					X	

* Vesication = blistering between a conformal coating and the underlying substrate.

TABLE 9.18 Typical Failure Modes/Failure Sites

Failure mode	Entire PWB	Conductor traces	PTHs/ vias	Spacing between conductors	Lands	Component leads	Conformal coating
Bow and twist	X						
Interfacial delamination	X				X		X
Loss of mechanical rigidity	X						
Creep						X	
Fatigue crack			X			X	
Brittle fracture			X			X	
Ductile fracture			X			X	
Dendritic growth				X			
Corrosion		X			X		
Vesication							X
Cross talk		X		X		X	X
Dielectric breakdown	X			X			X

9.6.4 Product Life Cycle

All too often the design effort has centered principally on the functional requirements (FRs); very often the life cycle requirements (LCRs) are considered only as an afterthought. The product life cycle extends through the following stages:

- Manufacturing
- Inspection and quality assurance
- Test and qualification
- Reliability
- Safety/environmental considerations
- Maintainability
- Procurement
- Storage
- Use in service environment
- Disposal

To ensure that the life cycle requirements are properly considered at the design/development stage is the chief purpose of the above suggestion regarding the formation of a concurrent engineering SMT team.

9.6.5 Life Cycle Stages

It is appropriate to apply the PoF approach over the entire anticipated life cycle usage profile of the product in question. The life cycle of a given entity can be discretely segmented into several distinct stages. These are given below.

The total life cycle environment of a system consists of several distinct stages, each of which constitutes a unique subenvironment. The ten life cycle stages (subenvironments) are

1. Storage, handling, and transport of components prior to printed wiring board (PWB) assembly
2. Fabrication and assembly of the surface mount PWAs
3. Storage, handling, and transport of the PWAs prior to integration into a suitable subsystem
4. Storage, handling, and transport of the various subsystems prior to final system assembly
5. Testing of the various subsystems prior to final system assembly
6. Subsystem and system assembly
7. Storage, handling, and transport of the system prior to system initiation and operation
8. Testing of the system prior to product initiation and operation
9. System operation from initiation to end
10. Disposal of the system after use

9.6.6 Life Cycle Load Exposures

For each life cycle stage or subenvironment, there are a number of critical parameters that must be characterized and quantified. These are the various thermal, electrical, magnetic, mechanical, chemical, and radiation loads to which the product is exposed over its entire life cycle. These loads are:

- Temperature range, ΔT
- Temperature rate of change, dT/dt
- Kinds of temperature cycle and number of temperature cycles.
- Mean temperature
- Humidity exposure
- Pressure conditions
- Vibration and shock
- Electromagnetic wave exposure (EMI)
- Electrostatic discharge (ESD) exposure
- Chemical exposures
- Radiation exposure
- Contamination/corrosion exposure

To best ascertain the effects of the different loads at each stage of the hardware life cycle, it is necessary to construct a life cycle stage/load matrix. This matrix quantifies the effects of each load at each stage of the product's life. This matrix should be completed by the

SMT concurrent engineering team. The suggested format of this matrix is shown in Table 9.19. Based on an assessment of this matrix, the next step is to consider possible failure sites, failure mechanisms and failure modes. The suggested format of this matrix is shown in Table 9.20.

TABLE 9.19   Suggested Life Cycle Stage/Load

Load \ Life cycle stage	Storage, handling, and transportation of components	Fabrication and assembly of surface mount PWAs	Storage, handling, and transportation of PWAs	Subsystem and system assembly	Storage, handling, and transportation of subsystems	Testing of subsystems prior to system assembly	Storage, handling, and transportation of system prior to use in service environment	Testing of system prior to use in service environment	System operation from beginning to end
Temperature range, ΔT									
Temperature rate of change, dT/dt									
Kind/number of temperature cycles									
Mean temperature									
Humidity exposure									
Pressure conditions									
Vibration and shock									
Electromagnetic exposure									
Electrostatic discharge exposure									
Chemical exposures									
Radiation exposures									
Contamination and corrosion exposure									

Tables 9.19 and 9.20 are to be used as checklists. They are the kind of PoF matrices that the concurrent engineering team should fill in for the product/system for life cycle stages 1–9. It is better to construct these two PoF matrices in conjunction with each other since various failure mechanisms may occur during different life cycle stages.

Note: After the system is retired from use, failure of the system is no longer a relevant concept. Hence, life cycle stage no. 10, disposal of the system after use, does not appear in Tables 9.19 and 9.20. However, the concurrent engineering team should decide up front how ultimate disposal is best to be performed so as to minimize all environmental burdens.

TABLE 9.20   Suggested Physics of Failure Matrix for Life Cycle Stages

Life cycle stage	Failure sites	Failure mechanisms	Failure modes
Storage, handling, and transportation of components			
Fabrication and assembly of surface mount PWAs			
Storage, handling, and transportation of PWAs			
Storage, handling, and transportation of subsystems			
Testing of subsystems prior to system assembly			
Subsystem and system assembly			
Storage, handling, and transportation of system prior to use in service environment			
Testing of system prior to use in service environment			
System operation from beginning to end			

9.7   GENERAL DESIGN CONSIDERATIONS

Closely connected with the concept of a concurrent engineering SMT team are two related concepts. These are

1. Chief design activities for SMT
2. Design for X (DfX) where X is a dummy variable representing the various life cycle requirements

Before considering these two issues, it is appropriate to set forth the generic design/development process for printed wiring assemblies.

9.7.1   PWA Generic Design and Development Process

Using the concepts of concurrent engineering for design and development of surface mount printed wiring assemblies, the general design/development process flow, departmental functions, and concurrent engineering interactions are set forth in Fig. 9.73. Because of the number of interactions, only the most important are shown in this figure.

9.7.2   Design Activities for Surface Mount PWAs

To understand design within its proper context, it is necessary to set forth the various design activities involved in creating a PWA. Each of these activities impacts other activities. It is important to understand that there is a hierarchy of activities, and the consequences that decisions at one level of this hierarchy have on lower levels. This concept of a hierarchy of activities will be extended into the section dealing with DfX. Almost without exception, the major design activities impact other activities, especially manufacturing and those dealing with life cycle requirements. Figure 9.73 shows the design/development process flow; it does not show the impact of design decisions further downstream.

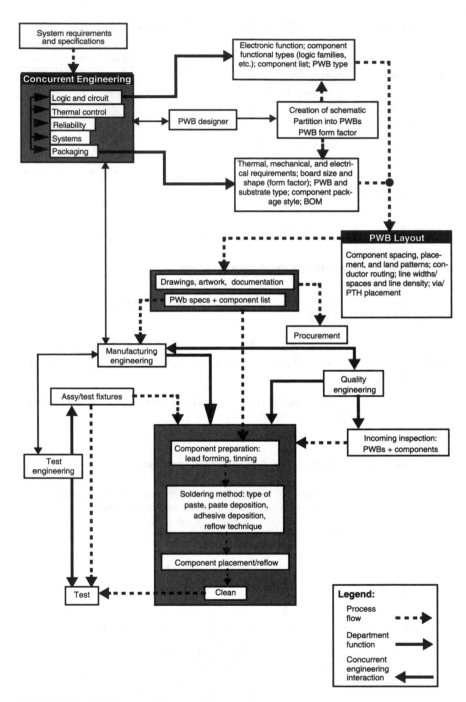

FIGURE 9.73 PWA design and development process flow.

Based on the system's functional requirements, a block diagram delineating the system's electrical functional requirements can be set forth. From this, the overall schematic diagram is defined and created. At this stage, the circuit design engineer should consult the PWB designer and the packaging engineer. This is no longer optional; it is mandatory to work as a concurrent engineering team. The PWB designer and the packaging engineer are instrumental in the creation of a physical entity from the schematic circuit design.

The overall schematic diagram is then partitioned into individual PWBs, and the functional type and number of components are determined. The selection of components based on the schematic will be called the component functional selection.This is to be distinguished from the component packaging style selection.

Normally, there is some give and take regarding the type and number of components per PWB. From the schematic and the partitioning of the schematic circuitry into PWBs, the individual PWB form factors (size/shape) are determined. This must be done in conjunction with the packaging engineer, who has the responsibility to ensure that the PWAs fit into a chassis or electronic box.

Note: Printed wiring board (bare board) manufacturers build panels, not individual PWBs. In the U.S.A., panel size used to be 18 in × 24 in. Note, though, that panel size can vary. Today, panels are often sized to specification. Nevertheless, effective utilization of the panel must be kept in mind when determining the PWB form factor to avoid unnecessary waste and panel underutilization. Check with the board manufacturer regarding the panel size prior to determining the final PWB form factor.

After component functional selection has been made, it is then necessary to select the component package styles. Component package style is often determined by component availability and whether the packaging engineer must work from an approved product list (APL).

At this point, a decision must be made whether one side or both sides of the PWB are to be populated with components. That is, a decision must be made whether the assembly is to be a Type 1 (components on one side only) or a Type 2 (components on both sides). This decision, in conjunction with component package style selection, determines the Type and Class of PWA (see Sec. 9.4.2).

In this chapter, the PWA will be either a Type 1B or a Type 2B (pure surface mount), or a Type 1C or Type 2C (mixed technology, with some surface mount components and some through hole components).

Component placement and component spacing on the surfaces of the PWB are next considered. The signal paths and power and ground busses (collectively known as traces) are then set forth and routed. The trace line widths and the spacing between adjacent traces must be determined as is the line density. The number and position of the signal and interconnect vias (signal vias and interstitial vias) and plated through holes if through hole components are to be used are also determined as are the diameters of the vias and PTHs. A CAD system is almost always used to accomplish these tasks. Component placement/spacing, in conjunction with trace routing, trace density, and via placement will determine if a multilayer PWB (MLB) is required to meet the design needs.

Component density refers to the number of components per unit area and follows directly from the component package styles, component placement, and component spacing.

Line density refers to the number of lines (traces) per channel when the lines run parallel to each other. Three different line density levels are shown in Fig. 9.74.

In conjunction with the above design activities, the material selections are also made. This includes the PWB material, solder mask material, and conformal coating material.

At this point, the PWB/PWA bill of materials (BOM) has been determined, and procurement of these items can then take place through the purchasing department.

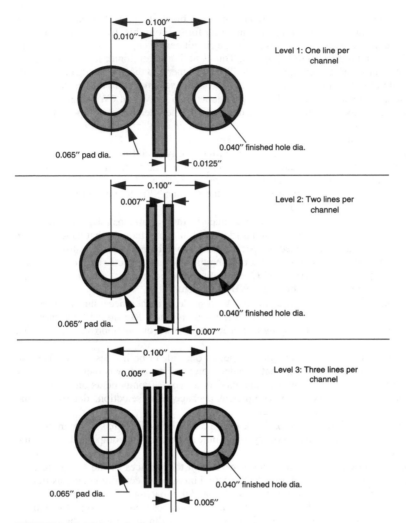

FIGURE 9.74 Line density levels.

9.7.3 Summary of Design Activities for PWAs

The above design activities can be conveniently summarized into 14 chief design activity categories. These activities affect others downstream in the design/development process (see Fig. 9.73). It is important to be aware of the impact of these major design activities. They are listed here in much the same order as they are invoked in designing a PWA.

- Schematic creation
- Schematic partitioned into PWBs
- Component functional selection—determination of the functional type and number of components

- Determination of the PWB form factor (PWB size/shape)
- Substrate material selection
- Component package style selection and determination of land pattern configurations
- Type of assembly
- Component placement on the PWB surfaces
- Component spacing on the PWB surfaces
- Determination of line widths/spaces and line density
- Routing of traces
- Placement of all electrical interconnect vias and PTHs on the PWB
- Solder mask selection
- Conformal coating selection

Given the above design activities for PWAs, some are quite generic for the design of any type of PWA. Among these, seven are herein considered generic.

- Schematic creation
- Schematic partitioned into PWBs
- Component functional selection
- Determination of PWB form factor
- Determination of line widths/spaces
- Routing of traces
- Placement of electrical interconnect vias and PTHs on the PWB

It is beyond the scope of this chapter to discuss these design activities in any detail since they are generic and not specific to SMT. If any has a specific impact on SMT, it will be discussed in an appropriate section.

In addition, there are two additional broad packaging design activities. These are

1. Thermal management considerations
2. Mechanical considerations

Again, these activities will not be specifically addressed in this chapter because they are generic. The other activities are more specific to the type and class of PWA, that is, SMT.

9.7.4 Chief Design Activities for SMT

Based on the above, the seven chief design activities for SMT that will be discussed in detail in this document are

1. Substrate material selection
2. Component package style selection and associated land pattern configurations
3. Type of assembly
4. Component placement on the PWB surfaces
5. Component spacing on the PWB surfaces
6. Solder mask selection
7. Conformal coating selection

9.7.5 Design Activities Interactions

When designing surface mount PWAs, there are a number of interactions among the different design activities. During component package style selection, it is assumed that some or all of the components are surface mount components (SMCs) (See Sec. 9.5.1). The decisions regarding component package styles, component placement, and the number of PWB sides utilized determine the type of surface mount printed wiring assembly (see Sec. 9.4.2). The decision regarding component package styles determines the corresponding land patterns and also impacts the material used for the substrate.

Caution: The packaging engineer must be especially concerned about the extent of mismatch between the components' and substrate material's coefficient of thermal expansion (CTE). This particular CTE mismatch is known as the global mismatch. The local CTE mismatch is that among the component lead material (copper, Alloy 42, and so on), the solder making up the solder joint fillet and land, and the PWB substrate material. See Fig. 9.75. See also DfR, below.

The type of component (component selection) combined with the component package style impacts the thermal management of the PWB, especially if some of the components dissipate large amounts of heat. This affects what design techniques are used for effective heat dissipation, e.g., thermal vias, thermal traces and other types of heat spreaders, conductive adhesive, internal power/ground layers used as heat sinks, the type of cooling techniques used, and so forth.

The determination of line widths/spaces, line density, and component density impact the decision regarding solder mask placement and conformal coating. The dielectric properties of the solder mask and conformal coating may also be an important consideration.

These design activity interactions are set forth in Table 9.21.

Global mismatch

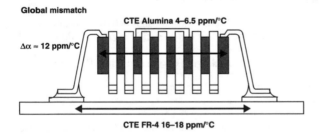

Local mismatch

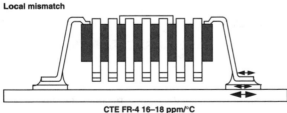

CTE Kovar leadframe = 5.3 ppm/°C
CTE Copper Leadframe = 17 ppm/°C
CTE Solder = 25 ppm/°C

FIGURE 9.75 Global versus local mismatch.

TABLE 9.21 Design Activity Interactions

Primary design activity / Secondary design activity	Component package style selection	Component placement on the PWB surfaces	Component spacing on the PWB surfaces	Line width/spacing	Line density	Vias/PTH placement
Substrate material selection	X					
Component land pattern configuration	X					
Type of assembly (e.g., Type 1B, etc.)	X	X				
Solder mask selection			X	X	X	X
Conformal coating selection			X			

A design activity having a direct impact on another design activity is a primary design activity. Conversely, a design activity affected by a primary design activity is a secondary design activity.

9.7.6 CAD/CAM/CAE

The paragraphs above delineated the design process for creating printed wiring assemblies (PWAs). In the past, the particular tasks were performed manually since there was no other way to accomplish them. Today suitable computer-assisted techniques exist whereby a large majority of these tasks can be automated. The acronyms, CAD/CAE/CAM, have existed for a long time. The meanings are

- Computer aided design or computer aided drafting (CAD)
- Computer aided engineering (CAE)
- Computer aided manufacturing (CAM)

CAD/CAE/CAM systems exist today that greatly facilitate the major design tasks discussed above. The principal components composing any CAD/CAE/CAM system are

- Hardware—computer and peripheral devices
- Software—both a suitable operating system (OS) for the computer and suitable CAD/CAE/CAM applications programs
- User interface—very important for usability
- A database (DB) and data base management system (DBMS)
- Communication

The CAD database (DB) typically contains libraries useful for CAD/CAM. For electronics packaging design, the libraries will generally have

- Standard symbols, such as electrical schematic symbols
- Standard components, with their names, part numbers, physical dimensions, and power requirements (if any)

- Standard drawing formats
- Manufacturing standards

In addition, the DB should contain all the necessary information to define electronic circuits and the way they function. The following information should be included:

- A graphical representation of the actual circuit (schematic diagram)
- The connectivity list (netlist)—this list shows how each component on the PWB is connected to every other component
- A list of all the parts used in the circuit
- The physical PWB layout (preferable if the DB contains this)
- The electronic simulation model—a CAE function enabling the designer to model the actual performance of the proposed electronic circuit using a suitable simulation technique

9.7.7 Application of CAD/CAE/CAM to SMT

The complexity of circuitry on surface mount PWAs is increasing to the point where it practically mandates the use of a suitable CAD/CAE/CAM system. Productivity goes up and costs are driven down with CAD/CAE/CAM. Some of the most notable items provided by a basic CAD/CAE/CAM system (see Fig. 9.76) are

- Generation of the schematic diagram (CAD)
- PWB layout (CAD)
- Design checking for errors (CAD)
- Netlist generation (CAD)
- Automated component placement (CAD)
- Automated trace routing (CAD)
- Automated circuit testing and simulation (CAE)
- Computer assisted process planning (CAPP/CAM)
- Documentation for manufacturing, such as artwork master, drill template master, phototools for solder mask and legending (CAM)

Advanced CAD/CAE/CAM systems are becoming available which provide many additional features. Among these are

- Automated thermal management calculations in conjunction with component placement (CAE)
- Automated mechanical calculations such as those for shock and vibration (CAE)
- Finite element modeling (FEM) for advanced thermal, mechanical, and reliability calculations (CAE)
- Design parameters impacting manufacturability(DfM/CAM)
- Design parameters impacting inspectability (DfI/CAM)
- Design parameters impacting testability (DfT/CAM)
- Design parameters impacting reliability (DfR/CAM)

9.7.8 Design for X

It is important that the packaging engineer understand the concept of design for X (DfX) where X is a dummy variable standing for various product life cycle requirements (LCRs). There are four principal DfX considerations. These are

1. Design for manufacturability (DfM)
2. Design for inspectability (DfI)
3. Design for testability (DfT)
4. Design for reliability (DfR)

To this list may be added a new life cycle requirement. This is to design with regard to environmental considerations. Hence, to the list above may be added the following additional DfX consideration:

5. Design for Environment (DfE)

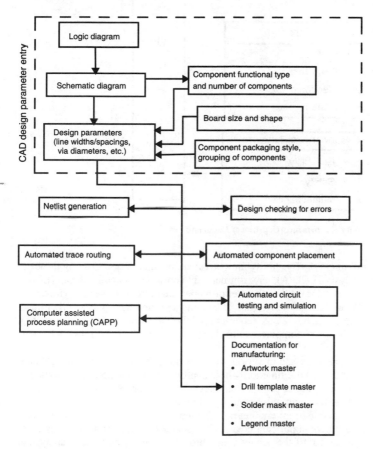

FIGURE 9.76   Basic CAD design features with output.

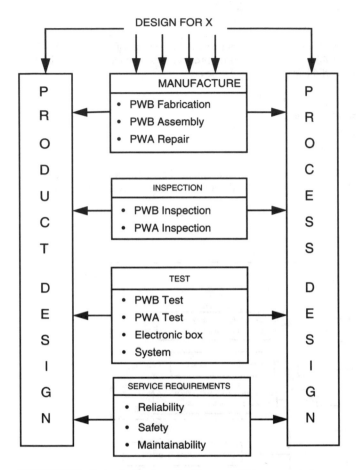

FIGURE 9.77 Design for X schematic. *(Courtesy of Edosomwan et al., 1989.)*

The design for environment (DfE) activity may be called by other names. Among these are life cycle design/analysis (LCD/A), environmentally-conscious manufacturing (ECM), and green manufacturing (GM). All of these expressions are more or less synonymous.

As mentioned above, the design activities of creating a printed wiring assembly impact these various downstream activities. A schematic of the DfX function is shown in Fig. 9.77. This diagram indicates the importance of designing the product and process concurrently.

As different topics are discussed in various sections throughout the rest of the chapter, attention will be called to the DfX function where appropriate. When attention is called to a particular DfX statement, the appropriate abbreviation will be placed in front of the statement, and the entire statement will be put in bold font.

Designing for X in a concurrent engineering fashion is a nontrivial issue since if performed correctly, it results in significant cost savings and a superior product. Costs escalate as the product passes out of the design stage through assembly and into final system integration. Figure 9.78 shows product costs versus stages of development.

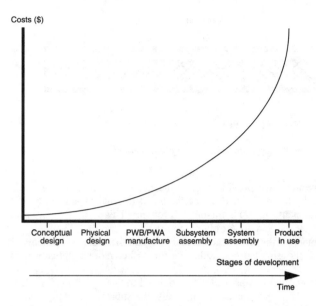

FIGURE 9.78   Product costs versus stages of development.

9.8   *DESIGN GUIDELINES FOR SMT*

9.8.1   SMT Packaging Design Topics Addressed

The scope outlined in the last section is directly relevant to this one. In addition, several other topics relevant to SMT design activities are also covered. The following nine packaging design topics are covered in more depth:

- Substrate selection
- Component land patterns, passives
- Component land patterns, actives
- Component placement and spacing
- Component lead forming
- Component packaging for shipping/handling
- Assembly types
- Solder mask selection
- Conformal coating selection

9.8.2   Substrate Selection

The choice of substrates for surface mount components (SMCs) is more complicated than for through hole components because in addition to providing good electrical interconnec-

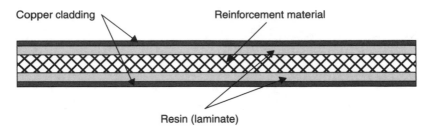

FIGURE 9.79   Substrate with two-sided copper cladding, copper clad laminate.

tion, the SMC solder joints must also provide a good mechanical interconnection. This issue becomes complicated by the difference in coefficient of thermal expansion (CTE) between substrate and component, the CTE mismatch problem. This issue is nontrivial, and may seriously compromise the final assembly reliability.

The substrate provides much of the electrical, mechanical, and thermal requirements of the completed SMT PWA so choosing a proper substrate for surface mount applications is very important.

The substrate typically comes with copper cladding bonded to one or both sides. Generally, two-sided copper-clad laminate is used. Structurally, the substrate is composed of two items, and the term "laminate" refers to the resin type making up the substrate along with the reinforcement material. Therefore the two constituents of a laminate (see Fig. 9.79) are

1. Resin type

2. Reinforcement material

There is another problem regarding substrates and their consequent reliability. Substrates are typically a composite of different materials, which affect their ability to expand in the X-Y or horizontal direction. They can often expand much more freely in the Z or vertical direction. Stated more formally, the CTE of a substrate can be segregated into a CTE/X-Y value and a CTE/Z value. This is reflected in Table 9.24.

Design for Reliability (DfR).   One of the most critical issues in surface mount technology is the reliability of solder joints and plated through hole vias. Their reliability is highly dependent on minimizing the difference among the various coefficients of thermal expansion. The global mismatch of CTEs, that is, the difference between the CTEs of substrate and components is, by far, the most important. The design engineer should be acutely aware of this ΔCTE and also the local CTE mismatch. The ΔCTE between the substrate and components, more than any other issue, seriously affects the reliability of the SMT PWA. Consult Table 9.24, which gives the CTEs of the materials used in PWA construction. In the case of component/substrate CTE mismatch, the substrate's CTE/X-Y value is critical. In the case of vias and plated through holes (PTHs), the substrate's CTE/Z value is the critical factor for through hole reliability. Both of these CTE issues must be carefully considered when picking a substrate; otherwise, serious reliability problems may occur.

In the case of substrates, one other parameter is of special importance. This is the substrate's glass transition temperature, or T_g. Below its glass transition temperature, a substrate made of a thermoset resin such as epoxy exhibits rigidity and good mechanical support properties. Above its glass transition temperature, the substrate loses a great deal of its rigidity and becomes more flexible and rubberlike. It can no longer adequately provide mechanical support when $T > T_g$. (See Fig. 9.80.)

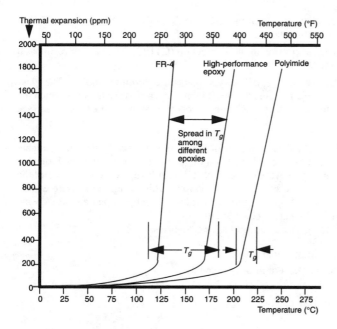

FIGURE 9.80   Glass transition temperatures of three resin types.

Design for Reliability (DfR).   Besides the CTE mismatch issue, the glass transition temperature, T_g, is another important property of the resin system. Above its T_g, a resin's mechanical properties rapidly change. The resin becomes markedly more rubbery, losing its ability to provide adequate mechanical support. See Fig. 9.80. Resin types having higher T_gs are required for service environments having large thermal excursions.

The circuit traces, land patterns, and via pads are created from the copper cladding by selective etching and plating processes. The resin type and reinforcement material contribute important mechanical and electrical properties to the finished PWA.

Along with the requirement for laminate material with reduced expansion in the Z direction, there exists the necessity for achieving higher packaging density. There are several ways to achieve higher packaging density. These are two of the ways:

1. Maximizing the area in the X-Y plane by utilizing finer traces and spaces and surface mount components.

2. Increasing the number of layers of the board to provide more layers of circuitry.

The first requires tighter control over the X-Y directions of the laminate material. The second requires tighter control over the Z direction.

Control over the X-Y direction is generally accomplished by using suitable reinforcement or constraining core materials. Control over the Z direction usually involves using a high-performance resin system having a high T_g.

It is possible to combine different resin types with different reinforcement materials. This is done chiefly to (1) control the CTE more carefully and/or (2) improve the thermal management characteristics of the substrate. Hence, the principal physical characteristics affected by carefully matching the resin type with a given reinforcement material are

- The coefficient of thermal expansion in the *X-Y* direction (CTE/*X-Y*). The type of reinforcement material has a pronounced effect on the CTE/*X-Y*
- The coefficient of thermal expansion in the *Z* direction (CTE/*Z*). The type of resin has a pronounced effect on the CTE/*Z*
- The glass transition temperature (T_g). The type of resin determines the T_g. The glass transition temperature is not a precise number. It represents a close range over which the resin's thermal expansion and other mechanical properties show a drastic change (see Fig. 9.80)

There is an alternative method to help control the CTE/*X-Y*. Specify that the laminate manufacturer should lay the fiberglass blankets on the bias. This means that each blanket is placed at a different angle from every other blanket during layup. This helps control the CTE/*X-Y* and makes it more uniform (i.e., less variation). This technique is more expensive, but it does allow the use of E-glass, keeping board processing simpler and less costly. Using this technique is a DfM and cost consideration.

Resin types are often categorized into six major categories.

1. Standard FR-4 epoxy (difunctional epoxy)

2. High-performance epoxy (tetra or multifunctional)

3. Polyimide

4. Bismaleimide triazine (BT)-epoxy

5. Cyanate ester

6. Polytetrafluoroethylene (PTFE) (Duroid®)

FR-4 epoxy is considered the standard workhorse of PWA manufacture. The resin systems 2 through 6 are considered high performance.

Table 9.22 shows some of the important physical characteristics of each of these resin types.

TABLE 9.22   Physical Characteristics of Laminate Resin Materials

Resin system	T_g, °C	Dielectric constant, ε_r resin only	Dielectric constant, ε_r resin w/E-glass	Propagation delay, ps/in	Percent of water absorption	Copper peel strength, lb/in² @ 25°C	Copper peel strength, lb/in² @ 200°C
Standard FR-4 epoxy	125	3.6	4.4	178	0.11	9.0	5.7
High-performance epoxy	150–180	2.9–3.6	3.9–4.4	168–175	0.04–0.20	6.0–9.0	6.5–7.9
Polyimide	220	3.4	4.3	176	0.35	6.5	8.0
BT-epoxy	195	3.1	4.0	170	0.40	8.7	5.2
Cyanate ester	240	2.8	3.7	163	0.39	8.0	6.3
PTFE	16	2.1	2.5	134	0.01	10.0	8.0

High-performance resin systems offer four major advantages.

1. High glass transition temperature, T_g
2. Lower Z-axis expansion rate
3. Reduced moisture absorption (but not in all cases)
4. Improved chemical resistance

There are five chief types of reinforcement materials.

1. E-glass
2. S-glass
3. Aramid fibers (Kevlar®) (woven or nonwoven)
4. D-glass (not used very often)
5. Quartz fibers (not used very often)

See Table 9.23 for some of the important properties of these reinforcement materials.

TABLE 9.23   Characteristics of Reinforcement Materials

Reinforcement	CTE/X-Y, −65°C to +125°C ppm/°C	Dielectric constant @ 1 MHz	Density, g/cm^3	Cost factor (app.)
Woven E-glass	5.5	6.5	2.54–2.58	1X
Woven S-glass	2.6	5.0	2.46–2.49	4X
Woven aramid	−5.0	3.8	1.44	5X
Nonwoven aramid	N/A	N/A	N/A	1.5X
Woven D-glass	2.7	4.1	2.14–2.16	10X
Woven quartz	0.5	3.8	2.50	30X
Metal core	2–6 (see Table 9.24)	N/A	8.33–8.43	10X

Note: Always assume that E-glass is the reinforcement material used in the laminate unless otherwise specified.

In addition, there are two other materials that can be added as reinforcement.

1. Copper-invar-copper (CIC)
2. Copper-molybdenum-copper (CMC)

These two materials are generally combined with a resin system to (1) control CTE and (2) for improved thermal management. Metal core as a reinforcement material is costly, creates many processing difficulties, and is high in mass. However, metal core can also act as a structural support and function as an electrical ground/power plane.

The percentage copper varies. For example, in the case of CIC, 20-60-20 (60 percent Invar) is common. The percent copper is varied to produced different CTE/X-Y values.

E-glass is currently the standard industry reinforcement material used in electronics. However, when matching the TCE (thermal coefficient of expansion) of substrate and component or when an improved dielectric constant are required, then alternate reinforce-ment materials should be considered.

For CTE matching, consider using Aramid (Kevlar®) or S-glass for improved dielectric constant properties.

There are five important types of substrate for surface mount applications. These are

1. Epoxy/fiberglass (FR-4).
2. Polyimide/fiberglass.
3. Epoxy/aramid (Kevlar®)—woven or nonwoven.
4. Polyimide/aramid (Kevlar®)—woven or nonwoven.
5. PTFE (nonwoven reinforcement—often called Duroid®)

The CTE and T_g of each of these materials is given in Table 9.24 along with other common PWA and component materials.

TABLE 9.24   CTE and T_g for Various Materials Using in PWA Construction

Material (fiberglass = E-glass)	CTE/X-Y, ppm/°C	CTE/Z, ppm/°C	T_g, °C
Epoxy/fiberglass (FR-4)	14–18	60–80	125–180
Polyimide/fiberglass	12–16	60	250
Epoxy/aramid (woven)	6–8	150	180
Epoxy/aramid (nonwoven)	6–8	110	180
Polyimide/aramid (woven)	5–8	60	230
Polyimide/aramid (nonwoven)	5–8	45	230
Epoxy/quartz	6–13	62	125–180
Polyimide/quartz	6–12	35	230
PTFE/fiberglass (woven)	10–25	55	N/A
PTFE/fiberglass (nonwoven)	20	50	N/A
PTFE/quartz	7.5–10	88	N/A
CIC (20/60/20)	5.3–5.5	16	N/A
CIC (12.5/75/12.5)	4.4	16	N/A
CMC	2–6	—	N/A
Copper/beryllium (lead frame material)	17.8	—	N/A
Copper (lead frame material)	16.7	—	N/A
Kovar (lead frame material)	5.2–5.9	—	N/A
Alloy 42 (lead frame material)	4.5	—	N/A
Eutectic and near-eutectic tin-lead solder	23–25	—	N/A
Alumina (ceramic component case)	4–6.5	—	N/A
Epoxy (plastic component case)	20–23	—	175
Silicon (die)	2.3–2.6	—	N/A
Gallium arsenide (die)	5.7	—	N/A

9.8.2.1 FR-4 Epoxy/Fiberglass. This is by far the most commonly used substrate material for a wide variety of applications. It is found in many commercial electronics products. However, a careful analysis of the reliability issues must be performed prior to using it for high reliability applications. The overall product/system requirements and intended service environment will determine whether one is justified using this material.

9.8.2.2 Polyimide/Fiberglass. This material has superior properties to FR-4, especially for high temperature applications. It possesses a much higher T_g than epoxy/fiberglass. Its T_g is ~230°C. However, it still has considerable CTE values, both CTE/X-Y and CTE/Z (see Table 9.24). It is more difficult to drill polyimide, and this decreases drill bit life. It is also more difficult to desmear and plate than FR-4. This means it costs more to process it than FR-4. In addition, it has poorer moisture absorption characteristics than FR-4. If polyimide is the substrate of choice, make sure that the PWB fabricator's processes are optimized for this material and not FR-4.

Design for Manufacturability (DfM). Polyimide has become the material of choice for use environments involving high temperature excursions. Polyimide/fiberglass (E-glass) is often quite satisfactory as opposed to using a more exotic material such as quartz or aramid.

9.8.2.3 Epoxy/Aramid and Polyimide/Aramid. If leadless chip carriers (LCCs) are going to be used extensively on the surface mount PWA, these materials should be considered since they give improved control over the CTE/X-Y. Aramid (Kevlar®) is not particularly good regarding its CTE/Z property, however, as seen from Table 9.24. Also, during processing, microcracks sometimes show up in the aramid (Kevlar®).

9.8.2.4 Polytetrafluoroethylene (PTFE). This material has a very low dielectric constant, making it the best material for high frequency applications. It is costly and difficult to process.

There are several common failure modes for substrate materials. These are presented in Table 9.25.

TABLE 9.25   Some Common Failure Modes in Laminates

Failure mode	Manifestation (detailed description)
Delamination	Separation of the layers making up the laminate—the laminate literally comes apart
Blistering	Localized delamination—observable as white spots at various places on the board surface
Mealing or crazing	Separation of the reinforcement fibers at weave intersections—observable as white crosses
Haloing	Delamination around drilled holes—observable as white rings
Measling or vesication	With a solder mask and/or conformal coating—small blisters under the coating
Solvent attack	During cleaning operations, partial dissolution of the laminate material by a solvent cleaning agent

9.8.3 Component Land Patterns

The component dimensions of each particular package style of component were set forth in Secs. 9.5.2 and 9.5.3. Typical land pattern formulas for each type and style of component are set forth herein. The first focus will first be on passive components, followed by active components.

The components with their attendant land patterns addressed in this section are

- Chip resistors and capacitors
- Molded plastic capacitors
- Cylindrical surface mount components
- Small outline transistors (SOTs)
- Small outline integrated circuit (SOICs)
- Leadless chip carriers (LCCs) and J-lead leaded chip carriers
- Gull wing leaded chip carriers, also known as quad flat packs (QFPs)
- Area arrays and ball grid arrays (BGAs)

9.8.3.1 Chip Resistor and Capacitor Land Pattern Parameters.   Refer to Fig. 9.23, showing the top and side view of a chip component. Figure 9.81 gives the land pattern parameters for chip components. For heuristic purposes, one can generate the following formulas to generate a set of land patterns for chip resistors and capacitors:

$$C = L(max) + 0.030 \text{ in}$$
$$A = L(min) - 2 \times T(max)$$
$$X = W(max) + 0.010 \text{ in}$$
$$Y = (C - A)/2$$

Note: For almost all the land patterns given here, the following relations hold:

$$C = A + 2Y$$
$$C = B + Y$$

9.8.3.2 Molded Plastic Capacitor Land Pattern Parameters.   Refer to Fig. 9.30, showing the front and side view of a molded plastic capacitor. Figure 9.82 gives the land pattern parameters for molded plastic capacitors. For heuristic purposes, one can generate the following formulas to generate a set of land patterns for molded plastic capacitors:

$$C = L(max) + 0.040 \text{ in}$$
$$A = L(min) - 2 \times L_T(max)$$
$$X = W_T(max) + 0.010 \text{ in}$$
$$Y = (C - A)/2$$

9.8.3.3 Cylindrical SMC Land Pattern Parameters.   Refer to Fig. 9.25, showing the front and side view of a MELF. Figure 9.83 gives the land pattern parameters for MELFs. For heuristic purposes, one can generate the following formulas to generate a set of land patterns for MELFs:

$$C = L(max) + 0.030 \text{ in}$$
$$A = L(min) - 2 \times T(max)$$
$$X = W(max) + 0.010 \text{ in}$$
$$Y = (C - A)/2$$
$$N = 0.012 \text{ in}$$

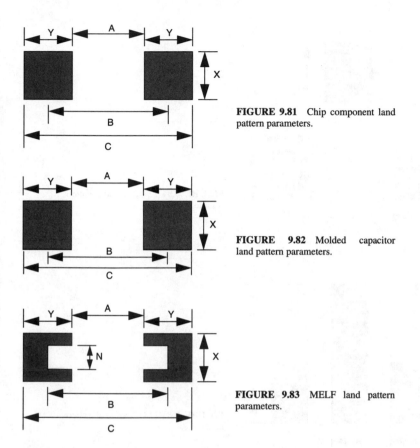

FIGURE 9.81 Chip component land pattern parameters.

FIGURE 9.82 Molded capacitor land pattern parameters.

FIGURE 9.83 MELF land pattern parameters.

9.8.3.4 Small Outline Transistor Land Pattern Parameters. Refer to the drawings for the SOT 23, SOT 89, SOT 143, and DPAK, Figs. 9.33–9.36, showing (1) the top and side view of an SOT 23, (2) the bottom and side view of an SOT 89, (3) the top view of an SOT 143, and (4) the top and side view of a DPAK. Figures 9.84–9.87 gives the land pattern parameters for the SOT 23, SOT 89, SOT 143, and the DPAK, respectively. Typical SOT and DPAK land pattern dimensions are given in Table 9.26.

9.8.3.5 Small Outline Integrated Circuit Land Pattern Parameters

Gull Wing SOIC Land Pattern Parameters. Refer to Fig. 9.38, showing the top and side view of a gull wing SOIC. Figure 9.88 gives the land pattern parameters for gull wing SOICs. For heuristic purposes, one can generate the following formulas to generate a set of land patterns for gull wing SOICs:

$C = A + 2Y$
$A = 0.140$ in if $W = 0.150$ in
$A = 0.300$ in if $W = 0.300$ in, i.e., SOLIC
$P = 0.050$ in

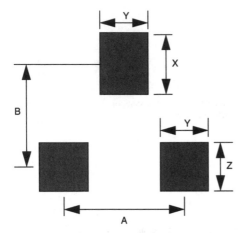

FIGURE 9.84 SOT 23 land pattern parameters.

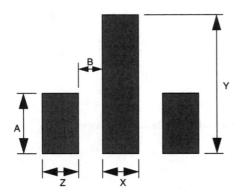

FIGURE 9.85 SOT 89 land pattern parameters.

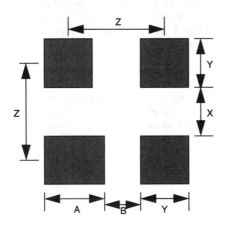

FIGURE 9.86 SOT 143 land pattern parameters.

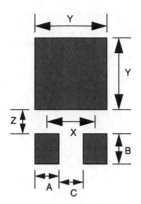

FIGURE 9.87 DPAK land pattern parameters.

$D = (N - 1)P$ where N = number of lands on one side

$X = 0.025$ in

$Y = 0.080$ in

A similar set of land pattern parameters can be derived for the shrink small outline SOICs. For SSOICs, $P = 0.025$ in.

TABLE 9.26 Typical SOT and DPAK Land Pattern Dimensions, inches (mm)

Component	X	Y	Z	A	B	C
SOT 23	0.055	0.040	0.048	0.080	0.088	—
	(1.4)	(1.0)	(1.2)	(2.0)	(2.2)	
SOT 89	0.040	0.213	0.040	0.055	0.020	—
	(1.0)	(5.4)	(1.0)	(1.4)	(0.5)	
SOT 143	0.040	0.040	0.080	0.050	0.030	—
	(1.0)	(1.0)	(2.0)	(1.25)	(0.75)	
DPAK	0.173	0.255	0.055	0.055	0.115	0.120
	(4.4)	(6.5)	(1.4)	(1.4)	(2.9)	(3.0)

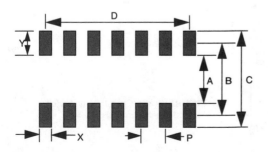

FIGURE 9.88 SOIC/SOJ land pattern parameters.

9.8.3.6 J-Lead SOIC Land Pattern Parameters. Refer to the reference drawing, Fig. 9.43, showing the top and side view of a J-lead SOIC, that is, an SOJIC. Figure 9.88 also gives the land pattern parameters for J-lead SOICs. For heuristic purposes, one can generate the following formulas to generate a set of land patterns for J-lead SOICs:

$C = A + 2Y$
$A = 0.235$ in if $W = 0.300$ in
$A = 0.285$ in if $W = 0.350$ in, i.e., SOLIC
$P = 0.050$ in
$D = (N - 1)P$ where N = number of lands on one side
$X = 0.040$ in
$Y = 0.080$ in

9.8.3.7 Leadless and J-lead Chip Carrier Land Pattern Parameters. Refer to Figs. 9.48 and 9.49, showing (1) the top, bottom, and side view of a leadless chip carrier (LCC), and (2) the top and side view of a J-lead leaded chip carrier. Figure 9.89 gives the land pattern parameters for leadless chip carriers and J-lead leaded chip carriers. For heuristic purposes, one can generate the following formulas to generate a set of land patterns for leadless chip carriers and J-lead leaded chip carriers:

$C = A + 2Y = L(max) + 0.050$ in
$A = C - 2Y$
$P = 0.050$ in
$D = (N - 1)P$ where N = number of lands on one side
$X = 0.030$ in
$Y = 0.080$ in
$Y' = 0.100$ in (pin 1)

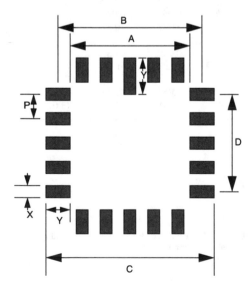

FIGURE 9.89 CLCC/J-lead leaded CC land pattern parameters.

9.8.3.8 Gull Wing Lead Chip Carrier Land Pattern Parameters. Refer to Fig. 9.54, showing the top and side view of a gull wing QFP. Figure 9.90 gives the land pattern parameters for gull wing QFPs. For heuristic purposes, one can generate the following formulas to generate a set of land patterns for gull wing QFPs:

$C = A + 2Y = W_{T-T}(max) + 0.050$ in

$A = C - 2Y$

$P = 0.050$ in (number of leads = 28 to 100)

$P = 0.025$ in (number of leads = 132 to 196)

$P = 0.020$ in (number of leads = 256)

$D = (N - 1)P$ where N = number of lands on one side

$X = 0.030$ in ($P = 0.050$ in); $X = W_L + 0.005$ in ($P = 0.025$ in or less)

$Y = 0.080$ in

9.8.3.9 Area Array and Ball Grid Array Land Pattern Parameters. Figure 9.91 gives the land pattern parameters for array array/ball grid arrays. For heuristic purposes, one can generate the following formulas to generate a set of land patterns for area array/BGAs:

$P = 0.060$ in

$P = 0.050$ in

$P = 0.040$ in

$D = (N - 1)P$ where N = number of lands on one side

$D' = 0.025$ in–0.035 in

Note: The above formulae for generating component land patterns are for heuristic purposes only. It is best to consult IPC-SM-782A, *Surface Mount Design and Land Pattern Standard,* for more specific details on land pattern design.

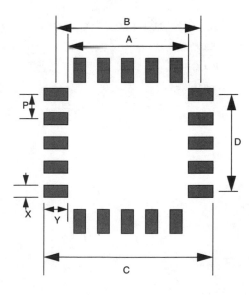

FIGURE 9.90 QFP land pattern parameters.

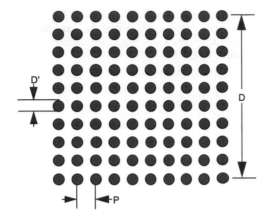

FIGURE 9.91 Ball grid array land pattern parameters.

9.8.4 Component Placement and Spacing

Component placement on the PWB surface is directly related to the component density. From the standpoint of maximizing component density, the designer wishes to place components as near to each other as possible. However, there are drawbacks to placing components too near together.

One of the drawbacks of placing components too close together is that it makes visual inspection difficult. It also makes rework/repair more difficult if the components are too closely spaced. Tables 9.27 and 9.28 gives the best recommended spacings for various components. Several examples are given herein to illustrate the use of these two tables.

TABLE 9.27   Component Type

A	B	C	D	E	F
0402	1812	SOIC	SOJ	PLCC 52	Edge of CCA[*]
0603	1825	SOLIC	PLCC 18	PLCC 68	
0805	SOT 89	SOP	PLCC 20	PLCC 84	
1206	Brick capacitor	TSOP	PLCC 28	Connector	
1210	Inductor		PLCC 32		
MML34			PLCC 44		
MML41			QFP		
SOT 23			DIP		
SOT 143			Axial THC		
DPAK			Radial THC		
			LCC		
			BGA		

* Strictly speaking, this is not a component type.

TABLE 9.28 Typical Component Spacing Dimensions, in inches

	A	B	C	D	E	F
A	0.025	0.040	0.050	0.100	0.150	0.250
B	0.040	0.040	0.050	0.100	0.150	0.250
C	0.050	0.050	0.050	0.100	0.150	0.250
D	0.100	0.100	0.100	0.100	0.100	0.250
E	0.150	0.150	0.150	0.150	0.150	0.250

When using Tables 9.27 and 9.28, if the component has extending leads, then the distance is from the end of its corresponding lands. If it has no extending leads, e.g., a chip component or an LCC, then the distance is from the component body.

Example 1. What is the minimum acceptable distance between an SOT 89 component and SOIC? According to Table 9.27, an SOT 89 is a B category component; an SOIC is a C category component. Note that Table 9.28 is set up in a matrix format. Beginning with a B category (row B) and moving to the C column, the minimum acceptable distance is 0.050 in. The minimum acceptable distance between two SOT 89s is 0.040 in.

Example 2. What is the minimum acceptable distance between a 1206 chip component and a PLCC 84? Again, by Table 9.27, a 1206 chip component is an A category component and a PLCC 84 is an E category. Using Table 9.28, beginning with an A category (row A) and moving to the E column, the minimum acceptable distance is 0.150 in. The minimum acceptable distance between two 1206 chip components is 0.025 in. The minimum acceptable distance between two PLCC 84s is 0.150 in.

A user may wish to modify Table 9.28 so that the values more closely conform with individual practical experience.

Design for Inspectability (DfI). The intent of Tables 9.27 and 9.28 is twofold. Following its recommendations is intended to make visual inspection after soldering easier. In addition, rework and repair, if required, should also be considerably easier.

For proper component placement for wave soldering, see Fig. 9.135.

9.8.5 Oomponent Lead Forming—CQFPs

Almost all CQFPs come with a carrier ring attached to the component body (see Fig. 9.55). Gull wing leads are very fragile and are easily damaged. The use of a carrier ring protects these delicate leads from damage. However, the carrier ring must be excised using a suitable tool and the gull wing leads formed. This may or may not be accomplished by the same tool.

One method used to form the gull wing leads of CQFPs is to use a universal trim and form device. This equipment first excises the carrier ring, then it forms the gull wing leads. Such a device normally operates on one side of the component at a time. Thus, the complete operation is: Excise the carrier ring on all four sides, one side at a time, then form the gull wing leads on all four sides, one side at a time. Such a device is shown in Fig. 9.92.

Another method is to use a dedicated die for each component part type. The die is loaded into an anvil and with one operation the carrier ring is excised and the gull wing leads formed simultaneously on all four sides. A device for accomplishing this is shown in Fig. 9.93. This method greatly cuts down on the variability of the operation.

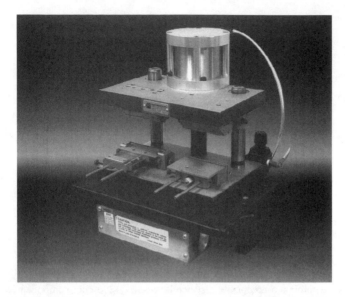

FIGURE 9.92 Universal trim and form die. *(Courtesy of Honeywell Space Systems.)*

FIGURE 9.93 Floating anvil press. *(Courtesy of Honeywell Space Systems.)*

Note: A dedicated die for a particular component part type is very expensive, approximately $8,000–$15,000 per die. Obviously, this method is not advisable for prototyping or for very small production runs. On the other hand, many ASICs are also quite expensive, so investing in a dedicated die may, in the long run, be cost effective.

When forming gull wings on a CQFP, the recommended parameters are given in Fig. 9.94 and Table 9.29. The gull wing leads should all be the same dimensions.

TABLE 9.29   Typical Values of Parameters during Lead Forming, inches

L_1	L_2	R_1, R_2, R_3	S	W_{T-T}
0.025 min.	0.035–0.040 (0.050 in pitch)	$2W_L$ minimum or 0.010, whichever is greater	0.010 min.	C – 0.020 See Fig. 6.36
0.050 max.	0.040–0.045 (0.025 in pitch)			
	0.045–0.050 (0.020 in pitch)			

9.8.6   Component Packaging for Shipping/Handling

There are four principal ways that surface mount components are packaged for shipping and handling prior to assembly processing. This kind of packaging is designated as the component carrier package. The four ways are

1. Bulk
2. Tape and reel
3. Tube
4. Matrix tray (waffle pack)

Tape and reel is by far the most common method for packaging components. It is the preferred choice for high-volume applications. Tubes are used for lower-volume applications.

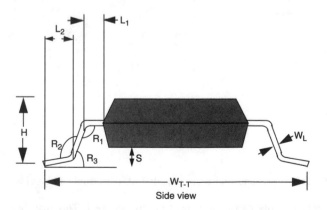

Side view

FIGURE 9.94   Parameters for the gull wing lead forming operation.

Tubes are sometimes called *sticks* or *magazines*. Large gull wing integrated circuit components such as quad flat packs are packaged in matrix trays. Table 9.30 indicates the different packaging methods and those most amenable for automated placement techniques.

TABLE 9.30   Component Packaging Methods

Components	Bulk	Tape and reel	Tube	Matrix tray
Resistors	X	A*		
Capacitors	X	A		
Diodes	X	A	X	
Transistors	X	A	X	
Inductors	X	A		
SOICs		X	A	
SOJICs		X	A	
PLCCs			A	
CLCCs			A	A
CQFPs				A
LCCs	X			A
PQFPs			A	A
TSOPs				A

* A indicates the method is more suitable for automated placement equipment.

Carrier tape is generally embossed plastic having a cavity for the component. There is a
cover tape over the carrier tape to keep the components from falling out. The entire ensemble is then mounted on a reel. There are seven carrier tape widths.

1. 8 mm
2. 12 mm
3. 16 mm
4. 42 mm
5. 32 mm
6. 44 mm
7. 56 mm

There are two reel sizes.

1. 7-inch.
2. 13-inch

Note: For more specific details on tape and reel, consult EIA (Electronic Industries
Association) RS 481A.

Many smaller components are only carrier packaged in tape and reel. This may be
many more components than are actually required for the job at hand. Provision should be

made to store the unused components and account for all components, both used and unused.

9.8.7 Assembly Types

The various assembly types have been covered in Secs. 9.4.2 and 9.4.4. For convenience, this classification scheme and a brief description of each type is reproduced in Table 9.31. The purpose of this section is to review important design considerations regarding the various assembly types. The caveat against manufacturing Type 2C (VC) is also repeated.

There are various advantages and disadvantages regarding each surface mount PWA type. It is appropriate to display these advantages and disadvantages in table format. This is done in Tables 9.32–9.36.

TABLE 9.31   Classification Scheme of Surface Mount PWAs

No.	PWA Classification	(1) Description + (2) Soldering method
1	Type 1B	(1) Pure surface mount/single side only (2) Reflow
2	Type 2B (no adhesive)	(1) Pure surface mount/both sides (2) Reflow
3	Type 2B (adhesive)	(1) Pure surface mount/both sides (2) Reflow
4	Type 1C	(1) Surface mount + through hole/single side only (2) SMCs: below; THTs: wave
5	Type 2C (S)	(1) Through hole on the top side + surface mount (small SMCs) on the bottom side (2) Wave (dual jet)
6	Type 2C (C)	(1) Surface mount + through hole on the top side + surface mount (small SMCs) on the bottom side (2) SMCs on top side: reflow; THTs + small SMCs on the bottom side: wave (dual jet)
7	Type 2C (VC) [This is a Type 2B + Type 2C (S) combined in one assembly]	(1) Surface mount + through hole on the top side + surface mount (small SMCs + large SMCs) on the bottom (2) SMCs: reflow; THTs + small SMCs: wave (dual jet)

9.8.8   Solder Mask Selection

The purpose of solder mask is to protect the various conductor traces during the soldering process. As the spacing between traces decreases, the chance for bridging and shorting due to solder balls greatly increases. Solder mask helps prevent this by covering the traces with an organic material prior to reflow. Because solder mask is typically an organic coating, it can also function as a dielectric on controlled impedance boards. This effect should not be forgotten.

TABLE 9.32 Advantages and Disadvantages of Type 1B

Description and main features	Advantages	Disadvantages
SMCs only	Straightforward process—3W304, Type 1B Process Flow	Component availability
Top side mount only	Single soldering process: reflow	TCE mismatch concerns—especially if LCCs are used
See Table 9.31	High reliability obtainable with controlled processing	Learning curve involved—mastery of new materials (e.g., solder paste) and new processing equipment (e.g., solder reflow equipment)
	PWB size reduction—improved miniaturization	Infrastructure required to support pure SMT
	High density possible	High density may make in-circuit testing (ICT) difficult

TABLE 9.33 Advantages and Disadvantages of Type 2B

Description and Main features	Advantages	Disadvantages
SMCs only	Straightforward process—see 3W305, Type 2B Process Flow	Component availability
Top side and bottom side mount	Single soldering process: reflow	TCE mismatch concerns—especially if LCCs are used
See Table 9.31	High reliability obtainable with controlled processing	Learning curve involved—mastery of new materials (e.g., solder paste) and new processing equipment (e.g., solder reflow equipment)
	PWB size reduction—greatly improved miniaturization	Infrastructure required to support pure SMT
	Very high density possible	Double side processing is more difficult
		High density may make in-circuit testing (ICT) difficult—this is especially true on a board populated on both sides

There are three principal types of solder mask. These are

1. Screened solder mask

2. Liquid photoimageable (LPI) solder mask

3. Dry film photoimageable solder mask

The first type of solder mask is applied using a screening technique similar to applying solder paste. For lines and spaces equal to or less than 10 mil, this form of solder mask is

TABLE 9.34 Advantages and Disadvantages of Type 1C

Description and main features	Advantages	Disadvantages
THCs on top side; SMCs on the top side	Don't have to worry about component availability	Two different soldering processes required; SMCs: reflow; THCs: wave
Mixed technology (THCs + SMCs)	Only have to worry about populating top side of the board	Infrastructure may be more difficult to obtain
See Table 9.31		

TABLE 9.35 Advantages and Disadvantages of Type 2C (S)

Description and main features	Advantages	Disadvantages
THCs on top side; small, discrete SMCs on the bottom side	Straightforward process— see 3W307, Type 2C Process Flow	Dual-wave wave soldering capability is required
Mixed technology (THCs + SMCs)	Single soldering process: wave	Must use adhesives to attach small SMCs to bottom before wave soldering
See Table 9.31	Only small SMCs required— resistors, capacitor diodes, SOTs	Ceramic SMCs subject to cracking during wave soldering
	Infrastructure to support THT + wave soldering widely available	
	Can be made easily testable	

not acceptable due to its inherent inaccuracy. Either liquid photoimageable or dry film photoimageable solder mask is recommended for SMT applications. Liquid photoimageable solder mask affords excellent coverage, but the application equipment is expensive. Liquid solder mask generally has a thickness after curing of 0.001–0.002 in. If fine line width/spacings are found on the PWB, liquid photoimageable solder mask will generally give superior coverage than will dry film photoimageable solder mask. The thickness of cured dry film solder mask is 0.003—0.004 in. One advantage photoimageable solder masks have is their ability to be placed very closely to features on the board. A setback from the lands is not necessary.

If conductor traces are found on the outer layers of the PWB, solder mask should be applied. If the outer layers are pads only, solder mask may not be necessary unless one wishes to define the pads with solder mask (solder mask defined [S/MD] pads).

Cleaning the assembly directly prior to the application of a solder mask material should be performed, and a suitable level of cleanliness should be achieved. See Secs. 9.9.13 and 9.9.14 regarding suitable cleaning agents and methods for testing for cleanliness.

TABLE 9.36 Advantages and Disadvantages of Type 2C (C)

Description and main features	Advantages	Disadvantages
THCs and SMCs on top side; small, discrete SMCs on the bottom side	Wide variety of components placed on assembly—promotes effective use of board real estate	Two different soldering processes required; SMCs: reflow; THCs: wave
Mixed technology (THCs + SMCs)	If functionality not available in SMT, THT available	Must use adhesives to attach small SMCs to bottom before wave soldering
See Table 9.31		Dual jet wave soldering capability is required
		Ceramic SMCs subject to cracking during wave soldering
		High density may make in-circuit testing (ICT) difficult—this is especially true of this type of board populated on both sides

If the solder mask tents over plated through holes and/or vias, the possibility of the entrapment of contamination residues and other debris within the PTHs exists. Dry film solder mask more easily tents holes than does liquid photoimageable solder mask.

9.8.9 Conformal Coating Selection

There are five principal types of conformal coatings. These are

1. Epoxy
2. Polyurethane
3. Silicone
4. Acrylic
5. Parylene® (polyparaxylylene)

Except for parylene, the other conformal coating materials consists of organic materials that polymerize to form a coating over the assembly. These materials may be a two-part system or a single-part system. Many of these materials also contain a suitable solvent as a carrier vehicle. Epoxies, polyurethanes, and silicones are thermoset plastics, meaning that during curing considerable crosslinking between molecules takes place. Thermoset plastics generally exhibit outstanding electrical, thermal, and mechanical properties. Acrylics, on the other hand, are thermoplastic. The coating forms when the solvent in which the acrylic material is dissolved evaporates. For the various types of conformal coating, excepting parylene, the normal coating thickness is 0.001 in (nominal). The nature of parylene is explained below.

Design for Environment (DfE). Many conformal coating materials are used with a solvent vehicle carrier. These solvents may (1) contribute to smog formation in the troposphere or (2) be ozone depleting substances (ODSs), e.g., 1,1,1-trichloroethane. The first class constitutes so-called volatile organic compounds (VOCs). In the case of ODSs, their manufacture is being phased out. In the case of VOCs, the current trend is to reduce as much as possible the quantity of solvent used in conjunction with the coating material.

The purpose of a conformal coating is to provide protection to the printed wiring board and components. Specifically, conformal coatings help protect and preserve the electrical performance of the assembly and mitigate the effects of environmental stresses. By encapsulating the surface of the PWB and the components, conformal coatings help ensure that moisture will not reach the surface and react with any contaminants found on the surface. Excessive moisture will lower the insulation resistance between circuit traces, accelerate high-voltage breakdown, and induce corrosion if suitable conditions are present. To ensure that as little contamination as possible remains on the assembly surface, it is necessary to clean. And if a conformal coating is applied, cleaning the assembly directly prior to the application of a conformal coating should be performed, and suitable level of cleanliness should be achieved.

Except for parylene, conformal coatings can be applied by dipping, spraying, flow coating, or brushing. Generally, dipping or spraying is the application method of choice, and it is possible to utilize both semiautomated and automated equipment to accomplish this task. Spraying is the easiest method to automate. However, spraying may result in little or no coating actually being applied underneath components and on smaller components that are shadowed by larger ones.

To apply paralene, a special piece of equipment is required. Essentially, the paralene is applied using chemical vapor deposition (CVD). The material (p-xylylene) exists as a dimer. By heating it under reduced pressure, it vaporizes. When heated further under further reduced pressure, it forms a free radical monomer. When the monomers condense on a suitable surface, they polymerize, forming the paralene conformal coating. See Fig. 9.95. The normal coating thickness of parylene is 0.0002–0.0003 in (nominal). Since parylene is

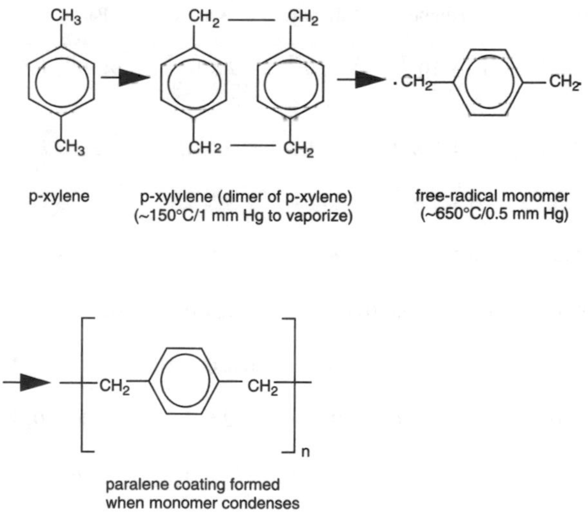

FIGURE 9.95 Formation of parylene conformal coating from p-xylylene.

applied by a vapor deposition method, the parylene coating covers everything—including under components, even components with very small standoff. But note that parylene, if it must be removed for repair, is very difficult to remove.

Choosing a suitable conformal coating depends on what properties the designer is seeking. The chief properties that the designer is typically interested in are

- Electrical properties
- Thermal properties
- Mechanical properties
- Resistance to humidity
- Resistance to chemicals and fungi

Manufacturing personnel are generally interested in

- Shelf life
- Pot life
- Ease of curing, both a room temperature and at elevated temperatures
- Ease of removal (during rework/repair)

Some of the specific electrical and thermal characteristics of these five materials is found in Table 9.37. A comparison of the above basic properties among the five types of coatings is found in Table 9.38. Table 9.38 is qualitative; it is to aid in the selection process of a suitable conformal coating.

TABLE 9.37 Typical Electrical/Thermal Properties of Conformal Coating Materials

Characteristic	Epoxy	Poly-urethane	Silicone	Acrylic	Parylene
Volume resistivity, Ω/cm (50% RH, 23°C)	10^{12}–10^{17}	1.1×10^{15}	2×10^{15}	2×10^{15}	2×10^{16}–10^{17}
Dielectric constant, 60 Hz	3.5–5.0	5.4–7.6	2.7–3.1	3.0–4.0	2.6–3.2
Dielectric constant, 10^3 Hz	3.5–4.5	5.5–7.6	—	2.5–3.5	2.6–3.1
Dielectric constant, 10^6 Hz	3.5–4.0	4.2–5.1	2.6–2.7	2.2–3.2	2.6–3.0
Dissipation factor, 60 Hz	0.002–0.010	0.015–0.048	0.007–0.010	0.020–0.040	0.0002–0.020
Dissipation factor, 10^3 Hz	0.002–0.020	0.043–0.060	—	0.020–0.040	0.0002–0.019
Dissipation factor, 10^6 Hz	0.030–0.050	0.050–0.070	0.001–0.002	2.5–3.5	0.0006–0.013
Thermal conductivity, 10^{-4} cal/s °C	4–5	1.7–7.4	3.5–7.5	3–6	~3
CTE (linear), ppm/°C	45–65	100–200	69	59	35–69

TABLE 9.38 Basic Properties of Conformal Coating Materials

Basic property	Epoxy	Polyurethane	Silicone	Acrylic	Parylene
Ease of application	G	G	G	E	E*
Chemical removal (solvents)	P	F	P	E	P
Removal by burn through	F	G	P	E	F
Abrasion resistance	E	E	G	F	E
Mechanical strength	E	E	F	F	E
Temperature resistance	F	F	E	F	E
Humidity resistance	G	E	E	E	E
Humidity resistance (extended periods)	F	E	E	G	E
Resistance to chemicals and fungi	E	G	G	F	E
Shelf life	F	F	G	G	E
Pot life	P	G	F	E	N/A
Room-temperature curing	G	G	G	E	N/A
Elevated-temperature curing	E	E	G	E	N/A
Avoidance of shadowing (when sprayed)	F	F	F	F	E

* Requires special processing.
E = excellent, G = good, F = fair, P = poor.

9.9 SURFACE MOUNT MATERIALS AND PROCESSES

9.9.1 General Process Considerations

In Sec. 9.4.4, the general process steps for producing surface mount PWAs, or SMAs, was outlined. It is appropriate to review the general process now.

Surface mount components (SMCs) are placed directly on the substrate surface. To create a metallurgical connection between the SMC and the substrate, solder paste is first deposited on the component lands. The SMCs are then mounted on the lands using a suitable placement method. The PWBs plus SMCs are then reflow soldered, forming a surface mount PWA, or SMA. The PWAs are then cleaned and tested. For producing surface mount PWAs with components on both sides, the application of a suitable adhesive may be involved. For producing mixed technology (some THCs and some SMCs) PWAs, wave soldering is involved. In a nutshell, the major steps for producing surface mount PWAs are

- Solder paste application
- Adhesive deposition (not always required)
- Component preparation (if required)
- Component placement
- Soldering
- Cleaning
- Inspection
- Rework (when necessary) and hand assembly

- Clean prior to conformal coat
- Conformal coat
- Test

This section will address all of these issues except test.

9.9.2   Solder, Fluxes, and Solder Paste

The type of solder most commonly used in electronic joining applications is eutectic tin-lead, 63 percent tin-37 percent lead, or near eutectic tin-lead such as 60 percent tin-40 percent lead. Almost all electronics soldering applications will involve one or the other of these solders. A phase diagram of eutectic tin-lead is shown in Fig. 9.96. A solder consisting of two elements, such as tin and lead, is referred to as a binary system. One consisting of three elements—such as tin, lead, and silver—is a ternary system. For reflow applications, a ternary solder is sometimes used, consisting of 62 percent tin-36 percent lead-2 percent silver, as detailed below.

Note: Metallurgically speaking, eutectic tin-lead solder is actually composed of 61.9 percent tin (Sn) and 38.1 percent lead (Pb) by weight. Nevertheless, much of the literature cites the eutectic composition to be 63 percent tin and 37 percent lead by weight. This practice will be adhered to in this chapter.

The specific purpose of solder is to form a metallic interconnect between the land pads on the printed wiring board surface and the components' terminations or leads. The process of producing a metallurgical bond between the component and the printed wiring board constitutes the most important interconnect method in printed wiring assembly manufacturing. The solder is the joining material. During the soldering process only the

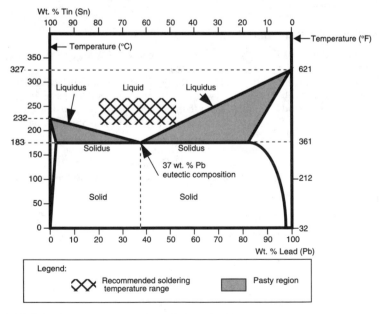

FIGURE 9.96   Phase diagram of eutectic tin-lead solder.

joining material melts; the base materials do not. Solders having melting points below 450°C are considered soft solders. Eutectic tin-lead solder has a melting point of 183°C.

Figure 9.96 depicts a phase diagram of eutectic tin-lead solder. There are several things to note about this diagram. The eutectic composition at 63 percent tin-37 percent lead by weight is a unique point in the diagram, meaning that this composition has a single unique melting point (m.p.) of 183°C (361°F). This is the only composition having a unique m.p. Note also that the eutectic melting point (183°C) is lower than the m.p. of pure tin at 232°C (450°F) and the m.p. of pure lead at 327°C (621°F). The m.p. of pure tin is at the left of the phase diagram, and the m.p. of pure lead is at the right.

The pasty range gives the melting temperature range of noneutectic tin-lead solders. For example, 60 percent tin-40 percent lead has a melting temperature range of 183°C to 191°C, that is, ΔTm.p. = 8°C for this particular noneutectic tin-lead formulation. The recommended soldering temperature is between 220°C (428°F) to 265°C (509°F).

There are a number of metallic impurities affecting the performance of solder. Some of these are described below.

Copper. Copper forms two intermetallic compounds (IMCs) with tin. These are: Cu_3Sn and Cu_6Sn_5 where Cu is the elemental symbol for copper (Latin: *cuprum*) and Sn the elemental symbol for tin (Latin: *stannum*). Although these IMCs can be considered as impurities in solder, they are inevitably formed during the soldering operation where copper traces (sometimes overplated with tin-lead or with tin-lead deposited by the HASL process) are found on a printed wiring board surface. In fact, some IMC formation is responsible for good solder joint formation.

Gold. Gold dissolves very easily in solder. Gold in solder joints causes embrittlement due to the formation of a number of gold-tin IMCs, $AuSn_4$ being the chief one.

Zinc. Zinc is very bad for solder because even a small amount of zinc will cause excessive graininess, poor adhesion, and poor solder joint performance. For different contaminants in solder and their recommended limits, see Table 9.39.

TABLE 9.39   Common Contaminants in Tin-Lead Solder

Contaminant	QQ-S-571-E, wt %[*]	New solder, wt %[†]	Contamination limits, wt %
Aluminum (Al)	0.005	0.003	0.006
Antimony (Sb)	0.2–0.5	0.3	—
Arsenic (As)	0.03	0.02	0.03
Bismuth (Bi)	0.25	0.006	0.25
Cadmium (Cd)	—	0.001	0.005
Copper (Cu)[‡]	0.08	0.010	0.25
Gold (Au)[‡]	—	0.001	0.08
Iron (Fe)	0.02	0.001	0.02
Silver (Ag)	—	0.002	0.01
Zinc (Zn)	0.005	0.001	0.005
Others	0.08	0.01	0.08

[*] Limits established by federal specification QQ-S-571-E for acceptable contamination levels for various elements.
[†] Levels of contamination generally found in new solder prior to use.
[‡] Copper and gold combined not more than 0.300.

If the contamination limits in Table 9.39 are reached, the solder should be replaced. A suitable method should be employed to determine the contamination level, and the results should be documented in a suitable way. The level of contaminants should be checked on a periodic basis.

When molten solder spreads over a surface, it is very important to consider the angle that the solder makes with the surface. The angle molten solder makes with the surface over which it is spreading is called the contact angle or the dihedral angle—symbolized by the Greek letter θ. The solderability of a surface is a function of the value of θ (see Fig. 9.97). The relation of the contact angle and a finished solder fillet is depicted in Fig. 9.98. The type of solder fillet shown in Fig. 9.98 is that formed from a gull wing lead configuration.

Surfaces to be soldered often have barrier layers on their surfaces such as oxides, hydroxides, and sulfides. Tin oxide is a very common barrier layer. If these barrier layers are not removed prior to the soldering operation, the solder will not wet the surface. This means the contact angle, θ, is greater than 90° (see Figs. 9.97 and 9.98). If this happens, the metallurgical bond will fail to form because the barrier layer inhibits the wetting action of the molten solder. It is important to determine the solderability of the solder being used. Figure 9.114 shows a wetting balance, a device used to measure the solderability of various items. It is recommended that, prior to the actual soldering operation, a use test be performed using a setup board to determine the solderability of the solder paste.

The purpose of a flux is to remove oxide and other barriers, such as sulfides, from a metallic surface to prepare that surface for soldering. This action renders the metallic surface active and in a state where it will readily combine with the molten solder in the soldering operation. During the formation of a good metallurgical joint, IMCs between the tin and copper are formed. Fluxing also reduces the surface tension between the solder and the surfaces to be soldered. This improves the surface wettability, facilitating the spread of the solder and promoting a good solder interconnect. Another function fluxes perform is to provide an inert atmosphere during the soldering operation whereby atmospheric oxygen is prevented from reaching the active base metals, tin-lead and copper, and reacting with them to reform surface oxides.

There are several distinct types of fluxes. The broad classifications are

- Rosin-based fluxes
- Synthetic activated (SA) fluxes
- Water-soluble (WS) fluxes, sometimes referred to as organic acid (OA) fluxes
- Low solids fluxes—so-called "no clean" fluxes

Table 9.40 gives the general categories of constituents of fluxes. The items given in this table are given as examples only.

Rosin-based fluxes today are complex. Flux manufacturers modify the organic acids contained in naturally occurring rosin to produce modified rosin fluxes. This is done to enhance certain properties of the flux, e.g., to render it more resistant to heat to lessen or

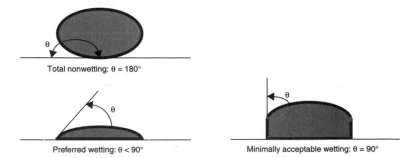

Total nonwetting: θ = 180°

Preferred wetting: θ < 90°   Minimally acceptable wetting: θ = 90°

FIGURE 9.97   Relationship between the contact angle and degree of wetting.

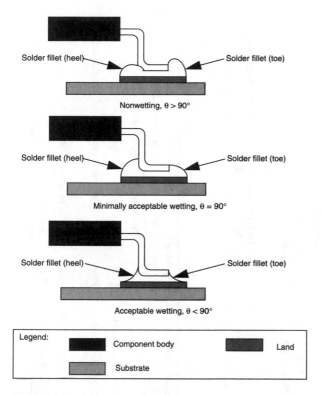

FIGURE 9.98 Degree of wetting for a gull wing solder fillet.

prevent the thermal degradation of the flux and subsequent rosin polymerization. Rosin-based fluxes can themselves be further categorized. The chief method of classifying them has been MIL-F-14256. This classification method is depicted in Table 9.41.

An alternative flux classification scheme has been proposed by the Institute of Inter-connecting and Packaging Electronic Circuits (IPC) as an alternative to MIL-F-14256. Many believe that MIL-F-14256 no longer reflects the present realities in flux formulation and development. The IPC classification scheme suggests that flux types be categorized as low (L), medium (M), or high (H) (see Table 9.42).

A solder paste is a material already containing the solder in powder form. Usually the percent solder powder by weight is 85–94 percent, with 88–90 percent being the most common in electronics soldering applications. In addition to solder powder, a solder paste contains flux, solvent, activators, and thickening agents. The thickening agents are added to impart the proper viscosity for the dispensing operation and to control slump. These thickening agents are sometimes called rheological control agents. A solder paste, then, contains the following ingredients:

- Solder powder, typically 88–94 percent by weight
- Vehicle (In the case of rosin-based pastes, rosin is the vehicle.)
- Activators

TABLE 9.40 General Categories of Fluxes and their Constituents

Flux type	Typical vehicle	Typical solvent (carrier)	Typical activator system
Rosin and modified rosin—amine activated	Rosin or modified rosin acids	Isopropyl alcohol (IPA) or glycol ether	Amine hydrohalide such as dimethylamine hydrochloride and trimethylamine hydrochloride
Rosin and modified rosin—nonamine activated	Rosin or modified rosin acids	Isopropyl alcohol (IPA) or glycol ether	Organic acid such as succinic or adipic acid
Synthetic activated (SA)	Alkyl acid phosphate Mono- and di-isooctyl phosphates	IPA	Sulfonic acid Alkylamine hydrohalide
Water soluble (WS)	Polyethyleneoxide Polypropyleneoxide (polyglycols)	Water or IPA	Citric acid
Low solids (LS)	Rosin or modified rosin acids Pentaerythritol tetrabenzoate (PETB)	IPA	Similar activators as in the other categories depending on the chemistry used

TABLE 9.41 Classification of Rosin-based Fluxes

Rosin-based flux	Resistivity of water extract, Ω-cm	Silver chromate paper test	Copper mirror test
R (rosin non-activated)	>500,000	Pass	No evidence of copper removal
RMA (rosin mildly activated)	>100,000 <500,000	Pass	No evidence of copper removal
RA-MIL (rosin activated meeting the requirements of MIL-F-14256	>50,000 <100,000	Pass	No evidence of copper removal or partial removal
RA (rosin activated)	<50,000	Fail	Partial to full removal
RSA (rosin superactivated)	<50,000	Fail	Full removal

- Solvents, used to control the tack time and working life of the paste
- Rheological control agents or thickening agents, used to control slump and other deposition characteristics

TABLE 9.42   Alternate (IPC) Flux Classification Scheme

Flux type (flux and flux residue activity level)	Test	Requirement
L	Copper mirror	No evidence of copper removal or partial removal of the copper mirror indicated by white background showing through in part of the fixed area.*
	Silver chromate or halide	Pass silver chromate test[†] or ≤ 0.5 percent halides.
	Corrosion[‡]	There shall be no evidence of corrosion.
M	Copper mirror	Partial or complete removal of the copper mirror in the entire area where the flux is located.
	Halide	≤ 2.0 percent halides.
	Corrosion	Minor corrosion is acceptable provided the flux or flux residue can pass the copper mirror or halide requirements of this category. Evidence of major corrosion places the sample or flux into the H category.
H	Copper mirror	Complete removal of the copper mirror in the entire area where the flux is located.
	Halide	> 2.0 percent halides.
	Corrosion	Evidence of major corrosion.

* Discoloration of the copper or reduction in the copper film thickness must not be sufficient to remove the flux from this category.
† Failure to pass the silver chromate test requires performing and meeting the requirements of the halide test to be considered for category L.
‡ The metal used for the corrosion test is copper.

Solder pastes are classified in the same manner as fluxes. RMA rosin-based solder pastes are the most commonly used in SMT, but WS and "no-clean" pastes have greatly increased in popularity, especially since the phaseout of ozone depleting solvents such as CFC-113 and 1,1,1-trichloroethane. Previously, these solvents were widely used, especially in the U.S.A., to clean PWAs after the flux/soldering operation. See Sec. 9.9.15 for further details about CFC-113 and 1,1,1-trichloroethane.

A number of critical parameters relate to depositing the solder paste on the lands prior to component placement and reflow. These are

- Paste viscosity, measured in units of centipoise (cps) or kilocentipoise (kcps) (see Table 9.43)
- Percent metal—the weight percent of the solder powder, generally between 88 and 94 percent by weight (see Table 9.43)
- Solder powder size (see Table 9.44)

- Solder powder shape—preferably a spherical shape with no satellites
- Slump—the property of the solder paste after deposition to spread out beyond the area on which it was deposited (Cold slump refers to this taking place at ambient temperature, whereas hot slump refers to this happening during the reflow operation.)

TABLE 9.43 Typical Viscosity Range and Metal Load for Types of Pastes

Solder paste type	Viscosity (kcps)	Metal load (wt %)
Needle dispensing	200–500	Up to 88
Screen printing	400–800	Up to 92
Stencil printing	600–1,400	Up to 94

TABLE 9.44 Solder Powder Sizes

Application	Size designation	Powder size	Percent
Standard pitch surface mount components (SP SMCs)	–200/+325* (Type 2)	75 μm 75–45 μm 45 μm	1% or less minimum 80% maximum 10%
Fine pitch surface mount components (FP SMCs)	–325/+400 (Type 3)	45 μm 45–20 μm 20 μm	1% or less minimum 80% maximum 10%
Fine pitch surface mount components (FP SMCs)	–400/+500 (Type 4)	38 μm 38–20 μm 20 μm	1% or less minimum 80% maximum 10%

* –200/+325 indicates that the powder will pass through 200 stainless steel mesh wires making up a screen but will not pass through 325 mesh wires.

- Tack time—the time the solder paste remains tacky (Tackiness is a desirable quality as it aids in holding the SMCs in place prior to reflow.)
- A setup board (see above—can be used to determine slump and tacktime
- Activator activation time—refers to the time-temperature profile of the activator during reflow (Ideally, the activator protects the solder powder from oxidation prior to use and is totally expended at the end of reflow.)

The solder powder size is directly related to its ability to successfully pass through the apertures (holes) of the screen or stencil during deposition (see Table 9.44). If the powder size is too large, it will not pass through. If it is too small, it may cause printing problems.

9.9.2.1 Lead-Free Solders and Alternatives to Lead-Bearing Solders. As pointed out above, eutectic and near-eutectic solder have been widely used in electronic applications. There is a vast amount of data in electronics dealing with this type of solder. However, both eutectic and near-eutectic solder contain the element lead (chemical symbol: Pb). This metal is quite toxic, and its toxic affects are most pernicious toward children. Lead is also toxic toward many other life forms. There have been moves within recent years, both by environmentalists and the U.S. Congress, to completely ban lead in numerous commercial products, including electronics products. Although the attempts so far have failed, and although the amount of lead consumed by the electronics industry is minuscule compared with its use in other industries, it has been estimated that lead use in electronics amounts to less than 1 percent of all lead usage by industry—numerous electronics companies and consortia have been actively pursuing alternatives to lead-bearing solders.

The task has not been easy. Some alternatives have emerged, but it is highly probable that no exact drop-in replacement for eutectic solder will emerge soon, if at all. What is more likely is that several solders will emerge whose range of application is more limited than eutectic and near-eutectic solder. Among the non-lead-bearing solders, the following are the most noteworthy at this time:

- Tin-silver solders, generally alloyed with a small amount of other ingredients
- Tin-indium solders, generally with a small amount of silver added
- Tin-bismuth solders, generally with a small amount of other ingredients

The tin-silver solders have higher melting temperatures than eutectic tin-lead solder. The range for tin-silver solders is roughly 208–223°C as opposed to eutectic tin-lead solder (183°C). The tin-indium and tin-bismuth solders have melting points either around 183°C or even lower. However, a series of tests run by different groups supports the contention that, all things considered, the tin-silver solders generally give the best application performance when compared directly to eutectic tin-lead solder. However, it is also necessary to investigate the long-term reliability of solder joints fabricated from these different materials.

Another intensive area of research is the elimination of metal solders altogether and utilizing a conductive adhesive (CA) instead. A CA consists of metallic particles, almost always silver in flake form, dispersed through an organic binder, generally an epoxy. There are two distinct types of CA. The first is called an isotropic conductive adhesive (ICA). The typical metal loading for an ICA is 70–80 percent by weight (25–30 percent by volume). In an ICA, the metal particles conduct electricity in all directions. The other type of CA is called an anisotropic conductive adhesive (ACA). The typical metal loading for an ACA is 5–20 volume percent; ACAs do not conduct electricity in all directions, only in one direction.

As with other adhesives, CAs can be either single stage or two stage. The single stage is activated by heat and normally requires higher temperatures to cure it properly. Two stage adhesives generally cure at lower temperatures, but they typically have shorter pot lives. CAs can be screen printed onto the lands of a PWB, but they do not wet nor spread as solders do and are not cured using reflow soldering equipment. CAs have low surface tension and do not have self-aligning characteristics like solder. CAs do not require the use of fluxes; hence, cleaning may be eliminated.

There are some problems with CA-formed joints. Epoxy tends to grow brittle as it ages, due to oxidation. The silver filler, under conditions of high temperature and humidity, can migrate (the well known silver migration problem) causing shorts. Rework is cumbersome. But the big drawback of CAs currently is that they do not conduct electricity nearly as well as metal solders do. It also seems that joints formed from CAs will not be as reliable as those formed from eutectic solder. However, for many noncritical electronic applications in which high reliability is not demanded, CAs may fulfill a niche.

Alternatives to HASL. Closely connected with the issue of discontinuing lead is the issue of finding alternatives to hot air solder leveling (HASL). HASL is a process that has been widely used in producing bare PWBs, especially solder mask over bare copper (SMOBC) PWBs. The PWB panel is immersed in molten solder and the solder is subsequently blown off. In addition to the use of lead, another disadvantage of the HASL process is that it creates solder lands convex in shape on top (= lack of land coplanarity). With standard pitch SMCs, this was not particularly a problem, but with fine and ultrafine pitch gull wing

QFPs, it is. So there has been a large amount of work to finding an alternative to HASL. Some of the chief alternatives are

- Organic solderability preservatives (OSPs)
- Electroless nickel/immersion gold
- Matte tin

Organic solderability preservatives (OSPs). The two most significant OSPs are

1. Benzotriazole
2. Substituted benzimidazole

These organic materials are anti-tarnishing agents. The more recently developed substituted benzimidazole compounds have several distinct advantages. The coating thickness of these compounds is of the order of 5–20 μin (0.1–0.5 μm). The substituted benzimidazoles can be either formic acid-based or acetic acid-based, meaning that either formic acid or acetic acid acts as the solvent carrier. More precise coating thicknesses are reported with acetic acid-based products. Such coatings can withstand multiple heat cycles typically found in SMT process technology. OSP does not involve heating, and the use of OSPs results in very flat lands since no solder is added to the lands. OSPs can be applied by dipping, flood immersion, and spraying at ambient temperature.

Electroless nickel/immersion gold. Depositing electroless nickel is accomplished using a process similar to depositing electroless copper. The process is an autocatalytic, self-limiting process. Deposits are typically 100 to 200 μin. Applying immersion gold then involves placing the part on which the electroless nickel deposit is found into a suitable gold solution. A small amount of the nickel dissolves and is replaced by a very thin coating of gold. An upper limit of gold deposit is typically 8 μm. The advantages of electroless nickel, immersion gold, besides eliminating the use of lead, are

- Excellent solderability for ultrafine pitch SMCs
- Corrosion resistance
- Suitability of directly wire bonding die onto the substrate (DCA)

For high reliability applications, the use of gold may entail gold embrittlement of the SJs (see Sec. 9.9.2).

Be aware though that not solder masks stand up to the plating bath chemicals.

Matte tin. In the case of matte tin, this refers to plating with tin. Matte tin plating is very controllable. Generally a thickness of tin of ≈300 to 400 μin is used. A nickel barrier layer should be used between the copper and the tin. Again, no lead is used during the process.

9.9.3 Solder Paste Application

Examining the process flow chart for surface mount assemblies (SMAs), except for Type 2C (S) involving wave soldering only, the first step is to apply solder paste to the component land patterns. There are two major techniques for applying paste to the lands. These are

1. Pattern printing
2. Needle dispensing

Pattern printing can be divided into the following two types:

1. Mesh screen

2. Stencil

When using a printer (see Figs. 9.99, 9.101, and 9.104) to apply solder paste to the component mounting lands on the PWB surface, there are three principal considerations to take into account during the printing operation. These are

1. Capabilities of the printing machine

2. Characteristics of the solder paste used

3. Design and characteristics of the stencil

However, each one of these principal considerations can be broken down into numerous further variables affecting the printing process. In particular, see Table 9.45. Common defects associated with this technique are found in Table 9.46.

The stencil printing machine is sometimes also called a screen printing machine or screen printer. It performs two chief functions. These are

- Registration of the PWB to the stencil (or vice versa)
- Squeegee control

The squeegee is a thin blade; its function is to sweep across the top of the stencil while at the same time pushing the solder paste in front of it. This sweeping action pushes the

FIGURE 9.99 Stencil printer in operation. *(Courtesy of the EMPF.)*

TABLE 9.45   Variables Affecting Solder Paste Printing

Capabilities of the printer	Solder paste characteristics	Stencil design and characteristics
Snapoff distance—if any	Viscosity	Thickness
Stencil/board registration	Solder particle size	Material of construction
Registration repeatability	Solder particle shape	Stepped or uniform thickness
Unidirectional or dual-directional printing	Solder metal composition	Aperture size
Squeegee pressure	Percent metal content, solder	Aperture shape
Squeegee speed	Vehicle rheology	Aperture aspect ratio (AR)[*]
Squeegee angle	Thixotropic index	Aperture arrangement
Squeegee material of construction	Susceptibility to breaking down when worked repeatedly	Sidewalls of aperture electropolished?
Squeegee hardness	Susceptibility to slumping	Stencil etched from one side or both sides?
Squeegee blade cross section	Solvent evaporation	Stencil cleaning frequency
	Susceptibility to spattering (solder ball formation)	
Ambient humidity (applies the whole system)		
Ambient temperature (applies the whole system)		

[*] Aperture aspect ratio (AR) = stencil thickness (T):aperture width (W) (see Figure 9.102).

TABLE 9.46   Common Defects Associated with Solder Paste Application

Printing defects	Causes of printing defects	Dispensing defects	Cause(s) of dispensing defects
Misregistration	Improper alignment of stencil with PWB	Misregistration	Improper alignment of the syringe with the PWB
Slump	Improper rheology characteristics of solder paste	Slump	Improper rheology characteristics of solder paste
Solder paste scooped out in the middle	Too much squeegee pressure, squeegee blade too soft, aperture too large	Dot diameter too large	Improper needle diameter, too much pressure during paste squeeze out
Too much solder paste	Incorrect squeegee blade height adjustment and improper squeegee pressure	Too much solder paste	Improper needle diameter, too much pressure during paste squeeze out
Solder paste at an angle	Excessive squeegee pressure	Solder paste spreads over side	Needle tip too close to the PWB surface

solder paste through the aperture holes of the stencil onto the component mounting lands. Both registration of the stencil with its apertures to the lands and squeegee control are of paramount importance for depositing the right amount of solder paste in the right place. When printing using a stencil, the snapoff distance is zero. Figure 9.100 shows a squeegee pushing solder paste in front of it and through an aperture of the stencil onto a mounting land.

Squeegees can be made from two different types of materials.

1. Synthetic such as polyurethane (for solvent resistance)

2. Metal—either brass or stainless steel

If the squeegee is made from a synthetic material, its hardness is measured in durometer units using the shore A hardness scale. If the squeegee used in the printing operation is made of a synthetic material, a blade of 80 to 90 durometer hardness is recommended. Metal squeegees can only be used for unidirectional printing. They are much more durable, and good results have been reported for them. If a metal squeegee is employed, one made of stainless steel is recommended.

A separate stencil must be used in the printer for each specific kind of printed wiring assembly being produced. For standard pitch surface mount PWAs, the stencil thickness should be 0.008 to 0.012 in. For fine pitch surface mount PWAs, the stencil thickness should be 0.006 in. If a board contains a mixture of standard pitch SMCs and fine pitch SMCs, a stepped stencil may be used. A stepped stencil is one having several (normally two) distinct thicknesses in different regions of the stencil in order to handle a mixture of standard pitch and fine pitch parts on the PWB surface (see Fig. 9.102). The number and arrangement of the apertures on the stencil must match exactly the component land patterns on the PWB surface. Extra squeegee pressure must be employed to ensure the correct volume of solder paste.

If a board contains a mixture of standard pitch SMCs and fine and ultrafine pitch SMCs, instead of using a stepped stencil, the apertures of the stencil may be adjusted in size to render the proper amount of solder paste for each type of SMC. Consult with the stencil manufacturer.

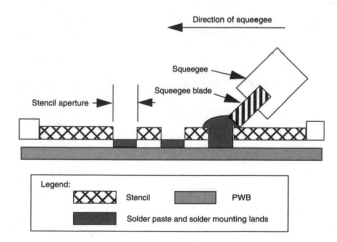

FIGURE 9.100 Squeegee motion depositing solder paste onto mounting lands.

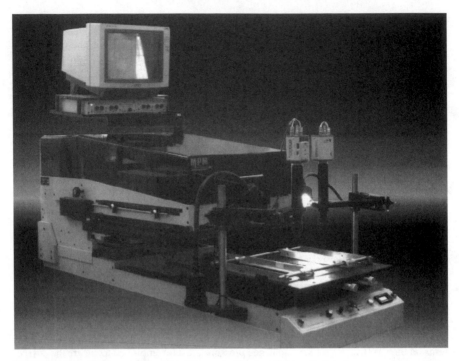

FIGURE 9.101 Semiautomated printer. *(Courtesy of Honeywell Space Systems.)*

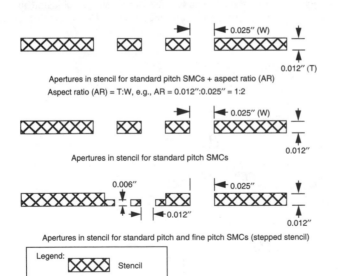

FIGURE 9.102 Standard versus stepped stencil + stencil aperture aspect ratio (AR).

Another issue becoming more important is stencil design and fabrication. There are several different ways that stencils can be made. The methods used today to produce stencils are

- Chemical milling or etching
- Laser cutting
- Electroforming

For standard pitch SMCs, etched stencils are generally used. They are relatively easy to produce. The etching can either be done from one side or from two sides. Obviously, when a stencil is produced by etching, the etching takes place not only in the Z-direction but also in the X-Y direction. Because of the etch factor of the solution performing the etching, the etched stencil will have either a larger opening in the middle of the stencil (assuming two-sided etching is used) or a larger opening at the bottom (trapezoidal shape). For the very precise deposition of solder paste required when fine and ultrafine pitch SMCs are used, etched stencils are not recommended.

A laser beam can be used to excise the stencil material, thus creating the required aperture. A laser-cut stencil has a much greater pattern accuracy than an etched stencil. Further, a laser-cut stencil has much straighter walls than an etched stencil because there is no removal of material in the X-Y direction as there is during chemical etching. Laser-cut stencils are more expensive than etched stencils. Also, the aperture walls tend to be rough. This roughness can be avoided by nickel plating the inside walls of the apertures. Of course, allowance must be made for the plating thickness buildup in the aperture walls.

Electroformed stencils are made using a photosensitive dielectric film. The film is coated over a metal substrate, then a suitable mask is placed over the film. The film is subsequently exposed and developed, leaving openings which are then plated, generally with nickel. Then the remaining film is stripped, leaving the aperture openings. The electroformed stencil is then removed from the metal substrate. The position and dimensional accuracy of electroformed stencils is very good, and the aperture walls are smooth. In addition, it is easy to produce stencils of varying thicknesses.

When a stencil is used in solder paste applications, it is important that it be cleaned on a regular basis. Figure 9.103 shows a cleaning machine dedicated specifically to stencil cleaning.

The solder paste must be applied uniformly on all the component mounting lands; further, it is very important to apply the correct volume of paste. A suitable method for determining solder paste volume should be employed and utilized to verify the correct solder paste volume for each assembly produced, and a record of the solder paste volume of each assembly should be kept. See Fig. 9.105.

Another technique for applying solder paste to the component mounting lands is to use a dispensing system. See Fig. 9.106. A solder paste dispensing system makes use of a syringe to dispense dots of solder paste on the component mounting lands. Precise control is achievable using this method. Common defects using this technique are found in Table 9.46. The three chief drawbacks are

1. The operation is slow. Printing from a stencil is a mass application technique; dispensing dots of solder paste is not a mass application technique.
2. Care must be taken to keep the syringe/dispensing portion of the machine in good operating condition as the syringes tend to become easily plugged.
3. Needle dispensing requires a lower viscosity solder paste, which may have poor slump characteristics.

FIGURE 9.103   Stencil cleaner. *(Courtesy of Honeywell Space Systems.)*

9.9.4   Adhesives and Adhesive Application

Adhesives in SMT assembly operations are used to secure components firmly to the PWB surface. This is often done with some Type 2B assemblies and Type 2C/simple assemblies. (See Sec. 9.4.4.) If an adhesive is used, it should be demonstrated that its use will not affect the solder joint reliability in the service environment.

Adhesives can be classified as either heat curable or ultraviolet (UV) light curable. If a UV curable adhesive is used and the body of the component obscures the adhesive dot, the UV light will not properly expose the adhesive since it is a line-of-sight application. One technique to ameliorate this is to apply two dots of adhesive, one on each side of the component with a majority of the dot not covered by the component. In this way, most of the dot will be directly exposed to the UV light. Heating for a final cure of the portion of the dots covered by the component may still be necessary. It may be preferable to use a heat curable adhesive in this case. The adhesives are generally epoxy resin. Follow the manufacturers recommended curing process.

Adhesives are generally applied using a syringe dispensing machine similar to the ones used to dispense solder paste. (See Sec. 9.9.3 and Table 9.46.)

If an adhesive is used for producing a Type 2C (S) surface mount PWA, the adhesive should be applied to the board surface following the recommendation given in Fig. 9.107.

FIGURE 9.104 Board being examined after printing. *(Courtesy of the EMPF.)*

9.9.5 Component Preparation

It is very important that the surface mount components (SMCs) used to producing a surface mount PWA be solderable. However, sometimes SMCs are placed in storage prior to use and may remain in storage for a fair amount of time. If their leads have been gold plated, this will prevent oxidation. If SMCs are kept in storage, it is recommended that they be stored in a nonoxidizing atmosphere, e.g., nitrogen gas (N_2), if at all possible.

To ensure good solderability, the SMCs may go through a tinning operation. All SMCs having gold-plated leads or terminations, e.g., LCCs, should have the gold removed directly prior to soldering since gold forms very brittle intermetallic compounds with tin, causing embrittled solder joints. Generally, this tinning operation consists of five steps.

1. Wetting the SMC leads with a suitable flux
2. Preheat (120–140°C)
3. First dip into a tin or solder bath to remove leads protected with gold plating
4. Second dip into a tin or solder bath to hot tin dip the leads
5. Clean off flux residues from the leads

The tinning operation may be manual, semiautomated, or automated. A fully automated tinning machine is shown in Fig. 9.108.

FIGURE 9.105   Station for determining solder volume deposition. *(Courtesy of JPL.)*

FIGURE 9.106   Solder paste needle dispensing unit. *(Courtesy of JPL.)*

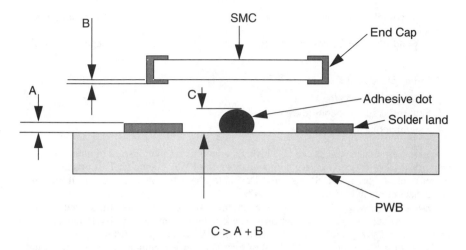

FIGURE 9.107   Adhesive dot placement before soldering. *(Courtesy of Traister, 1990.)*

9.9.6   Component Placement

Once a suitable solder paste has been deposited on the component mounting lands and the components suitably prepared, the next step is to mount them onto the mounting pads.

FIGURE 9.108   Automated tinning machine. *(Courtesy of JPL.)*

This is accomplished by means of placement machines, commonly called pick-and-place machines.

Pick-and-place machines can be conveniently placed into four categories (see Fig. 9.109).

1. In-line placement
2. Sequential placement
3. Sequential/simultaneous placement
4. Simultaneous placement

Pick-and-place machines can also be categorized as semiautomated or automated. It is also possible, of course, to place components on the PWB manually. This practice should be restricted to prototype SMAs only. Desirable characteristics shared by all pick-and-place machines can be found in Table 9.47.

In-line placement. This machine configuration has individual placement stations. Each placement station utilizes its own type of component. As the PWB moves past a given station, its respective components are mounted on the PWB surface.

Sequential placement. In this kind of pick-and-place machine, components are individually placed in a specific order. Either a preprogrammed placement head moves back

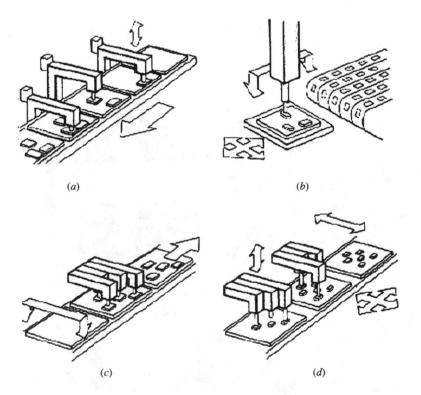

(a)   (b)

(c)   (d)

FIGURE 9.109 SMT placement machine categories; (a) in-line, (b) sequential, (c) simultaneous, and (d) sequential/simultaneous. *(Courtesy of Traister, 1990.)*

TABLE 9.47 Important Considerations regarding an SMT Placement Machine

Important considerations for a pick-and-place machine
Accuracy—ability to correctly place SMCs
Repeatability—ability to correctly place SMCs at all times
Reliability—machine performs as intended upon demand
Safety—safe for humans to use
Good vision system—very important when SMC pitch ≤ 0.025 in
Software control—good software control very important
Upgrade—ease of upgrade
Integration—ease of integration into existing SMT line
Gerber file—ease of downloading Gerber files
Maintainability—easy to maintain (modular approach is best)
Number of components—ability to place a wide variety of component types
Speed—not as important for high-reliability SMT PWAs
Adhesive dispenser—capability of also dispensing adhesive
Component testing—capability of testing CLR for chip components
Cost—probably the most expensive item on an SMT line

and forth, placing components on the PWB surface or else a moving X-Y table containing the PWB moves back and forth under a stationary placement head, which mounts components on the PWB surface. Figure 9.110 shows a semiautomated sequential placement machine, Fig. 9.111 shows an automated sequential placement machine, and Fig. 9.112 shows a component in a placement head about to be placed on the PWB surface—the machine has a movable X-Y table.

Sequential/simultaneous placement. This machine incorporates multiple placement heads and a moving X-Y table.

Simultaneous placement. All of the components are placed on the PWB surface in a single operation.

Note: Sequential/simultaneous and simultaneous placement machines are generally dedicated to high to very high-volume placement applications. Large sequential/simultaneous and simultaneous placement machines dedicated to the extremely rapid and accurate placement of chip components are commonly called chip-shooters.

Another important consideration regarding SMC placement machines is the component feeders. The feeders are directly based on the component carrier package type. There are four component carrier packaging formats (see Sec. 9.8.6). The formats are

1. Bulk
2. Tape and reel
3. Tube or stick
4. Matrix tray or waffle pack

FIGURE 9.110   Semiautomated SMC placement machine. *(Courtesy of JPL.)*

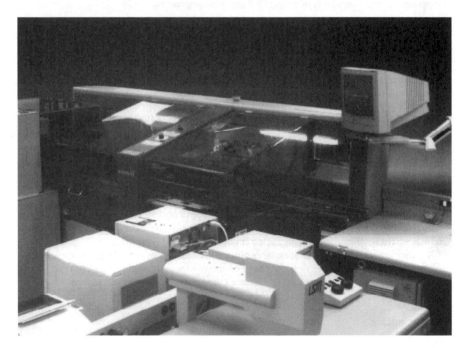

FIGURE 9.111   Automated SMT placement machine. *(Courtesy of JPL.)*

FIGURE 9.112   Component being placed on a PWA. *(Courtesy of the EMPF.)*

The use of fiducials in conjunction with an automated vision system (AVS) is very important for achieving both accuracy and repeatability when placing SMCs (see Fig. 9.113). A fiducial is a fixed location point on the PWB surface used by the machine's vision system to orient itself for accurate component placement. That is, the fiducial is an alignment target for the vision system. Many different kinds of symbols can be used as a fiducial. A circle is a common fiducial; a crosshair is another common fiducial. It is important, however, to use a fiducial symbol that is easily recognized by the machine's automated vision system. Global fiducials are associated with the PWB surface as a whole; local fiducials are associated with the footprint (land pattern) of a particular surface mount component. For standard pitch SMC placement, it is recommended that a minimum of three global fiducial targets be found on the PWB surface. For fine pitch SMCs (pitch $\leq$ 0.025 in), it is recommended that a minimum of two local fiducial targets per component shall be found on the PWB surface. Three or four is preferable. Refer to Fig. 9.113.

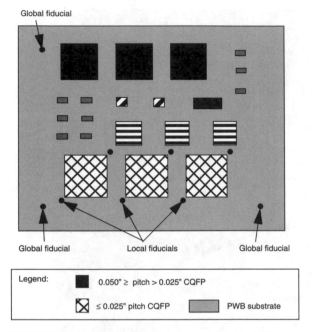

FIGURE 9.113   Global local fiducials on a surface mount PWA.

9.9.7   The Soldering Process

Once the components have been placed on their lands to which solder paste has been applied, the next step is to reflow the solder. There are three common methods used to reflow solder surface mount PWAs. These are

1. Vapor phase reflow soldering
2. Infrared reflow soldering
3. Forced air convection reflow soldering

Before soldering, it is important that the SMCs, PWB, and solder paste are solderable. Several pieces of equipment are now available to aid in determining solderability of SMCs (see Figs. 9.114 and 9.115). It is important that a suitable technique be chosen to determine the solderability of the printed wiring board and components prior to their being soldered to create the surface mount PWA. In addition, prior to the actual soldering operation, a suitable technique should be chosen to determine whether the solder paste is suitable for the application at hand. Again, a setup PWA is very useful for accomplishing this.

Reflow soldering means exposing solder paste deposited on land areas to sufficient heat to cause the solder in the solder paste to melt and become liquid. The flux in the paste is activated. It strips off surface contaminant layers (oxides, hydroxides, sulfides, etc.) and creates a virgin metallic surface. This metallic surface then combines with the solder in the paste to create a metallurgical bond. If the termination of a component becomes attached to a land via a bridge of solder, the resulting interconnect is known as a solder fillet. See Fig. 9.116.

FIGURE 9.114 Wetting balance. *(Courtesy of the EMPF.)*

Prior to the reflow operation, it may be necessary to bake the PWB on which solder paste has been deposited and on which the components have been placed. This should definitely be done if the reflow process does not have a sufficient preheat phase built into the machine. The purpose of this bake operation is to drive off the volatile materials in the solder paste to avoid spattering the paste during the reflow operation. If spattering occurs, solder balls may form, possibly leading to shorts. If baked, the board should be reflowed within ten (10) minutes of the bake operation. Always bake the PWB in an inert atmosphere, e.g., nitrogen gas, (N_2), if possible. The recommended bake schedule is:

Temperature 80°C±5°C, duration 30±5 min, oven humidity < 50 percent RH

By far, the predominant solder used in electronic applications is eutectic solder, that is, 63 percent tin (Sn)-37 percent lead (Pb) by weight.

To produce a high-quality solder joint, it is important to understand and control all the variables involved. Broadly speaking, there are five general variables governing any process. These are

- Materials (see Sec. 9.9.2)
- Process—In addition to the general variables, there are specific process parameters related to producing good products. For example, for any soldering operation, controlling the time-temperature profile is very important.
- Humans—refers to the so-called "operator effect." For manual and semiautomatic operations, the human input can be controlled to a great extent through proper training.

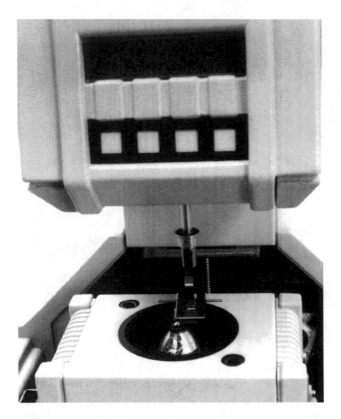

FIGURE 9.115 Details of a component being wetted. *(Courtesy of the EMPF.)*

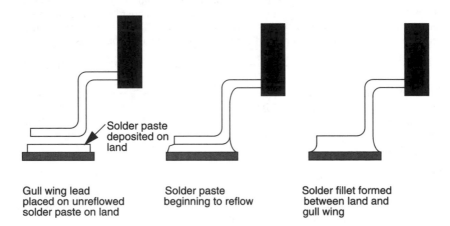

Solder paste
deposited on
land

Gull wing lead
placed on unreflowed
solder paste on land

Solder paste
beginning to reflow

Solder fillet formed
between land and
gull wing

FIGURE 9.116 Formation of a lap joint solder fillet during reflow.

- Ambience or environment—both external and internal. During the soldering process control of the ambient temperature and humidity is very important. Too much moisture adversely affects solder paste and possibly components too. The internal ambience inside the equipment can also have an affect on the final product being produced. The use of a nonoxidative internal atmosphere within the soldering machine can often ensure improved solder joints. See Sec. 9.9.11.

- Equipment—the equipment used to make product is described in the relevant sections below. The equipment is described in a general sense for each process. No particular piece of equipment from a given manufacturer is discussed.

Time and temperature are very important parameters during reflow soldering (see Table 9.48).

TABLE 9.48   Time-Temperature Effects Matrix

	Time	Temperature
Low	—	Poor solder joint formed, e.g., cold solder joint
High	—	Material damage such as burnt boards; excessive solder balls and solder spattering due to too rapid heating of the solder paste; cracked and/or damaged components
Long	Material failures and damage; thermal degradation; similar to high temperature exposure	—
Short	Inadequate flux action resulting in poor solder joints; similar to low temperature exposure	—

There are several distinct phases during reflow. All of these phases are part of the time-temperature profile. These phases take place over time and temperature. Figure 9.117 shows the different phases of the reflow process. The time-temperature profile in Fig. 9.117 is representative of a rosin-based paste. Water soluble and low solids ("no clean") pastes have a somewhat different time-temperature profile. Always consult with the solder paste manufacturer regarding the optimum time-temperature profile of a particular paste and its associated process window. These different phases are described below.

- *Phase 1. Preheat.*   Unless this phase is conducted in an outside oven, it should be accomplished rapidly, typically within 30–40 seconds. During preheat, the volatiles in the paste should start to be driven out, thus reducing the tendency to form solder balls. The temperature at the end of this phase should be ~100°C. The most critical parameter during this phase is the temperature ramp-up rate, which should not exceed 2.5°C/s. Other-wise, thermal shock to both the PWB and the components may result.

- *Phase 2. Flux activation.*   This phases lasts about 40–50 seconds. During this phase the flux begins to do its job. The activators begin to remove the surface contaminant layers (oxides, hydroxides, sulfides, etc.) rendering the metal underneath in an suitably active state. The flux also helps prevent reoxidation of the virgin metal surface during this stage because the volatiles in the solder paste start to be driven off due to intake of heat and chemical changes within the flux. The recommended temperature ramp-up rate during this phase should not exceed 2°–3°C/s. If the temperature rises too quickly, the paste

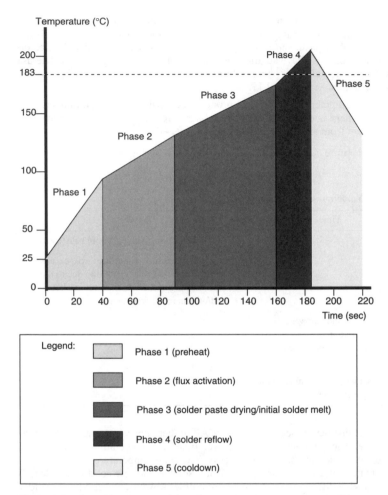

FIGURE 9.117 Different phases of the reflow process (time-temperature profile).

may spatter due to volatiles within the paste exiting too fast. This can results in voids and blow holes within the solder joints and excessive solder ball formation.

- *Phase 3. Solder paste drying/initial solder melt.*   This phase lasts 60–80 seconds. It is a bridge between Phase 2 and Phase 4. It is during this phase that the solder paste begins to melt and spread over the virgin surface of the metal replacing the flux in the process. During this phase, the flux and remaining portions of solvent are driven off, and the solder begins to melt and coalesce into a fillet. Nonuniformities in the temperature during this phase can result in nonuniform wetting of the solder and wicking of the solder. Solder tends to wick to a metallic surface having a higher temperature than the solder.

- *Phase 4. Solder reflow.*   This phase lasts 20–40 seconds. During this phase the metallurgical bond is formed, thus resulting in a solder joint. The wetting action of the solder causes it to spread over the metal in which it has contact. Sufficient time-temperature is needed to form the necessary intermetallic compounds (IMCs) required to create a good solder joint. Too long a time and/or too high a temperature will result in excessive IMC formation. This can result in a brittle joint. In addition, if excessive IMCs are formed or if the oxide layer has not been properly removed by the action of the activators in the paste, the solder will not properly wet the land and/or component lead. This can result in a dewetted or nonwetted solder joint.

- *Phase 5. Cooldown.* This phase can last several minutes. Longer cooldown times results in larger grain sizes in the solder whereas shorter cooldown times result in smaller grain sizes. The proper cooldown rate is still open to dispute. Experimental evidence exists indicating that a more rapid cooldown rate results in a finer grain size, which is considered preferable from the standpoint of solder joint life. However, over time the grain size in solder increases anyway. This area is still under investigation.

Regardless of the reflow soldering method used, a time-temperature profile for the process should be established and documented using a suitable time-temperature profiling technique. It may be necessary to redocument the profile if the process change is made. Use setup PWAs that resemble production PWAs as closely as possible to establish the time-temperature profile.

Every soldering process has an associated process window. This process window refers to the acceptable time-temperature range to which it can be subjected and still yield acceptable solder joints. The process window is determined by the requirements of several items, as illustrated in Fig. 9.118.

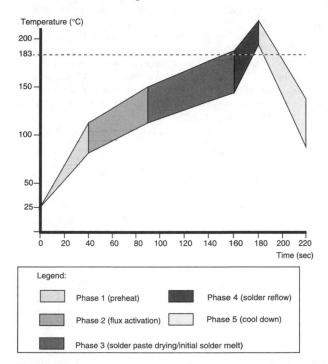

FIGURE 9.118 Process window of a time-temperature profile for a solder paste.

- The solder paste used
- The printed wiring board
- The components mounted on the printed wiring board

Note: It is important to determine the process window for each type of surface mount PWA before making a production run. To repeat, setup PWAs are well suited for this purpose.

Another issue of importance in reflow soldering regards the heat control in a reflow machine. For infrared soldering (IRS) machines (lamp and panel) and convection reflow machines, it is possible to independently control the top heaters and the bottom heaters. See Figs. 9.124 through 9.126 and Fig. 9.129. It is possible, in theory, to use a single-pass reflow operation to produce Type 2B and Type 2C (C) surface mount printed wiring assemblies because the top and bottom heaters can be controlled separately. However, in practice, a single pass generally does not work well. Some process development may be necessary to optimize the reflow process for producing Type 2B and Type 2C SMAs. Bear in mind that Type 2B and Type 2C assemblies, except for Type 2C (S), are more difficult to produce. (See Sec. 9.4.3.)

There are two common types of conveyor systems in in-line reflow machines.

1. Mesh conveyors

2. Edge conveyors

Both mesh conveyors and edge conveyors are found in in-line equipment—reflow soldering and cleaning equipment. Wave solder machines have only edge connectors with fingers to carry the board or a chain system used in conjunction with pallets. The boards are placed in the pallets. (See Sec. 9.9.7.5.)

Edge conveyors act as heat sinks on the edge of the product (the surface mount PWA); edge heating zones will become important when edge conveyors are used. A front zone of edge heat and a back zone will aid in compensating for the heat sink effect of edge conveyors. These zones are unnecessary if a mesh conveyor is used since the conveyor width can be chosen to comfortably encompass the product width.

The temperature rise of an object subject to heating can be expressed by Eq. (9.1).

$$\Delta T = q/C_p m \tag{9.1}$$

where ΔT = temperature rise of object (K)

q = heat absorbed (J or cal)

C_p = object's specific heat (J $\bullet$ kg^{-1} $\bullet$ K^{-1})

m = object's mass (kg)

The symbol K refers to the Kelvin (absolute) temperature, K = 273 + °C. For example, if the temperature is 27°C (81°F), the Kelvin temperature is 273 + 27 = 300K.

The specific heat (C_p) times the mass (m) is called the thermal mass. The greater the thermal mass, the more heat is required to achieve a given temperature rise. The thermal mass is important when considering heating effects on a PWB with different components. For example, several large CQFPs will have a much higher thermal mass than a number of small ceramic chip components.

9.9.7.1 Vapor Phase Reflow Soldering. This reflow method involves using the heat of condensation of a fluid to perform the reflow soldering. It is also referred to as condensation soldering. The boiling points of the various fluids used in this process typically fall between 215–235°C. The printed wiring assemblies to be reflow soldered are exposed to the vapors of this primary fluid. As the vapors condense on the board surface, the heat of condensation causes the solder to reflow thus creating a metallurgical bond. The heat transfer from vapor to workpiece is given by Eq. (9.2).

$$Q = hA(T_v - T_s) \qquad (9.2)$$

where Q = the heat transfer rate from the vapor to work piece—dq/dt $(J \cdot s^{-1})$

h = the heat transfer coefficient $(W \cdot m^{-2} \cdot K^{-1})$

A = work piece surface area (m^2)

T_v = saturated vapor temperature (K)

T_s = workpiece surface temperature (K)

Note: The workpiece = the PWB + suitably deposited solder paste on the mounting lands + components placed on the paste/lands. This workpiece will sometimes also be referred to as the PWB/solder paste/components.

It is evident from Eq. (9.2) that the driver is the difference in temperature between the primary vapor, T_v, and the workpiece, T_s. This difference can be expressed as ΔT. In vapor phase, or condensation, soldering the workpiece surface is heated first as the primary vapor condenses on its surface and reverts back to the primary liquid, releasing the heat of condensation as it does so. When the workpiece surface temperature (T_s) reaches the saturated vapor temperature (T_v) = b.p. of the liquid, $T_v - T_s = 0$ and no more heat is transferred. The heat transfer for vapor to workpiece surface is very rapid. The rate is typically 6–10°C/s. For this reason, it is best to preheat the workpiece (either externally in an oven or internally in a heat tunnel in the machine) to avoid thermal shock, tombstoning, and solder wicking.

There are several advantages of vapor phase soldering (VPS).

- VPS is the easiest soldering method to learn and use.

- Fewer profiling runs with setup boards are required with VPS to optimize the process.

- Process variables are minimal because the primary fluid has a fixed boiling point and a given heat of condensation that does not vary.

- VPS is an equilibrium soldering technique. The temperature between the workpiece and the primary vapor rapidly approaches zero, that is, $\Delta T \to 0°$.

- The VPS reflow system can not exceed the temperature of the primary vapor. Under normal operating conditions, this temperature remains constant.

- When solder reflow takes place, the workpiece is immersed in the primary vapor, thus preventing reoxidation from occurring. That is, the primary vapor acts as an effective blanket preventing reoxidation of the surfaces deoxidized by the flux.

There are three types of VPS.

1. Dual vapor batch VPS. [A dual vapor batch vapor phase reflow system often results in too rapid heating of the workpiece unless the workpiece is preheated (see Fig. 9.122)]
2. Single vapor batch VPS with IR preheat zone
3. Single vapor in-line VPS with IR preheat zone

The equipment used is different. A dual vapor batch VPS machine is very similar to a conventional vapor degreaser. (See Fig. 9.119, showing a dual vapor batch VPS unit.) The

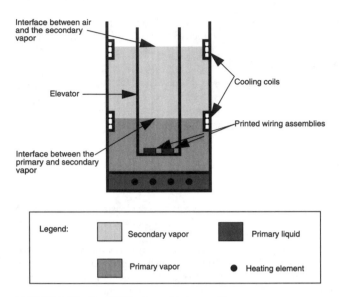

FIGURE 9.119   Dual VPS reflow soldering machine.

dual vapor batch system does not have a heat tunnel where the workpiece can be preheated prior to being reflowed. The printed wiring board/solder paste/components to be reflow soldered in the dual vapor VPS unit is put on an elevator that is operated mechanically. Two fluids are used in batch VPS: (1) a primary fluid and (2) a secondary fluid. The secondary fluid forms a vapor blanket over the primary vapor, thus keeping down unacceptable loss of primary fluid.

The primary fluid in any kind of VPS is a perfluorinated compound, or PFC. A PFC is a hydrocarbon in which all the hydrogen (H) atoms have been replaced by the element fluorine (F). PFCs are extremely stable chemically. They do not easily break down or decompose. They are typically very expensive, costing $U.S. 600–700/gal. In dual vapor batch VPS units, a secondary fluid is also employed. For many years, the secondary fluid of choice was straight CFC-113. However, because CFC-113 is an ozone depleting chemical, its production was discontinued at the end of 1995 (see Sec. 9.9.15). Newer PFCs have been developed to replace CFC-113 as the secondary fluid.

Caution: In certain instances, PFCs break down to form toxic substances. If breakdown occurs, hydrofluoric acid (HF) is generally formed. This substance is very toxic and hazardous. Some PFCs breakdown to form, in addition to HF, perfluoroisobutylene (PFIB). This material is also very toxic and hazardous. Always consult the material safely data sheet (MSDS) of the PFC and/or consult the PFC vendor.

It is very important to keep the VPS's heating elements clean because material such as flux can form deposits on them causing overheating of the PFC with subsequent breakdown. PFCs are not ozone depleting chemicals (ODCs) given that they contain no chlorine (Cl) in their molecules. They are, however, potential greenhouse warming chemicals (GWCs). Their use may be severely restricted or eliminated entirely at some time in the future.

A single vapor batch or in-line VPS consists of a conveyorized machine having some kind of mesh-wire belt or a pallet on a transport system. The machine transfers the workpieces to be soldered from the front end of the machine to the back end. In-line machines are used for higher throughput. In addition, they do not use a secondary fluid, thus eliminating the problem cited above regarding CFC-113 as the secondary fluid. In-line VPS machines are generally more costly than a batch unit. Figure 9.120 depicts in diagrammatic form an in-line VPS unit. Figure 9.121 shows a single vapor batch VPS with an IR heat tunnel.

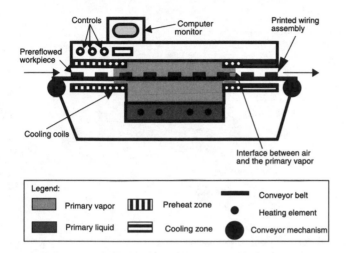

FIGURE 9.120   Single vapor in-line VPS reflow soldering machine.

FIGURE 9.121   Single vapor VPS reflow soldering machine. *(Courtesy of JPL.)*

As pointed out above, VPS systems supply heat very rapidly to workpieces. A typical time-temperature profile for VPS is shown in Fig. 9.122. Too rapid a rise in temperature causes the temperature of the lead to be higher than the land temperature (see Fig. 9.123). When this happens, the solder will tend to wick up the lead, resulting in an electrical open in the case of leaded components and tombstoning in the case of small, discrete chip com-

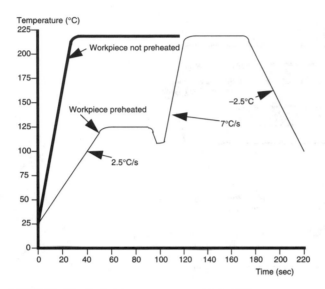

FIGURE 9.122   Typical time-temperature profile for VPS.

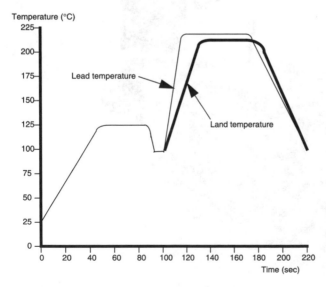

FIGURE 9.123   Typical time-temperature profile for VPS.

ponents. Make sure that the VPS machine has a good preheat section—generally accomplished using IR (see the next section).

Vapor phase reflow soldering has some drawbacks. These are

- VPS results in significantly fast temperature rises due to rapid heat transfer. This can cause wicking of the solder leading to insufficient solder in the fillet, electrical opens, and tombstoning of chip components.
- The primary fluid, the PFC, is very expensive. The equipment must be carefully maintained to avoid loss of this expensive material. It is also important to ensure that the heating elements are kept free from deposits since this can lead to undesired breakdown of the PFC.

9.9.7.2 Infrared Reflow Soldering.

This reflow soldering method involves using heating elements that emit infrared radiation to heat the workpiece. Infrared radiation consists of electromagnetic waves having wavelengths between 0.72 and 1000 micrometers (μm), although the wavelengths of the radiation used in infrared soldering (IRS) is between 0.7 to 8.5 μm. For the infrared (IR) radiation used in IRS, the wavelengths are classified as:

- 0.7 to 1.5 μm—near IR
- 1.5 to 5 μm—middle IR
- 5 to 8.5 μm—far IR

IRS depends on the workpiece absorbing the IR energy, which is converted into heat. The more IR energy absorbed, the hotter the workpiece grows. The IR waves are reflected from metallic surfaces; they are absorbed by organic materials. Thus, IRS depends on the absorption of IR energy by the organic materials making up the workpiece and the subsequent conversion of this energy into heat. The heat is responsible for the workpiece reaching sufficient temperature for reflow to take place. The absorption of IR energy depends on

- The wavelength of the incident IR energy
- The absorptivity and/or reflectivity of the material in contact and/or penetrated by the IR radiation

The equation governing the heat transfer rate in infrared reflow is given by Eq. (9.3).

$$Q = \varepsilon_1 \varepsilon_2 \sigma A_e (T_e^4 - T_a^4) \qquad (9.3)$$

where Q = the heat transfer rate from the infrared emitter to the workpiece ($J \cdot s^{-1}$)

ε_1 = emissivity factor for emitter (1 for a perfect blackbody; generally 0.8 to 0.9 for an IR machine)

ε_2 = geometric factor (amount of IR energy on the workpiece surface; generally 0.95)

σ = Stefan-Boltzmann constant ($5.67 \times 10\text{-}8 \ W \cdot m^{-2} \cdot K^{-4}$)

A_e = emitter surface area (m^2)

T_e = temperature (K) of the infrared emitter

T_a = temperature (K) of the energy absorber

The chief distinction in IRS is between lamp emitter units and panel emitter units. Lamp emitter units rely almost exclusively on infrared (IR) radiation to heat the workpiece; panel emitters make use of a combination of infrared and convection heating to heat the workpiece. The IR radiation used in lamp emitter units is between 0.7 to about 2.5 μm, and it is classified as near IR energy. The shorter the wavelength the deeper the penetration. Hence lamp emitter IRS units rely on a somewhat different reflow technique than panel emitter IRS units where the energy is confined more to the surface.

The lamps used in lamp emitter IRS are generally made of a tungsten filament encapsulated in a quartz tube and filled with an inert gas. An external reflector is often used which acts as a secondary emitter. Tungsten, which excited, emits IR radiation in the near wavelength range (0.7–1.5 μm—near IR). IR lamps operate at very high temperatures. A temperature of 1,100°C (2,012°F) is typical.

If a reflector is used as a secondary radiator, the radiation tends to fall predominantly in the middle IR region. Gases like air are transparent to near IR radiation, meaning that air, or any other gas, in such an IRS unit does not absorb near IR energy. Hence there is very little energy of convection transferred to the workpiece in a lamp emitter IRS unit.

The chief advantages of lamp emitter IRS are as follows:

- The system is generally easy to maintain.

- Because the energy in a lamp emitter IRS unit is either almost all absorbed or reflected, it is very easy to profile for different workpieces.

- Because lamp emitter IRS units are easy to profile for different workpieces, they are inherently flexible; they also have fast response times.

Panel emitter IRS units are area emitters, as opposed to lamps, which are source emitters. Generally, IR panels consist of three layers. See Fig. 9.124, which is exaggerated in size for illustrative purposes. The primary emitter in the panel generally contains a wound resistive element. The panel face can be either metallic (stainless steel, aluminum) or nonmetallic (glass, ceramic). If the face is metal, it can absorb some of the IR energy of the primary emitter and thus function as a secondary emitter. The insulation is normally some refractory material. Panel emitter IRS systems operate at much lower temperatures than lamp emitter IRS; 300 to 400°C (572 to 752°F).

In panel IRS units typically 40–60 percent of the heat transfer is accomplished by convection heating. The equation governing heat transfer by convection is Eq. (9.4) [same as Eq. 9.2)].

$$Q = hA\Delta T \tag{9.4}$$

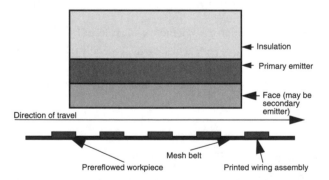

FIGURE 9.124   Infrared panel, exaggerated in size for illustrative purposes.

where Q = heat transfer rate from atmosphere to work piece

h = heat transfer coefficient

A = workpiece area to which energy is being transferred

ΔT = temperature difference between atmosphere and workpiece

The principal advantages of panel emitter IRS are

- The system is generally easy to maintain.
- Most of the material used in printed wiring assemblies absorb IR energy in the mid to far end.
- Panel emitter IRS systems are less load sensitive than lamp emitter IRS systems.
- Panel emitter IRS systems are less color sensitive than lamp emitter systems.

IRS units, either lamp emitter or panel emitter, are in-line machines. They all have several distinct zones based on the distinct phases of the time-temperature profile involved in reflow soldering (see Fig. 9.117). Figure 9.125 shows in diagrammatic form an in-line lamp emitter IRS unit, and an in-line panel emitter IRS unit in diagrammatic form is shown in Fig. 9.126. A typical time-temperature profile for IRS is shown in Fig. 9.127. Figure 9.128 shows the difference in temperature (ΔT) between a low thermal mass area on the work piece and a high mass area.

Infrared reflow soldering has some drawbacks. These are:

- It is generally necessary to profile the unit for each type of workpiece. With a vapor phase unit with an IR preheater, this is also the case. This means that a separate time-temperature profile must be determined for each type of workpiece processed through the unit.
- IRS is a nonequilibrium process. This can result in uneven heating of the workpiece with corresponding poor solder fillet formation.
- Lamp emitter IRS is color-sensitive. That is, different colors absorb IR energy differently. This can give rise to poor solder fillet formation.
- Lamp emitter IRS is also prone to a "shadowing" effect. Radiation reaches the substrate via line-of-sight. If a component lies in this line-of-sight, the component may absorb

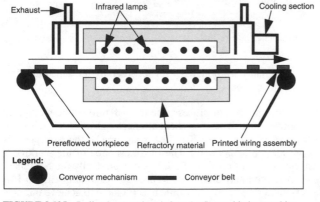

FIGURE 9.125 In-line lamp emitter infrared reflow soldering machine.

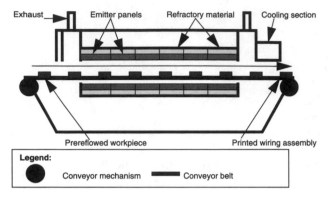

FIGURE 9.126   In-line panel emitter infrared reflow soldering machine.

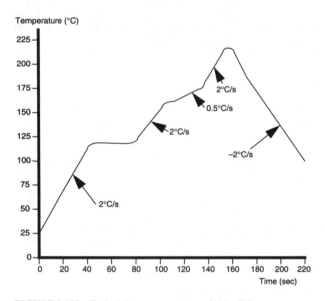

FIGURE 9.127   Typical time-temperature profile for IRS.

some of the IR radiation that normally would have been absorbed by the substrate. This also can give rise to poor fillet formation.

Caution: Because IRS is a nonequilibrium process, the temperature of the workpiece never reaches the emitter's temperature. There is unequal energy absorption. Different parts of the workpiece heat up at different rates. The low mass area of the workpiece is generally the hottest; the high mass area is cooler. The low mass area of the workpiece includes the corners and edges; the center portion of the board is generally the high mass area. See Fig. 9.128. This phenomenon may contribute to poor solder joint quality. If IRS is used, the component density should be balanced as much as possible to achieve uniform solder joint formation.

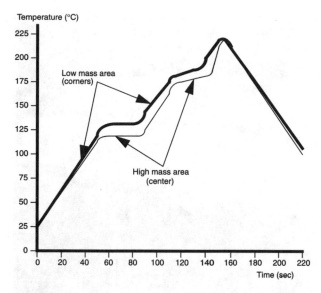

FIGURE 9.128   ΔT between low-mass area and high-mass area (IRS).

9.9.7.3 Convection Reflow Soldering.

Using pure convection heating to reflow solder is a more recent technique. The equation governing the transfer of heat energy by convection has already been presented as Eq. (9.4).

A convection reflow soldering (CRS) machine is shown in diagrammatic form in Fig. 9.129. Figure 9.130 shows an actual machine.

In convection heating, the rate of heat transferred to the workpiece, Q, can be changed either by changing h, the convective film coefficient, and/or ΔT, the temperature differ-

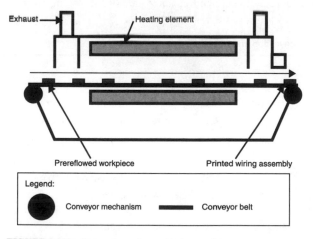

FIGURE 9.129   Convection reflow soldering machine.

FIGURE 9.130 Convection reflow soldering machine. *(Courtesy of Heller Industries.)*

ence between the workpiece and internal ambient temperature. The value of h is related to such factors as velocity of the air (or other gas in the heating tunnel) flow and angle of flow. Increasing the flow velocity will increase the value of h. Too high a velocity flow of the gas in the heating tunnel can cause components to shift and move in an undesirable manner.

The direction of flow also has an impact on Q. Perpendicular flow direction is superior to a flow direction parallel to the board surface because in perpendicular flow there is no boundary layer to impede the transfer of heat energy from the air to the workpiece.

There are a number of pros and cons to pure convection to pure IRS. These are given in Table 9.49.

TABLE 9.49 Pros and Cons of Convection and IR Soldering

Convection	IR
Pros (+)	
ΔT between source and workpiece is small	High efficiency (60–70%)
More uniform heating	Rate of temperature rise easy to control
Less sensitive to shading	Source temperature only controlled
Cons (−)	
Inefficient (efficiency 20–30%)	ΔT between source and workpiece is large
Air has low thermal mass and requires high flow rates to effectively transfer heat	Much more sensitive to shading
Leads get heated first—increasing chance of solder wicking	Less uniform high mass differences

9.9.7.4 *Controlled Atmosphere Soldering.*

Controlled atmosphere soldering does not refer to a new reflow soldering process; rather, it refers to using a machine internal atmosphere other than air while the reflow soldering process is taking place. In the case of vapor phase soldering (VPS) the atmosphere is provided by the vapors of the perfluorinated compound (PFC). In conventional infrared soldering (IRS), with or without convection, and in convection soldering, the internal atmosphere of the machine has conventionally been ambient air.

There are distinct disadvantages to conducting the reflow soldering process in an ambient air atmosphere. Air contains approximately 21 percent oxygen; hence, it is an oxidizing atmosphere. During reflow the flux must provide protection to the active metal surfaces in the form of nonoxidative vapors; otherwise, a passivating oxygen layer can reform on the active metal surfaces. If this happens, dewetting or nonwetting will occur. This results in a poor and/or malformed solder fillet whose reliability is compromised.

It is becoming more common to use high purity nitrogen gas (N_2) as the internal machine atmosphere of choice during reflow. Nitrogen is essentially a nonreactive gas. If the level of oxygen gas (O_2) is kept at low levels, typically superior results are obtained during the reflow soldering process (see Table 9.50).

TABLE 9.50 Affect of Nitrogen Purity on the Soldering Process

Description	Purity level of nitrogen gas (N_2)
Significant reduction in board discoloration	10,000 ppm
Ease of cleaning rosin-based flux residues (e.g., RMA flux)	500 ppm
Reduction in solder balling	200 ppm
Reflow of low-residue solder paste	< 100 ppm

The increasing use of "no clean" fluxes and pastes makes using a nonoxidizing internal machine atmosphere such as nitrogen gas even more important. This is because the so-called "no clean" fluxes and pastes all contain low solids. Hence, during the soldering operation, they provide even less of the necessary protective, nonoxidizing vapors than do regular fluxes and pastes.

There are other distinct advantages to using a nonoxidizing atmosphere during soldering. These are:

- Less solder ball formation
- Cleaning (if required) is not as difficult

In connection with wave soldering also, a nonoxidizing atmosphere such a nitrogen gas is becoming more prevalent. Again, in many cases the driver to such an inert atmosphere has been the use of the low solids, "no clean" fluxes.

The purity of the nitrogen gas can, in principle, have an affect on a number of important process parameters of the reflow soldering operation. Among these are

- Wetting force
- Wetting time
- Spread of flux
- Solder ball formation

- Amount of flux residue
- Microstructure of solder

Economic drawbacks to the use of nitrogen gas (or any other controlled atmosphere inside the soldering machine) are the extra expenses incurred. Machines built to successfully contain an inert atmosphere such as nitrogen are generally more expensive than machines lacking this feature. In addition to the one-time capital expenditure, there is the recurring operating cost of the nitrogen.

It is still an open issue whether the use of an inert, or nonoxidizing, atmosphere, such as nitrogen gas (N_2), truly produces solder joints that are significantly better and/or more reliable than those produced without using this process.

9.9.7.5 Wave Soldering. Wave soldering does not make use of a solder paste deposited directly onto land patterns; it involves applying the solder residing in a solder pot by the formation of a wave of solder generated mechanically by a pumping mechanism. The flux is applied separately to the bottom (secondary side) of the PWB prior to exposing the board to the wave of molten solder. Either eutectic solder or 60 percent tin-40 percent lead solder is normally used. The PWB is transported from the flux unit to the solder unit by means of a suitable conveyor system. Wave soldering is the mass soldering method of choice for pure through hole technology (THT) PWAs (Type 1A).

A wave soldering machine is composed of a number of distinct units or systems. These are

- Flux unit
- Flux density control unit (optional but recommended)
- Preheat unit
- Conveyor unit
- Solder unit (single wave or dual wave)
- Air knives
- Machine control unit

The process flow chart for the wave soldering operation is shown in Fig. 9.131; Fig. 9.132 shows a pictorial diagram of a wave soldering machine.

Two time-temperature profiles for wave soldering are shown in Fig. 9.133 and in Fig. 9.134. Figure 9.133 is for a single-wave machine and Fig. 9.134 for a dual-wave machine. As in any other soldering process, controlling the time-temperature profile is very important. Multilayer ceramic capacitors (MLCs) mounted on the bottom side of the PWB are easily prone to cracking when wave soldered. It is not advisable to solder them to the PWB using this technique. If this type of SMC is wave soldered, very careful control of the preheat is necessary to avoid cracking the component. In summary, it is best to restrict the SMC types to be wave soldered (on the bottom of the board) to chip resistors, small SOTs, and possibly small SOICs. These various components must be oriented properly to avoid shadowing and solder bridging. See Fig. 9.135.

Note: To successfully wave solder small, discrete SMCs on the bottom side of the PWB, make sure the bottom side (secondary side) reaches a minimum temperature of 150°C (300°F) prior to entering the solder wave(s) to avoid cracking the ceramic chip components.

Leaded SMCs should not be mounted on the bottom side of the PWB because a great deal of solder bridging may result. This is the most common solder joint defect of wave soldering. It is not advisable to solder them to the PWB using this technique.

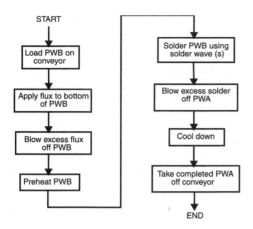

FIGURE 9.131   Flow chart for wave soldering process.

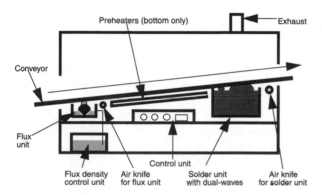

FIGURE 9.132   Wave soldering machine.

For Type 1C (some through hole components on the top side) and Type 2C (S), it is feasible to use wave soldering equipment to create the metallurgical solder connection. In the cases of Type 2C (S) and 2C (C), there are typically through hole components on the primary side of the printed wiring board and surface mount components on the secondary side. The SMCs are attached to the secondary side using adhesive so that they do not fall off when the board is inverted prior to the wave soldering operation. A minimum solder wave temperature of 260°C (500°F) is preferable to the maximum temperature of 275°C (527°F).

The flux is applied to the PWB prior to exposing the board to the solder wave. There are three principal methods for applying flux to the bottom side of the printed wiring board.

1. Foam (probably the most common in THT)
2. Wave
3. Spray (an external spray fluxer is very commonly used for low solids fluxes [LSFs = "no clean" fluxes])

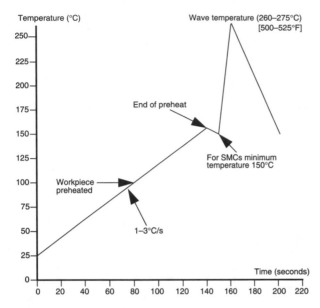

FIGURE 9.133   Typical time-temperature profile for wave solder (single-wave).

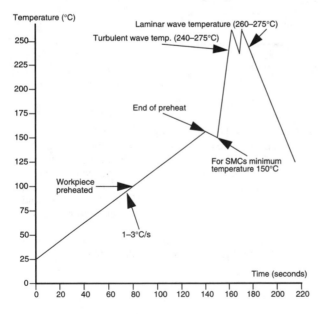

FIGURE 9.134   Typical time-temperature profile for wave solder (dual-wave).

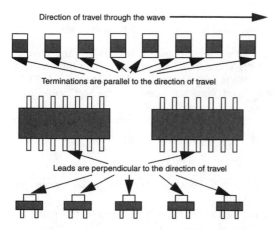

FIGURE 9.135 Orientation of SMCs for wave soldering.

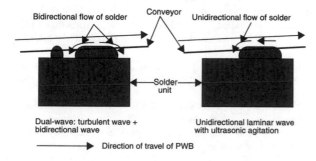

FIGURE 9.136 Dual-wave versus single unidirectional laminar wave.

After being fluxed on the bottom side, the printed wiring board to be wave soldered passes through a preheat zone. The function of preheat is the same as that of reflow solder- ing: to activate the flux and to prevent thermal shock when the board enters the solder wave (typically maintained at 260–275°C). In a wave soldering machine, the preheaters are generally on the bottom; they may be either calrod emitters or lamp emitters. It is important to adjust the preheaters so that the temperature of the board top side (primary side) is maintained between 200 and 220°C. A higher temperature is generally needed the thicker the substrate.

The angle of entry into the wave is also important. For any PWB, the solder of the wave should come to one-third to one-half of the laminate thickness. For a multilayer PWB, the solder of the wave should come to one-half of the laminate thickness. After pre- heat, the board enters the solder wave or waves if a dual-wave machine is used. For surface mount printed wiring assemblies, either a dual-wave or a single laminar wave with ultra- sonic agitation should be used.

There are two different approaches regarding the solder wave to solder mixed technol- ogy surface mount PWAs (see Fig. 9.136). Referring to Fig. 9.136, it can be seen that the dual-wave system consists of two waves.

1. Turbulent wave, the first wave to which the PWB is exposed

2. Smooth bidirectional wave, the second wave to which the PWB is exposed

The turbulent wave creates the basic solder fillet. Because it is turbulent and directed almost normally to the bottom of the PWB, it creates a uniform solder fillet for the small, discrete SMCs on the bottom side of the PWB with no shadow effect. The top side THCs are soldered with the smooth wave; the smooth wave also completes the SMC solder fillet formation.

In the case of the laminar wave, which is unidirectional, the solder fillets of the SMCs on the bottom side are created with ultrasonic agitation. This wave also creates the solder fillets of the top side THCs.

There are two types of conveyor unit for wave soldering machines.

1. Finger or edge conveyor—the PWB is held between finger-like metal appendages. Several types of finger styles are available.

2. Chain conveyor—the PWB is loaded into a suitable pallet or carrier. The pallet has built-in pins that fit into sockets on the chain.

Either type of conveyor must be maintained carefully. Keeping the fingers or pallet clean is very important.

A word must be said about dross. Dross refers to oxidized solder. Typically, it forms on the top of the solder pot and must be periodically skimmed off the top. Sometimes it is beneficial to utilize a thin layer of oil on the surface of the solder pot to reduce the amount of dross formation.

Caution: Because dross contains lead oxides, it is considered to be a hazardous material and must be disposed of properly. Check with the solder supplier. They typically provide a disposal service for recycling the dross.

It is possible to purchase controlled atmosphere wave soldering machines. The primary use of such machines is in conjunction with low solids fluxes. By using an inert atmosphere such as nitrogen gas during the soldering process, it is possible to employ fluxes containing much less activator and vehicle such as a low solids flux (LSF).

9.9.7.6 Other Soldering Techniques/Manual Soldering.

Several other reflow soldering processes exist; however, they are not used that often. Among these are

- Laser soldering
- Hot bar or thermode soldering
- Conductive belt soldering
- Heated platen soldering

Laser soldering is highly accurate. However, the throughput is low because each joint is individually soldered. It is not a mass soldering technique.

Hot bar (thermode) soldering is sometimes used to solder individual components such as quad flat packs. The bar simultaneously forms the leads and solders them to solder-paste-deposited pads on the PWB surface. It should be noted that hot bar soldering often sets up inherent stresses within the solder joints—thus it may compromise reliability. It is not a mass soldering technique.

Conductive belt soldering can only be used in conjunction with Type 1 assemblies having components on the top side. For ceramic substrates where the ceramic conducts heat well, this method of soldering is satisfactory. Substrates based on organic polymeric materials such as FR-4 do not conduct heat very well; hence, conductive belt soldering is generally not satisfactory for soldering organic-based surface mount assemblies.

What was stated above regarding conductive belt soldering also applies to heated platen soldering.

It is possible to solder all the solder attachments manually using a soldering iron. However, even if the operator is carefully trained and certified to perform hand soldering, because of the nature of this type of operation, it is not possible to consistently make each solder attachment identical to the next. That is, any manual operation is likely to experience greater variations than one performed on a machine.

Note: It is not recommended that hand soldering be performed to produce the solder attachments of surface mount PWAs. Remember, the idea is to reduce process variation as much as possible to ensure a more reliable product. Also remember that the cost of producing a hand soldered joint is much greater than producing one by a mass soldering method.

Hand soldering should be restricted to

- Touch up/rework operations
- Special components that can not be soldered by any other technique

9.9.8 Rework and Touch-Up

During the process of producing printed wiring assemblies, defects sometimes occur. If these defects are not too serious, the decision may be made to rework the defects rather than scrap the entire assembly. A list of common defects is shown below.

If rework is necessary, it generally indicates that the soldering process was not in control. A list of all defects that are repaired should be kept on a daily basis and should also be reviewed on a daily basis to be used for reducing process variation. Continuous process improvement (CPI) techniques should be employed to upgrade processes so that touch up/rework is not required. However, note that to perfect processes to the point that rework is not required may lead to exorbitant costs.

Ten common defects requiring rework are

1. Insufficient solder
2. Excess solder
3. Component misalignment
4. Solder balls/solder bridging
5. Rosin inclusion in the solder joints
6. Residue left after cleaning
7. Component leads (gull wing) require straightening before mounting
8. Incorrect component placement
9. Tombstoning of component (generally with small SMCs)
10. Solder dewetting/nonwetting

For rework and repair, two items are of especial importance. These are

- Have the proper piece of equipment available for performing the operation.
- Make sure that personnel performing the operation are properly trained.

Figures 9.137, 9.138, and 9.140 show rework/repair stations. Figure 9.139 shows an oblique viewing station. Such a station can be used for rework, but it is also very useful for

FIGURE 9.137 SMT rework/repair station. *(Courtesy of JPL.)*

FIGURE 9.138 Rework station with hot gas reflow. *(Courtesy of Honeywell Space Systems.)*

FIGURE 9.139   Oblique viewing system. *(Courtesy of Honeywell Space Systems.)*

FIGURE 9.140   SMT rework/repair station. *(Courtesy of the EMPF.)*

inspection. A rework/repair station ought to (1) have an adequate vision system and (2) utilize a heated, inert gas, e.g., nitrogen gas (N_2), for performing the reflow operation if components need to be removed and subsequently resoldered.

If a component needs to be removed, the following five steps should be performed:

1. Locate the component that needs to be removed. Either inspection or testing may be utilized to accomplish this.

2. Apply sufficient heat to reflow the solder joints. A hot gas (preferably N_2) reflow rework tool is best for accomplishing this task.

3. Remove the component and its solder. All the component's leads must be heated and reflowed at the same time to avoid damaging the component and its leads. Some kind of vacuum pickup tool is best.

4. Reapply solder paste or flux as required.

5. Position the replacement component and resolder, preferably using a reflow technique. Reinspect.

9.9.9 Cleaning

Cleaning refers to the removal or extraction of contaminants from the surface of the printed wiring board or the printed wiring assembly. If cleaning takes place very soon after the soldering operation, it is often referred to as defluxing.

During the manufacturing process of making bare printed wiring boards and printed wiring board assemblies, a wide number of possible contaminants can end up on the board surface. Table 9.51 summarizes the three categories of printed wiring contaminants. The three categories can be summarized as follows:

1. Category 1: particulate material

2. Category 2: polar/ionic, may be organic or inorganic

3. Category 3: nonpolar/nonionic, typically organic but may be inorganic

The preferred method for cleaning high reliability surface mount assemblies was to employ a suitable solvent in either a batch cleaning or an in-line cleaning machine. The solvent of choice was very often CFC-113 (1,1,2-trichloro-1,2,2-trifluoroethane: $CCl_2FCClF2$). CFC-113 is in that class of materials known as chlorofluorocarbons (CFCs). In some cases, 1,1,1-trichloroethane (TCA: CCl_3CH_3 = methyl chloroform) was used. TCA, or 1,1,1-trichloroethane is a more aggressive solvent than CFC-113; so its use was more restricted within the electronics industry. Generally, the CFC-113 was combined with another solvent(s) to improve its cleaning performance. For defluxing printed wiring assemblies, the CFC-113 was often combined with a low molecular weight alcohol such as methyl alcohol (CH_3OH) or ethyl alcohol (CH_3CH_2OH). Since alcohols are more water-like, their addition to the CFC-113 rendered the cleaning solvent capable of removing both ionic and nonionic materials quite well. Such an improved solvent was known as a bipolar solvent, and the cleaning operation was referred to as bipolar cleaning.

However, there is now a concern that compounds containing chlorine (Cl) atoms in their molecules and that are also relatively stable lead to the destruction of stratospheric ozone (O_3). Such substances are referred to as ozone depleting chemicals (ODCs) or ozone depleting substances (ODSs). In 1987, different countries around the world voluntarily agreed to first severely restrict, then ban, the production of ODCs. In the U.S.A. and in other more developed countries (MDCs), production for ODCs was discontinued at the end of 1995. In the case of electronics and SMT, this includes all solvents containing CFC-

TABLE 9.51   Chief Types of Contaminants on Printed Wiring by Category*

Category 1: Particulates	Category 2: Ionic	Category 3: Nonionic
Resin and fiberglass debris from drilling and/or punching operations	Flux activators and flux activator residues	Flux rosin
Metal and/or plastic chips from machining and/or trimming operations	Soldering salts	Flux resin (refers to synthetic resins)
Dust	Handling soils (especially sodium and potassium chlorides)	Oils
Handling soils	Residual plating salts	Grease
Lint	Residual etching salts	Wax
Insulation	Neutralizers	Synthetic polymers
Hair/skin	Ionic surfactants	Soldering oils
		Handling soils
		Hand creams and lubricants
		Silicones
		Nonionic surfactants

* *Source:* ANSI/IPC SC-60 (1987).

113 and all solvents containing 1,1,1-trichloroethane. This ban was also extended to certain hydrochlorofluorocarbons (HCFCs). The principal HCFC affected by the ban is HCFC-141b (1,1-dichloro-1-fluoroethane: CCl_2FCH_3). It turns out that its ozone depletion potential (ODP) is almost identical to TCA's. Hence, it is no longer considered an acceptable substitute for CFC-113 and TCA. Table 9.52 shows the boiling points (b.p.) of these three solvents along with the ODPs.

TABLE 9.52   Boiling Points and Ozone Depletion Potentials of Three Materials

Solvent material	b.p., °C	b.p., °F	ODP
CFC-113	47.6	117.6	0.8
1,1,1-trichloroethane	74.1	165.4	0.15
HCFC-141b	32.0	89.7	0.12

With the disappearance of ODCs, several cleaning options are open for cleaning surface mount PWAs. There are currently three cleaning methods that are being used in the electronics industry. They are

1. Semi-aqueous cleaning

2. Solvent cleaning

3. Aqueous cleaning, with or without saponification

Note: Solvent cleaning can entail use of a non-ODC solvent such as trichlorethylene. Several newer non-ODC solvents are coming to market. Most prominent among these is

HFC-43-10 mee, which has a 0 ODP and does form azeotropes. Solvents based on HCFC-225 are also finding use, but HCFC-225 does have a small ODP, namely, 0.002.

In addition to the above cleaning methods, the cleaning operation itself can sometimes be rendered unnecessary. The two principal ways this can be accomplished are

1. Using low solids (LS) fluxes and pastes, sometimes referred to as "no clean"

2. Controlled atmosphere soldering machines used in conjunction with LS fluxes/pastes

Semi-aqueous cleaning refers to a two-step operation. The first step involves immersing the workpiece in a suitable organic solvent to perform the solubilization process of contaminants on the workpiece surface. There are currently many semi-aqueous organic solvents to choose from. In general, these materials may be divided into two broad classes:

1. Terpene-based organic solvents

2. Nonterpene-based organic solvents

All semi-aqueous solvents in use so far are classified as combustible materials; a combustible material has a closed-cup flash point equal to or greater than 100°F (38°C). All semi-aqueous organic solvents, after performing the solubilization process on the workpiece to remove contaminants, must be rinsed off in a suitable manner. Generally, the rinsing solvent is water. Hence, the name of the process: semi-aqueous.

Semi-aqueous materials, both the solvent and the rinsing agent, can be utilized in either batch cleaning equipment or in-line equipment. Generally, batch equipment is perfectly suitable for small to medium production lots whereas in-line equipment is used in high production situations. Figure 9.141 shows a working semi-aqueous batch machine in which a technician is loading surface mount PWAs for cleaning. Another semi-aqueous batch cleaning system is shown in Fig. 9.142. This system consists of the semi-aqueous cleaning unit (seen on the right) and an isopropyl alcohol/water rinsing unit (seen on the left). Figure 9.143 shows a working semi-aqueous in-line cleaning machine.

Caution: If the semi-aqueous cleaning process is the cleaning method chosen, be sure to carefully investigate the equipment required. Although manufacturers of semi-aqueous materials sometimes claim that the cleaning equipment (either batch or in-line) used in conjunction with CFC-113-based solvents or with 1,1,1-trichloroethane can be used with the new semi-aqueous materials, this claim is not always true.

The term "semi-aqueous" is possibly not the most judicious term given that systems are coming on the market utilizing other materials as rinsing agents in place of water. For example, one company is proposing a perfluorinated compound (PFC) as the rinsing agent. Other alternatives to using a PFC are an HFC (hydrofluorocarbon) or an HFE (hydrofluoroether). Using water as a rinsing agent entails high energy input for drying. In addition, water is very difficult to dry adequately in tight places although it is possible to utilize a vacuum oven to accomplish this more easily. It is more desirable in these applications to use a material with a lower surface tension than water and one where less energy needs to be expended for drying. For these reasons, the use of a PFC as rinsing agent has arisen. If suitable precautions are taken, one of the lower molecular weight alcohols, such as isopropyl alcohol, can also be used as a rinsing agent. Alcohols are flammable; a flammable material has a closed-cup flash point less than 100°F. The equipment must be designed to handle such materials.

Note: Using a semi-aqueous organic material followed by rinsing with a PFC or an HFC, or an HFE in a suitable piece of batch equipment is called *co-solvent cleaning*.

There are still solvents that can be used to clean surface mount SMAs. At present, the use of HCFC-225 as a cleaning solvent is approved. This material, although it possesses

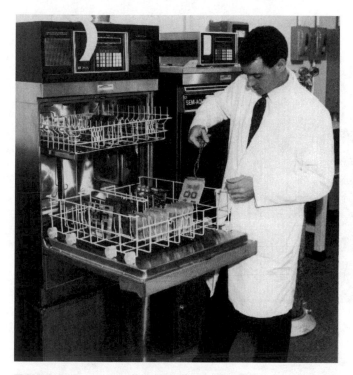

FIGURE 9.141 Semi-aqueous batch cleaning machine. *(Courtesy of the EMPF.)*

an ODP, has a very low one. In addition, its boiling point is very similar to that of CFC-113, and it can be combined with low molecular weight alcohols to enhance its cleaning ability. Using very near zero emission vapor degreasers, some people have even gone back to using trichloroethylene, although this usage is predominantly for precision cleaning rather than defluxing PWAs. Flammable solvents (closed-cup flash point < 100°F) such as isopropyl alcohol are being used, rather widely in Europe, as it turns out. Equipment can be constructed having suitable safety features so that liquids with low flash points such as the alcohols can be used.

Note: The batch or in-line equipment used in conjunction with CFC-113 can sometimes be used with little or no modification with some solvents if they are inflammable, such as HCFC-225 or solvents based on HFC-43-10 mee. To play it safe, however, consult with the equipment manufacturer.

Aqueous cleaning involves, obviously, cleaning with water. If water soluble fluxes and/or pastes are used, these easily lend themselves to aqueous cleaning. If rosin-based fluxes and/or pastes are used, then a suitable saponification agent must be added to the rinse tank. The purpose of the saponification agent is to render the rosin products of defluxing soluble in water by turning them into soluble ionic species. Saponification agents are very basic chemically, and their basicity is what functions to render rosin acids water soluble. Essentially, the reaction is as follows:

$$C_{19}H_{29}COOH \text{ (rosin acid)} + OH^- \text{ (saponifer)} \rightarrow C_{19}H_{29}COO^- \text{ (water soluble)} + H_2O$$

FIGURE 9.142 Batch semi-aqueous cleaning system. *(Courtesy of JPL.)*

FIGURE 9.143 Semi-aqueous in-line cleaning machine. *(Courtesy of the EMPF.)*

Sometimes a small amount of saponifying agent is used in conjunction with WS fluxes and pastes to aid in their efficient removal. Consult with the WS flux and/or paste manufacturer.

Bear in mind that aqueous cleaning is very energy intensive. In addition, the rinse water (effluent) may not be able to be disposed of by simply flushing it down the drain. It may be necessary to clean it up first. Rules differ from state to state and from municipality to municipality. Also, if aqueous cleaning is used, it is generally necessary to dry the part, and this operation too is energy intensive.

It is important that the cleaning operation be performed as soon as possible after the reflow operation. This is especially important in the case of rosin since this material tends to "set up." Generally, it is a good idea to clean within 30 minutes after the soldering operation.

9.9.9.1 Cleanliness Determination.

There are several widely recognized techniques for determining the cleanliness of printed wiring assemblies. For methods of cleaning, see above. Cleanliness is most generally checked after the defluxing operation. The trend in PWA technology is toward finer line widths/spaces and components with greater pin count and smaller pitch. All of these trends make cleaning a more difficult task. In addition, the component standoff is an important parameter in determining cleanliness. Along with these trends, the need for more sophisticated cleanliness determination techniques are required.

The solvent extract resistivity test is one of the oldest and most important methods of determining contamination levels. This test goes back to the early 1970s, when increasing electronics assembly complexity led to failures due to flux activator residues remaining on the assembly surface. This test method began as a manual extraction technique using a rinse solution containing 75 percent isopropanol (IPA)/25 percent water by volume. This rinse solution was determined to be the best for dissolving the flux activator residues of rosin-based fluxes. Only ionic species, which are charged atoms or portions of molecules, are detected in this test. Since ionic species remaining on the assembly surface are the most harmful, a convenient way of determining the gross level of these contaminants is important.

The solvent extract resistivity is often called ionic contamination testing. It is one of the oldest methods for testing for contamination, and it is still widely performed. A result equal to or greater than 10 $\mu g/in^2$ (1.55 $\mu g/cm^2$) of sodium chloride or equivalent is normally the cutoff point. Above this point, the assembly is considered to show too much ionic contamination. In this test, when calculating the board area, both sides of the board are counted. If a board shows an ionic contamination level $\geq$ 10 $\mu g/in^2$, it is recleaned until it shows a level < 10 $\mu g/in^2$.

In this test, an extract is made by rinsing the PWA surface with a known quantity of the IPA/water rinse solution. It is assumed that the rinse solution dissolves all ionic contamination off the PWA surface. Then the resistivity of the IPA/water solution is determined using suitable detection equipment. In practice, the conductance of the rinse solution is determined rather than the resistivity because the conductance is linearly related to the amount of ionic species. Today there are several well recognized ways of making this test semiautomatic. Figure 9.144 shows two common types of ionic contamination testers. The equipment can also address the specific issue of testing for ionics on surface mount PWAs.

Although ionic contamination testing is the most common method for detecting and quantifying ionic residues on PWA surfaces, it does not distinguish among different ionic species. A test having the capability of distinguishing among different ionic species and quantifying the amount of each is a more effective method for enhanced process control. Ion chromatography (IC) is such a technique. All forms of chromatography are essentially

FIGURE 9.144 Two common ionic contamination testers. *(Courtesy of the EMPF.)*

separation techniques. In IC, special columns separate the different ionic species in a suitable extract. If a detector such as a conductivity detector is used in conjunction with the columns, it is possible to detect and quantify the separate ionic species in a solution. Table 9.53 lists some common ionic species that can be detected by IC.

Note: Ionic species can be subdivided into cations and anions. Cations bear a net positive charge (+) and anions a net negative charge (–).

In reference to Table 9.53, amine hydrohalides are often found as activators in fluxes and pastes. Triethylamine hydrochloride $(CH_3CH_2)_3NHCl$ is a common amine hydrohalide. A halide is an anion formed from one of the halogens: fluorine (F), chlorine (Cl), bromine (Br), or iodine (I). The symbols R, R′, R″, and R‴ refer to the organic portion of the amine. In the example cited, R = CH_3CH_2 (ethyl group), R′ = CH_3CH_2 (ethyl group), and R″ = CH_3CH_2 (ethyl group). There is no requirement that R = R′ = R‴. In the case of triethylamine hydrochloride, this merely happens to be the case.

Ion chromatography can be used for detecting and quantifying ionic species in extracts made from printed wiring assemblies. As electronic technology is driven toward even greater miniaturization, such a technique may find increasing importance as a cleanliness determination method since determining the kind of ionic species and the amount of each species will greatly aid in tighter process control.

Ion chromatography and also liquid chromatography (LC or sometimes HPLC) can be used if further information is required to determine the various kinds and quantities of contaminants found on the PWA surface. The principal use of ion chromatography is to determine the types and quantities of ionic species. Liquid chromatography can be used to determine the types and quantities of nonionic species.

Another useful technique for inspecting defects and determining types of contamination found on a surface is that of scanning electron microscopy/energy dispersive analysis by X-rays (SEM/EDX) (see Fig. 9.145). Two types of images are possible with SEM.

TABLE 9.53 Some Common Ionic Species on PWA Surfaces

Name of species	Cations	Anions
Sodium	Na^+	
Potassium	K^+	
Calcium	Ca^{2+}	
Magnesium	Mg^{2+}	
Chloride		Cl^-
Bromide		Br^-
Fluoride		F^-
Sulfate		SO_4^{-2}
Phosphate		PO_4^{-3}
Primary amine	RNH_3^+	
Secondary amine	$RR'NH_2^+$	
Tertiary amine	$RR'R''NH^+$	
Quaternary amine	$RR'R''R'''N^+$	

1. Secondary electron images
2. Backscattered electron images

Secondary electron images are high resolution images and are very useful for determining the detailed morphological features of a surface. Magnification from 10× to 100,000× is possible. However, the usual range is 50× to 5,000×. Backscattered electron images emphasize elemental identification and are the preferred mode of detection for different regions having different compositions, e.g., two different phase regions. In a typical backscattered micrograph, dark regions indicate a region of low atomic number and light regions one of high atomic number. EDX is an elemental method of analysis and is commonly used to detect elements with atomic number equal to or greater than 11 (sodium: Na) although so-called thin-window EDX can detect elements with an atomic number of 5 or greater (boron: B). SEM/EDX is especially helpful for failure analysis. SEM/EDX is generally a form of destructive physical analysis (DPA).

9.9.10 Conformal Coating Application

The different kinds of conformal coatings have been covered in Sec. 9.8.9. Please refer to that section for guidance in choosing the most appropriate material.

There are several different techniques for applying conformal coatings to PWAs. In general, the methods can be divided into two broad categories.

1. Manual application methods
2. Automated application methods

FIGURE 9.145   SEM/EDX. *(Courtesy of the EMPF.)*

Manual methods consist of

- Brushing
- Spraying
- Dipping

The first two, brushing and spraying, are the most common manual methods for applying conformal coatings.

Automated methods consist of

- Spraying
- Dipping

Spraying is the most common automated method for applying conformal coatings.

Note: If Parylene® conformal coatings are used, a special piece of semiautomated equipment must be employed. The application method is one of chemical vapor deposition (CVD). Suitable dimers are volatilized which first form monomers and subsequently upon being deposited on the PWA surface polymerize to form the conformal coating (see Fig. 9.95).

Caution: The resins of almost all conformal coatings are still formulated with volatile solvents, considered to be volatile organic compounds (VOCs), and their use may be severely restricted in various geographical areas. In general, the wastes of conformal coatings are considered to be hazardous wastes and must be disposed of accordingly.

9.10 SMT INSPECTION AND QUALITY ASSURANCE

Rather than merely set forth quality assurance in its traditional role as inspection after the fact, the purpose of this section is to discuss the new role that quality assurance is assuming in many organizations. This new role is closely connected with such concepts as total quality management (TQM), total quality commitment (TQC), continuous process improvement (CPI), and ISO 9000. Today, quality is no longer equated with inspection. Rather, an infrastructure should be in place within any organization that fosters the customers' needs and goals above anything else.

9.10.1 Product-Driven versus Consumer-Driven Organizations

In today's marketplace, several new types of production operations are arising. The older type of organization can be called the product-driven organization. This type of organization is characterized by several key features:

- Inwardly oriented
- Makes products then seeks customers through marketing
- Quality philosophy: Inspect quality into finished product
- Internal result of quality philosophy (Quality is at odds with manufacturing; manufacturing is at odds with design.)
- External result: loss of customer satisfaction (Customers don't get what they want.)

The newer type of organization emerging in its place can be called the consumer-driven organization. As with the older type of organization, this new type can also be characterized by several key features:

- Outwardly oriented
- Marketing defines target markets
- Quality philosophy (Quality and customer satisfaction are the responsibility of everyone in the organization.)
- Internal result of quality philosophy (Concurrent engineering and total quality management (TQM) are very prevalent throughout the organization.)
- External result: satisfied customer (Customers get what they want.)

There is currently a migration away from the product-driven organization toward the consumer-driven type. This migration is extremely important for companies wishing to stay competitive in the 1990s and beyond. One clear indicator of an organization that is well on its way to becoming consumer-driven is a marked emphasis on continuous process improvement.

9.10.2 Continuous Process Improvement

Once new processes are being developed to meet the needs of a new product, a mind set of continual process improvement must be fostered. Again, the driving force should be the team leader. There are four recognized stages to continual process improvement, but the first stage can easily be bypassed once the mind-set of continual improvement has sunk in. Even the second stage will come almost automatically after this has taken place. The four stages are

1. The informal process
2. The formal process
3. Process capability determination and improvements
4. Continuous variation reduction improvements through design of experiments (DoE)

The first stage, the informal process, simply refers to the process before any thought has been given to controlling it or forming a basis of understanding it. This is the stage where the process engineer will try a trial-and-error (band-aid) approach when the process goes out of control, and this will occur on a regular basis. Little thought is invested or expended in truly understanding the process and finding out what is making it go out of control (i.e., the root causes). Many facilities are still operating at this first, rudimentary stage of process understanding, which is one reason why the quality of their product varies so much.

The second stage is that of initial control over the process. It is characterized by the application, generally in large doses, of statistical techniques. Generally, these techniques are subsumed under the heading of statistical process control, or SPC. There are a number of techniques in SPC; one of the chief ones is the control chart.

The control chart is not without its merits. Its systematic use will correctly identify when a process is getting out of control. Techniques like Ishikawa diagrams, or cause-and-effect diagrams, are also helpful in identifying possible causes of the process going out of control. But SPC techniques, by themselves, constitute only control methods; (1) they do not identify root causes, and (2) they can not reduce the inherent variation of the process. The real issue in understanding and controlling the process is to reduce as much as possible process variation (i.e., reduce the value of s as much as possible) and identify root causes when the process goes out of control. Eliminating the root causes for good ensures that the process will never go out of control, for the root causes, by definition, are the causes of the process going out of control.

The next stage is process capability improvements. After SPC has been initiated, the next stage is quantifying the process capability using two well defined parameters for accomplishing this task: C_P and C_{Pk}. C_P is defined as:

$$C_P = USL - LSL/6\sigma$$

where USL = upper specification limit

$\quad LSL$ = lower specification limit

$\quad \sigma$ = standard deviation

$$C_{Pk} = C_P(1 - k)$$

where $k = \dfrac{|T - \mu|}{(USL - LSL)/2}$

$\quad T$ = specification target

$\quad \mu$ = process average

Further details regarding these two process capability indices can be found in most statistics textbooks (see References).

The final stage, continuous variation reduction improvements, involves the concept of design of experiments (DoE). DoE is the key to variation reduction and the correct identification of the root causes behind the process going out of control. A question frequently asked is, When does a person know that he/she has correctly identified a root cause? The answer is, When the problem can be turned on or off by adding or removing the cause. DoE is built upon a rigorous statistical foundation. It is just the opposite to stage one, namely, the trial-and-error approach. To summarize, the true role of SPC is to follow process variation by charts and keep it within set bounds once it is reduced. This is the only correct use of control charts—as a maintenance tool, not as a problem-solving tool. DoE is the problem-solving tool. DoE is the method used to ensure that variation stays reduced at all times.

9.10.3 Quality and SMT

Printed wiring assembly has traditionally relied heavily on visual inspection. One important development going on in quality is to reexamine product requirements to determine which ones make sense to retain (they DO impact the final product) and which ones can be safely deleted (they don't impact the final product). There are a number of traditional inspection points in SMT. These are

- Incoming inspection of the bare board
  - opens and shorts
  - solderability of bare board
- Inspection of solder paste
  - verification of viscosity
  - metal content
  - tendency to spatter
  - slump (cold and hot)
  - tack time
- Inspection of stencil
  - correct hole pattern
  - aperture thickness and aspect ratio (thickness to diameter)
- Components
  - solderability
  - coplanarity (for leaded components)
- Solder paste characterization
  - solder paste deposition
  - solder paste volume
  - solder slump (hot and cold)
- Component placement
  - components placed on the board correctly
  - reflow time-temperature profile
- Inspection after cleaning
- Cleanliness determination

- Component defects
- Inspection before conformal coating
  - Solder joints
  - Leads, etc.
  - Board surface
- Inspection after conformal coating

There are a number of common defects in SMT. The most important are

- Solder balls/spattering
- Dewetting of solder
- Non-wetting of solder
- Inclusions in solder (microsection + SEM analysis)
- Voids in solder (microsection + SEM analysis)
- Icicles/projections on solder
- Insufficient solder
- Excess solder
- No fillet (electrical open)
- Lead overhang (may not be considered a defect in some organizations)
- Contamination on board and/or solder
- Light/moderate/heavy stress lines on solder
- Crack initiation/propagation/complete open
- Solder bridges
- Grainy solder (may not be considered a defect in some organizations)
- Lumpy solder—more pronounced than grainy (may not be considered a defect in some organizations)
- Deformed leads
- Damaged solder joints
- Contamination in solder (microsection + SEM analysis)

9.10.4   Development of Suitable Vendor Relations

Incoming inspection can be drastically reduced if good relations are established with suppliers. Qualified vendors with whom good relations are established can make much incoming inspection unnecessary. It is critical for an organization to set up vendor qualification programs and establish a close working relationship with its vendors. Incoming inspection by itself no longer suffices to keep material costs within bounds. Continuous process improvement (CPI) and total quality commitment (TQC) may very well have to be developed and implemented in a vendor's manufacturing facility since the vendor will often have problems similar to those the assembler has.

A procurement quality operating system, or PQOS, set up by the purchasing organization and established between itself and its vendors would undoubtedly prove beneficial to all parties. An action team should be organized within the organization to set up and control the PQOSs, and a corresponding team should be set up in the vendors' facilities. The essence of the PQOS is to institute the same kind of new productive system in the vendor's facility or at least facilitate such a change. The setting up of PQOSs and establishing a good working relationship between an organization and each of its vendors becomes the key to ensuring better quality parts and components while keeping material costs at a minimum.

9.11 TESTING IN SMT

After the assembly has been processed through paste deposition, component placement, reflow, and cleaning, the last step is typically testing. Planning for testing must take place at the design level because design factors can adversely affect testability. Testing of SMT PWAs must take into account higher component density, finer line widths and spacings, and reduced component pitch. In addition, testing both sides of the PWA may be required. The item being tested is designated the unit-under-test (UUT).

9.11.1 In-Circuit and Functional Testing

There are two chief testing methods for the finished PWAs. These are

1. In-circuit
2. Functional

In-circuit testing (ICT) generally checks for the following items:

- Electrical continuity such as shorts and opens
- Component verification/functionality

ICT may be used to check bare boards for electrical continuity, and it may also be used to check the completed PWA. Ideally, the test engineer would like to test every circuit node, a node being a junction where two components come together electrically speaking. To accomplish this, a bed-of-nails test fixture having numerous test probes is employed. The individual test probes make contact at various points on the PWA surface. The following points should be noted:

- The design engineer must allow suitable access to as many test nodes as possible. This entails design with testability (DfT) in mind.
- In SMT, the designer must take into account higher component densities, finer line widths/spacings, closer component spacings, and reduced component pitch.
- For Type 2B and 2C SMAs, testing both sides of the assembly may be required.
- Fine pitch and ultrafine pitch components greatly increase the difficulty of testing.

In functional testing, the PWA is checked to ascertain whether it is functioning electrically as it was intended. A functional test system applies various electronic stimuli to the UUT and then measures the resulting output. In functional testing, the UUT is normally plugged into the test system using a connector, e.g., an edge connector or some other kind of connector.

The functional tester normally employs one of two methods to check the response of the UUT.

- Stored response
- Signature analysis

9.11.2   Environmental Stress Screening

Other types of testing are also possible. For example, thermal cycling test PWAs and monitoring them for electrical opens is a way of testing them to determine solder joint reliability. One of the most common tests for revealing infant mortality failures is environmental stress screening (ESS).

Environmental stress screening (ESS) is another type of testing often performed on SMT products. Infant mortality failures are typically screened out by ESS. Among these failures are:

- Parameter drift
- PWB opens and/or shorts
- Parts incorrectly installed
- Contaminated parts
- Hermetic seal failures
- Defective solder joints
- Defective parts

ESS can be used to precipitate latent defects; it can also be used for product qualification. Among the different types of ESS tests, there are

- Acceleration
- Humidity
- Temperature and humidity
- Mechanical shock
- Vibration
- Thermal shock
- Thermal cycling
- Electrical stress
- Electrical stress and vibration

For precipitating infant mortality defects and for product qualification—the following tests are recommended:

- Temperature cycling
- Random vibration

For qualifying new soldering and/or cleaning processes—the following test is recommended:

- Humidity and temperature cycling with periodic application of electrical stress and measurement of the resulting insulation resistance [surface insulation resistance (SIR)].

9.12 SMT RELIABILITY ISSUES

Regarding the reliability of surface mount PWAs, failure is most closely associated with the solder joints forming the metallurgical interconnection between the components and the PWB substrate. Unlike through hole technology, solder joints in surface mount technology must not only function as the electrical interconnect but also as the mechanical interconnect of the components to the substrate. SMT solder joints are expected to function in a wide variety of service environments and under a wide variety of conditions. Further, both high quality and high reliability are expected from electronic products.

9.12.1 Fatigue Properties of Solder

Eutectic tin-lead solder, the principal material used for creating the electrical/mechanical interconnection, is a very unique and useful material, but it is characterized by a low melting point (183°C/361°F). Unlike most metals, eutectic solder experiences operating temperatures roughly near its melting point. This concept is succinctly expressed by the metallurgical concept of homologous temperature, namely, T/T_m where T is the operating temperature and T_m is the melting point, both temperatures expressed in degrees Kelvin (K). For eutectic or near-eutectic solder at normal room temperature (27°C/80°F), the homologous temperature is 300K/456K = 0.66. For any metal, if its homologous temperature > 0.5, almost all the damage is fatigue damage due to creep and stress relaxation. Obviously, eutectic and near-eutectic solder is a prime candidate for this type of fatigue failure. The primary drivers in the fatigue failure of solder are

- Difference in the coefficient of thermal expansion of the component and the substrate, ΔCTE or $\Delta\alpha$.
- Difference in temperature between the lowest operating temperature and the highest operating temperature, ΔT.
- The contribution to ΔT from power cycling and heat generated from the components (some large surface mount IC packages can generate up to 20 W of heat).
- Distance from neutral point of the component, L_D.
- The solder joint height, h.
- The stiffness ration of the combined system (PWB/PWA) to that of the solder joint, $k = K/K_s$ where K = system stiffness $(1/K_b + 1/K_s) - 1$, K_b = board/assembly stiffness and K_s = stiffness of the solder joint.

9.12.2 Solder Joint Life Prediction Methods (SJLPMs)

It is in the interest of everyone concerned with SMT to have a model that would reliably and accurately predict the mechanical behavior of solder under a variety of conditions. Numerous approaches have been developed for attempting to predict the expected life of solder under a variety of use conditions. These approaches range from almost purely mechanical-theoretical models all the way through models derived solely from empirical correlations. Some of the most important approaches to solder joint life prediction modeling (SJLPM) are as follows:

- Coffin-Manson-based
  - Engelmaier
  - Figures of merit

- Strain range partitioning
- Energy partitioning (chiefly mechanical-theoretical)
- Fracture mechanics (chiefly mechanical-theoretical)
- Finite element analysis/finite element modeling (FEA/FEM)
- Solder microstructure approach (chiefly metallurgical)

Of these various methods, the ones based on the Coffin-Manson are the most practical from a simple calculation point of view. The Engelmaier equations can be dealt with using algebraic techniques as can the figures of merit (FOMs). The advantage of the FOM approach is that it readily lends itself to a GO/NO-GO analysis based on the above solder joint reliability drivers. From a theoretical point of view, energy partitioning where the energy of fatigue is partitioned into its various types and each type is dealt with theoretically offers the most promise, especially when it is combined with microstructural knowledge. (What is happening to the solder at the microstructural level?) The FEA/FEM approach is also widely used, but an adequate model must still be employed, and this approach generally requires a lot of computer power.

9.12.3   Coffin-Manson-Based Calculation of Solder Joint Life

To get some feel for a calculation based on an algebraic formula, consider the following example. Consider a leadless ceramic chip carrier (LCC) on an FR-4 PWB. The component is a 68-termination LCC measuring 0.95 in on a side. The solder joint height, h, was determined to be 0.005 in. The total assembly is cycled from 25°C (room temperature) to 80°C. The component itself dissipates heat such that its final temperature is 75°C. Find the number of cycles to failure.

Starting with the Coffin-Manson approach, which assumes all the fatigue damage is strain related, the governing formula is Eq. (9.5).

$$N_f = \frac{1}{2}\left[\frac{\Delta\varepsilon}{\varepsilon_f}\right]^{\frac{1}{c}} \tag{9.5}$$

In this equation, c, the fatigue ductility exponent, < 0. N_f is the median (50 percent) number of cycles to failure, $\Delta\varepsilon$ is the strain range and ε_f is the fatigue ductility coefficient.

For an LCC, $\Delta\varepsilon \approx \Delta\gamma$ where $\Delta\gamma$ = the shear strain range. For eutectic and near-eutectic solder, $\varepsilon_f \approx 0.65$.

Now, by Eq. (9.6), $\Delta\gamma$ is

$$\Delta\gamma = \frac{L_D}{h} \cdot \Delta(\alpha\Delta T) \tag{9.6}$$

In this equation LD is the distance from neutral point of the component, h = solder joint height, and α = CTE. By Eq. (9.7), $\Delta(\alpha\Delta T)$ is

$$\Delta(\alpha\Delta T) = \Delta\alpha\Delta T_e \tag{9.7}$$

and by Equation 9.8, ΔT_e is

$$\Delta T_e = \frac{|\alpha_s\Delta T_s - \alpha_c\Delta T_c|}{\Delta\alpha} \tag{9.8}$$

In this equation, α_s is the CTE of the substrate, in this case the FR-4 PWB, ΔT_s is the temperature excursion of the substrate, α_c is the CTE of the component, ΔT_c is the temperature excursion of the component, and $\Delta\alpha = \alpha_s - \alpha_c$.

For eutectic and near-eutectic solder, $c \approx -0.5$ from temperatures ranging from about 0°C up to 120°C.

Note: There are more precise formulae for calculating c based on other parameters such as the estimated number of thermal cycles that the SMA will be exposed to and the mean solder joint temperature, TSJ. Consult the literature. For example, see J. H. Lau (1991).

By Eq. (9.9), L_D is

$$L_D = \frac{1}{2} \cdot (L^2 + W^2)^{\frac{1}{2}} \tag{9.9}$$

Using the values stated at the beginning of the example and consulting Table 9.24, $\alpha_s = 16 \times 10^{-6}$, $\alpha_c = 6 \times 10^{-6}$, $h = 5 \times 10^{-3}$, $L_D = 0.67$ [by Eq. (9.9)], $\varepsilon_f \approx 0.65$, $\Delta T_s = 55$, $\Delta T_c = 50$, $\Delta\alpha = 10 \times 10^{-6}$, $\Delta T_e = 58$ [by Eq. (9.8)], $\Delta\gamma = 77.7 \times 10^{-3}$ [by Eq. (9.6)]. Therefore, $N_f = 35$ cycles [by Eq. (9.5)].

Several researchers have refined the modified Coffin-Manson approach and some of the parameters that enter into the equation. Otherwise, the Coffin-Manson equation [Eq. (9.5)] is far too conservative for making accurate estimations. Another example will suffice. This one is drawn from a real case. Consider 20 dummy (no internal IC) 68-pin LCC packages. For each sample, one 68-pin LCC was reflow soldered to an FR-4 substrate. These samples were then thermal cycled between −55°C to +100°C every 246 min (4 h and 6 min = 1 cycle). In this case, L_D and $\Delta\alpha$ are the same as in the above illustration, namely, $L_D = 0.67$ and $\Delta\alpha = 10 \times 10^{-6}$ ppm/°C. Since the packages are nonfunctional, there is no Joule heat. That is, the package produces no internal heat, so $\Delta T_e = \Delta T = 155$°C. Using $c = -0.5$ and a solder layer height, h_L, of 0.005 in in conjunction with Eqs. (9.5) and (9.6), it can be shown that $N_f = 4.9 \approx 5$ cycles.

Actual experimental data found that the median (50 percent) number of cycles to failure, N_f, as detected by an Anatech® device (detects electrical opens) was 90 cycles. Instead of using the solder layer height, h_L, as h in Eq. (9.6), a better parameter is perhaps the effective solder fillet height, $\overline{h}_f$. (See Fig. 9.146.) Using $\overline{h}_f$ in conjunction with Eqs. (9.6) and (9.5), and using the experimentally measured value of $\overline{h}_f = 0.021$ in for the 68-pin LCCs, it can be shown that $N_f = 86.4 \approx 87$ cycles. A more precise theoretically calculated value of c, that is, $c = -0.464$, yields $N_f = 5.8 \approx 6$ cycles when $h = h_L = 0.005$ in, and $N_f = 128.8 \approx 129$ cycles when $h = \overline{h}_f = 0.021$ in. In this case, the experimental evidence suggests that $c \approx -0.5$ rather than −0.464.

Caution must be employed when using the Coffin-Manson equation. However, its judicious use can nevertheless illustrate the importance of the governing parameters driving the fatigue life and also enable one to perform a sensitivity analysis.

Another item of interest is calculating when the first solder joint failure will occur rather than calculating the median number of cycles to failure, N_f. Various modifications to the Coffin-Manson equation must be made in order to accomplish this. Again, caution must be used. Consult the literature. Of course, where does all this leave the ordinary design engineer who is not particularly interested in wading through all the literature on the subject and who does not wish to spend hours, maybe even days, in utilizing an FEM to calculate solder joint life. One important researcher in the field of SJLPM had this to say:

> Of all the necessary future work nothing is more important than the systemization of a design approach that will incorporate all the factors discussed above [issues in determining

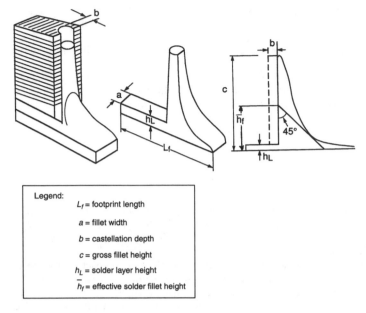

Legend:

L_f = footprint length

a = fillet width

b = castellation depth

c = gross fillet height

h_L = solder layer height

$\overline{h}_f$ = effective solder fillet height

FIGURE 9.146   68-pin LCC, SJ configuration. *(Courtesy of A. Wen, JPL.)*

thermal and mechanical fatigue life] in a manner that can be used by the average design engineer. [See Lau (1991), p. 448].

9.12.4 Reliability of Plated Through Hole Vias

One other item bears mentioning. This is the plated through hole (PTH). Although pure SMT assemblies do not require PTHs for component insertion, a number of SMAs do have PTHs in the form of through hole vias. Regarding the reliability of PTH vias, it is important to keep in mind that there are two important parameters.

- Ductility of the copper in the PTH barrel
- Tensile strength of the copper in the PTH barrel

This concept is illustrated in Fig. 9.147. This figure was derived from an FEA model. However, it was found to correlate closely with experimental findings. Bear in mind that the former governing military document for PWBs, MIL-P-55110, required a minimum of 6 percent elongation and 36 kpsi tensile strength. In addition to copper ductility and copper tensile strength, several other factors are also worth noting.

An investigation of the reliability of PTHs demonstrated a number of additional important factors in PTH reliability.

See, for example, IPC Technical Report, 1988, no. 579, Sept. This investigation revealed that the following factors are also of importance:

- Hole diameter—PTH reliability decreases as the PTH diameter decreases.

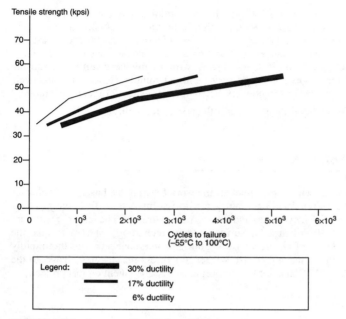

FIGURE 9.147   Copper quality effect on PTH fatigue life.

- PWB thickness—PTH reliability decreases as the PWB thickness increases. The aspect ratio (AR) of a PWB/PTH is the ratio of the PWB thickness to the drilled (unplated) hole diameter. For example, if a PWB is 0.062 in thick (standard thickness) and a drilled hole is 0.020 in, the AR is 3:1. High AR PTHs (AR > 5) exhibit less reliability than low to moderate AR PTHs.

- Copper thickness in the hole—PTHs with thin copper, either because the copper was plated thin or because of plating deficiencies, have a higher failure rate. PTHs with thicker copper that is uniformly deposited demonstrate better thermal performance.

- There does not appear to be any relationship between PTH reliability and the type of acid copper plating bath (e.g., acid copper sulfate, acid copper pyrophosphate) used to electrolytically plate the PTH.

- Standard electrolytic plating processes are not adequate for high AR PTHs in high reliability applications in severe service environments, even if optimum process controls are used on the plating baths.

- Functional or nonfunctional inner layer lands appear to improve reliability.

- In the case of epoxy resins, higher T_g values appeared to increase PTH reliability.

- Cross-sectional analysis revealed that most failures occur in the resin rich regions (as opposed to the copper rich regions).

- Differences in CTE and percent resin showed no affect on PTH reliability.

- The majority of failures observed in the –65°C to +125°C cycle were recorded during temperature transitions rather than at the high or low temperature dwells. This ΔT run for 400 cycles provides an excellent test for detecting potential PTH reliability problems. It revealed both infant mortality failures as well as wearout failures.

- The above thermal cycle (the MIL-T-Cycle) has a 30 minute dwell at each temperature extreme ($-55°C$ and $+125°C$; $\Delta T = 180°C$) with a 38 minute ($\sim5°C/min$) transition time between temperature extremes. One total cycle lasts 136 minutes; 400 cycles last 907 hours = 38 days. A more rapid cycle was discovered than correlated well with the MIL-T-cycle. This cycle goes from $-35°C$ to $+125°C$ with 15 minute dwell times and the transition between the two extremes was very rapid, ~1 minute. Hence, one total cycle lasts 32 minutes as opposed to 136 minutes. Then 400 cycles last 213 hours = 9 days.

- To summarize, PTH quality is the principal influence on reliability.

ACKNOWLEDGEMENTS

The author wishes to acknowledge that much of the work forming the basis of this chapter was performed at the Jet Propulsion Laboratory, California Institute of Technology, under a contract with the National Aeronautics and Space Administration. The research was conducted to support an infrastructure in surface mount technology and to further the advancement of knowledge of electronic packaging and interconnection in this rapidly advancing field. In addition, the author also wishes to express his sincerest thanks to the NASA sponsor, Dr. John W. Evans, for his support and encouragement on this project.

REFERENCES

1. Adams, J., and D. Malloy. 1989. "Discussion of 3-D X-Ray Laminography as a Provider of Process Control Data for Real-Time Quality Monitoring." IPC Technical Paper, *IPC-TP-851*.

2. Albert, T., and R. Albert. 1991. "New X-Ray Imaging Technology for Inspecting Fine Pitch Assemblies." IPC Technical Paper, *IPC-TP-978*.

3. Amick, P. J., and J. E. Mazurek. 1989. "Power Cycling of Kevlar® and Copper-Invar-Copper Printed Wiring Assemblies." IPC Technical Paper, *IPC-TP-802*.

4. Anderson, S., and W. C. Roman. 1991. "Life Testing of Surface Mount Power Assemblies Cleaned with CFCs and Semi-Aqueous Processes." IPC Technical Paper, *IPC-TP-967*.

5. Anjard, R. P. 1986. "The Rheology of Solder Pastes." IPC Technical Paper, *IPC-TP-614*.

6. _____ 1985. "Solder Paste Applications for Surface Mount Technology Used in PC Manufacture." IPC Technical Paper, IPC-TP-594.

7. Bakerjian, R. ed. 1992. *Tool and Manufacturing Engineering Handbook: Design For Manufacturability*, 1st ed., Vol. VI. Dearborn, MI: Society of Manufacturing Engineers.

8. Benzschawel, A., and L. Hymes. 1989. "A Real Time Approach to Process Control in SMT." IPC Technical Paper, *IPC-TP-838*.

9. Bergstrom, R. P. 1988. "Portrait of a Profession in Change." *Mfg. Eng.*, December, pp. 44–51.

10. Bhandarkar, S. M., A. Dasgupta, D. Barker, and M. Pecht. 1990. "Effects of Voids in Solder-Filled Plated-Through Holes." IPC Technical Paper, *IPC-TP-863*.

11. Bhote, K. R. 1988. *World Class Quality*. New York: American Management Association.

12. Bloomfield, D. 1990. "The Rheology of Solder Pastes." IPC Technical Paper, *IPC-TP-888*.

13. Buck, A. 1990. "Replacing the Missing Link between CAD and CAM." IPC Technical Paper, *IPC-TP-875*.

14. Capillo, C. 1990. *Surface Mount Technology: Materials, Processes, and Equipment*. New York: McGraw-Hill.

15. _____ 1985. "How to Design Reliability into Surface-Mount Assemblies." *Elec. Pkg. & Prod.*, July.

16. Carpenter, R. H. 1990. "Evolution of Ionic Contamination Testing." IPC Technical Paper, *IPC-TP-896.*

17. Clark, R. H. 1985. *Handbook of Printed Circuit Manufacturing.* New York: Van Nostrand Reinhold.

18. Classon, F. 1993. *Surface Mount Technology for Concurrent Engineering and Manufacturing.* New York: McGraw-Hill.

19. Coombs, C. F. ed. 1988. *Printed Circuits Handbook,* 3rd ed. New York: McGraw-Hill.

20. _____ 1979. *Printed Circuits Handbook,* 2nd ed. New York: McGraw-Hill.

21. Corbett, J., M. Dooner, J. Meleka, and C. Pym. 1991. *Design For Manufacture: Strategies, Principles and Techniques.* Reading, MA: Addison-Wesley.

22. Curtis, D. A. 1991. "Changing PCB Cleaning Operations to Accommodate the Phase-Out of CFCs." SME Technical Paper, *EMR91-19.*

23. Danielsson, H. 1995. *Surface Mount Technology with Fine Pitch Components: The Manufacturing Issues.* London, UK: Chapman & Hall.

24. Deram, B. 1989. "Technical Considerations for Use of No-Clean Fluxes." IPC Technical Paper, *IPC-TP-769.*

25. DeVore, J. A. 1990. "Can the Dip And Look Test Be Relied upon to Predict Good Soldering Performance?" IPC Technical Paper, *IPC-TP-915.*

26. _____ 1989. "The Metallurgy and Mechanisms of Failure in Surface Mount Solder Joints." IPC Technical Paper, *IPC-TP-847.*

27. _____ 1984. "The Mechanisms of Solderability and Solderability Related Failures." IPC Technical Paper, *WCIII-43.*

28. Dukes, J. M. C. 1961. *Printed Circuits: Their Design and Application.* London: MacDonald.

29. Edosomwan, J. A., and A. Ballakur. eds. 1989. *Productivity and Quality Improvement in Electronics Assembly.* New York: McGraw-Hill.

30. Ellis, B. N. 1986. *Cleaning and Contamination of Electronics Components and Assemblies.* Ayr, Scotland: Electrochemical Publications.

31. Engelmaier, W. 1991. *Designing for Surface Mount Solder Joint Reliability.* Dearborn, MI: Society of Manufacturing Engineers.

32. _____ 1987. "Is Present-Day Accelerated Cycling Adequate for Surface Mount Attachment Reliability Evaluation?" IPC Technical Paper, *IPC-TP-653.*

33. Ephraim, A. G. 1990. "Solder Paste: 'The Weaker Link'." IPC Technical Paper, *IPC-TP-906.*

34. Evans, J. W. 1994. "Fatigue and reliability in Near Eutectic Solder Alloys for Spacecraft Applications." *Ph.D. Dissertation.* The Johns Hopkins University. Baltimore, MD.

35. Fowlkes, W. Y, and C. M. Creveling. 1995. *Engineering Methods for Robust Product Design.* Reading, MA: Addison-Wesley.

36. Frear, D. R. 1987. "Microstructural Observations of the Sn-Pb Solder/Cu System and Thermal Fatigue of the Solder Joint." *Ph.D. Dissertation.* University of California at Berkeley. Berkeley, CA.

37. Frear, D. R., W. B. Jones, K. R. Kinsman. eds. 1990. *Solder Mechanics: A State of the Art Assessment.* Warrendale, PA: The Three M Society.

38. Froehlich, R. 1987. "Quality Control for Surface-Mount Technology Manufacturing." *Elec. Pkg. & Prod.,* August, pp. 69–71.

39. Garrett, R. W. 1990. "Eight Steps to Simultaneous Engineering." *Mfg. Eng.,* November, pp. 41–47.

40. Getty, H. and T. Barrett, T. 1989. "Quantitative Non-Ionic Cleanliness Measurement by HPLC." IPC Technical; Paper, *IPC-TP-796.*

41. Giglio, J. 1991. "Simultaneous Engineering for PCB Design." IPC Technical Paper, *IPC-TP-976.*

42. Gill, M. E. 1986. "Surface Analytical Techniques Applied to Printed Boards and Printed Board Assemblies." IPC Technical Paper, *IPC-TP-609.*

43. Ginsberg, G. 1991. *Printed Circuits Design.* New York: McGraw-Hill.

44. Haasen, P. 1978. *Physical Metallurgy.* Cambridge: Cambridge University Press.

45. Harper, C. A. ed. 1991. *Electronic Packaging & Interconnection Handbook.* New York: McGraw-Hill.

46. _____ 1969. *Handbook of Electronic Packaging.* New York: McGraw-Hill.

47. Harry, M. J. 1990. "The Nature of Six Sigma Quality." IPC Technical Paper, *IPC-TP-926.*

48. Hartley, N. 1988. "Yield Problems in SMT and Development of Process Equipment." IPC Technical Paper, *IPC-TP-723.*

49. Hastings, D. W. 1988. "A Military Airborne Application of LCCS on Cu/Invar/Cu and Kevlar Based PWBs." IPC Technical Paper, *IPC-TP-706.*

50. Hendrix, B. C. 1989. "The Interaction of Creep and Fatigue in Lead-Tin Solders." *Ph.D. Dissertation.* Columbia University. New York City, NY.

51. Hirakawa, T., H. Watanabe, and K. Nishimura. 1989. "A New Aramid Base Material for Advanced SMT." IPC Technical Paper, *IPC-TP-804.*

52. Hollander, D. 1988. "DPAK: A Surface Mount Package for Discrete Power Devices." Motorola Technical Paper, *AR319/D.*

53. Hollomon, J. 1989. *Surface-Mount Technology for PC Board Design,.* Indianapolis, IN: Howard W. Sams.

54. Hutchins, C. L. 1993. *Understanding and Using SMT & Fine Pitch Technology.* Raleigh, NC: C. Hutchins & Associates.

55. _____ 1990. *Surface Mount Technology—How To Get Started.* Raleigh, NC: C. Hutchins & Associates.

56. Hwang, J. S. 1989. *Solder Paste in Electronics Packaging.* New York: Van Nostrand Reinhold.

57. Hymes, L. ed. 1991. *Cleaning Printed Wiring Assemblies in Today's Environment.* New York: Van Nostrand Reinhold.

58. _____ 1986. "Designing For Manufacturability in Printed Wiring Assemblies." IPC Technical Paper, *IPC-TP-607.*

59. Kamezos, M. 1991. "Technology Trends in Electronic Packaging in the 90s." IPC Technical Paper, *IPC-TP-962.*

60. Kear, F. W. 1987. *Printed Circuit Assembly Manufacturing.* New York and Basel: Marcel Dekker.

61. Keller, J. n. d. *Surface Mount Fundamentals Handbook.* Plantation, FL: Joe Keller Associates.

62. Kevra, J., C. C. and Johnson. 1989. *Solder Paste Technology: Principles & Applications.* Blue Ridge Summit, PA: TAB Professional & Reference Books.

63. Kiko, T. 1984. *Printed Circuit Board Basics,* 1st ed. Alpharetta, GA: PMS industries.

64. Klein-Wassink, R. J. 1989. *Soldering in Electronics,* 2nd ed. Ayr, Scotland: Electrochemical Publications.

65. Lambert, L. P. 1988. *Soldering for Electronic Assemblies.* New York and Basel: Marcel Dekker.

66. Lau, J. H. ed. 1995. *Ball Grid Array Technology.* New York: McGraw-Hill.

67. _____ 1993a. *Handbook of Fine Pitch Surface Mount Technology.* New York: Van Nostrand Reinhold.

68. _____ 1993b. *Thermal Stress and Strain in Microelectronics Packaging.* New York: Van Nostrand Reinhold.

69. _____ 1991. *Solder Joint Reliability: Theory and Applications*. New York: Van Nostrand Reinhold.

70. Lazar, C. 1990. "How to Solve Problems with SEM/EDX." IPC Technical Paper, IPC-TP-882.

71. Lea, C. 1988. *A Scientific Guide to Surface Mount Technology*. Ayr, Scotland: Electrochemical Publications.

72. Leonida, G. 1981. *Handbook of Printed Circuit Design, Manufacture, Components & Assembly*. Ayr, Scotland: Electrochemical Publications.

73. Licari, J.J. 1970. *Plastic Coatings for Electronics*. New York: McGraw-Hill.

74. Livesay, B. R., and M. D. Nagarkar. eds. 1990. *New Technology in Electronic Packaging*. Materials Park, OH: ASM International.

75. Lozano, B. 1991. "AOI and SPC: The Implementation Challenge," IPC Technical Paper, *IPC-TP-958.*

76. Maki, J., D. Rubinic. 1988. *A Look at Surface Mount Technology*. Ridgecrest, CA: EMPF.

77. Mangin, C.-H., and S. McClelland. 1987. *Surface Mount Technology: The Future of Electronics Assembly*. Berlin and New York: Springer-Verlag.

78. Manko, H. H. 1992. *Solders and Soldering: Materials, Design, Production and Analysis for Reliable Bonding*, 3rd ed. New York: McGraw-Hill.

79. _____ 1986. *Soldering Handbook for Printed Circuits and Surface Mounting*. New York: Van Nostrand Reinhold.

80. _____ 1979. *Solders and Soldering*, 2nd ed. New York: McGraw-Hill.

81. Marcoux, P. P. 1992. *Fine Pitch SMT: Quality, Design, and Manufacturing Techniques*. New York: Van Nostrand Reinhold.

82. _____ 1991. *SMT Inspection, Rework and Repair Techniques*. Dearborn, MI: Society of Manufacturing Engineers.

83. _____ 1989. *Surface Mount Technology: Design For Manufacturability*. Dearborn, MI: Society of Manufacturing Engineers.

84. Mindel, M. J. 1989. "Solder Corrosion with 'Leave-On' Fluxes." IPC Technical Paper, *IPC-TP-765.*

85. Morris, J. E. ed. 1991. *Electronics Packaging Forum*, Vol. 2. New York: Van Nostrand Reinhold.

86. Munson, T. 1991. "Cleanliness Testing for the 90s by Ion Chromatography." IPC Technical Paper, *IPC-TP-938.*

87. Pattison, W. L. 1990. "In-Plane CTE Determinations of High Performance Materials within the Manufacturing Technology for Advanced Data Signal Processing Program." IPC Technical Paper, *IPC-TP-879.*

88. Pearne, N. K. 1987. "Packaging and Interconnection Developments into the 90's and the Future of the PCB." IPC Technical Paper, *WCIV-83,* June.

89. Pecht, M. G., L. T. Nguyen, and E. B. Hakim. eds. 1995. *Plastic-Encapsulated Microelectronics: Materials, Processes, Quality, Reliability, and Applications*. New York: John Wiley & Sons.

90. Pecht, M., A. Dasgupta, J. W. Evans, and J. Y. Evans. eds. 1994. *Quality Conformance and Qualification of Microelectronic Packages and Interconnects*. New York: John Wiley & Sons.

91. Pecht, M. ed. 1991. *Handbook of Electronic Package Design*. New York and Basel: Marcel Dekker.

92. Prasad, R. P. 1989. *Surface Mount Technology: Principles and Practice*. New York: Van Nostrand Reinhold.

93. Pshaenich, A., and D. Hollander. 1988. "Managing Heat Dissipation in DPAK Surface-Mount Power Packages." Motorola Technical Paper, *AR323/D.*

94. Reed, J. R., N. E. Thompson, and R. K. Pond. 1990. "Steam Aging, A Capability Study for PWBs." IPC Technical Paper, *IPC-TP-921.*

95. Rowe, G. L. 1985. "Setting Up a Surface Mount Facility." *Hybrid Circuits,* 6, pp. 27–34.

96. Rowland, R. J., and P. Belangia. 1992. *Applied Surface Mount Assembly: A Guide to Surface Mount Materials and Processes.* New York: Van Nostrand Reinhold.

97. Rua, R. 1988. *Surface Mounting Technology: (Professional Advancement Course).* Big Rapids, MI: Ray Rua Associates.

98. Scarlett, J. A. 1985. *The Multilayer Printed Circuit Board Handbook.* Ayr, Scotland: Electrochemical Publications.

99. _____ 1980. *Printed Circuit Boards for Microelectronics.* Ayr, Scotland: Electrochemical Publications.

100. Schonberger, R. J. 1986. *World-Class Manufacturing:* The Lessons of Simplicity Applied. New York: The Free Press.

101. _____ 1982. *Japanese Manufacturing Techniques: Nine Hidden Lessons in Simplicity.* New York: The Free Press.

102. Seraphim, D. P., R. C. Lasky, and C.-Y. Li. 1989. *Principles of Electronic Packaging.* New York: McGraw-Hill.

103. Shina, S. G. 1991. *Concurrent Engineering and Design for Manufacture of Electronics Products.* New York: Van Nostrand Reinhold.

104. Sinnadurai, F. N. ed. 1985. *Handbook of Microelectronics Packaging and Interconnection Technologies.* Ayr, Scotland: Electrochemical Publications.

105. Solberg, V. 1990. *Design Guidelines for Surface Mount Technology.* Blue Ridge Summit, PA: TAB Professional and Reference Books.

106. Stover, A. C. 1984. *ATE: Automatic Test Equipment.* New York: McGraw-Hill.

107. Strauss, R. 1995. *Surface Mount Technology.* Newton, MA: Butterworth-Heinemann.

108. Tautscher, C. J. 1991. *Contamination Effects on Electronic Products.* New York and Basel: Marcel Dekker.

109. _____ 1976. *The Contamination of Printed Wiring Boards and Assemblies.* Bothell, WA: Omega Scientific Services.

110. Traister, J. E. 1990. *Design Guidelines for Surface Mount Technology.* New York: Academic Press.

111. Tummala, R. R., and E. J. and Rymaszewski. eds. 1989. *Microelectronics Packaging Handbook.* New York: Van Nostrand Reinhold.

112. Turbini, L., M. Hook, D. Pauls, R. Sellers, H. Getty, and R. Lamoureux. 1991. "Phase 3 CFC Elimination Plan Water Soluble Flux Test Results—Part I." IPC Technical Paper, *IPC-TP-971.*

113. Turino, J. 1990. *Design-To-Test,* 2nd ed. New York: Van Nostrand Reinhold.

114. Vaughan, D. A., and E. Gorondy. 1986. "Flux-Solder Mask Interactions during Wave Soldering." IPC Technical Paper, *IPC-TP-615.*

115. Vardaman, J. 1993. *Surface Mount Technology—Recent Japanese Developments.* New York: IEEE Press.

116. Veilleux, R. F., and L. W. Petro. eds. 1988. *Tool and Manufacturing Engineering Handbook: Manufacturing Management,* 4th ed., Vol. V. Dearborn, MI: Society of Manufacturing Engineers.

117. Veilleux, R. F., and C. Wick. eds. 1987. *Tool and Manufacturing Engineering Handbook: Quality Control and Assembly,* 4th ed., Vol. IV. Dearborn, MI: Society of Manufacturing Engineers.

118. Vernon, G. L., and W. A. Collar. 1991. "A Comparison of the Physical Properties of Various Heat Sink Materials, and Related Processing Considerations." IPC Technical Paper, *IPC-TP-961.*

119. Ward, D. 1985. "High Volume Surface Mount Technology: How-To Advice based on Delco's Experience." *Circuits Mfg.,* March, pp. 58–71.

120. Walton, R. S. 1985. *Surface Mount Technology: The Connection to the Future.* Scottsdale, AZ: ICE.

121. Watanabe, Y. 1989. "Automated SMT Visual Inspection." IPC Technical Paper, *IPC-TP-792.*

122. Watson, R. M. 1991. "Statistical Process Control Implementation in a Printed Wiring Board Facility." IPC Technical Paper, *IPC-TP- 940.*

123. Wojslaw, C. F. 1978. *Integrated Circuits: Theory and Applications.* Reston, VA: Reston Publishing Co.

124. Wolfe, G. 1990. "Electronic Packaging Issues in the 1990s." *Elec. Pkg. & Prod.,* Oct.

125. Woodgate, R. 1988. *The Handbook of Machine Soldering,* 2nd ed. New York: John Wiley & Sons.

126. Yoo, C. 1990. "New Concepts and Methods in Surface Mount Technology." IPC Technical Paper, *IPC-TP-912.*

127. Zado, F. M., B. C. Chung, G. M. Wenger, and G. C. Munie. 1991. "Non-Rosin Organic Residue Characterization and Simple Quantitative Detection Methods." IPC Technical Paper, *IPC-TP-948.*

128. Zimmerman, J. 1991. "Solderability Improvement of Solder Mask Over Bare Copper Circuit Boards by Application of Taguchi Methods." IPC Technical Paper, *IPC-TP-941.*

129. *Component Solderability Guide.* n.d. Washington, DC: EIA.

130. *Summary of NASA's Surface Mount Technology (SMT) Survey (Findings, Results and Conclusions)* 1993. Jet Propulsion Laboratory RTOP Interim Significant Results (RISR) Report. Pasadena, CA.

131. *Surface Mount Technology—Status of the Technology/Industry Activities and Action Plan.* 1995. Lincolnwood, IL: Surface Mount Council (IPC & EIA).

132. *Surface Mount Technology—Status of the Technology/Industry Activities and Action Plan.* 1994. Lincolnwood, IL: Surface Mount Council (IPC & EIA).

133. *Surface Mount Technology—Status of the Technology/Industry Activities and Action Plan.* 1993. Lincolnwood, IL: Surface Mount Council (IPC & EIA).

134. *Surface Mount Technology—Status of the Technology/Industry Activities and Action Plan.* 1992. Lincolnwood, IL: Surface Mount Council (IPC & EIA).

135. *Surface Mount Technology—Status of the Technology/Industry Activities and Action Plan.* 1991. Lincolnwood, IL: Surface Mount Council (IPC & EIA).

136. *Surface Mount Technology—Status of the Technology/Industry Activities and Action Plan.* 1990. Lincolnwood, IL: Surface Mount Council (IPC & EIA).

137. "Changing PCB Cleaning Operations to Accommodate the Phase-Out of CFCs." 1991. Society of Manufacturing Engineers *Technical Report EMR91-19.*

138. UNEP *Solvents, Coatings, and Adhesives Technical Options Report: Montreal Protocol 1991 Assessment.* 1991. December.

139. *Mantech for Advanced Data/Signal Processing: Volume II—Final Report for Task II: VHSIC PWB Fabrication.* n.d. Wright Patterson Air Force Base Mantech Technology Transfer Center. WPAFB, OH.

140. *Mantech for Advanced Data/Signal Processing: Volume III—Final Report for Task III: Solder Process Controls.* n.d. Wright Patterson Air Force Base Mantech Technology Transfer Center. WPAFB, OH.

141. *Mantech for Advanced Data/Signal Processing: Volume IV—Final Report for Task IV: VHSIC IC, PWB, and Solder Process Integration.* n.d. Wright Patterson Air Force Base Mantech Technology Transfer Center. WPAFB, OH.

142. *Mantech for Advanced Data/Signal Processing: Volume VI—Handbooks.* n.d. Wright Patterson Air Force Base Mantech Technology Transfer Center. WPAFB, OH.

143. "Assembling Fine Pitch Surface Mount PCBs." Special Supplement. 1991, *Elec. Pkg. & Prod.,* May.

144. *IBM Surface Mount Technology.* 1991. Seminar. Austin, TX: IBM.

145. *Soldering Technology for Surface Mount Devices.* 1990. Dearborn, MI: Society of Manufacturing Engineers.

146. "Round Robin Reliability Evaluation of Small Diameter Plated Through Holes in Printed Wiring Boards." 1988. IPC Technical Report, *IPC-TR-579.*

147. *How To Use Surface Mount Technology.* 1984. Dallas, TX: Texas Instruments.

148. *Surface Mount Technology.* 1984. Silver Spring, MD: International Society for Hybrid Microelectronics.

149. *Microelectronics Interconnection and Packaging.* 1982. New York: McGraw-Hill Seminar Center.

150. *Electronic Packaging Handbook.* n.d. Lincolnwood, IL: IPC.

CHAPTER 10

ADVANCED ELECTRONIC PACKAGING

COB, BGA, FLIP-CHIP, TAB, AND HIGH-DENSITY INTERCONNECTIONS

Donald P. Schnorr
Wyndmoor, Pennsylvania

10.1 INTRODUCTION

Electronic packaging, which for many years was only an afterthought in the design and manufacture of electronic systems, increasingly is being recognized as the critical factor in both cost and performance. As the functional density of devices and systems increases, the role of electronic packaging and interconnection necessarily becomes more important. In spite of the revolutionary advances in semiconductor device technology, these improvements are insignificant when considering the constraining influence of the package in the system—the great disparity between the performance of the chip and performance outside the package.

Current electronic systems employ a packaging technology which limits system performance because of a number of factors. As the circuit density on a chip goes up, the speed of functions it performs increases, but it must connect the other chips in the system. Stage delay must necessarily be reduced, placing a burden on the interconnection technique. Retention of signal integrity is another consideration. The power needed to run the chips generates significant amounts of heat which must necessarily be removed—an important requirement.

10.1.1 Electronic Packaging Driving Forces

A number of driving forces influence the selection of an interconnection medium: mechanical, electrical, and thermal, to mention a few. The selection of material and the choice of processing are dependent on the needs of these driving forces.

10.1.1.1 Mechanical Requirements. The mechanical driving force strives to attain technology advancements to keep pace with the considerable improvements being made by the continued reduction (per function) in the integrated circuit package size.

10.1.1.2 Electrical Requirements.   The electrical driving force is bringing new technology to the fabricator of conventional printed wiring boards (PWBs), with expressions such as controlled impedance, dielectric constant, crosstalk control, and insulation resistance. Improvements in dimensional stability and registration accuracy and the use of new (more expensive) materials are the result of attempting to satisfy these needs.

10.1.1.3 Thermal Requirements.   Thermal driving forces are threefold. In one case, it is giving credence to the use of materials and processes that can satisfy the higher assembly processing temperatures to which the printed wiring boards must be subjected, that is, materials with a higher glass transition temperature, Tg. Another instance concerns coefficient of thermal expansion (CTE) "tailoring," which has resulted in some applications using reinforcing metal planes or constraining metal cores In the third instance, the printed board becomes an active element in the thermal management of the assemblies cooling system.

10.1.2 Cost Driving Factors

As a general case, the impact of new technologies on printed wiring board processing can possibly affect lead times, material costs, fabrication yields, and board performance, to name a few. The underlying cost drivers are critical operations of features that significantly increase expense or reduce yield. Board real estate use, board size and shape, laminate material, plating choices, and panel size can affect manufacturability. Conductor spreading improves board manufacturibility by closely centering copper between lands and vias, straightening "stairstepped" conductors and eliminating sharp bends. Mechanical and electrical test requirements are also cost drivers.

10.2 ADVANCED PRINTED WIRING TECHNOLOGY

Characteristic with all maturing technologies, there is considerable interest in advanced surface mounting and the new implementation technologies to support it. The majority of these advances support the use of high pin count, fine-pitch [0.065 mm (0.003 in) or less] IC packages as well as pad grid array packages of all types.

In any case, this advanced surface mount or fine-pitch technology is worth investigating to determine its impact on printed wiring board processing in the 1990s and beyond. (Table 10.1). As a surface mount technique that increases the maximum lead count to over 1000 and doubles the printed wiring board density, fine-pitch technology packages are also considerably less expensive (per terminal) than comparable surface-mount packages. However, there is a trade-off. Fine-pitch technology also increases the printed wiring board cost per unit area by requiring smaller line widths, and spaces, smaller pads, and in improved accuracy and dimensional stability.

Fine-pitch conductor widths and spaces are presently around 0.02 mm (0.0008 in), with the long-term preference being in the 0.13-mm (0.005-in) category or less. Fine-pitch technology also requires that the board fabrication to artwork registration have higher tolerances than the 0.05 to 0.1-mm (0.002 to 0.004-in) value common with today's processing equipment.

10.2.1 General Processing Trends

The design and fabrication of multilayer boards has been the prime mover in interconnection technology, forcing the search for smaller conductor widths, spaces, smaller holes and

TABLE 10.1 Typical Parameters for Printed-Board Technologies

	Through-hole technology	Conventional surface-mount technology	Fine-pitch technology
Process steps			
Package selection	0.100-in spacing Dual-in-line packages Axial-lead discretes Radial-lead discretes	0.050-in and less spacing Small-outline devices Chip carriers Leadless discretes	0.020-in and less spacing Quad flat packs Tape-automated bonding Chip carriers
Solder application	Printed-board fabrication	Printed-board fabrication, assembler applies paste	Printed-board fabrication, assembler applies paste and fuses
Component application	Assembler inserts manually and automatically	Assembler places manually and automatically	Assembler places manually and automatically
Soldering leads	Wave soldering	Wave soldering and reflow soldering	Reflow soldering
Cleaning	Water or solvent cleaning	Solvent cleaning	Water, solvent, or no cleaning
Printed-board features			
Vias	Same as lead holes	Separated from lands by 0.010 to 0.025 in	Small filled holes in land pattern (less than 0.018 in in diameter)
Conductor widths and spaces	Greater than 0.010 in	From 0.005 to 0.010 in	From 0.002 to 0.008 in
Land pitch	Greater than 0.050 in	From 0.020 to 0.050 in	From 0.004 to 0.020 in
Artwork generation	Manual or CAD layout, film artwork masters	Manual or CAD layout, film or magnetic tape artwork masters	CAD layout, magnetic tape or glass artwork masters
Testing methods			
In-circuit/postassembly	Test to land or test to node	Test to node or test to cluster	Test to cluster or self-test

10.3

pad sizes, improved materials, and innovations in manufacturing processes. This quest is expected to continue over the coming years as electronic equipment increases in performance capability and semiconductor circuits increase in density, thereby maintaining the pressure for using an efficient high-density interconnection system such as multilayer boards.

As with all rapidly advancing technologies, certain key elements provide the impetus for progress. For example, high-density conductor patterns, high-aspect ratio plated through holes, and sequential lamination will become more readily achievable with further advancements and refinements in drilling, plating, imaging, and etching technology.

Despite their advantages, multilayer boards are not simple to fabricate. In fact, fabricating them gives a good indication of the difficulties that arise when the drive to shrink dimensions runs up against physical processing constraints, particularly when yields, and ultimately costs, are involved. Advanced technology for its own sake, however, has limited use, and its wider application demands that it be reproducible with an emphasis on producibility. Therefore, this section explains some of the developments that are taking place in printed wiring board processing technology satisfy these requirements, with the understanding that only those that can be implemented in a cost-effective fashion will survive to become part of the next-generation state of the art.

10.2.2 Fine-Line Technology

For some time, the printed wiring board industry has been debating the need to manufacture fine-line features. Throughout this debate, the definition of fine-line has changed. Several years ago, conductor widths of 0.25 mm (0.010 in) or less were considered to be in the fine-line category, while today these conductors are felt to be normal practice. In the last few years, a consensus has emerged that categorizes fine lines to be 0.100 mm (0.0040 in) and less. For many shops, "fine" is the smallest conductor with the shop has ever made. However, there are always economic impacts from any decision to build a board that does not fit into the routine category.

Therefore, it is not enough to find the smallest geometry technically manufacturable; it is also important, in the actual factory environment, to find technology to match the requirements of the product to the practical reality of economically viable production.

Present processes can manufacture 0.1 to 0.125-mm (0.004 to 0.005-in) conductors in production quantities. However, in general, this requires tighter process control, improved cleanliness of the environment, and more care in handling than is presently being used in most facilities. Even with these restrictions, the yield, quality, and economics of producing these feature sizes are not completely known.

10.2.2.1 Primary Imaging.   The demand for competitively priced printed wiring boards with greater density and improved quality is driving the need for improved processing throughout the manufacturing procedure. This need is especially evident with regard to imaging, where improvements include the use of more automated techniques. Computer-aided design (CAD) and laser photo plotting have already revolutionized the front end of the imaging process. Improvement issues now center around better methods of transferring the artwork image to the fabrication panel.

After artwork generation, the basic equipment used to create fine-line boards is composed of the board preparation, resist deposition, developing, etching, and stripping systems that are common to most board fabrication processes. Complementing the equipment, and equally important, are the chemical systems, the resist material and particulate management.

In the synergistic and harmonious use of these ingredients, some critical process parameters subject to close scrutiny and management include board surface cleanliness and topography, light-source collimation, phototool stability, contaminant control, consistency of chemical systems, spray pressure, temperatures, residence time, and speed through various baths.

At one end of the spectrum, there are complete environmentally controlled panel lamination, exposure, and development systems (Fig. 10.1) which seek to expedite the handling of conventional film phototools. At the other end, there is direct imaging, which promises to do away with phototools altogether. In the middle, is glass phototooling, which would appear to be the most cost-effective near-term imaging improvement available to most fine-line printed wiring board fabricators.

Panel Preparation.   A critical operation at the beginning of the imaging process is panel surface preparation. The topography (cleanliness and flatness) of the substrate can influence fine-line imaging results dramatically.

Before resist is applied to the panel, the surface must be clean of contamination. The contamination may be either chemical or particulate. Unfortunately, the surface topography created by the typical laminate brush scrubber is detrimental to fine-line imaging. Non-planar surfaces created by uneven panel plating also inhibit fine-line imaging.

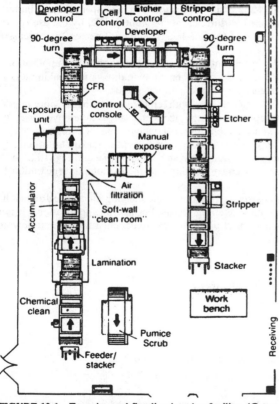

FIGURE 10.1   Experimental fine-line imaging facility. (*Courtesy of E.I. DuPont de Nemours & Co.*)

For fine features then, chemical cleaning is preferred. However, some chemical cleaning leaves an undesirable surface texture, which can ultimately produce a short or open circuit, depending upon the imaging process used.

Phototooling. The demands on phototool performance will continue to grow as board tolerances tighten. Photolab automation will eventually help reduce many defects that now result from the current method of producing phototools.

However, until significant cost-effective automation is in place, manufacturers need to develop procedures and to find products to reduce fine-line defects caused by artwork and phototools. This can be best achieved through improved size holding, phototool handling, and cleanliness. At the same time, manufacturers can begin to explore the increased use of robotics.

The effect of line quality on the resulting resist and plated sidewalls should not be overlooked. The absolute conductor size must be considered. A 0.01-mm (0.0004-in) fringe on a 0.25-mm (0.010-in) conductor, for instance, would have some effect on the plated conductor width. That same fringe, however, on a 0.1-mm (0.004-in) conductor would cause significant problems. Fringe can be caused by the imaging technique used to make the artwork or phototool, the development conditions, or the film or glass choice. In selecting a phototool, the board manufacturer has three basic choices: conventional photoresist film, high stability film, and glass (Table 10.2).

For fine-line imaging, conventional film, the mainstay of the printed wiring industry, has a major disadvantage in that it changes size with variations in temperature and humidity. While temperature can be confined within acceptable limits($\pm 2^\circ$C) by the use of air-conditioning equipment, humidity control for fine-line imaging (± 3 percent) is expensive and difficult to attain.

High-stability phototooling film has recently been developed for fine-line applications. These new films can hold size for longer periods when the humidity varies, making it possible to work within tight size tolerances for longer periods of time.

Glass changes size about half as much as film with temperature changes. More important, however, is the fact that glass does not change size with changes in relative humidity. But the use of glass has other drawbacks, including cost, weight, and, of course, breakage. To aid in keeping the entire glass phototool process in house for maximum efficiency, there are glass registration, hinging, and assembly systems as illustrated in Fig. 10.2. Such systems consist of separate alignment and mounting capable of holding registration to ± 0.013 mm (± 0.0005 in).

Plating and Etching Imaging Resists. To optimize the imaging process, care must be taken in choosing appropriate plating or etching resists. For fine-line applications, better resolution, artwork reproduction, and resist performance are obtained when the following conditions are fulfilled:

- The resist is a bulk polymerizable type.
- Thicker resists are used, consistent with other restraints.
- Photo printer collimation is improved.
- Exposure intensity is improved.
- The total exposure is controlled optimally controlled.
- Developing chemistry, temperature, and time are well controlled.
- Developing is followed by a good rinsing and complete air drying.

While numerous improvements have been made to image and process latitude, resist manufacturers are still upgrading the performance of their products whenever possible.

TABLE 10.2 Basic Phototool Alternatives

	Conventional phototooling films	"Dimension master" high-stability phototooling films	Glass plates
Size holding	Greatest change in response to changes in temperature and relative humidity	Holds size 3 to 5 times longer than conventional films under changing relative humidity conditions; equal to conventional film for temperature variation	Very stable for relative humidity changes; half the temperature response of conventional film
Durability	Base is most susceptible to kinks, dimples, and tooling hole wear	Thicker, stiffer base minimizes kinks, dimples, and tooling hole wear	Breakable, but does not kink or dimple
Handling, storage, and shipping	Flexible and lightweight; needs only normal handling procedures for photographic films; easy to ship and store		Heavier than film; extra caution required to prevent breakage; extra space and weight considerations for storage and shipping
Preconditioning (individual sheets)	≈4 h	10 to 12 h	Not normal practice
Use in standard exposure equipment	Suitable for all step-and-repeat and contact equipment		Weight and thickness require adjustment or special equipment
Plotters	Suitable for laser and vector plotters		May require adjustment for thickness; can only be used on flatbed
Processing	Suitable for standard automatic processors		Tray or special plate processors
Price	Least expensive	Moderate	Most expensive

FIGURE 10.2   Glass phototool registration system. *(Courtesy of Precision Art Coordinators.)*

Automated processing of the resist applications and imaging steps is being promoted as one of the major improvements that a printed wiring board manufacturer should implement toward achieving fine-line imaging results.

Another consideration relates to stricter waste-treatment regulations and higher recycling costs of using various solvents. Automated processing of the resist applications and imaging steps is being touted as one of the major improvements that a printed-board manufacturer should implement in trying to achieve fine-line imaging results.

Liquid Photoresist Imaging.   Over the past few years, several improvements in the field of liquid photoresists have made their use suitable for fine-line applications. The most important improvement responsible for this includes the development of negative-acting, fully aqueous processible, liquid photoresist products. These materials are capable of producing coating that yield high-resolution images for conductor widths and spaces of 0.04 mm (0.0015 in) and less. Several viable coating processes are readily available for using liquid photo resists materials in printed wiring board production.

Dry-Film Photoresist.   Dry-film photo resists provide a reliable fine-line imaging technology for high-density printed wiring boards. However, fabricating high-quality conductors of 0.05 to 0.15 mm (0.002 to 0.006 in) widths requires good process control and the use of techniques that are not usually required for medium- and low-density products.

Laser Direct Imaging.   Laser direct imaging employs a laser to expose the image of the printed wiring pattern directly from the CAD data onto the photoresist-coated substrate. This technique bypasses the traditional artwork and phototooling steps (Fig. 10.3).

For fine-line imaging applications, this can typically result in 0.06 mm (0.0024 in) conductor widths with an imaging accuracy on the order of ±0.03 mm (±0.001 in) and a repeatability of ±0.006 mm (±0.00025 in) or better. The reasons for these improvements over conventional imaging technology are that there are fewer steps in the process, less risk of error due to machine or operator error, and no phototool damage. Also, there is no panel damage due to contact exposure, and less contamination in the exposure process.

The throughput of laser direct imaging systems is the greatest impediment to creating a full-scale production system. The factors affecting throughput are the mechanical speed of

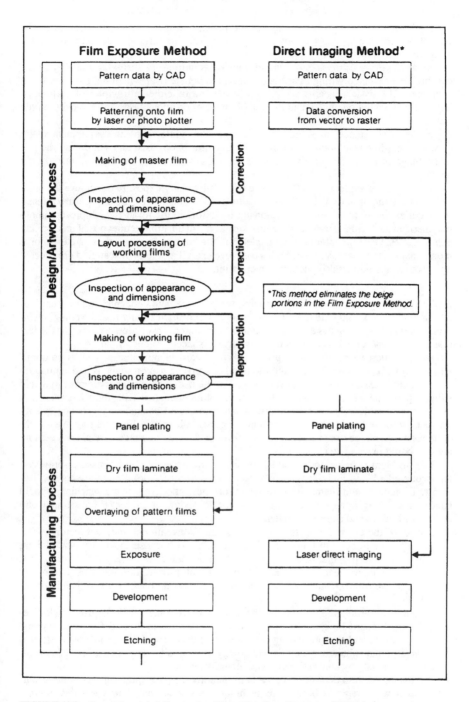

FIGURE 10.2 Conventional and direct printed-board imaging. *(Courtesy of Nikon.)*

the machine, the speed (or sensitivity) of the resist, and the data transmission rate. The cost of direct imaging resist is obviously another very important issue, particularly for high-volume production.

Photo-Exposure Systems. As circuit densities continue to shrink and challenge the capabilities of printed wiring board fabricators, the debate over which light source to use continues. The basic concerns are related to the use of either ultraviolet of collimated light sources on the one hand, and the use of either high-intensity or low-intensity on the other.

In the first instance, most manufacturers of ultra-violet exposure equipment claim excellent fine-line resolution performance on dry-film photoresist as used during the primary image process. Conversely, there are several advocates for the use collimated light sources.

There is less disagreement with the use of high-intensity exposure to achieve the resolution that is important for fine-line conductor widths and spaces. Low exposure intensities requires longer exposure times, allowing inhibitors in the resist (which prevent photo polymerization during storage and lamination) to diffuse from the unexposed areas to the exposed sites. Once the inhibitors are depleted in the unexposed areas, these areas become more sensitive to ultraviolet light. In this case, even small amounts of scattered or reflected light can initiate unintended photo polymerization.

10.2.2.2 Etching. Although the chemistry, temperature, and concentration of a particular material are normally the main factors involved in printed wiring board etching operations, conductor spacing has also been found to be a factor. This is due primarily to the fact that etchant flow is restricted when imaging closely spaced conductor patterns.

With the increase in multilayer printed wiring board manufacturing and its predominance of inner conductive layers, and the trend toward the use of aqueous and semiaqueous imaging resists, there is a renewed interest in the use of acid copper chloride and acid sulfuric peroxide etching systems. However, alkaline enchants, with their inherent high production capability, good economics, and the ability to consistently produce quality products, will continue to fill a critical and important role in fine-line imaging. New technologies based on the use of nitric acid and nitrogen dioxide as well as dry plasma etching are also being investigated.

This problem is being addressed by the use of conveyerized high-pressure spray systems coupled with in-line rinses, which are claimed to meet fine-line processing requirements. To achieve and maintain process control, these modern etching machines should have an oscillating spray, efficient pump and nozzle design, proper nozzle placement, and an integrally designed conveyor system.

Control of the etching process is also important. Conventional undercut factors of 0.3 to 0.5 will not suffice with fine-line conductors. This is because most chemical etching processes are isotropic; that is, the vertical and lateral etch rates are about the same. Thus attempts are being made to achieve almost straight-wall profiles through the use of newly developed anisotropic etchants. Anisotropic etchants have a greater vertical than horizontal etch rate.

With this being taken into account, there is a renewed interest in the use of additive-printed wiring board imaging technologies, which have been around for several years. Since, obviously, additive processing does not get involved with the problems associated with conventional subtractive etching technology, it is possible that its use will be considered more seriously by conventional board manufacturers.

The use of vacuum deposition or vacuum metallizing (i.e., sputtering) is being evaluated. Much work remains before this technique is operational, but it remains another option for fine-line manufacturers to consider.

10.2.3 Small-Hole Technology

One of the major conclusions of IPC-TR-579 is that the reliability of small-diameter plated-through holes [drilled holes of 0.5 mm (0.020 in) or less] is strongly influenced by the hole diameter, board thickness, thickness of plating, and copper metallurgical quality (ductility versus tensile strength). Specifically, the standard states:

Reliability decreases as the hole diameter decreases.

Reliability decreases as the board thickness increases.

Reliability decreases as the thickness-to-diameter (aspect ratio) goes above 3:1.

Holes with thin plated copper, whether plated thin or as the result of localized plating deficiencies, have a markedly higher failure rate.

Thicker copper plating that is deposited uniformly demonstrates better thermal cycling performance.

Solder coatings do not improve reliability.

Cross-sectional analysis reveals that most failures occur at the resin-rich regions.

Functional or nonfunctional internal lands appear to improve reliability.

With these considerations in mind, several companies are approaching small plated-hole fabrication with an increased emphasis on understanding, controlling, and improving the drilling, cleaning, and plating processes that play an important part in achieving reliable results.

10.2.3.1 Drilling.   Drilling smaller holes is not risk free. As the holes become smaller, the problems associated with the aspect ratio also increase. This change in emphasis has resulted in several problems for the printed wiring board manufacturer, including the following:

Developing process controls to prevent drill bits from breaking during drilling and handling

Compensating for the lower productivity because of slower "hit" rates and shorter stack heights

Higher board scrap rates resulting from drill breakage during drilling

Increased tool costs due to fewer hits per drill bit, more process variables, and less chance of resharpening

Small holes can be drilled successfully using conventional equipment and available drill bits, but with poor yields and profitability. To be successful, different approaches have to be taken with respect to machine design, operation maintenance, tool bit selection, and so on (Fig. 10.4). There are basically seven main variables in the drilling process:

The drilling machine

Drilling parameters

Drill bits

Stacking and pinning

Entry materials

Backup materials

The first six are controlled directly and determined by the drilling operation. The last, the laminate material, is controlled by the design. With the use of new laminate reinforcing

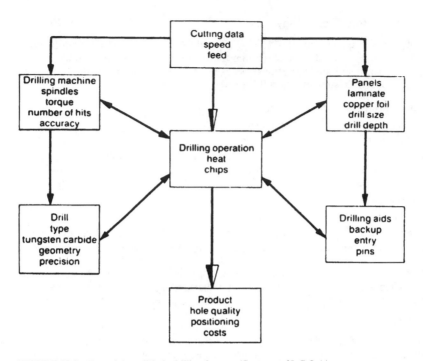

FIGURE 10.2   Essential small-hole drilling factors. *(Courtesy of L.C.O.A.)*

materials, such as aramid fibers and quartz, as well as sophisticated constraining cores for thermal management, this variable plays a more significant part in achieving successful small-hole drilling than with conventional printed wiring board technologies.

Each variable can be subdivided into a number of subvariables. When these are known and controlled properly, small-hole drilling can be accomplished with consistent and acceptable quality. Proper control not only implies verifying that all of the variables are correct, but also requires maintaining them consistently.

The ultimate quality of a drilled hole is measured by its compatibility with subsequent processing. Although the process chemistries following drilling, such as desmear and plating, are designed to prepare and improve the quality of the drilled hole, these operations cannot fix the defects created by improper drilling. Thus, with the use of smaller holes, hole drilling is receiving much more attention by board fabricators.

10.2.3.2 Hole Cleaning.   Once a small high-aspect-ratio hole is drilled properly, the traditional sequence of hole preparation prior to copper plating must be followed. However, limitations that the hole geometry imposes must be overcome by the selection of a desmear and plating process specifically formulated to address these limitations.

The desmear process utilized must remove resin smear completely and, according to the mechanism of adhesion, provide a clean, cohesively strong substrate for subsequent plating. But in case of smear removal of small high-aspect -ratio holes, the process chosen must provide complete initial solution contact with the hole wall, and must impart a hydrophilic surface on the hole wall to ensure adequate solution contact in subsequent wet processing steps.

The key to successful small-hole plating is the ability to move working solutions and rinse waters in and out of the holes effectively. This demands that enough energy be supplied into the hole to move sufficient volumes of working solution through the hole and that this energy be supplied uniformly across the panel.

For holes as small as 0.35 mm (0.014 in) in diameter with aspect ratios as high as 5:1, many classical smear-removal methods are falling under attack, both from a performance standpoint and because of increasing environmental concerns. Thus, current concentrated sulfuric acid smear removal methods are being challenged by the use of permanganate-based chemical processing and dry plasma processing.

Methods for solution agitation are also being evaluated. As an extreme example, with holes less than 0.23 mm (0.009 in) in diameter, mechanical or vigorous air agitation cannot reliably supply enough energy through the hole to dislodge tenacious particles of epoxy from the hole wall. Spray processing can give good results; however, there are several considerations that must be taken into account.

Ultrasonic agitation can be an effective means of supplying sufficient energy into a small hole to give good solution flow-through, debris removal, and hole-wall cleaning and rinsing. Ultrasonics is particularly effective when used in conjunction with good mechanical agitation.

Thus, the wet processing of small high-aspect ratio holes can be a straightforward proposition as long as allowance is made at each and every step for the reduction in solution flow-through caused by the smaller geometries. It also illustrates that proper rinsing may indeed be the most important step in through-hole processing.

10.2.3.3 Plating. The need for uniform plated through holes in high-aspect-ratio multilayer boards is well understood in the printed wiring board industry. However, the trend toward the increased use of even smaller holes in thicker boards does require the electroplater to devise new ways to obtain acceptable results. Most important is the need to relate much of the results of the research in this area to practical implementation.

The successful plating of these holes can only be achieved if the equipment, chemistry, and control methods are designed properly and fully integrated. Therefore, for these sophisticated boards, every step in the process is critical and must be monitored and controlled carefully. Table 10.3 lists some of the problems caused by the unique features of this type of multilayer board.

Proper surface preparation in the desmear or etchback process is only one of the important prerequisites for complete adherent electroless copper coverage when plating small high-aspect-ratio holes. The plating process must also be designed to circumvent the shortcomings of diminished fluid dynamics, as noted for the hole-cleaning process. For these applications, special plating processes have been developed that give excellent throwing power, deposit mechanical properties, and level at relatively high plating speeds.

10.2.4 Multilayer Technology

Supplemental to concerns about the use of fine lines and small holes, process development for high-density multilayer boards is also concerned with issues relating to the increased complexity of the product. In particular, the dimensional accuracy in the laminated layers is affected by the stress inherent in lamination because the internal board material is usually thinner and thus has lower strength than that in conventional multilayer boards.

Therefore, the usual processing techniques often fail to satisfy requirements for registration accuracy. The complexity factor for multilayer boards increases disproportionately

TABLE 10.3 Small High-Aspect Plating Problems

Feature of board	Problem	Solution
Many small-diameter holes	Difficult to rinse; drag-in can affect other processes	Thorough rinsing and rugged, forgiving chemistry
	Hole-plugging by debris	Clean environment, ultra filtration; must have excellent hole desmear and cleaning process
	Air entrapment and gas generation during processing (hole desmear, electroless copper)	Rack enters solution at an angle; bump agitation removes gases
Panels are expensive	Require 100 percent yields	Tight controls on all processes; institute statistical quality control, statistical process control; hole preparation and electroless copper must provide void-free copper with excellent hole wall and interconnect adhesion
Panels are often large (30 by 30 in)	Panel agitation can vary from top to bottom in a nonrigid system, causing nonuniformity in processing	Panel rack support must not deflect
	Improper cell and rack design can cause wide deposit thickness variations	Racks must provide uniform current distribution to panel and provide proper thieving; cell and rectifier are designed to minimize primary current variations; plating additive helps even out remaining variations

with the layer count, and in many cases it increases exponentially. Therefore, the increase in the number of process control elements results in a potential nightmare for the board fabricator.

An additional complication is the fact that multilayer boards are not only used to achieve higher interconnection densities but also to improve the electrical and thermal characteristics of this type of packaging and interconnection structure. As an example, this brings into play controlled characteristic impedance requirements that place stringent dimensional tolerances on the board's features; thermal management issues sometimes necessitate the use of new board reinforcements or constraining metal cores.

10.2.4.1 Lamination. Dimensional variations in the course of lamination appear to be decreased by pressure, the shrinking of the resin caused by hardening, and the high shear stress that accompanies the flowing of the resin in the prepreg. Thus, the selection of an appropriate lamination system and proper processing conditions is important to minimize the stress. In particular, the use of vacuum lamination under the appropriate time and temperature conditions is often ideally suited for the fabrication of complex multilayer boards.

On the minus side, it should be noted that the use of low-pressure vacuum lamination is sensitive to tooling setups and prepreg properties. Thus, compared with controlled conventional high-pressure lamination, the vacuum process may neither guarantee achieving high-pressure lamination nor guarantee achieving an equally toleranced multilayer board. In fact, lower pressures may result in greater resin retention, a condition inconsistent with having reinforcement-influenced regions that assist in controlling board thickness uniformity.

Vacuum Hydraulic Lamination. The vacuum lamination technique involves forming a vacuum assembly that is placed in each of the openings of the static hydraulic press. In its simplest form, and lamination lay-up is placed on a carrier plate and a flexible membrane is draped over the package and sealed. A vacuum is drawn on this package and pressure is then applied to the assembly.

Vacuum Autoclave Lamination. The vacuum autoclave lamination technique uses a pressurized vessel to perform the lamination procedure. In its most basic form, the lamination lay-up is placed on a carrier tray equipped with vacuum connections prior to its insertion into the pressure vessel. This forms the vacuum assembly.

The most common method of vacuum autoclave lamination consists of a flexible membrane placed over the lay-up and sealed at the edges. (Fig. 10.5). The vacuum assembly is then transferred into the pressure vessel and the door is sealed. The vessel is then pressurized with inert gas. Following completion of pressurization, enclosed heaters are used to heat the gas to the desired lamination temperature. To maintain temperature, the gas is circulated throughout the vessel by means of an enclosed fan. At the end of the cycle, the chamber is depressurized and the package removed.

10.2.4.2 Dimensional Stability. Dimensional stability is dependent on several process and raw material properties. Thus, a system approach must be taken to achieve acceptable results. Understanding that no one variable by itself can cause all dimensional stability

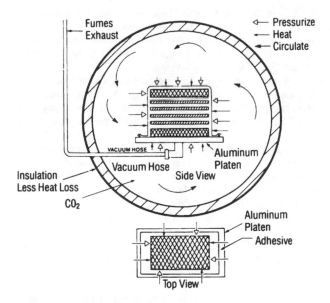

FIGURE 10.2 Vacuum autoclave lamination. *(From Bourdon.[45])*

problems is key to success. One way to control material shift is through proper material selection. Some common laminate materials will shift more than others, and only rigorous testing will reveal how much shift is acceptable for a particular board design and which materials will produce the least amount of shift.

Another method of controlling material shift is to break the manufacturing process into smaller steps. The use of sequential lamination permits controlling the shift between adjacent layers during processing. The process begins with the lamination of critical layers into a subassembly and then pressing subassemblies together into a final board. This technique is said to permit better control of misregistration, or shift, than pressing all the layers together at once.

A manufacturing process for controlled impedance boards must consistently meet tight tolerances. Unfortunately, a small population of test runs is frequently misleading. Also, by doing this, the board fabricator is left with only a very general estimate to account for the effects of manufacturing variability.

Fortunately, the manufacturing capability of controlled-impedance boards can be predicted in advance by utilizing a computer simulation. A simulation can allow a number of variables fluctuate in a random fashion, just as they might occur in a production process but rarely occur in a test sample run. By using simulation, the production costs can be determined quickly and accurately. Moreover, alternative designs and alternative processes can be tested in advance on the computer to determine the most economical result.

The problem of dimensional stability is very complex, involving many variables, and results in the "marrying" of materials with entirely different chemical and physical properties. In addition to the processing concerns already mentioned, one important control that fabricators are learning to do is to compensate (size) photo tooling to allow for dimensional changes in the laminate. The resizing factor, which normally allows for shrinkage of the laminate after the lamination process, is being determined from historical data on a particular prepreg and inner-layer core combination.

10.2.4.3 Registration. Every step in the manufacturing process that requires position of dimensional accuracy adds to the ultimate misregistration of the inner layers of a multilayer board. However, imaging and lamination tooling can only be maintained to certain positional tolerances, while the materials themselves change dimensionally due to uneven thermal expansion and compression forces in the lamination press.

If these and other errors add up evenly across the fabrication panel and appear consistently for each layer, the relative land positions will remain constant in their vertical orientation. Unfortunately, more often registration errors occur randomly and may or may not cancel out at some locations, while at other locations they build up to a worst-case deviation from their true position.

The use of real-time x-rays provides a means for the board fabricator to overcome registration problems and thereby improve processing yields. Using this technique, registration error compensation is achieved by determining the optimum "centroid" of lands for a plated through hole in the actual multilayer board after lamination. The drill pattern is then modified to provide the best fit for the lands on each fabrication panel.

In addition, real-time x-ray inspection is a very powerful engineering tool that can be used to aid in process development, quality verification, and product rework. For instance, it can be used to monitor conductor width etching consistency, detect out-of-tolerance drilling conditions, and localize the position of open and short circuits determined by continuity testing.

10.2.4.4 Process Control. The impact of new technology on printed wiring board processing is being felt most by the increasing need for improved process controls, throughout the manufacturing facility. These demands are being satisfied by the use of statistical and real-time process control methodologies.

The beginning of an effective statistical process control (SPC) program begins at the highest levels within a company, with an understanding that only by using methods that will ensure continuous process improvements can the manufacturer attain the required quality levels and cost and productivity performance necessary to satisfy customer requirements. This understanding must then be communicated down through the organization and a consensus to act arrived at. After this has been accomplished, a plan to implement SPC must be decided upon, delineated, reviewed, accepted, and implemented. Then its performance must be rigorously monitored.

Real-time process control (RTPC) systems are emerging, which integrate manufacturing planning, control, and implementation functions into a closed-loop computer-integrated manufacturing (CIM) system. These systems are supported by computers that offer data processing power together with advances in system technology and the availability of intelligent processing equipment. The basic functional requirements fulfilled by and RTPC system include system tracing, manufacturing floor data collection and reporting, and automatic machine setup and processing. It also provides integration with other manufacturing facilities, and short-term planning assistance.

For example, as mentioned previously, x-ray inspection equipment can be used in a real-time mode to alter hole-drilling locations to be actual, rather than theoretical, land-pattern locations. The information gathered can also be used to provide data to facilitate making modifications to consistently out-of-tolerance fabrication processes.

10.2.5 State-of-the-Art Capabilities

There are several important observations that can be made concerning the impact of new technology on printed wiring board processing:

- It is possible to produce fine-line conductor widths and spaces of 0.13 mm (0.005 in) or less with today's production technology.
- It is possible to drill and plate through small-diameter [0.5 mm (0.020 in) or less high-aspect-ratio holes] with today's production technology.
- It is possible to produce dimensionally stable, tight-tolerance multilayer boards with today's production technology.

However, true manufacturing control over the processes used to produce these products requires improved material and equipment controls in order to achieve an industry wide state of the art for wide-window comfortable high-yield production. Thus, continued product development will be required in these areas.

The significant concern for the printed wiring board manufacturer will not only be the ability to meet these requirements but also to allow for the progression into future technology that will be even more demanding.

The level of performance that can be expected from existing processing equipment, materials, and procedures must be carefully assessed along with an analysis of which improvements are most likely to yield the most effective results. However, action plans must be made to improve the state of technical capability without unduly taxing the resource foundation of the existing facilities.

Fine-line small-hole printed wiring boards will certainly not be inexpensive to either make or buy, but they will be required to meet the future challenges of high-performance electronic products. The processes are available to allow for the manufacturing of these boards and, while not in common use, they are being used more and more by an ever-increasing number of companies that have an eye to the future.

10.3 CHIP-ON-BOARD

With the drive for smaller and more densely populated printed wiring boards, packaging design engineers are specifying components without the traditional package. Direct attach, or chip-on-board (COB), requires a process different from that used for packaged components because it uses printed wiring boards as the interconnection substrate. In chip-on-board packaging, the IC chip is bonded directly on an intererconnecting substrate—either a printed wiring board or hybrid—and protected with a "glob" top encapsulant. Therefore, it merges levels 1 and 2 connections to realize maximum space conservation. For years, COB technology has been used in consumer electronics for such applications as digital watches, pocket calculators, video games, and consumer electronic products.

Since there is no package involved, COB applications rely on a glob top, or encapsulation of some kind, for protection against moisture and to support the fragile leads. The encapsulants are materials such as silicone or epoxy which, although they do not provide a true hermetic seal, do provide very good moisture sealing in applications where high reliability is required. Thermal performance is poor, since both the materials above and below the components are not good thermal conductors.

Connection to the chip is traditionally by wire bonding, but a variety of other methods can be used, as discussed below.

The factors that lie behind the increasing interest in COB include the following:

Size.   The benefits of using COB all stem from the absence of the IC package. A wire-bonded chip occupies about 1/4 the area of a dual-in-line package (DIP), and is also more space efficient than leadless chip carrier (LCC) packages. Also, with a lower profile, COB can be used in applications not possible with other packaging methods (e.g., in "smart" credit cards).

Cost.   Lower overall system cost is another factor behind the selection of COB technology. A related cost-saving consideration is due to the fact that cost for making circuit connections is much less when made directly on the IC chip, as opposed to making them at a higher packaging level. Therefore, the trend has been toward more sophisticated IC chips.

Semiconductor technology.   Trends in semiconductor technology have also paved the way to COB technology. Lower-power CMOS devices, already taking a major share of the IC markets, are ideal for the power-limited COB technology. Also, the trend toward custom and semicustom ICs should provide additional importance to COB.

Encapsulation.   Reliability has improved with the development of encapsulation materials that more closely match thermal expansion properties of conventional printed wiring board materials. In the past, the expansion mismatch between the encapsulant (usually a "glob" of epoxy) and the board would excessively stress the joint between the chip and the board. However, recent developments with improved sealants have reduced the significance of this problem, thereby making COB attractive for use in even more applications.

Automation.   Companies that have had only limited results competing in labor-intensive assembly markets may pay more attention to COB, since many, if not all, of the processes involved can be automated.

10.3.1 Printed Wiring Board Considerations

Printed wiring board layout for COB applications, as with normal boards, is initiated with the positioning of components and the routing of conductors. Special considerations for the use of COB is required in layout of the PWB. A prime factor to remember is that exact placement of components is not always possible. A layout for production should be tolerant to variations component placement. Ideally, the board design should allow for positional discrete parts dislocation of up to 0.010 in and up to 10° of rotation.

Board substrate materials for COB application depends a good deal on the cost tradeoffs associated with the end-product application and the type of COB technology being used. The resin system chosen for a COB construction is the most significant concern. For instance, for many low-cost applications where thermocompression wire bonding is not used, conventional printed wiring board materials may be used. If thermocompression bonding is used, with its accompanying high temperatures, consideration must be given to using special high-temperature materials such as epoxy/glass or polyimide/glass board materials. Also thermosonic wire bonding is done with a heated stage, hot enough to soften epoxy, making bonding more difficult. This is normally remedied by using a higher temperature resin such as bismalimide triazine(BT) or polyimide.

Care should be taken to ensure the board flatness, since COB is especially sensitive to out-of-flat surfaces. While industry standards allow board warpage of up to 0.010 in per inch of length (measured across the greatest dimension) for standard 0.062 in thick board, this is not satisfactory for COB mounting. It is also important to recognize that the design of the board can affect flatness.

10.3.2 Chip Termination Techniques

COB applications are accomplished using one of the three basic chip attachment and termination techniques, or derivatives thereof, i.e., wirebonding (chip-and-wire), tape-automated bonding (TAB), and controlled-collapse (flip-chip) bonding, all described below.

10.3.2.1 Wire Bonding. There are three types of wire bonding (chip-and-wire) techniques used in COB, namely, thermocompression, ultrasonic, and thermosonic. They derive their names from the method the energy source employs to terminate very small diameter gold or aluminum wire, approximately 0.18 to 0.25 mm (0.0007 to 0.010 in) for COB applications, between the die and the substrate.

Thermocompression Wire Bonding. Thermocompression wire bonding is the most frequently used wire bonding process The basic method actually encompasses three different bonding processes, i.e., wedge, ball, and stitch. The principle is to join two metals using heat and pressure but without melting. The elevated temperature maintains the metals in an annealing state as they join in a molecular metallurgical bond,. The process is quite involved (see Fig. 10.6), but generally speaking, the softer the metals, the more readily the bonding takes place.

Ultrasonic Wire Bonding. Ultrasonic bonding (see Fig. 10.7) is a different concept of bonding that uses pressure in addition to a rapid scrubbing or wiping motion to achieve a molecular bond. The scrubbing action effectively removes any oxide films that may be present. A slightly larger area of contact is necessary because of the scrubbing action in involved in the operation. Extreme care must be taken so that the chip is not damaged during the ultrasonic wire-bonding action. Since ultrasonic bonding can create bonds between a wide variety of materials, it is an extremely flexible process. Therefore, both gold and

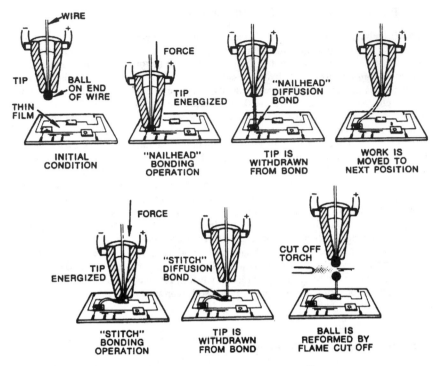

FIGURE 10.6   Mechanics of pulse-heated thermocompression bonding.[8]

aluminum wire are compatible with this technique. Void-free junctions are produced by ultrasonic wire bonding with relatively few foreign-body inclusions, making it a desirable means of creating high-quality, low-resistance electrical junctions. Also, bonding dissimilar metals at low temperatures eliminates or greatly decreases the formation of intermetallic compounds and allows bonds to be made in the immediate vicinity of temperature-sensitive components without adverse effects.

Thermosonic Wire Bonding.   Ultrasonic energy is used with thermocompression wire bonding methods, resulting in a technique known as *thermosonic wire bonding*. The technique depends on vibrations created by ultrasonic action to scrub the bond area to remove any oxide layers and also to create the heat necessary for wire bonding. Since this can be accomplished at temperatures in the vicinity of 120 to 150°C, thermosonic wire bonding can be done with low-temperature materials, including gold and aluminum wire.

10.3.3   Chip Protection

Increasingly, many applications are requiring that the COB assembly protect the chips from the atmosphere. The means for doing this fall basically into two categories: glob-top coatings and lids.

In selecting the technique to be used, several important properties must be considered, namely:

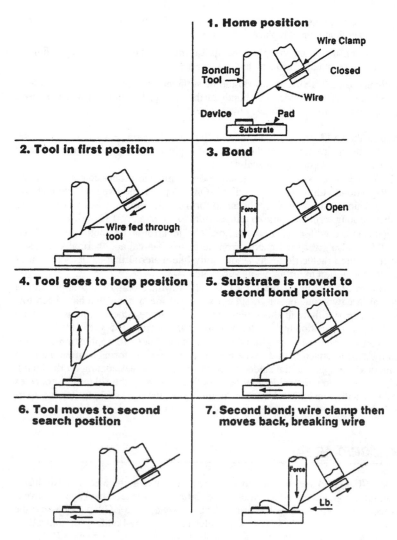

FIGURE 10.7 Mechanics of pulse-heated thermocompression bonding.[8]

- *Sealing temperature and time.* The technique chosen must be able to form a seal at a low enough temperature and in a short enough time to minimize the heating effects on the chip and board components.

- *Thermal expansion.* The thermal expansion of the sealing material/device should closely match that of the board to minimize the thermal stresses and thus maintain the integrity of the seal.

- *Hermeticity.* The seal must provide the degree of hermeticity required and maintain this level when exposed to the operating and storage environment of the equipment.

- *Cost.* The sealing technique must be cost-effective, not only with respect to material cost but also in application and replacement costs.

- *Repairability.* If it is necessary to replace a chip it is important that the sealing technique be such that it can be readily "broken" and replaced.

- *Stability.* When coatings are used, they should be sufficiently stable that they do not tend to put excessive stress on the wire/lead bonds or the die attachment bond when exposed to the equipment operating and storage environments.

The final step in the COB process is to protect the bare die from environmental abuse. Depending upon the environment and the degree of sophistication needed, this is usually done with either protective coatings or sealing lids.

Typical RTV dispersion coatings are one-component silicone-rubber coatings supplied as a xylene dispersion, with no mixing required. Once applied and exposed, the material vulcanizes by reaction to moisture from the air to form a soft, resilient elastomeric coating that will withstand long-term exposure to temperatures as high as 250°C (482°F).

Silicone gel is also used for die encapsulation. The gel affords the nonflowable permanence of a solid but also gives the freedom from large mechanical and thermal stresses of a fluid. The fully cured dielectric gel is a soft, jelly-like material that exhibits tenacious pressure-sensitive adhesion to virtually any substrate.

Epoxy coatings are also available for COB self-crowning or glob-top applications. Typical materials are two-component, liquid epoxy/anhydride systems that have been formulated for their superior thermal shock performance, substrate adhesion, moisture resistance, as well as glass transition temperatures in the range of from 165°C to 180°C.

Metal lids can also be used to seal individual chip sites. For COB applications, soldering is generally used to attach the lid to the board. The use of preforms is often the most convenient method for applying the solder, in addition to the solder coating on the board, The heat required to reflow the solder can be supplied by one of the soldering processes that is associated with conventional·surface-mounting technology, such as IR, or vapor-phase soldering.

10.4 BALL GRID ARRAY

Ball grid array (BGA) packages are rapidly gaining acceptance in the electronics industry as a low-cost, higher yielding leadless alternative to fine-pitch leaded packages. Conventional BGA technology emphasizes soldering directly between the BGA package and the interconnecting substrate. Thus, it is also often referred to as face-bonding, or controlled-collapse soldering. This is identical to the controlled-collapse chip connection (C4) process promoted by IBM more than 30 years ago. The BGA takes advantage of the previously unused under-chip real estate, increasing both the number of I/Os and pitch, eliminating perimeter leads, and reducing the handling problems associated with high pin count devices.

The key characteristic for identifying any BGA package is that the leads—actually solder bumps configured in an array—are on the bottom instead of on the perimeter. The BGA is a leadless package that is not susceptible to bent or skewed leads and thus can be easily handled.

JEDEC standards for bump pitch include 1.5, 1.27, and 1.0 mm (0.058, 0.050, and 0.040 in). This wide spacing of ball-shaped contacts allows up to 12-mil misregistration during placement of lighter plastic BGAs, since surface tensions act to self-align the part during reflow.

Among the many advantages of using BGAs is the fact that they offer the maximum in board space efficiency. They have a lower profile and have vastly improved pin count capability. BGAs have excellent thermal and electrical performance The electrical improvements are due to the shorter wire lengths that reduce path delay and inductance. There is maximum thermal dissipation by direct heat-sink bonding to the die. There are no coplanarity problems when using BGAs, no bent component leads, and good package yield. In production, BGAs are compatible with existing surface mount, test, and handling equipment.

An important limitation to using BGA is the inability to inspect the solder joint, since they are hidden. Industry users maintain that the best approach is to use good process controls such as paste inspection and reflow profiling to ensure good solder joints. With low level of solder defects noted, electrical test alone is a sufficient indicator of yield and reliability. Removal of BGAs is not considered difficult. Removed BGAs are unusable, but the high soldering yields obtained allow scrapping some number of packages in deference to costly rework.

There are potential routing problems in using BGAs. Various techniques are being investigated to improve routability by using staggered pitches and placing I/Os close to edges, thereby leaving a larger route area in the middle of the footprint. Many BGAs, however, need a multilayer substrate to handle the dense routing required.

This section, then, presents BGA technology as a generic process, with the C4 process as a subset of a particular application. The concepts of design of the BGA connection configuration are stressed, along with the material choices for the connection medium, options in implementing BGA, and processing issues in light of process limitations and existing infrastructure.

10.4.1 BGA Packages

BGA packages offer a significant size advantage over conventional surface mount packages: they allow the highest I/O-to-package body size. For example, a 1.27 mm BGA package offers more that 350 I/Os on a 25 mm body size package compared to 304 I/Os on a 40 mm body size QFP (quad flat pack) of 0.5 mm lead pitch. BGAs are available in several configurations, as described below.

10.4.1.1 Plastic Ball Grid Array.

In the plastic ball grid array (PBGA), a die is mounted to the top side of a single-layer, double-sided bismaleimide triazine (BT) resin epoxy PWB (see Fig. 10.8). The standard core thickness of this two-layer substrate is typically 0.2 mm (0.008 in) with 18 µm (half-ounce or 0.7 mil) copper on each side. A 2 mil thick (thickness over epoxy glass) dry or dual-pass wet film solder mask is currently used to ensure that all the substrate vias are completely tented. The silicon chip containing the integrated circuit is die bonded to the gold-plated top surface of the substrate with a silver epoxy.

Interconnection of signal and power lines between the die and the PW-board contact points is through thermosonic gold wire bonding. From there, copper traces are routed to an array of metal pads on the bottom of the printed wiring board. Vias in the form of plated through holes run from the top surface to the bottom and are usually around the periphery of the board. The bottom side pads are arranged on a square or rectangular grid with either a constant 1.5 mm (0.058 in) or 1.27 mm (0.05 in) pitch. A "glob-top" encapsulation or overmold is then performed to completely cover the chip, wires and substrate bond pads. Solder balls typically composed of 62 percent Sn, 36 percent Pb, and 2 percent Ag are partially reflowed to the metal pads to form the package leads.

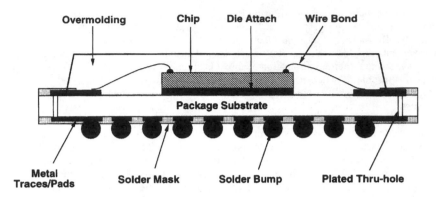

FIGURE 10.8   Plastic ball grid array package.

PBGA advantages include:

- lower system cost—no new equipment or processes
- thermal—thermal enhancement via detachable heat sink
- small package size—smaller foot print for high I/O devices
- electrical—shorter delay, less noise
- reduced manufacturing cycle time
- ease of assembly
- depopulation (removal of inner contacts) of array does not affect thermal cycle life

PBGA limitations include:

- possible solder joint fatigue
- higher package height
- possible package warp
- board routing problems—optimized board layout to reduce layer count
- moisture sensitivity—susceptible to "popcorning"

10.4.1.2 Ceramic Ball Grid Array. Ceramic ball grid array (CBGA) packages use a three- to five-layer ceramic substrate with wire bond or C4 die attach (see Fig 10.9). CBGAs use high-temperature solder balls (10 percent Sn, 90 percent Pb,) that are pre-attached to the substrate with a eutectic tin/lead solder paste. The principle construction of the CBGA package begins with a multilayer cofired ceramic substrate where tungsten of molybendum refractory metallization has been screen printed with desired circuit patterns including die attach pad, wire bond or C4 pads and back-side solder bump pads.

The substrate is electrolytically nickel plated, plated with a gold strike and then a thin layer of gold on the solder bumps and C4 pads. It is then electrolytically gold plated with standard thickness on wire bond pads. Subsequently, solder paste is screened on the back side, reflowed, then solder balls are attached either by pick-and-place methods or with a stainless steel template.

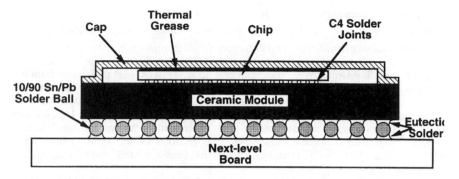

FIGURE 10.9 Ceramic ball grid array package.

To relieve stresses due to the large coefficient of thermal expansion (CTE) mismatch between the ceramic substrate and the printed wiring board substrate, the 10/90 Sn/Pb solder ball that is used does not melt during reflow soldering. The eutectic solder joins the ball to the package and to the board. The high solder ball ensures a 0.035 in standoff to facilitate cleaning and improve thermal cycle life. Cofired ceramic chip carriers have inherent excellent thermal and electrical characteristics, including built-in reference and power planes, low inductive power connections with multi-power paths. CBGAs can support all types of chip interconnection (wirebond, flip-chip, and/or TAB), plus a variety of lid sealing or encapsulation techniques.

CBGA packages have excellent thermal performance due to the thermal conductivity of the ceramic substrate. In addition to the thermal path through the solder balls and off the ceramic substrate, flip-chip CBGA packages can dissipate heat from the back side of the chip through a thermally conductive compound to a directly attached aluminum cap.

CBGA packages provide short signal paths from the chip to the card, reducing electrical parasitics that impact signal time of flight. As the number of package interconnections increases, the importance of electrical performance increases. The package benefits from the multilayer ceramic structure. Molydenum metallurgy on each ceramic layer carries the signals and the multilayer structure can provide stripline or controlled impedance signal conditions. Multiple power and ground planes can provide both impedance control for signal lines, as well as reduce the inductance in the power paths.

CBGA advantages include:

- It exhibits excellent thermal characteristics.
- It is ideally suited to chip attachment.
- Provides a short electrical path, low parasitic with flip-chip.
- Has low warpage.
- It can accommodate multi-chip, including mixed (flip-chip and wire bond) technology.
- The package can be hermetically sealed.
- Components can be removed after testing and can be tested and rebelled for reuse.
- CBGA is not sensitive to moisture.
- High melt balls provide standoff during assembly.

CBGA limitations include:

- Ceramic may have high mass, making reflow profiling difficult.
- Depopulation of array reduces thermal cycle life of remaining interconnects.
- PCB to ceramic thermal expansion difference limits thermal cycle.

10.4.1.3 Tape Ball Grid Array. Tape ball grid arrays (TBGAs) get their name from the tape automated bonding (TAB) type frame that connects the chip with the next level board, Figure 10.10 is a schematic representation of a TBGA tape frame is typically a two-conductor tape with a 2-mil thick dielectric layer sandwiched between them. Usually, one conductor layer contains the signal traces and the other contains the ground plane. The two conductor layers are connected using plated-through holes set on 50-mil centers. High melting point solder balls (10/90 Sn/Pb), 25 mils in diameter, are reflowed into each plated through hole.

The chip attachment to the tape may be accomplished using a standard TAB inner lead bonding (ILB) process, or a solder attached ILB method. The chip is subsequently encapsulated with a filled epoxy that is dispensed around the chip edges. A heat sink may be bonded to the back side of the die to provide a high-performance package. The TBGA is reported to easily dissipate 10 to 15 W with a heat sink attached.

Current TBGA packages range from 21 to 40 mm (0.82 to 1.57 in) body size with 192 to 736 I/O connections. Since the TBGA provides a CTE that is matched to the next-level board, the size of the package is not limited by the maximum distance from neutral point (DP) for temperature cycling tests.

TBGA advantages include:

- It is thin in size.
- It is low in weight.
- It offers long thermal/power cycle life.
- Package CTE matches board.
- It provides improved system-level performance with high speeds and low noise.

TBGA limitations include:

- Repair is difficult, touch-up impossible.
- Cost is high.

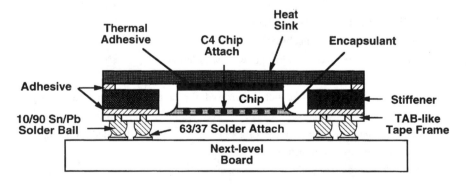

FIGURE 10.10   Tape ball grid array package.

- Solder joints are low in reliability.
- Escape routing impact is a negative factor.

10.4.2 Assembly Issues

10.4.2.1 Mother Board Design.

Board Design.   One of the outstanding advantages of BGA package technology is that they can typically be placed on printed wiring boards and assembled using existing surface mount equipment.

The package, with its balls in an array configuration, presents some odd challenges to overcome with regard to mother board routing when compared to other devices, such as peripherally leaded devices. None of the advantages of using BGAs would have any significance without the ability properly interconnect the devices.

Various techniques are used to improve the routability by using staggered pitches and placing I/Os close to edges, thereby leaving a larger route area in the middle of the footprint. However, with a large number of I/Os, BGA designs will invariably use multilayer substrates.

Fortunately, technology exists to produce fine line multilayer boards with adequate density capability to accommodate the requirements dictated by the space requirements of BGAs. The use of chip carriers and surface mounting allows for the use of thin multilayer boards, with small holes, dictated only by the substrate thickness and the via hole technology used.

Escape Routing.   A perceived drawback of using BGAs is the challenge of routing all the required signal, power, and ground pins to the system board without increasing the printed wiring board complexity, and therefore cost. This task is overcome by thoughtful pin assignment and device configurations (pitch, ball count, ball depopulation methods) in conjunction with the choice of solder pad geometry and board technology (number of line/space widths).

If signal pin assignments are made too deeply within the BGA matrix, board-level escape using conventional 0.008 in printed wiring board fabrication technology becomes difficult for large matrices. Although current PWB technology with 8 mil lines and spaces does not incur any additional cost, the configuration of the packages used may require the employment of a denser interconnection technology, forcing the move to 5 mil lines and spaces.

As pin count increases and matrix size goes up, it may be necessary to make signal assignments on more than one buried row on each of the outer layers to be escaped on a given PWB layer to maintain a two signal layers. If this takes place, it may be necessary to decrease further from the standard line/space standards in the design. Table 10.4 shows how many traces can be routed between two pads for given line and space technologies down to three and three and all the current standard BGA pad diameters.

Footprint Geometry.   Proper design of the board-level solder pad is important for the successful implementation of BGA technology. This is due to its influence on the controlled collapse and self-alignment of the solder ball interconnections. The design of a standard mother board solder pad is shown in Fig. 10.11. The solder pad on the mother board is defined by an opening in the solder resist. The diameter of this opening in the solder resist will determine the amount of collapse of the package during the reflow process. For instance, use of a 0.63-mm (0.024-in) diameter solder pad opening at the board and

TABLE 10.4 Number of Escape Traces that Can Be Routed between PBGA Pads Given Device Pitch, Pad Diameter, and Board Line/Space Widths

Board technology (line/space widths in mil)	No. of escape traces routed between pads given ball pitch and pad diameter (mm/mil)			
	0.5/31	1.5/35	1.27/31	1.27/35
3/3	4	3	2	2
4/4	2	2	1	1
5/5	2	1	1	1
6/6	1	1	1	
8/8	1	1		

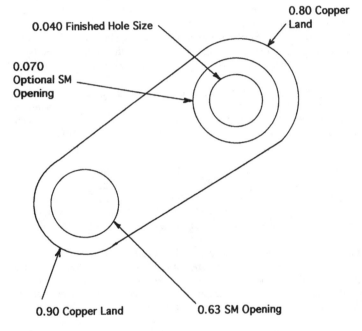

FIGURE 10.11 BGA substrate pad design.

package level, and a 0.76-mm (0.030-in) diameter solder ball will produce a 0.2-mm (0.008-in) solder ball collapse during the reflow process.

As shown in Fig. 10.11, the escape vias next to the board level solder pad must be designed so that solder cannot run from the solder pad into the via during the solder ball reflow process. Allowing reflowed solder to run into the escape vias can result in an uncontrolled solder ball collapse condition whereby the interconnections have varying volumes of solder.

The solder pad finish is dependent on the specification of the end user. Several metalizations have been used successfully: bare copper with an antioxidant coating, hot air solder leveled copper, and electroless gold. Board-level solder mask selection is based on the same criteria as used for other SMT applications.

10.4.2.2 Solder Assembly Processing. BGAs offer great solder assembly advantages over conventional leaded packages. The primary advantage is that solder-bumped BGAs can be IR reflowed in a production environment with extremely low solder defect levels. The high solder assembly yield of BGAs is due to the use of a eutectic tin/lead solder bump technology, which diminishes the impact of the three factors responsible for yield loss of conventional leaded plastic packages: (1) opens due to lack of adequate lead coplanarity or skew, (2) shorting between terminations due to solder paste bridging, and (3) opens or shorts due to package misplacement.

Lead coplanarity and skew problems are eliminated by the addition of leads and the addition of a collapsible solder bump technology. The solder balls of a BGA package are designed to collapse approximately 0.2 mm during the solder assembly process, eliminating any yield loss due to coplanarity or solder ball size variation. Finally, BGA placement tolerances are not as critical as other fine-pitch devices due to their large pad pitch and self-centering capability during reflow.

The self-centering of BGAs during solder reflow is due to wetting forces that align the package on the solder lands of the mother board. This phenomenon increases the allowable positional tolerance for placement of the package on the substrate.

The surface mount assembly process for BGAs is virtually identical to other surface mountable packages. The BGA is picked from a JEDEC/EIAJ sized tray or tape and reel by a standard pick-and-place machine and then placed into prescreened solder paste on the mother board. Solder paste selection for BGA soldering is not critical. End users report that they have implemented BGAs into production using the same solder paste materials and processes (including solder past thickness) that they have been using for fine-pitch leaded plastic package soldering.

After placement of other surface mount components, the mother board is transported into an IR or IR/forced convection furnace for mass reflow of the BGAs and other components. After soldering, the package is cleaned using standard cleaning processes. Selection of a cleaning method and material is dependent almost entirely on the solder paste used in the existing SMT process. Concerns about cleaning the flux from the middle of the package appear to be unwarranted due to the 0.040-mm (0.0015-in) standoff of the package from the mother board after soldering.

Most companies that are in production with BGAs are not performing any type of post-reflow solder joint inspection due to the extremely high solder assembly yields. However, it is understood that, during BGA implementation, some manufacturers will want to perform inspection of BGA solder joints. Due to the high density of solder, x-ray inspection of BGA solder joints is easily accomplished. Inspection of solder assembled BGAs can be performed with conventional x-ray or 3-D laminography systems.

10.5   FLIP-CHIP

Conventional flip-chip technology employs soldering directly between the integrated circuit die "face" and the interconnecting substrate. Thus, it is often referred to as face bonding, controlled-collapse soldering, or controlled-collapse chip connection.

The solder-bump interconnection of flip-chips has been practiced for 30 years. Face-down soldering of silicon devices to alumina substrates was introduced by IBM in 1964 on solid logic technology (SLT) hybrid modules in System 360 computers.

The original flip-chip concept employed small, solder-coated copper balls sandwiched between the chip termination lands and the appropriate lands on the interconnecting substrate. The resultant solder joints were made when the unit was exposed to an elevated temperature. However, the handling and placement of the small-diameter balls was extremely difficult, and the operation was costly. In a more advanced technique, a raised metallic bump or lump, usually solder, is provided on the chip land. This is fabricated on lands of all the chips, in wafer form.

The individual chip is then aligned to the appropriate circuitry on the substrate and bonded in place using reflow soldering techniques. In this way, the interconnection bonds between the chip and the substrate are made simultaneously, reducing fabrication costs. An alternative procedure is to place the bump on the printed wiring board pad, rather than as a part of the chip. In this way, the COB assembler can use any of the devices that are available in bare-chip form.

Since flip-chip bonding does not use bonding wires or beam leads to land patterns outside the die's perimeter, it can be used to achieve the highest ratio of active silicon surface to substrate area. In this way it can be used to achieve a very high number of fine-pitch I/O terminations per chip.

The original flip-chip concept used small solder-coated copper balls sandwiched between the chip termination lands and the appropriate lands on the interconnecting substrate. The resulting solder joints were made by solder reflow.

A more-advanced flip-chip technique (Fig. 10.12) employs a solder bump at all of the die-bonding pad locations. This is usually done when the die are still in the uncut wafer or on the substrate prior to final assembly.

Flip-chip technology is extendible to meet the requirements of future high performance chips. Since the connections are arranged over an area, many more I/O terminals can be accommodated in comparison to wire bonding or TAB, which rely on peripheral connections.

10.5.1 Advantages

Compared to wire-bonding and TAB technologies the flip-chip technology has several advantages. Flip-chip is the only connection configuration that permits assembled active chips to assume the form for which they were created, namely, a wafer. The chip is capable of handling a higher number of I/Os because solder bumps can be arranged in an area array rather than being restricted to the chip's periphery. Due to surface tension factors

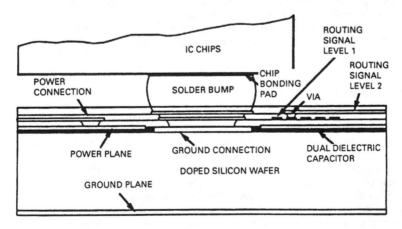

FIGURE 10.12   Cross section of flip-chip termination. *(Courtesy of AT&T Bell Laboratories.)*

related to the solder, this technique has a self-aligning capability during bonding. The actual bonding can be carried out under reflow conditions. The flip-chip has the best circuit performance, since it has the shortest interconnection distance, providing superior electrical and thermal performance. A separate die-attach operation is also not required for flip-chip bonding.

10.5.2 Disadvantages

The disadvantages of flip-chip are related to infrastructure and not technical issues. First and foremost is the lack of chips available with solder bump array connections. Another disadvantage to flip-chip technology is the inability to inspect the assembled joints. Flux removal is difficult, and there is a need for underfill of flip-chips to maintain connection reliability.

10.5.3 Manufacturing Flip-Chips

The manufacturing steps for conventional integrated circuits and flip-chips are similar through most of the processing steps until they reach the passivation step. Integrated circuits are normally passivated, that is, coated with a protective silicon nitride layer, and then readied for testing and encapsulation. Flip-chips require more processing that includes and additional coating, back-side marking, and solder-bump formation steps.

10.5.3.1 Back-side Marking. Back-side part numbers and orientation markings can be applied to nitride-coated silicon wafers by sputtering on a pattern of titanium tungsten. The markings can also be coated with a thin layer of silicon nitride to protect them and to enhance their contrast with the underlying silicon.

10.5.3.2 Bumping. Flip-chip uses a solder bump to connect the bond pads to the substrate. The flip-chip assembly consist of some method of patterning the balls, called pad limited metallurgy (PLM) on the chip bond sites, solder bump, and top surface of the bond pads. PLM is a multilayed structure with a barrier layer, and adhesive layer, and a bonding layer. Common materials for the PLM structure include titanium and chromium for the adhesion layer, copper, palladium, nickel, and platinum for the barrier layer, and gold for the bonding layer. Common materials for the solder bumps include 50Pb/50Sn and 95Pb/55Sn solders (Fig. 10.13).

Solder bump diameters usually range from 100 µm (4 mils) to 250 µm (10 mils). Typical solder bump heights range from 50 µm (2 mils) to 200 µm (8 mils). Minimum I/O pitches achieved using flip-chip interconnects are generally greater than achievable by using TAB or wire bonding. The maximum number of I/Os bonded using flip-chip interconnects are in the vicinity of 700 or greater. Typical values of mutual and self-inductance for flip-chip interconnects are in the range of 1 pH and 0.05—0.01 nH, respectively. Flip-chip interconnects are ideally suited for use with area array bonds.

The bumps, which are composed of a series of metal layers, are formed over the integrated circuit aluminum bonding I/O pads. In a typical application, a layer of titanium is first sputtered over the I/O pads to enhance bump adhesion.

This is generally followed by a sputtered copper layer that acts as a transition layer for the solder. This layer is etched to form a pedestal upon which the solder bump rests. The pedestal shape distributes bump stress so that the adjacent silicon nitride protective coating does not crack. Another layer of electroplated copper is added, and this is followed by an

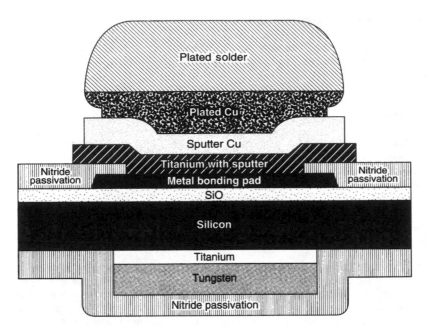

FIGURE 10.13   Cross section of a typical solder bump.

electroplated layer of tin-lead solder. Finally, the "bumped" wafers are heated to promote solder reflow. the final result is the formation of controlled height solder bumps over each I/O bonding pad (Fig. 10.13).

10.5.4   Substrate Preparation

The substrate must be prepared as shown in Fig 10.14 prior to mounting of the flip-chip. In a typical process, the substrate and its conductive pattern are first precleaned, then a solder paste is stenciled or screened onto the substrate's land pattern at locations corresponding to the associated locations of the solder bumps on the flip-chip. It is critical to deposit the proper amount of solder paste on the lands. Too much solder paste can result in bridging during reflow, and too little will compromise the termination. It is often recommended that the solder paste should contain a mildly activated rosin-based flux (RMA) and 45 to 75 μm-diameter spherical particles of 10 percent tin, 88 percent lead, and 2 percent silver alloy.

10.5.5   Flip-Chip Mounting

Flip-chips may be mounted on the substrate either manually or automatically. This operation is often the gating operation in the entire assembly process. The flip-chips are usually supplied by the IC manufacturer in waffle packs. They also have some sort of identification marking and mounting orientation designators. If the assembly is being done manually, the operator simply removes the individual die from the waffle pack using a vacuum

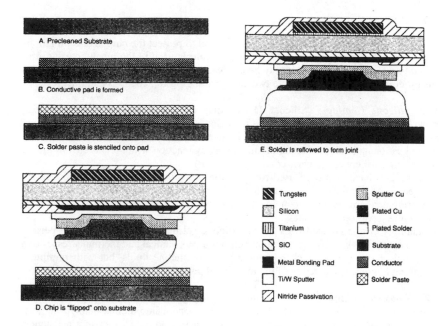

A. Precleaned Substrate

B. Conductive pad is formed

C. Solder paste is stenciled onto pad

D. Chip is "flipped" onto substrate

E. Solder is reflowed to form joint

	Tungsten		Sputter Cu
	Silicon		Plated Cu
	Titanium		Plated Solder
	SiO		Substrate
	Metal Bonding Pad		Conductor
	Ti/W Sputter		Solder Paste
	Nitride Passivation		

FIGURE 10.14 Typical flip-chip assembly process.

pencil. Each die is then positioned over the substrate land pattern, which was previously covered with a tacky solder paste flux. The solder paste will hold the chip in place until the soldering reflow operation.

For automated assembly, there are several pick-and-place machine options. If the bare dice are provided in precision waffle packs, the assembly machine can either move the individual dice and place them directly onto the substrate, or they can be transferred automatically to a holder prior to final component mounting.

Regardless whether the dice are placed manually or automatically, their position with respect to their land patterns may be off by as much as slightly less than one-half the pad pitch. This is true because the surface tension created during the reflow operation will be sufficient to force the proper alignment between flip-chip pads and the substrate lands.

10.5.6 Reflow Soldering

Once all the dice are positioned, the entire substrate assembly is placed in an oven to drive off any flux residues. The assembly can move through an infrared reflow soldering machine with and inert nitrogen atmosphere, or a vapor-phase soldering machine.

The thermal profile of the reflow soldering oven will vary on the equipment being used. The critical factors are the magnitude of the peak temperatures and the duration of time the assembly is exposed to it in the reflow soldering oven. If the solder reflow has been accomplished correctly, the flip-chip solder joints will be columnar, smooth, and shiny. The adhesion of the solder bump can then be determined quantitatively using a shear tester.

10.5.7 Chip Protection

Bare chips need protection from the atmosphere, and to protect the chip and terminations from mechanical shock, moisture, and various chemicals used in the manufacturing process, and to which the end product may be exposed to during its functional lifetime. The means for doing this fall basically into two categories: glob-top coatings and lids.

10.5.7.1 Silicone Coatings. Silicone (RTV) dispersion coatings are one-component, room-temperature vulcanizing rubber coatings that are supplied as a xylene dispersion. The curing process uses a cross-linking mechanism that generates methanol during cure. After application and cure, the material vulcanizes by reaction with moisture from the air to form a soft, resilient elastomeric coating.

Silicone gels have also been used for bare-die protection applications. Silicones are formulated especially for use with integrated circuits and are widely used in that applications. Curing noncorrosively, silicon gels bond satisfactorily to the surface of bare silicon die, as well as to various substrates.

Technology can be controlled to yield a silicone gel that affords nonflowable permanence of a solid, but also give the freedom from large mechanical and thermal stresses of a fluid. A typical silicone gel is chemically similar to silicone fluids, but with minimum cross-linking to prevent separation of the individual polymer chains.

10.5.7.2 Epoxy Coatings. Epoxy coatings are also applicable for chip protection or glob-top applications. Typical materials are two-component liquid epoxy-anhydride systems that have been formulated for their superior substrate adhesion, moisture resistance, and thermal shock performance. Epoxy is also used to relieve the stress in the solder bump connections. The die and the substrate have different CTEs and, as the temperature of the die increases, the stress on the solder bumps due to the higher strain on the substrate causes failure of the solder joints.

10.5.7.3 Sealing Lids. Metal lids can be used to seal individual chip sites. For most applications, soldering is generally used to attach the lid to the substrate. The use of preforms is the most convenient method for applying the solder, in addition to the solder coating on the board. The heat required to reflow the solder can be applied using one of the soldering processes that is associated with conventional surface-mounting technology.

10.6 TAPE AUTOMATED BONDING

The continuing trend toward increasing the complexity of an integrated circuit die to its ultimate limit by maximizing the number of circuit elements and putting as many interconnections as possible on the die where they cost the less is creating a cost/density gap that conventional wire-bonding technology has difficulty meeting. The use of smaller die feature sizes and the resulting increase in I/O terminals have created further pressures on packaging that can be satisfied by the use of a cost-effective, high-speed/high-density packaging technology such as tape automated bonding (TAB).

The most visible aspect of the TAB system is the carrier tape, which is stored on reels similar to movie film, in widths from 8 to 70 mm. Windows are punched in the tape at specific locations and a thin, often 0.036-mm (0.0014-in) thick, conductive foil is bonded to the tape, which is usually either mylar or polyimide. A conductive pattern is then etched in the foil to give the desired interconnection circuitry with "beam-type" leads extending over the windows in the tape.

In subsequent processing, the chips are first bonded to the tape (ILB, or inner lead bonding), with each pad on the chip connected to a beam lead by thermocompression bonding. The entire inner lead bonded unit is then removed from the tape and attached to the printed wiring board using conventional die-bonding techniques.

The exact location of the chip with respect to the sprocket holes on the tape carrier permits automated tape-handling equipment to be used that accurately positions the individual chips to printed wiring boards for the subsequent processing operations. The last step in the operation is to bond the outer lead of the beam leads to the substrate pad using heat and pressure. The outer lead bond is usually a gang or single-point (one lead at a time) in a thermocompression or reflow soldering operation.

10.6.1 TAB Process Overview

TAB is an electronic integrated circuit die fine-pitch packaging technology that uses fine-line conductors on a flexible printed circuit or on bare copper. It is a connection method for connecting integrated circuit die to a variety of packages. Connection is made by bonding a patterned conductor to corresponding pads on the IC die. TAB parts can be assembled in three basic configurations: spider mount, inverse mount, and flat mount (see Fig 10.15).

Spider mount TAB involves mounting the chip with the back side attached to the substrate. Leads are then formed to connect form the chip to the package or substrate. Inverse mount TAB consists of mounting the chip with the back side or active side facing the sub-

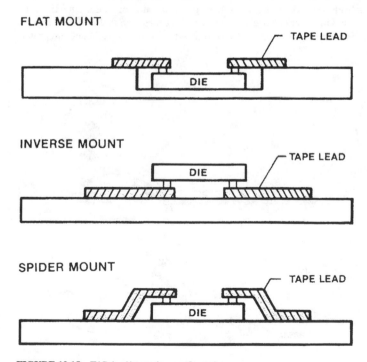

FIGURE 10.15   TAB lead/mounting configurations.

strate. The bonding pitches on the die can be the same as those on the substrate, with no leadforming required. Flat mount TAB involves placing the chip in a cavity in the substrate or package with its back side attached to same.

The most common TAB construction is a tape/carrier interconnect product that has special features to facilitate reel to reel processing (see Fig. 10.16). The tape used is similar to photographic film in format. The tapes come in standard widths ranging from 8 to 70 mm (0.314 to 2.75 in). Many applications use a two layer (copper-polyimide) or three-layer (copper-adhesive-polyimide) combination.

However, unlike conventional flexible printed circuits, TAB tapes have precision sprocket holes along their edges. This provides them with a feature to accommodated the reel-to-reel processing. An opening, appropriately called a *window*, is formed near the center of the conductor array in the dielectric base film if one is present. (One-layer TAB tape has no dielectric base.) The window permits the etched conductor leads to extend over the opening to create the essential beam-type interconnect array.

A process not unique to TAB, known as "bumping," is performed to provide a raised surface at the die-bonding sites, The bump can be formed either on the semiconductor wafer (as in flip-chip), or on the tape carrier. The integrated circuit wafer, with or without bumps, is scribed or sawed into individual dice. In the initial bonding process, the TAB tape unwinds from one reel and is taken up by the another.

A die is selected and positioned under the window in the tape. A die can be picked up either from waffle packages or from sawed wafers on expanded tape while the tape can be fed either from reel-to-reel or in the slide carrier format. The chip-to-tape connections are accomplished using an inner-lead bonding (ILB) machine. The ILB machine precisely aligns the patterned tape to the die and then descends to the die. The ILB tool mass "gang" bonds the TAB tape leads to the die either by a thermocompression or solder/eutectic technique. This results in having a sprocket-driven tape with bonded die that can be mounted directly onto a printed board or other substrate, or readied for use as a packaged integrated circuit component.

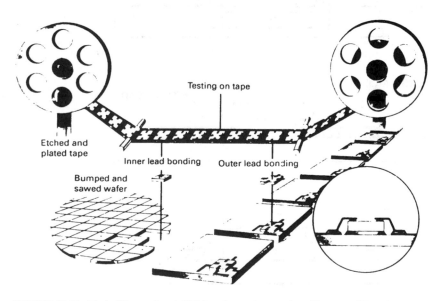

FIGURE 10.16   Typical "bumped-wafer" TAB reel-to-reel processing. *(Courtesy of Farco.)*

After the plating busses have been disconnected from the individual leads, testing of the bonded die can be performed while it is still on the TAB tape. This affords the opportunity to remove poorly bonded die from the TAB tape so that they do not proceed further along the assembly process, thereby enhancing the reliability and yield of the finished product.

The bonded die is then usually protected by applying an organic potting compound, or "glob." To perform the final bonding of the tape to the printed board, an outer-lead bonding (OLB) process is used. OLB requires that the TAB interconnection area be "blanked" or excised from the tape. The excised TAB device is bonded to the surface of the printed board or substrate by aligning the TAB outer-lead frame with corresponding sites on the land pattern and then applying energy. The OLB is usually a gang or single-point (one lead at a time) thermocompression or reflow soldering operation.

10.6.1.1 Advantages. Advanced TAB technology provides IC printed wiring board assembly, hybrid microcircuit, and multichip module (MCM) packaging engineers with options that can possibly enhance the cost effectiveness and performance of the end product being developed. The following sections reflect some of the reasons for this.

Density. TAB die bonding patterns can be up to one-half the size and pitch of those associated with wire bonding, namely, 0.1-mm (0.004-in) pads on 0.2-mm (0.008-in) centers. Thus, smaller die can be used, or more I/Os can be provided on the same die size. The packaged devices will also be significantly smaller using TAB.

Electrical Performance. As compared with wire bonding, TAB is characterized by shorter lead lengths and lower impedances, which can result in less stage delay and smaller signal distortion. The rectangular beam leads of the TAB tape provide superior high-speed circuit electrical performance at operating speeds above 40 MHz. Power and ground leads can be larger, thereby lowering the already low resistance inherent in the TAB interconnection.

Reliability. The die bonding pad areas on each TAB device are sealed by a thin-film metallization. This, plus the use of "bumped" die, provides a high degree of protection against contamination from moisture and chemicals to achieve longer life and improved end-product reliability.

Productivity. Through the use of "gang" or "mass" bonding, TAB provides a higher assembly throughput. Also, by reducing the number of connections between the IC die and the interconnecting substrate from three to two (for wire bonding), TAB can significantly increase first-pass yield. TAB eliminates the IC production problem of "lead wash" in bare-die assembly processes involving cleaning and coating high-lead-count devices, wherein wire bond leads tend to be moved so as to touch (short out) adjacent wires.

Automation. TAB's use of reel-to-reel or reel-to-carrier principles makes it highly suitable for integration into a manufacturing facility that emphasizes the use of automation. TAB reel-to-reel tape and component carrier packages are also compatible with many surface-mount pick-and-place assembly systems.

Low Profile. TAB can provide a packaging profile that is as low as 0.08 mm (0.003 in) or less since it does not have to contend with the strain-relief service loop associated with wire bonding.

Testing and Burn-In. TAB devices, usually in packaged carriers, can be tested and burned in prior to excising the die for next-level assembly. Thus, when required, more reliable devices can be attached to interconnecting substrate.

Thermal Management. In some applications, TAB provides another thermal management option over wire bonding through the use of the increased mass or conductivity of copper TAB leads.

10.6.1.2 Disadvantages.   Use of TAB is not without its disadvantages. The following considerations must also be taken into account.

Equipment and Tooling.   A large capital investment is required in order to set up a TAB production facility, both for the equipment, and for personnel training. Also, equipment availability is limited to a small number of suppliers, which are located throughout the world, and each TAB product requires new tooling. Conversely, many companies have extensive experience in wire bonding and have the necessary equipment in place at both the component packaging and the chip-on-board (COB) level.

Die Availability.   Semiconductor manufacturers are sometimes unwilling to invest the necessary resources to accommodate an unpackaged-device TAB market. They are also generally not prepared to supply the special metallization and assembly features associated with the use of bumped die.

Planarity.   To achieve successful TAB gang bonding with "hard" metallurgical systems, very tight planarity must be maintained consistently, so that the large forces involved are distributed evenly to each bonding site. However, single-point tape automated bonding can be used to overcome this problem. Also, to dissipate heat properly, solid contact is generally required between the TABed die and a substrate acting as a heat sink. This is particularly critical with large silicon die that are more sensitive to temperature and thermal management.

Handling.   Blank etched TAB tape must be handled with care in order to avoid mechanical displacement of the delicate leads. If care is not taken, stress problems can occur whenever silicon die are in contact with other materials that have different coefficients of thermal expansion.

Cost Effectiveness.   All things considered, TAB is presently an expensive technology with costs that are often difficult to quantify unless all of the device and assembly-level cost savings are measured.

10.6.2   TAB Tape Considerations

The TAB tape carrier provides several essential functions to the TAB production process. Initially, it provides the interconnection leads and supports the attached die after inner-lead bonding (ILB). The next step depends on the application. For unpacked component applications, the TAB tape usually carries the die to an optional encapsulation station prior to excising and outer-lead bonding (OLB) to the interconnecting substrate. If a packaged component is being processed, the TAB tape carries the attached die to an optional encapsulation station that creates the TAB component carrier package for subsequent processing by the substrate assembler. (Special consideration is given to die to be burned in or tested while it is still attached to the TAB tape. However, this is readily done with the packaged TAB component carrier package.)

To provide adequate physical support for the die, the TAB tape carrier must be capable of maintaining critical die location during repeated movements through a variety of sprocketed transport production equipment. Additionally, the tape must supply a circuit pattern, usually etched copper foil, with lands to accommodate ILB and OLB for probing for the testing and burn-in process.

Several different TAB tape widths (Table 10.5) are available to satisfy the primary requirements for various TAB applications. Depending primarily on die size and tape width, a range of lead I/O counts can be supplied, which may vary from a very few (Fig. 10.17) to over 200 (Fig. 10.18).

TAB tape consists of a patterned conductor, either laminated to a dielectric layer, or free standing. The pattern fits the I/O patterns on the die and provides for the connection of

TABLE 10.5 Wire-Bond Versus TAB Component Packaging

Lead type	Wire-bond packaging*						TAB packaging†			
	40-Lead DIP		44-Lead PLCC		132-Lead PQFP		40-Lead		132-Lead	
	Long	Short	Long	Short	Long	Short	Long	Short	Long	Short
Lead length, in	0.99	0.13	0.20	0.15	0.38	0.26	0.12	0.06	0.31	0.23
Resistance, mΩ	125	123	98	98	102	101	3.6	2.2	11	8.2
Inductance, nH	22	3.9	4.6	3.3	10	7.2	2.1	0.8	6.7	5.1
Capacitance, pF	0.68	0.12	0.12	0.16	0.21	0.15	0.04	0.02	0.11	0.08

*DIP—dual-in-line package; PLCC—plastic leaded chip carrier; PQFP—plastic quad flat pack.
†National Semiconductor Tapepak.

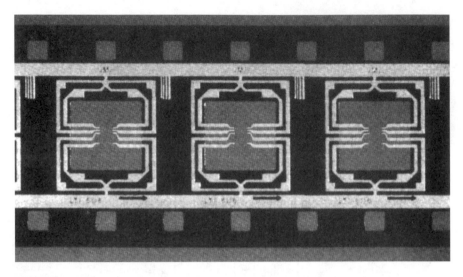

FIGURE 10.17   16-mm TAB tape with 8 leads. *(Courtesy of International Micro Industries.)*

FIGURE 10.18   70-mm TAB tape with 224 leads. *(Courtesy of International Micro Industries.)*

the die to the package. In general, the three most commonly used types of TAB tape carriers have either single-layer, two-layer, or three-layer construction. The single-layer construction consists solely of a metal foil; two-layer construction contains the metal foil and a base dielectric; and three-layer TAB tape consists of a metal foil, a base dielectric, and an intervening adhesive layer.

Each TAB tape construction offers certain advantages over the others, and each has its limitations. It remains for the packaging designer to choose the construction (and material) best adapted to a specific application in order to obtain the best results. An understanding of the choices available in base material, conductors, plating, adhesives, and construction will help the designer to choose judiciously.

10.6.2.1 Standards.   The Joint Electronic Device Council (JEDEC) has developed TAB tape carrier standards to help the manufacturer or user to purchase die on tape from a variety of suppliers who use identical interconnecting substrate land pattern and the same tooling for OLB on all of them, without the need for (or expense of) custom assembly tooling. The standards also facilitate the testing and burn-in of die on the tape. The following features are included.

OLB Windows.   A fixed number of OLB windows are defined to allow for a wide range of lead counts with a minimum number of lead forming and excising tools.

Lead Pitches.   Three standard lead pitches documented are on 0.25-mm (0.010-in), 0.375-mm (0.015-in), and 0.5-mm (0.020-in) centers.

Test Pads.   Standard test pad locations and sizes for each OLB window and lead pitch are covered in order to allow standard sockets and carriers to be tooled.

Superformat.   Standardization was made on a superformat of 1.4 by 1.4 mm (0.056 by 0.056 in) (Table 10.6) so that the maximum lead count for each type was allowed.

Alignment Holes.   Standard alignment holes to facilitate accurate positioning were defined.

Dimensioning and Tolerances.   Dimensions and tolerances were defined to accommodate existing two-and three-layer tapes and future formats.

10.6.2.2 TAB Tape Materials.   Tab tape materials and the metallurgy of the die bonding site are critical to the overall reliability of the TAB assembly. Therefore, developments

TABLE 10.6 Superformat Window Sizes and Lead Counts

Outer/inner window size	OLB center-to-center distance		
	0.010-in pitch	0.015-in pitch	0.020-in pitch
S35 mm			
0.410 in	124	84	60
0.610 in	204	132	100
0.810 in	284	188	140
0.810 × 1.620 in	444	296	220
S48 mm			
1.010 in	364	244	180
1.310 in	484	324	244
S70 mm			
1.510 in	564	372	284
1.710 in	644	428	324
1.910 in	724	484	364
2.110 in	804	532	404

have been undertaken to improve tape materials and bump metallurgies to afford maximum bond reliability and product manufacturability. Attempts have also been made to tightly control the material systems so that process repeatability is optimized.

Base Dielectric.   Polyimide is by far the most widely used two-layer and three-layer TAB base dielectric material. Characteristically, the dielectric should be mechanically stable, resistant to high temperatures, and have high resistance to moisture. Although polyimide is the material of choice, some applications use less expensive materials such as polyester and epoxy-glass dielectrics. However, polyimide is the material of choice since it has proven dimensional stability, is not flammable, has controllable shrinkage, withstands high temperatures, and has excellent electrical properties.

Polyimide film is typically used in a 0.05-mm (0.002-in) thickness for two-layer TAB constructions, where it must be etched, and in a 0.125-mm (0.005-in) thickness, where it must be etched, and in a 0.125-mm (0.005-in) thickness for three-layer constructions because of its mechanical strength and controlled-environment dimensional stability.

The potential drawbacks to using polyimide film, in addition to expense, include the possibility of pinholing and dimensional instability when the material is temperature and humidity stressed, Also, high-speed controlled-impedance applications may be better suited to the lower dielectric constant materials that have been developed for stripline and microstrip configurations, such as teflon.

Finishes.   The plating on the copper leads furnishes the desired bond metallurgy and corrosion protection to the copper. Typical finishes for TAB tape products include solder, gold, or immersion tin. A nickel barrier plate between the copper and the gold is sometimes included as a metallurgical barrier. In the case of solder and gold, the tapes must be electrically bussed so that electroplating may be used, and this in the continuous reel-to-reel manner.

Immersion, or electroless tin has also been a successful finish for TAB tape. In any case, the substrate materials used must be capable of withstanding the plating and chemical operations involved without exhibiting delamination or contamination of any type.

Adhesives.   In the three-layer tape, the adhesive is a critical element. It must bond the conductor to the base dielectric material without deterioration throughout all of the processing, including solvents and chemical baths involved in stripping photoresist, plating, and cleaning.

The major adhesive groups used in the manufacture of TAB tape are acrylic, epoxy-amide, phenolic butyral, and polyester, usually with proprietary formulations. The acrylics and epoxy-amides exhibit good insulation resistance, but have a high moisture absorption and relatively low temperature resistance. The phenolic butyrals are extremely low in moisture absorption, and because they are thermosetting in nature, they resist softening at bonding temperatures.

The polyesters are generally lower in desirable qualities than the others, but they serve well when used with polyester TAB tape carrier materials. The use of a polyimide adhesive offers very high-temperature capabilities.

10.6.2.3 Single-Layer Construction.   The single-layer ("all copper") TAB tape construction consists of a copper tape into which the interconnection pattern and support structures have been etched. A rolled or electrodeposited copper foil is normally used for this product in thicknesses of either 0.07 mm (0.0028 in), that is, 1 oz, or 0.05 mm (0.002 in).

In a continuous processing operation (Fig. 10.19) the copper foil is coated with photoresist (one or both sides), imaged, and etched by the processes normally used in the manufacture of printed wiring boards. However, the etching is a much more difficult task than is usually met because of the fine lines involved and because the TAB beam leads are unsupported. Typically, beam-lead widths of 0.0754 mm (0.003 in) are utilized, which also require careful handling and inspection.

Single-layer TAB tape is the least expensive TAB construction and can be used with standard die if the tape is bumped. Its main drawback is that all of the leads are electrically interconnected by supporting structure for plating, thus prohibiting device test or burn-in while the die is in the tape. However, its use is highly suitable for making fine-pitch TAB component carrier packages as the output of the semiconductor manufacturing process.

10.6.2.4 Two-Layer Construction. Two-layer TAB tape construction employs a base dielectric material to support the metal conductive pattern. The most common combination of materials for this type of construction is obtained either by casting a polyimide film upon copper foil, usually of the same thickness as used in single-layer tape, or by additively depositing copper directly onto the polyimide film.

The two-layer TAB tape is manufactured in a manner similar to that for single-layer construction, but with added steps for etching the base dielectric material (Fig. 10.20). In addition, careful indexing between the conductive pattern and the base material is required.

Since the features in the polyimide are formed by chemical etching, a support ring can be left inside the window containing the free-standing TAB beam leads. The use of this

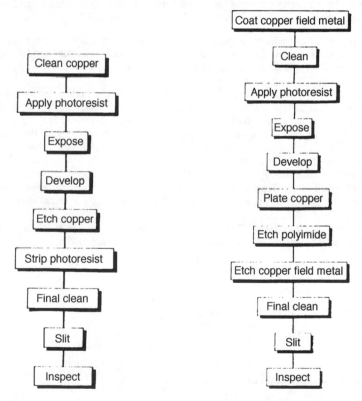

FIGURE 10.19 Single-layer TAB tape construction. *(Courtesy of 3M Electronic Products.)*

FIGURE 10.20 Two-layer TAB tape construction. *(Courtesy of 3M Electronic Products.)*

support ring is essential for high-lead-count devices. A further advantage of using the two-layer construction, as compared to using the three-layer type, is that no adhesive is used between conductive and dielectric materials. This means that if part of the dielectric base material structure, such as the supporting ring, is to be left in the end product, no organic adhesive material will be present.

Proponents of using the additive two-layer TAB tape construction process assert that it offers significant advantages over the subtractive copper-foil two-layer and three-layer types. First, the lead shape can be closely controlled, rather than having the undercut side-walls associated with subtractive copper-foil etching processes. Also, the additive process permits the fabrication of high-aspect-ratio leads, that is, higher than they are wide, which permits increased current carrying capabilities without compromising lead spacing.

10.6.2.5 Three-Layer Construction. Three-layer TAB tape construction is the original TAB technique that has been utilized since the 1970s. This construction (Fig. 10.21) features an adhesive-coated base dielectric material which is first mechanically punched and sprocket-holed. Copper foil is then laminated to the film-adhesive combination. Last, the TAB tape is imaged and etched.

Reported advantages of using the three-layer TAB tape construction include having a higher bond strength between the base material and the copper foil, good insulation resistance, and low moisture absorption as a consequence of using the adhesive layer. As a consequence of the expensive nature of the hard-tooled punching equipment, three-layer TAB tape is generally use only in the 35-mm (1.38-in) and 75-mm (2.95-in) widths.

10.6.2.6 Other Constructions. Multilayer TAB tape construction is an extension of TAB technology that allow for multiple-die attachment to a single substrate for increased packaging density. It consists of a base dielectric material with a conductive layer on both sides that are connected by means of vias.

Multilayer TAB tape is manufactured using a process that produces an "adhesiveless" metal-dielectric-metal construction. Since this process does not use an adhesive in the manner that three-layer tape does, vias interconnecting the conductive layers can be formed by chemical processing rather then by mechanical means. Because the conductor-via patterns are photoimaged onto the tape, the registration of the layer is superior to that achievable by mechanical processes.

Dice are mounted directly to the tape and coated with a polymer for protection. Then the tape is bonded to a stiffener in order to increase rigidity. In this respect, multilayer tape differs from the other constructions, since most TAB applications are used in conjunction with other kinds of packaging.

A patented "multiple-layer" TAB tape construction (Fig. 10.22) is also available for multiple die-attachment by having from two to four tapes assembled back-to-back onto a

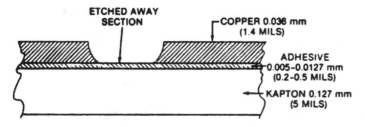

FIGURE 10.21   Typical three-layer TAB tape construction. *(Courtesy of International Micro Industries.)*

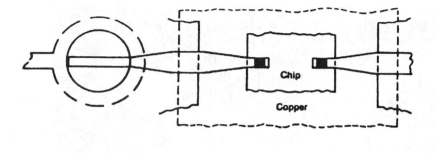

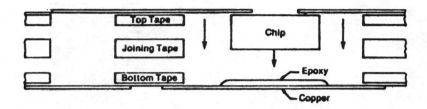

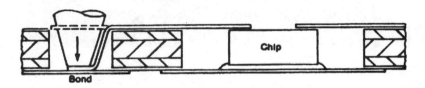

FIGURE 10.22 Multiple-level TAB tape construction. *(Courtesy of International Micro Industries.)*

middle layer of film. The top tape contains the bonding land patterns; the bottom tape(s) provide a conductive pattern to facilitate die interconnections. The center film's thickness is dimensioned to control die seating and thermal requirements.

Interconnections between conductive layers are accomplished by single-point bonding of cantilevered leads through punched holes. Therefore the entire process is done without the use of plated-through holes.

10.6.2.7 Area-Array Tapes.

As die lead counts continue to increase, it is inevitable that I/O pads within the perimeter of the die will be used to supplement those traditionally used on its periphery. To accommodate this trend, TAB technologists have devised a scheme called area-array TAB (Fig. 10.23).

A single- or two-layer TAB tape that interconnects to two peripheral rows of bonding pads fits this need. However, more generally, an area-array tape consists of multiple layers of metal sandwiched between layers of dielectric material.

FIGURE 10.23 Area-array structure.

10.6.3 Wafer Bumping

In addition to choosing the TAB tape construction, the TAB user must decide what kind of bumping is most appropriate for the application. Usually, small metal standoffs or bumps are added to the I/O pads on an integrated circuit. This is a relatively thick metal bump added either to the bonding pad sides of the individual die while they are still together on the IC fabrication wafer, or to the ILB sites on the land patterns of the TAB tape. This is to facilitate the ILB process—not only do the bumps provide a standoff to prevent shorting the lead to the edge of the die, but they also to provide corrosion protection to the die pad conductor. A typical bump configuration is shown in Fig. 10.24.

10.6.3.1 Wafer Bumping.   The bumping of die bonding sites on a semiconductor manufacturing wafer requires the fabrication of bonding pads that are raised above the planar surface of the wafer. Sometimes called BTAB, these bumps facilitate inner lead bonding without having the TAB tape touch the surface of the die.

The initial stage of bumped wafer preparation is almost identical to that used for wire-bonding die. A pinhole-free silicon nitride passivation layer, silicon dioxide, or polyimide

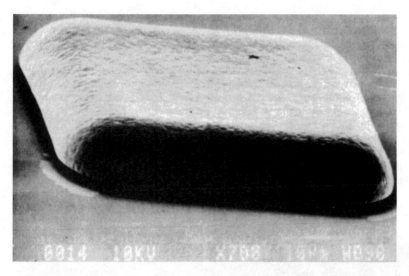

FIGURE 10.24　Mushroom-shape wafer bump. *(Courtesy of International Micro Industries.)*

in some cases, is deposited at low temperatures. The passivation is selectively removed to expose a good portion of the aluminum die bonding pads.

At this step in the preparation of the IC wafer, a barrier metal, such as titanium-tungsten, is deposited over the exposed aluminum and over the passivation on the periphery of the pad. These steps, plus the addition of 0.025 mm (0.001 in) of gold or copper bumps and subsequent electroplating, hermetically seal the die to help ensure the reliability of the ILB terminations.

The want of commercially available wafers in low quantities, reduced yield, and the additional cost of wafer bumping are often cited as disadvantages associated with the use of TAB technology. However, the wafer bumping technology is relatively straightforward, and good wafer bumping equipment is becoming readily available for barrier-metal deposition and bump plating for both prototype and production applications.

10.6.3.2 TAB Tape Bumping.　　Different techniques are available to create the desired ILB bump, sometimes called BTAB, on the TAB tape. The most common method is to subtractively etch a relatively thick copper foil on the tape so as to expose the bumps in the land pattern areas of the beam leads. However, this approach suffers from variations in the etching process and from misregistration owing to the fact that at least two photolithic stages are used in the BTAB manufacture.

An additional approach to achieving a testable BTAB tape system uses semiadditive fabrication techniques which overcome most of the shortcomings of the subtractive etching method. This approach begins with a relatively thin copper foil to which the bump is added by electrodeposition. By making the bump of a more compliant metallurgy, one of the potential major difficulties of the BTAB system, i.e., low bond reliability, is overcome.

Another radically different method first forms the bumps on a substrate by means of gold plating. Then the bumps are transferred from the substrate onto each corresponding land-pattern bonding site. The bump transfer is carried out by the application of pressure at elevated temperature.

10.6.4 Inner-Lead Bonding

The initial TAB assembly process begins with the ILB—where the patterned TAB tape frame is interconnected to associated I/O bond pads on the die. The plated TAB leads combine with pad or bump metallurgy to form the outer lead bond. As applicable, ILB is followed by burn-in, testing, and encapsulation or molding of the attached semiconductor die (Fig. 10.25). The type of bond produced depends primarily on the metallurgies at the tape-die interface and the ILB method used. For instance, thermocompression bonding may be used with copper tape and copper bumps and can be bonded by both gang and single point techniques. Common combinations are gold-to-gold, gold-to-tin, and copper-to-copper.

Each ILB combination offers different parameters with respect to cost, density capability, ease of processing, bonding force, and so on, which must be matched to each particular application. For example, solder bumps offer a low-cost low-bonding-force alternative, but they cannot be spaced as closely together as gold bumps due to the needed volume of solder to produce a reliable connection (Table 10.7).

FIGURE 10.25   Inner-lead bonding terminations. *(Courtesy of Siemens.)*

TABLE 10.7   Typical Minimum Inner-Lead Bonding Pitch

		Lead pitch	
Type of ILB metallurgy	Bonding method	in	mm
Gold-plated leads bonded to gold pads	Thermocompression	0.25	0.010
Tin-plated leads bonded to gold pads	Eutectic	0.30	0.012
Gold-plated leads bonded to tin pads	Eutectic	0.30	0.012
Tin-plated leads bonded to tin pads	Reflow	0.50	0.020
Tin- or gold-plated leads bonded to solder pads	Reflow	0.63	0.025

The termination process is based on the use of either thermocompression, eutectic, or reflow soldering principles, ILB may also be done with all of the leads terminated simultaneously by using mass (gang) bonders, or one lead at a time with single-point bonders.

10.6.4.1 Thermocompression Bonding. Thermocompression inner lead bonding (ILB) is a popular TAB assembly technique. It involves the application of pressure and heat to form the bond between the pad and bump metallurgy. A diamond or synthetic diamond-tipped steady-state thermode is the most widely used bonding tool for thermocompression mass-bonding technique (Fig. 10.26). The diamond has been chosen because of its heat uniformity. The system produces a very strong and reliable bond in approximately 200 to 250 ms.

The metallurgy generally consists of electroless gold on copper bumps joined to bare-copper pads. A second metallurgy that is rapidly increasing in use is the joining of gold-plated leads to gold bumps.

The advantage of thermocompression bonding is the ability to join similar metals. This results in excellent bonds due to the possibility that intermetallics are eliminated. The joining method is basically simple, with an established process and a wide process window. Furthermore, when gold is used, the total metallurgy system is inert and unaffected by the environment. (This is not true with all-copper metallurgies. Thus, they must be overmolded or protected due to copper's inherent oxidizing and corrosive properties in an open environment.)

The disadvantages include the relatively high bonding temperature and pressure required, the latter of which may damage the integrated circuit die if proper bump design is not employed. Another disadvantage is that the gold on the TAB leads requires electroplating. Thus, the lead fingers initially must be electrically interconnected (short-circuited), and a means must be provided for their separation prior to electrical testing.

10.6.4.2 Reflow Soldering. Solder-bumped die with electroless-tin-plated TAB tape have been used successfully for ILB. In these reflow soldering processes, the ILB has been

FIGURE 10.26 Mass inner-lead bonder. *(Courtesy of International Micro Industries.)*

made using a pulsed-tip thermode. Due to the molten solder interface during the bonding cycle, the reflow soldering tool and die planarity considerations are not major concerns. Unfortunately, the bonding time cycle for this method is approximately 4 to 6 seconds, considerably longer than that required for the thermocompression ILB technique.

It is recommended that the solder-bumped die have a pedestal beneath the solder. The pedestal keeps the TAB tape beam leads from coming into contact with the die surface in order to avoid cracking of the passivation during the bonding operation. An excellent metallurgy for the pedestal is a chrome-copper sputtered layer for adhesion and an electroplated copper to define the pedestal.

Solder used for the ILB reflow soldering operation may be any of the commonly used reflow bonding solders. However, consideration must be given to the OLB method used when selecting the appropriate solder. Most users choose an ILB solder that has a higher melting temperature than that required for the OLB operation. A solder flux is also typically used for all solder reflow operations and is usually applied to the die prior to reflow.

10.6.4.3 Eutectic Soldering.   Eutectic bonding requires a gold-tin eutectic or solder configuration of some type. The gold-tin eutectic ILB process is quite often done with a pulsed thermode rather than with a constantly heated thermode, as utilized with thermocompression bonding the gold-tin eutectic bond occurs at a lower temperature than does the gold-to-gold thermocompression bond. Although the bond itself is usually considered to have a less reliable metallurgy, it has the advantage of requiring less heating of the die and can be done with lower applied pressures.

10.6.4.4 Reflow and Thermocompression Bonding.   Another inner lead bonding method is the combination is the combination of reflow soldering and thermocompression bonding. It uses electroless-tin-plated TAB tape beam leads and gold-bumped die. The resulting bond is achieved most successfully using a heated steady-state thermode.

Pulsed-tip thermodes have been used, but the prolonged heating during a pulsed reflow soldering operation causes heating of the surfaces away from the bonding area. Also, heating of electroless tin in a non-reducing atmosphere causes tin oxides to be generated, which later cause problems with the OLB operation. However, there have been few problems associated with using the pulsed-tip thermode when the tin has been reflowed prior to bonding.

This combination bonding system is an excellent ILB technique when all of the parameters are controlled properly and the end product is not placed in a humid environment without proper protection from corrosion.

10.6.4.5 Mass and Single-Point Bonding.   Mass gang bonding has been the traditional method for implementing the ILB technique with any of the processes just described. This has been primarily because of the higher production rated achieved by the simultaneous bonding of all the TAB tape beam leads. Unfortunately, with mass gang bonding, the flatness of the surfaces to be joined and the bonding pressures are critical. Also, the die holder, thermode, and clamp are exclusive for each die. Thus, the advent of larger die with high I/O counts has created applications that are better suited to a single-point inner lead bonder.

Single-point bonding is a one-at-a-time process that uses a small wedgelike tool (Fig. 10.27). It can be used equally well for making both inner-lead and outer-lead terminations. The following features are unique to single-point bonders.

- The height of each lead is taken into account during the bonding process in order to eliminate the need for having a high degree of planarity between the bonding tool and the IC die.

FIGURE 10.27   Single-point inner-lead bonding. *(Courtesy of International Micro Industries.)*

- A consistent force is applied to each lead.
- Individual leads can be reworked.
- A variety of ILB and OLB configurations can be terminated readily by the same machine without significant tooling modifications.

10.6.5 Die Protection

An important development in TAB technology involves coating or molding the die for environmental protection. This protection can be provided on automatic reel-to-reel processing machines.

General practice for bare-die TAB assemblies is to apply a liquid coating to the die prior to curing the material. When a TAB component carrier package is the end product of the IC fabrication process (Fig. 10.28), molding is used to encapsulate the bonded die and to provide the protective handling ring around the exposed outer TAB leads.

Common die-protection materials used include epoxies, silicones, and polyimides. The criteria for selecting the materials are dictated by the specific application, especially with respect to solvent resistance, cure schedule, ease of application, level of ionic contaminants, and mechanical properties.

For low-volume production, the die protection can be done semiautomatically using a positive displacement dispenser that controls the volume of the material applied. Fully automatic coating machines are also available, which typically combine automated dispenser with integrated curing ovens.

10.6.6 Final Assembly

The final TAB assembly process begins with an excise-and-form operation that removes the die from the TAB tape. This is followed by outer lead bonding, which attaches the TAB leads to the interconnecting substrate. The OLB process can done successfully on a variety of substrates including glass epoxy, silicon, and ceramic.

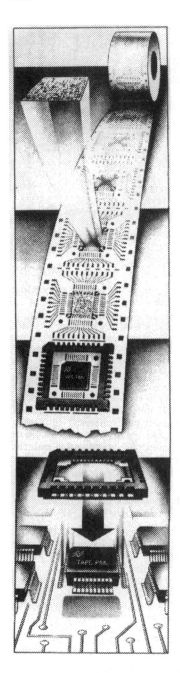

FIGURE 10.28 TAB component carrier fabrication. *(Courtesy of National Semiconductor Corp.)*

10.6.6.1 Excising and Forming. Excising and forming refers to excising, or punching out of the inner-lead bonded Integrated circuit and bending (forming) its outer leads. The special configuration, usually gull-wing, shape provides a compliant lead that can withstand thermal and mechanical stresses and still allow for a low overall assembled height.

10.6.6.2 Outer-Lead Bonding. The actual outer lead bonding process is accomplished using either mass or gang bonder, or a single point bonder (Fig. 10.29). Mass OLB processes use vapor phase and reflow solder operations common in conventional assembly. OLB equipment is often customized because of the wide variety of substrates involved, for example, hybrid circuit ceramics and printed wiring boards. However, for a given class of substrates, only the excising tool, lead-forming tool (when required), and work fixtures need to be customized.

10.6.7 Quality Assurance

Unfortunately the fine-pitch advantages of using TAB are somewhat offset by the basic inability to ensure the quality of the end product. Thus, without a suitable test, the long-term reliability of the electrical connection and IC die performance may not be assured. Fortunately, improvements have been made in the use of burn-in, functional and continuity testing, and laser-based inspection technology for TAB applications.

10.6.7.1 Burn-In and Testing. Burn-in and testing are done by environmentally cycling and electrically contacting test lands on the TAB tape. The test lands are produced by the same procedure that is used to define the functional lands on the tape.

Testing can also be performed with the TAB tape in a reel-to-reel mode if provisions are made to do so. Also, "singulated" die, which are either in individual die slide carriers or in TAB component carrier packages that hold one TAB tape component, can be used if they are compatible with either socket of common test handlers.

10.6.7.2 Scanning Laser Acoustic Microscopy. Reliable visual inspection and nondestructive tests on the bonded TAB beam leads are difficult, if not impossible, to perform due to the tight

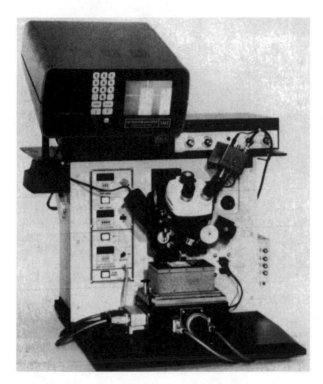

FIGURE 10.29   Single-point outer-lead bonder. *(Courtesy of International Micro Industries.)*

geometries involved. Fortunately, the use of scanning laser acoustic microscopy (SLAM) has the potential for determining the bond quality of both inner and outer-lead TAB bonds. Acoustic microscopy is the term applied to high-resolution, high-frequency (100-MHz) ultrasonic inspection techniques that produce images of features beneath the surface of the sample. Because ultrasonic energy requires continuity of materials to propagate, internal defects (such as voids, delaminations and cracks) interfere with the transmission or reflection of ultrasound signals.

SLAM has been used to image nondestructively gold-to-gold thermocompression TAB bond interfaces. The resulting data correlated with the pull-strength of values that were obtained subsequently through conventional destructive testing. Thus SLAM appears to be an attractive inspection method for TAB applications.

10.6.8 TAB Applications

The predominant packaging technology in the 1980s for electronic applications was conventional surface mounting. TAB is likely to have a similar impact on the packaging applications of the 1990s. The marketplace for TAB will consist of product applications utilizing one or more of its benefits to achieve low profile, low inductance, high speed, high lead count, high reliability, package compactness, increased density, and greater use of automated assembly processes.

Using these attributes, the markets and products benefiting from the use of TAB packaging will include products for communications (pocket telephones, modems), consumers (smart cards, watches calculators, video cameras, portable audio and compact discs), computer (lap-top portable, memory storage), military, medical (hearing aids, pacemakers, monitors), and hand-held industrial equipment.

10.6.8.1 Unpackaged Bare-Die Assemblies. TAB will find wide usage in the ever-increasing unpackaged bare-die applications. Depending upon the end-product requirements, these applications will incorporate TAB-based CAB, hybrid microcircuits, and MCM technologies. One typical example of how this is being done is in pocket calculators (Fig. 10.30). Low-profile pocket calculators contain elaborate displays with an increasing number of display elements having high lead counts. To meet these packaging requirements, several companies have switched from surface-mount packaged components to unpackaged TAB parts.

In one such application, a customized piece of equipment at the front end of the assembly line excises the part from the TAB tape, forms the leads, and presents the ILB-terminated die to a robot in a fixture that registers its position. Prior to pickup of the part, the robot searches each substrate for a fiducial mark using an optical sensor. Next, the robot carries the part to a dispenser for a drop of glue and then places the die on the interconnecting substrate. The use of solder paste and reflow soldering complete the TAB assembly operation.

10.6.8.2 TAB Component Carrier Packages. Using inert-molding technology, the previously used single-layer TAB construction has been crafted into a low-cost, easy-to-test IC TAB component carrier package (Fig. 10.28) that is being used by several semiconduc-

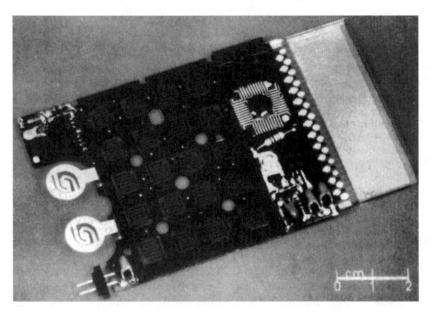

FIGURE 10.30 TAB device in a pocket calculator assembly. *(Courtesy of International Micro Industries.)*

tor manufacturers. As supplied by the IC manufacturer, the tested molded die is in its own carrier frame and is ready for excising, lead forming, and OLB by the interconnecting-substrate assembler.

The use of expensive polyimide dielectrics and gold plating has been avoided in order to bring the package cost down to where it can compete with conventional dual-in-line packages (DIPs). The TAB component carrier package is projected to provide fine-pitch packages with lead counts ranging from as low as 20 to over 360, with four standard package body sizes. The 40-lead package in a 7-mm (0.286-in) square body has been selected for automotive production products. The 132-lead package has a 20-mm (0.805-in) square body.

10.7 CHIP-SCALE PACKAGING

Chip-scale technology is categorized as semiconductor chip structures that have been rug-gedized to facilitate ease of chip handling, testing and chip assembly. The chip-scale tech-nologies have common attributes of minimum size, no more that 1.2 × the volume of the original die size, and being direct surface mountable as opposed to the use of wire bond-ing. Chip-scale packages are generally fully packaged chips in which the chips represent more than two-thirds of the package volume. Chip-scale packages (CSPs) are actually a variety of packaging and interconnecting schemes that allow the user to go directly from the bare silicon chips to second-level interconnects, such as printed wiring boards (PWBs) with packages virtually no larger that the die themselves. Packages are close to or the same size of the semiconductor die they encompass.

CSPs are poised between fully packaged die at one end of the C-packaging spectrum and flip-chips, multichip modules (MCMs), and chip-on-board at the other. CSPs provide the size and performance benefits of flip chips without their drawbacks. Mechanically robust, CSPs can be handled, tested, reworked, and easily integrated into surface-mount assembly processes. These pretested, pre-speed sorted, and prepackaged die require no specialized additional testing or other processing common to bare die. Chip-scale com-bines most of the benefits of bare dice without giving up the easy handling and testing afforded by packaged parts.

A typical chip-scale package (Fig. 10.31) consists of the silicon die with a small amount of packaging material supporting a connecting medium including flexible leads,

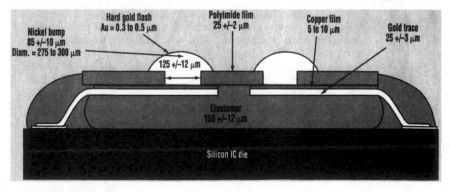

FIGURE 10.31   Construction of a compliant μ BGA.

a copper interconnecting structure, and bumped leads for connection to the next-level substrate. The compliant leads, which can be attached using standard techniques to standard printed wiring board substrates, decouple the thermal expansion of the die from that of the substrate without overly stressing the solder joints. Also, as with any SMT device, the typical chip-scale device such as the FBGA is reworkable using hot air to desolder it. The package provides the performance and form-factor advantages of flip-chip assembly. Because of the package's extremely short leads, the typical power/ ground inductance is on the order of 0.5 nH. In addition, a direct back side thermal path is very efficient.

A standard in chip-size packages does not exist. Chip-size packages come in many forms (Fig. 10.32). These include flex-circuit interposers, rigid substrate interposers, transfer molded packages, custom lead-frame packages, wafer-level assembled packages, and TCP lead-frame packages.

The factors in the increased interest in chip-scale include the following:

Size.   The benefits of using chip-scale stem in part from the absence of the IC package. Packages are about 1.2 times the size of the chip. The thin outline enables use in low-profile applications. Packages are available in pinouts ranging from 120 to 600 leads.

Reliability.   Reliability has improved with the development of encapsulation materials which more closely match the thermal expansion properties of conventional printed wiring boards.The result is a robust, solid package. Flexible internal leads allow for stress reduction between the die and the package connection. Chip protection is assured by package structure. The ability to test the die in the final package obviates the need for known good die and assures a reliable device.

Performance.   The electrical performance of a chip-scale package is enhanced by the short on-package trace length. The average trace length between the bonding leads and the bumps on the package is approximately 1.5 mm.This results in low stage delay, and low power/ground inductance, and greatly reduced on-package parasitics. A direct back side thermal path provides for good thermal performance.

CHIP-SCALE-PACKAGING SCORECARD TAKING SHAPE
CSP-based products expected to reach full production by 1998

Example	CSP type	Companies
	Flex circuit interposer	General Electric, NEC, Nitto Denko, Tessera
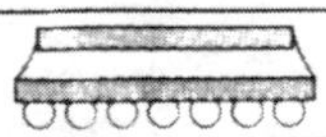	Rigid substrate interposer	IBM, Matsushita, Motorola, Toshiba
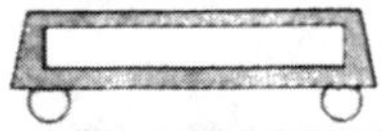	Transfer molded	Mitsubishi Electric
	Custom lead frame	Fujitsu, LG Semicon, Hitachi Cable
	Wafer-level assembly	ChipScale, ShellCase
	TCP lead frame	Rohm

SOURCE: TECHSEARCH INTERNATIONAL INC

FIGURE 10.32   Chip-scale packages.

Cost.   Lower overall system cost is another factor behind the selection of chip-scale technology. The device is cheaper and easier to test. The known good die issue is solved, since the devices can be tested in the package. Present indications are that CSPs will be price competitive with quad flat packs.

Manufacturability.   Chip-scale devices are easier to handle and easier to burn in and test and can be machine attached. Current automation setups can easily accommodate CSPs. Standard reflow soldering processes can be used for manufacturing; hot air-gas and vapor soldering methods are preferred.

Applications.   There is no doubt that the package's appeal to manufacture of portable equipment, disk drives, computers, telecommunications gear, and PC and smart cards, where small size is a continuing marketing edge. Early adoption is expected in consumer products such as PCMCIA cards, camcorders, mobile phones, and personal digital assistants. Applications in automotive electronics, portable computers, wireless equipment, entertainment boxes, and appliances can also be expected.

Chip-scale packages, such as the FBGA, offer solutions not available to the ordinary market. These new package types provide the user/assembler many of the advantages of bare dice but do not require the procuring and packaging of the known good dice procured from the IC manufacturer. In this way, chip-scale packages—the newer form of packaged die—are meant to be used as today's commercial packaged parts because they will be made on standard pad formats, statically and dynamically tested, and burned in prior to assembly. This development should promote the popularity of chip-scale technology—a new branch of electronic packaging.

REFERENCES

1. G.L. Ginsberg, "High Pin Counts Power Advanced Surface Mounting," *Electronic Packaging and Production*, pp. 36–39, Nov. 1988

2. R. Daniels, "Cost: The Complexity Factor, "*Printed Circuit Fabrication*, pp. 90–94, Sept. 1989

3. G. Hendrickson and L. Neitling, "Surface Mount Success: Redesigning the Process," *Printed Circuit Fabrication*, pp. 90–94, Sept. 1989.

4. P.P. Marcoux, "Surface Mount Turns to Fine Pitch," *Electronic Engineering Times*, p. 51, Mar. 6, 1989.

5. R.D. Rust, "Introduction to the General Problems in the Production of Fine Lines," Technical Paper IPC-TP-799, IPC *Technical Review,* July 1989.

6. S.O. Huntress, "Glass Phototooling: Sorting Out the Options, "*Printed Circuit Fabrication*, pp. 97–102 Sept. 1989.

7. S.L. Spitz, "Fine Lines: Tweaking the Process or a Quantum Leap," *Electronic Packaging and Production*, pp. 58–60, Jan. 1988.

8. L. Smith-Vargo, Process Windows Widen on Aqueous Resists," *Electronic Packaging and Production*, pp. 65–67, May 1987.

9. P.J. Oleske and D.B. Jones, "The Revival of Liquid Resist," *Printed Circuit Fabrication*, pp. 40–50, May 1988.

10. R.C. Linsdell, Dry Film Demands, "*Printed Circuit Fabrication*, pp. 24–36, Jun 1989.

11. D.Cook, "Printed Wiring Board Direct Imaging," *Printed Circuit Fabrication*, Jan. 1988.

12. S.C. Miller "Collimated Light from a User's Viewpoint," *Printed Circuit Fabrication*, pp. 81–85, June 1989.

13. S. Sugasawara, "Using Automated Optical Inspection Techniques for Inner/Outer Layer PCB Testing, "*Electronic Manufacturing*, pp. 8–11, Sept. 1988.

14. A. Angstenberger, "Optical Inspection of Printed Circuits: Requirements Placed on AOI in the Manufacturing Process," *Circuit World,* Vol 13, No. 3, pp. 8–10, 1987.

15. G.L.Ginsberg, "Automation is the Key to Small-Hole Processing," *Electronic Packaging and Production,* pp. 44–46, Aug. 1988.

16. J.J. D'Ambrisi, J.L. Cordani, P.E. Kukanskis, and R.A. Letize, "The Chemistry of Plating Small-Diameter Holes, Part 1," *Printed Circuit Fabrication,* pp. 78–80, 85–90, Apr. 1989.

17. H. Vandervelde, "The Game is Small Hole Drilling- The Rules are Strict," *Printed Circuit Fabrication,* pp. 18–22, 102–103, Mar. 1988.

18. D. Goulet, "Small Hole Drilling…The Future is Now," *Electronic Manufacturing,* pp. 22–24, Apr. 1988.

19. 19.R.E. Haslow, "Cleaning and Plating of Sub-9 Mil Holes," *Printed Circuit Fabrication,* pp. 78–90, Feb. 1989.

20. J. Wynshenk, "Update of PCB Technology for Surface Mount," *Printed Circuit Fabrication,* pp. 106–116, Jan. 1989.

21. S.I Amadi, "Plating High Aspect Ratio Multilayer Boards," Printed *Circuit Fabrication,* pp. 85–94, Oct. 1987.

22. K. Otoh, "Japan's PWBs: Small Features for a Big Business," *Electronic Packaging and Production,* pp. 70–72, Oct. 1987.

23. "The Dimensional Stability Debate," *Printed Circuit Fabrication,* pp. 50–62, July 1988.

24. L. Petroff, "Controlling Material Shift in Fine Line PWB Fabrication," *Electri-Optics,* pp. 11–12, Apr. 1987.

25. W.,G. McQuarrie, "Predicting the Manufacturing Capability for Controlled Impedance Multilayer Boards," *Proceedings, National Electronic Packaging and Production Conference West,* Mar. 1989.

26. K. Bieneman, "Drill Position Optimization for MLB Process Control," *Printed Circuit Fabrication,* pp. 41–50, Mar. 1988.

27. R. F. Benson and J.W. Wilent, "X-Rays Go on Line at Predrill Tooling," *Printed Circuit Fabrication,* pp. 72–80, Mar. 1987.

28. R.C. Marrs and G. Olachea, "BGAs for MCMs," *Advanced Packaging,* pp. 48–52, Sept/Oct 1994.

29. R.E. Sigliano, "Using BGA Packages," *Advanced Packaging,* pp. 36–45, Mar/Apr 1994.

30. E. Zamborsky, "BGAs in the Assembly Process," *Circuits Assembly,* pp. 60–64, Feb 1995.

31. B.Freyman, J. Briar, and R,.Marrs, "Surface Mount Process Technology for Ball Grid Array Packaging," *Journal of the SMT.* pp. 27–31, July 1994.

32. B. Grove, "Infrastructure Issues and Future Packaging Technologies," *Circuits Assembly,* pp. 34–38, Jun 1995.

33. J. Pokavian, "Ball Grid Array Technology, *Proceedings,* IPC, pp. S4–4, Oct. 1995.

34. V. Solberg, 'A Practical Guide to Ball Grid Array Assembly," *Proceedings,* IPC October 1995 pp. S4-3-1 to -10.

35. O. Vaz, "Outer Lead Bonding," *Advanced Packaging,* pp. 100–103 July/Aug 1994.

36. S. Tostado and P. Hoffman, "Solder Ball Attach Processing for Metal BGAs," *Proceedings,* 1995 IEPS Conference, pp. 559–564.

37. J. Fjelstad, "Designing Manufacturable COB Circuits," *Electronic Packaging and Production,* pp. 74–77, Feb. 1991.

38. G.L. Ginsberg, "Chip-on-Board Profits from TAB and Flip-Chip Technology," *Electronic Packaging and Production,* pp. 140–143, Sept. 1985.

39. A.J. Barbiarz, "Chip Size Packaging," *Advanced Packaging,* pp. 30–32, May/June 1995.

40. E. Jan Vardaman, "International Activity in Flip-chip on Board Technology," *NepCon West Proceedings* 1995, pp. 345–350.

41. D. Maliniak, "Chip-Scale Packages Bridge the Gap Between Bare Die and BGAs, *Electronic Design,* pp. 65–73, Aug. 21, 1995.

42. M. Willett, "Multichip Carrier Modules," *Advanced packaging,* pp. 12–14, Mar/Apr 1994.

43. R. Heitmann, "A Direct Attach Evolution," *Advanced Packaging,* pp. 95–99.Jul/Aug 1994.

44. "Guidelines for Chip on Board Technology Implementation," IPC-SM-784, *Institute of Interconnecting and Packaging Electronic Circuits,* 1989.

45. G. Ginsberg and D. Schnorr, "Multichip Modules and Related Technologies," McGraw-Hill, 1994.

46. A.J. Babiarz, "Dispensing Process in Chip Assembly for Direct Chip Attach," Electronic Packaging and Production, pp. 42–46, April 1995.

47. J.U. Knickerbocker and M.S. Cole, "Ceramic BGA," *Advanced Packaging,* pp. 20–25, Jan/Feb 1995.

48. J. Houghten, "Ball Grid Arrays: The Next Surface- Mount Package of Choice," *Electronic Products,* pp. 33–36, Jun 1995.

49. J. Ewanich, H. Tagaguchi, M. Kobayashi, N. Oie, "Development of a TAB (TCP) Ball Grid Array Package."*Proceedings,* 1995 IEPS Conference, pp. 558–594.

50. R.D. Schueller, "Design Considerations for a Reliable Low Cost Tape Ball Grid Array Package," *Proceedings, 1995 IEPS Conference,* pp. 595–605.

51. P. Mescher and G. Phellan, "A Practical Comparison of Surface Mount Assembly for Ball Grid Array Components," *Journal of SMT,* pp. 24–29, April 1995.

52. G.H. Michaud, "Fine-Pitch Wire Bond for COB," *SMT,* pp. 66–69, Oct. 1994.

53. A. Pecht, D. Barker, and A. Dasgupta, "An Approach to the Development of Package Design Guidelines," *Proceedings,* ISHM, 1991, pp. 217–240.

54. J. Partridge and P. Viswanadham, "Organic Carrier Requirements for Flip-chip Assemblies," *Journal of SMT,* pp. 15–19, July 1994.

55. G.L.Ginsberg, "Chip and Wire Technology, The Ultimate in Surface Mounting," *Electronic Packaging and Production,* pp. 78–83, Aug. 1985.

56. J.Morris, "An Introduction to Tape Automated Bonding (TAB), the Next Generation High Density Packaging," *Proceedings, International Electronic Packaging Conference,* Oct. 1987 pp. 360–366.

57. W. Schroen, "Chip Packages Enter the 21st Century, "*Machine Design,* pp. 137–143, Feb 11, 1988.

58. P.W. Rima, "TAB Technology: An Historic Overview of Lessons Learned m, Relating to Current Technology and Material Issues," in Proceedings, *National Electronic Packaging and Production Conference West,* (Feb. 1987), pp. 587–590.

59. K. Gallen, "Future Directions for Tape Automated Bonding," *Connection Technology,* pp. 16–20, Mar. 1988.

60. J. Walker, "Molded versus Direct Mount TAB: A Comparison," in Proceedings, *National Electronic Packaging and Production Conference West* (Feb. 1988), pp. 775–784.

61. H.W. Markstein, "TAB Rebounds as I/Os Increase," *Electronic Packaging and Production,* Aug. 1986.

62. P. Burggraaf, "TAB Rebounds as I/Os Increase," *Semiconductor International,* pp. 72–77, June 1988.

63. A. Oscilowski, "Tape Automated Bonding for VHSIC Parts," *Electronic Engineering Times,* July 14, 1986.

64. S.T.Holsinger, "Material Choices for Tape Automated Bonding (TAB)," in *Proceedings, National Electronic Packaging and Production Conference,* (Feb. 1988), pp. 325–331.

65. D.A. Behm and B. Parizek, "Beam Tape Carriers-A New Design Guide," in *Proceedings, International Symposium on Microelectronics,* (9 Nov. 1985).

66. J.S. Sallo, "An Overview of TAB Applications and Techniques," in *Proceedings, National Electronic Packaging and Production Conference West* (Feb. 1985), pp. 18–24)

67. M. Eleccion, "Tape-Automated Bonding Pushes in New Directions," *Electronics*, pp. 90–92, Sept. 3, 1987.

68. P. Hoffman, "TAB Implementation and Trends," *Solid State Technology*, pp. 85–88, June 1988.

69. I. Milosevic, A Perret, E. Losert, and P. Schlenkktich, "Polyimide Enables High Lead Count TAB." *Semiconductor International*, pp. 122–124, Oct. 1988.

70. B. Sharenow, 'Package Design Considerations Using Tape Automated Bonding Materials," in *Proceedings National Electronic Packaging and Production Conference*, (Feb. 1987), pp. 123–126.

71. G. Silvlenger, "Single-Point Tape Automated Bonding-A Versitile, Efficient Interconnection Technique," *Microelectronic Manufacturing and Test*, pp. 35–37, Feb. 1988.

72. A.C. Taub, "Non-Destructive Evaluation of TAB Bonds." Vanzetti Systems Co.

73. J. Walker, "It's Time for Better High-Density Packaging," *Electronic System Design*, pp. 49–53, Oct. 1987.

74. K. Skidore, "Package Trends for VLSI Devices, "*Semiconductor International*, pp. 60–65, June 1988.

75. N. Griffin "High Lead Count TAB in a Multi-Chip Environment," In *Proceedings, National Electronic Packaging and Production Conference West*, (Feb. 1987), pp. 569–575.

76. P.W. Rima, "TAB Gains Momentum," *Connection Technology*, pp. 28–32, Aug. 1987.

77. C.J. Bartlett, "Advanced Packaging for VLSI," *Solid State Technology*, pp. 119–123, June 1986.

78. G. Messner, "Cost-Density Analysis of Interconnections, "*Proceedings, Printed Circuit 'world Conference*, IV (Tokyo, Japan, June1989).

79. A.H. Mones, and R.K. Spielberger, "Interconnecting and Packaging VLSI Chips," *Solid State Technology*, pp. 119–122, Jan. 1984.

80. D. Walshak, H. Hashemi, "BGA Technology," *Advancing Microelectronics*, pp. 14–15, Jan/Feb 1995.

81. J. Liu, H. Berg, Y. Wen, S. Mullgaonker, and R. Bowlby, "PBGA: Pros and Cons," *Advancing Microelectronics*, pp. 16–17, Jan/Feb 1995

82. M.S. Cole, and T. Caulfield, "Ceramic Ball Grid Array Packaging," *Advancing Microelectronics*, pp. 18–19, Jan/Feb 1995.

83. W. Bernier, B.T. Ma, P. Mescher, A. Trivedi, and E. Vytlacil, "BGA vs. QFP," Circuits Assembly, pp. 38–40, Mar. 1995.

84. M.L. Barton, "Ball Grid Array Technology: Is it the Answer to Defect-Free Assembly," SMT, pp. 128–132, Aug. 1994.

85. B. Freyman and M. Petrucci, "High-Pincount PBGAs," *Advanced Packaging*, pp. 44–46, May/Jun 1995.

86. B. Levine, "Chip-Scale Packaging Gains at SMI," *Electronic News*, p. 62, Sept. 4, 1995.

87. G. Derman, "Chip-Size Packages Proliferate," *Electronic Engineering Times*, p. 18, Dec. 15, 1995.

88. A. Bindra, "Chip-Scale Packages Poised for Growth," *Electronic Engineering Times*, p. 24, June 19, 1995.

89. T. Costlow, "Chip-Scale Breaks from the Pack," *Electronic Engineering Times*, pp. 24–26, Jun. 19, 1995.

90. J.H. Lau, *Flip Chip Technologies*, McGraw-Hill, 1996.

SYSTEM PACKAGING TECHNOLOGIES

CHAPTER 11

PACKAGING OF HIGH-SPEED AND MICROWAVE ELECTRONIC SYSTEMS[*]

Stephen G. Konsowski

Associate
Technology Seminars, Inc.
Lutherville, Maryland

11.1 INTRODUCTION

Microwave systems have unique requirements for packaging technology as a result of the relatively short wavelength of the electromagnetic energy, the functions performed by the systems, and the type of devices used. The microwave band usually refers to the centimeter wavelength range of the spectrum (300 MHz to 30 GHz).

The major applications of microwaves are communication systems, including point-to-point line-of-sight microwave radio, satellite relay systems, point-to-multipoint radio for cellular mobile systems, avionics systems, navigation systems, landing systems, weather radar; defense systems, including ground-based and airborne radar, space-tracking radar, guidance systems, industrial systems, including drying, identification, speed measurement, and consumer equipment, including intrusion alarms, cooking equipment, and home satellite receivers.

A major difference between microwave and lower-frequency systems is that circuit components must be considered as distributed elements rather than as lumped elements due to the relatively short wavelengths. There are exceptions when very small integrated circuit (IC) components are used. The latter case is discussed in this chapter. The design of packaging and interconnects for microwave and high-speed digital systems requires extreme care to maintain impedance matching or low return loss to minimize detrimental effects on performance. Furthermore, since typical circuit configurations approach a significant fraction of a wavelength, or are several wavelengths in physical size, radiation of electromagnetic energy may occur under certain conditions, which leads to increased loss and crosstalk.

The major packaging issues that must be addressed in any microwave and high-speed digital system are coupling and interconnects.

---

*With credits to *Packaging of Microwave Electronic Systems*, by Harold Sobol, *Packaging of High Speed Digital Electronic Systems*, by Robert E. Canright, James C. Chamblin, William L. Pattison, and James A. Root, and *Electronic Packaging and Interconnection Handbook*, 1st ed., McGraw-Hill.

Coupling is the mechanism that enables microwave energy to be extracted from or added to a device or circuit. The coupling may be accomplished using a direct connection, or as the result of coupled fields with no physical connection. A direct connection is used for coupling to a device such as a transistor or for joining an input or output coaxial connector to the package. Field coupling is used when a probe connected to an input or output waveguide or coaxial connector is used to couple to the fields within a microwave resonator or cavity. The coupling must provide a low-loss connection between a package and a source or load, or between a device and its circuit, with the appropriate bandwidth and impedance level with minimum parasitic effects. The input and output ports for microwave band packages at power levels up to 5 W utilize miniature coaxial connectors. The ports for packages in the 60-GHz and higher frequencies utilize rectangular waveguide.

The interconnect is the medium that transports microwave energy within the package. It also functions as circuit elements used for resonators, tuners, dc blockers, and chokes. The circuit elements define the range and bandwidth of operation of the active devices, including microwave transistors and diodes. Theoretically any form of microwave waveguide, including coaxial line, rectangular or circular waveguide, or planar waveguide, may be used for interconnects. However, modern microwave circuits with solid-state devices generally utilize planar waveguides. These planar waveguides have the appropriate size and shape factors to ensure good coupling with excellent reliability and performance for moderate power levels. Planar waveguides have higher loss than rectangular waveguides and are therefore best suited for broadband circuit applications. An additional advantage of the planar media is low manufacturing cost through the use of photolithographic fabrication.

This chapter presents design information for modern microwave and high-speed digital packaging techniques. Microwave circuitry differs from dc and low frequency circuits in that the signals carried in both the conductors *and* in the dielectrics surrounding the conductors are influenced strongly by the electrical properties of the dielectric at high frequencies. For this reason the dielectric properties of packages which contain microwave circuits are quite important. A brief discussion of the development of materials for microwave circuits is presented here.

Early microwave printed wiring technology was derived from microwave power dividers for antennas and was then developed into flat coaxial configurations. In the 1940s, a configuration was developed that had a single conductor and a single ground plane with a solid dielectric. This was called *microstrip*. Later a central conductor, two solid dielectric, two ground plane system emerged which was called tri-plate. This configuration has emerged into what is known today as stripline, although the original name *stripline* referred to an air dielectric arrangement.

The dielectric materials which were used at that time were fiberglass followed by unreinforced lower dielectric constant and lower loss plastics that were simply matrix materials. Reinforcement in the form of woven glass fabric provided mechanical strength but, at the same time, detracted somewhat from the quality of the dielectric. Less lossy reinforcement (quartz and randomly oriented short glass fibers) were introduced to restore the performance while still providing mechanical strength. This increased the design flexibility considerably for a wide variety of applications (microwave components such as couplers, filters, dividers, and combiners) and power distribution manifolding in phased array antennas. The component manufacturing community continued to seek higher performance materials and began to experiment with microstrip dielectrics such as quartz, aluminum oxide, sapphire, beryllium oxide, and some titanates as well. These materials not being organic, printed circuit boards required a technique to deposit and adhere a conductor system to them. Thin-film technology was the most convenient since it was being developed and refined to produce integrated circuits and hybrid components and was well characterized. Sapphire because of its original cost and the extensive machining required to render a usable substrate finish for thin-film metallizations did not survive as a major substrate material. Quartz had similar machining costs and found limited though continued use. Aluminum oxide, being easily produced in reasonable purity levels, emerged as a major material of choice. Beryllium oxide, because of its high thermal conductivity and availability, was also used extensively.

As dielectric technology continued to be refined and a commercial market for microwave components developed in the communications field, considerable effort was spent on reducing the costs of both the materials and the processes to produce microwave circuit components. This required a different approach to metallizing the substrates. While an industry was under development for the production of analog and digital hybrid circuits, technologies emerged that began to be applied to microwave circuits. Thick-film metallizations that could be applied by stenciling techniques to alumina or beryllia substrate surfaces and subsequently baked/fired into the ceramic had potential use in the microwave arena, but the electrical conductivities of the metallizations were generally too low for most microwave applications. This issue was resolved with the introduction of high electrical conductivity thick film pastes by DuPont, EMCA, ESL, and others in the late 1960s. At this point, microwave circuitry moved into the realm of affordable high performance.

Solid state microwave receivers and transmitters of low power (less than 500 W) had hardware made up of many microwave components such as low noise receivers, circulators, amplifiers, mixers, and phase shifters connected to each other with miniature coaxial conductors, referred to as coax or "hardlines." This arrangement was bulky and costly to fabricate. A better integration of the components was required. Systems producers began to combine various subfunctions into the same package rather than interconnect individually packaged subfunctions with coax. This resulted in more compact electronics equipment that became lighter in weight and easier to install and maintain.

Subfunctions were generally constructed on ceramic substrates and placed together into large packages where they were interconnected together by ribbons of gold or by gold wire bonds. Solder and diffusion bonding were the methods of interconnecting the substrates together. Although this approach was economical there were subfunction designs that needed to be larger than the standard ceramic sizes in vogue. Larger ceramic plates were available, but handling and assembly issues dictated 2 in on a side to be an optimum size. A solution to the size limitation of ceramic substrates was offered by the 3M Company with their Epsilam 10, which was a double sided copper cladding on a proprietary dielectric composed of a high dielectric constant (K) filler dispersed within a lower K matrix with a combined effective K of approximately 10. The value of 10 was chosen to simulate the K of 99+ percent alumina, so the Epsilam 10 could be substituted directly for it. This clad material was available in sheets 10×10 in, making it possible to construct large integrated microwave assemblies because the copper cladding could be patterned into the desired conductor configurations through photoetching. The copper cladding on the bottom side served as the ground plane. This material did not have the handling and breakage characteristics associated with alumina and therefore offered more versatility. Irregular shapes and sizes could be achieved by the user through the use of printed circuit technology, namely routing. Etching and plating was possible since the dielectric was compatible with standard processes. Several other manufacturers provided materials with similar characteristics later.

The previous discussion related primarily to microstrip circuitry. Since the ceramic substrates were only double sided, that is, printed with a backplane on one side and a conductor pattern on the other they conformed to the definition of microstrip. Continued microwave circuit materials development made it possible in the mid 1980s for designs to include several layers of conductors and dielectrics in the same structure, much as a printed wiring board is configured. This development concerns ceramic-like materials known as low temperature cofired ceramics (LTCC). The cofiring of conductors and dielectrics at less than 1000°C allows the use of high electrical conductivity metals for top surface and buried layers as well. This means that microstrip configurations can be constructed and used in conjunction with lower frequency circuitry and digital circuits, all within the same substrate. Such flexibility offers the designer opportunities to combine, within the same substrate, circuit functions that should be located in close proximity to each other. It also allows more compactness and has the potential to improve circuit performance because of fewer interconnections. With a wide variety of dielectric properties, the LTCC materials are opening new vistas for their application. This includes components such as resistors and capacitors printed onto buried layers dielectric layers. Certain passive components can be included within the multilayer structure to increase overall component density because the materials that constitute resistors and low value capacitors have demonstrated compatibility with the cofired LTCC systems.

11.2 TYPES OF CIRCUIT MEDIA FOR MODERN MICROWAVE PACKAGING

A wide variety of circuit media exist and are used for microwave packaging in modern microwave ICs. The media include microstrip transmission line, coplanar waveguide slot line, inverted microstrip, dielectric waveguide, and lumped-element circuits. Examples and brief descriptions of the more popular circuit types follow.

11.2.1 Microstrip Transmission Line

The microstrip line shown in Fig. 11.1 consists of a strip conductor on one surface of a dielectric or high-resistivity semiconductor substrate with a conducting ground plane on the opposite surface of the substrate. This waveguide is the most popular form used for microwave packaging today. Figure 11.2 illustrates direct coupling of a packaged microwave device to a microstrip, Fig. 11.3 illustrates direct coupling of a microwave field effect transistor to a microstrip line on a semiconductor substrate,[1] and Fig. 11.4 shows the coupling of a microstrip waveguide to an external-port coaxial conductor.[2] Typical parameter sizes for a 50-Ω characteristic impedance line are shown in Fig. 11.1. Various distributed circuit forms, including resonators, tuning lines, couplers, and so on, can be easily

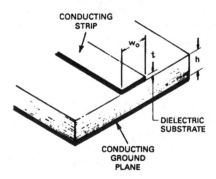

FIGURE 11.1 Microstrip transmission line. Typical values: $h = 0.025$ in; $w = 0.0245$ in; $t = 0.00025$ in; $\varepsilon_r = 9.6$ (Al$_2$O$_3$); $Z_0 = 50\ \Omega$.

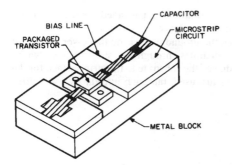

FIGURE 11.2 Packaged transistor in hybrid microstrip package.

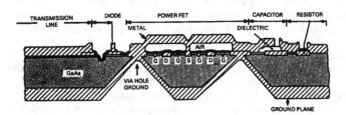

FIGURE 11.3 Coupling in monolithic microwave circuits. *(From Pucel.[1] © 1981 IEEE.)*

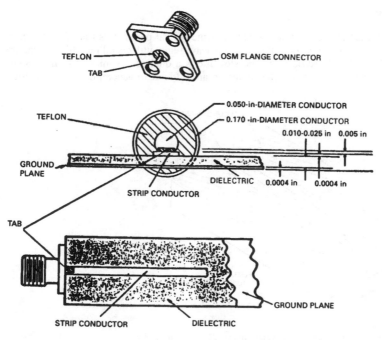

FIGURE 11.4 In-line coaxial-to-microstrip transition. *(From Caulton et al.[2])*

defined using microstrip as the interconnect medium. The packaged device mounting scheme shown in Fig. 11.2 provides low thermal impedance.

A disadvantage of the microstrip line is that the ground plane is on the lower surface of the substrate, thus requiring metallized vias in holes through the substrate, or "wrap-around" to extend the ground plane over an edge of the substrate to the strip surface for the mounting of shunt elements. Both approaches are used, but each introduces some difficulty in the processing of circuits.

11.2.2 Coplanar Waveguide

Figure 11.5 illustrates a coplanar waveguide[3] and shows typical parameters for a 50-Ω characteristic impedance. The waveguide consists of a strip conductor on a dielectric substrate with the ground plane located on the same surface as the strip and symmetrically about the strip.

The coplanar waveguide overcomes the problem of ground connections and shunt mounting that the microstrip line has, since both the strip and the ground plane are on one surface. However, the electrical performance of the coplanar guide is somewhat inferior to that of the microstrip. Figure 11.6 illustrates direct coupling of a device to the coplanar waveguide. Figure 11.7 shows coupling of the coplanar guide to a microstrip through field coupling, and Fig. 11.8 shows coupling of the coplanar guide directly to a coaxial connector.

11.2.3 Slot Line

Figure 11.9[4] illustrates a slot line and shows some typical parameters for a 50-Ω characteristic line. The slot line, as does the coplanar waveguide, utilizes all metallic conductors on one surface of a dielectric slab. The slot line is a dual of the microstrip line. Shunt mounting is fairly simple, but series elements are difficult to accomplish. The electrical

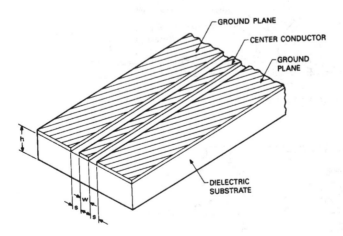

FIGURE 11.5 Coplanar waveguide. Typical values: $H = 0.025$ in; $w = s = 0.025$ in; $\varepsilon_r = 16$.

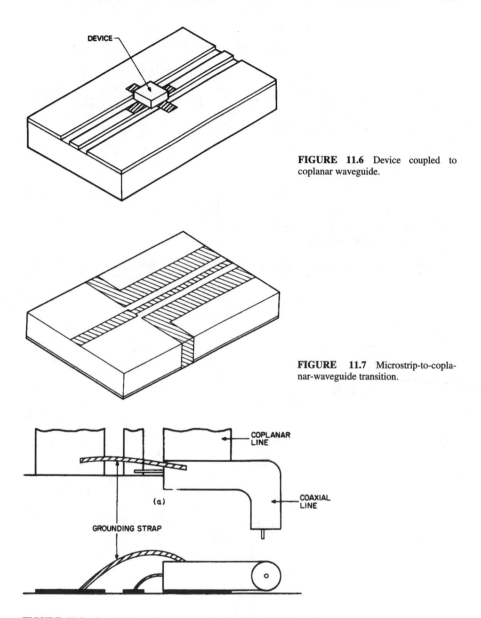

FIGURE 11.6 Device coupled to coplanar waveguide.

FIGURE 11.7 Microstrip-to-coplanar-waveguide transition.

FIGURE 11.8 Coaxial-to-coplanar-waveguide transition; (a) top view, (b) end view.

performance is not as good as for a microstrip. The slot line is used in conjunction with a coplanar waveguide to fabricate circuit elements such as resonators or couplers. Figure 11.10 shows the coupling of a coaxial conductor to the slot line, and Fig. 11.11 shows the combination of coplanar waveguide and slot line to implement a resonator.

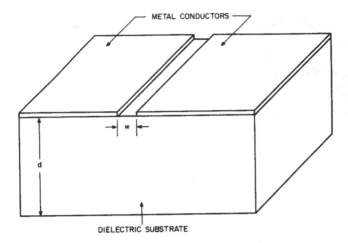

FIGURE 11.9   Slot-line waveguide. Typical values: $d = 0.125$ in; $w = 0.010$ in; $\varepsilon_r = 9.6$.

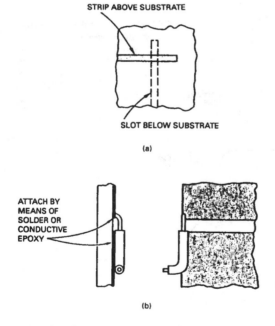

FIGURE 11.10   Coupling to slot line; (a) simple transition between slot line and microstrip, (b) broadband transition between slot line and miniature semirigid coaxial line. *(From Cohn.*[4] *© 1960, IEEE.)*

FIGURE 11.11   Resonant slots in slot line. *(From Cohn.[4] © 1960, IEEE.)*

11.2.4   Lumped-Element Circuits

Miniature inductors, capacitors, and resistors can be fabricated using photolithographic technology.[6] These components, which are on the order of 100 μm in size, are small enough compared to a wavelength at microwave frequencies to truly appear as lumped elements. Because of their small size, they are ideally suited for monolithic microwave ICs. However, the elements are used in certain applications in hybrid circuits as well. Figure 11.12 shows various forms of lumped circuits.[7] In monolithic circuits the lumped elements are usually produced simultaneously with the device metallization. Certain capacitors and feedthroughs in multiturn special inductors require multiple layers of metal and dielectric layers.

11.3   LIMITATIONS OF MICROWAVE INTEGRATED CIRCUITS

The IC technology has been applied to a very wide range of uses, yet there are some basic considerations that limit the scope of application. This section discusses the limits of microwave IC technologies. The constraining factors are loss, peak field strength, thermal characteristics, tunability and open structures.

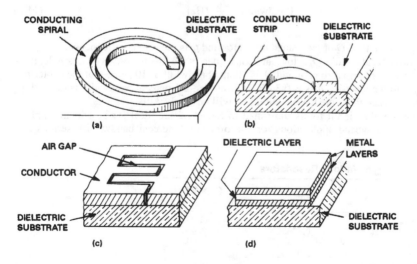

FIGURE 11.12   Lumped-element components; (*a*) spiral inductor, (*b*) strip inductor, (*c*) interdigital capacitor, and (*d*) metal-oxide-metal capacitor. *(From Caulton.[7])*

11.3.1 Loss

The loss in propagating structures or resonators results from metallic losses of the conductors and dielectric loss due to the imaginary component of the dielectric constant. In addition losses will result from radiation of open structures and excitation of higher-order modes. The primary losses in the structures discussed in Sec. 11.2 are metallic losses that result from the small size of the transmission structures and high current density.

The unloaded quality factor Q_0 of a resonator is equal to the ratio of the stored energy to the energy dissipated per cycle in the unloaded resonator. For the conductor loss dominated resonator, the Q_0 is proportional to the ratio of the volume of the resonator to the surface area of metallic conductors. This latter ratio is very small for the circuit types described in Sec. 11.2 and has a value of approximately one-half the substrate thickness (typically 0.010 to 0.025 in), as compared to a spherical metallic resonator, which has a ratio equal to one-third the resonator radius. In the microwave band this will be several centimeters. Table 11.1 lists typical values of resonator Q_0 for various structures as a function of surface resistance. Clearly the Q_0 of the microstrip is a factor of 20 to 50 less than that of the larger metallic resonator structures. The loss of slot line, coplanar waveguide, and lumped-element circuits is higher than that of microstrip, and consequently the Q_0 will be even lower. The dielectric resonator is an alternate structure that is small but has no metallic conductors and consequently no metallic loss. With typical low-dielectric-loss materials, the Q_0 of this resonator approaches that of the full metallic cavities. A disadvantage of the dielectric resonator is that it usually cannot be fabricated directly in a monolithic microwave IC.

The impact of the low Q_0 on performance depends on the loaded bandwidth of the resonator. The loss and bandwidth of a singly tuned resonator are approximately

$$\text{Dissipation loss} \approx 8.7 \frac{Q_t}{Q_0} \text{ (dB)} \tag{11.1}$$

$$\text{Bandwidth} \approx \frac{f_0}{Q_t} \text{ (Hz)} \tag{11.2}$$

where Q_t is the total Q of the loaded resonator and f_0 is the resonant frequency.

For broadband circuits Q_t is low, and consequently the loss is low. As an example, if Q_0 = 300 for a microstrip resonator and Q_t is 10, the bandwidth is 10 percent of the center frequency, and the insertion loss is 0.3 dB. If, however, a 1 percent bandwidth is desired, Q_t must be 100, and therefore the insertion loss will be nearly 3 dB.

Therefore, the circuit forms illustrated in Sec. 11.2 are ideal for broadband applications. For narrowband applications (on the order of 1 percent bandwidth), serious loss

TABLE 11.1 Q_0 **of Various Resonators**[*]

Resonator	Q_0	Q_0 at 9 GHz $(r_s = 2.5 \times 10^{-2})$
Cube metal	$0.74\eta/r_s$	11,200
Cylindrical metal	$0.47\eta/r_s$	6,000
Spherical metal	η/r_s	15,000
Microstrip	$0.02\eta/r_s$	300
Dielectric resonator	—	10,000

[*]r_s—surface resistance, $\Omega/\square$; $\eta = 377\ \Omega$.

problems arise. These can be overcome by the use of dielectric resonators in place of microstrip, or slot-line, and similar resonators. Obviously, the loss problems for the lumped-element devices, slot line, and coplanar waveguide are more severe than those for microstrip since the Q_0 values are lower.

11.3.2 Peak Field Strength

Peak field strengths are an additional limitation in microwave IC packaging. The dielectric breakdown of the substrate or the environment above the substrate is a matter of concern because of the relatively close spacing of conductors and the very thin substrates, on the order of 0.010 to 0.025 in.

Circuits using typical solid-state sources producing 5 to 10 W of microwave power will normally not encounter peak field problems; however, circuits such as T-R switches or limiters used in conjunction with high- or medium-power tubes may be subject to dielectric breakdown.

The dielectric breakdown field of a bulk alumina substrate is 250 kV/in which translates to a peak voltage limitation of 6.3 kV for a 0.025-in-thick substrate with an applied uniform field. Because of the distorted fields at the edge of a microstrip conductor, the effective peak voltage is reduced to 1 to 2 kV, which for a 50-Ω line is a peak power of 10 to 40 kW.

Using the effective value of maximum peak voltage, adjacent conductors should be separated by at least 0.025 in to prevent voltage breakdown in the air space above the conductors.

Operation near peak breakdown strengths should only be considered for narrow-pulse, low-duty-cycle operation since fairly high average power operation will lead to high temperature rises as a result of loss in the planar waveguide structures that could lead to severe damage.

11.3.3 Thermal Characteristics

The temperature rise of a microwave package must be controlled to meet the system maximum temperature requirements. A thermal analysis should always be performed to ensure that the devices operate reliably over a long period. Generally the packages are connected to the system thermal ground, and temperature rises are measured with respect to the temperature of the thermal ground.

The small size of the IC packages and the topology of typical circuits (components on an insulating or semiconductor substrate) can result in relatively high thermal impedance. As an example, the thermal impedance presented to a circular heat source of radius R_0 mounted on the top surface of a substrate that is thermally grounded on the opposite surface as shown in Fig. 11.13 is given approximately by

$$R_{th} \approx \frac{1}{KA} \frac{h}{1 + h/R_0} \quad (°C/W) \tag{11.3}$$

where $A = \pi R_0^2$.

Therefore a 0.025-in-diameter heat source on a 0.025-in-thick alumina substrate would have a thermal resistance of $R_{th} = 566°C/W$. This example illustrates the problem of mounting heat sources on the substrate. The solution is to mount the heat source directly on the thermal ground. Figure 11.2 shows a suitable arrangement for mounting a power

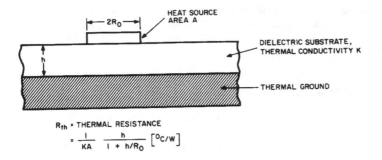

$$R_{th} = \text{THERMAL RESISTANCE}$$
$$= \frac{1}{KA} \cdot \frac{h}{1 + h/R_0} \left[{}^\circ C/W \right]$$

FIGURE 11.13 Simplified thermal model.

transistor on thermal ground. More detailed information on the thermal characteristics of a microstrip is given in Gupta et al.[8]

11.3.4 Tunability

Variations in device characteristics, parasitic reactances, placement of devices, and final packaging often result in the need for fine tuning of the final product in manufacturing runs. In discrete microwave circuits, this is accomplished with tuning screws or slugs. However, the problem is difficult in microwave ICs since their physical size is so small.

Monolithic circuits minimize some of the problems since parasitics within the chip are controlled and are exactly the same from run to run. Circuits requiring matched devices are subject to considerably less variation with monolithic technology since the devices are fabricated in close proximity on the same wafer.

The need for tuning can be somewhat minimized by careful design using simulation programs that contain accurate models of the processes involved in the circuit fabrication. The designer should then use the program to find a circuit solution that minimizes sensitivity to expected process variations (including length of bond wires, for example). Final tuning of hybrid circuits can be accomplished using pads, as illustrated in Fig. 11.14. The placement of the pads should be tested in simulation runs or during early manufacturing. The trimming is accomplished by wire bonding pads to the microstrip circuit to achieve the required performance.

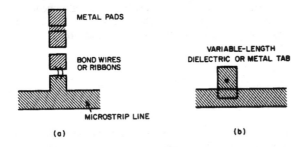

FIGURE 11.14 Methods for manual tuning of hybrid microwave circuits.

11.3.5 Open Structures

The waveguide media described in Sec. 11.2 are open structures, as compared to metallic rectangular waveguides or coaxial lines that are closed structures and contain the electromagnetic fields within the structures. The open structure fields may result in radiating modes, surface modes on the dielectric-air boundary, as well as bound fields that extend beyond metallic boundaries. As a result the media for microwave ICs are subject to crosstalk and coupling between functional units, which can produce oscillations, intermodulation products, and generally degrade circuit performance. To circumvent problems associated with these effects, functions or groups of functions are compartmentalized with metallic barriers to limit coupling. The compartments may be totally or partially closed. In many cases the compartmented circuits are partitioned to be stable units that can be manufacturing qualified in advance of being assembled into the final system. When partially enclosed compartments are used, microwave absorbing materials may be placed on the subsystem or the system package lid to minimize the effect of coupled fields.

11.4 TECHNOLOGY OF MICROWAVE INTEGRATED CIRCUITS

Various technologies are used in packaging microwave functions. Both hybrid circuits using passive substrates with deposited metallic conductors for circuits and interconnects and with attached devices and components, and monolithic circuits with semiconductor substrates that contain grown devices and components and metallized circuits and interconnects are utilized. In addition dielectric resonators, microwave ferrites and low-frequency ICs must be added to form complete circuit functions. This section describes the major technologies and the requirements for substrates and conductors used in microwave ICs.

11.4.1 Hybrid Circuits

Hybrid technology is a more mature technology than monolithic technology and now is the more popular approach. Hybrid circuits are generally fabricated by depositing interconnects and circuits on passive substrates and attaching active devices and certain components, such as capacitors, to the conductor pattern defined on the surface of the substrate. Die bonding, soldering, epoxy bonding, and wire or ribbon bonding are used in the attachment process.

A major advantage of the hybrid circuits is that a wide variety of device types made from different semiconductor materials can be used. The devices can be tested in advance of mounting to ensure good yield. The device may be used in packaged or unpackaged form. The hybrid circuit utilizes the semiconductor area only for active devices. In-circuit tuning is possible in hybrid circuits during production.

Disadvantages of hybrid circuits include high labor content in production assembly, variation in parasitics due to the manual operations in die mounting and wire bonding, increased parasitics due to the wire bonding, and somewhat decreased reliability due to bonding and mounting. Hybrid circuits tend to be larger than monolithic circuits, which may be a disadvantage in certain applications. The thermal management of hybrid circuits is more difficult than that of monolithic circuits because of the wide variety of materials utilized. Additive or subtractive approaches may be used in defining circuit patterns on substrates. The subtractive approach, or thin-film process, incorporates photolithographic definition of patterns in metal layers that were deposited by electron beam deposition, sputtering, or resistive heating evaporation from crucibles. The additive, or thick-film, process applies metal patterns by the printing of conductors through a screen and the subsequent firing of the inks.

The technological issues that must be considered are the selection of appropriate substrate and conductor materials, the surface finish of the substrates, the metal-deposition process to ensure good adherence to the substrates, microwave performance, and long-term stability of the metal system.

11.4.2 Monolithic Circuits

Monolithic microwave circuits are fabricated by growing and depositing active devices and passive circuit elements on a semiconductor wafer. The major technology differences between microwave monolithic circuits and conventional ICs are (1) the use of gallium arsenide as the primary semiconductor for microwaves, with silicon used for lower frequencies, and (2) the use of a relatively high percentage of the wafer area for passive circuitry such as microstrip lines or lumped elements, compared to a much lower fraction for passive circuitry on conventional circuits.

The advantages of monolithic circuits are that, ultimately, the production cost should be low because the assembly labor content is minimal, the control of parasitics is excellent, and device characteristics in a circuit are matched because they are fabricated simultaneously in close proximity on the same wafer. The monolithic circuits are small, which leads to simpler enclosure problems and less weight and size in system applications.

The disadvantages of monolithic circuits are that the current yield of complex circuits is quite low, virtually no trimming is possible, which leads to the requirement for extremely accurate computer models and process control, a significant area is devoted to passive circuitry, the devices are limited to those that can be grown in the starting wafer, crosstalk is more prevalent than in hybrid circuits, and the unloaded Q_0 is generally lower than in hybrid circuits.

Relatively simple circuits, such as wideband amplifiers, power amplifier stages, switches, and attenuators, are currently in production and are being produced at acceptable yield levels.

The primary processes used for fabricating monolithic circuits are ion implantation in epitaxial layers grown on semi-insulating gallium arsenide wafers. Multilevel metallization is used for devices and passive components. The metals used must be consistent with the semiconductor processing. Research work is under way on the use of other compound semiconductors, including indium phosphide and other materials and heterogeneous technology growth of silicon on gallium arsenide substrates and vice versa.

11.4.3 Substrates

The basic requirements for substrates used in microwave ICs are listed in Table 11.2. The properties of substrates that have been used for hybrid ICs[9] are listed in Table 11.3. The most popular material is alumina or aluminum oxide.

Permittivity. Two very important features of materials that affect electromagnetic waves are permittivity and permeability. Permittivity relates to electric fields and permeability relates to magnetic fields. Permittivity is a term which describes how a material under consideration allows itself to become polarized by an electric field imposed upon it from outside itself. Polarization refers to the realignment of charges within the dielectric along the lines of the imposed electric field. The polarizability of a given material is expressed in relation to the polarizability of a vacuum or air. (For practical purposes the values of per-

TABLE 11.2   General Requirements for Microwave IC Substrates

Low loss at microwave frequencies
Good adherence for conductors
Polished surfaces with 1–5 μin finishes
Stable properties
Little variation of properties at temperatures up to 150°C
No deformation over processing temperatures
Uniform dielectric constant from batch to batch
Low cost
Possible to fabricate various shapes and drill holes
Good thermal conductivity

mittivity and permeability of air are the same as for a vacuum or what is commonly referred to as "free space.") This relationship is given as

$$\varepsilon_r = \varepsilon_o \mathbf{k}$$

or stated another way, $\mathbf{k} = \varepsilon_r / \varepsilon_o$ where $\mathbf{k}$ is known as the dielectric constant of a material. The value of $\mathbf{k} = 1$ for air. The value of ε_o is 8.85×10^{-12} farads/meter (F/m). The subscript r indicates that the value of the polarizability of a material is related to the polarizability of air or a vacuum or free space.

In similar fashion, permeability is a measure of the magnetic property of a material that allows a magnetic field to be set up within itself by another magnetic field imposed upon it from outside. The relationship of the relative permeability of a material to the permeability of air is

$$\mu = \mu_r \mu_o$$

where μ_r is the relative permeability of a given material. The relative permeability of air is 1, so the value of μ for air is μ_o which is 4×10^{-7} henrys per meter (H/m). It should be stated here that the dielectric constants of insulating materials can vary a great deal but their relative permeabilities are about the same, namely 1. The relative permeabilities of most metals can be assumed to be approximately 1, except for iron, nickel, and cobalt, which can be much greater than 1. These three metals are said to be ferromagnetic because they tend to behave magnetically very much like iron.

Typical substrates are usually 0.010 to 0.025 in thick. The width of the strip conductor of microstrip is approximately inversely proportional to that of the substrate for a given impedance level and substrate thickness. The substrate thickness for the onset of higher-order modes of propagation is inversely proportional to ε_r. Taking these factors into account, a relative dielectric constant ε_r of about 10 and a substrate thickness of 0.025 in are suitable for circuits up to 10 GHz. Circuits in the 20- to 30-GHz range and above will typically use substrates with a lower relative dielectric constant, on the order of 4, and a substrate thickness of 0.010 to 0.025 in, which allows the use of practical-width conductors and results in acceptable loss levels. The analytical relationships between these parameters are discussed in Sec. 11.5.

Analysis of the effect of the surface roughness of the substrate on loss shows that rms. surface finishes of 1 to 5 μin are required to achieve conductors with a negligible increase in loss at microwave frequencies. Figure 11.15 shows the increase in effective surface resistivity due to substrate roughness, and Figure 11.16 shows the increase in loss, in dB/cm, as a function of rms. surface roughness with frequency as the parameter.[10] For Figs.

TABLE 11.3 Properties of Substrates Used in Hybrid Microwave ICs

Material	ϵ_r (approximate)	$\tan \delta \times 10^4$ at 10 GHz	Surface roughness, μm	Thermal conductivity K, W/(cm·°C)	Remarks and applications
RT-duroid 5880	2.16–2.24	5–15	0.75–1.0* 4.25–8.75†	0.0026	Cu-plating, flexible/stripline, microstrip
RT-duroid 6010	10.2–10.7	10–60	0.75–1.0* 4.25–8.75†	0.0041	Cu-plating, flexible/stripline at L band, microstrip
Epsilam-10	10–13	20	—	0.0037	Cu-plating, flexible/stripline at L band, microstrip
Alumina 99.5%	9.6–10.4	0.5–3	0.05–0.25	0.37	Cr-Au layer/microstrip, microstriplike lines, slot line, coplanar lines
Fused quartz 99.9%	3.75–3.8	1	0.006–0.025	0.01	Cr-Au layer, optical finish/microstriplike lines at millimeter wave frequencies
Beryllia (BeO)	6.6	1	0.05–1.25	2.5	High conductivity/compound substrate
Glass	5	20	0.025	0.01	Lossy/lumped element
Kapton	3–3.5	—	—	—	Flexible substrate/stripline
Cu-flon	2.1	4.5	—	—	Flexible substrate/stripline, microstrip
Rutile (TiO_2)	100	4	0.25–2.5	0.02	High dielectric constant/microstrip, slot line, coplanar lines
Ferrite/garnet	13–16	2	0.25	0.03	Porous/nonreciprocal devices in slot line, coplanar lines
Sapphire (single crystal)	$\epsilon_{r\perp} = 9.4$ $\epsilon_{r\parallel} = 11.6$	0.4–0.7	0.005–0.025	0.4	Well-defined and repeatable electrical properties, anisotropic/microstrip
Pyrolytic boron nitride	$\epsilon_{r\perp} = 3.4$ $\epsilon_{r\parallel} = 5.12$	—	—	—	Anisotropic/microstrip, suspended strip line

*Average peak-to-valley difference in height (rolled copper).
†Average peak-to-valley difference in height (electrodeposited copper).
Source: Bhat and Koul.[10]

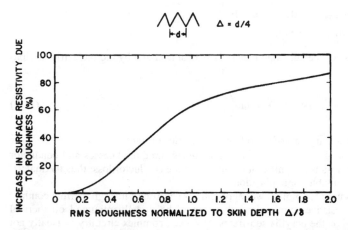

FIGURE 11.15 Percent increase in surface resistance as a function of surface roughness normalized to skin depth. *(From Sobol and Caulton.[10])*

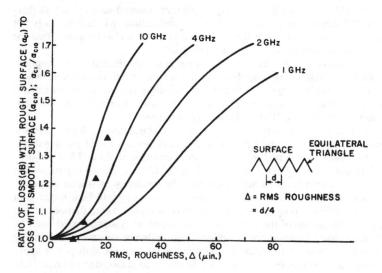

FIGURE 11.16 Loss as a function of substrate roughness with frequency as parameter. ▲—measured data, 9 GHz. *(From Sobol and Caulton.[10])*

11.15 and 11.16, the roughness is modeled as an equilateral-triangle serrated surface. The rms value of this roughness is one-fourth of the side of the equilateral triangle.

The skin depth of a conductor is given as

$$\delta = \sqrt{\frac{2}{\omega\mu_0\sigma}} \tag{11.4}$$

where $\omega = 2\pi$ frequency

μ_0 = free-space permeability = $4\pi \times 10^{-7}$

σ = dc conductivity, $(\Omega \cdot m)^{-1}$

The dielectric loss of the substrate materials is characterized by the dielectric dissipation factor D,

$$D = \tan\left(\frac{\sigma'}{\omega\varepsilon}\right) \qquad (11.5)$$

where σ'/ω is the imaginary part of the relative dielectric constant.

For values of D on the order of 10^{-4} and typical substrate thicknesses and conductor widths, the dielectric loss is insignificant compared to the conductor loss; thus the materials of Table 11.3 are suitable for application.

Beryllia (BeO) is used for high-power applications because of its high thermal conductivity. Devices and resistive terminations that cannot be directly mounted on thermal grounds are mounted on the beryllia substrates. However, complex circuitry is usually not placed on the beryllia substrate; instead it is placed on adjacent alumina, Duroid^{TM}, quartz, or sapphire substrates and interconnected by wirebonds or ribbons to the simple transmission line on the beryllia.

Aluminum nitride(AlN) is rapidly becoming a substrate material selected for its thermal conductivity (2.5 times greater than alumina) for applications where beryllia is not allowed because of its hazardous properties. Silicon carbide may also be substituted for beryllia if the thermal analysis determines its use acceptable.

Duroid is an alternative to alumina and is used for many distributed-circuit applications. Since its thermal conductivity is not high, its use is restricted to circuit areas where the dissipation is low. Circuits requiring active device mounting are usually not fabricated on Duroid. Low-microwave-frequency circuits are occasionally fabricated on the high-dielectric-constant materials ($\varepsilon_r > 10$) to achieve size reduction.

Monolithic structures are usually fabricated on semi-insulating gallium arsenide (GaAs) substrates since the majority of microwave devices are fabricated using gallium arsenide. These substrates have a real relative dielectric constant of about 13 at frequencies up to 100 GHz and resistivity in the range of 10^6 to 10^7 $\Omega \cdot$ cm. Silicon can be used to fabricate low-microwave-frequency devices. However, the resistivity of silicon is usually limited to the range of 10^2 to 10^3 $\Omega \cdot$ cm. The substrate resistivity will be a function of temperature and will normally drop as the temperature is increased due to thermal excitation of carriers. The properties of the particular semiconductor substrates used must be known and any variations accounted for in design. Semiconductor substrates are usually sliced to thicknesses of about 0.010 in, although during processing for high-power applications, the substrate may be thinned in the area of the device to allow adequate heat sinking of the device.

Figure 11.17 shows the relative loss of 50 Ω microstrip on three types of substrates. Clearly the lowest loss and consequently the highest Q_0 circuits can be fabricated on 0.025-in-thick ceramic substrates. Circuits on 0.010-in-thick gallium arsenide substrates have considerably higher Q_0 than circuits on comparable silicon substrates.

11.4.4 Conductors

The basic requirements for conductors for microwave ICs are given in Table 11.4. The material systems used for hybrid and monolithic circuits differ in that the conductors for

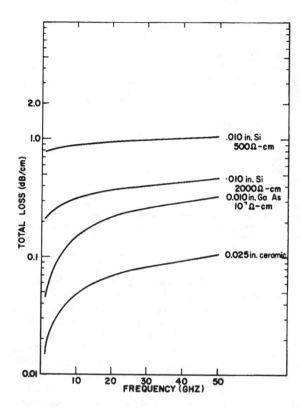

FIGURE 11.17   Loss, in dB/cm, as a function of frequency for 50-
Ω microstrip lines on silicon, gallium arsenide, and ceramic substrates. *(From Sobol and Caulton.*[10]*)*

TABLE 11.4   General Requirements for Microwave IC Conductors

Low loss at microwave frequencies
Good adherence to substrates
Corrosion-free over specified range of environmental conditions
Good bonding properties for die mounting and wire bonding
Compatibility with other materials used in fabrication

hybrids must meet only the requirements listed in Table 11.4, while the metal systems for monolithic circuits must also serve as resistive contacts to heavily doped semiconductors, Schottky contacts, and diode contacts. Table 11.5 lists the characteristics for metals used in hybrid thin-film ICs.[10] The metals fit into several classes, including conductors with low dc resistivity and poor adherence to substrates, higher-resistivity materials with good adherence to substrates, and moderate-resistivity materials that act as buffers between metallic systems.

The poor conducting materials achieve good adherence through the formation of chemical bonds with the substrate. Low-loss metallization is achieved by using a thin layer of

TABLE 11.5   Conductor Characteristics for Microwave ICs

Material		DC resistivity ρ (relative to Cu)	Skin depth δ at 2 GHz, μm	Thermal expansion α_T, /°C × 10^6	Adherence to dielectrics	Typical deposition technique*
I	Ag	0.95	1.4	21	Poor	RB
	Cu	1.0	1.5	18	Poor	RB, P
	Au	1.36	1.7	15	Poor	RB, P
	Al	1.6	1.9	26	Poor	RB
II	Cr	7.6	4.0	8.5	Good	RB
	Ta	9.1	4.5	6.6	Good	Sp, EB
	Ti	27.7	7.8	9.0	Good	RB, Sp
III	Mo	3.3	2.7	4.6	Fair	Sp, EB
	W	3.2	2.6	6.0	Fair	Sp, V
IV	Pt	6.2	3.6	9	—	Sp, EB
	Pd	6.2	3.6	11	—	Rb, Sp, Ps

*EB—electron-beam evaporation; P—electroplating; Ps—electroless plating; RB—resistance boat evaporation; Sp—sputtering; V—chemical vapor deposition.
Source: Sobol and Caulton.[10]

the poor conductor at the substrate surface followed by a thicker layer of the high-conductivity material. Some metal systems require an intervening layer of a buffer metal to prevent alloying effects over time and temperature that degrade the conductivity of the thick layer of good conductor. Table 11.6 lists some thin-film metal systems used for hybrid microwave ICs.[10] The chromium-based systems are not stable over temperature and time and should be avoided for high-reliability production units. The titanium-based systems, with buffer layers, particularly the titanium-palladium-gold system, are very stable, highly reliable metal systems. The titanium layer is usually 500 to 1000 Å, and the palladium layer is 1000 Å. The gold layer, or main conductivity strip, is 5 to 10 skin depths thick. Skin depth as a function of frequency is given by Eq. (11.4) and plotted for various materials in Fig. 11.18. The resistivity for very thin films may be considerably higher than the bulk value. Very thin film resistivity is highly dependent on substrate condition, temperature, and rate of growth. Typically the bulk resistivity may be reached at a thickness of 600 to 800 Å. The resistivity at 100-Å thickness is at least double the bulk value.

The thin resistive metal layers are usually produced by electron-beam or resistive heating evaporation or by sputtering. The thickness must be controlled carefully to ensure that the films do not exceed 1000 Å. The high-conductivity metal film is required to be 5 to 10 skin depths thick, or about 5 to 10 μm in thickness. This film is fabricated by deposition of an evaporated or sputtered film of the metal to a thickness of 1 μm. The remaining thickness is achieved by electroplating through a photoresist mask.

TABLE 11.6   Thin-Film Metal Systems for Microwave ICs

Chromium-based*	Titanium-based (or other)†
Cr-Cu	Ti-Au
Cr-Cu-Au	Ti-Pd-Au
Cr-Cu-[group (IV)]-Au	Ti-Pt-Au
Cr-Au	Mo-Au
Cr-Cu-Cr	Ti: W-Au

*Cr layer 100–300 Å; Cu or Au 2–10 μm.
†Ti, Pd, Pt layers 500–1000 Å.
Source: Sobol and Caulton.[10]

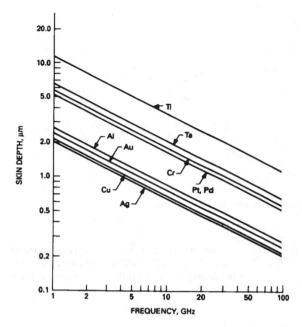

FIGURE 11.18   Skin depth as a function of frequency for conductors.

11.5   PLANAR WAVEGUIDES

Planar waveguides, the most popular circuit types used in both hybrid and monolithic packaging, were introduced in Sec. 11.2. Microwave propagation in these waveguides requires the presence of six field components: E_x, E_y, E_z, and H_x, H_y, H_z, to meet the boundary conditions. However, for some of the waveguides very useful results at low microwave frequencies may be obtained by assuming the presence of only transverse fields, or quasi-TEM propagation. Higher microwave frequencies must be treated using full-wave analyses of the structures.

The important parameters required for circuit design are characteristic impedance, guide wavelength, and loss. This section presents design information for several planar waveguides.

11.5.1   Microstrip Waveguides

The microstrip waveguide shown in Fig. 11.19 is a very simple physical structure, yet due to the dielectric discontinuity at the surface of the substrate that supports the strip conductor, the field analysis of this waveguide is extremely complex. The discontinuity requires the presence of both a longitudinal electric field and a magnetic field, and as a consequence the microstrip does not support true TEM propagation, as does the coaxial line and the symmetric strip line.

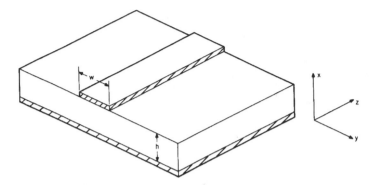

FIGURE 11.19   Microstrip waveguide. ε_r—relative dielectric constant of the substrate, t—conductor thickness.

For typical configurations of microstrip in microwave ICs, a TEM approximation to propagation can be used to accurately determine the characteristics for frequencies below 5 GHz. For frequencies above 10 GHz, it is necessary to use the full-wave solution considering the six field components to obtain accurate design parameters.

11.5.2 Low-Frequency Microstrip Approximations

The low-frequency or quasi-TEM approximation to propagation is determined using conformal mapping.[8,11–13] The mapping is carried out in two steps, the first step to determine the properties of the waveguide with the dielectric substrate replaced by free space, and the second step, which includes the effect of the dielectric substrate. In the second step, the use of a filling factor q that arises in the conformal mapping and an effective relative dielectric constant are very useful in the computations.

The low-frequency approximation to characteristic impedance of the microstrip waveguide is

$$Z_0 = \frac{1}{v_a\sqrt{C_a C}} = Z_{0_a}\sqrt{\frac{C_a}{C}} \tag{11.6}$$

where v_a = velocity of light in free space, 3×10^8 m/s

C_a = capacitance per unit length of microstrip with dielectric substrate replaced by free space

C = capacitance per unit length of microstrip with dielectric substrate present

Z_0 = characteristic impedance of microstrip waveguide

Z_{0_a} = characteristic impedance of microstrip with dielectric substrate replaced by free space

The effective relative dielectric constant $\varepsilon_{r_{eff}}$ and the filling factor q are related by

$$\varepsilon_{r_{eff}} = 1 + q(\varepsilon_r - 1) = \frac{C}{C_a} \tag{11.7}$$

The dielectric filling factor q is shown in Fig. 11.20.[13] The guide wavelength of the microstrip is given as

$$\frac{\lambda_g}{\lambda_{TEM}} = \sqrt{\frac{\varepsilon_r}{\varepsilon_{r_{eff}}}} = \sqrt{\frac{\varepsilon_r}{1 + q(\varepsilon_r - 1)}} \tag{11.8}$$

where

$$\lambda_{TEM} = \frac{\lambda_a}{\sqrt{\varepsilon_r}} \tag{11.9}$$

and λ_a is the free-space wavelength.

The low-frequency approximation to wavelength is plotted in Fig. 11.21 for several values of substrate dielectric constant.[2]

The characteristic impedance of the low-frequency microstrip is approximated for wide strips ($w/h > 1$) and narrow strips ($w/h < 1$). The impedance of the wide-strip case with the substrate replaced by free space is given as

$$Z_{0_a} = 189\left[\frac{1}{2}\frac{w}{h} + \frac{1}{\pi}\ln 2\pi\varepsilon\left(\frac{w}{2h} + 0.92\right)\right]^{-1} \qquad \frac{w}{h} > 1 \tag{11.10}$$

where ε is the base of the natural logarithm. For narrow strips,

$$Z_{0_a} = \frac{189}{\pi}\left[\ln\frac{8h}{w} + \frac{1}{8}\left(\frac{w}{2h}\right)^2\right] \qquad \frac{w}{h} < 1 \tag{11.11}$$

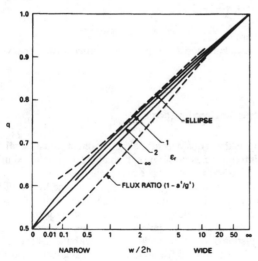

FIGURE 11.20 Microstrip effective filling fraction q. *(From Wheeler.[12] © 1965, IEEE.)*

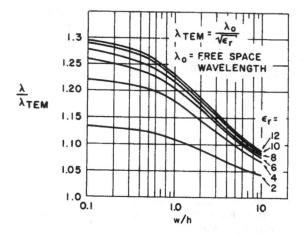

FIGURE 11.21 Microstrip low-frequency approximation. Normalized guide wavelength as a function of conductor width to substrate thickness. *(From Caulton et al.[2])*

The characteristic impedance of the waveguide with the dielectric substrate present is given as

$$Z_0 = \frac{Z_{0_a}}{\sqrt{1 + q(\varepsilon_r - 1)}} \qquad (11.12)$$

where Z_{0_a} is found from Eqs. (11.10) or (11.11).

Figure 11.22 shows the low-frequency characteristic impedance of microstrips for several values of dielectric constant.[2]

11.5.3   High-Frequency Microstrip Approximations

As mentioned earlier, a full-wave analysis that considers all field components must be utilized to treat the microstrip line at high frequencies. Several such analyses are published in the literature. Both shielded and open microstrip models have been analyzed.[8,13]

The study of the shielded microstrip line shows a logarithmic growth of the longitudinal electric field as a function of frequency, as illustrated in Fig. 11.23. The results of several analyses of the calculation of the guide wavelength as a function of frequency are shown in Fig. 11.24 for various values of the substrate dielectric constant. A comparison with the experimental results shown illustrates the excellent agreement with the Mittra and Itoh analysis. The guide wavelength thus determined from the full-wave analysis is close to the low-frequency (quasi-TEM) solution at frequencies up to about 5 GHz. At higher frequencies results from the wave analysis are required to achieve accurate predictions.

An example of the variation of the characteristic impedance with frequency is shown in Fig. 11.25.[17] Again, below 5 GHz the impedance is fairly constant at the level predicted by the quasi-TEM analysis, but above 5 GHz, impedance errors of several percent will result if low-frequency predictions are used.

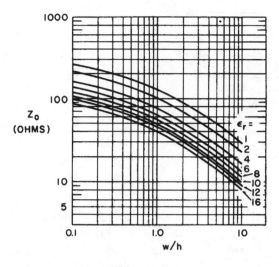

FIGURE 11.22 Microstrip low-frequency approximation. impedance as a function of conductor width to substrate thickness. *(From Caulton et al.[2])*

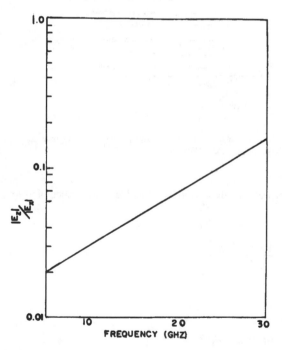

FIGURE 11.23 Microstrip full-wave analysis. Relative magnitudes of E_x and E_z components as a function of frequency. *(From Mitra and Itoh.[13])*

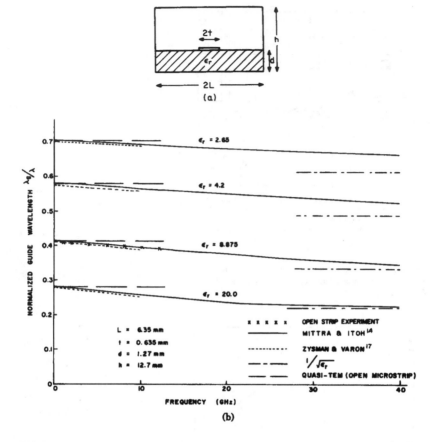

FIGURE 11.24 Shielded microstrip, full-wave analysis; (a) model and (b) variation of guide wavelength as a function of frequency for dominant mode. *(From Mitra and Itoh.[13])*

An excellent empirical form of wavelength derived by Schneider[17] as a function of frequency is given by

$$\frac{\lambda_g}{\lambda_0} = \frac{1}{(\varepsilon_r \varepsilon_{eff})^{1/2}} \frac{(\varepsilon_{eff})^{1/2} f_n^2 + \varepsilon_r^{1/2}}{f_n^2 + 1} \tag{11.13}$$

where $f_n = 4d(\varepsilon_r - 1)^{1/2}/\lambda_0$

λ_0 = free-space wavelength

$\varepsilon_{eff} = [(\varepsilon_r + 1)/2] + [(\varepsilon_r - 1)/2][1/1 + 5d/t)^{1/2}]$

d = substrate thickness

t = one-half strip width

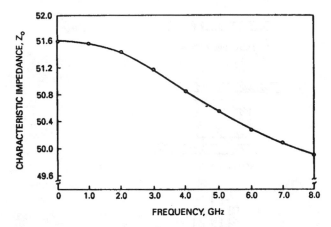

FIGURE 11.25 Microstrip characteristic impedance as a function of frequency. $\varepsilon_r = 15.87$; $w/h = 0.543$; $h = 0.1016$ cm. *(From Denlinger.[16] © 1981, IEEE.)*

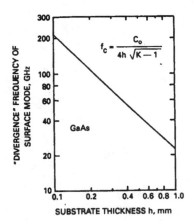

FIGURE 11.26 Frequency of onset of lowest-order TE surface wave on gallium arsenide substrate as a function of substrate thickness. *(From Pucel.[1] © 1981, IEEE.)*

The onset of the lowest-order surface mode on a gallium arsenide monolithic IC substrate is shown in Fig. 11.26.[1] Various forms of circuit configurations used in monolithic and hybrid ICs are shown in Fig. 11.27. An excellent discussion of the effects of discontinuities on microstrip propagation is given in Gupta et al.[8]

11.5.4 Coplanar Waveguides

A schematic drawing of the coplanar waveguide structure is shown in Fig. 11.28. Conformal mapping has been used to analyze the propagation of electromagnetic energy in this waveguide.[3] Figure 11.29 shows the characteristic impedance as a function of strip conductor width to slot width for a wide range of dielectric constants.

Useful approximate analytical expressions for the coplanar waveguide are as follows.

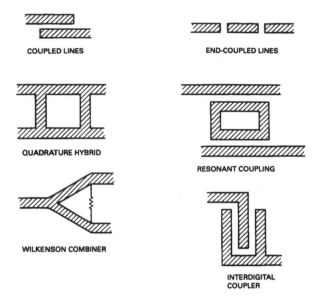

FIGURE 11.27   Microstrip circuit elements.

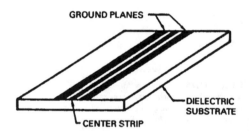

FIGURE 11.28   Coplanar waveguide, a surface strip transmission line.

1. Capacitance per unit length,

$$C = (\varepsilon_r + 1)\varepsilon_0 \frac{2a}{b} \tag{11.14}$$

2. Effective relative dielectric constant,

$$\varepsilon_{r_{eff}} \cong \frac{\varepsilon_r + 1}{2} \tag{11.15}$$

3. Phase velocity,

$$v_{ph} = \left(\frac{2}{\varepsilon_r + 1}\right)^{1/2} v_a \tag{11.16}$$

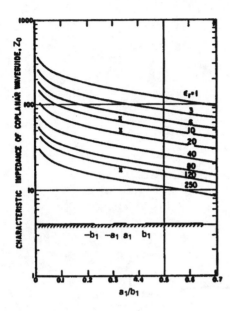

FIGURE 11.29 Characteristic imped-ance Z_0 of coplanar waveguide as a function of a_1/b_1 with relative dielectric constant ε_r as parameter. x—measured values. *(From Wen.[3] © 1969, IEEE.)*

where $v_a = 3 \times 10^8$ m/s.

4. Characteristic impedance,

$$Z_0 = \frac{1}{Cv_{ph}}$$

(11.17)

11.5.5 Slot Line

The slot-line waveguide is illustrated in Fig. 11.30. The characteristic impedance and the guide wavelength as determined in Mariani et al.[18] are shown in Fig. 11.31 for a dielectric constant of 9.6 (alumina). Comparisons of the behavior of slot line with microstrip charac-teristic impedance and the effective dielectric constant are presented in Figs. 11.32 and 11.33, respectively. A general comparison of the characteristics of the various waveguides considered is given in Table 11.7.

11.6 LUMPED-ELEMENT CIRCUITS FOR MICROWAVE PACKAGES

Microwave ICs can utilize extremely small passive components, produced photolitho-graphically, which exhibit lumped rather than distributed behavior. The requirement that a circuit element behave in a lumped manner is that the size be much smaller than a wave-length and that it have negligible phase shift across any dimension. These circuit elements can be used with frequencies as high as 12 GHz. They are the lowest-Q devices, the most difficult to process, but most useful in monolithic circuits where size is always a consider-ation. Hybrid circuits also utilize lumped circuits extensively.

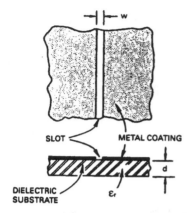

FIGURE 11.30 Slot line on dielectric substrate.

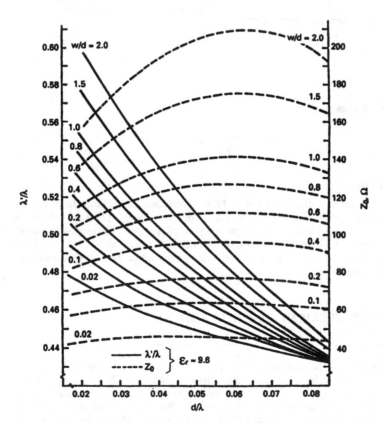

FIGURE 11.31 Slot-line characteristics for relative dielectric constant $\varepsilon_r = 9.6$. (*From Mariani et al.*[18] *© 1969, IEEE.*)

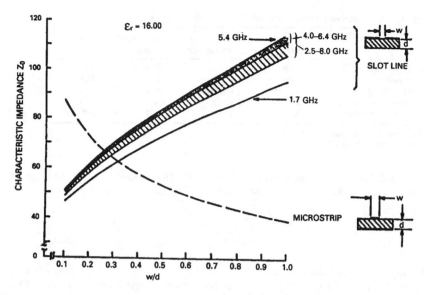

FIGURE 11.32 Comparison of slot-line and microstrip characteristic impedance. *(From Mariani et al.[18] © 1969, IEEE.)*

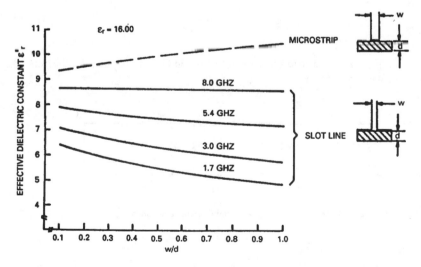

FIGURE 11.33 Comparison of slot-line and microstrip effective dielectric constant. *(From Mariani et al.[18] © 1969, IEEE.)*

TABLE 11.7 Qualitative Comparison of Propagation Modes

	Microstrip	Coplanar waveguide	Coplanar strips	Slot line
Attenuation loss	Low	Medium	Medium	High
Dispersion	Low	Medium	Medium	High
Impedance range, Ω	10–100	25–125*	40–250*	High
Connect shunt elements	Difficult	Easy	Easy	Easy
Connect series elements	Easy	Easy	Easy	Difficult

*Infinitely thick substrate.
Source: Pucel.[1] © 1981 IEEE.

The model for a general line is shown in Fig. 11.34. The elements L, r, G, and C are all per unit length. Z_l is the terminating load on the line. The characteristic impedance is:

$$Z_0 = \sqrt{\frac{r + j\omega L}{G + j\omega C}} \tag{11.18}$$

A lumped-element inductor can be represented by using a short-circuit load ($Z_l = 0$) and taking the limit of the driving-point impedance as the length l approaches zero. The resultant equivalent circuit is shown in Fig. 11.35, and the expressions for driving-point impedance and Q are given by the equations

$$Z_{in} \approx rl + j\omega Ll \tag{11.19}$$

Similarly, if the load is an open circuit, the driving-point impedance for very-short-length lines approaches

$$Z_{in} \approx \frac{1}{3}rl + \frac{Gl}{(\omega Cl)^2} - \frac{j}{\omega Cl} \tag{11.20}$$

which is a capacitor with the equivalent circuit shown in Fig. 11.36. The Q for the lumped-element capacitor is given by

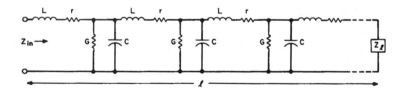

FIGURE 11.34   General transmission line. L, r, G, and C are per unit length.

FIGURE 11.35   Equivalent circuit for lumped-element inductor; r and L are per unit length.

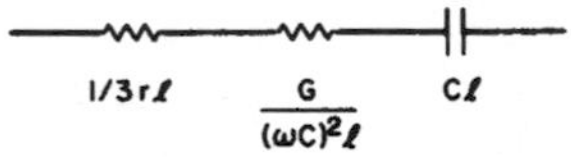

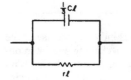

$1/3r\ell$ $\dfrac{G}{(\omega C)^2\ell}$ $C\ell$

FIGURE 11.36 Equivalent circuit for lumped-element capacitor; r, G, and C are per unit length.

$$\frac{1}{Q} = \frac{1}{Q_c} + \frac{1}{Q_d}$$

$$Q_c = \frac{3}{\omega C r l^2}$$

$$Q_d = \frac{\omega C}{G}$$

$$(11.21)$$

A lumped-element resistor may be represented by assuming $r \gg \omega L$, $G \to 0$, $Z_l = 0$, and taking the limit as $l \to 0$. The equivalent circuit is shown in Fig. 11.37, and the driving-point impedance is given by

$$Z_{in} \approx \frac{rl}{1 + jr\omega C l^2/3} \qquad (11.22)$$

Inductors for microwave circuits can be made as ribbons, spirals, or lengths of wire. The spiral inductor is shown in Fig. 11.38.

Several candidate forms for capacitors are shown in Fig. 11.39. Broadside-coupled, end-coupled, and interdigitated capacitors are simpler to fabricate since they do not require dielectric films. Capacitors that require dielectric films are subject to short circuits and must be fabricated very carefully to avoid this problem. Some that may be considered

$\frac{1}{3}C\ell$

FIGURE 11.37 Equivalent circuit for limped-element resistor; r and C are per unit length.

$r\ell$

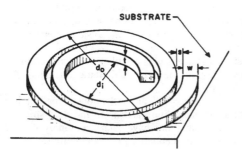

SUBSTRATE

FIGURE 11.38 Spiral inductor.

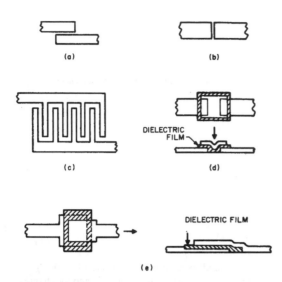

FIGURE 11.39 Planar capacitor designs; (a) broadside-coupled, (b) end-coupled, (c) interdigitated, (d) end-coupled overlay, and (e) overlay. (*From Pucel.*[1] © *1981, IEEE.*)

dielectric films are listed in Table 11.8.[1] The process system selected must be compatible with the total processing scheme used.

Resistors can be fabricated in various shapes, depending on the circuit configuration. There are a variety of materials that can be applied; some are listed in Table 11.9. Once again, the choice is dependent on the overall material system selected.

11.7 HIGH-SPEED DIGITAL PACKAGING

As the requirements for higher signal speeds and higher data rates increase, the use of the associated high-speed logic devices demands that the high-speed packaging and interconnection design be considered on a system basis, not merely on a logic-device basis. With device rise times decreasing into the subnanosecond region and operating frequencies increasing beyond the 500-MHz region, transmission-line characteristics must be considered in all designs involving printed wiring boards (PWB) and other interconnect media. Some critical factors include electrical noise, crosstalk, reflections, interconnection and line impedance, transient current, and others. Insulating material and system interconnection design factors must be properly selected.

The principal starting point for such electrical analysis is established at the definition of an electrical noise budget that is compatible with the individual characteristics of the circuit device elements being used in the design.

The two key contributors to electrical noise are crosstalk and reflections. Crosstalk has been examined for both vertical and lateral adjacent line patterns in the dual-stripline PWB lay-up configuration. These two crosstalk situations have each been analyzed and modeled to determine their impact using newly derived equations for calculations.

TABLE 11.8   Properties of Applicable Dielectric Films

Dielectric	K	Temperature coefficient of capacitance, ppm/°C	C/A,* pF/mm^2	$(C/A) \cdot Q_d$	$(C/A) \cdot V_b$	Comments
SiO	4.5–6.8	100–500	300	Low	Medium	Evaporated
SiO$_2$	4–5	50	200	Medium	Medium	Evaporated, CVD, or sputtered
Si$_3$N$_4$	6–7	25–35	300	High	High	Sputtered or CVD
Ta$_2$O$_5$	20–25	0–200	1100	Medium	High	Sputtered and anodized
Al$_2$O$_3$	6–9	300–500	400	High	High	CVD, anodic oxidation, sputtering
Schottky-barrier junction	12.9	—	550	Very low	High	Evaporated metal on GaAs
Polyimide	3–4.5	–500	35	High	—	Spun and cured organic film

*Film thickness assumed 2000 Å, except for polyimide, 10,000 Å.
Source: Pucel.[1] © 1981 IEEE.

TABLE 11.9   Properties of Some Resistive Films

Material	Resistivity, $\mu\Omega \cdot$ cm	Temperature coefficient of resistance, ppm/°C ·	Method of deposition	Stability	Comments
Cr	13 (bulk)	+3000 (bulk)	Evaporated sputtered	Good to excellent	Excellent adherence to GaAs
Ti	55–135	+2500	Evaporated sputtered	Good to excellent	Excellent adherence to GaAs
Ta	180–220	–100–+500	Sputtered	Excellent	Can be anodized
NiCr	60–600	200	Evaporated (300°C) sputtered	Good to excellent	Stabilized by slow anneal at 300°C
TaN	280	–180––300	Reactively sputtered	Good	Cannot be anodized
Ta$_2$N	300	–50––110	Reactively sputtered	Excellent	Can be anodized
Bulk GaAs	3–100Ω/□	+3000	Epitaxy or implantation	Excellent	Nonlinear at high current densities

Source: Pucel.[1] © 1981 IEEE.

The second critical contributor to electrical noise is signal reflections. Such reflections occur when the characteristic impedance of a transmission line is interrupted or changed significantly.

The remainder of this chapter establishes a baseline starting point to specify the physical geometry limitations for line length, line width, line spacing, and multilayer lay-up configurations required to meet a specific calculated noise budget. Once the layout is under way, these limitations may lead to conflicting requirements in terms of reducing the number of signal layers to achieve the design. These guidelines also provide for examining and resolving such conflicts or for adjusting the requirements. The trade-offs of their impact on electrical performance are described in depth.

Other electrical characteristics, such as transmission-line impedance, rise and fall times, and propagation delay, were modeled and are included as a part of these guidelines.

11.7.1 THE DESIGN PROCEDURE

First, it is assumed that a netlist has been finalized for the PWB to be designed. The choice of whether to use leaded or leadless packages has been made based on the pin count of each chip function. Package footprints and their tolerances have been noted, and the packages have been laid out in a suitable grid. The physical PWB layout should be done in the most efficient way to accomplish the electrical function and meet the thermal requirements (Fig. 11.40).

Initial considerations for designing a VHSIC-type PWB require that the characteristic line impedance Z_0, line width, line thickness, and spacing between lines be defined before layout can begin. First a PWB material is selected and its dielectric layer thickness and dielectric constant are determined. When the margin for the allowable crosstalk has been defined and the conductor designed for Z_0, the line spacing (vertical and horizontal) can be found.

Once the conductor design, line width, and spacing have been completed and the initial component layout is done, a visual inspection of the PWB layers can be conducted to detect potential manufacturing problems. If the layer limit is exceeded, it may be possible to use diagonal routing or move components to help improve layout efficiency. Reassessing the noise budget to allow for closer line spacing or eliminating ground planes may also help to reduce the layer count.

Computer programs can be used to design PWBs. One program will determine the line width given the desired Z_0 and other related PWB design parameters. The other program

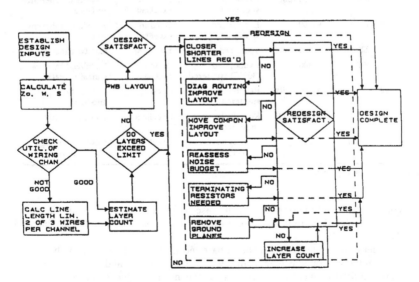

FIGURE 11.40 Design guidelines—process flow.

will determine the conductor spacing necessary to achieve the specified noise budget for lateral crosstalk.

11.8   ELECTRICAL DESIGN INPUTS

This section describes digital device characteristics that are useful for transmission-line design. Designers should use the most recent information available to them. Device output impedances are important for the analysis of reflections. Input impedance values are always high, so the package capacitance of the load is the dominant input characteristic. Input and output logic threshold levels are important for determining the dc noise margin. The device characteristics for several technologies are discussed in the following subsections.

11.8.1   TTL Device Characteristics

Various TTL versions exist for this popular type of logic. Most of the earlier versions were generally not suited for driving long transmission lines. The device characteristics were not symmetrical with respect to the power supply voltage levels for the device states. However, improvements in device performance and specifications resulted in the present devices, which are capable of having design characteristics for a wide variety of applications. The following data are based on the specifications and descriptions published for the FAST family by Fairchild Semiconductor.[20] Specification conditions are a temperature range of –55 to + 125°C and a power supply voltage V_{cc} of 5.0 ± 10 percent.

TTL switches current, so significant current must be driven to maintain the low state. The maximum input current is specified at 0.6 mA in the low state, with a maximum leakage current of 20 μA in the high state for typical FAST devices. The input capacitance is not normally specified. The FAST device schematic shows input clamp diodes to ground to limit excursions below ground, but no such limitation with respect to V_{cc}.

The output impedance is not specified directly. However, the current-limiting resistor shown on the FAST device schematic largely determines the high-state impedance; approximately 50 Ω is typical. Short-circuit current is specified at 60 to 150 mA for most nonbuffer devices; this corresponds to 75 to 36 Ω for short-circuit impedance in the high state with respect to V_{cc}. Some FAST line drivers are also specified at V_{OH} = 2.0 V minimum at 12 mA to provide an indication of operation in a transmission-line environment. This results in an output impedance of approximately 200 Ω with respect to V_{cc} = 4.5 V. However, the device is nonlinear; impedance with respect to ground or the equivalent open-circuit voltage is lower.

The output impedance in the low state is not specified directly. The schematic indicates that the forward (above ground level) impedance is based on the conducting Schottky transistor to ground, and the reverse (below ground) impedance is based on the clamp diode to ground. Therefore the actual dynamic impedance in the conduction region is expected to be low (possibly near 10 Ω). The maximum output voltage of 0.5 V at 20 mA implies an impedance with respect to ground of 25 Ω maximum, but actual characteristics are nonlinear. Although the diode impedance is not specified, the absolute maximum ratings for reverse input voltage and current of 0.5 V and 30 mA might imply an impedance of 16 Ω.

The output impedance in the OFF (high-Z) state is usually specified at 50-μA maximum leakage current. The schematic indicates that three-state output devices include a series diode to ensure isolation from V_{cc}. The absolute maximum ratings indicate that the three-state outputs may be raised to +5.5 V when $V_{cc} = 0$ (power off) without damage. The schematic does indicate a clamp diode to limit excursions below ground for all outputs.

TTL families are not generally specified for operations in a controlled-impedance long-transmission-line environment. The basic transient response control is provided by reasonably limiting the source current in the high state and by limiting the load-voltage overshoot in the low state.

TTL families are specified for inputs and outputs over the temperature and power supply voltage range. The computed noise margins for the FAST family, which include the worst-case power supply differential, are taken from the *FAST Data Book*[20] as follows:

Low-state V_{IL}	0.8 V maximum
Low-state V_{OL}	0.5 V maximum @ 20 mA
Low-state noise margin	0.3 V minimum

High-state V_{IL}	2.5 V maximum
High-state V_{OL}	2.0 V maximum @ 20 mA
High-state noise margin	0.5 V minimum

If the 0.3-V noise margin is compared to a typical signal swing of 4 V, the noise margin is 7.5 percent of the signal swing.

Some of the TTL families specify V_{OH} with respect to the power supply V_{cc}. For example, Texas Instruments, in its *TTL Data Book*,[21] specifies V_{OH} at V_{cc} −2.0 at 0.4 mA for advanced lower power Schottky TTL and at 2 mA for advanced Schottky TTL. Since this also corresponds to the $V_{OH} = 2.5$ V minimum at a minimum power supply of 4.5 V for the FAST specifications, it may be reasonable to imply that FAST (and possibly other Schottky TTL) will drive high-state outputs to within 2 V of the power supply voltage.

Some of the TTL families are also specified for high-state current at approximately $(V_{cc} - 1)/2$, which is similar to the $V_{OH} = 2.0$ V data for FAST line drivers and provides an indication of operation in a transmission-line environment. This output current for typical advanced Schottky devices is specified at 30 mA minimum and 112 mA maximum at 5.5 V power supply, according to Texas Instruments.[21] This indicates a high-state impedance of 20 to 75 Ω with respect to $V_{cc} = 1.0$ V.

11.8.2 Other Device Characteristics

This section discusses characteristics of four different families of devices. Emitter-coupled logic (ECL) is intended for high-speed use in controlled-impedance terminated applications. It has traditionally been well specified for a transmission-line environment. The devices are tested for performance specifications driving terminated coaxial lines. Both minimum and maximum rise times under terminated coaxial-line loading are specified for most devices. Unterminated operation is allowed if the line length is less than the lengths recommended in the Motorola handbook.[22] The impedance of the output transistor is given as 7 Ω, so long lines require external termination. The slower 10K family is specified at approximately twice the rise time (3.5 ns).

The ECL 10KH family is specified to have a minimum calculated noise margin of 0.15 V over a temperature range of 0 to +75°C and a power supply range of 5.2 V +5/−10 percent. Since the signal swing is typically only 1 V, the noise margin is about 15 percent of the signal swing; this is better than for TTL. However, the ECL noise margin applies for a limited temperature range. Loading and power supply specifications are different for the full temperature range. Device characteristics ensure that unconnected (open) inputs achieve a definite state, so that ambiguous results are prevented. The types of performance characteristics specified for ECL suggest that a more complete specification of VHSIC devices is required to characterize performance.

FACT is a high-performance Fairchild advanced CMOS technology family. The input thresholds are specified at 30 percent of power supply for the low state and at 70 percent of power supply for the high state. The output levels are specified both at very low current loads (20 times CMOS input current) to be consistent with CMOS-CMOS interfaces and at higher current loads to indicate compatibility with TTL and open-ended transmission lines. The following results are taken from the *FACT Data Book:*[23]

Low-state V_{IL}	1.35 V maximum
Low-state V_{OL}	0.40 V maximum @ 20 mA
Low-state dc noise margin	0.95 V minimum @ 24 mA
	1.25 V minimum @ 20μA

High-state V_{OH}	4.40 V maximum @ 4.5 V_{cc}, 20 μA
	3.70 V minimum @ 4.5 V_{cc}, 24 mA
High-state V_{IH}	3.85 V maximum @ 20 mA
High-state dc noise margin	0.70 V minimum
Worst-case V_{cc} differential	Negative @ 24 mA

High-state V_{IH}	3.15 V minimum @ 4.5 V_{cc}
High-state dc noise margin	1.25 V minimum @ 20 μA
Equal power supply voltage V_{cc}	0.55 V minimum @ 24 mA

It should be noted that although the 24-mA loading may seem large compared to steady-state CMOS input loads, it is not sufficient to characterize dynamic operation in a controlled-impedance transmission-line environment adequately. A 24-mA current change into a 50-Ω line results in only a 1.2-V signal change. Conversely, a 5.0-V signal change into a 50-Ω line requires 100 mA of current change. Even if we consider the signal change to be only the minimum difference between the worst-case signal levels of 3.7 V and 0.4 V, namely, 3.3 V, the line impedance required for this signal change with 24 mA is greater than 3.3 V/24 mA = 137.5 Ω. This transmission-line impedance is far greater than the practical range. However, if two devices are paralleled for 48 mA of current, this results in an initial voltage change of 2.4 V. If this doubles when the signal arrives at an open-circuited end of the line, a full signal amplitude will occur at the load end.

Another potential solution to both the power supply differential and TTL compatibility is to specify input thresholds that are equivalent to standard TTL thresholds. The Fairchild FACT family includes a family called ACT, which specifies input thresholds at the standard TTL levels of 2.0 V for the high state and 0.8 V for the low state over the full military temperature range and ±10 percent power supply voltage range. Since the output characteristics are unchanged, this results in the following worst-case dc noise margins, which are based on the *FACT Data Book:*[23]

Low-state V_{IL}	0.80 V
Low-state V_{OL}	0.40 V maximum @ 24 mA
	0.10 V maximum @ 20 μA
Low-state dc noise margin	0.40 V minimum @ 24 mA
	0.70 V minimum @ 20μA

High-state V_{OH}	4.40 V maximum @ 4.5 V_{cc}, 20 µA
	3.70 V minimum @ 4.5 V_{cc}, 24 mA
High-state V_{IH}	2.00 V maximum @ 5.5 V_{cc}
High-state dc noise margin	2.40 V minimum @ 20 µA
	1.70 V minimum @ 24 mA

Even if the high-state output is from a TTL device, such as the FAST described in Sec. 11.8.1, and $V_{OH} = 2.5$ V minimum at 1 mA, the high-state dc noise margin is a minimum of 0.50 V. This is approximately equivalent to the low-state noise margin. Therefore, both TTL compatibility and power supply differential immunity are achieved by TTL input thresholds, but at the cost of a reduced noise margin in the low state.

Additional interface characteristics of high-speed CMOS devices can be obtained from other sources. For example, the Cypress Semiconductor *CMOS Data Book*[24] discusses CMOS interface characteristics. Their device inputs are not diode clamped, but the information provided is not sufficient to completely characterize operation in a controlled-impedance transmission-line environment.

VTC Inc.'s *Data Book and Applications Manual*[25] includes an extensive discussion of CMOS power dissipation and output as a function of input voltage. Their TTL-compatible inputs have a typical power supply current of 1 mA at the threshold level of approximately 1.4 V. They increase noise margins with some input hysteresis, but guarantee the 0.8- to 2.0-V threshold region common to TTL. Typical curves for the clamp diodes to both ground and V_{cc} are given, with 1.2 V at 18 mA guaranteed. Outputs are specified at within 0.10 V of the power rails for 20 µA as well as at 2.4 V at 24 mA for the high state and at 0.5 V at 32 mA for the low state. The output impedance is much lower at low temperature, resulting in much higher currents, which reduces propagation delay.

11.9   NETLIST ANALYSIS

This section discusses the different signal types in a digital design. The signal types are divided into two segments, based on the required signal quality of each.

11.9.1   Signal Quality

When dealing with digital signals of a synchronous system, the signal quality must be examined. In the discussion of signal quality, device performance must be recognized. Each device family has a defined voltage range over which the device will operate. In addition, each device has a range of signal input and output voltages, which are recognized as specific logic states, and also a tolerance to the distortion of these waveforms. In the remainder of this section, signal quality is discussed.

11.9.2   Critical Signals

Critical signals are signals that are constantly examined or sampled to determine the control of logic elements in a digital system. Some of the best examples of critical signals are clock lines, write enables, certain control and address lines, and asynchronous resets. Further, any signal in a critical timing path is considered critical. The signal quality of these lines needs to be controlled very precisely. Since lock lines are present in all synchronous systems, the effect of the signal quality on clock lines is used as an example.

In synchronous systems, clock transitions define intervals at which data are sampled and presented. False triggering of these transitions can cause data to be processed before they have stabilized, interrupt a timing sequence, or cause other problems. False triggering can occur any time the voltage transitions from the range of a logic state to an undefined state, or returns to the original state. In addition, to prevent false triggers, voltage transitions through the indeterminate voltage region (the region between the minimum logic high voltage and the maximum logic low voltage) must not pass the same voltage point more than once on each transition. Examples of false triggering are given in Figs. 11.41 and 11.42.

In working with digital devices, one must assume that the transition from one logic state to another can take place at any voltage point in the indeterminate region. Passing any voltage point in the indeterminate region more than once can produce an undesired state transition. To prevent these transitions, all signals must control the amount of reflection, crosstalk, and loading that can cause the transitions. Note that in Fig. 11.43 the signal does have a fluctuating voltage for a short period of time, but that the transition is controlled to be within the voltage limits for each logic state.

In addition to the transition discussed, signal quality must be controlled for other reasons. The devices that transition need to achieve a steady state before the next transition begins. The devices are designed to transition from a given ready state. When a device

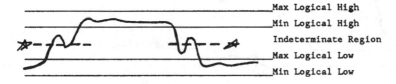

FIGURE 11.41   False triggering of critical line that has exceeded voltage limits for logic states. Drawing of load end of line.

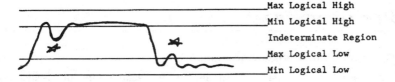

FIGURE 11.42   False triggering of critical line that has passed the same indeterminate voltage point more than once. Drawing of load end of line.

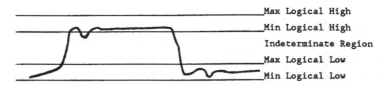

FIGURE 11.43   Valid critical-line transition. Voltage does not enter indeterminate region except during transition. In addition, during transition, voltage does not achieve the same voltage more than once per transition. Drawing of load end of line.

begins to transition before it has achieved steady state, it could be driven into regions of voltage and current requirements for which it was not designed. Operations outside the design specification can either damage the device or provide false information.

11.9.3 Noncritical Signals

Noncritical signals are signals that are sampled on a known interval. The state of the signals is only important near the sampling time. These signals include the majority of signals on the PWB. Noncritical signals include most data and control signals on a synchronous digital design. Like critical signals, these signals must have achieved a steady state before the next transition period. These signals can enter the indeterminate region and transition through the region in any fashion as long as they achieve and maintain a valid logic state before the sampling time. In the past, in TTL-type designs these signals were not generally examined for crosstalk and reflection. Because of high edge speeds and high-speed designs, reflections and crosstalk must be examined to ensure that the signal is within limits at the sampling time, and that the signal has achieved steady state before the next transition. In addition, examination of the signal must ensure that the maximum signal limits have not been exceeded at any time. Examples of noncritical lines are illustrated in Figs. 11.44 and 11.45.

11.10 DESIGN RESTRICTIONS

11.10.1 Optimum Characteristic Impedance

The Z_0 chosen for a specific interconnection is influenced by several factors, such as attenuation, reflections, signal distortion, crosstalk, power dissipation, and PWB producibility.

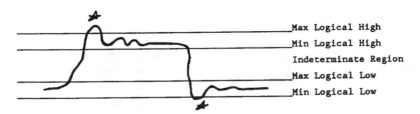

FIGURE 11.44   For noncritical lines, a major concern is exceeding the maximum device voltage. (This is usually equal to the maximum logical high voltage.)

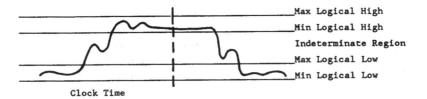

FIGURE 11.45   For noncritical lines, as long as signal is within tolerance at sampling time and stable before transition, it is acceptable.

Many of these factors result in contradictory objectives, so that a reasonably optimum result is achieved when counteracting factors are balanced. Producibility requires not only that a cost-effective design center value be chosen, but also that a certain tolerance be allowed to improve PWB production yields.

Published textbook analysis indicates that an optimum Z_0 exists that minimizes coaxial-cable losses and, therefore, attenuation on long lines. Depending on the specific situation considered, the minimum losses result when Z_0 is approximately in the 60- to 80-Ω range. Applications where attenuation is significant, such as closed-circuit television, tend to standardize on 75-Ω coaxial cable. Other applications, such as microwave and ECL interconnections, have tended to standardize on 50 Ω due to consideration of other factors.

Increasing or decreasing Z_0 affects several performance factors, which are summarized qualitatively. The negative impacts of lower-signal Z_0 include overload of source drivers, increase in switching current noise (ground upset and V_{cc} collapse), increase in switching power CV^2f, because C is inversely proportional to Z_0), and reduction of signal amplitude.

The negative impacts of higher-signal Z_0 include increased sensitivity to crosstalk and load capacitance. Therefore electrical performance is reasonably optimum when Z_0 is above the value that causes significant source driver loading, but sufficiently low that crosstalk and load-capacitance sensitivity are not excessive.

Lumped effects such as crossovers in adjacent planes, plated-through holes, and buried via connections, stub lines, and capacitance of loads connected to the signal tend to decrease the effective Z_0 from the unloaded value. Therefore the unloaded Z_0 should be chosen sufficiently high so that the loading effects result in a Z_0 within the lower limits.

11.10.2 Loading Effects

The basic models for describing the Z_0 of a line are based on a uniform line of significant length with intrinsic inductance and capacitance. The normalized signal propagation (based on the TEM mode) and the dynamic impedance Z_0 are defined by the intrinsic inductance L_0 and capacitance C_0 as follows:

$$Z_0 = \left(\frac{L_0}{C_0}\right)^{0.5}$$

and the time per unit length is

$$t_0 = (L_0 C_0)^{0.5} = Z_0 C_0$$

However, if lumped loading is added to the intrinsic line, both the effective Z_0 and the propagation delay t_{pd} are changed. If the loading is modeled as additional capacitance, Z_0 is decreased and the normalized propagation time increased. If the length of the line to which the lumped loading is added is relatively short, then it is reasonably accurate to model the lumped loading capacitance as effectively added to the intrinsic capacitance for modeling purposes. The lumped capacitance is divided by line length and added to C_0. The definition of "short" depends on frequency, which is a function of the signal rise time. The definition of the cutoff length l_{co} used for defining long lines comes from equating twice the propagation delay t_{pd} of the line to the rise time t_r,

$$2t_0 l_{co} = t_r$$

$$l_{co} = 0.5 v t_r$$

where $v = 1/t_0$ is the signal velocity.

For a rise time of 2 ns and a dielectric constant of 4, the cutoff length is approximately 6 in. It is convenient to regard less than one-half of the cutoff length as being "short."

Based on the long-length condition, adding lumped capacitance to the Z_0 and t_{pd} models results in the following:

$$Z_{0 \text{ loaded}} = [L_0(C_0 + C_D)]^{0.5} = \left[\frac{L_0}{C_0\left(l + \dfrac{C_D}{C_0}\right)} \right]^{0.5} = \frac{Z_0}{(l + C_D/C_0)^{0.5}}$$

$$t_{0 \text{ loaded}} = [L_0(C_0 + C_D)]^{0.5} = \left[L_0 C_0\left(l + \frac{C_D}{C_0}\right)\right]^{0.5} = t_0\left(l + \frac{C_D}{C_0}\right)^{0.5}$$

where C_L = lumped capacitance from package(s)

$\quad C_D$ = distributed package capacitance = C_L/L

$\quad l$ = line length

These equations show that Z_0 and the propagation delay are changed by a factor depending on the ratio of distributed load capacitance C_D to intrinsic capacitance C_0 in a short length of line.

11.10.3   Delay

The total delay of a signal may be considered as consisting of several parts. The basic speed of light in a vacuum is reduced by a factor determined by the actual dielectric material through which the wave propagates. Conductor losses and package capacitance cause the signal rise time to increase as the wave propagates. The wave shape may be further distorted by lumped loading at the end of the line or by reflections caused by nonperfect termination at the end of the line. No matter how many loads are on a line, the propagation delay of the signal measured near the zero percent amplitude value is the same. Capacitive loading slows the rise time and the 50 percent amplitude ahead, thereby appearing to increase device propagation delay.

The effective delay per unit length may be computed as shown in the following example:

Propagation delay in vacuum	1.02 ns/ft = 0.085 ns/in
× 2 for relative dielectric of 4	0.17 ns/in
× 1.25 for maximum lumped loading	0.21 ns/in
× 2 for rise-time delay to final value	0.42 ns/in

However, reflections may cause the signal to ring after the first incident wave. The magnitude and polarity of the reflection are defined by the reflection coefficient, which is given as $p = (R - Z_0)/(R + Z_0)$, where R is the impedance at any discontinuity, such as the end of a line, and Z_0 is the impedance of a long line. If a line is not long, then the effect of the reflection coefficient is reduced by the short-line scaling factor q, which is the ratio of actual line length to cutoff length, $q = 2lt_{pd}/t_r$ for $q < 1$, l being the line length.

It can be shown that the load voltage after the full linear rise time of the Nth incident wave (toward the load) is of the form $V(N) = V_{ss}(1 - K^N)$, where V_{ss} is the steady-state voltage change and K is the line response characteristic. $K = q^2 P_s P_l$, where q is the short-line scaling factor defined above, P_s is the reflection coefficient at the source, and P_l is the reflection coefficient at the load. For long lines with an open circuit at the load, $K = P_s$. Settling time must be allowed for ringing signals until the undershoot (which occurs on even-numbered incident waves) is within allowed limits. The signal is initially considered valid when the previous odd-numbered incident wave arrived. The overshoot and ringing are caused by P_s having a negative value. Additional rise-time delay is usually ignored since signal excursions are small.

For example, if $3 \times$ underdamped mismatch at the source is allowed ($3R = Z_0$), the first incident-wave overshoot for long open-circuit lines is 50 percent and the second incident-wave undershoot is 25 percent, which clearly exceeds reasonable noise budget allocations. The fourth incident-wave undershoot is 6.25 percent. If this is acceptable, then the signal is valid after the third incident wave, which requires 5 times the one-way delay. However, if an undershoot less than 1.56 percent is required, then the sixth incident wave is required and the time needed for reflections to settle is $T = [2(6) - 1]t_{pd} = 11$ times the one-way line delay. These settling times are as follows. The distorted rise is assumed to be less than or equal to the one-way line delay, $t_r = lt_0$.

One-way propagation delay for relative dielectric of 4	0.17 ns/in
$\times$ 1.25 for maximum lumped loading	0.21 ns/in
$\times$ 6 for third incident wave plus 1 rise time	1.28 ns/in
or	
$\times$ 12 for sixth incident wave plus 1 rise time	2.52 ns/in

The resultant propagation delay times shown for third and sixth incident waves include allowance for rise time, propagation, and settling time delay. The total delay time for a comparison with available time (generally the clock interval) must also include the device delay (including an allowance for capacitive loading conditions), clock skew tolerance, and a timing design margin for such factors as modeling uncertainty and device performance degradation.

11.10.4   Noise Budget

The noise budget is the allocation of the noise margin to account for various possible electrical disturbances seen on the interconnection between parts.

Each family of digital devices has a specified range over which signals can be generated or received. In the logic low, or 0 state, an output will drive the output pin of the device, when loaded within specifications, to a specified voltage. A logic low is guaranteed to provide an output voltage within some voltage range for all the devices made by a specific manufacturer and for a given type of technology. Many manufacturers use the standard ranges for well-defined families, such as TTL, FAST, ECL, and CMOS at the other end of the wire providing the logic connection is the receiver. A properly matched receiver is guaranteed to accept a minimum range of voltages as a logic low, which include the full range of the driver voltages and some additional tolerances. The amount of voltage difference between the driver output being at an extreme voltage limit for a given logic state, compared to the receiver being at the corresponding extreme voltage limit for the same logic state, defines the noise margin. An example of well-matched devices is shown in Table 11.10.

TABLE 11.10 Noise Margins

Driving device	Voltage levels*	Receiving device
		Logic 1, maximum voltage
	(1)	
Logic 1, maximum voltage		
	Driver tolerance region	
Logic 1, minimum voltage		
	(2)	
		Logic 1, minimum voltage
	Indeterminate region of devices	
		Logic 0, maximum voltage
	(3)	
Logic 0, maximum voltage		
	Driver tolerance region	
Logic 0, minimum voltage		
	(4)	
		Logic 0, minimum voltage

*Four tolerances are shown: (1) Receiver logic high maximum voltage—driver logic high maximum voltage; (2) driver logic high minimum voltage—receiver logic high minimum voltage; (3) receiver logic low maximum voltage—driver logic low maximum voltage; (4) driver logic low minimum voltage—receiver logic low minimum voltage.

The smallest number in Table 11.10 defines the logic connection noise margin. If any of the values are negative, the devices do not work together. Usually the second and third numbers in Table 11.10 place the limit on the noise margin. The first number is the high overshoot margin, and the fourth number is the low overshoot margin. The interconnect noise margin is the smallest of the terms defined. If a PWB uses several different types of devices, and uses interconnections with different types of devices, separate noise margins may need to be defined for each of the device configurations.

A short list of noise margins and their associated families is given in Table 11.11.

TABLE 11.11 Noise Margins and Their Associated Families

Device family	Noise margin
Texas Instruments Schottky TTL	0.30 V
Motorola 10KH	0.15 V
Fairchild FAST	0.30 V
Fairchild FACT	0.55 V @ 5.0 V V_{cc}
Fairchild FACT	0.30 V @ 3.3 V V_{cc}
Westinghouse VHSIC	1.00 V @ 5.0 V V_{cc}

11.10.5 System Requirements

The operation of the system depends on the voltages on the lines being within the margins of the logic devices, as explained. The exact allocation of the noise margin to define the

noise budget for each type of noise in the electric system varies with the system requirements. In a high-speed design the requirements might be as follows:

1. 2- to 3-ns rise times
2. 25- to 40-MHz operation
3. high-density PWBs (10-layer multilayer, 10-mil-diameter buried vias)
4. lines to be terminated only where absolutely critical
5. design must utilize worst-case analysis
6. 5-mil line width and 5-mil space minimum

These constraints define many of the boundaries of design. Designing with worst-case
analysis guarantees that the board will function as long as the parts being used have better
performance characteristics (shorter clock to output delay, larger noise margins, and so on)
than the guaranteed operation specification defined by the manufacturer. In the past many
designs have simply derated the signal propagation speed of a part to allow settling time
for all electrical phenomena discussed in the noise budget. This method provided fairly
accurate results since the noise was small (due to the slow edge speeds of the devices and
the less dense boards) and would disappear in a small settling time. In most cases the settling time was achieved by the derating factor and the speed gained from devices operating
faster than guaranteed.

However, as the rise times become faster, and the line density of the PWB gets higher,
these electrical phenomena must be accounted for directly. The trade-offs involved in
obtaining the noise budget need to be made using knowledge of electrical, mechanical,
and manufacturing tolerances. The major areas of noise budget allocations are listed in
Table 11.12, which is typical of a budget developed for high-speed PWBS.

TABLE 11.12 Noise Budget Allocations

Reflection on line	10%
Crosstalk from orthogonal layers	2%
Reflections at ends of line and crosstalk	38%
Ground upsets	25%
Other	25%
Total	100%

11.10.5.1 Examination of Noise Budget.   Reflection is a term used to describe the condition that causes part of a signal traveling along a transmission line in one direction to
travel in the opposite direction. A signal may reflect for a significant period, until the signal is dampened to its final state. The magnitude and polarity of the reflection are defined
by the reflection coefficient, which is here.

Reflections at the ends of the lines due to impedance mismatches between the devices
and the controlled-impedance lines are a major concern in high-speed systems. Associated
with the high-speed signal transition is crosstalk. Furthermore, crosstalk is linked with
reflections, for there exist reflections of crosstalk and crosstalk of reflections. It is difficult
to deal with crosstalk without considering the effects of reflections.

Crosstalk is one of the major areas of concern in any high-speed design and layout.
Earlier papers on crosstalk deal with crosstalk where the driven lines are terminated with
matched impedances. However, some devices have a crosstalk component associated with
each reflection of the driven line and a reflection with each imposed crosstalk wave. The
devices have this additional reflection of crosstalk, and crosstalk of reflection characteris-

tic because they do not have matched impedance between the board and the inputs and outputs of the devices.

Simply stated, crosstalk is the induction of voltage and/or current between interconnection media which are not physically connected. Any two wires that pass in proximity to one another induce some small electrical effect on each other. This electrical effect, due to mutual capacitance and induction between the wires, is known as crosstalk. For worst-case design, plan on the crosstalk portion of the noise budget accounting for noise from two adjacent lines, one to each side. Crosstalk can be divided into two types, forward and backward. Forward (far-end) crosstalk is usually very small in magnitude and travels in the direction of the wave inducing the crosstalk. Theory says that there is zero forward crosstalk when the dielectric is homogeneous, as in stripline designs. Backward (near-end) crosstalk is usually the only significant voltage transient due to electrical coupling and travels opposite to the direction of propagation of the signal inducing the crosstalk.

For the analysis of crosstalk that occurs on any PWB using fast-rise-time components that are not terminated in a matched impedance, it is important to remember that the crosstalk will reflect, just as a normal signal that is not properly terminated will reflect. In addition, the reflections of the driven line inducing the original crosstalk induce additional crosstalk caused on the passive lines. In the classic crosstalk analysis, the backward crosstalk, which is larger in magnitude than forward crosstalk, lasts only for a time equal to twice the line delay, $t = 2lt_0$. The near-end reflections are the principal concern for backward (near-end) crosstalk.

Ground upsets are dynamic changes in the voltage of the ground plane and voltage plane due to current transients. When several devices change states, large amounts of current can move across the PWB or in the devices themselves, affecting the references the devices use to operate. One of the worst effects that could be seen by loss of the reference is false triggering of the clock of asynchronous lines. This ground upset transient will decrease the amount of noise budget available for other items.

Reflections on the line are the reflections that will occur due to changes in the line impedance along the length of the line due to manufacturing changes (such as etching width differences) and geometry (such as a line passing close to a hole, vias, or buried vias). This does not include reflections due to stubs or other loads connected along the length of the line.

Crosstalk of orthogonal layers is the coupling of voltage from one signal to another during transition periods. The coupling takes place over the length of the parallelism between lines. To prevent large coupling from taking place between vertical signal layers, orthogonal signal layers are required. Further, the dual-stripline construction of controlled-impedance PWBs can be chosen to eliminate as much vertical crosstalk as possible (only orthogonal layers exist between ac ground planes). The budget for this section can account for noise from the small crossover areas between orthogonal conductors (see Sec. 11.10.6).

Other allocations include items such as temperature variance, as well as a derating factor to ensure operation through hostile environments and device degradation.

A complete evaluation of parts, supply voltage, line widths, line spaces, and impedance tolerances should be performed for worst-case design for each family of parts and for the interface of different families. The voltage at the output device should also be examined for overvoltage in order to prevent device damage.

11.10.6 "Zero" Vertical Crosstalk

Due to the large mutual coupling between circuit lines arranged in a horizontal or vertical fashion (top to bottom or side to side when observed in the plan view), lines paralleling these on adjacent layers are not allowed. This eliminates the strong effects of broadside coupling.

An exception is small amounts of parallelism between adjacent layers in a dual-strip-line configuration which can be accepted if the parallelism is short. This value of crosstalk must be included when allocating a noise budget. To allow for a small amount of vertical parallelism, a noise budget for vertical parallelism should be established with a maximum amount of vertical parallelism defined.

11.10.7 Critical-Line Design

Critical lines are those that carry critical signals as described previously (Sec. 11.9.2). These signals are required to be valid signals at all times, with well-behaved transitions when the signal changes state. The clock lines in a synchronous system are a common example of critical lines. Clock-signal quality must be maintained so that the clock samples the data signals at only the proper time, with one unambiguous sample per clock cycle.

Two invalid conditions to be avoided were described and illustrated in Sec. 11.9.2. The first is undershoot ringing, which results in a signal that returns to the indeterminate region at a time other than the intended transition time. The second is an ambiguous transition through the indeterminate region. Both of these could cause multiple samples per clock cycle.

The traditional technique for controlling ringing is to parallel terminate the load with resistors. In this way, overshoot and ringing are minimized if the termination matches the PWB impedance. However, in the case of low power CMOS devices, parallel resistive termination is usually an unacceptable alternative. The termination resistor will consume additional power on the board. In addition, some signal degradation will occur due to the additional signal local current caused by the termination resistor.

The use of nonlinear loads can also control ringing. Nonlinear loads (such its a clamp diode) connected to input signals can control the ringing by limiting overshoot. It is important to incorporate these clamping devices into the load device to prevent adverse effects on the layout density, interconnection length, and line capacitance. Clamping characteristics must be selected that are capable of controlling the voltage and current of the signal arriving at the load in order to dampen the line transients to quickly establish steady-state conditions.

Another effective technique for controlling overshoot and ringing at the load is to series-terminate the source driver. Common series-termination components include resistance and composite elements such as ferrite beads. This technique limits the initial signal amplitude out of the driver and stretches the signal rise time. Overshoot and ringing result when the source driver impedance is less than Z_0 line impedance. This is why components are sometimes added in series to increase the source impedance.

The other condition to be avoided as illustrated in Sec. 11.92 is a nonuniform transition through the indeterminate region. The signal transition should be monotonic (that is, it must not reverse direction while in the indeterminate region). Furthermore, the best clock-timing precision is achieved when the transition through the indeterminate region is as fast as practical. Unfortunately, several common conditions tend to cause the signal transition through the indeterminate region to "hesitate" or even reverse. Lumped loads such as stubs and input capacitance cause a negative reflection to be superimposed on the initial signal transition. Source driver impedance causes the initial voltage to divide according to the ratio of line impedance to source impedance. Most source drivers are not capable of launching a full signal swing into a transmission line, so the initial level tends to remain until the reflection returns from the open-circuit end. (If the far end is terminated, the initial level is also the final level.) Only the loads located sufficiently close to the open-circuit end of the unterminated line will see a reflected signal, which is double the initial signal with no hesitation or step at the intermediate level.

Recommended guidelines for PWB layout of critical lines are given in the following sections. It should be noted that these guidelines may not be appropriate for all applications and that careful attention to all design characteristics is required for an effective, conservative, worst-case design of critical signals. In many applications these guidelines may be compromised or even ignored without obvious adverse results. However, the noise margin and other characteristics may be reduced to the extent that the correct operation is uncertain or sensitive to the operating conditions.

11.10.7.1 Point-to-Point Routing.

All critical lines shall interconnect the loads in a series, with the source driver at one end of the line. No branching is allowed. If a single driver circuit is not specified to be capable of providing at least one-half-amplitude signal, additional drivers should be paralleled until this amplitude can be ensured. An exception to the one-half-amplitude signal requirement for critical lines is allowed if the total signal-line length is within the maximum stub length rule for noncritical lines (approximately 3.0 in for the conditions generally considered).

The starting level is the signal level farthest from the indeterminate region that could result from the settling of the previous transition. The worst case starting level is generally the best-case previous signal level.

The halfway intermediate level is usually considered to be the midpoint of the indeterminate region, or the average of the minimum logic high and the maximum logic low input threshold levels.

Although conventional maximum loading (fan-out) derating rules are not generally applied to transient loading, conservative design procedures indicate that some derating should be allowed for such factors as modeling uncertainty and device performance degradation.

11.10.7.2 Load Placement and Maximum Stub Length.

All loads must be placed sufficiently close to the end of the line so that the reflection effectively eliminates the intermediate step, and full signal amplitude (with margin) is achieved quickly. Full signal margin after the transition is important for clock lines because the clock transition causes simultaneous transitions on the data lines which, in turn, results in ground upsets and V_{cc} collapse. The general recommendation for "sufficiently close" to the end of the line is the maximum stub length rule for noncritical lines, which is one-half of the cutoff length for 2-ns edges of the 500-MHz quarter-wave length (approximately 3.0 in for the conditions generally considered, dielectric constant $\varepsilon_r = 4$. Therefore all loads are to be within a stub length of the end of the line.

11.10.7.3 Series Termination.

The layout should allow for a series component to be inserted in the signal at the source end. This component should be located as close to the driver as reasonably practical, but no more than a stub length from the driver. This component may be needed either to slow the rise time (this "shortens" the electrical length of the line by increasing the cutoff length), or limit the initial signal amplitude launch into the line (which limits overshoot), control the ringing, or compensate for capacitive loading. The series components providing these characteristics are generally resistive/inductive in nature (series capacitance effects are to be avoided in general). Small resistor values (less than Z_0 minus the driver impedance) or ferrite beads are commonly used.

11.10.7.4 Minimum Stub Length.   Since the total interconnect length between loads is limited to within the maximum stub length for noncritical lines, the length of individual stubs to each load is necessarily limited. The number of loads is also limited, since some length is unavoidable due to package leads. Each load interconnection (especially those farthest from the end of the line) should have minimum stub length, preferably a small fraction of the maximum stub length for noncritical lines.

11.10.7.5 Ground Distribution.   The ground distribution is crucial to the critical signal lines because source, interconnection, and load signal are all with respect to ground for these nondifferential interconnections. Furthermore, the clock transitions cause transitions in the clocked signals. Ground and power planes must be as continuous as possible, with minimal cutouts. The placement and number of ground connections to devices are critical in most applications. The device kind PWB interface ground connections must be connected as directly to the ground plane as possible to minimize the inductance resulting from the lead. It is good practice to have as many connections to power and ground as possible.

All critical signal lines must be run with continuous proper spacing (to maintain constant Z_0) to the ground plane or a parallel ground line at all times. This is particularly important if nondifferential critical signals are interconnected between different (but connected) ground planes, such as between PWBs or multiple ground-plane layers.

V_{cc} power distribution is generally considered part of the ground distribution, but the V_{cc} differential requirements are not intended to be as crucial as ground upset. Larger systems may tend to have different power supplies, and some fault-tolerant and continuously maintained systems must continue to function with power loss in some portions of the system. Wherever drivers attempt to change signal states at a high rate (especially when loaded with long transmission lines), the result is a tendency for V_{cc} collapse. This is counteracted by the stored energy in the V_{cc} distribution, which is available for transient loading after the effects of inductance and distance have been considered. Therefore energy stored nearest the switching devices and with the least inductance is most effective in reducing transients in the V_{cc} power supply. Discrete capacitors are superior to closely spaced V_{cc} and ground planes to control this noise. Total V_{cc} capacitance must be much greater than the total load capacitance of all signals and transmission lines. A ratio of 100:1, or higher, is recommended for high-frequency bypass capacitance. Typically, large capacitors are near the connector and smaller-value capacitors are located around the PWB. For more information about noise on the V_{cc} line, see Canright.[26]

11.10.7.6 Minimum Crosstalk.   Crosstalk to critical signal lines must be strictly controlled. Critical lines must not be run parallel to other signals (laterally or vertically) for any significant distance without providing adequate spacing to ensure minimum crosstalk.

Increasing the conductor spacing around a critical signal line is recommended rather than grounding an interposed conductor (Fig. 11.46). Grounding the interposed conductor reduces crosstalk, but does not eliminate it because the ground conductor is not an effective shield. Increasing the spacing is recommended because of its simplicity.

11.10.8 Capacitive Loading

The propagation delay of a digital device is generally measured from the time the input reaches 50 percent to the time the output reaches 50 percent. Digital logic using CMOS technology has its propagation delay stretched out by the effective capacitance of line loading. Logic devices will have their maximum propagation delay specified with a 50-pF

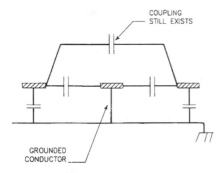

FIGURE 11.46 Ground interposed conductor.

load. The capacitive load on an output depends on the number of devices driven by the output and the total length of conductor driven by the conductor. The capacitance per unit length C_0 of the conductor is

$$C_0 = \frac{1}{vZ_0}$$

where v is the signal velocity, and

$$C_0 = \frac{\varepsilon_r^{0.5} \text{ pF/cm}}{0.03Z_0}$$

If the 50-pF maximum-capacitance load is exceeded (many devices have an input capacitance of 5 pF or more), the manufacturers' charts should be used that relate the excessive capacitance load to the increase in device propagation delay.

11.11 PWB LAYOUT CONSIDERATIONS

For anyone unfamiliar with PWB interconnection or lay-up, this section provides a description.

11.11.1 Lay-up Configurations

Lay-up configurations can vary from buried microstrip (Fig. 11.47) to dual stripline (Fig. 11.48).

A combination of the two configurations, such as the one shown in Fig. 11.49 is a possible compromise that meets most electrical and package design requirements. The lower section of this configuration uses the dual stripline and the upper section uses the buried microstrip. The major difference is in adding layers to the configuration. To add a layer pair to the dual stripline requires the addition of two C stages and four B stages of PWB material. But the same layer pair added to buried microstrip requires the addition of one C

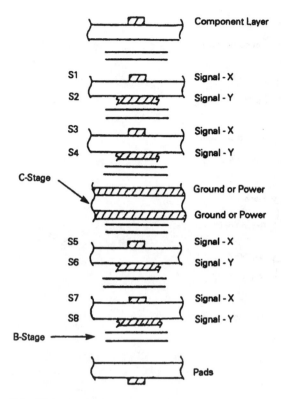

FIGURE 11.47 Buried-microstrip lay-up.

stage and two B stages. This affects the total thickness of the PWB. The buried microstrip is more efficient in terms of total layer thickness but much less effective in electrical performance. In order to compare the total thickness of the PWB to the configuration, the following assumptions apply:

1. Power/ground and signal layers without buried vias,
   Cu thickness = 0.0014 in

2. Signal layers with buried vias,
   Cu thickness = 0.0018 in

3. External layers,
   Cu thickness = 0.0026 in
   Sn/Pb plate = 0.0004 in

4. C stage,
   Material thickness = 0.0045 in

5. B stage,
   Material thickness = 0.0025 in

Depending on the specific dielectric material used, these dimensions may vary a bit. Table 11.13 lists thicknesses for 6, 8, and 10 signal layers for both buried microstrip and stripline configurations.

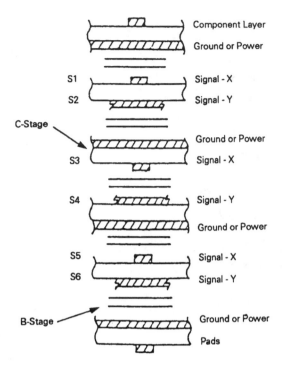

FIGURE 11.48   Dual-stripline lay-up.

TABLE 11.13 Signal Layer Thickness

Signal layers	6	8	10
Buried microstrip	0.0718	0.0847	0.0978
Dual stripline		0.0736	0.0967	0.1216

11.12 PREFERRED REDESIGN OPTIONS

Rules are established in the beginning of a PWB design effort that establish the lay-up of the PWB layers, the maximum number of layers permitted, line widths and spaces, and the length of permissible parallelism. The PWB design falters when there are connections that cannot be routed within the established restrictions, requiring an increase in the number of layers to continue routing the board. This section describes options available to continue routing the PWB without increasing the number of layers.

The need to reduce the number of layers in the PWB can be recognized at two places in the PWB design process—after the number of layers necessary for routing the PWB has been estimated and after the PWB layout has resulted in too many layers. The number of redesign options available in the former case is limited.

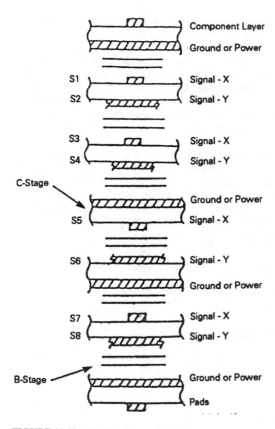

FIGURE 11.49   Combination configurations.

Only when an algorithm estimates the number of layers required to interconnect the components of the PWB can the design features be used to estimate the layer count and be modified to reduce the estimated number of layers. Removing components from the PWB is the most certain technique for reducing the estimated layer count, perhaps by moving to another PWB within the same electronic chassis. If the layer estimate is performed early enough in the product development cycle, then gate arrays, programmable logic arrays, application-specific integrated circuits (ASICS), or multichip modules (MCMs) could be used to reduce the number of components on the PWB. One need only to inspect the inside of a laptop PC to realize the effective use of board space provided by such alternatives, particularly MCMs.

If the PWB has a grid of plated-through holes (PTHs) and if the algorithm for estimating the layer count makes use of the PTH grid, then removing rows of PTHs from the grid will improve the routability of the PWB, but the layer-estimating algorithms may lack the finesse to assess the impact of the wider wiring channels accurately.

The noise budget can be altered to allow for closer line spacing if the algorithm for estimating the layer count takes this into account. Next, layers allocated for power and ground planes can be used instead for signal routing. This option is undesirable because it increases crosstalk. Finally, the PWB lay-up can be altered to provide a larger number of layers.

The guidelines recommend designing PWBs so as to completely avoid vertical crosstalk. This is the condition of electrical coupling to a line on a layer above or below. This restriction may be a contributor to ineffective conductor routing. If a conductor turns 90° and develops a segment parallel to a conductor on an adjacent layer, then this segment must be separated from the parallel line on the adjustment layer by a distance in the X-Y plane at least equal to the minimum specific line spacing in order to avoid vertical crosstalk. This rule requires the 90° segments to always lie over a conductor routing track on the adjacent layer. A congested layer might not allow 90° conductor turns on the adjacent layer, forcing the use of a PTH and another signal layer for routing the hard-to-place conductor segment.

Changing the PWB design rules to permit vertical crosstalk requires a reevaluation of the noise budget for crosstalk. If the longest length of parallelism for coplanar lines is actually less than the designed line length, then only a fraction q of the budgeted crosstalk is used. The remaining fraction $1 - q$ may be used for vertical crosstalk and the permitted amount of vertical crosstalk [equal to the crosstalk noise budget times $(1 - q)$] can be used to calculate the maximum length of parallelism for vertical crosstalk.

The vertical crosstalk permitted V_{VCTP} is

$$V_{VCTP} = (1 - q)V_{CTNB}$$

where q = fraction of crosstalk noise budget used by lateral crosstalk

V_{CTNB} = crosstalk noise budget

The maximum line length experiencing vertical crosstalk l_{VM} is

$$l_{VM} = l_{co}\frac{V_{VCTP}}{V_{VCTM}}$$

where l_{co} = cutoff line length

V_{VCTM} = maximum vertical crosstalk

If the noise budget for crosstalk is fully utilized by lateral crosstalk, then some other portion of the noise budget needs to be reassigned for vertical crosstalk. If the noise budget has a safety margin, this margin could be reduced in order to permit some amount of vertical crosstalk. If the noise budget has a portion of the noise margin allocated for reflections, then the use of terminating resistors would free this portion of the noise budget for vertical crosstalk.

A technique that avoids the problem of vertical crosstalk is to turn a conductor 45° from its principal direction instead of 90°. If one layer has the X direction (0°) as its principal direction, if all deviations from the principal direction on this layer are at 45°, if the next layer has the Y direction (90°) as its principal direction, and if all deviations from the principal direction on this layer are at 135°, then vertical crosstalk can be avoided in the case of dual-stripline configurations.

If a reference plane is removed from the PWB lay-up and replaced with a signal layer, then the PWB would have five signal layers between two reference planes. This alternative to exceeding the maximum number of layers for the PWB will cause a number of problems. The previous paragraph described using the 45° and 135° directions for minor deviations from a principal direction. With five layers between two reference planes, the principal directions can be 0, 45, 90, 135, and 0°, in that order. Minor deviations from the principal direction of a layer would then turn 90° from that principal direction. Vertical crosstalk, therefore, will occur between the two layers farthest away from each other and between a deviated segment and a principal-direction segment that are two layers away from each other. The CAD system used for routing the PWB might not perform automatic routing with four principal directions.

Other problems created by removing an inner layer reference plane (power or ground) are increased crosstalk and a loss of impedance control. When a reference plane in a dual-stripline configuration is replaced by a signal plane, the signal planes nearest the new signal plane will experience increased crosstalk if they maintain the same minimum spacing. Therefore removal of a reference plane might not permit increasing the minimum spacing used on the affected signal layers. Similarly, the signal layer replacing the reference layer will experience more crosstalk than the layers nearest the remaining reference planes. The signal layer nearest the remaining reference planes will also see an increase in crosstalk.

This case of five signal planes between two reference planes will require that the shortest conductor paths in the PWB be place on the center signal plane of the five, with some expectations. Only clocked signals should be placed on these five signal layers. No asynchronous lines should. be placed on these five layers because of the higher crosstalk. No inputs to unclocked logic; no reset, clear, clock, enable, or chip-select interconnections can be placed on these five signal layers.

If more layers are added without changing the individual layer thickness, then the overall PWB thickness increases. Mechanical design constraints may preclude a thicker PWB. An alternative is to use thinner layer materials. Thinner layers reduce the conductor-to-ground distance, which increases the conductor capacitance, decreases crosstalk, lowers characteristic impedance, and raises power consumption.

11.13 CAD INTERFACE

This section provides the circuit designer with a brief overview of the work of a PWB computer-aided engineering (CAE) group. It provides a list of inputs needed by the CAE group and a description of advanced features useful for computer-aided design (CAD) of high-speed PWBs, and it identifies some useful CAD system outputs.

11.13.1 CAD Inputs

The following inputs are required from CAE to design high-speed, high-density PWBs:

- parts list
- netlist
- schematic
- data sheets on parts (mechanical and electrical)
- physical board definition
- connectors and location
- critical component placement
- allowable I/O and/or chip swapping
- thermal characteristic
- critical line length definition
- line width and spacing

- parallelism constraints
- crossover constraints
- capacitor layout
- board material
- copper weight
- plated-through-hole locations/grid
- buried-via locations/grid

CAE must be aware of the fact that any constraints placed on the design will increase the cost and time required to design a PWB. Therefore CAE should review all design constraints for criticality.

11.13.2 CAD Capabilities and Functions

The following capabilities are desirable in the CAD design system to design high-speed, high-density PWBs:

1. Component autoplacement with weighting between component thermal characteristics and component affinity relative to minimum total connection length.
2. Buried and blind vias used by the autorouter. It might be possible to restrict buried vias to layer pairs sharing a common C-stage laminate.
3. An autorouter that will efficiently handle fine-line, high-density designs.
4. Ability to automatically spread apart closely spaced conductors after routing, where spacing permits, for manufacturability.
5. Identify sending and receiving ends of all device lines that are unidirectional and identify tristate lines. The CAD program will use this information when applying design rules.
6. Capability of handling transmission-line rules of any specific controlled impedance (30 to 130 Ω). Inputs to the CAD database include characteristic impedance as well as distance from conductor to nearest power and ground planes. The type of controlled-impedance structure being used must be identifiable, such as stripline or dual stripline. The CAD program will use an identification input such as stripline to select the appropriate subroutine for calculating the line width. When conductors change layers, the CAD system must be capable of assessing whether the original line width must change to maintain constant impedance. The CAD program should be capable of implementing design rules for changing the characteristic impedance at branch points to avoid reflections. The CAD program, as a minimum, should allow the user to label conductors and also write subroutines that the CAD program will invoke to select conductor widths and layer assignments.
7. Spacing between parallel conductors must be controlled by the CAD program. There are two options that should be available to users—to specify the separation between conductors (same layer or different layers) and to specify the length of parallelism permitted for the specified separation.
8. The PWB CAD program must be able to report the actual conductor length between each output-input pair of component pins. For example, U1-1 to U7-11 is 5.87 in. These data must be placed in a computer file that can be interfaced by other software programs. The goal is to use this line length information, along with knowledge of the

material dielectric constant, to calculate the propagation delay of the wiring and then to use this delay information when timing analysis is performed by another CAE program.

9. A PWB CAD program for high-speed designs must be capable of generating at least two and preferably two sheets of artwork for each layer of signal conductors. The two sheets of artwork are required in order to perform selective etching of double-clad PWB laminates. The two layers of clad materials have different electrical resistivities in order to build terminating resistors within the PWB by selective etching. (See Mahler and Schroeder.[27]) The CAD program must recognize a special reference designator for these planar resistors, such as RT instead of R, and refer to a design rule to select the aperture size and the length of the trace at this aperture size.

11.13.3   CAD Outputs

Besides the normal outputs to create tools needed to manufacture the PWB, there are reports needed to support electrical engineering. These reports will help ensure a workable design by checking outputs before building the first PWB. The reports required are on parallelism (vertical and lateral), line lengths, crossovers per line, line delay, and settling time. In addition to these reports, any change by the CAE group to the schematic from the engineering inputs (such as change in reference designators) must be used to update the engineering schematic.

REFERENCES

1. R. Pucel, "Design Considerations for Monolithic Microwave Circuits," *IEEE Trans. Microwave Theory Tech.,* vol. MTT-29, June 1981.

2. M. Caulton, J. J. Hughes, and H. Sobol, "Measurements on the Properties of Microstrip Transmission Lines for Microwave Integrated Circuits," *RCA Rev.,* vol. 27, Sept. 1966.

3. C. P. Wen, "Coplanar Waveguide: A Surface Strip Transmission Line Suitable for Nonreciprocal Gyromagnetic Device Applications," *IEEE Trans. Microwave Theory Techn.,* vol. MTT-17, Dec. 1969.

4. S. Cohn, "Slot Line on Dielectric Substrate," *IEEE Trans. Microwave Theory Techn.,* vol. MTT-17, Oct. 1969.

5. H. Sobol, "Microwave Integrated Circuits," in L. Young and H. Sobol (Eds.), *Advances in Microwaves, vol. 8,* Academic Press, New York, 1974.

6. H. Sobol, "Application of Integrated Circuit Technology to Microwave Frequencies," *Proc. IEEE,* vol. 59, Aug. 1971.

7. M. Caulton, "Lumped Elements in Microwave Integrated Circuits," in L. Young and H. Sobol (Eds.), *Advances in Microwaves, vol. 8,* Academic Press, New York, 1974.

8. K. C. Gupta, R. Garg, and I. J. Bahl, *Microstrip Lines and Slotlines,* Artech House, Dedham, Mass., 1979.

9. B. Bhat and S. K. Koul, *Stripline-Like Transmission Lines for Microwave Integrated Circuits,* Wiley, New York, 1989.

10. H. Sobol and M. Caulton, "The Technology of Microwave Integrated Circuits," in L. Young and H. Sobol (Eds.), *Advances in Microwaves, vol. 8, Academic Press,* New York, 1974.

11. H. A. Wheeler, "Transmission Line Properties of Parallel Wide Strips by a Conformal Mapping Approximation," *IEEE Trans. Microwave Theory Techn.,* vol. MTT-12, May 1964.

12. H. A. Wheeler, "Transmission Line Properties of Parallel Strips Separated by a Dielectric Sheet," *IEEE Trans. Microwave Theory Techn.*, vol. MTT-13, Mar. 1965.

13. R. Mittra and T. Itoh, "Analysis of Microstrip Transmission Lines," in L. Young and H. Sobol (Eds.), *Advances in Microwaves, vol. 8*, Academic Press, New York, 1974.

14. H. Sobol, "Extending Integrated Circuit Techniques to Microwave Equipment," *Electronics,* vol. 40, no. 6, Mar. 1967.

15. G. I. Zysman and D. Varon, "Wave Propagation in Microstrip Transmission Lines," in *Dig. 1969 IEEE Int. Symp. on Microwave Theory and Techniques*.

16. E. Denlinger, "A Frequency Dependent Solution for Microstrip Transmission Lines," vol. MTT-19, Jan. 1971.

17. M. Schneider, "Microstrip Dispersion," *Proc. IEEE, vol. 60*, Jan. 1972.

18. E. A. Mariani, C. Heinzman, J. Agrios, and S. B. Cohn, "Slot Line Characteristics," *IEEE Trans. Microwave Theory Techn.*, vol. MTT-17, Dec. 1969.

19. F. E. Terman, *Radio Engineers' Handbook*, McGraw-Hill, New York, 1943, p. 36, fig. 5.

20. *FAST Data Book*, Fairchild Semiconductor Corp., South Portland, Me., 1985.

21. *TTL Data Book*, vol. 3, Texas Instruments, Inc., Dallas, Tex., 1984.

22. *Motorola MECL System Design Handbook*, 4th ed., Motorola, Phoenix, Ariz., pp. 50–52, 1983.

23. *FACT Data Book*, Fairchild Semiconductor Corp., South Portland, Me., 1985.

24. *CMOS Data Book*, Cypress Semiconductor, Santa Clara, Calif., 1986, app. B.

25. *Data Book and Applications Manual*, VTC Inc., Bloomington, Minn., 1985.

26. R. Canright, "A Formula to Model Noise," in *Proc. 37th Electronic Components Conf.*, 1987, pp. 354–361.

27. B. Mahler and P. Schroeder, "Planar Resistor Technology for High-Speed Multilayer Boards," *Electronic. Packaging and. Production.*, pp. 150–154, Jan. 1986.

CHAPTER 12

PACKAGING OF HIGH-VOLTAGE SYSTEMS

William G. Dunbar

12.1 INTRODUCTION

A detailed knowledge of the application of electrical insulation and high-voltage techniques is essential if reliable high-voltage systems are to be designed and manufactured. Today's trend is to develop systems that operate at high voltages in compact, economical packages, which have been subjected to a multitude of diagnostic tests to assure that the customer will receive high-quality, reliable assemblies. To keep in the context of this chapter, high voltage will generally refer to assemblies or circuits that operate at voltages exceeding 150 Vac or 225 Vdc.

12.2 PROPERTIES OF INSULATING MATERIALS

Electrical insulation properties change when they are subjected to electric fields and temperature variations, mechanical stresses, pressure variations, and contact with rough electrode configurations. Long-term overstresses are contributors to voltage breakdown and failure. The designer must be aware of these changes and understand certain fundamental characteristics of insulating behavior before a long-life product can be assured. The basic theory of insulating gases, liquids, and solids is provided in the literature.[1-10] This section is devoted to application data useful in the packaging design of high-voltage equipment. Each insulation class is treated separately with respect to the electrical properties of the material when subjected to high-voltage applications.

12.2.1 Vacuum

Vacuum as an insulation is considered very good, provided the electrodes are completely isolated and the spacing between electrodes is small, less than 1.0 cm and the gas pressure is less than 10^{-3} Pa. Higher pressures and larger spacings can result in Paschen breakdown, explained in Sec. 12.2.2.1. For small electrode spacings, vacuum breakdown exceeds 150 kV/cm. For practical applications the breakdown voltage is much less, as shown in Fig. 12.1. Techniques to prevent voltage breakdown include the following.

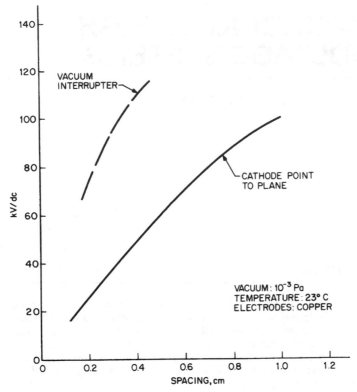

FIGURE 12.1 Dc breakdown voltage as a function of gap length at 23°C in vacuum.

- Well outgassed electrodes, that is, porous electrodes or electrodes with a large amount of absorbed gas, will have lower breakdown voltage because the gas will form on the electrode surface, increase the surface gas pressure, and result in lower breakdown voltage in the vicinity of the gas pocket.
- The electrodes should be polished and, if possible, made of a hardened metal.

  Magnetic-field interaction with the electric field may result in lower breakdown voltage. The magnetic field strength usually has to be greater than 0.5 tesla to be effective. Pulse power and dc field experiments at Texas Tech University[11,12] have shown that breakdown can be either increased or decreased, depending on the relationship of E × B, as shown in Fig. 12.2. A comparison of several experiments shows that the field is influenced by the smoothness of the material, the type of material, the size and shape of the material, and the orientation of the electric and magnetic fields.
- All forms of gases will lower the breakdown voltage because of materials outgassing, such as dust, materials spalling, oils and greases, and other forms of debris.
- Cryogenic temperatures have a tendency to raise the breakdown voltage slightly for some electrode configurations. Cold temperature for most cases does not improve the breakdown voltage.

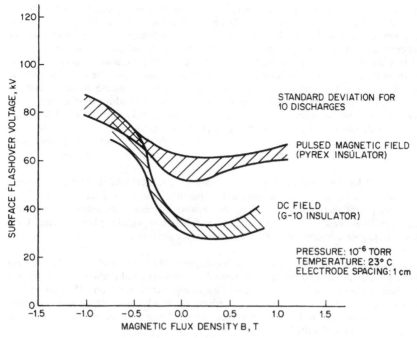

FIGURE 12.2 Flashover versus magnetic flux density using pulsed and dc magnetic field, where B is parallel to surface and perpendicular to E.

- Outer space, as between the earth and the moon, is not a good vacuum. There are many breakdown mechanisms in space, such as plasma, ultraviolet light, gamma and beta rays, and dust particles. All these, and others too numerous to mention, can lower the breakdown voltage.

12.2.2 Gases

For a detailed discussion on the insulating properties, the designer is encouraged to refer to the literature.[4-10] These references contain excellent data concerning voltage breakdown, prebreakdown currents, ionization, glow discharges, corona, and partial discharges. All these phenomena are properties leading to gaseous discharges. The designer must know how to use these data to develop robust equipment.

12.2.2.1 Paschen's Law. The breakdown voltage of a uniform-field gap in a gas can be plotted to relate the voltage to the product of gas density and gap length. This is known as Paschen's law curve.[8,9] The law may be written in the general form

$$V = f(pd) \qquad (12.1)$$

where p is the gas density and d is the distance between parallel plates. Usually the gas pressure p is used. This is valid, except at higher pressures, such as at several atmospheres in SF_6, where this simple proportionality no longer exists.[11]

Figure 12.3 shows typical V versus pd curves for several gases. [Note that 1 pascal (Pa) $= 7.5 \times 10^{-3}$ torr.] There is a minimum in this curve because as the density is reduced, the spacing between gas molecules becomes so large that although every electron collision produces ionization, it is not possible to achieve enough ionizations to sustain the chain reaction. Finally the pressure becomes so low that the average electron travels from one electrode to the other without colliding with a molecule. This is why the minimum breakdown voltage varies with gas density and spacing. For a 1-cm spacing at room temperature, this minimum occurs at about 100 to 300 Pa • cm, depending on the gas or gas mixture. A representative minimum for air at room temperature is 326 Vdc at 0.7 torr • cm. For an electrode spacing of 1 cm at standard atmospheric conditions, the breakdown voltage of air is 31 kV. The Paschen breakdown curves are valid except at very high pressures, as in SF_6, and at very low pressures, below 10^{-3} torr, or for very wide spacings.

12.2.2.2 Breakdown Strength of Gases.   Aerospace electrical and electronic equipment must be designed to operate at the maximum specified altitude and temperature. Table 12.1 lists the potentials required for voltage breakdown in gases at the minimum pressure-spacing condition (Paschen's law minimum) and between parallel plates spaced 1 cm apart. Of these gases, conditioned air is used whenever possible.

12.2.2.3 Gas-Density Relationship.   The breakdown voltage in a uniform field is always the same for a given gas and gas density, regardless of what combination of pressure and temperature produces that density for electrode spacings between 0.5 and 25 mm and temperatures below 500°C. For aerospace applications, the breakdown voltage may be related to temperature and altitude by using the U.S. standard atmosphere, which gives the air pressure as a function of altitude.

The conditions for simulating a given operating altitude and temperature can be calculated by using this relationship derived from the ideal gas law,

$$P_t = P_o \left(\frac{273 + t_t}{273 + t_0} \right) \quad \text{at constant volume} \tag{12.2}$$

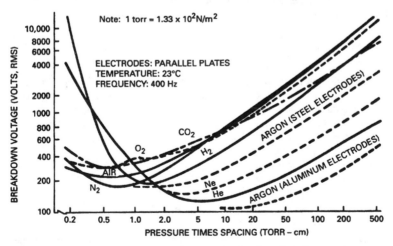

FIGURE 12.3   Voltage breakdown of pure gases as a function of pressure × spacing.

TABLE 12.1 Breakdown Voltage between Bare Electrodes Spaced 1 cm Apart

Gas	Minimum at critical pressure × spacing		Breakdown voltage at 1 atm	
	V ac rms	V dc	kV ac rms	kV dc
Air	223–230	315	23	33
Ammonia	320	400	18.5	26
Argon	196	280	3.4	4.8
Carbon dioxide	305	430	24	28
Freon 14	340	480	22.8	32
Freon 114	295	420	64	90
Freon 115	305	430	64	90
Freon 116	355	500	53	75
Freon C 138	320	450	59	83
Helium	132	189	1.3	1.63
Hydrogen	205	292	12	17
Nitrogen	187	265	22.8	32
Oxygen	310	440	—	—
Sulfur hexafluoride	365	520	67	95

where t_o = operating temperature, °C

t_t = test temperature (usually room temperature), °C

P_o = operating pressure, torr

P_t = test-chamber pressure, torr

The corona initiation voltage (CIV), as a function of gas pressure and temperature, for round nichrome wires will follow Paschen's law in the low-pressure region, decreasing to a minimum voltage at a predictable pressure, followed by a voltage increase as the pressure is further reduced. There is some variation in the minimum CIV at temperatures between 500 and 1100°C, but little or no change from 23 to 500°C.

Dielectric-coated electrodes can increase the breakdown voltage in gases compared with bare electrodes,[13] as shown in Fig. 12.4. This technique for increasing the breakdown voltage is not recommended unless the coating materials are given sufficient life testing and the coating process is held to a very tight tolerance. The application of coatings to the electrodes can be recommended for improving the safety margin. However, a coating that becomes unbonded may flake or blister, lowering the breakdown voltage to values below those of bare electrodes.

Metallic impurity particles can reduce the breakdown voltage in gas-insulated systems dramatically.[13-16] This effect becomes more pronounced with increasing pressure and particle length (Fig. 12.5). The particles become charged in the field and move between the electrodes. The field nonuniformity at the particle initiates the low-voltage breakdown.[14-17] Particle traps can be used to deactivate the particles.

When using nitrogen and SF_6 mixtures, it has been found that the addition of 20 to 30 percent SF_6 to nitrogen raises the dielectric breakdown strength of the pressurized mixture to over 80 percent of that of pure SF_6. These mixtures could be used to reduce the cost of the insulating gas in gas-insulated systems.

Curves showing the ratio of impulse voltage to steady-state breakdown voltage for three electrode configurations in air at 1-atm pressure are given in Fig. 12.6. These curves show that very fast single-pulse short-duration transients (less than 10-ns duration) may

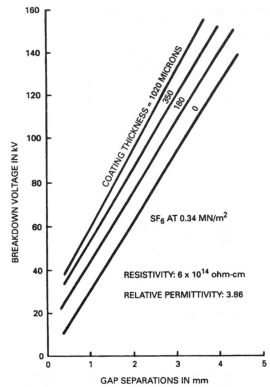

FIGURE 12.4   Effect on dc breakdown in SF$_6$ of unloaded polyurethane coating on electrodes.

not cause breakdown at normal operating voltage. Slow transients (less than 1-μs duration) require 105 to 110 percent of steady-state voltage for breakdown.

Some gases have very low breakdown characteristics and should not be considered. Helium is an example. Fluorocarbons are the preferred gases. Of these gases, SF$_6$ is generally the accepted gas because it is stable, electronegative, and easily contained, as shown in Fig. 12.7.

Electronegative gases are chemically inert and have good thermal stability, but can decompose chemically when exposed to partial discharges or arcs. The products of decomposition are often toxic and corrosive. In addition, a small quantity of water decomposes the SF$_6$ to form hydrofluoric acid when in the presence of a partial discharge or arc. Once formed, the hydrofluoric acid etches into crevices and requires special cleaning of all parts within the pressurized module.

12.2.3   Liquid Dielectrics

Liquids used as insulators include mineral oils, askarels, silicone oils, fluorocarbons, vegetable oils, organic esters including castor oil, synthetic hydrocarbons, and polebutenes (polyhydrocarbon oils). The characteristics of some typical liquids[18,19] used for high-volt-

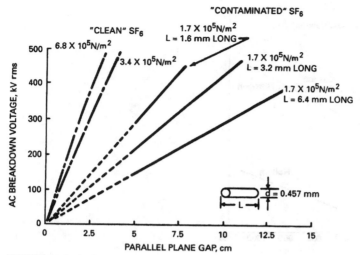

FIGURE 12.5 Ac breakdown voltage gap characteristics in SF_6 with copper particles of various lengths.

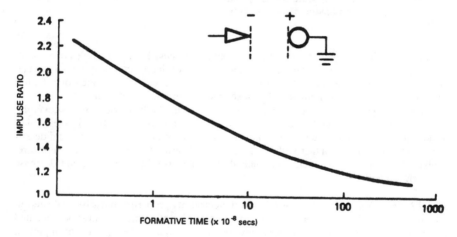

FIGURE 12.6 Relation between formative time and impulse ratio for various gap lengths and gas pressures in negative point-sphere gap in air.

age applications are given in Table 12.2. Mineral oils are most commonly used for transformer or high-voltage equipment insulation. Silicones are used for specialized applications when their fire resistance and high operating temperature characteristics are required. The other fluids are used with capacitor systems when their high cost is justified by the superior performance required for the high-voltage stress conditions typical in capacitors.

Advantages of liquid dielectrics are that they may be used as insulators and as heat transfer media. Often, liquid dielectrics are used as impregnants in conjunction with solid

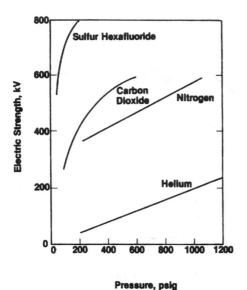

FIGURE 12.7   Effect of pressure on dc voltage strength of representative gases tested in uniform field.

insulations such as papers, films, and composite materials. By eliminating air or other gases, liquid dielectrics improve the dielectric strength of the insulation system. They can also be self-healing, in contrast to solid dielectrics, since the affected area of a failure caused by a breakdown is immediately reinsulated by fluid flow back to it.

Disadvantages, which always accompany the use of liquid dielectrics, are cost, weight, and relatively limited operation temperature with respect to many solids and gases. Other disadvantages with many liquids are combustibility, oxidation, contamination, and deterioration of materials in contact with the liquid. The deterioration may generate moisture, evolve gas, form corrosive acids, produce sludge, increase dielectric loss, and decrease dielectric strength.

12.2.3.1 Selection.   In selecting a liquid dielectric, its properties must be evaluated in relation to the application. The most important properties of a fluid that must be evaluated for an electrical application are temperature, moisture, and dissolved gas. Other important properties are dielectric strength, dielectric constant, conductivity, flammability, viscosity, thermal stability, purity, flash point, chemical stability, and compatibility with other materials of construction and the local atmosphere.

Temperature affects the solubility of gas, so heating can cause dissolved gas to evolve from the liquid. The same is true for changes in pressure. Corona will start within the gas bubbles when subjected to electric fields, leading to eventual dielectric breakdown. Thus liquids used as impregnates must have a low, stable gas content.

12.2.3.2 Mineral Oils.   Mineral oils are the mostly widely used of all liquid dielectrics. Typical characteristics of mineral oils used in common dielectric applications are listed in Table 12.3. The refining process must be evaluated to be certain it was designed to remove deleterious materials, such as sulfur and nitrogen, without removing or destroying the

TABLE 12.2 Properties of Dielectric Fluids

Property	Mineral oil	Dimethyl silicone at 50 cP	Hydrocarbon distillates	Isopropyl biphenyl	Isobutyl* monochloro biphenyl oxide	Di-2-ethylhexyl phthalate-25 without trichlorobenzene	Phenol xylyl ethane	Di-2-ethylhexyl phthalate
Relative dielectric constant at 60 Hz, 100°C	2.1	2.6	2.1	2.6	4.0	4.5	2.5	4.3
Dissipation factor at 60 Hz, 100°C, %	0.1	0.01	0.4	0.2	1.0	1.0	0.2	0.5
Dielectric strength, ASTM D-877, kV	35	35	38	60	35	35	55	42
Gas absorption coefficient, ASTM D-2300, μL/min	21	-13	-5	180	80	38	120	25
Viscosity at 100°F (37.8°C), cP	10	41	310	5.8	10.5	13	6.1	29
Pour point, °C	-55	-55	-20--30	-55	-45	-52	-48	-45
Fire point, °C	150	345	300	155	200	255	148	235
Use	Transformers, cables, circuit breakers	Fire-resistant transformers	Fire-resistant transformers	High-voltage capacitors	High-voltage capacitors	High-voltage capacitors	High-voltage capacitors	Low-voltage capacitors

*No longer available.

TABLE 12.3 Typical Characteristics of Mineral Insulating Oils

Property	For use in solid-type cables	For use in capacitors and hollow-core cables	For use in transformers, switches, and circuit breakers
Condition	Clear	Clear	Clear
Viscosity, cSt	100 (98.9°C)	100 (37.8°C)	58 SSU (37.8°C)
Specific gravity	0.930 (15.5/15.5°C)	0.885	0.885 (15.5/15.5°C)
Color	2.3 (NPA)	1 or less (NPA)	1 or less (NPA)
Neutralization number, mg KOH/g	0.2	0.02	0.02
Flash point (open cup), °C	235	165	135
Burn point (open cup), °C	280	185	148
Pour point, °C	−5	−45	−45
Free sulfur	Nil	Nil	Nil
Total (fixed) sulfur, %	0.35	0.15	0.1 or less
Evaporation (8 h at 100°C), %	—	—	8
Dielectric strength, kV	30	30	30
Specific heat at 30–35°C	—	0.412	0.4252
Power factors at 100°C	0.001	0.001	0.001
Chlorides and sulfates	Nil	Nil	Nil
Resistivity at 100°C, $\Omega \cdot cm$	$1\text{--}10 \times 10^{12}$	$50\text{--}100 \times 10^{12}$	$1\text{--}10 \times 10^{12}$
Coefficient of expansion, °C^{-1}	0.00075	—	0.00070
Specific optical dispersion	115–120	115–120	110–115
Thermal conductivity, cal/(s · cm · °C)	—	—	0.39
Refractive index at 25°C	—	—	1.4828
Aniline point, °F (°C)	—	—	169 (76)

crude-oil constituents which appear to be necessary for long life and stability, such as the aromatic hydrocarbons. Like inhibitors which are added during the manufacture of mineral oils, aromatic hydrocarbons slow down the rate of oxidation.

Oil color is an index of the degree of refinement for unused oils. It also gives a rough measure of the deterioration of oil in service. Cloudiness indicates the presence of moisture, sludge, particles of insulation, products of metal corrosion, or other undesirable suspended materials. Contaminants are introduced into mineral oils from:

Improper manufacturing and refining methods

Improper handling and shipping procedures

Oxidation of the oil

Soluble polar particles produced by moisture

Leakage from the materials of construction or other insulations

Solid components, which may or may not be used in contact with mineral oils for long periods of time, are shown in Table 12.4. The interfacial test is a sensitive detector of small concentrations of polar contaminants and oxides. This and other tests are necessary in specifying electrical insulating oils.[20] There are also new methods of accelerated testing and rapid measurement.[21]

12.2.3.3 Askarels.   Askarels are synthetic liquid dielectrics, which have been used primarily in capacitors and transformers. They are derived by chlorination. However, because they are now considered environmentally suspect, they can no longer be used.

12.2.3.4 Silicone Oils.   Silicone oils most commonly used as liquid dielectrics are dimethyl silicone polymers. These silicones are characterized by a relatively flat viscosity-

TABLE 12.4 Material Compatibility with Mineral Oils

Compatible materials	Incompatible materials
Alkyd resins	Acrylic plastics
Cellulose esters	Asphalt
Cork	Chloride flux
Epoxy resins	Copper (bare)
Masonite	Fiber board
Melamine resins	Greases
Nylon	Polyvinyl chloride resins
Phenol-formaldehyde resins	Rubber, natural and synthetic
Polyamide-imides	Saran resins
Polyester-imides	Silicone resins
Polyethylene terephthalate (Mylar)	Tars
Polyurethane	Waxes (petroleum)
Pressboard	
Shellac	
Silicone rubber	
Wood	

temperature relationship, resistance to oxidation, stability at high temperature, and excellent high-frequency characteristics. They are unique in two important characteristics. The viscosity of various silicones ranges from 1 to 1,000,000 centistokes (cSt), and they are stable for appreciable lengths of time at 150°C in air and at about 200°C in the absence of air.

Silicone liquids resist oxidation and do not form sludge as do mineral oils. Their stability in the presence of oxygen makes them low fire and explosion hazards, even at temperatures up to 200°C.

12.2.3.5 Miscellaneous Insulating Liquids. Other liquid dielectrics include fluorocarbons, vegetable oils, organic esters, and synthetic aromatics. These, among which there are new capacitor fluids (Table 12.2), are discussed in Clark[22] and *Insulation/Circuits.*[23]

12.2.3.6 Filtering and Outgassing. Oils used as liquid dielectrics should be filtered before use and outgassed when installed. Mineral oils, vegetable oils, and organic esters should be outgassed at 85°C at a pressure of 10 Pa (0.7 torr) for 4 h prior to application.

Where possible, mineral oils, while serving as high-voltage dielectrics, should be continuously circulated through sorbents, such as activated alumina, activated Fuller's earth, or activated bauxite, by thermosyphon action. Such filtering by controlling contamination limits the loss of dielectric strength.[24]

12.2.4 Solid Insulation

Potting or encapsulating material for an electrical/electronic component is selected on the basis of its physical, electrical, thermal, mechanical, and chemical properties. An ideal potting material should be a very lightweight material with low dielectric constant, dissipation factor, thermal expansion, viscosity, glass transition temperature, cure shrinkage, and moisture absorption. It should possess high volume resistivity, dielectric strength, surface resistivity, insulation resistance, dry arc resistance, tensile strength, lap shear strength, peel strength, tear strength, and coefficient of thermal conductivity. It should possess long pot and shelf lives and should cure overnight at room temperature with no exothermic reactions and no by-products. It should flow easily and uniformly and should fill spaces with no voids. It should stick to all the materials covering the surfaces of various components and connections present in the potted component but should be easily demolded, if necessary. It should cure into a hard material in order to physically support the components and connections, but should be soft enough to cut and remove if it is necessary to repair failed components and connections.

For high-quality aerospace applications, the critical properties and their acceptable ranges of values are given in Table 12.5. The acceptable range of values for each material property was obtained from data published in the military specification MIL-1-16923G[25] and from private communications with experienced high-voltage designers and analysts. These data are relevant to ASTM tasking procedures for fresh materials processed under ideal conditions.

The magnitude of most of these material properties depends on the thickness, dimensions, preparation, environment, and cure procedures used to make the samples. Contamination of the potting compound will affect its properties. For instance, removal of metallic particulate will improve the life of an epoxy, as shown in Fig. 12.8. The measured values will also be affected, sometimes extensively, by the environmental factors such as temperature and humidity. Likewise, some of the electrical properties are extensively influenced by the applied voltage and its frequency. Most of the values for the material properties can be obtained from the manufacturers' data sheets, but in some cases the data sheets do not provide adequate information on important variables such as sample preparation, dimen-

TABLE 12.5 Material Properties for Potting Materials

Properties	Range of values
Dielectric strength, 0.1-in-thick sample	≥ 350 V/mil
Dielectric constant, 50-Hz to 1-MHz range	< 5.0
Dissipation factor, 50-Hz to 1-MHz range	< 0.1
Volume resistivity	$> 1.0 \times 10^{13}$ $\Omega \cdot$ cm
Surface resistivity	$> 1.0 \times 10^{13}$ Ω
Dry arc resistance	> 120 s
Coefficient of linear thermal expansion	$< 10^{-4}/°C$
Coefficient of thermal conductivity	$> 5 \times 10^{-4}$ cal/(s $\cdot$ cm $\cdot$ °C)
Thermal shock	Pass -55 to 125°C for 10 cycles
Thermal stability	No reversion -55 to 125°C
Brittleness temperature	$< -60°C$
Tensile strength	> 500 lb/in^2
Lap shear strength	> 500 lb/in^2
Peel strength	> 10 lb/in
Tear strength	> 20
Viscosity, at 25°C	< 600 cP
Pot life	> 30 min
Shelf life	> 6 months
Cure shrinkage	$< 0.2\%$
Cure inhibitors	None
Moisture absorption	$< 0.2\%$ for 7 days
Repairability	Yes
Fungus resistance	Nonnutrient
Flammability	Self-extinguishing
Wettability	Yes
Compatibility	Yes

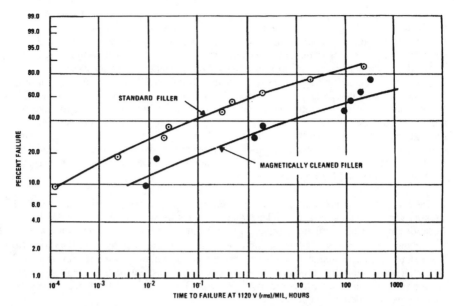

FIGURE 12.8 Comparison of high-voltage endurance of high-performance epoxy encapsulating materials using magnetically cleaned and standard filler.

sions and thickness, and the environmental conditions during the test. As an example, the short-term dielectric strength is influenced by the material thickness, as shown in Fig. 12.9. Sometimes it is difficult to get such information even from the manufacturer's technical assistance group. Certain properties, such as the glass transition temperature, can be measured by more than one technique and may yield differing values. There are some properties that have no standard definition or test procedure and, therefore, may have little reliability or repeatability for their measured values.

The data available from the product manufacturers' data sheets are based on statistical averages and do not claim very high degrees of repeatability or reliability. Therefore, such data should be used only as preliminary criteria toward characterizing the materials. For a proper comparison of the materials, some of the critical design properties should be defined and evaluated by independent or qualified testing laboratories using established procedures for sample preparation and calibrated equipment and standard procedures for measurements.

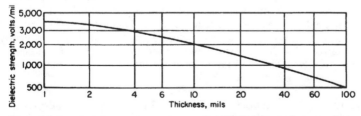

FIGURE 12.9 Short-term dielectric strength of TFE fluorocarbon as a function of thickness.

The data compiled for eight selected materials are listed in Table 12.6. They were collected from published data sheets and from telephone conversations with representatives of the manufacturers' technical support groups. Data on lap shear strength and peel strength, which are very critical in assessing the adhesive strength and wettability of the material with the substrates, are conspicuous by their absence on the data sheets.

Data on the relaxation frequencies of the materials due to polarization are also missing from the produce manufacturers' data sheets. If the relaxation frequency of the material is an odd multiple of the operating frequencies of the high-voltage power supply, the loss index will increase considerably and will result in excessive heating of the material. The polarization frequencies also have a tendency to decrease with increasing temperature and may remain well within the odd harmonics of the operating frequencies of the component. Therefore relevant data on the polarization frequencies of the material should be measured by independent testing laboratories.

12.2.4.1 Practical Applications.
In applications, the environmental and electrical stresses vary sporadically; some independently, others depending on each other. These variations may make it difficult to select an ideal insulation for a specific application.

For some of the materials that contain fillers, there is a tendency for the fillers to settle to the bottom of the container. In order to ensure that the mix ratios for such materials are correct, it is important that the fillers be brought back into suspension by stirring the materials in order to ensure that they are homogeneous. Both resins and hardeners should be stirred separately before weighing.

Degassing to eliminate voids is achieved by applying a vacuum of 27 to 29 in Hg (absolute pressure 25 to 75 torr) on the mixed material for a period of 2 to 10 min. The applied vacuum and its duration depend on the material and its mass. When the vacuum is applied, the entrapped air expands, the liquid froths, and the air bubbles rise to the surface of the material. To accommodate the expansion, the container should be about four to five times the volume of the material. To ensure proper degassing, it may be necessary to apply the vacuum more than once and let the trapped air escape from the material.

The effect of prolonged application of vacuum on mixed materials should be studied carefully. When a vacuum is applied too long, the catalyst may separate from the resin and leave the compound nonhomogeneous in nature. In such cases, the potting material may not cure properly and may not possess the properties critical for reliable potting. Therefore, the degassed compound should be checked visually for the presence of streaks denoting separation of the catalyst. It may also be necessary to measure the viscosity of the degassed compound to detect any changes in the physical properties of the material.

Composite and laminated insulations are used for terminal boards, and also for supports that separate the coils and wiring from the cores, structure, and containers. Typical properties of glass and nylon containing laminates are shown in Tables 12.7–12.9. More complete listings of materials can be found in the literature.[26–28]

It must be noted that the measurement of the dielectric strength of a laminated board is critical to the orientation of the electrodes with respect to the grain of the laminations. When the dielectric strength is measured perpendicular to the laminations, it may be as much as 20 times greater than when the same measurement is made parallel to the laminations. It is important to keep this characteristic in mind when using small blocks of laminated boards to separate high-voltage and low-voltage circuits or modules. The data in Table 12.9 give the perpendicular dielectric strength.

12.2.4.2 Imperfections and Life.
If voids, cracks, or imperfections are present, gas discharges will occur. If the void dimension, shape, position, and filling gas are known, then the applied voltage required for the onset of partial discharges in the gas inside the void

TABLE 12.6 Material Characterization of Potting Compounds

Material	STYCAST 2651-40	EN-2521	RTV 615	ECOSIL 4952	ECOSIL 5952	SILASTIC E	SYLGARD 170	DC X3-6121
Classification	Epoxy	Polyurethane	Silicone	Silicone	Silicone	Silicone	Silicone	Silicone
Manufacturer	Emerson Cuming	Conap	General Electric	Emerson Cuming	Emerson Cuming	Dow Corning	Dow Corning	Dow Corning
Catalyst/hardener	Catalyst 11/9	Part B	Component B	Catalyst 50	Part B	Curing agent	Part B	Curing agent
Primer	None	AD6,PR1167	GE SS4155	S-11	S-11	DC 1200	DC 1200	DC 1200
Shelf life, months	6	12	6	6	6	6	12	6
Physical properties								
Color	Black	Tan	Transparent	Red	Red	White	Dark gray	Translucent
Specific gravity	1.44	1.53	1.02	2.2	2.05	1.12	1.33	1.13
Viscosity at 25°C, cP	4000	4000	3500	35,000	40,000	100,000	2500	35,000
Pot life, h	4.0	1.3	4.0	0.5	1.6	2.0	0.5	2.0
Electrical properties								
Dielectric strength, V/mil	450	650	500	500	500	500	450	510
Sample thickness, in	1/10	0.0625	0.076923	0.1	0.1	0.076923	0.125	N/A
Dielectric constant	4.8; 4.6; 4.0	4.8; 4.3; 3.7	3.0; 2.83	5.0; 5.2	5.0	2.81; 2.82	3.15; 3.1	2.91; 2.91
Frequency, Hz (k=kilo; M=mega)	(60);(1k);(1M)	(100);(1k);(1M)	(100)(938M)	(1M);(8600M)	(1M)	(100);(100k)	(100);(100k)	(100);(100k)
Dissipation factor	0.02; 0.01; 0.02	0.072; 0.045; 0.016	0.001	0.01; 0.001	0.01	0.006; 0.002	0.008; 0.002	0.0003; 0.0002
Volume resistivity, 10 Ω·cm	14	1.2	10.0	2.5	1.0	5.0	10.0	15.0
Surface resistivity, 10 Ω	N/A	4.6	N/A	.044	1.0	N/A	N/A	N/A
Arc resistance, s	120	130	100	N/A	N/A	121	120	120
Thermal properties								
Coefficient of linear thermal expansion, 10°C	0.47	1.6	2.7	1.8	2.0	2.9	2.7	2.6
Thermal conductivity, 10 cal/s·cm·°C	8.97	6.5	4.5	26.0	26.0	4.35	7.5	N/A
Thermal stability, −55 to 125°C	Pass	Pass	Pass	Pass	Pass	Pass	Pass	Pass
Thermal shock, −55 to 125°C; 10 cycles	Pass	Pass	Pass	Pass	Pass	Pass	Pass	Pass
Glass transition (brittleness) temperature, °C	N/A	N/A	−60	−65	−65	−65	−65	−115
Mechanical properties								
Tensile strength, lb/in²	9200	1600	800	718	500	700	500	550
Lap shear strength, lb/in²	N/A	N/A	N/A	N/A	N/A	N/A	N/A	700
Peel (tear) strength, lb/in²	N/A	N/A	(25)	(25)	(20)	(90)	(25)	(55)
Miscellaneous								
Mixing ratio, base:curing agent	100:8.6	1:5	10:1	250:1	1:1	10:1	1:1	10:1
Degassing pressure, mm Hg	50-75	1-5	25	50-75	50-75	50	50	50
Time for room temperature curing, h	24	168	168	24	168	24	8	72
Elevated temperature, °C; curing, h	100;4	60;6	65;4	65;4	65;4	65;0.5	70;0.25	65;4
Cure shrinkage, %	0.2	0.7	0.2	0.55	0.22	0	0	0
Water absorption (7 days), %	0.1	0.55	N/A	0.21	0.1	0.1	0.1	N/A
Cure inhibitors	Si	Amines	S, Chl. Rubber	S, N	S, N	S, Org. Rubber	Organotin, S	Chl. Rubber
Repairable	No	No	Yes	Yes	Yes	Yes	Yes	Yes
Mold release	TFE-20	MR5002	Teflon spray	TFE-20	TFE-20	Soap sud	Soap sud	Soap sud

TABLE 12.7 Material Properties of Laminates and Compositions

NEMA grade	Base material	Resin	Specific gravity	Water absorption, %
G-7	Glass cloth	Silicone	1.68	0.55
G-9	Glass cloth	Melamine	1.9	0.8
G-10*	Glass cloth	Epoxy	1.75	0.25
G-11	Glass cloth	Epoxy	1.75	0.25
N-1	Nylon	Phenolic	1.15	0.6
FR-4*	Glass	Epoxy	1.75	0.25
FR-5	Glass	Epoxy	1.75	0.25

*Many NEMA G grades are being replaced with fire retardant FR grades. For instance, G-10 is equivalent to FR-4.

TABLE 12.8 Mechanical Properties of Laminates and Compositions

	Flexural strength, 1.6-mm thick, $10^8 \times$ N/m^2	Tensile strength, $10^8 \times$ N/m^2	Compressive strength, $10^8 \times$ N/m^2	Bond strength, kg	Rockwell hardness M scale
G-7	1.4	1.6	3.1	295	100
G-9	4.1	2.7	4.5	770	—
G-10*	4.1	2.4	4.8	900	110
G-11	4.1	2.4	4.8	725	110
N-1	0.7	0.6	1.9	450	105
FR-4*	4.1	2.4	4.8	900	110
FR-5	4.1	2.4	4.8	725	110

*Many NEMA G grades are being replaced with fire retardant FR grades. For instance, G-10 is equivalent to FR-4.

TABLE 12.9 Electrical Properties of Laminates and Compositions

	Dielectric constant at 1 MHz, 0.8 mm	Dissipation factor at 1 MHz, 0.8 mm	Dielectric strength at 0.8 mm, kV/mm	Volume resistivity, $\Omega \cdot$ cm	Surface resistance, MΩ	Arc resistance, s
G-7	4.2	0.003	11	—	—	180
G-9	7.5	0.018	10	—	—	180
G-10*	5.2	0.025	20	10^{12}	10^4	128
G-11	5.2	0.025	16	10^{12}	10^4	115
N-1	3.9	0.038	15	—	—	—
FR-4*	5.2	0.025	18	10^{12}	10^4	128
FR-5	5.2	0.025	18	10^{12}	10^4	128

*Many NEMA G grades are being replaced with fire retardant FR grades. For instance, G-10 is equivalent to FR-4.

can be calculated.[29,30] The partial-discharge onset voltage for gas-filled cavities will occur when the voltage across the cavity corresponds to the *pd* value for the cavity on the Paschen curve (Fig. 12.3). The continual presence of discharges may lead to complete failure after a period of time that depends on the resistance of the material to discharges. Figure 12.10, for example, shows the life of polyethylene film with and without discharges.

The erosion, or weakening, of insulation through internal discharge attack may be the result of several effects progressing simultaneously:

Thermal degradation caused by local heating from ionization, streamers and increased losses in surrounding solid materials

Degradation of solid material and reaction with gas in cavity

Degradation of gas and reaction with cavity surfaces

Partial breakdown in solid material (treeing)

Extensive studies have been made on the voltage-time endurance characteristics of epoxy and other materials.[31,32] Figure 12.11 shows the effect of electrode material and configuration.[31] These characteristics have been treated mathematically to determine the band of the curve using extreme-value statistics. It is of interest to note that in many cases no discharges above the background of 0.05 pC could be detected before the sudden onset of failure.

A recommended technique to determine the life of a material or dielectric system is to use the Arrhenius plot of the life as the reciprocal of absolute temperature. The endpoint for the life could be, for example, breakdown strength, specific loss of weight, elongation, mechanical strength, or evolution of gas. Figure 12.12 shows a typical Arrhenius plot for insulated magnet wire; similar plots can be made for cast epoxy, impregnated systems, composites, and all-solid dielectric material systems. The Arrhenius plot can, if its limits are recognized, allow extrapolation of the life of the material from elevated-temperature tests to the service-temperature conditions. A thermal-life design rule is that for every 8 to 12°C temperature rise, life is decreased by approximately 50 percent.

For accelerated life testing, higher frequencies are used, for example, testing at 1240 Hz for 60-Hz materials. This is usually valid. Tests have shown that the insulation life degrades inversely to the sinusoidal frequency, or

$$L \approx 1/f \tag{12.3}$$

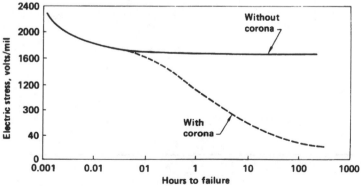

FIGURE 12.10 Dielectric life of polyethylene with and without corona.

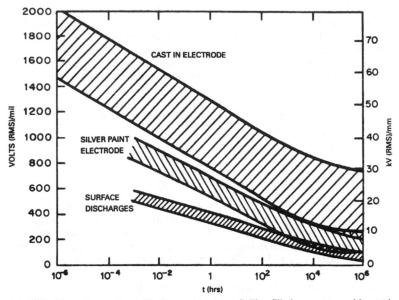

FIGURE 12.11 Comparison of voltage endurance of silica-filled cast epoxy with cast-in electrodes, silver paint electrodes, and surface discharges.

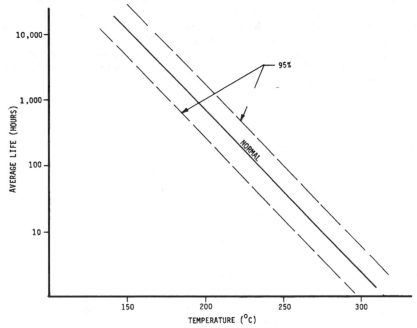

FIGURE 12.12 Arrhenius plot of insulation life versus temperature for class H insulation.

where L is life in hours and f is frequency in hertz. When frequency, temperature, and voltage stress derating factors are all combined, it is easily understood how an insulation with a 10,000-V test at 1 min can be reduced to less than a few hundred volts in application.[24,33,34]

Applying large ratios of test frequency to estimate the application life may be in error, especially for ratios greater than 25:1. A factor that may influence the life is that when high-pressure gases are formed in the partial discharges which cannot diffuse out effectively at the higher frequencies, the gas pressure in the void is higher and the partial-discharge onset voltage therefore is higher than at the lower frequencies.

As a rule, the three generic-type encapsulating materials for aerospace applications have maximum field stresses for 1 h of operation, as shown in Table 12.10. The lower values are for very thick insulations; higher values are for insulation less than 10 mil thick. Note that the data in Table 12.10 are for 1-h life, not 10,000 h. To obtain 10,000 h, the maximum stress must be decreased considerably. The voltage stress must be decreased by 8 to 10 percent for each order of magnitude increase in service life. For 10,000 h the stress should be decreased to about 65 percent of the values shown in Table 12.10. A generic curve that may help in design is presented in Fig. 12.13. This curve shows the voltage stress, in V/mil, as a function of life. This generalized curve has already included many insulation derating factors normal to electronic circuits. Not included are frequency, temperatures greater than 85°C, and units with volumes greater than 500 in³.

TABLE 12.10 One-Hour Life Field Stress

Materials	Maximum field strengths, V/mil
Epoxies	200–350
Silastics	300–600
Urethanes	250–500

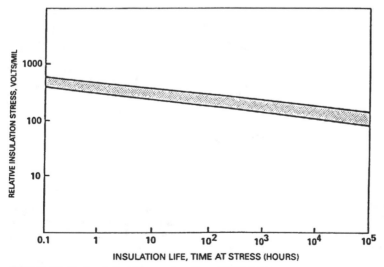

FIGURE 12.13 Insulation life as a function of field stress.

12.2.4.3 Treeing and Surface Flashover. Surface flashover along an insulator occurs when there is electrical breakdown in the gas at the surface. The type of flashover mechanism depends on the insulation surface conditions.[35–37] One technique that has helped to reduce carbon tracking is to coat or fill the insulating material with aluminum trihydrate.

Under dry gas and clean insulator conditions, the criterion for the gas-insulator flashover is usually one of the following conditions: (1) the field at the insulator surface or at an adjacent electrode reaches the critical breakdown field for the gas; (2) the interface of the insulator at an electrode is poor so that discharges there are sufficient to initiate flashover; or (3) particulate or moisture contamination on the insulator surface creates regions of high field, sufficient for ionization in the gas and leading to flashover.

The current flowing across the surface of an insulator, especially when slightly wetted and containing a conductive contaminant, may produce enough heat to generate a track of carbon, which becomes a conductive path, tending to reduce the capability of the insulator to resist the voltage. With some materials the surface erodes, but no track is produced. Fillers can effectively reduce the tracking tendency of organic materials. Eroding materials, such as acrylics, do not require filler protection. Tracking can also be controlled by reducing the applied stress on the surface. Petticoat insulation configurations lengthen the surface creepage path to reduce stresses tending to cause tracking.

Figure 12.14 shows the typical flashover voltages in air across uncoated epoxy bonded laminate under dry conditions. The test configuration was 1.95-cm diameter washers on both surfaces of the laminate, the spacing being varied up to 4 cm. The effect of higher frequency reducing the flashover voltage can be seen clearly and should be considered in any design application. All materials have lower flashover strength at higher frequencies. The example given in Fig. 12.15 illustrates the magnitude of the change.

High-dielectric-constant materials have much lower resistance to surface voltage creep than the low dielectric constant materials. Figure 12.16 illustrates the advantage in selecting insulation with the correct dielectric constant. The breakdown factor in the illustration represents the results of many measurements, showing how a decreasing flashover voltage can be expected across the dielectric when insulations with progressively higher dielectric constants are tested.

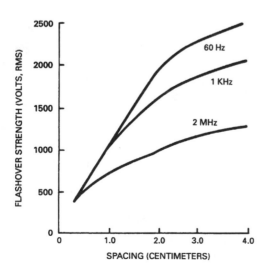

FIGURE 12.14 Effect of spacing on flashover voltage along epoxy surface in air.

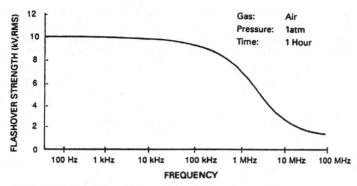

FIGURE 12.15 Effect of frequency on flashover strength.

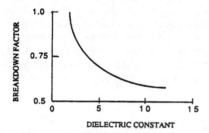

FIGURE 12.16 Variation of flashover voltage with changing insulation dielectric constant.

12.3 FIELD STRESS AND CONFIGURATIONS

Packaging for long life and high mean time between failure (MTBF) rates is a function of terminal, lead, and part configurations with respect to the electric field stresses within the materials. Several techniques for calculating the electric field stress between two or more electrodes are:

Analytic solution
Conformal mapping techniques
Finite-difference computer programs
Simulated-charge computer programs
Resistance-network analogs
Conducting-paper analogs
Curvilinear hand-plotting techniques

It is not necessary to calculate all the high-stress areas. The designer must have an intuitive insight toward focusing attention to those parts that have very small radii and small spacings. A small-radius part such as a wire or sharp-edged component could have a maximum electric stress 3 to 10 times that of the average stress, as will be shown. Once it is established which parts require field stress calculation or plotting, a technique familiar to the designer should be used to calculate the field stress. For quick field stress estimations, the utilization factors shown in Fig. 12.17 may be used. These factors are based on experimental data and have been shown to have less than 10 percent error. The utilization factor is numerically equal to the required voltage derating of a configuration. In equation form,

$$\eta = \frac{E}{E_m} < 1 \tag{12.4}$$

where E = average voltage stress between two electrodes spaced a unit apart, kV/mm

E_m = maximum voltage stress at conductor surface spaced a unit apart, kV/mm

12.3.1 Empirical Field Equations

An empirical field equation or formula is the shortened, simplified form of a rigorous equation. Rigorous equations, manageable with electronic calculators, are still difficult to use in everyday design work, especially if the design has to be assembled piecewise. Furthermore, maximum stress may be the only value needed, and the plotting of the complete field using a rigorous equation is not essential. Empirical equations for the maximum field stresses at the smaller electrodes are given in Table 12.11 for several electrode configurations. Electrical stresses calculated with these equations are within 10 percent of the values obtained with rigorous equations.

12.3.2 Configurations

Several electrode and insulation configurations that follow show the importance of maximum field stresses as compared to average field stresses.

12.3.2.1 High-Voltage Cable. High-voltage wire should be constructed with semiconducting layers of insulation around the stranded center conductor and over the insulation

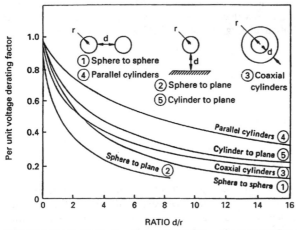

FIGURE 12.17 Utilization factor for various electrode configurations.

TABLE 12.11 Maximum Field Strength E with Potential Difference V between Electrodes for Various Electrode Configurations

	Configuration	Formula for E_{max}
	Two parallel plane plates	$\dfrac{V}{a}$
	Two concentric spheres	$\dfrac{V}{a} \cdot \dfrac{r+a}{r}$
	Sphere and plane plate	$0.9\dfrac{V}{a} \cdot \dfrac{r+a}{r}$
	Two spheres at distance a from each other	$0.9\dfrac{V}{a} \cdot \dfrac{r+a/2}{r}$
	Two coaxial cylinders	$\dfrac{V}{2.3r \log\left[\dfrac{r+a}{r}\right]}$
	Cylinder parallel to plane plate	$\dfrac{0.9V}{2.3r \log\left[\dfrac{r+a}{r}\right]}$
	Two parallel cylinders	$\dfrac{0.9V/2}{2.3r \log\left[\dfrac{r+a/2}{r}\right]}$
	Two perpendicular cylinders	$\dfrac{0.9V/2}{2.3r \log\left[\dfrac{r+a/2}{r}\right]}$
	Hemisphere on one of two parallel plane plates	$\dfrac{3V}{a}$, $a \gg r$
	Semicylinder on one of two parallel plane plates	$\dfrac{2V}{a}$, $a \gg r$
	Two dielectrics between plane plates, ε_1 ε_2	$\dfrac{V\varepsilon_1}{a_1\varepsilon_2 + a_2\varepsilon_1}$

media, as shown in Fig. 12.18. In this construction, the air trapped within the stranded center conductor is not electrically stressed and need not be eliminated. The insulation can be made advantageously of several layers, with the dielectric constant e of each layer being successively higher toward the center. The voltage gradient can then be maintained nearly constant from the inner to the outer conducting layer, rather than being much higher near the inner conducting layer (Fig. 12.19).

12.3.2.2 High-Voltage Connectors and Feedthroughs.   Connectors and feedthroughs must be designed to eliminate air voids between conducting surfaces. One successful method is to make one side of the mating interface from soft, pliable insulation (Fig. 12.20). When mated, the pliable insulation conforms closely to the opposite dielectric. The pliable insulation should first contact the molded insulation near the center conductor, then the contact should progress out to the shell without trapping air between the two contact surfaces.

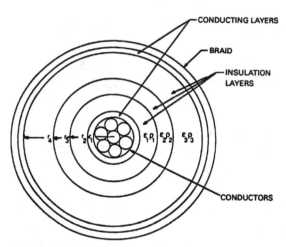

FIGURE 12.18   High-voltage wire.

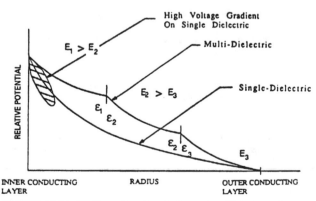

FIGURE 12.19   Field gradient for single- and multiple-layer dielectric.

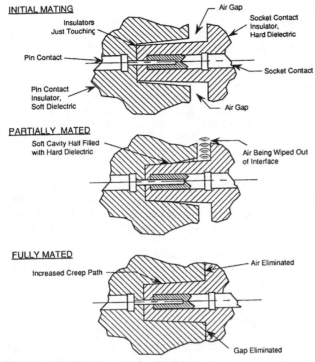

FIGURE 12.20 Proper connector mating sequence.

A thin layer of silicone grease may be applied to the insulation surfaces of some connectors to fill micropores in the insulation. Too much grease (more than 5 mil) tends to (1) prevent complete closure of the connector, (2) introduce air cavities, or (3) deform the pliable insulation. Therefore silicone or other additives are not recommended for properly constructed high-voltage connectors.

The contacts within a connector must float to allow for manufacturing tolerances and pin insertion and extraction. To do this, small spaces surround each contact. Furthermore, small air gaps will form about the wire insulation unless the insulation is pressed firmly against the contact. Many connectors are made with spring-loaded or crimp contacts, which leave an air gap above each contact.

Small air gaps around the pins are ideal for partial discharges and cannot be tolerated for high-voltage equipment designs and applications.The most common failure modes are (1) improper connector surface mating, leaving a small air gap along the surface of the mating insulator surfaces, and (2) burrs or strands protruding through the semiconducting layer, producing a point-to-ground highvoltage stress, which may result in treeing and voltage breakdown.

12.3.2.3 Solder Balls. Solder joints on circuit boards, especially printed-circuit boards, are designed to have minimum solder. This leaves the electrical post termination protruding through the solder on the circuit interconnection. Often the post terminal is clipped after soldering, leaving sharp edges protruding above the circuit metal surface. This is unacceptable for circuits with voltage exceeding 250 V peak. Solder balls must be made to cover the sharp tips to prevent corona. Some large components may have bolted high-voltage terminals. If the bolts do not have smooth rounded surfaces, they should have smooth rounded caps with spherical ends clamped over the nuts to prevent corona. In some cases the nuts may be placed inside larger U-shaped cups. The cups must have large rounded edges above the nuts.

12.3.2.4 Resistors.   Solid-core resistors should be used in aerospace applications throughout the highvoltage module. Hollow-core resistors, if sealed on the ends, will slowly depressurize at high altitude and arc internally from end to end along the hollow-core surface. Resistor coatings can be a problem. One supplier made resistors with light- or dark-blue coatings. The light-blue coatings were well bonded to the elements, the dark-blue coatings tended to debond. The resulting air gaps then depressurized and the resistors failed.

12.3.2.5 Capacitors.   Dielectrics used for high-voltage capacitors include liquids, liquid-impregnated paper or film, electrolytics, plastic film, paper-plastic combinations, mica, ceramics, glass, compressed gas, and vacuum. Unless special requirements, with respect to temperature, stability, radiation resistance, or packaging, are involved, liquid-impregnated paper or plastic film offer the best energy-space-cost combination and, consequently, are more widely available for high-voltage capacitors.

High-voltage and low-voltage capacitors are required for high-voltage power supplies. Electrolytic, filmwound (unimpregnated), and ceramic capacitors are used for both high- and low-voltage applications. High-voltage applications use impregnated film and reconstructed mica. These capacitors have proven performance for filters. Ceramic capacitors have excellent properties for voltage multipliers.

In most ceramic applications the metal plates are bonded to the ceramic wafer. The relationship between the thermal expansion coefficients of metal and ceramic often determines the success of the capacitor. The expansion coefficients need not be identical, but they must be well matched for optimum results. For instance, Kovar and nickel are a much better match than silver or aluminum for beryllia or magnesia ceramics.

High-frequency capacitors used in high-frequency circuits must have low inductance and very rapid charge and discharge characteristics. To obtain these goals, design and materials engineers look for better packaging and improved field enhancement at the foil edges, as well as higher-dielectric-strength materials. One application has the capacitor case biased to 25 percent of the working voltage to relieve the stress on the outer foil and negative lead.

Many high-voltage capacitors designed for small power supply applications have nonmetallic cases. Some of these capacitors have excellent film-plate construction, but less than desired end caps. In some cases the end caps have very sharp edges, which can lead to high field stress; therefore a small modification is recommended. Corona shields should be designed so that they can be placed over the high-voltage ends of the capacitors. The leads should be bent to penetrate through the center of the shields and soldered to make them a part of the design. The high-voltage leads can then be soldered to a more centralized location on the corona shield to improve the stress levels. This modification will greatly reduce the maximum field stress between the ends of the capacitor and another part or ground. It must also be remembered that a sharp point also results in a thermal short circuit. Better thermal control can be achieved by using the corona shields.

A high-voltage capacitor may be constructed using two or more sections. This means that the outer surface of a two-section capacitor is not at one equipotential, that is, half-voltage will exist over one-half of the outer surface. This potential must be considered when packaging the capacitor with many other highvoltage or low-voltage parts. Spacing between parts is very important.

12.3.2.6 Transformers. Transformer insulation life is based on two factors—overstress and time. Overstress must include two factors—the turn-to-turn overstress within the winding and the impressed input/output voltage magnitudes. An example is a 50-percent-overvoltage 1-μs transient. For many transformer designs, most of the transient overvoltage will appear across the initial turns of the winding. This could momentarily triple the stress between turns or between the turns in the affected layer and the adjacent layer. This type of overstress can cause puncture to occur between the inner or outer layers of a transformer designed with high-voltage stresses. At first it would be expected that the failure would occur between the end turns. The failure will occur in the vicinity of least insulation, that is, at insulation flaws such as voids, or places where impurities are embedded in the insulation. Thus the breakdown could occur in the center of the winding. Pulse transformers and very-high-frequency transformers are less prone to this type of failure.

12.3.2.7 Terminal Boards, Mounts, and Terminations. A terminal board for high potential should always be made from qualified insulation. The board may be flat if the voltage is less than 20 kV, provided the electrical stress is either less than 10 V/mil for long life (10 to 30 years) or less than 10 to 25 V/mil for short life (1 month to 1 year) with treated boards. In a dry, clean atmosphere of pure gas, the latter values can be increased by about a factor of 3.

Terminal boards operating at voltages greater than 20 kV should be contoured to increase the creepage paths. There are four basic methods of contouring.

1. cutting slots (gas-filled regions) between terminals
2. building barrier strips between terminals
3. mounting terminals on insulated stand-offs
4. ruffling the surface to increase creepage paths

Insulated stand-offs are a form of barrier strips. They are difficult to design because they must withstand the forces applied by the terminals, and the terminal anchor must be embedded in the top surface of the stand-off. The anchor must be contoured for minimum electrical stress.

12.3.2.8 High-Voltage Leads. Leads between high-voltage parts should be made of round, smooth-surface, polished metal tubing. Steel and nickel-plated metals are preferred, but other, softer metals are often used because they are easier to fabricate. The radius of curvature on all bends should be at least 2.5 times the conductor diameter to avoid flattening or crushing the tube at the bend. The ends of the tubes should be flattened as little as possible, but this becomes difficult for pieces other than straight sections. When the end of the tubing is flattened, the corona suppression shield should extend over the edges of the flattened end.

12.3.2.9 Solid-State Devices. Solid-state devices are referred to as *chips*. The actual solid-state device has very sharp corners, is very small, and may be made of a multitude of chips. This implies that, before potting, the assembled arrangement may appear as a group of needle points or razor-sharp edges. When coated, all these sharp points and edges are hidden, but the fields remain to emanate from the surface of the coating. These fields must be considered. The fields exist, and the problem is what are the dielectric properties of the coating material. A high-dielectric-constant coating material reduces the maximum field stress next to the chip within the coating. This implies that if an encapsulating material is selected with much lower dielectric constant, the chip coating material has little voltage drop, and a large voltage drop will exist at the insulation interface. Thus, the engineer must obtain the dielectric properties of the coating material and its thickness over the chip or multichip internal construction before recommending an insulation thickness.

12.3.2.10 Circuits. The high-voltage insulation design starts with a circuit diagram showing all parts and their anticipated design voltage levels. The parts are then arranged in a preliminary package which minimizes the net voltage stress between parts and the voltage drop across each part. In designing highvoltage assemblies it is important to avoid wire and part crossovers that place a lowvoltage surface on one part next to a very-high-voltage surface of another part. Circuits containing resistive or capacitive voltage dividers require careful design, especially if the resistor or resistor string is extended. For example, a resistor or group of resistors may form a voltage divider between the high-voltage terminal and local ground. The normal plan is to zigzag many resistors from the highvoltage terminal to the ground terminal, or to have one resistor with one end attached to the high-voltage terminal and the other end grounded. Sometimes other high-voltage parts near the center of the resistor or resistor chain may be at full voltage or at ground potential, stressing a zone that is not normally designed for voltage stress. This is poor design practice and must be avoided.

Low-voltage devices and their wiring must be well separated from the highvoltage circuits. Low-voltage conductor shielding has rough surfaces that look like multiple points, which enhance field gradients with respect to the high voltage. This will result in lowering the breakdown voltage between high-voltage parts, or conductors, and the low-voltage shields.

Shielding the low-voltage components and wiring should be adequate to hold the induced impulses to less than 750 V peak in common-mode and differential-mode circuits, and to less than 7500 V peak in the wiring. When properly filtered and protected, these limits will prevent the destruction of most hardened solid state devices, inductors, capacitors, and resistors that are commonly used in the control circuits. Many circuits have been evaluated for damage or malfunction by electromagnetic pulses.[38]

For large modules with several circuit boards and large parts, the designer may have to interconnect two or more components with a high-voltage flexible wire that has insulation inadequate to sustain the full electrical stress of the applied voltage. This can be done if (1) the diameter of the wire is increased with more insulation and (2) adequate and rigidly controlled spacing is provided between the wire and ground planes. With dc voltage stress, the low resistance of the insulation and near infinite resistance of the gas allow the surface of the wire insulation to charge to the conductor voltage level. This larger diameter lowers the voltage gradient in the highly stressed gas next to the conductor. With ac, the voltage at the surface of the wire is determined by the configuration and dielectric constants of the wire insulation and gas space.

Generally, extra flexible wire should be used only when the bending and placement of the tubing through the high-voltage volume are too difficult or will mechanically stress parts during installation. Terminations on extra flexible wire will not stay in place as they will with solid tubing. Therefore either the terminations must be keyed to a slot in the insulation barrier, or a special locking device must be developed for the termination or wire end.

12.3.2.11 Field Modeling. It is a good practice to test the electronic circuit continuity and electrical continuity before encapsulation. This can be done by immersing the prototype unit in a fluorocarbon fluid. A few things should be watched for.

Fluorocarbon fluid is an excellent heat conductor and will not give a desirable thermal profile.

Some fluorocarbons will take on water and other impurities, reducing their dielectric strength to unacceptable levels. This is especially true for voltages above 10,000 V.

The fluorocarbon must be completely dehydrated from the parts, modules, and wiring. This will take a long bake period. If not removed, the fluorocarbon could affect the dielectric properties of materials where it exists.

Fluorocarbon is a cleaning fluid. Any inks or unstable unbonded material, dirt, and debris will become loose and float about between some high-voltage parts and ground, causing problems.

12.4 AEROSPACE DESIGN CONSIDERATIONS

All the terrestrial environmental conditions exist for aerospace hardware. In addition, mechanical vibration, shock, and thermal variations, in many cases, are more severe in airplane and space hardware applications than for terrestrial applications. Thus the design requirements are more strenuous. In addition, high altitude airplanes and spacecraft may be subjected to severe temperature variations, space particles, and debris.

Unpressurized electrical systems and modules for high-altitude airplanes may use conventional design practices for voltages to 208 V rms or 270 Vdc. High frequency electronic hardware will require special attention to the insulation system for switching transients and voltages over 200 V rms.

12.4.1 Pressure

12.4.1.1 System.   Many aerospace vehicles are designed or will be designed to operate in near space or space where the low pressure (less than 10^{-4} Pa) makes the theoretical dielectric strength of the volume of gas greater than 5×10^5 V/cm, a value 16 times the dielectric strength of dry air at sea level. This is because there are few carriers and the mean free path exceeds the gap length between closely spaced electrodes. But in space or near space the electronic modules and circuits slowly outgas during ascent to altitude and for the remainder of the mission. For instance, during ascent to altitude, the vehicle's external pressure may decrease rapidly from sea-level pressure (1.013×10^5 Pa) to values approaching 1×10^{-5} Pa. This pressure change takes only a few minutes, but the pressure next to the outer surface and inside the aerospace vehicle remains at a higher level until the vehicle once again returns to the earth's surface because of the outgassing of various materials. The pressure decrease as a function of time for a large space vehicle varies in different compartments, as shown in one space vehicle application. The outgassing area was measured to be about 1 cm^2/L volume in one electronic box, the value recommended for adequate outgassing when high-voltage experiments or equipment are on board. The resulting pressures[39,40] are summarized in Fig. 12.21.

The data in Fig. 12.21 show that outgassing may be a lengthy process. When there is little outgassing and the box is next to or exposed to space, outgassing will be rapid. For interior boxes with many outgassing products, the process will take a long time after entering into orbit. Furthermore, outgassing products within a high-voltage module may keep the pressures much too high for a safe, reliable operation of some high-voltage circuits, making it advisable to delay their turn on. Likewise, the outgassing products of the vehicle and reaction control propellants increase the pressure in the vicinity of the vehicle. For very high power, high-voltage equipment it may become necessary to package in pressur-

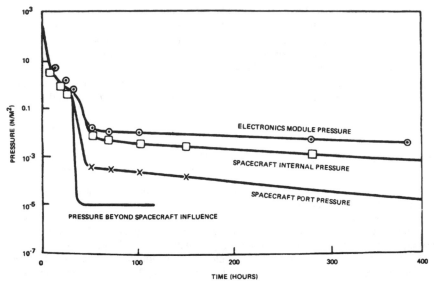

FIGURE 12.21   Gas pressure inside Apollo telescope mount.

ized gas or oil to prevent corona or voltage breakdown due to slow outgassing and external pressurization by propulsion exhaust gases.

12.4.1.2 Gas-Density Relationship.   It is possible to simulate any altitude and temperature in a room temperature chamber containing gas at an appropriate pressure. When calculating pressure-temperature relationships, the designer must use the temperature and pressure for simulating a given operating altitude and temperature, as derived from the ideal gas law, Eq. (12.2).

12.4.1.3 Helium Leak Test.   Helium, if used as part of the atmosphere, or if entrapped during leak detection, may reduce the breakdown voltage. If mechanical or electrical stressing should cause the insulation to crack internally, the smaller cracks may fill with helium rather than nitrogen or other pressurizing gas. When the helium partial pressure is between 13.3 Pa (1×10^{-1} torr) and 2.66×10^3 Pa (20 torr), it can ionize, generating partial discharges at very low voltage within the void, and possibly resulting in insulation failure. A comparison of helium and air is shown in Fig. 12.22.

12.4.2   Insulation Parameters

12.4.2.1 Insulated Conductors.   Most insulated conductor data are developed for terrestrial applications under normal sea-level operations. When the same insulated conductors are subjected to very high altitudes, the data may not apply because the gas between the conductors or between a conductor insulation and ground may go into corona, heat the insulation, and eventually cause the insulation to fail. When designing wiring systems, the voltage between components and wires must include the long field paths as well as the short paths. For example, the spacing between wire insulation and ground plane is very small or negligible, whereas the space between the upper surface of the insulation and

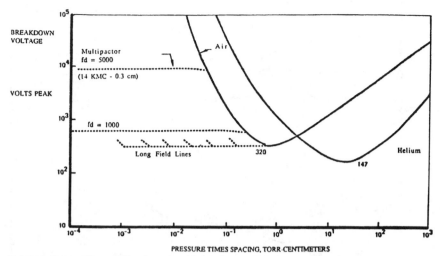

FIGURE 12.22 Theoretical voltage breakdown for parallel-plate configurations.

ground is very long compared to the insulation thickness. As gas pressure decreases, partial discharges are first formed in the narrow space between insulation and ground, as shown in Fig. 12.23, curve *A*. At that gas density, most of the voltage drop is across the insulation rather than in the air gap. As pressure is further decreased, the optimum mean free path for voltage breakdown becomes longer and the voltage gradient in the insulation decreases, as does the voltage, decreasing the breakdown potential between the conductors. This condition continues until the insulation contributes little to the voltage drop (curve *B*). As long as a glow discharge exists on the surface of the insulated conductor, the conductor is being degraded and will soon become carbonized, if an organic, and an arcover will follow.

Insulated conductors in flat-wire configurations are used in many low-voltage applications. They can be used at higher voltages, provided the pressure-spacing criteria are observed. For flat wires, the Paschen law curve for adjacent conductors in a cable has a U shape, as shown in Fig. 12.24. It must be remembered when designing with flat conductors that the end conductors have almost the same pressure-spacing characteristic as round insulated conductors.

12.4.2.2 High-Voltage Systems.   Power and electronic systems having voltages greater than 250 Vac or 350 Vdc require pressurized or sealed (potted) wiring and components. Wiring can be made corona- and partial-discharge-free by using coaxial power lines instead of the usual two-conductor system. The connectors must be corona-free, that is, the conductor pin cannot have an air pocket surrounding the pin, as in most military standard connectors. All relays, switches, disconnects, and connected components should be either pressurized or solidly potted.

12.4.2.3 Frequency and Wave Shape.   A generalized curve showing the breakdown of air as a function of frequency at the Paschen law minimum is shown in Fig. 12.25.[41,42] These data indicate little change in the Paschen law breakdown minimum for frequencies between 50 Hz and 2 kHz. Indeed, the curve is nearly flat. Then there is a gentle lowering of the breakdown voltage from 220 V above 2 kHz to 170 V at 2 MHz. Between 2 and 200

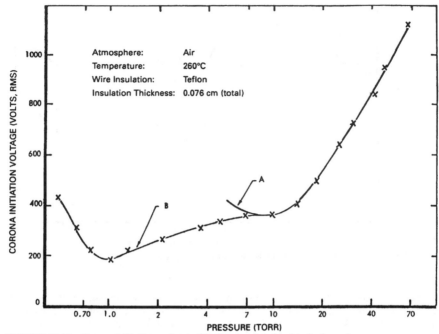

FIGURE 12.23   Corona initiation voltage of Teflon-insulated twisted wire pairs.

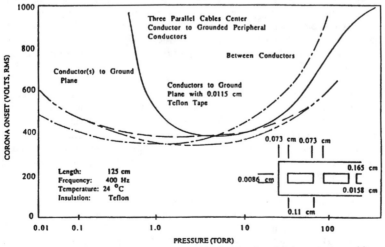

FIGURE 12.24   Corona initiation voltage of Teflon-insulated flat conductor cable.

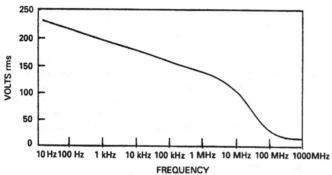

FIGURE 12.25 Lower breakdown voltage resulting from higher frequency between thin-film-coated parallel plates.

MHz the breakdown voltage drops drastically to above 50 V. Above 5 GHz, the breakdown voltage rises again.

The effect of a square wave is similar to that of adding an impulse to an ac voltage. Densley[43] developed a technique of analyzing square waves. He found that the leading edge of a square wave will have the same effect as an ac voltage with an impulse at the zero-voltage point on the sine wave. The impulse from the square wave will initiate partial discharges, which may continue throughout the waveform. Most of the discharges occur immediately after the impulse, with few or none at the end of the constant-voltage plateau. The number of discharges and their duration depend on the amplitude of the square wave, the reverse stress across the void or crack after the leading edge passes, and the frequency of the square waves.

12.4.3 Environment

12.4.3.1 Particulates.   Many future high-power space missions will involve large spacecraft to support the on-board loads.[44] Some of these spacecraft will require kilowatts to megawatts of electric power. Even though the amount of cosmic dust flux is very small in low earth orbit to geosynchronous orbits, the effects of cosmic dust on large spacecraft are significant enough to produce problems with high-voltage systems. A number of predictions on the effects of the space environment on large spacecraft have been based on experimental experience gathered from small satellites.[45–50]

Large spacecraft with large electric power systems, on the order of hundreds of kilowatts, are more subject to problems caused by particle debris than are smaller spacecraft because of the following characteristics, which are inherent in the design and the function of large spacecraft.

The high voltages and high currents of interconnectors, cables, and bus lines generate strong electric and magnetic fields, which strongly influence particulate debris-spacecraft electromagnetic interaction.

The large areas of dielectric materials in the high-power solar array structure produce strong electric fields because of surface charging and differential charging. This tends to collect particulate debris.

The use of low-density materials in the spacecraft structure yields some outgassing and fragmentation products.

The requirement for long life (10 to 20 years) will lead to cumulative effects of particulate debris impacts. These characteristics are barely evident in the design of small spacecraft; therefore their effects are not significant.

When the debris particles have entered the field of the conductor, they will become charged and polarized to form a seta (hairlike) growth on the conductors. Some of these charged particles will form bridges or "strings of pearls" between high-voltage and low-voltage or ground planes. The field stress at the ends of the seta is very high, so any plasma in that region can and will initiate an arc between the conductor and the ground plane or another conductor. Other problems resulting from particulate debris contamination include:

Short-circuiting of electrical elements

Arcing and surface damages over the main structure

Arcing on the main bus lines to communication antennas, which causes current irregularity and produces intolerable noise in addition to physical damage

Degradation of paints and surface finish

Excessive wear and binding of movable parts because of contamination of lubricants, seals, and sliding surfaces

12.4.3.2 Interactions with Space Electronics.   When an aerospace plane operates in orbit, it interacts with the naturally occurring space plasma. The low temperature ionospheric plasma density distribution is shown in Fig. 12.26[51] as a function of altitude for an equatorial orbit. The plasma density varies with the 11-year sun-spot cycle, and the figure provides maximum and minimum densities. Both maximum and minimum peak near 150 km, reach a plateau near 1100 km, and fall off drastically above 10,000 km.

The initiation voltage breakdown for bare contacts in a plasma environment is shown in Fig. 12.27. These data are based on laboratory tests[52-54] and flight tests.

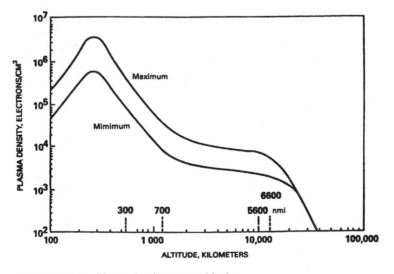

FIGURE 12.26 Plasma density versus altitude.

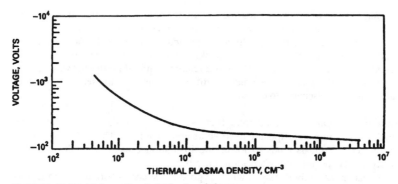

FIGURE 12.27 Voltage threshold for breakdown.

12.4.3.3 Surface Charging. The primary vehicle interaction is that of the air striking the vehicle. Differential charge buildup between various satellite elements can occur, which can result in arcing between neighboring surfaces. This arcing can affect component and system operation or cause significant damage to on-board components. Table 12.12 lists potential hazards due to arcing.

Differential charging on vehicle surfaces can result from at least two distinct processes—surface charge buildup and charge deposition in dielectrics. Dielectrics or isolated conducting surfaces can build up potentials, over long periods of time, that are well in excess of the breakdown-potential threshold for the various materials. Certain insulating materials are also known to store charge in the bulk of the dielectric material, which can be released by an external perturbation such as X radiation. This will initiate an internal discharge and "punch through" the insulator, resulting in a fault.

TABLE 12.12   Potential Spacecraft Hazards

Electromagnetic interference
   Upsetting of logic circuits, for example, false commands
   False telemetry signals
   Alteration of electrical component parameters
   Damage or burnout of electrical components

Thermal control degradation
   Burnout of ground straps
   Deposition of burnout materials on surfaces
   Alteration of surface absorption and emittance properties
   Insulator contamination or tracking

Optical sensor degradation
   False optical signals due to arc discharge flashes
   Deposition of arc materials on optical surfaces

Mechanical effects
   Coulomb forces due to electrostatic charging
   Coulomb forces due to electrostatic arc discharges (quick release of steady-state forces)
   Mechanical distortions due to long-term changes

12.4.3.4 Discharge Due to Improper Bonding.   Another place where the effect of vehicle charging can be detrimental is where the conducting sections of the vehicle are not bonded together; for example, a rocket vehicle is charged triboelectrically on the forward surfaces and discharged through corona from the skirt at the aft end. If the forward section is not electrically connected to the aft section, the charge acquired on the forward section cannot flow to the aft section unless the potential difference between the sections becomes large enough for a spark discharge to occur.

12.4.3.5 Radiation.   Other environmental factors affecting insulation are ultraviolet radiation, nuclear radiation, and exposure to solvents and chemicals. The addition of ultraviolet radiation and chemicals tends to lower the insulation breakdown voltage, either instantaneously or as exposure and deterioration progress.

Some materials are severely damaged when exposed to space radiation. Teflon suffers degradation to both its mechanical and its electrical properties at levels as low as 10^3 to 6×10^4 rads. Some organic materials with moderate radiation resistance are listed in Table 14.13.

To produce reliable radiation-hardened equipment, proper part selection, circuit design techniques, and low active-to-passive component-count ratios must be implemented. Some hardening techniques, as outlined in Bloom[55] and King,[56] follow.

Switching Circuity

For bipolar switching transistors, utilize maximum base drives and minimal resistances across base emitter junctions.

Implement switching transistors with the largest possible gain-bandwidth product.

Protective circuitry must be included to prevent overcurrent or secondary breakdown of the switching transistors.

For maximum transistor gain, higher operating temperatures should be used.

If automatic reset is not possible, do not use overvoltage or overcurrent silicone-controlled rectifier crowbars.

Control/Sensing Circuits

Implement relay control rather than transistor control.

Use transformer-coupled feedback, avoiding optocouplers or isolators. To minimize ionization-induced effects, use minimal impedance.

TABLE 12.13   Organic Materials with Moderate
Radiation Resistance

Material*	Radiation, rads
Polyethylene	10^7–10^8
Silicones	10^6
Polyamide	10^8
Polyolefin	10^8
Alkanex	10^7
Silicone-alkyd	10^7
Polypropylene	10^7

*Inorganic and fiberglass products generally can withstand much higher doses, up to 10^7 to 10^8 rads.

In high total-dose situations, MOS semiconductors cannot be used.

Do not assume that "matched pairs" of integrated diodes and transistors will have characteristics that track identically during and after irradiation.

Minimize fan-out from logic gates and ultra stable reference voltage requirements.

Implement dielectrically isolated ICs, negative feedback, and direct-coupled gain stages.

Filter Requirements

Use shielding, filtering, component selection, and circuit layout to reduce EMP susceptibility on input/output lines.

Paths between power sources and filter capacitors must be current-limited.

To ensure proper operation, circuitry must be designed to operate when capacitors (especially tantalum) undergo a 25 percent charge loss.

Use only the necessary amount of capacitance required for filtering.

12.5 DESIGN AND MANUFACTURING GUIDELINES

12.5.1 Design

For a high-voltage electronic module or system to be successful, the following criteria for long-life, trouble-free operation should be considered.

Design the circuit to minimize transients and coupling between high-voltage and low-voltage circuits and component parts.

Avoid electrical overstress between high-voltage parts, circuits, ground planes, and adjacent modules.

Select parts, materials, processes, and circuit designs that meet the life, voltage, thermal, and environmental requirements specified by the customer.

Select materials and spacings that have adequate thermal conductivity through the thick insulation systems required.

Analyze critical spacings and configurations for electrical overstress. Include frequency, peak voltages, spacing, and configuration of adjacent parts or circuits.

Select materials and processes that are known to be void-free and will not debond when environmentally tested for temperature extremes, mechanical vibration and shock, and high voltage.

Isolate sensitive circuits by using electromagnetic filters and shields. Solder-ball all high-voltage connections.

Improve spacing to reduce maximum electric field stress to less than 100 V/mil between high-voltage parts, conductors, and plates.

Eliminate sleeving and other void-entrapping materials within magnetic devices and circuits.

Upgrade parts to screened military standard parts. A few specialty items that do not have military designations must be screened and "burned in" prior to use.

Provide corona testing on magnetic devices and critical high-voltage modules before and after plotting.

Use encapsulants and processes found acceptable for the specified environment.

Include partial-discharge (corona) tests with high-voltage tests.

Obtain a list of high-failure-rate items from manufacturer or repair records.

Determine possible failure modes by analysis and testing of critical configurations.

Simulate power supply transient, steady-state, and ac small-signal response.

Evaluate worst-case component voltage, current, and power stress.

Review parts list for component upgrade where possible.

The following guidelines may be used to assess circuit and module designs.

Arrange conductors so that high- and low-voltage groups are separated. Lowvoltage-sensitive conductors should be placed close to the ground plane and separated from the high-voltage conductors.

Use at least 0.25-mm insulation between rounded electrodes and 1.25-mm insulation between flat surfaces.

Use semiconducting coatings on high-voltage surfaces where necessary and around stranded wiring to control uniformity of fields.

Design terminations so that mechanical stress points are minimized. This can be accomplished by molding the terminal in a solid insulating material that is attached to the board, or by placing metal spacers with flanges through the board.

High-voltage extra flexible wiring is acceptable when guided from terminal to terminal to eliminate the probability of the wire insulation intermittently touching other surfaces containing higher- or lower-voltage circuits.

Potting materials should not completely fill a box or module. Spacing is required for stress relief. A semiconducting layer or a ground plane may be used.

Prevent contamination such as salt spray and outgassing residues from other airborne equipment, structures, and controlled emissions.

Avoid corona, which may be enhanced by gases such as helium, argon, neon, and hydrogen if they should be mixed with the atmospheric gases.

Control electrical stress. Creepage and tracking can cause increased temperatures in localized areas and eventually lead to surface flashover.

Prevent voltage transients, which may cause surface flashovers. Between 6 and 10 flashovers will form tracking and eventually lead to voltage breakdown.

The following design and packaging problems need to be considered.

Corona resistance: withstand 1-pC discharge continuously; 10-pC discharge for 500 to 2500 h.

Compatibility with reinforcing fibers: demonstrate good adhesion and load transfer capability under severe environmental exposure.

Good adhesion to metals: 10-lb/in peel; 500- to 1000-lb/in^2 shear.

High dielectric strength: (1) 100 V/mil design, 2500 h operation; (2) 150 V/mil peak, 2500 h operation; (3) 50 to 100 V/mil, continuous operation.

Wide temperature range: usable from −65 to 160°C.

Ease of repair: capable of removal without damage to unfailed components.

Environment for spacecraft systems should be analyzed for the following.

Plasma interactions.

Open lines exceeding 300 V peak: (1) arcing, (2) insulation deterioration.

Debris collection by high-current lines: arcing.

Effluents from cryogenics, engines, magneto hydrodynamics, and controls: (1) voltage breakdown, (2) arcing.

Spacecraft charging: (1) composite structures; (2) isolation requirements.

Bonding and grounding: (1) composite structures, (2) ground returns, (3) shielding required for pulsers.

High-power RF system: multipactor at high frequency.

The remainder of this section contains suggested techniques that have worked successfully for many terrestrial and aerospace designs. Most of these techniques were developed for aerospace high-density high-voltage electronic packaging. The same considerations should be used for terrestrial designs because many parts and circuits are designed for high density to reduce coupling, size, and weight.

12.5.2 Parts

12.5.2.1 Resistors. Resistors, as a rule, are considered very reliable. Unfortunately some resistor characteristics must be considered for high-voltage designs.

When potted, some resistors are stressed mechanically by material shrinkage, temperature extremes, and aging. Resistors that are designed with minimum tolerances must be evaluated for these characteristics.

There is a voltage gradation along the surface of the resistor. Resistors placed next to a ground plane another higher- or lower-voltage part, or a wire must consider the field gradient between the surfaces.

All construction parts within the resistor must be examined for curvature.

The dielectric constant of the resistor coating material should be greater than that for the surrounding material for ac designs. Likewise, the insulation resistance should be lower than that of the surrounding material for dc designs.

Solid-core resistors should be used throughout the high-voltage module. Hollow-core resistors, if sealed on the ends, will slowly depressurize at high altitude and arc from end to end along the hollow-core surface.

12.5.2.2 Capacitors. There are many types of high-voltage capacitors, such as ceramics, electrolytics, foilwound, sealed liquid-filled, potted, or nonimpregnated. Very-high-voltage capacitors may be constructed with several sections connected in series. All these design characteristics must be considered when packaging a capacitor.

The dielectric of a ceramic capacitor should be constructed with nonchipped edges. If chips or small fissures are developed during assembly, the voids will not be filled with the outside potting material and the plate material will migrate into the cracks. The voids will be overstressed and fail the capacitor.

Wax-filled or coated outer surfaces should be avoided since they will not bond.

Many ceramic capacitors fail because the outer surface material is incompatible with the bulk potting material. Check for compatibility and cleanliness (Fig. 12.28).

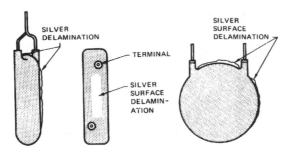

FIGURE 12.28 Delaminated high-voltage capacitor.

Foil-wound capacitors should be constructed with multiple film layers to avoid the probability of a pinhole between plates. It is mandatory that liquid-filled capacitors be sealed. A leaky capacitor will develop bubbles (voids) within the dielectric system and fail.

The capacitor conductors should have smooth plates with no sharp edges caused by shearing the foil.

High-voltage capacitors with multiple sections should be placed between parts with spacings in accordance with the voltage differentials.

All capacitors should be tested for dielectric properties and corona. The corona test should be with the correct type of voltage, ac or dc, at 1.25 to 1.5 times the operating voltage.

With liquid-impregnated capacitors, the container terminations and seals are important. All free space must be filled with liquid to exclude gas that can readily ionize. Rectangular or oval cases are to be designed with enough flexibility to permit the liquid to expand and contract as temperature and pressure change. Cylindrical or rigid-walled cases must be designed with enough flexibility to permit the liquid to expand and contract as temperature and pressure change.

The frequency spectrum of each film material should be determined as a function of temperature. A polarization at a critical pulse frequency will cause heating and loss of the dielectric. The frequency spectrum should be measured through 100 MHz.

Low inductance can be achieved using short leads.

Processes should be specified that will result in an excellent bond between foil and lead caps. A high resistance joint will result in overheating and capacitor failure.

Trade-offs between large reliable capacitors and lightweight new designs will be necessary. Doubling the number of lightweight capacitors to meet reliability values is a poor design practice when 20 percent more weight and volume for a single unit can achieve the same goal.

Capacitors with large insulators mounted between terminal and case should have a tapered insulator for higher stand-off voltage and a case having a smooth surface without exposed seal ridges. Dome-shaped ends are preferred.

12.5.2.3 Magnetic Dences. Magnetic devices placed next to cold plates must be insulated with a void-free material to prevent partial discharges between coil conductors and grounded surface. Field plots should be made to evaluate the stresses within enclosed magnetic devices or between conductors and core edges. Other winding, construction, and potting recommendations are as follows.

When coils and circuits are encapsulated, film-tape tabs such as Mylar adhesive may cause built-in gas pockets or voids. These voids may initiate partial discharges and eventual voltage breakdown. If tabs are required, they should be made of porous materials that are compatible with and easily wetted by the encapsulant. Tabs should be very small, preferably less than 1 cm square.

Creepage distances between layers must be large enough to prevent a carbon path from forming. The creepage voltage stress should be less than 36 V/mil for long-life designs.

The insulation between layers and between windings must be based on the maximum voltage calculated using the utilization factor with respect to wire diameter and insulation thickness. Wire insulation must be considered in the calculation. Maximum transients must be entered into the formulation to obtain the allowable stress levels for the insulation.

Potted transformers must be as void-free as possible. The corona test should be used with the following restrictions. (1) The frequency should be the transformer operating frequency. (2) If the operating frequency cannot be used, overvoltage should be used based on the operating frequency versus the test frequency. (3) The energy should be less than 1.0 pC per 5000-V test.

Windings with unfilled epoxy should be encapsulated.

Insulation should consist of wettable materials to eliminate voids.

Porous materials should be used for penetration by potting material.

Transformer should be potted with filled epoxy for better heat transfer.

Output leads must be stressed.

12.5.2.4 Solid-State Devices.   Many companies change from "old style" parts to "new" parts without notifying the customer of the changes. Some newer devices are much faster and do not meet the circuit characteristic requirements. This causes many failures.

Solid-state manufacturers often change device characteristics without changing the device nomenclature.

Some devices have become obsolete and units that need replacement are difficult or impossible to find in time to keep up with high-failure-rate electronics.

Diodes with low voltage drop will help the cooling problem.

Diodes should not be placed next to power resistors.

Matched-set solid-state devices must be replaced as a set. The replacement matched-set devices often are not properly characterized and matched.

Some solid-state devices appear to change characteristics with aging. These changes are severe enough to cause a failure in that device and in other devices in the board.

The solid-state chips or wafers have very sharp edges or are pointed. The coating dielectric should be determined so that the field stress of the insulation system can be evaluated accurately.

12.5.2.5 Mounts, Interconnects, and Surfaces.   One rule that all high-voltage designers should follow is that a good high-voltage mount, surface, termination, or ground surface should have a smooth rounded structure. Decreasing the maximum electric field stresses is the first step in a good thermal design. It is as important for the solid dielectric surfaces of components and parts to be rounded as it is for the metallic surfaces.

12.5.2.6 Terminal Boards and Supports.   Composite and laminated insulation may be used for terminal boards, and also for supports that separate the coils and wires from the cores, structures, and containers.

Laminated boards should not be overstressed mechanically. Overstressing may cause the board to delaminate; then creepage paths may form in the delaminated space, resulting in internal breakdown (Fig. 12.29).

Laminated boards should not be held together with screws. The screw threads will also cause small fissures near the threads and may result in voltage breakdown.

Adhesion tests should be performed to ensure that the materials used are compatible.

Bonded voltage barriers are not recommended. Debonding or incomplete bonding results in a creepage path between the two parts.

12.5.2.7 Insulated High-Voltage Wiring.   A designer may have to interconnect two or more components with a high-voltage flexible wire that has insulation inadequate to sustain the full electrical stress of the applied voltage. In such a case, the following has to be considered.

The diameter of the wire is increased with more insulation. With dc voltage stress, the low resistance of the insulation and near infinite resistance of the gas allow the surface

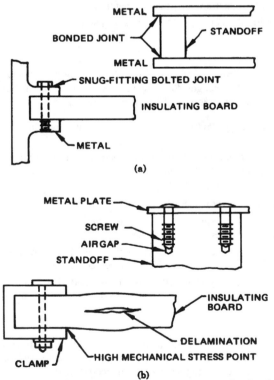

FIGURE 12.29 Methods of board attachments. (a) Good designs. (b) Poor designs.

of the wire insulation to charge to the conductor voltage level. This larger diameter lowers the voltage gradient in the highly stressed gas next to the conductor. With ac, the voltage at the surface of the wire is determined by the configuration and dielectric constants of the wire insulation and gas space.

Adequate and rigidly controlled spacing is provided between wire and ground planes.

The knots on wire ties should be either burnished or enameled. Otherwise the feathered tie ends will become points from which corona discharges can emanate.

In general extra flexible wire should be used in large electronic modules only when the bending and placement of tubing through the high-voltage volume are too difficult or will mechanically stress parts during installation and operation. Terminations on extra flexible wire will not stay in place as with solid tubing. Therefore either the terminations must be keyed to a slot in the insulation barrier, or a special locking device must be developed for the termination or wire end.

One method of preparing coaxial wire terminations is to cut the wire and shield to length and then place three or more turns of a small conductor or formed metal ring over the end of the wire shield and insulator. This method holds the shield end strands in place and forms a round conformal metal surface at the end of the shield.

Triple points between vacuum, metal, and insulation can be eliminated by making the coaxial termination appear as a doughnut.

The protruding insulation between doughnut and wire terminations must he sufficient to satisfy creepage requirements.

If Teflon insulation is used, it is necessary to etch the Teflon. This etching should be done within a few hours before encapsulation. If the etching material is placed on the Teflon for more than 24 h at ambient temperature before encapsulation, or if the Teflon is heated because of soldering, the Teflon will cold flow and ruin the usefulness of the etching. If the etching must be applied more than 24 h before encapsulation, it is recommended that the etched surfaces be kept cold to prevent cold flow.

Insulated stranded wire must be allowed to outgas in aerospace applications. Otherwise the insulation will expand and form gas pockets between insulation and conductor.

Closely spaced small-pin multipin connectors are not recommended for high-voltage unpressurized equipment, especially if there is a probability that the voltage between pins will exceed 450 V peak in the pressure region between 1000 and 10 Pa.

Connectors must also be designed to eliminate air voids between conducting surfaces. One successful method is to make one side of the mating interface from soft, pliable insulation.

Avoid wells around the pins. Filled wells may have voids at the bottom of the well, resulting in partial discharge, which disrupts sensitive circuits.

12.5.3 Materials

Insulation and insulation systems fall into two design categories—unpressurized, or open, construction and pressurized, or potted, construction.

12.5.3.1 Open Construction.   Open-construction design refers to modules with few or no solidly potted submodules. In this type of construction, circuit boards must be conformally coated to reduce breakage due to vibration and shock, inhibit corrosion and fungus during handling and storage, and reduce surface tracking along board surfaces between high- and low-voltage parts, terminations, or connections.

Before a board is conformally coated, it should be inspected for clearance, cleanliness, and spacing of parts and wires above the board so that conformal coating material will fill the void between the wire parts and the board. Then the following rules should apply.

All boards, conductors, wiring, and electrical components must be cleaned per appropriate specification before the unit is conformally coated. This includes solder flux, fingerprints, particles from the workbench, and dust.

The circuit boards should be conformally coated with at least three separate layers of a low-viscosity insulation. Application may be by either dipping or brushing, with each layer applied at right angles to the preceding layer. Three layers are recommended to eliminate the pinholes (continuous leakage path) and uncoated areas that normally occur in single or double coating processes. The completed process should be checked by an insulation test.

The final step in an electrical assembly is the joining of the printed-circuitboard assembly. If wired, the wire and solder joints must be cleaned and conformally coated with the same precaution as the electrical networks on the printed-circuit boards after the connections are made to the other board subassemblies within the module.

The conformally coated board materials must be compatible with the pressurizing gases or liquids and all parts, solder, terminations, traces, and wiring on the board.

12.5.3.2 Potted Construction. One method of preventing a gas-discharge voltage breakdown is to exclude gases from the high-voltage areas. This can be accomplished by encapsulating the high-voltage circuitry. Encapsulation provides the system with mechanical protection from external damage, gives structural support to the components against shock and vibration, and protects the high-voltage system from gas discharge damage.

12.5.3.3 Processed Materials. For processed materials the following conditions must be considered.

Mold release agents such as silicone products may contaminate certain epoxies, urethanes, and other insulating materials.

Coatings must be evaluated with proper materials under identical environmental and electrical stress conditions to be fully qualified.

Yield problems often result from variations between batches of potting compounds, mix-ratio variations from pour to pour, and a lack of adequate process detail.

Wettability is as important as viscosity.

Sleeving promotes air bubbles and must be kept short (Fig. 12.30).

Molecular sieving may separate the catalyst through fine mesh or fillers.

Degassing forces parts and wires apart; permanent stakes are necessary.

Thermal or mechanical stresses may break joints and connections during cooling.

12.5.4 Manufacturing

12.5.4.1 Manufacturing Controls. Regardless of the excellence in electronic design, material and process evaluation, and packaging of high-voltage systems, the electronics will not perform reliably if manufacturing controls are not maintained to ensure that the

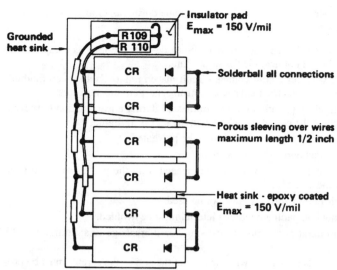

FIGURE 12.30 High-voltage bridge circuit mounted on grounded heat sink.

electronics are properly assembled, fabricated, and tested. The main considerations in proper manufacture are:

Contamination and cleanliness

Accuracy in all physical measurements and other aspects of processes, such as weight, volume, time, temperature, pressure, and material control

Thorough knowledge of all aspects of incoming process controls and materials processing

12.5.4.2 Tooling. Some problems associated with tooling for electrical insulating materials that must be overcome to increase efficiency include:

Mold release contamination

Chipping of materials as they are removed from the mold

Excessive grinding and smoothing required for encapsulated components because of tooling surface roughness

Wire breakage on small high-voltage transformers due to misalignment

Contaminated mixing crucibles (Do not use epoxies in the same crucibles or tools as silicones.)

Improper cleaning of tools

Equipment controls

Workmanship

Properly developed and designed tools are essential for the technical personnel to perform their duties so that high yield can be achieved. It is equally important to train each technician in the correct operation and use of these tools.

12.5.4.3 Assembly.   Many problems must be overcome during assembly. Typical problems include the following.

When solder balls are required, the ball must be made after the joint is complete and inspected. The ball must not degrade the solder joint.

Twisted wires and wires in a high-voltage circuit placed randomly between modules can be subjected to high electric field stress, causing short circuits and arcs.

Wires for high-altitude application must be outgassed during encapsulation to prevent air pressure buildup at the end of the wire when the unit is in flight.

Movement of wires during encapsulation must be prevented.

Waxed or contaminated parts must be deleted.

Contamination by grease and debris must be cleaned from the part as received from the supplier or by the assemblers.

Tabs that entrap air in transformers must not be used.

Use of tabs with noncompatible bonding material must be avoided.

Overtightening of circuit-board mounts, causing board breakage or splitting, must be prohibited.

Use of screws in boards or plastics with air gaps at the end of the screws must be prohibited.

Sharp-edged high-voltage parts must not be used next to low-voltage circuits and grounds.

Chassis with sharp edges or burrs must be smoothed.

These problems should be brought to the attention of quality control before potting of the components or immersion in an oil. It is difficult to inspect for these problems after potting in a colored or opaque material.

Quality control engineers should coordinate their efforts with the electronic and packaging design engineers to obtain a more complete technical knowledge of all engineering specifications. The design and packaging engineers must determine the minimum standards that can be achieved by quality control before the design is released to manufacturing. After a new development prototype assembly has been developed the high-voltage designers should spend time coordinating with quality control, detailing critical component inspection, outlining pass/reject criteria, and determining whom to call for immediate consultation.

12.5.4.4 Testing.   Yield and manufacturing efficiency can be improved with proper testing and test equipment. The following tests should be required for all designs.

Magnetic Devices

Continuity tests before and after encapsulation

Corona test after potting to determine insulation integrity between coils, between coils and cores (if available), and between turns within a winding

Parts burn-in where burn-in is required

Out-of-tolerance tests (High-voltage components may be in tolerance at low voltage but fail at high voltage. High-voltage testing is preferred.)

Assembly

Assembly continuity tests

Material characterization tests

Material viscosity tests to indicate aging or formulation changes

Final assembly tests to determine operation at ambient, altitude, and high and low temperatures

Electrical high-voltage tests—insulation resistance, dielectric withstanding voltage tests, and partial-discharge tests

Many inferior and out-of-tolerance parts have been found and rejected by imposing the proper test program to the parts and modules before final assembly. This results in better yield and lower cost.

12.5.5 Test Methods

Data for the material properties of the potting compounds are available from manufacturers' data sheets, research papers, discussions with researchers, and manufacturers' technical assistance groups. When some of the critical parameters are missing or not measured or documented, it is necessary to determine these properties using standard testing procedures and reliable equipment. It is also necessary to use standard testing methods, such as ASTM standards, to determine some of the critical properties for qualification test, quality conformance test, recertification test and quality control, as shown in Table 12.14. To facilitate a quick reference, the properties and the test methods may have to be modified slightly in order to hold the sample or to measure the property of the potting compound.

TABLE 12.14 Material Properties and Test Methods for Potting Materials

Properties	Test method
Dielectric strength	ASTM D-149-87
Dielectric constant	ASTM D-150-87
Dissipation factor	ASTM D-150-87
Volume/surface resistivity	ASTM D-257-78
Dry arc resistance	ASTM D-495-84
Viscosity	ASTM D-1824-83
Pot life	Correlate measured viscosity or flow rate
Coefficient of thermal expansion	ASTM E-831-86
Coefficient of thermal conductivity	ASTM D-1674-67
Brittleness temperature	ASTM D-746-79
Tensile strength	ASTM D-638-84
Lap shear strength	ASTM D-429-81
Peel strength	ASTM D-429-81
Tear strength	ASTM D-624-86
Thermal shock resistance (cracking)	ASTM D-1674-67
Thermal shock adhesion (peeling)	See Morel and Dung[24]
Miniwound transformer test (penetration)	See Morel and Dung[24] and Sec. 14.6.2
Thermal aging	ASTM D-2304-85
Cure shrinkage	ASTM D-1917-84
Moisture absorption	MIL-I-16923G
Specific gravity	MIL-I-16923G
Flammability	MIL-I-16923G

12.6 TESTS

12.6.1 Performance

Performance tests were developed to verify compliance with specified electrical performance criteria. Electronic units are generally tested to meet the applicable environmental tests.

A common test voltage for 28- and 120-V equipment is two times normal plus 1000 V. Some airborne equipment is tested with lower voltages, especially if short life and dense packaging are involved. Sometimes this equipment is designed with a dielectric withstanding voltage (DWV) that is less than 160 percent of the operating voltage, but it should be at least 160 percent for quality hardware. Some systems have large voltage transients generated by silicone controlled rectifiers, solid-state controls, or mechanical switches. The DWV test must exceed the highest of these transients by at least 20 percent.

In a high-potential test the intentional grounds of the component being tested should be disconnected and the voltage applied between mutually insulated elements of the electric equipment and between insulated elements and the frame or "ground," usually for 60 s. High-voltage insulation is tested to evaluate its physical and electrical properties and to predict its service life. Equipment tests should be designed to verify the quality of the insulation rather than to serve as a failure analysis tool.

12.6.2 Insulating-Material Tests

There are two categories of insulation testing—material-evaluation and component-insulation tests. Material evaluation tests include tests of the electrical and physical properties listed in Table 12.15. Physical properties include flexural strength, tensile strength, warp and twist, water absorption, linear and bulk coefficients of thermal expansion, heat capacity, chemical resistance, and flammability. Materials are usually evaluated in commercial testing laboratories and in laboratories operated by manufacturers of insulation.

The following sequence of testing could prevent high-voltage use of a material that is unsuitable for that application.

Visual inspection

Insulation resistance measurements, namely, volume resistivity and surface resistivity

High potential applied to solid insulation between two metal electrodes

Test for tracking susceptibility

Final insulation resistance measurement

Life test

12.6.3 High Potential Tests

The purpose of testing high-voltage parts and components is to determine their worthiness. The suggested order for these tests is insulation resistance, high potential dielectric withstand voltage (DWV), impulse or basic insulation level (BIL), and partial-discharge or corona test.

12.6.3.1 Insulation Resistance and DWV Tests. High insulation resistance by itself does not prove that the insulation of a component does not have cracks or other faults where insulation breakdown may start subsequently. Therefore an insulation-resistance test is not a substitute for high-potential tests, which should follow an acceptable insulation-resistance test. Insulation resistance should also be measured after high-potential tests because insulation damage from a high potential breakdown may otherwise be difficult to detect. Insulation resistance that is low because of moisture can usually be restored by baking.

12.6.3.2 BIL Tests. Impulse or basic insulation level (BIL) tests are required for components and equipment that will be used where electromagnetic pulses (EMP) or switching surges are expected. A BIL test subjects the insulation to a voltage pulse having a rise time of 1.5 µs to 90 percent of peak voltage and a 40-µs decay to 50 percent of peak voltage. The slowest EMP transients are essentially the same as the BIL standard transient. Factors that decrease the impulse level that an insulation can withstand are material aging, power system transients experienced, and the maintenance status of the equipment. Insulation impulse ratings are decreased to 0.75 to 0.85 of their original values by these phenomena.

Pulse tests, like DWV tests, can be destructive and must be planned and executed carefully. Some rules for the application of pulse and dielectric withstanding voltage tests follow.

TABLE 12.15 Tests of Electrical Properties of Insulation

Tested property	Test condition	Evaluated	Test method
Dielectric strength	dc/ac, 1/4-in electrodes	When received and following environmental stress	ASTM D-149-61 ASTM D-877 ASTM D-1816 IEEE-262
Tracking	dc/ac	Following environmental stress	ASTM D-495 ASTM D-2302 ASTM D-2303
Complex permittivity (dielectric constant and dissipation factor)	1 kHz	When received	ASTM D-150 ASTM D-2520 ASTM D-924
Volume resistivity	125 V	When received and following environmental stress	ASTM D-257-61 (modified)
Surface resistivity	dc	When received and following environmental stress	ASTM D-257-61 (modified)
Insulation resistance	dc	Following environmental stress	Based on 0.05-mF wound parallel-plate capacitor
Corona	dc/ac	Following environmental stress	ASTM D-3382-75
Life	dc/ac	Vacuum (plasma)	ASTM D-2304-64T (modified)
Voltage endurance (continued ac stress above partial discharge onset level until failure)			ASTM D-2275

The pulse-test peak voltage should not exceed 200 percent rated (peak) voltage.

Pulse and DWV tests should be limited to incoming inspection, component acceptance, and system or subsystem acceptance. Overtesting overstresses the dielectrics and some critical electric parts.

Tests beyond those outlined in the preceding step should be at 80 to 85 percent of the original voltage.

The DWV and pulse tests must be magnitude-limited, that is, the magnitude must be limited to coincide with the acceptable dielectric stress limits of the insulations within the insulation system of the item tested. For instance, capacitors in pulse-forming networks are rated near the electric-stress limit of the insulation system.

Components and equipment recommended for pulse tests are cable assemblies, capacitors, and inductors used in high-voltage, high-power equipment. Pulse tests are not recommended for delicate experiments or very-low-power equipment.

12.6.3.3 Corona Tests. Corona tests are used to seek out insulating material flaws by detecting partial discharges in spaces, cracks, and voids. Some techniques follow.

Electrical detectors record the current pulses due to partial discharges using capacitance-coupled devices, or by using a resistance or coil connected between the test object and ground. These can be calibrated to give the magnitude of the partial discharges. The circuitry of these detectors can be designed to eliminate or balance out background noise. Several techniques are available for detecting and measuring partial discharges.[5,58]

Acoustic systems detect the sonic signal from the partial discharge, for example, by using microphones and piezoelectric crystals. These can be mounted externally on the equipment case or internally. Recently acoustic waveguides have been used for mounting inside oil-insulated equipment to couple the external detectors very efficiently to the internal acoustic signals.

Visual detection permits viewing the discharges directly in some systems. Photo multipliers are used to increase the sensitivity. Some systems have selected the spectrum to inspect (such as ultraviolet or infrared) in order to increase the sensitivity, and some systems permit operation with ambient light.

Thermal detection uses thermal or infrared detectors.

The RF signals of radio frequency coils or loop antennas can be analyzed to determine the source.

Several partial-discharge test instruments are listed in Table 12.16. Many of these instruments may be used to evaluate their own electronic systems. For instance, a photo multiplier will easily detect its own partial discharges when the filters are bypassed between the suspect electronic circuit and the photo multiplier pickup input circuit. Partial discharges are often responsible for several types of disturbances which can affect sensitive circuitry.

With square waves, the detector will pick up the front and tail of each wave and display them as very large pulses, which look like partial discharges having hundreds of pico coulombs of energy. These pulses will, of course, have to be separated from true partial-discharge pulses in the subsequent processing of the signals.

Some requirements for improving the yields of the test facility and the test equipment are listed here.

Have complete test facilities to test electrical, physical, and environmental parameters.

TABLE 12.16 Corona Detection Categories

Detector	Input	Characteristics
Light sources		
Solar cells Photomultiplier Camera Television camera Solid-state detector	Measure light generated by gaseous ionization between open electrodes in darkened chamber	Sensitive to stray light Mobile Cannot measure discharge in voids or enclosures Very directional
Mechanical devices		
Accelerometer Ultrasonic	Measure mechanical vibration setup by gas pressure shock waves	Massive discharges are required Subject to external noise sources Light insensitive
Electromagnetic radiation		
Voltage standing-wave ratio Antenna Electrometer SATO probe Capacitor probe	Measure radio-frequency emanations generated by gaseous discharge	Light, temperature, and pressure insensitive Sensitive to outside radio frequency Semidirectional Unattached to test article Mobile
Electronic pickups		
Capacitor coupling Attached RF coil Series resistor	Measure high-frequency voltage and current impulses generated by corona partial discharges	Attached to corona-sensitive circuit Immobile Light, temperature, and pressure insensitive
Chemical detectors (gaseous)		
Mass spectrometer	Measure generated ozone and outgassing products	Must be located close to discharge High-voltage power supply required Nondirectional
Scientific instruments		
Geiger counter Curved plate analyzer Cerenkov detector Solid-state detector	Measure charged particles radiated by corona discharge	Located near discharge Mobile High cost Require special modification and instrumentation

Test-equipment sensitivity must have standardized calibration procedures to validate sensitivity.

Have separate procedures for parts, materials, submodules, and modules.

Define accept/reject criteria, as shown in Table 12.17, for 1000- to 5000-h life for parts and components. Lower values are required for long-life modules and parts.

Test specialists should make analyses, prepare failure reports, and verify test procedure correlation.

Fabricated assembled parts and circuits should be assembled per specification and tested through the temperature extremes with all circuits normally energized. Five to nine cycles are recommended. Pre-environmental and post-environmental tests should include corona, dissipation factor, insulation resistivity, and a visual inspection for breaks, tears, and deformation.

The corona initiation voltage (CIV) of an electrical apparatus can be determined when the design parameters and the applicable Paschen-law curve are known. The particular Paschen curve used depends on the type of gas in which the corona would occur, the temperature of the gas, and the configuration of the electrodes.

BIL and DWV tests may require limitation of the test voltage to prevent insulation overstress for high density packaged parts, components, and modules.

For gaseous insulation breakdown the Paschen-law minimum is affected by frequency. Likewise, the lifetime of an epoxy varies approximately inversely with the applied voltage frequency.3' With these data it can be readily established that partial-discharge and life testing of high-frequency components at 60 Hz gives poor or inconclusive results. Unfortunately most corona test equipment for partial-discharge testing is designed to operate at frequencies of 10 to 500 kHz, much lower than the aerospace electronic equipment frequencies of 40 kHz to I MHz. Recently, Eeman developed partial-discharge detection apparatus for spacecraft operating at the lower-power electronic frequencies.[59] The output from Eeman's equipment is similar to that obtained by tests by the author several years ago.[60] In those tests it was established that the partial-discharge initiation voltage is frequency-dependent. In the near future corona test equipment will be developed to test high-frequency commercial and aerospace high-voltage electronic equipment. The new test procedure should also establish partial-discharge limits. New test methods only say "test." Anyone can test, but the problem is how to interpret the test.

Some corona and partial-discharge detection equipment may have limited or wide frequency bandwidths, depending on use and design. One commercial detector is designed with several frequency bands. These bandwidths are all between the frequencies of 40 and 400 kHz. Experiments indicate that partial discharge and corona frequency spectra are attenuated at frequencies above 10 kHz in rarefied gases at pressures between 0.01 and 50 torr. This implies that corona discharge measurements made in altitude chambers would

TABLE 12.17 Partial-Discharge Test Values at Rated Voltage, in pC

Test article	ac/dc	Charge limit, pC/kV	Counts per minute above limit	Maximum charge, pC/kV
Dc tests				
HV modules	dc	1	2	5
Power supply outputs	dc	1	2	5
Filters	dc	1	2	5
Ac tests				
Transformers	ac	2	10	5
Inductors	ac	2	10	5

have greater peak values than those recorded by the corona-detecting equipment. More work is in progress on this topic and should be followed closely by engineers evaluating aerospace electronics for corona and partial discharges.

Signals with a charge of less than 1 pC should be measured in a well shielded screen room. High-frequency partial-discharge signals of less than 1-μV amplitude are easily lost when the background includes interfering signals of several microvolts. The power supply should also be appropriately isolated.

12.7 PROBLEMS AND SUGGESTED SOLUTIONS

12.7.1 Design

High-voltage, densely packaged electronics require great care for design, packaging, parts selection, field effects, manufacturing, and test than do low-voltage electronics. This fact is quickly brought to the attention of the design engineers when the failure analyses of both types of electronics are evaluated. The life of high-voltage electronics is more apt to fall short of the specified value when compared to that of low-voltage electronics. Some of the more frequent failure modes and a few recommendations are shown below.

Overheating is the greatest failure model. Integrated circuits fail because of poor heat transfer through silicones and other encapsulants.

A board mounted next to a high-temperature transformer is poor design and causes the solid-state devices to overheat. Better packaging procedures are required.

Board flexing during vibration causes solder and trace breakage. More support should be given to the boards.

Many parts have improper or no heat sinks and overheat. The solution to this design problem is obvious.

Electronic units often overheat because of poor cooling, such as power transistors without heat sinks, insufficient coolant, and cooling paths barred by thermal and electrical insulation.

Wire routing in many power supplies is very poor and inconsistent. Some wires are laid out in such a way that they cause high failure rates. A high-voltage wire should not be allowed to touch the ground plane intermittently during vibration.

Ties often have loose ends "sparking" to ground planes, causing noise.

Wires are often too short or too long. The short ones have high breakage and the long ones are doubled back in such a way that it makes repair and maintenance difficult.

Repair of potted units often results in poor bonding due to uncleanliness and incompatibility between new and old materials.

Fluorocarbon products such as FC-77 evaporate without leak identification. In leaky units, sludge has been caused by arcing within the unit.

Silicone wire in silicone-oil-filled units becomes spongy and fails.

An air pocket above the oil allows the air bubble to move about during the flight plan, uncovering some high-voltage parts and resulting in arcs and corona.

Cleaning materials should be specified for all potted or fluid-filled supplies to detect material incompatibility.

Low-voltage boards should not be encapsulated. This makes repairability and maintainability much simpler.

Some air-cooled units get inadequate air cooling at altitude, overheat, and fail.

Airflow paths are often eliminated by modifications and redesign.

Instruction should be given for the use of FC-77, compatibility, and leakage problems.

A high failure rate has been caused by water condensation on low-voltage connectors. Weatherproof connectors should be required on all sealed units. Air-cooled units have high failure rates due to condensation.

Modularized units are repaired easily. Parts are standardized and easy to replace.

Qualification units are often built by engineers and skilled technicians. Some failure modes introduced by the production program do not appear on the drawings.

Ceramic capacitor coatings often crack and fail the capacitor because of insufficient testing before application.

12.7.2 Parts

Parts problems are listed as follows.

All parts should be screened to meet a minimum specification.

Some capacitor and resistor values change with age. This throws the system out of tolerance.

Parts are crushed or broken due to insufficient stress relief. Solder pull and broken leads are also detected.

Parts are not all identical in characteristics or shape.

Some tantalum capacitors are overstressed when wave-soldered.

Capacitor circuits with tantalum-silver capacitors that were replaced with tantalum-tantalum capacitors have much improved MTBF.

Parts stored for long periods of time become corroded before being installed in a power supply.

12.7.3 Manufacturing

There are three more manufacturing problems that may exist during high-voltage electronic module manufacturing.

Epoxy-coated cards are often repaired during manufacture by burning off the epoxy with a hot iron. If the residues are not removed, they cause high failure rates for modified boards.

Vibration makes wires break at potted joints or connections if not properly laced to a support.

Workmanship appears to be fairly consistent for low-voltage electronics. For higher-voltage units workmanship and packaging defects produce high failure rates.

12.8 CONCLUSIONS

Packaging designs must minimize electrical stresses, both inside and along the surface of the insulation media. Due account must be given to the environment, such as temperature, pressure, humidity, and mechanical stresses that may be present. For reliable systems, care must be taken with the manufacturing processes to avoid contamination and incorrect assembly, which can directly affect the dielectric performance or which can initiate subsequent failure by, for example, starting partial discharges. Testing of the materials, components, and system is critical for reliable performance.

This chapter has only been able to describe briefly the dielectric characteristics of gaseous, liquid, and solid insulation systems and some of their applications. To obtain higher mean time between failures, manufacturing and test procedures must be developed to address the high stresses and life aspects of highvoltage electronics.

REFERENCES

1. W. R. Smythe, *Static and Dynamic Electricity,* McGraw-Hill, New York, 1968.

2. J. D. Stratton, *Electromagnetic Theory,* McGraw-Hill, New York, 1941.

3. E. Weber, *Electromagnetic Fields,* Wiley, New York, 1950.

4. E. A. R. von Hippel, *Dielectric Materials and Application,* Wiley, New York, 1954.

5. W. E. Greenfield, *Introduction to Dielectric Theory and Measurements,* College of Engineering, Washington State University, Pullman, Wash., 1972.

6. J. M. Meek and J. D. Craggs (Eds.), *Electrical Breakdown of Gases,* Wiley, New York, 1978.

7. L. B. Loeb, *Electrical Coronas,* University of California Press, Berkeley, 1965.

8. F. Llewellyn-Jones, *Ionization and Breakdown of Gases,* Wiley, New York, 1957.

9. F. D. Corbine, *Gaseous Conductors,* Dover, New York, 1958.

10. L. B. Loeb, *Basic Processes of Gaseous Electronics*, 2d ed., University of California Press, Berkeley, 1960.

11. M. Lehr et al., "Inhibiting Surface Flashover with Magnetic Fields," presented at the 6th Int. Symp. on High-Voltage Engineering (New Orleans, La., Aug. 28–Sept. 1, 1989).

12. R. Korzekwa et al., "Inhibiting Surface Flashover for Space Conditions Using Magnetic Fields," *IEEE Trans. Plasma Sci,* vol. 17, pp. 612–615, Aug. 1989.

13. D. J. Chee-Hing and K. D. Srivastava, "Insulation Performance of Dielectric-Coated Electrodes in Sulphur Hexafluoride Gas," *IEEE Trans. Elec. Insul.,* vol. EI-10, pp. 119–124, Dec. 1975.

14. A. H. Cookson and O. Farish, "Motion of Spherical Particles and ac Breakdown in Compressed SF_6," in *Conf. on Electrical Insulation and Dielectric Phenomena 1971 Annual Rep.,* National Academy of Sciences, Washington, D.C., pp. 129–135.

15. A. H. Cookson and R. E. Wootton, "Particle-Initiated ac and dc Breakdown in Compressed Nitrogen, SF_6, and Nitrogen-SF_6 Mixtures," in *Conf. on Electrical Insulation and Dielectric Phenomena 1973 Annual Rep.,* National Academy of Sciences, Washington, D.C., pp. 23o241.

16. W. Mosch, H. Hauschild, J. Speck, and S. Schierig, "Phenomena in SF_6 Insulation with Particles and Their Technical Evaluation," presented at the 3d Int. Symp. on High-Voltage Engineering (Milan, Italy, 1979), paper 32.01.

17. J. G. Trump, "Dust Precipitator," U.S. Patent 3515959, 1970.

18. R. N. Sampson, "Insulation, Electrical Properties and Materials," in R. E. Krik and D. F. Othmer (Eds.), *Encyclopedia of Chemical Technology,* vol. 13, 3d ed., Wiley, New York, 1981, pp. 53F564.

19. L. Mandelcorn et al., "High Voltage Capacitor Dielectrics Recent Developments," in *Proc. Symp. on High Energy Density Capacitors and Dielectric Materials* (Nat. Acad. Of Sciences, Washington, D.C., 1981), pp. 97–104.

20. E. Simo, "Large-Scale Dielectric Test of Transformer Oil with Uniform Field Electrodes," *IEEE Trans. Elec. Insul.,* vol. El-5, pp. 121–126, Dec. 1970.

21. R. Bartnikas, "Dielectric Loss in Insulating Liquids," *IEEE Trans. Elec. Insul.,* vol. El-2, pp. 3343, Apr. 1967.

22. F. M. Clark, Insulating *Materials for Design and Engineering Practice,* Wiley, New York, 1962.

23. *Insulation/Circuits,* Desk Manual, vol. 27, no. 7, Lake Publishing, Libertyville, Ill., June/July 1981.

24. J. F. Morel and P. N. Dung, "Thermal Agigob Bi-Axially Oriented PET Films; Relation between Structural Changes and Dielectric Behavior," *IEEE Trans. Elec. Insul.,* vol. EI-15, Aug. 1980.

25. MIL-I-16923G, "Insulating Compound—Electrical Embedding."

26. C. A. Harper, *Handbook of Plastics and Elastomers,* McGraw-Hill, New York, 1975.

27. A. Roth, *Hochspannungstechnik,* Springer, Vienna, 1959.

28. H. L. Samms and W. W. Pendleton, *Materials for Electrical Insulating and Dielectric Formations,* Hayden, Hasbrouck Heights, N.J., 1973.

29. R. Bannikas and E. J. McMahon (Eds.), *Engineering Dielectrics,* vol. 1: *Corona Measurements and Interpretation,* ASTM Spec. Publ. 669, Philadelphia, Pa., 1979.

30. F. H. Kreuger, *Discharge Detection in High Voltage Equipment,* American Elsevier, New York, 1965.

31. S. A. Studniarz and T. W. Dakin, "The Voltage Endurance of Cast Epoxy Resin—II," in *Conf. Rec. 1982 IEEE Int. Symp. on Electrical Insulation,* pp. 19–25, 1982.

32. P. Paloniemi, "Theory of Equalization of Thermal Aging Processes of Electrical Insulating Materials in Thermal Endurance Tests," *IEEE Trans. Elec. Insul.,* vol. El-16, pp. 1–30, 1981.

33. W. Nelson, "Survey of Methods for Planning and Analyzing Accelerated Tests," *IEEE Trans. Elec. Insul.,* vol. EI-9, Mar. 1974.

34. G. C. Mononori and G. Pattini, "Thermal Endurance, Evaluation of Insulating Materials: A Theoretical and Experimental Analysis," *IEEE Trans. Elec. Insul.,* vol. EI-21, Feb. 1986.

35. G. Alex, "Tracking Resistance of 1-V Insulating Materials," *IEEE Trans. Elec. Insul.,* vol. EI-1, Mar. 1965.

36. M. V. Bilings and K. W. Humphreys, "An Outdoor Tracking and Erosion Test of Some Epoxy Resins," *IEEE Trans. Elec. Insul.,* vol. EI-3, pp. 62–70, Aug. 1968.

37. M. J. Billings, A. Smith, and R. Wilkins, "Tracking in Polymeric Insulation," *IEEE Trans. Elec. Insul.,* vol. EI-2, pp. 131–136, Dec. 1967.

38. "Component Damage/Malfunction Levels," Tech. Memo. TM-75, Boeing Aerospace Co., Seattle, Wash., Dec. 1974.

39. W. G. Dunbar, "Skylab High Voltage System Corona Assessment," presented at the 11th Electrical /Electronics Insulation Conf. (Chicago, Ill., 1973).

40. J. F. Scannapieco, "The Effects of Outgassing Materials on Voltage Breakdown," in *Proc. 2d Workshop on Voltage Breakdown in Electronic Equipment at Low Air Pressure,* Tech. Rep. 33–447, Jet Propulsion Lab., Pasadena, Calif., 1969.

41. W. A. Prowse, "The Initiation of Breakdown Gases Subject to High Frequency Electric Fields," *J. IRE (British),* p. 333, Nov. 1950.

42. P. I. Nouye and E. R. Bunker, Jr., "High Voltage Electric Packaging Flight Equipment," Rep. DM505139, Rev. A, Code 23835, Jet Propulsion Lab.; California Institute of Technology, Pasadena, Nov. 24, 1971.

43. R. J. Densely, "Panial Discharges in Electrical Insulation under Combined Alternating and Impulse Stress," *IEEE Trans. Elec. Insul.,* vol. EI-5, pp. 96–103, Dec. 1970.

44. R. C. Finke, I. T. Meyers, F. Terdan, and N. J. Stevens, "Power Management and Control for Space Systems," Future Orbital Power Systems Technology Requirements, NASA Conf. Pub. 2058, 1978.

45. B. von Herzen, "Light Pressure and Solar Wind Perturbations to Payload Trajectories," presented at the 4th Princeton/AIAA Conf. on Space Manufacturing Facilities, 1979.

46. E. Miller, W. Fischbein, M. Stauber, and P. Suh, "Environmental Interaction Implications for Large Space System," Spacecraft Charging Technology, NASA-CP-2071, 1978.

47. J. W. Kaufman, "Terrestrial Environment (Climate) Criteria Guidelines for Use in Aerospace Vehicle Development," NASA Tech. Memo. 78118, rev., 1977.

48. G. S. West, Jr., J. J. Wright, and H. C. Euler, "Space and Planetary Environment Criteria Guidelines for Use in Space Vehicle Development," NASA Tech. Memo. 78119, 1977.

49. J. A. M. McDonnel, *Cosmic Dust,* Wiley, New York, 1978.

50. H. Fechtic, E. Grun, and G. Morfill, "Micrometeoroids with Ten Earth Radii," *Planet Space Sci,* vol. 27, p. 511, 1979.

51. "High Voltage Design Guide: Spacecraft," Air Force Tech. Rep. AFWAL-TR-822057, vol. 5, 1983.

52. K. C. Purvis, H. B. Garrett, A. C. Whittlesey, and N. J. Stevens, "Design Guidelines for Assessing and Controlling Spacecraft Charging Effects," NASA Tech. Paper 2361, Sept. 1984.

53. 53. NASA *Space Systems Technology Model,* vol. IIA: *Space Technology Trends and Forecasts,* 5th issue (preliminary draft), Aug. 1983.

54. 54. N. T. Grier and N. J. Stevens, "Plasma Interactions Experiment, PIX Flight Results," NASA Conf. Pub. 2071, 1978.

55. G. Bloom, "Designing Power Supplies against the Effects of Nuclear Radiation," in *Proc. Power Con 6,* 1979.

56. E. E. King, "Radiation-Hardening Static NMOS RAMS," *IEEE Trans. Nuc. Sci,* vol. NS-26, Dec. 1979.

57. "Detection and Measurement of Discharge (Corona) Pulses in Evaluation of Insulation Systems," ASTM D-1868, American Society for Testing and Materials, Philadelphia, Pa.

58. "Measurement of Energy and Integrated Charge Transfer Due to Partial Discharges (Corona) Using Bridge Techniques," ASTM D-3382, American Society for Testing and Materials, Philadelphia, Pa.

59. J. C. J. Eeman, "Discharge Detection in High Voltage Flight Hardware," in *Proc. 16th Ann. IEEE PESC,* ESA Session (University Paul Sabatier, Toulouse, France, June 2428, 1985), p. 217.

60. W. G. Dunbar, unpublished, 1967.

APPENDIX

SOME ACRONYMS AND SYMBOLS FOR ELECTRONIC PACKAGING AND INTERCONNECTING*

ABS	acrylonitrile-butadiene-styrene
ACPI	automated component placement and insertion
AgPd	silver palladium
AI	artificial intelligence
AIA	Aircraft Industries Association
AID	automatic insertion DIP
AIN	aluminum nitride
ALU	arithmetic logic unit
Al_2O_3	alumina, aluminum oxide
ANSI	American National Standards Institute
AP	arithmetic processor
APC	array processor controller
APE	asynchronous processing element
APIO	array processor input/output
AQL	acceptable quality level
As	arsenic
ASIC	application-specific integrated circuits
ASP	advanced signal processor

*This listing is, in part, adapted with permission from M. B. Miller, *Dictionary of Electronic Packaging, Microelectronic, and Interconnection Terms* (McGraw-Hill Companies, New York, N.Y.).

A.1

ASTM	American Society for Testing and Materials
ASW	antisubmarine warfare
ATAB	area-array tape-automated bonding
ATE	automatic test equipment
AuGe	gold-germanium
AuPt	gold-platinum
AuSi	gold-silicon
AuSn	gold-tin
AWG	American wire gauge
$BaTiO_3$	barium-titanate
BeO	beryllia, beryllium oxide (bromellite)
BGA	ball grid array
BIT	built-in test
BL	beam-lead
BMC	bulk molding compound
BN	boron nitride
BOPS	billion operations per second
BQFP	bumped quad flat pack
Brassboard	field-demonstrable electronic module
Breadboard	laboratory-demonstrable electronic module
BTAB	bumped tape-automated bonding
C	capacitance
C and W	chip and wire
CAD	computer-aided design
CAE	computer-aided engineering
CAM	computer-assisted manufacturing
CAT	computer-aided testing
CAVP	complex-arithmetic vector processor
CBGA	ceramic ball grid array
CC	chip carrier
CCB	controlled-collapse bonding
CCC	ceramic chip carrier
CCGA	ceramic column grid array
CDA	clean dry air
CDR	critical design review
CERDIP	ceramic dual-in-line package
CERQUAD	ceramic quad flat pack (pressed ceramic/glass sealed)
CFC	chlorinated fluorocarbon
CGA	configurable gate array
CIM	computer-integrated manufacturing

CLA	centerline average (a measure of substrate surface roughness)
CLCC	ceramic leaded chip carrier
CLDCC	ceramic leaded chip carrier (preferred)
CLLCC	ceramic leadless chip carrier (preferred)
CML	current-mode logic
CMOS	complementary metal-oxide semiconductor
COB	chip on board
CPGA	ceramic pin grid array (laminated cofired ceramic)
CPU	central processing unit
CQFP	ceramic quad flat pack
CQP	ceramic quad pack (laminated cofired ceramic)
CPU	central processing unit
CSA	Canadian Standards Association
CSIC	customer-specified integrated circuit
CSP	chip scale packaging
CTE	coefficient of thermal expansion
CVD	chemical vapor deposition
DAP	diallyl phthalate
dB	decibel
DCAS	Defense Contract Administration Service
DI	deionized water
DIP	dual-in-line package
DMA	dynamic mechanical analysis
DODISS	Department of Defense Index of Specifications and Standards
DPA	destructive physical analysis
DRAM	dynamic RAM
DSC	differential scanning calorimetry
DTA	differential thermal analysis
DUT	device under test
E	voltage, potential
EAPROM	electrically alterable programmable ROM
ECL	emitter-coupled logic
ECM	electronic countermeasures
EDS	energy dispersive spectroscopy
EEPROM	electrically erasable programmable ROM
EIA	Electronics Industries Association
EIAJ	Electronics Industries Association of Japan
EMC	electromagnetic compatibility
EMI	electromagnetic interference
EMP	electromagnetic pulse; electromagnetic potential

EMPF	Electronic Manufacturing Productivity Facility (Navy)
EO	electrooptic
EOS	electrical overstress
EPIC	environmentally protected integrated circuit (laminated plastic CC)
EPROM	electrically programmable ROM
ESCA	electron spectroscopy for chemical analysis
ESD	electrostatic discharge
ETPC	electrolytic tough pitch copper
EUT	equipment under test
EW	electronic warfare
f	frequency
FA	failure analysis
FACI	first-article configuration inspection
FC	flip chip
FCB	flip-chip bonding
FCC	flat conductor cable
FCFC	flat conductor flat cable
FEA	finite-element analysis
FEP	fluorinated ethylene propylene (Teflon)
FET	field effect transistor
FLIR	forward-looking infrared
FP	fine pitch; flat pack
FPAP	floating-point arithmetic processor
FPL	fine-pitch leaded
FPT	fine-pitch technology
FPW	flexible printed wiring
FQFP	fine-pitch quad flat pack (molded plastic)
FR	flammability rating; flame-retardant; failure rate
FR-4	NEMA designation for epoxy-glass laminate (*see* NEMA)
FRP	fiber-glass-reinforced plastics
GaAs	gallium arsenide
Ge	germanium
GFE	government-furnished equipment
Gnd	ground
GTAB	ground-plane tape-automated bonding
GW	gull wing (leads)
HASL	hot air solder leveling
HAST	highly accelerated stress testing
HAZMAT	hazardous material
HCC	hermetic chip carrier

HCMOS	high-density CMOS
HDCM	high-density ceramic module
HDI	high-density interconnect
HDPE	high-density polyethylene
HEMP	high-altitude electromagnetic pulse
HF	high frequency
HIC	hybrid integrated circuit
HMOS	high-performance MOS
HOL	high-order language
HTRB	high-temperature reverse bias
HV	high voltage
Hz	hertz, cycles per second (1 Hz = 1 cycle per second)
I	current
IAPU	image array processing unit
IC	integrated circuit; internal connection
IDC	insulation displacement connector; insulation displacement contact
IEC	International Electrotechnical Commission
IEEE	Institute of Electrical and Electronics Engineers
IEPS	International Electronics Packaging Society
ILB	inner-layer board; inner-layer bond (for TAB)
IMC	insertion-mounted component
I/O	input/output
IPC	Institute for Interconnecting and Packaging Electronic Circuits
IPS	instructions per second
IR	infrared
IRS	infrared scan
ISA	instruction set architecture; imaging sensor autoprocessor
ISHM	International Society for Hybrid Microelectronics
I/SMT	interconnect/surface-mount technology
ISO	International Standards Organization
JC	JEDEC Committee
JEDEC	Joint Electronic Devices Engineering Council
JIT	just-in-time
K	dielectric constant
KGD	known good die
L	inductance
LCC	leadless chip carrier
LCCC (LLCC)	leadless ceramic chip carrier (laminated cofired ceramic)
LCD	liquid-crystal display
LCP	liquid-crystal polymer

LDCC	leaded chip carrier
LED	light-emitting diode
LEMP	lightning electromagnetic pulse
LGA	land grid array
LIC	linear integrated circuit
LID	leadless inverted device
LLCC	leadless chip carrier (preferred)
LMCH	leadless multiple-chip hybrid
LPGA	leadless pad grid array
LPPQFP	low-profile plastic quad flat pack
LSI	larger-scale integration
LTPD	lot tolerance percent defective
MCC	multiple-chip carrier; miniature chip carrier
MCM	multichip module
MCP	multichip package
MCRPQFP	molded-carrier ring plastic quad flat pack
MED	microelectronic device
MELF	metal electrode face (bonded)
MESFET	metal-semiconductor field-effect transistor
MFD	microelectronic functional device
MIC	monolithic integrated circuit; microwave integrated circuit
micrometer	10^{-6} meters, micron (approximately 25 μm equals 1 mil)
mil	0.001 inch
MIL-STD	military standard
MIPS	million instructions per second
MIS	metal insulator semiconductor
MLB	multilayer board
MLC	multilayer ceramic (laminated cofired ceramic)
MMIC	monolithic microwave integrated circuit
MMPQFP	multilayer molded-plastic quad flat pack
MNOS	metal nitride-oxide semiconductor
MNS	metal nitride semiconductor
MOPS	million operations per second
MOS	metal-oxide semiconductor
MOSFET	metal-oxide semiconductor field-effect transistor
MQFP	metal quad flat pack
MRB	material review board
MSI	medium-scale integration
MTBF	mean time between failures
MTM	multiple-termination module

MTNS	metal thick nitride semiconductor
MTOS	metal thick oxide semiconductor
MTTF	mean time to failure
MTTR	mean time to repair
NDE	nondestructive evaluation
NDRO	nondestructive readout
NDT	nondestructive test
NEMA	National Electrical Manufacturers Association
NEMP	nuclear electromagnetic pulse
NMOS	n-channel metal-oxide silicon
OEM	original equipment manufacturer
OLB	outer-lead bond (for TAB)
OPS	operations per second
PAA	pad area array
PBGA	plastic ball grid array
PCB	printed-circuit board
PCBT	pressure-cooker bias test
PCT	pressure-cooker test
PCTFE	polychlorotrifluoroethylene
PDA	percent defective allowable
PDIP	plastic dual-in-line package (molded plastic)
PEL	picture element in display
PEM	plastic encapsulated microcircuit
PGA	pin grid array (laminated cofired ceramic); pad grid array
PIN	p-n junction with isolation region (diode)
PIND	particle impact noise detection test
PLCC	plastic leaded chip carrier (molded plastic)
PMP	premolded plastic
PMOS	p-channel metal-oxide semiconductor
POS	porcelain-on-steel (substrate)
PPGA	plastic pin grid array (laminated plastic)
ppm	parts per million
PQFP	plastic quad flat pack (molded plastic)
PROM	programmable read-only memory
PSG	phosphosilicate glass
PSP	programmable signal processor
PTF	polymer thick film
PTFE	polytetrafluoroethylene
PTH	plated-through hole
PVC	polyvinyl chloride

PVF_2	polyvinylidene fluoride
PWB	printed-wiring board
QFP	quad flat pack
QPL	qualified parts list; qualified products list
QUIP	quad-in-line package (molded plastic)
R	resistance
RAM	random-access memory
RF	radio frequency
RFI	radio-frequency interference
RGA	residual gas analysis
RH	relative humidity
RISC	reduced instructions set computing
RISP	rapid-impingement speed plating
RMS	root mean square
ROM	read-only memory
RTI	radiation transfer index
RTL	resistor-transistor logic
RTV	room-temperature vulcanizing
SAM	scanning acoustic microscope
SAW	surface acoustic wave
SB-DIP	side-brazed dual-in-line package (laminated cofired ceramic)
SCC	stress-corrosion cracking
SDI	strategic defense initiative
SEM	standard electroicn modules; scanning electron microscope
Si	silicon
SIM	single-in-line module (laminated cofired ceramic and laminated plastic)
SIMM	single-in-line memory module (laminated cofired ceramic and laminated plastic)
SIP	single-in-line package (laminated cofired ceramic and laminated plastic)
SLAM	single-layer alumina metallized; scanning laser acoustic microscopy
SM	surface mount
SMD	surface-mount device
SMOBC	solder mask over bare copper
SMP	surface-mount package
SMPGA	surface-mountable pin grid array (laminated cofired ceramic)
SMT	surface-mount technology
SO	small-outline package, gull wing leads (molded plastic)
SOIC (SOG)	small-outline integrated circuit package, gull wing leads (molded plastic)
SOJ	small-outline package, J leads (molded plastic)
SOP	small-outline package (molded plastic)

SOS	silicon-on-sapphire
SOT	small-outline transistor
SOW	statement of work
SP	signal processor
SPC	statistical process control
SPDT	single-pole double throw
SPICE	simulation program for integrated-circuit emphasis
SPS	systolic processing superchip
SPST	single-pole single-throw
SRAM	static RAM; short-range attack missile
SREMP	source region electromagnetic pulse
SSI	small-scale integration
SSWS	static safe work station
STL	strip line
TAB	tape-automated bonding
TC	thermocompression (bonding)
TCC	temperature coefficient of capacitance
TCE	temperature coefficient of expansion; same as CTE
TCR	temperature coefficient of resistance
TF	trim and form; thick film
TFE	tetrafluoroethylene
T_g	glass transition temperature
TGA	thermal gravimetric analysis
TH	through-hole mount
THB	temperature-humidity bias
TIC	tape-automated bond in cap
TMA	thermomechanical analysis
TO	transistor outline package
TOB	tape-automated bond on board
TQFP	tape quad flat pack
TS	thermosonic (bonding)
TTL	transistor-transistor logic
ULSI	ultralarge-scale integration
US	ultrasonic (bonding)
UV	ultraviolet
VCD	variable center distance
VHF	very high frequency
VHL	vacuum hydraulic lamination
VHSIC	very-high-speed integrated circuit
VIL	vertical-in-line

VLSI	very-large-scale integration
VPS	vapor-phase soldering
VQFP	very small quad flat pack (molded plastic)
VSOP	very-small-outline package (molded plastic)
WB	wire-bonded
WIP	work in progress
WSI	wafer-scale integration
X	reactance
X-MOS	high-speed MOS
YAG	yttrium aluminum garnet
Z	impedance
ZIF	zero insertion force
ZIP	zigzag-in-line package

INDEX

ABOUT THE EDITOR

Charles A. Harper is president of Technology Seminars, Inc., an organization devoted to presenting seminars on electronic packaging and related subjects to the electronics industry. A graduate of The Johns Hopkins University School of Engineering, Baltimore, Md., he is a member of the engineering faculty of The Johns Hopkins University and is active in numerous professional societies, including the International Electronics Packaging Society (of which he is a past president), the National Electronic Packaging Conference (NEPCON), and the Institute of Electrical and Electronics Engineers. He is series editor of McGraw-Hill's *Electronic Packaging and Interconnection Technology Series* and a member of the editorial advisory board of *Electronic Packaging and Production* magazine.